AF691527

Published by SciTech Publishing, an imprint of The Institution of Engineering and Technology, London, United Kingdom

The Institution of Engineering and Technology is registered as a Charity in England & Wales (no. 211014) and Scotland (no. SC038698).

© The Institution of Engineering and Technology 2019

First published 2018

This publication is copyright under the Berne Convention and the Universal Copyright Convention. All rights reserved. Apart from any fair dealing for the purposes of research or private study, or criticism or review, as permitted under the Copyright, Designs and Patents Act 1988, this publication may be reproduced, stored or transmitted, in any form or by any means, only with the prior permission in writing of the publishers, or in the case of reprographic reproduction in accordance with the terms of licences issued by the Copyright Licensing Agency. Enquiries concerning reproduction outside those terms should be sent to the publisher at the undermentioned address:

The Institution of Engineering and Technology
Michael Faraday House
Six Hills Way, Stevenage
Herts, SG1 2AY, United Kingdom

www.theiet.org

While the authors and publisher believe that the information and guidance given in this work are correct, all parties must rely upon their own skill and judgement when making use of them. Neither the authors nor publisher assumes any liability to anyone for any loss or damage caused by any error or omission in the work, whether such an error or omission is the result of negligence or any other cause. Any and all such liability is disclaimed.

The moral rights of the authors to be identified as authors of this work have been asserted by them in accordance with the Copyright, Designs and Patents Act 1988.

British Library Cataloguing in Publication Data
A catalogue record for this product is available from the British Library

ISBN 978-1-78561-357-9 (hardback)
ISBN 978-1-78561-358-6 (PDF)

Typeset in India by MPS Limited

Radar and Communication Spectrum Sharing

Edited by
Shannon D. Blunt and Erik S. Perrins

The Institution of Engineering and Technology

Related titles on radar:

Advances in Bistatic Radar Willis and Griffiths
Airborne Early Warning System Concepts, 3rd Edition Long
Bistatic Radar, 2nd Edition Willis
Design of Multi-Frequency CW Radars Jankiraman
Digital Techniques for Wideband Receivers, 2nd Edition Tsui
Electronic Warfare Pocket Guide Adamy
Foliage Penetration Radar: Detection and characterisation of objects under trees Davis
Fundamentals of Ground Radar for ATC Engineers and Technicians Bouwman
Fundamentals of Systems Engineering and Defense Systems Applications Jeffrey
Introduction to Electronic Warfare Modeling and Simulation Adamy
Introduction to Electronic Defense Systems Neri
Introduction to Sensors for Ranging and Imaging Brooker
Microwave Passive Direction Finding Lipsky
Microwave Receivers with Electronic Warfare Applications Tsui
Phased-Array Radar Design: Application of radar fundamentals Jeffrey
Pocket Radar Guide: Key facts, equations, and data Curry
Principles of Modern Radar, Volume 1: Basic principles Richards, Scheer and Holm
Principles of Modern Radar, Volume 2: Advanced techniques Melvin and Scheer
Principles of Modern Radar, Volume 3: Applications Scheer and Melvin
Principles of Waveform Diversity and Design Wicks *et al.*
Pulse Doppler Radar Alabaster
Radar Cross Section Measurements Knott
Radar Cross Section, 2nd Edition Knott *et al.*
Radar Design Principles: Signal processing and the environment, 2nd Edition Nathanson *et al.*
Radar Detection DiFranco and Ruby
Radar Essentials: A concise handbook for radar design and performance Curry
Radar Foundations for Imaging and Advanced Concepts Sullivan
Radar Principles for the Non-Specialist, 3rd Edition Toomay and Hannen
Test and Evaluation of Aircraft Avionics and Weapons Systems McShea
Understanding Radar Systems Kingsley and Quegan
Understanding Synthetic Aperture Radar Images Oliver and Quegan
Radar and Electronic Warfare Principles for the Non-Specialist, 4th Edition Hannen
Inverse Synthetic Aperture Radar Imaging: Principles, algorithms and applications Chen and Marotella
Stimson's Introduction to Airborne Radar, 3rd Edition Baker, Griffiths and Adamy
Test and Evaluation of Avionics and Weapon Systems, 2nd Edition McShea
Angle-of-Arrival Estimation Using Radar Interferometry: Methods and applications Holder
Biologically-Inspired Radar and Sonar: Lessons from Nature Lessons from Nature Balleri, Griffiths and Baker
The Impact of Cognition on Radar Technology Farina, De Maio and Haykin
Novel Radar Techniques and Applications, Volume 1: Real aperture array radar, imaging radar, and passive and multistatic radar Klemm, Nickel, Gierull, Lombardo, Griffiths and Koch
Novel Radar Techniques and Applications, Volume 2: Waveform diversity and cognitive radar, and target tracking and data fusion Klemm, Nickel, Gierull, Lombardo, Griffiths and Koch

Radar and Communication Spectrum Sharing

Contents

About the editor

Shannon D. Blunt received his Ph.D. degree in electrical engineering from the University of Missouri in 2002. From 2002 to 2005, he was with the Radar Division of the Naval Research Laboratory in Washington, D.C. Since 2005, he has been with the Department of Electrical Engineering and Computer Science at the University of Kansas (KU) where he is currently a professor. Since 2011, he has been the Director of the Radar Systems & Remote Sensing Lab (RSL) that was founded in 1964 by Richard Moore, who pioneered the field of radar remote sensing and established KU as one of the preeminent radar research institutions in the world.

In 2012, Prof. Blunt received the IEEE/AESS Nathanson Memorial Radar Award for "contributions to adaptive radar signal processing and waveform diversity" and in 2008 he received the AFOSR Young Investigator Award to explore radar-embedded communications. He has published and contributed over 140 refereed journals and conference publications and 14 patents/patents-pending. He also previously co-edited the book *Principles of Waveform Diversity & Design*, the first on the topic.

He is currently Chair of the IEEE/AESS Radar Systems Panel and the Nathanson Award Committee. He is on the Editorial Board for *IET Radar, Sonar & Navigation* and is an Associate Editor in the radar area for *IEEE Transactions on Aerospace & Electronic Systems*. He was General Chair of the 2011 *IEEE Radar Conference* held in Kansas City, Technical Co-Chair of the 2018 *IEEE Radar Conference* held in Oklahoma City, and is a member of the Program Committee for the *MSS Tri-Service Radar Symposium* series. He was Chair of the NATO SET-179 research task group on "Dynamic Waveform Diversity & Design" and a member of SET-182 on "Radar Spectrum Engineering & Management" and SET-227 on "Cognitive Radar". In 2016, he became a Fellow of the IEEE for "contributions to radar waveform diversity and design".

Erik S. Perrins received his B.S. (*magna cum laude*), M.S., and Ph.D. degrees from BYU, Provo, UT, in 1997, 1998, and 2005, respectively, all in electrical engineering. From 1998 to 2004, he was with Motorola, Inc., Schaumburg, IL, where he was engaged in research on land mobile radio products. Since 2005, he has been with the Department of Electrical Engineering and Computer Science, University of Kansas, Lawrence, where he is currently a professor. Since 2004, he has also been an industry consultant on communication problems such as reduced-complexity receiver design, receiver synchronization, and error control coding. His current research interests include digital communication theory, synchronization, channel coding, energy efficient communications, and complexity reduction in receivers.

Dr. Perrins is a member of the IEEE Communications Society and the IEEE Aerospace and Electronic Systems Society. He was an editor and an area editor for the *IEEE Transactions on Communications* from 2007 to 2018, and he is currently Chair of the Communication Theory Technical Committee within the IEEE Communications Society.

Preface

In light of the old adage "necessity is the mother of invention," one could make the argument that the driving force behind this book was not the publisher, or the editors, or the authors, but was instead the tremendous number of individuals (ourselves among them) who make use of the modern miracle that provides us near-instantaneous access to a mind-boggling amount of information via wireless communications and the globe-spanning internet. In other words, this capability that has so radically altered our world … has likewise introduced its own set of problems, not least of which is the increasing pressure upon other, equally important uses of the radio frequency spectrum. Of course, for those of us involved in research, problems such as these (or more precisely, the process of working towards viable solutions) do keep us gainfully employed. They are, as a colleague likes to say, "the coin of the realm".

The increasing army of cellular communication users/devices notwithstanding, the true heroes of our tale are the chapter authors who so graciously accepted our request to convert their individual knowledge and research contributions into book chapter form. It is no understatement to assert that the book simply would not exist if not for their efforts. Given the breadth of topics encompassed by spectrum sharing, we as editors learned quite a few new things during this endeavor, and our hope is that you the readers do as well.

Finally, on a more personal note, we would like to thank the many mentors we have had over the years, the fantastic students with which we have had the opportunity to work, and most importantly our families, for putting up with the long hours of obsessive perfectionism that should be the dictionary definition of editor. This book is dedicated to you.

All the best,
Shannon D. Blunt and Erik S. Perrins

Part I

The big picture

Chapter 1

The case for spectrum access

Shannon D. Blunt[1] and Erik S. Perrins[1]

1.1 Introduction

Fundamentally, this book seeks to capture the salient aspects of the exceedingly complex topic of radar and communication spectrum sharing, which is itself a menagerie of different problem spaces that depend on the particular goals one is attempting to achieve. The most prominent problem, and consequently the one most often first considered, is that of sharing between commercial communications and monolithic, stationary radars (e.g. for weather monitoring or air traffic control). However, the notion of spectrum sharing is actually far more diverse. For example, Table 1.1 illustrates a variety of characteristics that could possibly be used to classify a given spectrum-sharing arrangement, though it is a veritable certainty that one could envision situations that are still not considered here. Further complicating the issue are the physical attributes imposed by the transmitter/receiver hardware combined with the multitude of different radar and communication operating modes and requirements. Simply put, radar and communication spectrum sharing is anything but simple.

Due to the sheer breadth of this topic, we do not attempt to codify all the many ways in which spectrum sharing could be performed. In fact, with the notion of radar spectrum sharing only rather recently becoming manifest due to the confluence of exponentially growing spectrum demand and emerging software-driven radio capabilities, one could well contend that we are now only glimpsing what will later be considered the earliest stages of spectrum sharing innovations. To that end, the purpose of this book is three-fold.

First, it is vitally important for these separate communities to gain a better understanding of the other. In particular, for sharing arrangements to work, it is necessary for the communications community to have a better appreciation for the incontrovertible need for radar technology (which itself continues to grow and expand), the realistic constraints and propagation/phenomenological conditions under which radars operate, and the vast variety of radar modes and systems that are arguably the antithesis of the standardization drivers often encountered in communications (especially for military radar systems). As it will be discussed in Chapter 4, students in electrical engineering and related disciplines commonly receive instruction in communications,

[1]Electrical Engineering & Computer Science Department, University of Kansas, USA

Table 1.1 Some prospective characteristics of a spectrum-sharing arrangement

Radar: commercial or safety/defense application?
Communications: commercial or safety/defense application?
Radar has knowledge of communication operating parameters?
Communication system has knowledge of radar operating parameters?
Cohabitation or co-design?
If cohabitation: Interference avoidance or interference tolerance?
If co-design: separate systems or single multi-function system?
Long range or short range?
Spectrum access structured (e.g. primary/secondary) or unstructured?
Time-division access (e.g. alternating operation)?
Frequency-division access (e.g. spectrum sensing and notching)?
Code-division access (e.g. separability of waveforms)?
Space-division access (e.g. transmit nulling or aperture sharing)?
Stationary or moving platforms?
Radar modality (e.g. surveillance, imaging, tracking)?
Contiguous or disjoint spectral content?
Standardized (for efficiency/interoperability) or maneuver flexible (for robustness)?

and so it tends to be a more broadly understood topic, not to mention the fact that nearly all of us now carry around a smart phone that would have been considered a super-computer not all that many years ago. By comparison, applications involving human safety and defense contribute to a general reticence with regard to the dissemination of information on radar technology (often with good reason). Further, while nearly everyone has personally experienced the principles involved with echoes produced by their own voice in an enclosed space and the Doppler effect induced by a passing vehicle, the actual technical characteristics and requirements of an operational radar system can be rather less well understood. Consequently, Chapter 2 provides an overview of radar operation as well as a radar perspective on spectrum sharing. Likewise, Section 1.4 of this chapter considers the different domains in which radar and communications operate along with the associated similarities and differences between them in each domain. For a quite in-depth coverage of the vast diversity of radar operating attributes and modes, the reader is also directed to the *Principles of Modern Radar* [1–3], a three-volume set that provides an excellent treatment of the subject. Furthermore, the long-established website radartutorial.eu [4] provides a database of over 400 distinct radar systems that are publicly known.

A second important attribute of the book is to consider the spectrum-sharing problem space in all of its complexity. While it is most certainly useful for us as engineers to take a "divide and conquer" approach by considering individual components of a problem, the multi-level interactions involved with spectrum sharing may limit the efficacy of such a methodology. Beyond simply treating these systems as mutual sources of interference, joint consideration of the electromagnetics, radio frequency (RF) systems engineering, and signal-processing aspects of each modality may shed far more light on how actual radar and communication systems could achievably interact. Put another way, while it can be convenient to mathematically model these structures at a signal-processing level where theoretically optimal estimation and detection can be determined, it is imperative that the physical characteristics and limitations of the respective transmitters, receivers, antennas, and propagation phenomenology also be considered to realize meaningful results. As such, Chapters 5, 6, 9, 10, 11, and 12 address different attributes of these physical aspects and discuss relevant research ongoing in these areas that are pertinent to future spectrum-sharing capabilities. Chapters 3 and 4 likewise take a high-level perspective on the operational roll-out of radar/communication spectrum sharing to date and the valuable lessons that have been learned in the process. Chapter 13 subsequently considers how the associated radar protection regions are assessed and posits analyses that may ensure these regions are not excessively conservative.

Finally, the remainder of the book contains a broad sampling of spectrum-sharing research formulations, ranging from theoretical investigations that seek to understand the limits of various performance trade-spaces to experimental demonstrations of prospective spectrum-sharing capabilities. In some cases, the only common thread is that these approaches investigate new ways in which to exploit limited spectral resources, thereby highlighting the sheer breadth of the topic. It should be noted that, given the rapidly expanding body of work in this area, it is simply not possible to include all the present ongoing work (particularly if one wishes to be able to physically lift the book). However, we have sought to provide a representative cross section of many of the different lines of research being explored with an eye toward the identification of areas needing further investigation or emerging topics. Consequently, many of the chapters contain a "Looking Ahead" section at the end where the authors discuss where the research could go next. Thus, this book is not so much a treatise on what we as a research community have thus far accomplished but should be viewed instead as an early roadmap of the problem space and the subsequent innovations that could be realized in the future.

1.2 Why spectrum access is so vital

This ongoing conflict over finite resources in the RF spectrum is driving research into how all users can achieve their own individual goals while doing so collectively in a more efficient manner. From a radar perspective, performance is largely driven by *resolution* and *sensitivity*. The former speaks to why radar bands are so large, since bandwidth and range resolution are inversely proportional. For defense applications, large dedicated bands also provide a maneuver space in which to avoid interference that may be dynamically changing—a gap in coverage could have disastrous consequences. Note that Doppler and spatial resolution are also important factors, though they have a weaker dependence on spectral content. In contrast to resolution, radar sensitivity corresponds to the need to detect/image/identify/track noise-limited scattering and is the reason why these large bands are traditionally not shared, since in-band interference would inherently limit sensitivity. This attribute cannot be overstressed, because while a radar does indeed already take measures to avoid/suppress interference to attempt to reach a noise-limited condition, there are only so many available degrees of freedom with which to do so.

It is also important to keep in mind that radar comes in a wide variety of different forms depending on the propagation/scattering phenomenology, such as ground or foliage penetration and over-the-horizon (OTH) operation; the nature of the sensing mode being deployed, which includes various forms of surveillance, imaging, and tracking; the availability of and interplay between the various spatial, spectral, and temporal degrees of freedom such as occurs for synthetic aperture radar (SAR) imaging and space-time adaptive processing (STAP) for moving target indication (MTI); and the size, weight, power and processing limitations of the radar platform. Along with a general perspective on radar operation and the impact of interference to/from a radar, Chapter 2 provides a useful table that delineates *typical* radar operation according to frequency band, which relates directly to propagation/scattering effects. Of course this list should not be considered a complete and definitive inventory of the many different radar modes and applications. (See [1–4] for a more exhaustive delineation.)

In contrast, communications are largely driven by information capacity and the receiver error rate, though more recently notions of delay-sensitive operation, reliability, and energy efficiency have become greater concerns as well. Where radar is sometimes jokingly referred to as a "communication system that talks to itself"—which also speaks to the reason why radar enjoys greater operational diversity—communication systems require a higher degree of standardization so that individual units can actually exchange information in a meaningful and efficient manner. Consequently, communication systems tend to have agreed upon implementations regarding the structure of symbol modulation schemes, error correction coding, interleaving, synchronization procedures, and a variety of handshaking protocols to ensure the link is maintained at a sufficiently high quality of service [5].

Driven by the demand to achieve ever higher rates of bits/second/Hertz (which can also be understood as the amount of data per unit of time–bandwidth product), the communications community long ago moved from noise-limited operation into the interference-limited regime as a means to realize greater capacity. For instance, it is well known that equalization can be performed to compensate for intersymbol interference (ISI) that can arise when the data rate is high enough for channel multipath effects to be observable. In fact, the code-division multiple access (CDMA) scheme was devised for users to intentionally occupy the same band simultaneously, with separability achieved in the receiver by exploiting knowledge of the spreading code(s) of the desired user(s). Given the context of this book, it is interesting to note that one of the important technical challenges associated with the implementation of CDMA involves the "near/far" problem, whereby signals from mobile units at further distances from the base station are essentially overwhelmed by the higher received power of nearby users. This issue necessitated investigations into automated mechanisms for power control [6] and optimal/adaptive symbol estimation schemes generally referred to as *multiuser detection* [7]. Consider this arrangement, which at most involves perhaps 10–20 dB of dynamic range, relative to the transmit/receiver powers between radar and communication modes that could differ by many orders of magnitude.

For digital communications in particular, underlying enablers that support this general idea of interference-limited operation are that (1) a one-way link exists with parametrized channel characteristics that can be adequately predicted/estimated, so that (2) a communication signal can confidently be determined to be present, and thus (3) the problem becomes one of estimating the correct symbol from a constellation containing a finite set of known symbols. Of course, the practical problems of realizing delay, phase, and frequency synchronization and accurately estimating channel distortion effects of fading and multipath cannot be overstated. Emerging technologies envisioning the tremendous connection density of massive machine-type communications (also referred to as the Internet of Things) and ultra-reliable/low-latency communications will serve to drive the possibilities of interference-limited operation even further [8].

While communications has actively embraced interference-limited operation to achieve higher capacity, radar instead seeks noise-limited operation to maximize the achievable sensitivity, which is particularly important because the scattering response represents a continuum of possible values and is pervasively distributed over range, spatial angles, and portions of Doppler. Of course, that is not to say that radars actually operate in interference-free environments; quite the opposite in fact. For instance, radars that perform ground/foliage penetration [9,10] or take advantage of ionospheric propagation to peer beyond the horizon [11] tend to be limited to bands below 1 GHz (HF, VHF, and UHF) where many other spectrum users also exist. These systems must therefore take measures to identify and avoid such interference and/or employ cancellation techniques to reach a noise-limited condition, or at least as close as possible. Further, airborne/space-based radars whose function is to detect ground movers (e.g. vehicles, people) even have to contend with self-induced interference in the form of an undesired *clutter* response (i.e. scattering from everything else on the ground) that can be orders of magnitude higher than the desired target echoes.

Add to these effects the prospects of signal attenuation caused by two-way propagation and nondirectional scattering, the possibility of both neutral and intentional interference (i.e. *jamming*), and the need for ultra-sensitive radar receiver components that could rather easily become saturated, and it should become clear that spectrum sharing creates quite an operational dilemma for a radar to at least approach, much less maintain, the sensitivity enabling noise-limited condition. Indeed, despite these limiting factors and the continued push for increased spectrum sharing, it bears mentioning that the constantly evolving conditions in which radars are deployed are driving research into ways in which they could become even *more sensitive* to address an ever-growing expansion of operational requirements. Innovation alone will be the key to achieving these conflicting goals of enhancing both robustness to interference and sensitivity.

1.3 Tolerating versus cooperating

Consider for a moment, another type of relational arrangement. Many can remember growing up with a sibling who, at least while living together under the same roof, must have been the most annoying person on the planet. They used your stuff without asking, one of you was always in the shadow of the other, possibly living with hand-me-down clothes and toys after the other was finished with them, and they were generally just an all-around nuisance that never left you alone. It was perhaps only through sheer willpower, and the strong hand of watchful parents as necessary, that you managed to refrain from throttling them. Simply put, you tolerated one another, but you may not have liked one another all that much.

Continuing this analogy, with maturity may have come the selection of a partner with which to live in eternal bliss (admittedly the bliss part is a bit oversold, but you get the point). In so doing, you learned one another's idiosyncrasies, you became a team, and together became stronger than you could have been apart. In short, your collective good is put ahead of what would be most beneficial to either of you as an individual.

While it may seem a bit odd to read about sibling rivalry and tips for a healthy relationship in a book on radar and communications (the technology, not interpersonal), these descriptions are actually rather apt representations of both the early/present state of spectrum sharing and where it could potentially go in the future. Right now the older sibling (primary user) routinely tells the younger sibling (secondary user) to leave his/her room (spectrum allocation), though they may eventually settle into a reluctant acceptance of the other's presence. Of course, tensions may remain high and it could take little provocation (interference) before it becomes necessary for the parents (federal regulators) to step in and take action.

This state of affairs is particularly relevant within the context of radar systems providing for human safety (air traffic control and weather monitoring) and the demand for spectrum by commercial communications (cellular, Wi-Fi) that is largely being driven by the data requirements to provide wireless video access. When posed as a question between entertainment and potential loss of life, the decision is quite clear cut. That said, every one of us (the authors included) has at various times become frustrated when the download of a file/video/etc. to our wireless device is too slow or is just impossible in a given location. Consequently, there is a strong business case for higher data rates and improving wireless coverage based on market demand that cannot be ignored, particularly since it is such a potent driver of our evermore technology-focused economy.

1.3.1 Achieving greater tolerance

Few siblings continue to cohabitate into adulthood, yet we have every reason to believe that this manner of radar/communication spectrum sharing will not only continue, but ultimately will become a standard occurrence. However, this arrangement can only exist with the establishment of well-defined ground rules that dictate acceptable behavior. For example, Chapter 4 discusses the introduction of Unlicensed National Information Infrastructure (U-NII) wireless systems into the 5 GHz radar band. As part of this deployment, some users found that they could (unlawfully) modify these devices to enhance performance, albeit while (presumably) unwittingly causing serious interference to nearby terminal Doppler weather radars used for microburst warning at airports. Just like normal parents who cannot be perpetually present, it is not feasible to expect regulators to step in and solve every problem, especially when decisions and actions need to be made on operational time scales. Using the U-NII deployment as a rather pointed illustration, no amount of regulatory action could undo the damage of an air traffic incident. It is therefore necessary for technology to do what regulation cannot, which is to address these issues in real time to a degree that is sufficient to avoid catastrophic failure.

It stands to reason that the first line of defense in mitigating unacceptable communication-to-radar interference is to establish the operating rules by which these wireless devices are governed, and this approach is precisely what has been developed for the dynamic frequency selection (DFS) paradigm in the 5 GHz U-NII deployment [12] that is discussed in Chapters 3, 4, and 13. While not actually intended for radar, Chapter 8 likewise explores how the TV white space wireless regional area network (WRAN), as specified by the IEEE 802.22 standard [13], could have similar merit for radar spectrum sharing.

Whereas a good set of working rules should dictate the behavior of the constituents for a vast majority of the time, there still remains the nonzero probability of technical malfunctions and the prospect of unlawful modifications to be considered. From a pragmatic standpoint, these situations necessarily fall to the radar to address through some means of interference avoidance and/or cancellation in real-time. That said, given that many of these radar systems have already been in use for several years—and thus are based on the available technology at the time of deployment—and many will continue to be used for years to come due to high replacement cost, the question becomes a matter of how best to upgrade/modernize their capabilities within the constraints of the given baseline technology.

Some of the advances discussed in this book, or their technological descendants, may be potential candidates to facilitate these needed capabilities for legacy radar systems. In particular, Chapters 10, 11, and 12 address emerging RF transmitter and receiver architectures that provide greater operational flexibility, including the prospect of real-time reconfigurability. Likewise, Chapters 9, 14, 15, 16, 17, and 18 discuss a variety of different ways that a radar can proactively compensate for in-band interference, many of which rely on forms of cognitive radar that, at least in this context, have much in common with cognitive radio [14].

Another reason the radar will need to share a heavier burden in a spectrum-sharing arrangement with communications is the inherent *information asymmetry* that arises from the need to ensure secure radar operations. Simply put, if the structure of the radar emission scheme—often referred to as the *waveform*, and in this context broadly defined to include the modulating signal, the repetition rate, the polarization, and even the direction/beampattern of the illumination—was widely known, then the radar could be susceptible to malicious actions, with potentially disastrous consequences. Because any secondary wireless devices can therefore possess only limited predictive capability with regard to the primary (incumbent) radar, these devices must be robust with respect to the detection and discernment of a wide variety of possible radar emissions (see Chapter 4). While these devices must be compliance certified to perform this way in order to operate, the very notion of *maneuver space* combined with the continued evolution of radar signal structures may preclude guarantees of noninterference. As a result, the radar must be prepared to address interference that may occur in cases that these secondary systems do not detect its presence in a timely manner.

The information asymmetry condition also implies that there will remain some fundamental limit to how well this manner of spectrum sharing can ever be achieved in practice. For instance, Chapter 19 examines some of the theoretical attributes of radar/communication co-design in this context. Of course, a tenet of good engineering is to avoid designs that involve a single point of failure, whereby one action, or lack thereof, defeats the entire system. Indeed, the nonzero probability that any among a multitude of spectrum-sharing devices could malfunction suggests that this arrangement could very well possess many possible single points of failure, since any one of them could be highly detrimental. However, this effect can generally be avoided by incorporating some manner of redundancy into the overall architecture.

One candidate solution, which itself can loosely be viewed as a form of spectrum sharing, is the notion of *passive radar* [15]—or passive coherent location, passive bistatic radar, parasitic radar, etc. as it is also known—whereby illuminators of opportunity such as broadcast TV/radio are exploited to serve as the transmit component of the sensing function. As discussed in Chapters 7 and 8, passive radar does have notable limitations in comparison to standard radar operation, not least of which because the passive radar receiver has no control over whether the illuminating signal actually possesses good characteristics from a sensing perspective, i.e. range resolution, sidelobe levels, energy on target, spatial coverage, etc. That said, for radars operating in signal-rich environments where spectrum sharing is far more likely to be needed, the passive exploitation of these signals could possibly provide a useful failsafe mechanism.

1.3.2 Living in harmony

Returning to our relationship analogy, there are many forms of spectrum sharing one could envision beyond the somewhat contentious arrangement involving commercial communications (e.g., see Chapters 19–23). Indeed, that manner of spectrum sharing may even be considered a single special case among a far larger set of diverse possibilities. While avoiding any discussion of more exotic interpersonal arrangements (which are most certainly outside the scope of this book!), a central theme does stem from the simple adage that "the whole is greater than the sum of the parts," with the modes behaving in a completely cooperative manner. In other words, different forms of tightly coupled multi-function systems may facilitate as yet unrealized modes of operation, with prospective benefits such as more efficient usage of allocated spectrum (obviously), self-tagging mechanisms for co-registration of sensor data products, embedded control signals for distributed sensor "swarms," communication emissions designed to possess inherent attributes that make them more amenable for exploitation by passive radar, leveraging radar characteristics of a dual-mode emission to enhance communication synchronization, etc.

Take, for example, the myriad ways in which a communication signal could be incorporated into a radar emission, and the various trade-offs involved with doing so. First, the embedding of information necessitates the presence of uncertainty, which translates into the radar emission now possessing what can be viewed as a degree of randomness. This condition tends to be at odds with the vast majority of operating radar modes that conform to a strict structure (even if it is unknown beforehand to an external observer). If we presume the transmitter has an arbitrary waveform generation (AWG) capability, has dual-polarization capability, and can independently control the signals produced by the individual elements in its antenna array, then an information-bearing component could be integrated into any or all of these associated signal dimensions. More specifically, communication could be achieved by modulating the waveform on a slow-time (pulse-to-pulse) basis, on a fast-time (within each pulse) basis, by modulating the mainlobe and/or sidelobes of the antenna beampattern on a slow-time or fast-time basis, by manipulating transmit polarization in slow-time or fast-time, by altering the transmit center frequency, or many different combinations of these. When one also considers the myriad ways in which a signal can be modulated (AM, FM, PM, PSK, PAM, etc.), it quickly becomes clear that a rather wide variety of different emission schemes may exist.

While the multi-function system could be designed to have a separate communication mode, joint operation provides the communication function with the possibility of access to greater overall bandwidth, use of a high-power transmitter, and self-referential attributes known to the receiver (i.e. leverage knowledge of the radar waveform in the same manner as pilot symbols). From a radar perspective, the incorporation of an information-bearing signal represents a type of waveform agility not unlike a noise waveform [16–20] that can be leveraged to reduce sidelobes and/or extend the maximum unambiguous range. On that same note, however, waveform agility is also known to introduce a *range sidelobe modulation* effect that can limit clutter cancellation unless appropriately compensated [21]. In short, joint operation is expected to involve a trade-off between the expected optimal performance of either mode alone. That said, the increasingly complex environments in which these systems operate may eventually prove that the freedom to realize different joint modalities, particularly if such can be achieved on-the-fly as needed, may enable greater overall performance in practice.

A variety of fundamental and practical questions begin to arise when one considers the joint design of a dual-purpose radar/communication system. For example, what is/are the appropriate metric(s) to use to measure "goodness"? Chapters 5, 19, and 20 consider the role information theory may play in this context. Clearly, bit error rate is also important for communications, as are the related probabilities of detection (P_D) and false alarm (P_{FA}) for surveillance radar modes and position/velocity errors for tracking modes. In general, all of these metrics depend on the signal-to-interference-plus-noise ratio (SINR) of the received signal.

Alternatively, how does one design a waveform to perform the radar-sensing function when a communication attribute is either embedded in the same emission (see Chapters 9, 20, 22, and 23) or is known to reside in the same band (see Chapters 6, 12, 14, 15, 16, 17, and 18)? Among these, Chapters 9, 14, and 23 examine structures based on orthogonal frequency division multiplexing (OFDM), Chapters 16, 17, 18, 20, and 22 consider polyphase codes, and Chapters 6, 12, and 23 employ frequency-modulated (FM) waveforms. It is noteworthy that OFDM, which serves as the basis for fourth generation (4G) cellular communications, and continuous phase modulation (CPM), itself a form of digital FM used in aeronautical telemetry, deep-space communications, and the Bluetooth™ wireless standard, are signal structures presently being explored in various capacities for radar applications. Wholly new ways of representing communication symbols for these joint modes are also examined in Chapters 21 and 22. Further, Chapters 6 and 23 examine the incorporation of physical effects, such as transmitter distortion and spectral containment, into dual-mode emission designs. With new combinations and variations of these modes appearing in the literature on a regular basis, it can be stated with some confidence that we have still only scratched the surface of the diverse arrangements to come.

1.4 Scoping the spectrum-sharing problem space

In the remainder of this chapter, we provide some scope to the diverse problem space that is radar/communication spectrum sharing, as it is useful to understand the relationship between typical parameters and operating conditions that arise for each as a means to pose realistic engineering problems to be solved. Specifically, we consider perspectives in terms of spectral content, transmit and receive power levels, spatial arrangements, temporal attributes, Doppler effects, modulation and coding (i.e. waveform) characteristics, and polarization. These are then concluded with a brief discussion on the general requirements to achieve separability of mutually interfering signals. To the possible degree, we avoid detailing particular modes and instead attempt to paint a picture based on the utility of various design arrangements, with the hope of ultimately stimulating the reader to conceive of new and useful spectrum-sharing formulations.

1.4.1 Spectral perspective

It is generally understood that bandwidth directly relates to data rate for communications and range resolution for radar. (See Chapter 6 for a discussion of the more comprehensive notion of *spectral content*.) For instance, the long-term evolution (LTE) commercial wireless standard employs *resource blocks* of spectrum that occupy from just over 1 MHz up to 18 MHz of bandwidth (with the *channel bandwidth* about 11% wider) thereby realizing data rates up to 100 Megabits/second (Mbps). A single resource block spans 12 subcarriers with 15 kHz spacing, thus resulting in 180 kHz. The maximum occupied bandwidth of 18 MHz therefore corresponds to 100 resource blocks, or 1,200 subcarriers [22].

In contrast, with a 1 m range resolution being associated with a bandwidth of 150 MHz, depending on the particular mode typical radar bandwidths can be on the order of a few tens of MHz for surveillance applications up to a few GHz for imaging applications [1–3]. Of course, there are notable exceptions such as OTH radar that relies on ionospheric reflection in the HF band and employs bandwidth on the order of 10 kHz [11].

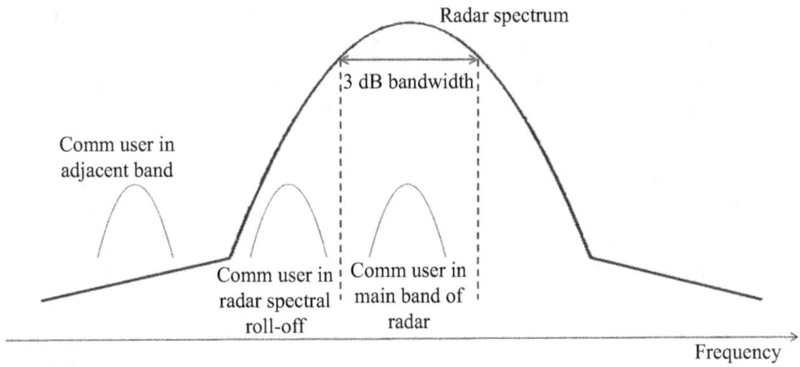

Figure 1.1 Spectral relationships between radar and communications

While there are clearly exceptions (e.g. the OTH mode above), contiguous radar bands generally tend to be wider than communication bands. Thus, Figure 1.1 illustrates a generic depiction of radar spectrum and three corresponding regions where a communication signal may reside. Obviously, the two signals may overlap within the radar main band, where each can be expected to impose the greatest amount of interference on the other. It is for this reason that many researchers have investigated radio frequency interference avoidance techniques for the radar (see Chapters 9, and 14–18). In contrast, a communication signal in an adjacent band to the radar would introduce far less interference, though the pulsed nature of many radar systems and nonlinear transmit effects produced by both systems means the interference is still not zero. Finally, a communication signal could inhabit the radar spectral roll-off region, where the degree of mutual interference would depend on the spectral shaping attributes of both signals. This latter arrangement is the focus of the radar-embedded communication paradigm discussed in Chapter 21.

It also bears mentioning that frequency hopping spread spectrum (FHSS) communication systems, such as those used for military radios to facilitate secure operation, rapidly hop around the spectrum they occupy. This hopping may be characterized as *slow hopping*, in which the hopping rate is lower than the symbol rate, or *fast hopping* in which the arrangement is reversed [23]. Many radars likewise employ frequency hopping as a mean to avoid interference or to achieve a form of synthetic wideband operation through the generation of piecewise narrowband emissions [24].

1.4.2 Power perspective

While it depends considerably on the operating range, communication transmit power varies from 200 mW or less for mobile units and femto-cells up to about 20 W for macro-cell base stations [25]. Factors affecting these power levels are the antenna gain, system-level power control (to reduce interference to other users), and the one-way propagation environment. Likewise, short-range radars such as those used in automotive applications have a peak power of 10 mW [26], while long-range radars can have peak powers on the order of hundreds of kilowatts to megawatts [1]. Radars require such higher transmit powers because of two-way propagation loss and the fact that only a fraction of the incident power density is scattered back towards the radar receiver.

On the receive side, the communication metric of received signal strength indicator (RSSI) provides a useful baseline measure. For Wi-Fi, RSSI values around 10^{-10} W are considered to be the minimum signal strength for reliable data packet delivery, while RSSI values of 10^{-12} W or lower are essentially considered unusable [27]. With 10^{-8} W considered to be a very strong received signal, the operational dynamic range is on the order of 40 dB. By comparison, using Sandia's Lynx SAR/MTI system [28] as a relevant example, surveillance radar modes like MTI may be expected to detect signals on the order of 10^{-14} W, while SAR imaging modes may be expected to capture signals below 10^{-18} W. Further, the dynamic range over which a radar receiver must operate can exceed 120 dB due to the need to represent without distortion the extremely different power levels encountered for large and small targets, clutter, and noise [1].

It is worth noting that multi-function approaches in which a communication signal is embedded into the radar emission (e.g. Chapters 9, 20, 22, and 23) would subsequently be transmitted at the higher power generally employed for radar. Likewise, with only the one-way path instead of the radar's full two-way path, these communication signals could be received at relatively high power levels at rather far distances.

1.4.3 Spatial perspective

To ensure adequate coverage of the serviced area, communication base stations tend to have relatively wide beams that are sufficiently depressed in elevation to focus the emitted power density over a broad swath of the ground where most mobile units reside. Improved spectral re-use can be achieved via 120° or 60° sectoring [5]. In more recent years, the use of multiple-input multiple-output (MIMO) configurations involving two or more transmit and receive antennas has emerged as a means to facilitate enhanced data rates via spatial multiplexing, which requires some knowledge of channel state information, or to improve signal diversity against fading at the receiver via space-time coding [29]. These techniques are now considered mature technology that are included in the IEEE 802.11 wireless standards and are in widespread usage.

Research on future 5G systems is exploring *multi-user MIMO*, itself an outgrowth of space-division multiple access, whereby separate spatial channels can be allocated to different users that are simultaneously sharing the same spectrum, thus increasing overall capacity [30]. In other words, broad coverage beams can be replaced by individual user-specific beams so that time-sharing the channel is not required. The notion of *massive MIMO* has also emerged from this paradigm in which multi-user MIMO is accomplished through the use of large-scale antenna arrays or distributed antennas [31]. The underlying goal of all of these methods is to maximize the number of users that can be supported with the maximum achievable data rates.

For radar applications, many legacy systems rely on narrow beams to achieve high directivity gain to meet the sensitivity requirement, with these fixed beams generated from a rotating antenna to provide 360° coverage. Depending on the particular usage, the antennas may form "pencil" beams that are narrow in both azimuth and elevation dimensions (e.g. for 3-D search) or "fan" beams that are in narrow in one dimension and wider in the other (e.g. for air traffic control) [1]. SAR systems may also employ wider beams as a means to ensure persistent illumination for a large number of pulses, the echoes from which produce a predictable phase history that can be used to facilitate finer resolution in the *along-track* direction than could feasibly be achieved with a real antenna (i.e. the synthetic aperture size may exceed the length of the platform) [32].

The MIMO concept has also been explored for radar [33,34]. For instance, it has been shown that the spatial diversity afforded by MIMO radar can be used for OTH radar to separate different ionospheric reflection paths for enhanced discrimination capability [35]. More broadly, MIMO is less mature and remains a topic of ongoing investigation to ascertain where (and if) it could be beneficial. In fact, there has been significant debate regarding the practical utility of MIMO for general radar applications [36]. That said, [34] does describe a production X-band MIMO radar, and Chapters 18, 22, and 23, respectively, illustrate how MIMO could be used to avoid in-band interference via space-time transmit notching, modulate the spatial sidelobes of a radar emission to embed communications, and simultaneously produce a standard quadrature phase-shift keying (QPSK)-modulated communication signal and a radar waveform from the same antenna aperture.

Finally, while communication between a transmitter and receiver is really only meaningful when they are separated, most modern radars involve a *monostatic* arrangement where the transmitter and receiver are collocated. However, many early radars were bistatic, such as the famed Chain Home radar that provided the UK with an early warning of in-bound bombers during World War II. As discussed in Chapters 7 and 8, and in greater detail in [37], bistatic and the generalization to multistatic radar configurations provide a more expansive spatial perspective to the sensing problem that may be beneficial given that the scattering from an object could change by 30 dB or more according to the *aspect angle* from which it is observed [1].

1.4.4 Temporal perspective

Again using LTE as a basis for comparison, the fundamental communication symbol duration is 66.7 μs, from which larger signal structures are organized. Specifically, 7 symbols plus the addition of individual *cyclic prefixes* comprise a 0.5 ms *time slot*, 2 time slots make up a 1 ms *subframe*, and 10 subframes constitute a 10 ms *radio frame*. The cyclic prefix for each OFDM symbol provides a guard interval to avoid ISI and facilitates a circular convolution model that permits use of simple discrete Fourier transform processing to address frequency-selective multipath effects. While there are certainly other temporal structures that a communication signal could take, LTE provides a useful benchmark because it was designed for backwards compatibility with the 3G universal mobile telecommunications system (UMTS) that is based on wideband CDMA, while future 5G systems are expected to be a further expansion of the 4G LTE architecture.

Of course, there are many other forms of RF communication besides commercial cellular. FHSS systems can hop on orders of microseconds (fast hop) up to 100s of milliseconds (slow hop). Alternatively, the CPM framework [38] is used when both power and spectral efficiency are of utmost importance, with Chapter 6 discussing the particular CPM standards used in aeronautical telemetry applications. Further, for embedded communication schemes the timing would presumably take on that of the host radar, such as described in Chapters 21, 22, and 23.

Where communication timing structures tend to be very strictly defined and standardized, the "talking to itself" nature of radar means that far greater diversity is observed from a temporal perspective. At one extreme, magnetron-based radars represent some of the very earliest system concepts and they may have pulsewidths on the order of 100 nanoseconds. The lower peak power of solid-state transmitters requires longer pulsewidths of 100s of microseconds to achieve commensurate *energy on target*. For pulsed instantiations the pulses often repeat, which is referred to as the *pulse repetition interval* (PRI), on the order of a few microseconds up to a few milliseconds. There is a fundamental trade-off involved with setting the PRI, or its inverse the *pulse repetition frequency* (PRF), between the maximum unambiguous range that can be achieved and the maximum unambiguous Doppler. While there are signal-processing techniques to address these limitations, a system is generally considered to be *low PRF* if it is unambiguous in range for all target ranges of interest, *high PRF* if it is unambiguous in Doppler for all target Dopplers of interest, and *medium PRF* if both range and Doppler ambiguities exist.

In contrast to pulsed modes, continuous wave (CW) systems emit a signal containing no breaks. A prominent example is frequency-modulated continuous wave (FMCW) in which a repeated chirping signal is generated that could repeat on the order of microseconds up to milliseconds depending on the application. These systems tend to be used for shorter-range applications than their pulsed counterparts and must contend with the fact that the receiver is operating at the same time as the transmitter, thus necessitates very good isolation between them. It is particularly noteworthy that FMCW is the mode of choice for automotive radar and is being increasingly employed for collision avoidance and blind-spot detection. FMCW radar is likewise expected to play a role in collision avoidance for unmanned aircraft systems [39].

Given the above typical temporal values for communications and radar, it is interesting to consider the different regimes in which these systems may interact. At one extreme, Figure 1.2 illustrates a short to moderate radar pulse width relative to the duration of a generic single-stream communication signal corresponding to a data rate on the order of Mbps as well as the symbol duration for the LTE standard. In these cases, as long as the communication system has sufficient receiver front-end protection (e.g. power limiter), the radar pulse would appear like shot noise that could be readily addressed using forward error correction (FEC) and interleaving, both of which tend to be standard attributes of modern communications. However, this manner of radar operation may at most have a time–bandwidth (*BT*) product on the order of 100 or so (perhaps much less), thus necessitating reliance on spatial

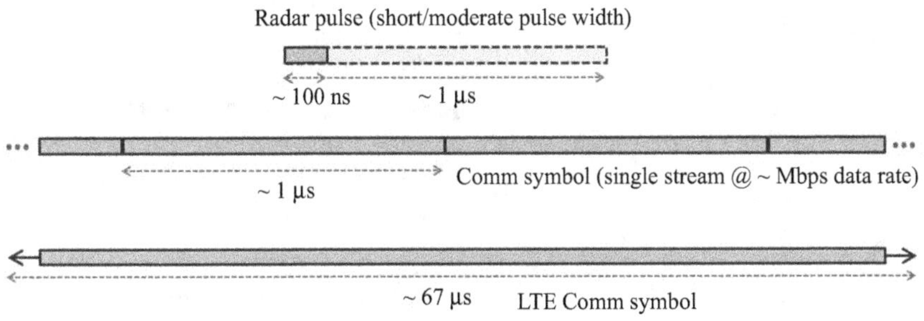

Figure 1.2 Temporal relationships between communications and short/moderate pulse radar

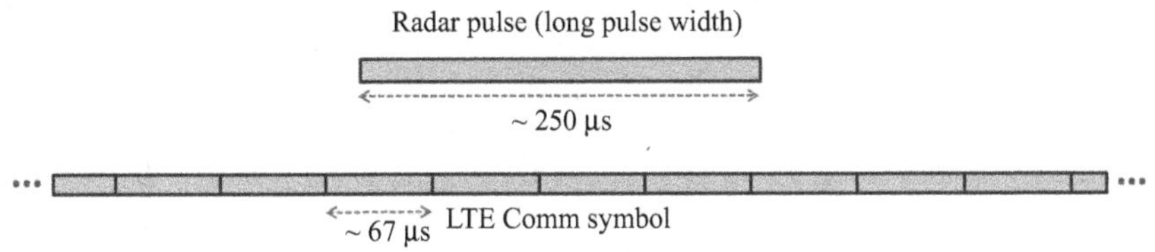

Figure 1.3 Temporal relationships between communications and long pulse radar

separability and arrangements such as DFS to ensure that interfering systems avoid the radar.

Alternatively, Figure 1.3 depicts the relationship between a long radar pulse on the order of 100s of microseconds (typical when high-range resolution is required) and an LTE symbol; the much shorter symbol intervals of the ~Mbps signal from Figure 1.2 are now too short to meaningfully show. For these modes, the radar tends to employ a much higher *BT* that could be well into the thousands, thereby potentially aiding separability in the radar receiver. While the peak transmit power for the radar also tends to be lower when such longer pulses are employed, the fact that the radar emission now overlaps a few (for LTE) to many (for the Mbps signal) symbols could hinder the efficacy of FEC and interleaving. Hence, spread spectrum communications like CDMA may be better suited to this manner of spectrum sharing.

The last relationship to consider is illustrated in Figure 1.4 and involves FMCW radar. Because it is "always on," the radar provides persistent interference to the communication system, though such radars also tend to operate at a much lower peak power (relative to pulsed radars) for the same reason. Consequently, this arrangement could prove to be the most manageable and robust form of spectrum sharing as long as both the radar and communication systems involved possess sufficient dimensionality to facilitate separability; again suggesting spread spectrum schemes like CDMA.

Another temporal factor for communications is the *delay spread*, which is a measure of the time interval over which a received signal arrives due to the presence of multipath in the channel and is typically on the order of a few microseconds for

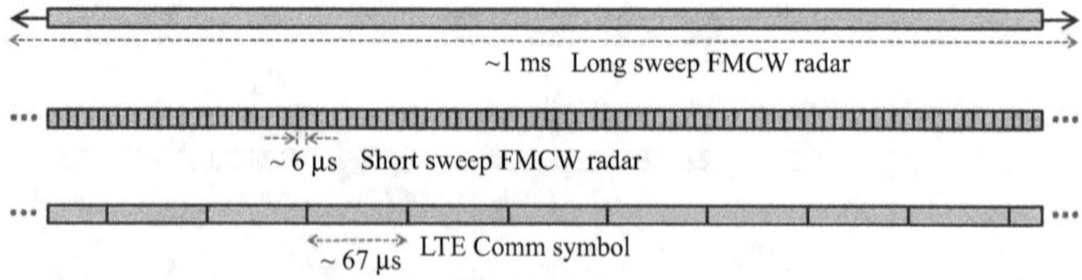

Figure 1.4 Temporal relationships between communications and long-sweep/short-sweep FMCW radar

macro-cells and 100s of nanoseconds for the much smaller pico-cells. The delay spread is also inversely proportional to the *coherence bandwidth* that is used to determine whether a signal experiences flat fading or frequency selective fading (the latter occurs if the signal bandwidth sufficiently exceeds the coherence bandwidth). It should be noted that from a phenomenological perspective, multipath is essentially the same scattering that radar measures. Of course, the particular geometry of the communication transmitter and receiver, with multipath in between, is referred to as the *forward scatter* regime in radar and implies a bistatic arrangement where little to no range or Doppler information are available (see Chapter 7). In contrast, most radars operate in the backscatter regime.

A final comment on the temporal perspective concerns the notion of latency. Along with reliability, constraining latency is becoming an important topic of research in communications for applications in which response lag could be detrimental to the user experience [8]. Perhaps the most prominent example is massively multiplayer online gaming, though the emergence of telemedicine and the impact it could have on the developing world is a far more serious matter. One would certainly not want to be on the operating table when the surgeon half-a-world away experiences latency problems.

With the increasingly congested and complex environments in which radars operate, they too are seeking to address latency issues in the context of cognitive radar. For instance, if the disposition of in-band interference sources change, the radar may need to rapidly reallocate the locations of spectral notches to avoid severe SINR degradation [40]. Further discussion of the impact of latency on cognitive radar can be found in Chapter 15.

1.4.5 Doppler perspective

In communications, Doppler effects due to radial motion between the transmitter and receiver are generally viewed as a nuisance that impacts the rate of fading. The *Doppler spread*, which is the difference between the minimum and maximum possible Doppler values over the collection of multipath components, is inversely proportional to the *coherence time* over which the channel fading can be considered invariant. For OFDM systems like LTE, the Doppler frequency shift can also cause interchannel interference. For example, the LTE standard was established to support speeds up to 300 km/h to enable operation in the "High-Speed Train" scenario.

For radar, Doppler can be viewed as both a nuisance and a useful discriminant. The former generally refers to *fast-time* Doppler because it occurs during a single pulse (or FMCW sweep). For radars employing pulse compression, which essentially all do except for those employing very short pulses or unmodulated CW, the first processing stage in the receiver is the matched filter to the transmitted waveform. Sufficient radial motion between the transmitter and receiver imposes a Doppler frequency shift onto the reflected waveform that, after matched filtering, introduces a mismatch loss that translates into lower signal-to-noise ratio (SNR) and thus degraded sensitivity to the given target. While a bank of Doppler-tuned matched filters could conceivably be employed, the implementation complexity increases accordingly, which may not be feasible. This effect plays a contributing role to the widespread deployment of linear frequency-modulated (LFM) waveforms, which are *Doppler tolerant* in that there is a graceful degradation in terms of mismatch loss as the magnitude of Doppler increases.

In contrast, the notion of *slow-time* Doppler that is observable over a set of pulses (or FMCW sweeps) denoted as the coherent processing interval (CPI) serves as a fundamental discriminant for certain radar modes. For example, MTI modes rely on slow-time Doppler to differentiate moving targets from stationary background clutter, where scattering from the latter may be several orders of magnitude larger, thus necessitating the use of clutter cancellation techniques on receive. Alternatively, SAR is a high-resolution imaging mode that achieves fine range resolution in the *cross-track* (radial) direction by emitting wideband waveforms (100s of MHz up to several GHz) and in the *along-track* direction (e.g. platform flight path) by forming a synthetic aperture that could be considerable longer than an antenna that could feasibly be mounted on the platform. Relative to a single point in the illuminated field of view, the Doppler changes as the platform traverses past, thereby enabling the construction of a *Doppler phase history* response that can be used to coherently process the received data (after pulse compression) to achieve the fine spatial resolution corresponding to the long synthetic aperture.

One of the most important evaluation tools for radar is the delay-Doppler ambiguity function proposed by Woodward in 1953 [41], which describes the matched filter response as a function of delay (relative to a matched reference point at a delay of 0) and Doppler shift of the reflected version of the emitted waveform captured at the receiver. Various measures of this function such as peak sidelobe level (PSL) and integrated sidelobe level (ISL) serve as the predominant means of optimizing radar waveforms, noting the important attribute that the volume under the surface of the normalized ambiguity function (in its squared embodiment [42]) remains a constant, thus realizing a *conservation of ambiguity* [43]. There are also bistatic [37] and multistatic [44] instantiations of the ambiguity function that necessitate inclusion of spatial geometry as well.

1.4.6 Modulation and coding perspective

In general, communication systems have allocated resources to users according to three schemes: (1) frequency-division multiple access (FDMA) in which users are apportioned separate frequency channels, (2) time-division multiple access (TDMA) where users are assigned separate time slots, and (3) CDMA for which unique spreading codes are allocated to different users that are sufficiently separable from other codes so that receive separability can be achieved. Of course, the variations and combinations of these arrangements are considerably more diverse. For instance, the OFDM framework that is the basis for LTE can be viewed as the extreme limit of packing together signals within an FDMA context, thereby realizing high capacity. Alternatively, CPM is essentially a form of digital FM that can be viewed as an extreme case with regard to achieving high power efficiency (it maintains constant amplitude) and very good spectral containment for a single user.

Focusing attention on modern digital communications, the information-bearing symbols can be organized in a variety of constellations that can be classified as amplitude-shift keying (ASK), frequency-shift keying (FSK), and phase-shift keying (PSK), or some combination thereof. For instance, 3G and earlier standards employed forms of PSK comprised of two symbols (binary or BPSK), four symbols (quadrature or QPSK), or eight symbols (8PSK), with a differential form denoted as differential phase-shift keying (DPSK) commonly used to avoid phase ambiguities. While the higher number of symbols realizes a higher data rate, the closer proximity of adjacent symbols in terms of Euclidean distance also produces a higher rate of symbol decision errors. In contrast to these constant amplitude constellations, 4G standards have adopted quadrature amplitude modulation (QAM) schemes, with 4, 16, and 64 symbols being the customary arrangements for LTE. For applications in which control of peak to average power ratio (PAPR) is a more pressing concern, amplitude/phase-shift keying (APSK) that is comprised of concentric circles of symbols is presently being explored [45] as a means to attain higher capacity.

Beyond the notion of individual symbols, the idea of *information coding* envisioned by Shannon [46] established the ground work for a variety of schemes that fundamentally provide a trade-off between the probability of receive error and the achievable data rate, with examples including linear block codes, cyclic redundancy check codes, and convolutional codes. As opposed to the Euclidean distance that is desired to be maximized between symbols, it is the Hamming distance that is maximized between *code words*. Generalizations of these structures denoted as space-time codes have likewise been developed for MIMO implementations.

Given the complexity of communication signal structures, radar waveforms might at first glance seem almost simplistic by comparison, particularly since the single most widely used radar waveform is the LFM *chirp* that involves a linear sweep between some start and end frequencies. However, it is important to understand that LFM possesses a useful Doppler tolerance property (as discussed above), is constant amplitude and thus can be produced at high power, is relatively easy to generate for very wide bandwidths (several GHz), and is inherently amenable to receiver *stretch processing* that permits sampling at a rate far below what Nyquist would require for the bandwidth of the received signal (can be thought of as a waveform-dependent precursor to *compressive sampling*). While LFM realizes relatively poor sidelobes following receiver matched filtering (the peak sidelobe is only about 13 dB below the mainlobe), it is also easy to apply receive tapering that reduces sidelobes by tens of dB in trade for some degradation in range resolution and SNR.

Further, while LFM may be widely used for the reasons stated above, nonlinear FM (NLFM) represents an attractive alternative to LFM that can realize the low sidelobes achieved by receive tapering without suffering the attendant SNR loss, though without quite the degree of Doppler tolerance achieved by LFM. Binary codes implemented with derivative PSK [47] or the biphase-to-quadriphase transformation [48] are also commonly deployed. Their generalization to polyphase codes have been the subject of considerable research, though their widespread deployment has previously been limited by the lack of an implementation scheme that permits their use in high-power transmitters. That said, it has recently been shown that a form of CPM denoted as polyphase coded FM (PCFM) can be used to optimize physically realizable FM waveforms via an underlying polyphase coding [49,50].

In fact, under the umbrella of what is referred to as *waveform diversity* [43, 51–53], the digital revolution that so transformed communications is likewise doing so for radar through high-fidelity AWG capabilities, real-time spectrum analyzers, the emergence of high-performance parallel computing, and the availability of high-capacity memory storage. A whole host of new sensing schemes such as waveform agility, cognitive radar, and MIMO radar are the direct result of these enabling technologies, as are many of the radar/communication spectrum-sharing approaches described in this book.

1.4.7 Polarization perspective

Figure 1.5 shows the Poincaré sphere, which represents all possible polarization states as points on or within the sphere. Of particular note is that orthogonal polarization states are diametrically opposed on the sphere, thus making it a useful tool to characterize modes that are separable.

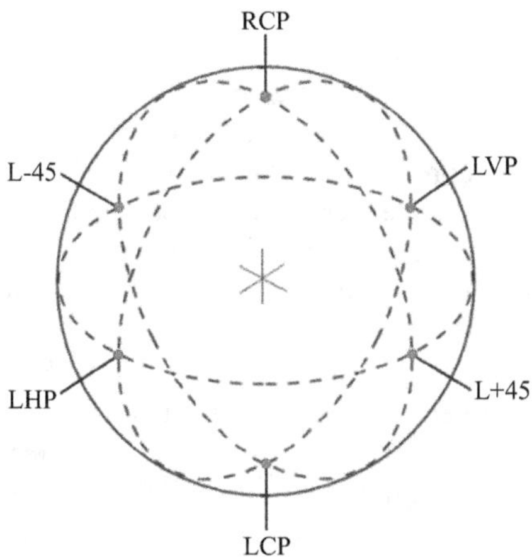

Figure 1.5 Poincaré sphere to represent polarization state. RCP: right-hand circular, LCP: left-hand circular, LHP: linear horizontal, LVP: linear vertical, L-45: linear with −45° tilt, L+45: linear with +45° tilt

Using these antipodal representations, some communication schemes leverage the concept of polarization multiplexing similar to the notion of spatial multiplexing that arises in MIMO communications as a means to improve spectral efficiency. Likewise, polarimetric radar exploits the diverse phenomenology captured by dual-polarized operation for a variety of remote sensing applications.

From a transmitter-centric perspective, it would initially stand to reason that polarization may be a useful dimension in which shared-spectrum radar and communication could be separated. However, ideal polarization is not possible, and thus the imperfect *polarization purity* would induce some leakage interference between these different modes. Further, both radar backscattering and communication multipath impose random polarization transformations, thereby further limiting separability. Consequently, while it is always dangerous to predict that something cannot be done, it may be a relatively safe bet to say that radar and communication spectrum sharing based solely or even significantly upon the polarization dimension is unlikely.

1.4.8 Separability

Finally, no discussion of spectrum sharing would be complete without consideration of separability. While co-design schemes such as in Chapters 21, 22, and 23 involve a multi-function arrangement where some mechanism for separation may be intentionally incorporated, the sharing of spectrum by completely independent systems requires some means to isolate the desired signal from the other, which may simply be treated as interference. From a general standpoint, the notion of separability can be thought of in terms of three main characteristics: dimensionality, fidelity, and dynamic range.

Dimensionality refers to the inherent complexity of a signal as well as the available degrees of freedom with which one can discriminate one signal from another. As noted in Section 1.4.4 of this chapter, radar waveforms that possess a high time-bandwidth (BT) product should be more easily separable from shared spectrum interference than those with a low BT. The simple reason for this difference is that BT directly corresponds to the coherent processing gain of the waveform. A similar argument can be made for beamforming using a higher number of antenna array elements or Doppler processing using more pulses. In fact, these different dimensions can also be combined to obtain multiplicative increases in dimensionality. A prominent example in radar is STAP from a moving platform where the spatial and Doppler channels are combined to facilitate cancellation of the clutter, which is likewise coupled in spatial angle and Doppler due to the platform motion [54].

More than just the attendant higher gain, though, is the associated increase in degrees of freedom that higher signal dimensionality provides so that adaptive receive filtering can suppress interference. Of course, doing so requires either some statistical (usually second-order) or structural knowledge of the interfering signal. The former is generally obtained from other received data and is the mechanism by which many adaptive interference cancellers like STAP operate. In contrast, the use of structural knowledge of interference signals does not tend to be as widely useful, though given the inherent standardization of commercial cellular it could conceivably be applicable to some spectrum sharing arrangements [55].

A last note on dimensionality pertains to determining the utility of a given signal domain as a means of separation. In other words, one or both of the signals must possess some discriminating feature that allows them to be separated. Returning again to the STAP example, the desired echoes from moving targets can be extracted from multiple orders-of-magnitude greater interference produced by stationary ground clutter via adaptive processing by exploiting their different angle-Doppler signal structures.

Even with an identified discriminant, signals may not be separable if the receiver representations of those signals do not possess sufficient *fidelity*. While high dimensionality can provide significant coherent integration gain to enhance SNR and provide modest separability from another shared-spectrum signal, it also tends to be rather forgiving with respect to small errors in the matching filter structure, where at most a modest mismatch loss penalty may be incurred. In contrast to coherent integration, the process of cancellation tends to require a more accurate representation of interference to be rejected. For example, it is well known that successive interference cancellers that estimate and subsequently remove successively smaller sources of interference experience an effect known as *error propagation* in which residual interference from previous stages was not perfectly removed [7]. Of course, it is not possible in practice to realize a perfect model of any interference effect; it is just a matter of representing it with as high a fidelity as possible.

Lastly, given adequate discriminating dimensionality and sufficient fidelity to enable interference suppression, good separation still may not be possible if the receiver does not possess enough *dynamic range* to capture both the desired signal and interference without saturating. The general rule based on quantization noise is that each additional receive digitization bit provides roughly 6 dB of increased dynamic range. Further, with the most significant quantization bit set to correspond to the highest amplitude-received signal, the least significant bit should be set enough lower than the expected noise floor that the thermal noise dominates the quantization noise [56–58]. Doing so permits coherent integration to extract signals out of the interference-plus-noise to the degree possible by the available dimensionality and fidelity.

1.5 Conclusions

The above introduction serves as a starting point from which to delve deeper into spectrum sharing. Given the vast diversity of the topic, it is without doubt that some fundamental characteristics, important configurations, and/or sharing arrangements have been omitted. With the remainder of this book as a (hopefully) useful starting point, we leave it to the reader to discover these as yet unknown innovations.

Acknowledgments

The authors would like to sincerely thank Prof. Christopher Allen of the University of Kansas for his great help and feedback in establishing the scope of some of the radar parameters.

References

[1] M.A. Richards, J.A. Scheer, and W.A. Holm, *Principles of Modern Radar, vol. I: Basic Principles*, Edison, NJ: SciTech Publishing, 2010.

[2] W.L. Melvin and J.A. Scheer, *Principles of Modern Radar, vol. II: Advanced Techniques*, Edison, NJ: SciTech Publishing, 2013.

[3] W.L. Melvin and J.A. Scheer, *Principles of Modern Radar, vol. III: Radar Applications*, Edison, NJ: SciTech Publishing, 2014.

[4] http://www.radartutorial.eu/19.kartei/ka01.en.html

[5] J. Proakis and M. Salehi, *Digital Communications, 5th ed.*, New York, NY: McGraw-Hill, 2007.

[6] A.J. Viterbi, *CDMA: Principles of Spread Spectrum Communication*, Boston, MA: Addison-Wesley, 1995.

[7] S. Verdu, *Multiuser Detection*, Cambridge: Cambridge University Press, 1998.

[8] J.G. Andrews, S. Buzzi, W. Choi, *et al.*, "What will 5G be?," *IEEE Journal of Selected Areas in Communications*, vol. 32, no. 6, pp. 1065–1082, June 2014.

[9] D.J. Daniels, *Ground Penetrating Radar*, 2nd ed., London: The Institution of Engineering and Technology, 2007.

[10] M.E. Davis, *Foliage Penetration Radar: Detection and Characterization of Objects Under Trees*, Edison, NJ: SciTech, 2012.

[11] G. Fabrizio, *High Frequency Over-the-Horizon Radar: Fundamental Principles, Signal Processing, and Practical Applications*, New York, NY: McGraw-Hill, 2013.

[12] Recommendation ITU-R M.1652-1, "Dynamic frequency selection in wireless access systems including radio local area networks for the purpose of protecting the radiodetermination service in the 5 GHz band," May 2011.

[13] IEEE Standard 802.22.2-2012, "Telecommunications and information exchange between systems: wireless regional area networks (WRAN)," September 2012.

[14] A.M. Wyglinski, M. Nekovee, and T. Hou, *Cognitive Radio Communications and Networks*, Burlington, MA: Elsevier, 2010.

[15] H.D. Griffiths and C.J. Baker, *An Introduction to Passive Radar*, Norwood, MA: Artech House, 2017.

[16] B.M. Horton, "Noise-modulated distance measuring systems," *Proceedings of the IRE*, vol. 47, no. 5, pp. 821–828, May 1959.

[17] S.R.J. Axelsson, "Random noise radar/sodar with ultrawideband waveforms," *IEEE Transactions on Geoscience & Remote Sensing*, vol. 45, no. 5, pp. 1099–1114, May 2007.

[18] M. Malanowski and K. Kulpa, "Detection of moving targets with continuous-wave noise radar: theory and measurements," *IEEE Transactions on Geoscience & Remote Sensing*, vol. 50, no. 9, pp. 3502–3509, September 2012.

[19] J. Jakabosky, S.D. Blunt, and B. Himed, "Waveform design and receive processing for nonrecurrent nonlinear FMCW radar," *IEEE International Radar Conference*, Arlington, VA, May 2015.

[20] J. Jakabosky, S.D. Blunt, and B. Himed, "Spectral-shape optimized FM noise radar for pulse agility," *IEEE Radar Conference*, Philadelphia, PA, May 2016.

[21] S.D. Blunt, M.R. Cook, and J. Stiles, "Embedding information into radar emissions via waveform implementation," *International Waveform Diversity and Design Conference*, August 2010.

[22] http://www.3gpp.org/ftp/Specs/html-info/36-series.htm

[23] D. Torrieri, *Principles of Spread-Spectrum Communication Systems*, New York, NY: Springer, 2018.

[24] D.R. Wehner, *High-Resolution Radar*, 2nd ed., Norwood, MA: Artech House, 1995.

[25] A. Agrawal, "Heterogeneous networks: a new paradigm for increasing cellular capacity," Stanford seminar, January 2009.

[26] S. Yamano, H. Higashida, M. Shono, *et al.*, "76 GHz millimeter wave automobile radar using single chip MMIC," *Fujitsu Ten Technical Journal*, no. 23, 2004.

[27] https://www.metageek.com/training/resources/understanding-rssi.html

[28] S.I. Tsunoda, F. Pace, J. Stence, *et al.*, "Lynx: a high-resolution synthetic aperture radar," *IEEE Aerospace Conference*, Big Sky, MT, USA, March 2000.

[29] B. Vucetic and J. Yuan, *Space-Time Coding*, Hoboken, NJ: John Wiley & Sons, 2003.

[30] P. Viswanath and D. Tse, "Sum capacity of the vector Gaussian broadcast channel and uplink–downlink duality," *IEEE Transactions on Information Theory*, vol. 49, no. 8, pp. 1912–1921, Aug. 2003.

[31] E.G. Larsson, O. Edfors, F. Tufvesson, and T.L. Marzetta, "Massive MIMO for next generation wireless systems," *IEEE Communications Magazine*, vol. 52, no. 2, pp. 186–195, February 2014.

[32] I.G. Cumming and F.H. Wong, *Digital Processing of Synthetic Aperture Radar Data: Algorithms and Implementation*, Norwood, MA: Artech House, 2005.

[33] J. Li and P. Stoica, *MIMO Radar Signal Processing*, Hoboken, NJ: John Wiley & Sons, 2009.

[34] J. Bergin and J.R. Guerci, *MIMO Radar: Theory and Application*, Norwood, MA: Artech House, 2018.

[35] G.J. Frazer, "Experimental results for MIMO methods applied in over-the-horizon radar," *IEEE Aerospace & Electronic Systems Magazine*, vol. 32, no. 12, pp. 52–69, December 2017.

[36] F. Daum and J. Huang, "MIMO radar: snake oil or good idea?," *IEEE Aerospace & Electronic Systems Magazine*, vol. 24, no. 5, pp. 8–12, May 2009.

[37] N.J. Willis and H.D. Griffiths, *Advances in Bistatic Radar*, Raleigh, NC: SciTech Publishing, 2007.

[38] J.B. Anderson, T. Aulin, and C.-E. Sundberg, *Digital Phase Modulation*, New York, NY: Springer, 1986.

[39] L. Shi, C. Allen, M. Ewing, *et al.*, "Multichannel sense-and-avoid radar for small UAVs," *IEEE/AIAA Digital Avionics Systems Conf.*, Syracuse, NY, USA, October 2013.

[40] J.W. Owen, B. Ravenscroft, B.H. Kirk, *et al.*, "Experimental demonstration of cognitive spectrum sensing & notching for radar," *IEEE Radar Conference*, Oklahoma City, OK, USA, April 2018.

[41] P.M. Woodward, *Probability and Information Theory, with Applications to Radar*, Pergamon Press, Oxford, 1953.

[42] N. Levanon and E. Mozeson, *Radar Signals*, Hoboken, NJ: Wiley-IEEE Press, 2004.

[43] S.D. Blunt and E.L. Mokole, "An overview of radar waveform diversity," *IEEE Aerospace & Electronic Systems Magazine*, vol. 31, no. 11, pp. 2–42, November 2016.

[44] V.S. Chernyak, *Fundamentals of Multisite Radar Systems*: *Multistatic Radars and Multiradar Systems*, Amsterdam, The Netherlands: Gordon and Breach Science Publishers, 1998.

[45] S. Pathak and E. Perrins, "LDPC-coded APSK for aeronautical telemetry,"' *International Telemetering Conference*, Las Vegas, NV, USA, October 2017.

[46] C.E. Shannon, "A Mathematical Theory of Communication," *Bell System Technical Journal*, vol. 27, no. 3, pp. 379–423, July 1948.

[47] H.H. Faust, B. Connolly, T.M. Firestone, R.C. Chen, B.H. Cantrell, and E. Mokole, "A spectrally clean transmitting system for solid-state phased-array radars," *IEEE Radar Conference*, Philadelphia, PA, April 2004.

[48] J.W. Taylor and H.J. Blinchikoff, "Quadriphase code—A radar pulse compression signal with unique characteristics," *IEEE Transactions on Aerospace & Electronic Systems*, vol. 24, no. 2, pp. 156–170, May 1988.

[49] S.D. Blunt, M. Cook, J. Jakabosky, J. de Graaf, and E. Perrins, "Polyphase-coded FM (PCFM) radar waveforms, part I: Implementation," *IEEE Transactions on Aerospace & Electronic Systems*, vol. 50, no. 3, pp. 2218–2229, July 2014.

[50] S.D. Blunt, J. Jakabosky, M. Cook, J. Stiles, S. Seguin, and E.L. Mokole, "Polyphase-coded FM (PCFM) radar waveforms, part II: Optimization," *IEEE Transactions on Aerospace & Electronic Systems*, vol. 50, no. 3, pp. 2230–2241, July 2014.

[51] M. Wicks, E. Mokole, S.D. Blunt, V. Amuso, and R. Schneible, eds., *Principles of Waveform Diversity & Design*, Raleigh, NC: SciTech Publishing, 2010.

[52] S. Pillai, K.Y. Li, I. Selesnick, and B. Himed, *Waveform Diversity: Theory & Applications*, New York, NY: McGraw-Hill, 2011.

[53] F. Gini, A. De Maio, and L.K. Patton, *Waveform Design and Diversity for Advanced Radar Systems*, London, UK: IET Press, 2012.

[54] W.L. Melvin, "A STAP overview," *IEEE Aerospace & Electronic Systems Magazine*, vol. 19, no. 1, pp. 19–35, January 2004.

[55] H. Griffiths, S. Blunt, L. Cohen, and L. Savy, "Challenge problems in spectrum engineering and waveform diversity," *IEEE Radar Conference*, Ottawa, Canada, April/May 2013.

[56] M. Bartz, "Large-scale dithering enhances ADC dynamic range," *Microwave RF*, vol. 32, pp. 192–198, May 1993.

[57] B.C. Brock, "The role of noise in analog-to-digital converters," *Sandia National Laboratories Report*, SAND96-2697, November 1996.

[58] C.T. Allen, S.N. Mozaffar, and T.L. Akins. "Suppressing coherent noise in radar applications with long dwell times," *IEEE Geoscience & Remote Sensing Letters*, vol. 2, no. 3, pp. 284–286, July 2005.

Chapter 2
The spectrum crunch – a radar perspective

Hugh Griffiths[1]

2.1 Introduction

When the greatest strides in the development of radar were made, around the middle of the twentieth century, there was little concept of microwave frequency allocation. An International Telecommunications Union (ITU) document from 1927 [1] discusses frequency allocations of 56 to 60 MHz for 'amateurs and experiments' and above 60 MHz as 'not reserved'. By 1937 [2] the maximum frequency covered by the allocations had risen to 300 MHz 'for experimental purposes'.

As radar was developed, the radar frequency bands were dictated to a large extent not only by the applications but also by the dimensions of the waveguides employed – for example, the WG16 waveguide has internal dimensions of 0.9 inch by 0.4 inch, and is still widely used at X-band. These radar bands were designated by letters, partly for convenience and partly to avoid having to reveal exact radar frequencies, and that convention has stood ever since. Table 2.1 shows the radar bands [3], their letter designations, and typical applications in each case.

As other uses of the microwave spectrum developed rapidly in the post-war years, there was a need to allocate specific microwave frequency bands to specific applications as had been done with lower radio frequencies, and this allocation was essentially straightforward. Although there were occasional instances of interference of one service with another, particularly under conditions of anomalous propagation, these were rare. However, the last decade of the twentieth century and the first decades of the twenty-first century have witnessed phenomenal growth in use of the RF and microwave spectrum for radio and TV broadcasting, cell phones, broadband Wi-Fi, and radionavigation. Whereas previously a small number of terrestrial UHF TV channels was considered adequate, now there is demand for hundreds of high-definition channels, streamable to mobile devices. It is these factors that have led to the problem that is described as the 'spectrum crunch'.

Compared to the very simple frequency allocations of the 1920s, the situation has now become enormously more complicated, both because of the much greater number and types of users and signals, and because the allocations now extend through

[1]Department of Electronic and Electrical Engineering, University College London, UK

Table 2.1 Standard radar frequency letter bands and radar-operating modes [3]

Radar frequency bands		
IEEE Band designation	**Frequency range**	**Typical usage**
HF	3–30 MHz	Over-the-horizon surveillance (ionospheric propagation); *long range and low resolution*
VHF	50–330 MHz	Long-range (line of sight) surveillance, foliage penetration (FOPEN), counter-stealth, ground penetrating; *low/medium resolution*
UHF	300–1000 MHz	Long-range surveillance, FOPEN; *low/medium resolution*
L	1–2 GHz	Long-range surveillance, long-range air traffic control; *medium resolution and small weather effects*
S	2–4 GHz	Moderate-range surveillance, terminal air traffic control, long-range weather observation, airborne early warning (AEW); *moderate weather effects in heavy precipitation*
C	4–8 GHz	Long-range tracking, weather observation, weapon location, long-range tracking; *increased weather effects in light/medium rain*
X	8–12 GHz	Short-range tracking, missile guidance, mapping marine radar, airborne intercept, battlefield surveillance, weapon location; *reduced to short range operation in rain*
K_u	12–18 GHz	High-resolution mapping, satellite altimetry, man-portable/unmanned air vehicle (UAV) radar; *short range due to water vapour absorption*
K	18–27 GHz	Police radar; *very limited use due to high water vapour absorption*
K_a	27–40 GHz	Short-range very high-resolution mapping, airport surveillance; *short range due to water vapour absorption*
V	40–75 GHz	Scientific remote sensing, *high water vapour absorption*
W	75–110 GHz	Automotive radar (77 GHz), missile seekers, very high-resolution imaging (94 GHz), *high water vapour absorption elsewhere in band*
mm	110–300 GHz	Experimental; *limited to short range due to high water vapour absorption*

microwave, millimetre-wave and even to Terahertz frequencies. The process of spectrum allocation continues to be regulated by the ITU, and this allocation is continually reviewed at an international level by the World Radiocommunication Conference (WRC), held every three to four years.

A chart that shows the frequency allocations for the USA, based on this plan, can be found on a U.S. government website [4]. This chart is detailed and complex, and worth careful study. It shows the different services (Aeronautical Mobile, Aeronautical Mobile Satellite, Aeronautical Radionavigation, Amateur,…, Space Research, Standard Frequency and Time Signal, Standard Frequency and Time Signal Satellite), all

Table 2.2 Prices paid by telecommunication companies when 3G and 4G spectrum was sold off (Figures from GSM Association – other sources vary)

3G (The UK)	£22.5.2bn (2000)
4G (The UK)	£25bn
4G (Germany)	£2.9bn
4G (Italy)	£4.1bn
4G (Spain)	£1.1bn
4G (France)	£2.9bn
4G (USA)	£12bn

colour coded. Some are allocated on an exclusive basis, and some are shared. The chart also shows which allocations are government exclusive, which are non-government exclusive, and which are government/non-government shared.

It is also instructive to consider the prices paid by telecommunications companies for spectrum in different countries (see Table 2.2). These show that spectrum has a definite – and substantial – monetary value, of something like £20 million per MHz at the current market rate.

The purpose of this chapter is to present the spectrum problem from a radar point of view, and hence to set the scene for the rest of the book. It draws on a paper published in 2015 in *IEEE Proceedings* [5], and also on the work of the NATO SET-182 Task Group on Radar Spectrum Engineering & Management which reported in 2016 [6]. Section 2.2 elaborates on the nature of the radar spectrum environment, and then Section 2.3 discusses the spectra of radar and other emissions and their imperfections. Section 2.4 explains how the spectrum is currently allocated, while Section 2.5 and 2.6 discuss, respectively, the nature of radar interference to other users, and the interference other users impose on radar, since these relationships are crucial in framing effective regulations and in developing novel technical approaches. A number of these technical approaches are introduced in Section 2.7, and are covered in much greater detail in individual chapters throughout the rest of the book. Finally, Section 2.8 draws some conclusions.

2.2 The radar spectrum environment

Radar represents a fundamentally important use of the electromagnetic (EM) spectrum, and is used for a wide range of purposes, including air traffic control, geophysical remote sensing from space of land, sea, and polar regions, automotive safety, meteorology, and surveillance and tracking for defence and security. As a sensor, it has the merits of allowing day/night all-weather operation (at frequencies below about 10 GHz) and of providing information such as target range, bearing, elevation and velocity, atmospheric measurements, on-board altimetry, long-range imaging capabilities and collision avoidance.

A specific radar will not usually occupy the total frequency allocation in any particular radar band. The bandwidth of a radar determines its ability to resolve targets in range, so the range resolution ΔR is related to the radar signal bandwidth B by $\Delta R = c/2B$, where c is the velocity of propagation (3×10^8 m/s). The bandwidth that an individual radar occupies depends on the range resolution required, as well as on the need to reduce the potential for interference with other radars in the same band by operating on different frequencies. In some situations there may be a large number of radars operating in close vicinity. As an example, in a busy shipping lane within, say, a 20 nmi radius, there may be many tens of large ships, fishing vessels, and recreational craft, each with a radar operating in S-band or X-band. There might be a civil aviation flight path above, with each aircraft operating a weather radar in X-band. It is easy to imagine that in such a scenario there might be more than one hundred emitters simultaneously detected by a wideband receiver.

The higher frequency bands provide some advantages to radar. For a fixed fractional bandwidth (bandwidth divided by centre frequency, expressed as a percentage) increasing the operating frequency subsequently increases the achievable bandwidth, thus providing finer range resolution. Further, for a fixed angular beamwidth, the antenna size decreases as the wavelength is reduced (antenna azimuth beamwidth is proportional to λ/D, where λ is the RF wavelength and D is the antenna aperture width). This relation is particularly important for airborne radars, where the physical space available for antennas may be severely restricted.

However, at higher radar frequencies the effect of weather becomes more significant, so radar sensing in the higher bands tends to be limited to shorter-range applications such as airport surface movement surveillance, automotive collision avoidance, and police speed radar. There are also some specific advantages associated with lower frequency bands, such as long-range over-the-horizon propagation via ionospheric reflection (at HF), foliage penetration (at VHF and UHF), and long-range terrestrial applications such as surveillance and tracking, air traffic control, and meteorological monitoring (at L-, S- and C-band). The choice of radar frequency therefore involves a trade-off between a number of factors.

For airborne radars, the space available for antennas, particularly in fast jets, is quite limited. The use of X-band therefore represents a compromise between a reasonable antenna size on the one hand, and on the other the ability to operate in poor weather conditions at long ranges [7].

In the HF (short-wave) bands, the primary applications are communications and broadcasting, and the radar allocations are interleaved or shared. These frequencies provide very long range operation (thousands of kilometres), but the propagation conditions do vary with the time of day, time of year, and time in the sunspot cycle, so the choice of frequency needs to be made adaptively. There have previously been no specific frequency allocations for HF radars, with HF radar operation designated as a secondary user, having to avoid interference with primary users. However, the 2012 World Radiocommunication Conference did propose a number of primary allocations for the radiolocation service between 3 and 50 MHz to support high-frequency oceanographic radar operations [8].

At VHF and UHF frequencies the spectrum is mostly used for communications and broadcasting, and is also very crowded. Again, radar operation in these bands is generally on a secondary basis. At microwave frequencies (L-band and above), there are radar allocations as a primary user, but there is an increasing number of applications putting pressure on the available bandwidth. Typical applications around the radar L- and S-bands include mobile telecommunications, in particular, and also wireless LAN, Bluetooth and GPS. At higher microwave frequencies, up to about 30 GHz, other uses include radio astronomy, microwave communication links, satellite television broadcasting and communication satellites. These radar applications in the various frequency bands have been summarised in Table 2.1.

2.3 Signal spectra

2.3.1 Spectra of practical emissions

Academic textbooks tend to suggest that signals – of all kinds – are ideal, clean, and spectrally pure. In practice this is not the case, both because the modulation sidebands extend well beyond the nominal bandwidth, and because of distortion products generated in the transmitter. Indeed, the classical tool used to show the performance of a radar waveform – the ambiguity function [9] – says nothing about its spectral properties.

A somewhat extreme example is illustrated in Figure 2.1, which shows the measured emission spectrum from an X-band magnetron transmitter used in a marine navigation radar. The pulse length is 100 ns, giving a nominal bandwidth of 10 MHz. However, the figure clearly shows that the emitted energy extends well beyond this bandwidth, and in fact there is radiation over the whole of X-band. In dealing with the spectrum congestion problem there is therefore great scope for techniques that can generate spectrally cleaner signals of all kinds.

2.3.2 Radar emissions

The emissions from radar systems are modulated to permit the measurement of target range and Doppler shift, and to resolve distinct scatterers. The nature of this modulation depends on the requirements of range resolution, Doppler resolution, maximum unambiguous range and separation of Doppler ambiguities. Most radars emit a regular sequence of pulses at a repetition rate known as the pulse repetition frequency (PRF), and the choice of pulse length and PRF depends on a number of factors. Some important relations here are:

Range resolution
$$\Delta R = \frac{c\tau}{2} = \frac{c}{2B} \tag{2.1}$$

Maximum unambiguous range
$$R_{\text{max}} = \frac{c}{2 \times \text{PRF}} \tag{2.2}$$

Pulse energy
$$E = P_{\text{T}}\tau \tag{2.3}$$

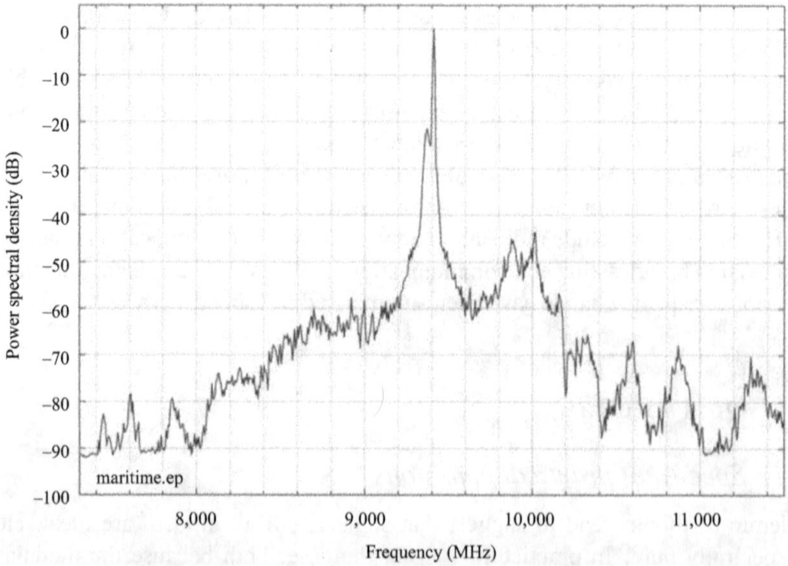

Figure 2.1 Power spectral density of the measured emission from a radar using a pulsed magnetron (Furono Model 1953C X-band maritime surface radar), with a pulse length of 100 ns giving a nominal bandwidth of 10 MHz. Reproduced, with permission, from [5]

$$\text{Average transmitted power} \qquad P_{av} = P_T\tau \times \text{PRF} \tag{2.4}$$

$$\text{Echo Doppler shift} \qquad f_D = \frac{2vf_0}{c} \tag{2.5}$$

$$\text{Spacing of Doppler ambiguities} \qquad f_{max} = \frac{1}{\text{PRI}} \tag{2.6}$$

where τ is the pulse duration, v is the radial velocity of the target with respect to the radar, f_0 is the radar carrier frequency and PRI is the pulse repetition interval (the time from the beginning of one pulse to the beginning of the next pulse).

The maximum unambiguous range occurs because any echo that arrives after the transmission of the next pulse cannot unambiguously be associated with the original pulse. Doppler ambiguities occur because of the sampling of the Doppler shift at the PRF, so if the one-way range to the target changes by a multiple of a half-wavelength during one PRI, the phase of the echo stays constant and the target appears to be stationary. Decreasing the PRF does increase the maximum unambiguous range, but it also reduces the spacing of Doppler ambiguities.

The pulse duration τ may be as short as 100 ns to longer than 100 μs depending on the application, and the PRF may likewise vary from around 300 Hz or even lower to 100 kHz or more. The duty cycle (the ratio of pulse duration τ to the PRI) can be much lower than 1% to as high as 30%, which means that the radar spends most of the PRI 'listening' for echoes from targets, during which time it requires exclusive access to the relevant part of the spectrum in time, frequency and space.

Pulse compression is frequently used in modern radars, transmitting a longer pulse to increase the pulse energy, and hence the maximum detection range of the radar. Pulse compression operates by modulating the radar pulse, either by linear frequency modulation ('chirp') or by digital coding [10], followed by appropriate matched or mismatched filtering of the receive echoes to achieve a range resolution much finer than the pulse duration as well as a signal-to-noise ratio (SNR) processing gain.

The other class of radar emission is known as continuous wave (CW). Here, the transmitter radiates continuously, so the duty cycle is 100%. This type of radar requires separate transmit and receive antennas with good isolation between them. The range information is obtained by sweeping the frequency of the radiated waveform, usually linearly with time. A typical frequency-modulated continuous wave radar of this kind might sweep over a bandwidth of 100 MHz in a repetition period of 2 ms. This operation can be compared to a pulsed radar that might sweep frequency over the same bandwidth within the time duration of a pulse, which might be 5 μs, for example.

Radar transmitters usually radiate a constant-power envelope (i.e. no amplitude modulation) so as to maximise the energy incident upon, and subsequently reflected from, the illuminated objects and to thereby maximise the achievable sensitivity. The use of constant envelope pulses is also motivated by the need to drive power amplifiers in saturation to obtain the best power efficiency. However, driving the amplifiers in this way also tends to cause a high degree of non-linearity that subsequently produces a broadening of the transmitted spectrum.

2.3.3 Radar transmitters

The radar transmitter is the subsystem that generates and amplifies the radar signal to the level necessary to detect target echoes at long range. The final device may use either vacuum tube or solid state technology, and the peak radiated power may range from milliwatts to megawatts, depending on the application. The simplest, and one of the earliest, high-power transmitter types is the magnetron tube, which is a high-powered pulsed oscillator. Although the magnetron is simple and inexpensive, its spectrum can be broad and 'dirty' (Figure 2.1), so is far from ideal.

Virtually all other types of radar transmitter use separate waveform generation and amplification stages that provide better control over the radiated signal. Many modern radars use digital techniques to generate the radar waveform, thus enabling precise, wide-bandwidth signals to be synthesised. The digital techniques even permit the waveform to be varied on a pulse-to-pulse basis as well as pre-distorted to compensate for the distortion introduced by power amplifier stages. This highly flexible capability is the basis for the field of *waveform diversity*, which is discussed further in Section 2.7.2 of this chapter, as well as nearly all the chapters in the second half of the book.

Commonly used radar power amplifiers based on tube technology include the klystron, travelling wave tube (TWT), and crossed-field amplifier. Klystrons can generate megawatts of peak power but are limited in bandwidth due to restrictions on their resonant cavities. TWTs provide peak powers on the order of 0.1 to 50 kW, and typically have much broader bandwidths than klystrons (up to two or three octaves). While the current work seeks to improve the spectral purity of these tube devices, the reality is that legacy systems, particularly for defence applications, will be in abundance for the next 50 years due to the long acquisition cycle for such systems and the enormous costs involved with building large modern radar systems.

Solid-state power amplifiers have also been used for several decades, principally in distributed modules as part of active electronically scanned arrays (AESAs). At frequencies below 3 GHz, silicon bipolar transistors may be used with duty cycles less than 10% to generate peak powers on the order of 100 W. At higher frequencies – usefully up to about 30 GHz – gallium arsenide devices are used. A transmit/receive module with two or more devices in parallel in an AESA radar at X-band might generate a peak power of 10 W. A small AESA may have 1,000 transmit elements yielding a total peak transmitter power of 10 kW, with duty cycles up to 20%.

As a general rule, solid-state amplifiers cannot provide the high peak power of tubes but they are usually able to sustain a much higher duty cycle, which consequently causes solid-state radars to rely on waveforms with large duty cycles to provide commensurate 'energy on target'. It should be kept in mind that, currently, solid-state amplifiers represent a small minority of the total number of operating radar systems. They do, however, offer much scope for improved spectral control.

Table 2.3 lists a number of transmitter types and properties for various radar applications. A detailed discussion on radar transmitters can be found in Chapter 11.

Table 2.3 Examples of generic radar types and their transmitters. Reproduced, with permission, from [5]

Radar band	Radar role	Transmitter type	Peak power/ Mean power	Pulse bandwidth	PRF	Duty cycle
X	Airborne maritime reconnaissance	Travelling wave tube (TWT)	50 kW/ 500 W	10–100 MHz	300 Hz–2 kHz	0.01%
Ku	Small airborne synthetic aperture radar (SAR) or ground-moving target indication radar	TWT or solid state microwave power	100 W/ 25 W	10 MHz– >1 GHz	1–5 kHz	0.25%
S	Air traffic management	Solid state	15 kW/ 1.2 kW	1 MHz	1 kHz	0.08%
X	Marine radar	Magnetron	10 kW/10 W	10 MHz	1–2 kHz	0.001%
X	Fast-jet radar	Active electronically scanned array (AESA)	5 kW/1 kW	1–100 MHz	1–10 kHz	0.1%–0.2%

2.4 Spectrum allocation

2.4.1 Competition for spectrum

The RF electromagnetic spectrum is becoming increasingly more congested as a result of rapid expansion by the commercial wireless industry and other RF applications. This growing congestion continues to apply ever-greater pressure on the parts of the spectrum formerly reserved exclusively for radar.

Since its inception in the 1950s, HF radar has always competed for spectrum with the primary users of HF communications and the amateur radio world. Since the late 1970s, the world's communication industries have shown greater interest in the UHF part of the spectrum and in 1979, the World Administrative Radio Conference (WARC) took the decision to downgrade the primacy of radar in portions of the UHF band, specifically 420–430 MHz and 440–450 MHz, to secondary status. In the language of spectrum management, downgrading to secondary status means that radars can operate only as long as they do not interfere with primary users. Over the last 10 years, wireless industries have lobbied their member nations within the ITU to downgrade radar in the 3.4–3.7 GHz band to secondary status as well. Currently, the big competitor for the 3.4–3.7 GHz band is 4G wireless communications (Worldwide interoperability for microwave access (WiMAX) or long-term evolution (LTE), though all indications are that the latter will dominate).

2.4.2 Spectrum regulation

A key part of spectrum management is how frequency use and emissions are regulated (see [11] for a broad perspective on spectrum regulation). For radar, regulation is particularly complex due to the variety of different radar modes, their necessary power outputs (which dictates the nature of the specific transmitter) and spectral emissions. Many, but not all, countries adopt ITU emission standards. In the United States, emission standards are determined by two organisations: the National Telecommunications and Information Administration (NTIA), the governing body for all U.S. Federal Government spectrum use; and the U.S. Federal Communications Commission (FCC), the regulatory authority on spectrum use by non-federal entities such as the commercial broadcasting industry. Today, the regulations cover the frequency range from 9 kHz to 1000 GHz for 40 different radio services, including radar (which falls under the classification of radiodetermination or radiolocation services), in a 1,000+ page publication [12]. These regulations can only be changed by the WRC.

Within the ITU guidelines, provisions are made for the computation of spectral masks that delineate out-of-band emissions, thereby demarcating the spectral bounds on what a real transmitter can emit. For example, Figure 2.2 shows a typical emission mask that might be applied to a radar transmission. Broadly, there is a band over which the radar is designed to transmit that is fixed in frequency, with the edges decreasing to −40 dB from the peak. Outside of this band, at lower emission levels, various 'roll-off rates' could be applied. Figure 2.2 shows two examples, one at a roll-off rate of −20 dB per decade of frequency and the other at −30 dB per decade. A roll-off

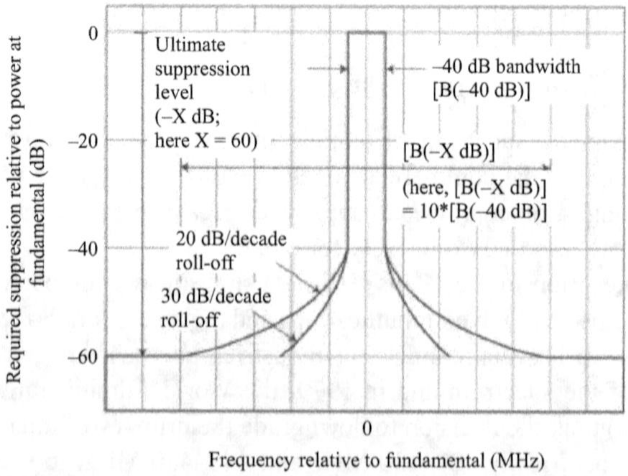

Figure 2.2 Graph of a generic ITU spectral mask showing permissible regions of operation for differing rates of spectral roll-off [11]

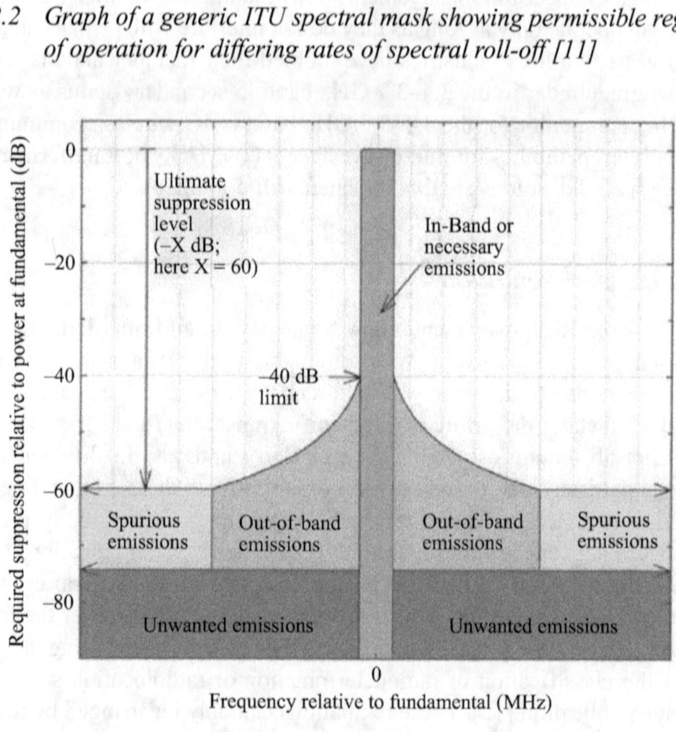

Figure 2.3 RSEC emissions mask showing the signal domains and their permitted relative levels. Reproduced, with permission, from [13]

rate of −20 dB per decade is the current standard while 30 and 40 dB per decade is presently under consideration.

The radar emissions have to reside within this mask whose power versus frequency shape is determined by the regulatory bodies. The shape of the mask has

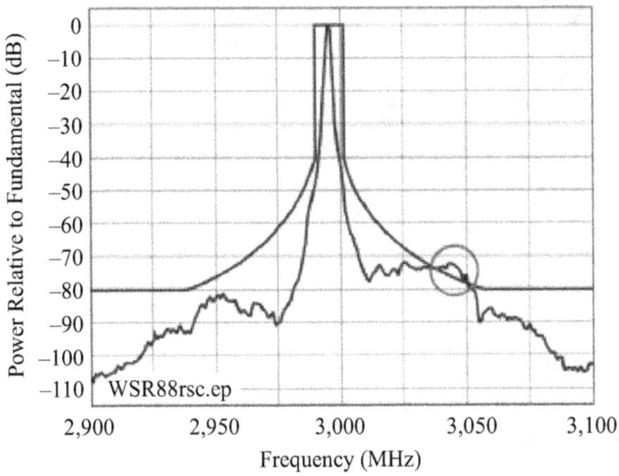

Figure 2.4 Example showing the measured emission of an unmodulated pulse from a weather surveillance radar relative the RSEC mask. At around 3,050 MHz, the emission exceeds the allowable limit for the signal domain. Reproduced, with permission, from [13]

significant implications for radar design and performance. While lessening the potential for adjacent band interference to other services, a −40 dB per decade roll-off rate poses extreme challenges to the radar designer due to the intrinsic spectral spreading that results from pulsed operation combined with the requirement for high transmit power, which tends to drive the use of tube technology and its subsequent limitations (i.e. non-linear distortion in the form of intermodulation products).

In the United States, the NTIA publishes a guide: *Manual of Regulations and Procedures for Federal Radio Frequency Management*, better known as the *Red Book* [12]. Of most relevance here is Chapter 5.5, in which the Radio Spectrum Engineering Criteria (RSEC) are defined. In the RSEC, radars are divided into five classes, A to E. This partition considers such factors as frequency coverage, peak power output, type of waveform (pulsed versus non-pulsed), and functionality (wind profiler, etc.). The RSEC determines a spectral mask based on a 40 dB bandwidth with roll-off rates that are calculated with equations according to the criteria specified in the five class designations. Figure 2.3 shows a generic example of an RSEC mask. The desired in-band radar emissions are contained within the 40 dB bandwidth as shown, with a subsequent allowable roll-off at today's −20 dB per decade down to a lower limit requirement of −60 dB at all other frequencies. The radar transmissions should not exceed the limits implied by this mask, but unwanted emissions do often occur.

Unwanted emissions from a radar transmitter are composed of out-of-band (OOB) and spurious components that are generated by non-linearities that occur within the transmitter together with the steep rise and fall times of the radar pulses. Figure 2.4 shows an example of a radar emission relative to the RSEC mask for the current roll-off rate of 20 dB per decade. The radar operates in S-band with a transmitted signal that is designed to emit between 2990 MHz and 3000 MHz only. Note that the radar emission is only marginally out of compliance in the upper sideband, i.e. it exceeds the value set by the mask. This condition necessitates that modifications be made to the transmitter so that it can meet the requirements demanded by the mask. There are potential methods to better control radar emissions and these are discussed in more detail in Section 2.6. However, the effects of such a transgression on the performance of, say, a communication system operating in an adjacent band are unclear, but will be a function of waveform and signal processing.

2.5 Radar interference to other users

2.5.1 Radar interference to other radars

The majority of radars currently in service operate in a pulsed mode, with rotating antennas having narrow transmit and receive beamwidths and low spatial sidelobes. These features help to protect against interference from other radars. Any significant interference tends to occur when two radar beams are aligned, which is usually only for a short period of time. The pulsed nature of the systems also means that unless the PRFs of the two radars are synchronised, the interference will be suppressed in the receive signal processing. Given these features and the relatively small number of radars in a local area, it has in the past usually been possible to minimise radar-to-radar interference by careful allocation of operating frequencies and control of geographical location.

For those radars employing traditional plan position indicator (PPI) detection displays, human operators are very good at identifying and addressing interference issues. Examples of interference observed on the PPI display of a non-coherent airborne radar are shown in Figure 2.5. The plots on the PPI are an overlay of relatively unprocessed radar returns (the 'raw' radar returns), which show the presence of clutter and interference before the main radar detection processing is applied. In this case, the radar was using pulse-to-pulse frequency agility. In Figure 2.5(a), some high duty cycle or CW interference can clearly be seen. This interference is manifested as distinct spokes, as only certain frequency channels in the radar receiver are affected. The interference is also only seen in the main antenna beam and near sidelobes when the radar scans past the interfering signal. Some similar interference can also been seen in Figure 2.5(b). Also observed in this image is a spiral of pulse-like returns that arose due to the presence of another pulsed radar with a low-duty cycle and a similar PRF. In this case, the interference must have been entering the radar through the antenna sidelobes, because it is visible at all azimuth angles over the display.

For a modern high-performance radar with advanced adaptive detection processing, such interference actually might not result in false detections because the radar would be able to discern the presence of interference and raise its thresholds accordingly. However, this in turn would reduce the sensitivity of the radar in the vicinity of the interference, making it difficult to detect smaller targets. Such effects can be quite insidious, with the radar losing performance without it being obvious to the operator.

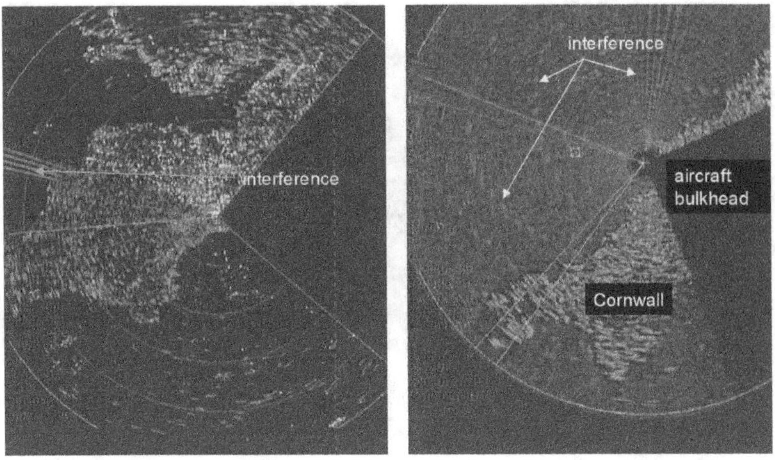

(a) (b)

Figure 2.5 Interference to an X-band airborne radar using pulse-to-pulse frequency agility: (a) high-duty cycle or CW interference; (b) high- and low-duty cycle interferences. Reproduced, with permission, from [5]

2.5.2 Radar interference to other systems

Various cases of interference by radars to commercial electronic equipment have been reported. Examples include cashpoint machines (automatic teller machines), satellite and terrestrial television receivers, and hearing aids. Sometimes this occurrence is a problem of poor electromagnetic compatibility (EMC) design where improved screening and filtering within the victim equipment would mitigate the interference. However, even with the best EMC design, some problems may be unavoidable and such interference may well increase in the future without careful spectrum planning and advances in spectrum-sharing technology.

2.5.3 WiMAX and LTE communication systems

Wireless communications, particularly the push for 4G (and eventually 5G) systems, are driving much of the worldwide pressure for spectrum [14]. The 4G systems, based on either the LTE or WiMAX standards, rely on orthogonal frequency division multiplexing (OFDM) and have variable modulation types (4, 16 and 64 quadrature amplitude modulation (QAM)) depending on channel conditions and required data rates. While the growing number of mobile users plays an important role in this demand for spectrum, it is the proliferation of bandwidth-hungry streaming video that is currently the driving factor. The LTE and WiMAX systems operate in some spectrum currently used predominantly by radar. With the availability of inexpensive RF and signal processing chip sets in S-band (specifically 3.4–3.7 GHz), the wireless community has become a competitor to radar in this band. The potential for radar to interfere with LTE and WiMAX is significant. Figure 2.6 shows a 64 QAM WiMAX

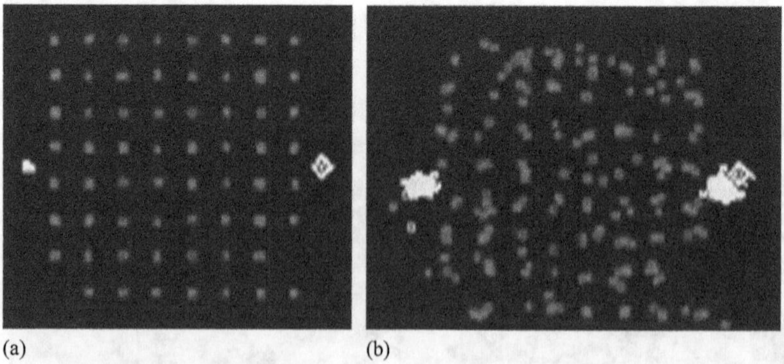

Figure 2.6 (a) 64 QAM constellation with no interference; (b) with interference from an S-band radar [15]

constellation before (Figure 2.6(a)) and during (Figure 2.6(b)) interference from an S-band radar, measured in a bench test [15]. It is clear that in the presence of radar interference, the ability of the WiMAX receiver to accurately recover transmitted symbols is severely degraded.

2.6 Interference to radar by other users

A typical radar, such as a primary surveillance radar used for air traffic control, may have a peak pulse power from the transmitter of 20 kW and an antenna with a directive gain of 30 dB, yielding an effective radiated power of +73 dBW. However, the power levels detected in the radar receiver are expected to be noise-limited, where the echo power from a distant target may be on the order of −110 dBW. In order to achieve the maximum detection performance for a given transmitter power, the radar will aim to achieve the best possible receiver sensitivity. It is usually more cost-effective to improve the receiver sensitivity than to attempt to increase the transmitter power by the same margin. For this reason, long-range radars usually employ ultra-low noise amplifiers at the receiver front end.

For many applications, radar receivers have traditionally been designed on the assumption that the radar is the exclusive user of its allocated frequency spectrum and that performance will be limited by thermal noise in the receiver. It was assumed that interference from other radars in the same band could usually be managed by careful frequency planning and site selection for ground-based radars (see also Section 2.5.1). With increasing demand for spectrum, there is growing potential for interference from other users, either those sharing the same frequency allocation or from those occupying adjacent frequencies. The effect of interference from different sources on radar performance can be quite complex to define. For example, a study of this problem was reported in [16]. This report concluded that interference at low duty cycles, such as from other radars, can often be sustained at interference-to-noise (I/N) levels as high as +30 dB to +60 dB without degrading receiver performance (i.e. its ability to detect small targets). An example of the effects of such interference on a radar display can be seen in Figure 2.5. In contrast, interference at higher duty cycles (above 1% to 3%), such as is the case with most communication signals, can cause target detection losses to begin at I/N levels between −10 dB and −6 dB, depending on the type of radar. It was noted in [16] that the loss in performance may be very insidious. No obvious effects may be visible on the radar display, but nevertheless target detections may be lost.

An area of current concern to radar users is the allocation of spectrum to communication systems such as WiMAX and LTE in the band of 2500 MHz to 2690 MHz. This band is immediately adjacent to radar allocations for air traffic control (ATC) radar at 2700 MHz to 2900 MHz, and maritime radars at 2900 MHz to 3100 MHz. Also, the U.S. National Weather Service (NWS) Next Generation Weather Radars (NEXRAD) operate in the band from 2700 MHz to 3000 MHz. It has been found that OOB signals (i.e. the spectral sidebands) from WiMAX base station transmitters can at times cause co-channel interference to radar users in adjacent bands. Equally, the OOB response of typical ATC and maritime radar receivers, which had not been designed to cope with adjacent band signals of this sort, may also render them susceptible to interference. These problems have been found to occur even though both the radars and WiMAX base stations are compliant with the appropriate national EMC standards.

This problem has been widely investigated by national regulatory authorities. In the United Kingdom, Ofcom has undertaken its own studies and commissioned independent research, including the effects of radar transmissions on adjacent WiMAX systems. This study has resulted in a proposed S-band remediation plan [17]. One outcome from this work has been to show that front-end filtering (e.g. a low-loss passive bandpass filter in the main antenna feed) in radar receivers can considerably reduce the effects of high power adjacent communication systems without discernibly degrading radar performance.

In the United States, studies have been undertaken by the National Telecommunications and Information Administration/Institute for Telecommunications Sciences (NTIA/ITS) [18]. These studies investigated sources of electromagnetic interference (EMI), how EMI is exhibited within the radar receiver, and what technical solutions could be taken to mitigate the effects. In particular, it was found that the NEXRAD radars were susceptible to interference from WiMAX. The NTIA investigation quantified power levels of the interference in NEXRAD receivers under different conditions. It also assessed the amount of de-coupling required to reduce the observed interference levels to a non-interference level, together with the amounts of decoupling required to mitigate the interference for various frequency and spatial separations. Because the interference is partly co-channel, the application of additional filtering (see discussion above) within the radar receiver does not totally eliminate the WiMAX EMI. Further mitigating procedures include careful local frequency planning (with a local frequency coordinator or manager to maximise the frequency differences between WiMAX transmitters and radars), management of the vertical beam angles of WiMAX base stations and control of WiMAX antenna locations and heights.

In short, radars will increasingly have to co-exist with other users. This necessity will require improvements to radar designs to minimise the effects of interference, together with careful frequency and site planning.

2.7 Technology developments

This section provides an introduction to a number of technology developments that might be used to address the spectrum congestion problem, and which are developed in subsequent chapters in this book.

2.7.1 Passive bistatic radar

Passive bistatic radar (PBR) [19] may be defined as a set of radar techniques that exploit existing signals, such as broadcast, communication, or radionavigation emissions as their transmitting source (Figure 2.7). PBR has a number of attractive qualities, but in the context of the spectrum problem it has the particular advantage of needing no addition spectrum because it makes use of existing transmissions. As such, it also therefore has no spectrum licensing issues. For these reasons, PBR techniques have been referred to as 'Green Radar'. Also, the more useful illuminators of opportunity, as they are often referred, are usually sited on hills or tall buildings to provide broad coverage, an attribute that is likewise useful from a radar standpoint.

Some particular issues for PBR, however, are that the waveforms of such transmissions are not explicitly designed for radar purposes and so have less-than-optimum

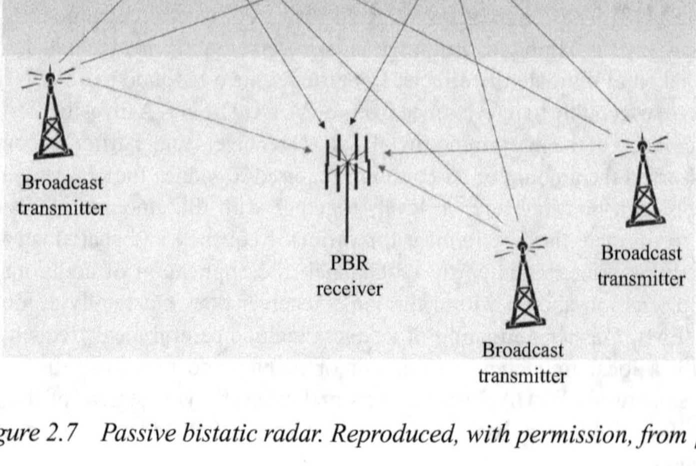

Figure 2.7 Passive bistatic radar. Reproduced, with permission, from [5]

properties with respect to resolution and range sidelobe behaviour. Digital modulation schemes are found to be better in this regard. Also, their coverage may not be ideal, since (for example) the elevation-plane radiation patterns may be tailored to avoid wasting power above the horizon. Care must be taken, therefore, to choose the right transmissions, in terms of both waveforms and coverage.

Taking this further, it may also be desirable in the future to design the signals of PBR illuminators so that they not only fulfil their primary function but also have favourable waveform properties for radar purposes. This idea has been termed 'commensal radar' – or literally 'at the same table' – and is an example of the sort of approaches that will be necessary as the spectrum crunch problem becomes worse.

Although the principles of PBR have been known for many decades – and indeed, some of the very first radar experiments exploited broadcast transmissions – it is only in the past decade or so that the subject has attained maturity. PBR and its applications to the spectrum problem are described in greater detail in Chapter 7, with further variants considered in Chapter 8.

2.7.2 Waveform diversity

It is possible to use emissions far more efficiently than is currently the case. Modern digital technology means that it is now feasible to generate precise, sophisticated wide-bandwidth radar waveforms that can be varied, potentially even on a pulse-by-pulse basis. This capability forms the foundation for what has become known as 'waveform diversity' [20], which includes the optimisation of waveforms based on mission requirements and prior knowledge of the spectral environment, greater exploitation of available degrees-of-freedom for both transmit waveform design and receive filtering, and the general convergence of electromagnetic, systems engineering, and signal processing requirements and capabilities.

For example, it is possible to design radar waveforms that have spectral nulls at particular frequencies or frequency bands [21,22], and potentially even to adapt these waveforms dynamically in response to a changing interference environment (the latter being an example of a cognitive system). Likewise, recent schemes [23] for the implementation and optimisation of polyphase-coded FM waveforms based on the continuous phase modulation (CPM) framework demonstrate new avenues for the design of radar emissions with enhanced spectral containment. These latter techniques also highlight the prospective benefits to incorporating system-level effects such as transmitter distortion directly into the waveform design process (as opposed to performing pre-distortion after the fact). Such hardware-in-the-loop design schemes enable a holistic view of the radar so as to address jointly the waveform performance and its actual emitted spectral footprint.

A particular open problem with regard to radar emission control is the spectral spreading induced by the rapid rise-time and fall-time of each transmitted pulse. In this context, the important distinction can be drawn between a *code*, which is an idealised set of phase values, a *waveform*, which is the result of modulating a code onto a carrier, and taking into account the transitions between one code element and the

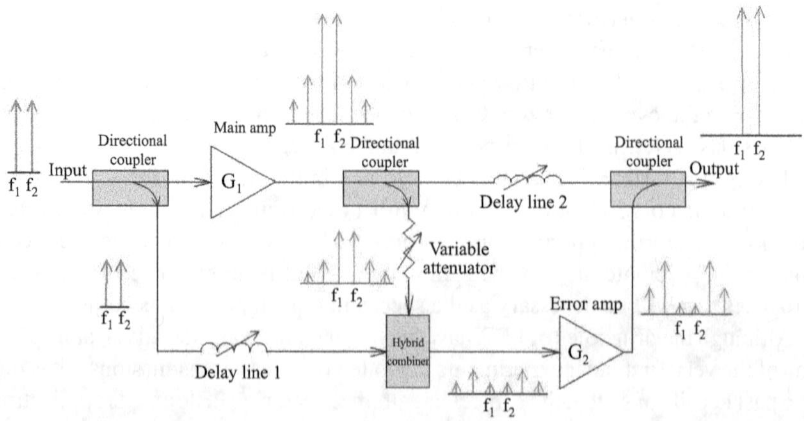

Figure 2.8 Open-loop feedforward power amplifier linearisation technique [27]

next, and an *emission*, which is the result of passing the waveform through practical upconversion and power amplification processes that embody the transmitter, including the distortion imposed by those processes [24]. High-power transmitters do not presently have a means to 'slow down' the rise/fall-time, as they behave more like a switch that is either on or off. Furthermore, any action taken to alter the spectral content of the radar emissions also directly impacts the waveform structure and, by extension, radar sensitivity. Within this context, hardware-in-the-loop design through the use of linearisation techniques [25,26] may potentially provide a means to slow down the rise/fall time (see Chapter 6).

As an example of power amplifier linearisation, Figure 2.8 shows an open-loop feedforward technique [27]. To understand its operation, consider an input signal consisting of two tones. The main amplifier produces an amplified version of these, plus distortion (intermodulation) products. Directional couplers are used to sample the input signal and the output signal, and with suitable weighting these are combined to provide the distortion signal alone. The distortion signal is subsequently amplified and added back into the main signal path, adjusted in amplitude and phase to cancel the distortion in the main signal path. Of course, this manner of operation would require careful adjustment in setting up, and the performance would be likely to degrade as a function of time due to drift. Closed-loop architectures would yield better performance in this respect, but would also be more complex.

Additional approaches to perform transmitter linearisation include different forms of adaptive predistortion [28] as well as variation of the load impedance that is connected to the power amplifier. The latter is discussed in detail in Chapter 12, where the well-known Smith Chart is also extended to realise the Smith Tube concept [29–31], with the ultimate goal of achieving joint transmitter/waveform optimisation. Simply put, the waveform and transmitter – particularly for high-power operation – are intrinsically linked, and to properly address radar spectrum issues it is necessary to consider them in that way.

2.7.3 Bio-inspired approaches

There are numerous examples of sensing in nature – for example, bats and dolphins use echolocation to navigate and to detect and intercept their prey [32]. It has been found that bats use very sophisticated waveforms in an adaptive and intelligent manner. Also, large numbers of bats are able to fly and navigate without apparently causing interference to one another.

Natural systems of this kind will have benefitted from optimisation through the process of evolution and natural selection [33]. By studying and understanding the processing techniques used by these natural systems, it is expected that we can use this knowledge in the design of our own systems to provide improved capability and to avoid mutual interference.

2.7.4 Cognitive radar

The notion of cognitive radar can be viewed from two different perspectives: as the evolution of bio-inspired control systems to higher level decision-making [34,35] or as the natural out-growth of knowledge-aided sensor signal processing [36]. Regardless of its roots, in the most general sense cognitive radar is essentially the application of Bayesian learning, through the use of prior knowledge and feedback, to facilitate the development of autonomous decision-making within the radar.

If prior knowledge of the spectral environment exists, it can also be exploited. This approach has enabled cognitive radio to make great strides in recent years. However, cognitive radio has tended to concentrate only on radio communication rather than considering the problem in its entirety. A more comprehensive approach would be to map out spectrum usage in terms of spectral, temporal, and spatial occupancy of all emitters and exploit this total 'spectral landscape' in cognitive type approaches. This perspective would enable cognitive approaches to embrace all emitters in an intelligent fashion. For example, most radar systems scan at a rate of less than one rotation per second. Most power is concentrated in the main beam, whose width may only be a few degrees. Thus at any one time the vast majority of the swept volume (typically 90%) is not being used by the radar. As this operation is fully determinable in advance, there is considerable opportunity for further improving spectrum usage and possibly spectrum sharing. This form of approach clearly offers efficiency gains in spectrum use without sacrificing performance, thus making it an attractive topic of future study.

Existing radar procedures such as automated frequency agility to avoid other spectral users and dynamic time-division resource allocation to enable different sensing (and possibly other) modes to share the same antenna [37] can be considered as early examples of cognitive systems. However, on-going research is also exploring more radical modifications, such as by leveraging the burgeoning work in waveform diversity to enable the radar to design waveforms 'on the fly' according to the observed spectral environment and mission requirements (e.g. [38–40]) through the use of complex feedback mechanisms and automated decision-making. In other words, by viewing active sensing as a 'question and answer' exercise, how can we enable the radar to select the best questions (i.e. waveforms) so as to obtain the best answers (given the spectral usage constraints) in real time?

Besides the sensor-centric area of waveform diversity, cognitive radar research is also building from previous work in cognitive psychology and artificial intelligence to mimic our own attributes of learning, memory, attention and intelligence [41], all with the goal of making the radar 'smarter'. Compared to the relative ease with which many animals can sense and interact with their environments, it is clear that we are only just beginning to realise the potential of artificial cognitive sensor systems, though continued research is necessary to quantify the likely gains that could be accrued.

2.8 Conclusions

This chapter has provided an introduction to the spectrum crunch problem from a radar perspective. At one level it can be said that the problem is already severe. There is ever-greater competition for a resource that is strictly finite, and radar is only one voice among many pressing their case.

All users have a need for greater bandwidth, and the only thing that can be said with certainty is that the problem is only going to get worse. Yet if spectrum usage were measured at a given point as a function of frequency, time, space and polarisation, it would certainly be found that the spectrum is currently not being used that efficiently. There is therefore great potential for approaches aimed at using the spectrum in a more efficient and dynamically controlled manner.

The regulatory framework has thus far taken a relatively conservative approach. However, it is important to have a proper quantitative understanding of the effect of interference of one service upon another in order to adopt appropriate regulation measures, rather than taking the view that no service should ever occupy the same part of the spectrum as any other.

To date, a number of novel radar technology approaches have been introduced, including improvements to the spectral purity of transmitters, intelligent, cognitive approaches to dynamic frequency allocation, passive sensing based on the emissions of other RF applications, and even through learning to mimic the behaviour of echo-locating animals. These topics and others are developed in the chapters that follow, and give some cause for optimism that a solution can be found.

Acknowledgements

I express my thanks to the many people who have helped me over many years with the work described in this chapter. These include Chris Baker, Alessio Balleri, Shannon Blunt, Larry Cohen, Eric Mokole, Simon Watts and Mike Wicks. I am also grateful to the organisations who have funded work on the subject, including THALES UK, the Royal Academy of Engineering, the UK Engineering and Physical Sciences Research Council (EPSRC), the IET A.F. Harvey Research Prize and the Office of Naval Research (Global).

References

[1] General Regulations annexed to the International Radiotelegraph Convention (Washington,1927) http://search.itu.int/history/HistoryDigitalCollection DocLibrary/1.4.48.en.100.pdf#search=frequency%20allocation, accessed 27 June 2017.
[2] Documents of the Moscow Telecommunications Conference, 1946, http://search.itu.int/history/HistoryDigitalCollectionDocLibrary/4.6.51.en.100.pdf#search=frequency%20allocation, accessed 27 June 2017.
[3] IEEE standard letter designations for radar-frequency bands, IEEE Std 521-2002.
[4] United States Frequency Allocations, http://www.ntia.doc.gov/files/ntia/publications/spectrum_wall_chart_aug2011.pdf
[5] Griffiths, H.D., Cohen, L., Watts, *et al.*, "Radar spectrum engineering and management: technical and regulatory approaches," *IEEE Proceedings*, vol. 103, no. 1, pp. 85–102, January 2015.
[6] *Radar Spectrum Engineering and Management*, NATO SET-182 report, TR-SET-182, October 2016.
[7] Griffiths, H.D., Baker C.J., and Adamy, D., *Stimson's Introduction to Airborne Radar, 3rd Edition*, SciTech Publishing, Raleigh NC, 2014.
[8] [http://www.itu.int/dms_pub/itu-r/oth/0C/04/R0C040000070001PDFE.pdf
[9] Woodward, P.M., *Probability and Information Theory, with Applications to Radar*, Pergamon Press, London, UK, 1953; reprinted by Artech House, 1980.
[10] Levanon, N. and Mozeson, E., *Radar Signals*, Wiley, Hoboken, NJ, USA, 2004.
[11] ITU Manual of Radio Regulations, 2012; http://www.itu.int/pub/R-REG-RR-2012.
[12] Manual of Regulations and Procedures for Federal Radio Frequency Management (Red Book), NTIA, May 2013.
[13] Richards, M.A., Scheer, J.A., and Holm, W.A., *Principles of Modern Radar*, IET, Stevenage, 2010.
[14] McQueen, D., "The momentum behind LTE adoption," *IEEE Communications Magazine*, vol. 47, no. 2, pp. 44–45, February 2009.
[15] Cohen, L., Daly, E., DeGraaf, J., and Sheff, K., "Mitigation of radar interference with WiMax systems," *International Waveform Diversity and Design Conference*, Niagara Falls, Canada, August 2010.
[16] Saunders, F.H., Sole, R.L., Bedford, B.L., Franc, D., and Pawlowitz, T., "Effects of RF interference on radar receivers," NTIA Report TR-06-444, September 2006.
[17] Ofcom Information Update, "Coexistence of S band radar systems and adjacent future services," 11 December 2009.
[18] Sanders, F.H., Sole, R.L., Carroll, J.E. Secrest, G.S., and Allmon, T.L., "Analysis and resolution of RF interference to radars operating in the band 2700–2900 MHz from broadband communication transmitters," NTIA Technical Report TR-13-490, US Department of Commerce, Washington DC, October 2012.

[19] Griffiths, H.D. and Baker, C.J., *An Introduction to Passive Radar*, Artech House, Norwood MA, March 2017.
[20] Wicks, M., Mokole, E., Blunt, S., Amuso, and V. Schneible, R. (eds), *Principles of Waveform Diversity and Design*, SciTech Publishing Inc., Raleigh, NC, 2010.
[21] Gerlach, K., Frey, M.R., Steiner, M.J., and Shackelford, A., "Spectral nulling on transmit via nonlinear FM radar waveforms," *IEEE Transactions on Aerospace and Electronic Systems*, vol. 47, no. 2, pp. 1507–1515, April 2011.
[22] Patton, L.K., Frost, S.W., and Rigling, B.D., "Efficient design of radar waveforms for optimized detection in coloured noise," *IET Radar, Sonar & Navigation*, vol. 6, no. 1, pp. 21–29, January 2012.
[23] Jakabosky, J., Blunt, S.D., Cook, M.R., Stiles J., and Seguin, S.A., "Transmitter-in-the-loop optimization of physical radar emissions," *IEEE Radar Conference*, Atlanta, GA, May 2012.
[24] Blunt, S.D., Cook, M., Jakabosky, J., de Graaf, J., and Perrins, E., "Polyphase-coded FM (PCFM) radar waveforms, part I: implementation," *IEEE Transactions on Aerospace & Electronic Systems*, vol. 50, no. 3, pp. 2218–2229, July 2014.
[25] Choffrut, A., Van Veen, B.D., and Booske, J.H., "Minimizing spectral leakage of nonideal LINC transmitters by analysis of component impairments," *IEEE Transactions on Vehicular Technology*, vol. 56, no. 2, pp. 445–458, March 2007.
[26] Ryan, L., Jakabosky, J., Blunt, S.D., Allen, C., and Cohen, L., "Optimizing polyphase-coded FM waveforms within a LINC transmit architecture," *IEEE Radar Conference*, Cincinnati, OH, May 2014.
[27] Griffiths, H.D., "Introduction to Waveform Diversity and Cognitive Radar," in *Novel Radar Techniques and Applications*, vol. 2, IET, Stevenage, UK, October 2017.
[28] Raab, F.H., Asbeck, P., Cripps, S., *et al.*, "Power amplifiers and transmitters for RF and microwave," *IEEE Transactions on Microwave Theory & Techniques*, vol. 50, no. 3, pp. 814–826, March 2002.
[29] Baylis, C., Moldovan, M., Wang, L., and Martin, J., "LiNC power amplifiers for reducing out-of-band spectral re-growth: a comparative study," *IEEE Wireless and Microwave Technology Conference*, Melbourne, FL, April 2010.
[30] Baylis, C. Wang, L., Moldovan, M., *et al.*, "Designing transmitters for spectral conformity: power amplifier design issues and strategies," *IET Radar, Sonar & Navigation*, vol. 5, no. 6, pp. 681–685, July 2011.
[31] Baylis, C., Fellows, M., Cohen, L., and Marks, R.J. II, "Solving the spectrum crisis: intelligent, reconfigurable microwave transmitter amplifiers for cognitive radar," *IEEE Microwave Magazine*, vol. 15, no. 5, pp. 94–107, July 2014.
[32] Balleri, A., Griffiths, H.D., and Baker, C.J. (eds), *Biologically Inspired Radar and Sonar: Lessons from Nature,* IET, Stevenage, UK, July 2017.

[33] Darwin, C., *On the Origin of Species*, John Murray, Albemarle Street, London, 1859, https://en.wikisource.org/wiki/On_the_Origin_of_Species_(1859)

[34] Haykin, S., "Cognitive dynamic systems: radar, control, and radio," *Proceedings of the IEEE*, vol. 100, no. 7, pp. 2095–2013, July 2012.

[35] Haykin, S., *Cognitive Dynamic Systems: Perception-Action Cycle, Radar, and Radio*, Cambridge University Press, Cambridge, UK, 2012.

[36] Guerci, J.R., *Cognitive Radar: The Knowledge-Aided Fully Adaptive Approach*, Artech House, Norwood, MA, 2010.

[37] Tavik, G.C., Hilterbrick, C.L., Evins, J.B., *et al.*, "The advanced multifunction RF concept," *IEEE Transactions on Microwave Theory & Techniques*, vol. 53, no. 3, pp. 1009–1020, March 2009.

[38] Bell, M.R., "Information theory and radar waveform design," *IEEE Transactions on Information Theory*, vol. 39, no. 5, pp. 1578–1597, September 1993.

[39] Pillai, S.U., Oh, H.S., Youla, D.C., and Guerci, J.R., "Optimum transmit-receive design in the presence of signal-dependent interference and channel noise," *IEEE Transactions on Information Theory*, vol. 46, no. 2, pp. 577–584, March 2000.

[40] Goodman, N.A., Venkata, P.R., and Neifeld, M.A., "Adaptive waveform design and sequential hypothesis testing for target recognition with active sensors," *IEEE Journal of Selected Topics in Signal Processing*, vol. 1, no. 1, pp. 105–113, June 2007.

[41] Stevenson, R., "A driver's sixth sense," *IEEE Spectrum*, vol. 48, no. 10, pp. 50–55, October 2011.

Chapter 3

Spectrum sharing between radar and small cells

Luiz DaSilva[1], *Francisco Paisana*[1], *and Nicola Marchetti*[1]

The escalating interest on the topic of coexistence between radar systems and broadband communication devices is a direct consequence of the significant portion of the international radio spectrum currently allocated to radar systems. However, studies show that their spectrum occupancy is low in the spatial, temporal, and frequency domains [1–3].

The most promising radar bands for shared use are the L, S, and C bands located in the 960–1400 MHz, 2.7–3.6 GHz, and 5.0–5.85 GHz frequency ranges, respectively. These frequencies are sufficiently low to avoid high power consumption and the usage of highly directional antennas, and sufficiently high to offer considerable bandwidth to commercial services. Furthermore, they are close to the cellular and ISM bands used for 2G/3G/4G and wireless local area networks (WLAN), respectively, facilitating the production of low-cost devices capable of using all these frequencies.

3.1 Radar systems—the incumbents

Radar systems find application in almost any area or service that requires object detection, localization, and recognition, in particular aeronautical and maritime radionavigation, weather forecast, and radiolocation. Despite the basic operating principles being the same, this wide range of applications makes radar specifications very diverse, even for systems within the same band of operation [4].

As incumbent features have a strong influence on how dynamic spectrum access (DSA) systems are implemented and deployed, we start by providing in this section a brief classification of radar systems commonly found in L, S, and C bands in terms of operating principles, applications, waveform properties, and beam/scanning patterns. We then follow this classification with a qualitative discussion of these bands' spectrum occupancy characteristics, and how these may affect spectrum sharing feasibility.

[1]Electronic & Electrical Engineering, Trinity College Dublin, College Green, Ireland

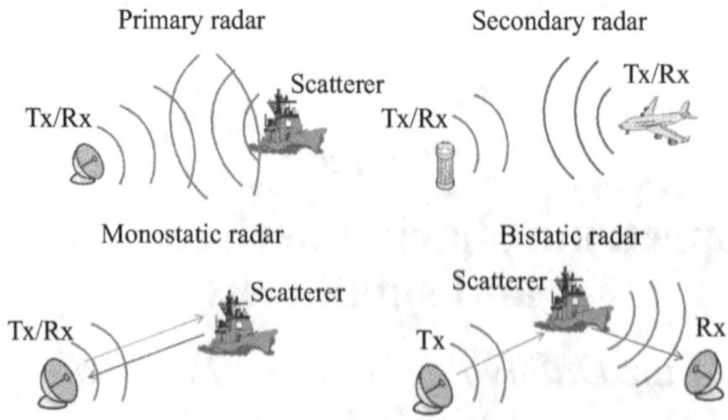

Figure 3.1 Examples of types of radar

3.1.1 Classification of radar systems

3.1.1.1 By operating principle

Depending on its technology, a radar can be classified as (see Figure 3.1):

- Imaging/Nonimaging: Imaging radars form a picture or map, while nonimaging radars produce a one-dimensional representation of the observed object or area. Typical implementations of a nonimaging radar system are speed gauges, radar altimeters, and scatterometers, while synthetic aperture radar (SAR) and Doppler weather radars are examples of imaging radars.
- Primary/Secondary: In primary systems, the picture of an object or area is formed using the echoes of the transmitted signal, while in secondary systems it is formed through a two-way communication between an interrogator and a transponder. Secondary radar systems are commonly used in air traffic control and can acquire more information, such as the altitude and identification code of aircraft, than primary systems.
- Monostatic/Bistatic: In monostatic primary systems, the transmitter and receiver are colocated, while in the bistatic case they are separated.
- Static/Mobile: Radar systems may be static or mobile, depending on whether the radar emitter or receiver is placed on mobile platforms, such as ships and aircraft, or in towers at a fixed location.

3.1.1.2 By application

Radar systems can be employed in numerous different services. We highlight the following:

- Air-traffic control (ATC) – Several different types of radar systems are encompassed by this service. ATC equipment can be either airborne, placed at airports or in en-route positions to assist operators with the landing, take-off, and guidance of civilian and military aircraft. Secondary radar is preferred in most ATC operations, as it provides identification information.

- Air surveillance – These systems are used to detect, identify, and determine the position, course, and speed of air targets in a given area. Air surveillance equipment can detect foe/hostile aircraft, drones, and cruise missiles. For this reason, it plays an essential role in many types of military operations, including battle, air-policing, and missile tracking. Surveillance systems can be further subcategorized into 2D and 3D types, depending on whether they offer height information. Their bearing coverage is generally 360°, and their range may exceed 460 km.
- Aeronautical radionavigation service (ARNS) – These systems ensure the correct and safe operation of aircraft, providing pilots with essential tools for airborne weather and obstacle avoidance, windshear detection, and other safety services.
- Maritime radionavigation services – Ship helmsmen rely mostly on visual information for navigation in restricted waters. However, in nonclear weather conditions visual piloting can become impossible, and radionavigation may provide the safest method to ensure ship safety. These radar systems can map the surrounding terrain and determine the position of ships in the vicinity.
- Meteorological aids – Applications of these services include weather forecast, in particular precipitation and wind measurements. These systems perform an essential role in early warning and emergency management to mitigate potential loss of life and damage to property. The weather data may be shared with other sectors such as transport, defense, aviation, and agriculture.
- Military missile control – Equipment that provides information to guide missiles to hostile targets can be placed on mobile platforms, including ships and ground vehicles.
- Miscellaneous – In this chapter, we mostly focus on the previously described applications, which are predominant in the 200–6000 MHz frequency ranges. However, radar can find application in a wide range of other areas, including vehicle cruise control, speed gauges, subsurface investigation, motion detection, and material testing.

3.1.1.3 By waveform

From the perspective of its waveform features, a radar system can be classified on the basis of the following properties:

- Constant wave (CW)/Pulsed waveform: CW radar signals are continuous in time, while pulsed radar (PR) signals can be represented by a train of short pulses. Both types, as well as other defining waveform features, are illustrated in Figure 3.2.
- Intra-pulse modulation (IPM): Pulses can take the form of a simple pulsed carrier (PC), or be shaped through compression techniques, such as linear frequency modulation (LFM), nonlinear frequency modulation (NLFM), and phase modulation (PM), to increase imaging resolution without sacrificing maximum range. Some examples of these waveforms are illustrated in Figure 3.3.
- Pulse repetition frequency (PRF): The PRF of a PR system can be low, medium, or high, depending on the range of the radar and on whether target velocity is of interest. The PRF or pulse repetition interval (PRI) may not remain constant in some cases, e.g., in staggered and jittered PRF systems.

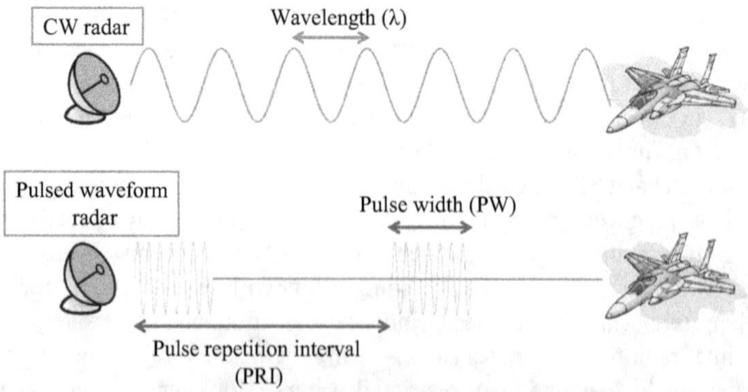

Figure 3.2 Different types of pulse waveforms utilized by radar systems, and some of their temporal features

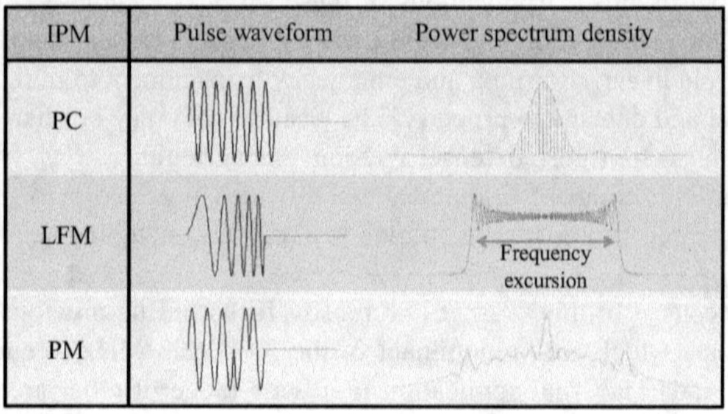

Figure 3.3 Different types of pulse modulation waveforms utilized by radar systems

- Pulse width (PW): The duration of each pulse in a pulsed waveform system, in conjunction with the modulation bandwidth, defines the range resolution of the radar system.
- Frequency agility: To avoid enemy jamming, some military radar systems may hop from channel to channel in a pseudo-random fashion.

3.1.1.4 By beam and scanning pattern

To scan their environment, radar systems steer their antennas either electronically by virtue of adaptive multiantenna techniques, or mechanically using a highly directive antenna supported on a rotating mast. In recent years, the more sophisticated electronically scanned array (ESA) systems, in particular active electronically scanned array (AESA), have seen increased use in specific military systems such as fighter jets. The major advantages of AESA include high jamming resistance and superior detection performance. However, their high manufacturing cost has limited their wide adoption

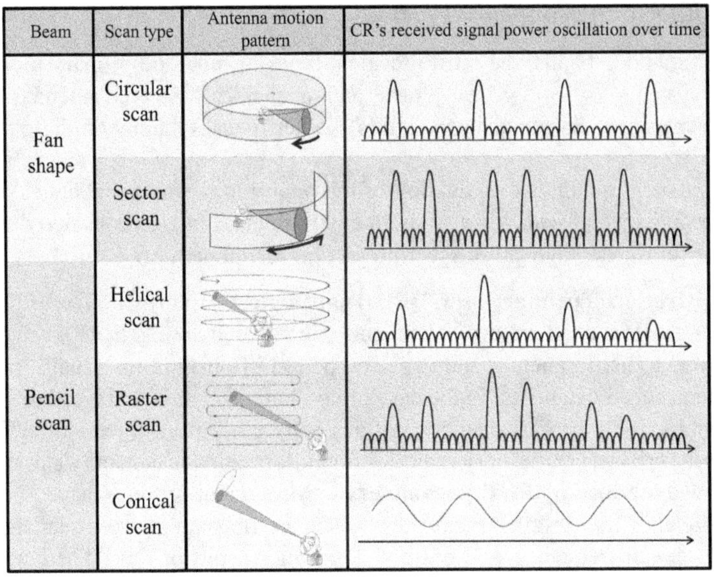

Figure 3.4 Types of mechanical antenna scan patterns and their signal level variation over time from a remote receiver's point of view

in other markets. Mechanically steered radars (MSRs) still constitute the majority of systems in radar bands worldwide, and considering their high acquisition cost and low cost of operation, replacing them by more advanced technologies may take several decades [5].

MSR antenna radiation patterns can be divided into two general types according to their shape: fan and pencil-shaped. Fan beams are characterized as being very wide in one direction and very narrow in the orthogonal direction and, therefore, are only used for either azimuth or elevation estimation. They are found in most military, maritime and aeronautical surveillance and radionavigation systems, which employ circular or sector scanning for detecting targets in 2-D [6]. Fan-shaped beam systems usually employ a relatively short antenna scan period (ASP), between 2.5 and 10 s, and their dwell times (i.e. the time their beam spends illuminating a particular direction) is around 7–30 ms. Pencil-shaped beams are directional in both elevation and azimuth and are thus generally associated with raster, helical, or spiral scanning patterns that are found in tracking, airborne intercept, and weather forecast radars. Scanning the environment in one extra dimension makes most of these systems part of the slow rotation class, with antenna revolutions that can take several minutes. These systems can also vary their antenna rotation speed over time. For instance, weather forecast radars change their ASP, reacting to the presence of rain or snow on the horizon. Tracking radars, on the other hand, may lock their beam on a missile or aircraft they locate in the sky. Some of these scanning motion trajectories and resulting signal energy variation from the perspective of a hypothetical remote receiver are illustrated in Figure 3.4.

3.1.2 Radar environmental factors

In the L, S, and C bands, the most prevalent types of radar are primary monostatic imaging systems, with pulsed waveforms and mechanically steered antennas. Therefore, the spectrum sharing analysis in this chapter focuses mainly on these types of systems.

To pursue a qualitative evaluation of the opportunities and challenges of DSA in radar bands, we provide a list of radio environmental factors considered to be of relevance for the deployment of spectrum-access techniques.

1. **Registration of primary users' (PUs) parameters** – Information regarding each radar station's position, equivalent isotropic radiated power (EIRP), waveforms, scan patterns, frequency, and expected period of operation is usually available in centralized databases. In the case of civilian navigation and weather forecast services, the positions of these stations can be accessed in real time through websites. However, military radiolocation information is generally classified.
2. **Time dynamics of PUs' parameters** – Most radar systems have fixed and predictable transmitting parameters, such as transmit power, scan patterns, PRF, center frequency, and position, for long periods of time. There are some exceptions, though, such as
    i. Military systems that employ electronic counter-countermeasures like frequency hopping when subject to enemy jamming.
    ii. Secondary radar transponders, whose emissions are not periodic and only occur after the reception of an interrogation message.
    iii. Some tracking systems that have irregular antenna scan patterns, designed to detect and track air targets.
    iv. Weather radar systems that display quasi-periodic scan patterns, varying their rotation speed several times a day [7].
3. **Mobility** – ATC and weather forecast systems are generally ground-based, and therefore have static positions. In contrast, radionavigation systems tend to be placed on mobile platforms, such as ships and aircraft. Surveillance and imaging systems can be either stationary or mobile.
4. **Diversity of incumbents** – Despite the same fundamental operating principles, radar systems may have very distinct features depending on their application. In particular, they may employ different waveforms, scan and radiation patterns, frequency hopping techniques, transmit power levels, and be on various stationary or mobile platforms. This diversity makes it challenging to design a universal, efficient spectrum sharing scheme for radar bands.
5. **Density of incumbent systems** – ATC, surveillance, and weather forecast radars are generally sparsely deployed to decrease redundancy of gathered information and minimize interference. For the case of airborne and shipborne radionavigation services, where distances between systems cannot be effectively controlled, spectrum can become quite congested in specific areas, such as harbors and airports.

    To limit cochannel interference and increase frequency reuse, radionavigation and some generations of weather and radiolocation systems employ interference suppression techniques [8–10]. It is therefore not unusual for secondary users (SUs) in these radio environments to observe more than one radar operating simultaneously.

6. **Planning** – Radar operators for services such as ATC and weather monitoring tend to coordinate among themselves to limit radio frequency (RF) interference. Hence, it is not unusual for stations' positions and parameter specifications to be planned ahead. In the case of quasi-periodic weather radar systems, changes in antenna rotation speed tend to be scheduled in advance and could easily be made available to other spectrum users [7]. Of course, severe weather would necessitate deviating from the schedule.
7. **Duty cycle (DC)** – With the exception of CW systems, radars emit pulsed signals with duty cycles that can be as low as 0.1% and as high as 25%, depending on the pulsewidth and PRF for the particular mode. Directive antennas and long rotation periods can contribute to an apparent reduction of the radar's duty cycle from the perspective of a remote receiver. For example, radar systems used for weather forecasting can take several minutes to make a full antenna revolution [11].
8. **Robustness to interference** – In a 2006 technical report [12], the US National Telecommunications and Information Administration (NTIA) comprehensively examined the impact of RF interference caused by communication systems on radar receivers' performance. Several conclusions were drawn from these tests, including
   i. Communication-type devices degrade radar performance at interference to noise ratio (INR) levels between −9 and −2 dB.
   ii. Radar systems can withstand low-duty cycle asynchronous interference caused by other radars up to INR levels as high as +63 dB. This robustness is mainly achieved through the use of suppression techniques [8].
   iii. At low levels, interference at radar receivers is manifested through a loss of detection for desired targets. This loss is not directly related to the distance between the targets and radar stations.
   iv. The effects of low-level interference are "insidious," meaning that there is no indication to radar operators that target losses are occurring.
9. **Public safety concerns** – The importance of radar systems to public safety varies with the application. We highlight ARNS and weather forecast as safety critical services that perform an essential role in early warning and emergency management systems to mitigate potential loss of life and damage to property [13]. As a consequence, every possible precaution must be taken to ensure their protection.
10. **Susceptibility to fading** – Ground-based and shipborne radar stations operate outdoors and are generally close to coastal areas and airports. Considering their highly directive antennas and very high peak transmit powers, their signals can reach radio receivers at very long distances, sometimes even beyond the horizon. For instance, a radar station with 10 kW of transmit power, 3.5 GHz center frequency, and 35 dBi antenna gain could, in free-space propagation conditions, reach a radio receiver at a distance of 1,920 km with a relatively high signal strength of −64 dBm. This distance, however, is well beyond the radar's horizon of 25 km. Taking such long ranges into account, one can infer that spatial sharing

between radar systems and SUs will predominantly occur in non-line-of-sight (NLOS) arrangements, i.e. where Rayleigh fading, shadowing, and high path loss exponents occur.

For the case of airborne systems, their operation at high altitudes reduces the likelihood of obstructions from buildings, terrain, or the Earth's curvature. Hence, line-of-sight (LOS) sharing conditions may take place more frequently.

11. **Channel dynamics** – Due to the use of rotating antennas and the presence of reflections in the environment, the interference channel between a radar and a remote terminal suffers rapid variations over time in amplitude, power delay profile, and angle of arrival. The coherence time of this channel will be proportional to the respective radar antenna's dwell time or time on target, which is on the order of a few milliseconds. For instance, the dwell time of a radar with ASP of 4 s and antenna beamwidth of 1° would be approximately 11 ms. We denote these rapid changes caused by the radar antenna rotation as *intra-scan channel decoherence.*

    During a measurement campaign at Cork naval base in Ireland [14], the impact of LOS versus NLOS on the signal amplitude from a shipborne radar was observed. The received signal power as a function of time for each case is shown in Figure 3.5. Clearly the effect of path reflections is far more noticeable in the NLOS case, while in the LOS case the times when the rotating radar beam is pointed at the receiver is quite evident. Another observable feature is the cyclic structure, with the period coinciding with the radar ASP of 2.6 s. This phenomenon is expected in reasonably static environments like the one in which these measurements were made. In such contexts, the channel has a low *inter-scan decoherence.*

12. **Incumbents' exclusion zones** – Radar systems have very large coverage areas, in some cases with a radius over 200 km due to high radiated powers on the order of kilowatts to megawatts and high antenna directivities.

    The highly directive antennas of ground-based radar systems are generally tilted upward to avoid ground reflections. As a consequence, remote radio receivers may notice strong variations in received signal strength as a function of elevation (e.g. street level versus a building rooftop). In technical report [11], the European Telecommunications Standards Institute (ETSI) depicted the footprint of a weather radar in the 5 GHz band as a highly irregular shape, whose contour distance from the radar station can vary from 35 to more than 100 km. The authors in [11] justified that this situation, far from the idealized footprint of a spherical shape for Earth's curvature, mainly resulted from the actual elevation profile of the terrain.

13. **Hidden receiver** – Most primary radar systems are monostatic, with colocated transmit (Tx) and receiver (Rx) chains, using the same antenna for both. For the case of secondary radar systems, both interrogators and transponders transmit and can likewise be detected.

14. **Recognizable features/hidden periodicities** – The intrinsic periodicities created by virtue of constant PRFs, pulsewidths, waveforms, and antenna rotation periods can be recognized through time–frequency analysis or matched filtering.

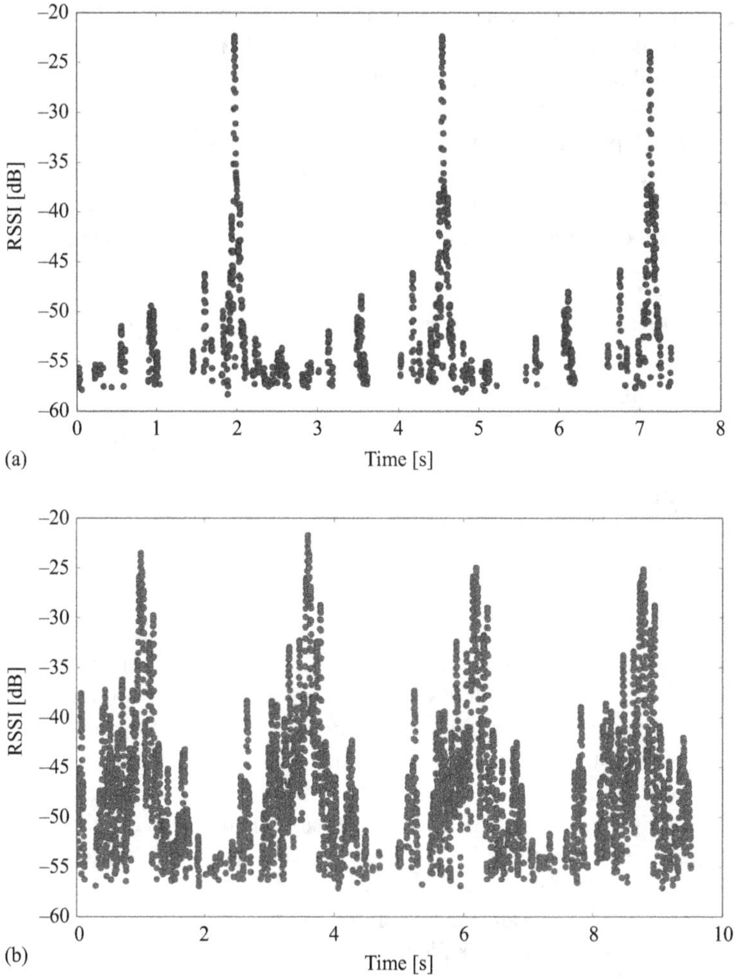

Figure 3.5 Measured power from a shipborne radar at different locations. Each dot corresponds to a detected pulse with a specific time of arrival (TOA) and pulse amplitude (PA). (a) LOS environment and (b) NLOS environment

Of course, military systems tend to avoid such predictable behaviors by utilizing complex or variable scan motions, waveforms, and frequency hopping.

15. **Uplink (UL)/Downlink (DL) bands separation** – Primary radar systems use the same channel for transmission and reception. Therefore, one can often make the assumption of channel reciprocity when sharing their spectrum. On the other hand, secondary radar systems such as secondary surveillance radar and distance measuring equipment use two separate bands, one for the transponder and another for the interrogator.

16. **Aggregate interference margin** – Radar systems tend to have long ranges and many are deployed close to areas of high population density such as coastal regions and airports, which make them more prone to be affected by a large number of SUs. On the other hand, the authors in [15] show the positive effect that the high path loss exponent of an urban environment has on the reduction of the aggregate interference caused by a specific density of SUs. In addition, the high directivity of most radar systems' antennas makes the interference caused to them to be dominated by the SUs situated inside their main lobe.

3.2 Current regulation on radar spectrum sharing

3.2.1 5150–5925 MHz band

Several bands in the 5 GHz spectrum, referred to by the U.S. Federal Communications Commission (FCC) as Unlicensed National Information Infrastructure (U-NII) bands, have already been made available in many countries for license-exempt radio frequency devices called wireless access systems (WASs) [11,16]. These bands include 5.15–5.25 GHz (U-NII-1), 5.25–5.35 GHz (U-NII-2A), 5.35–5.47 GHz (U-NII-2B), 5.47–5.725 GHz (U-NII-2C), 5.725–5.825 GHz (U-NII-3), and 5.85–5.925 GHz (U-NII-4). WLAN has been the predominant type of WAS, which is intended for low power, indoor, short-distance, broadband communication between traditional business and personal products such as computers, phones, and networking equipment. Recently, Qualcomm, Ericsson, the 3rd Generation Partnership Program (3GPP), and other groups have also started considering the deployment of long term evolution (LTE) networks in the 5 GHz spectrum as a way to boost coverage of current cellular networks [17]. To reduce the risk of interference to incumbents or other unlicensed devices operating in the same band, WASs typically employ DSA techniques and listen-before-talk (LBT) protocols.

The set of applied rules and nomenclatures for the use of U-NII bands varies from country to country, making it challenging to provide a comprehensive worldwide regulatory view. For this reason, in this section, we focus on the current U.S. and European regulation frameworks and frequency allocations.

3.2.1.1 Radio environment

The U.S. and European 5150–5925 MHz frequency allocations currently accommodate the following incumbent services:

- Radiolocation services. These systems may perform several different functions, including sea and air surveillance, tracking space launch and aeronautical vehicles undergoing operational testing, Earth imaging, and national defense.
- Aeronautical radionavigation systems. Applications include airborne weather avoidance, windshear detection, and other safety services.
- Maritime radionavigation services.
- Meteorological services in the 5600–5650 MHz band. Typical operational characteristics of these systems include the use of volume scanning patterns (e.g. helical),

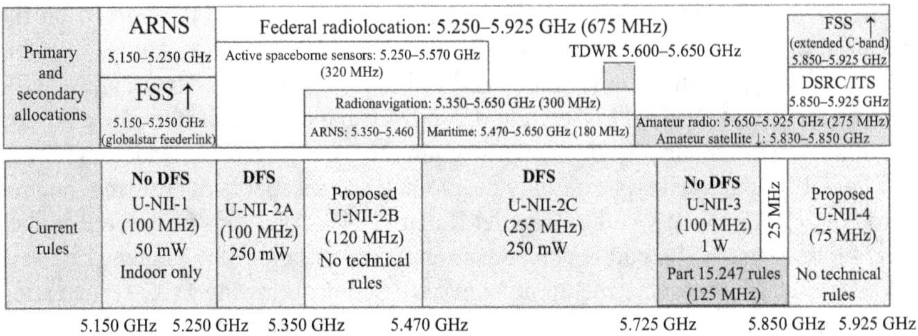

Figure 3.6 Chart summarizing the frequency bands and rules for U-NII device operation in United States [taken from the FCC's Notice of Proposed Rulemaking (NPRM) [16]]

antenna gains of 45 dBi, peak powers of 250 kW, pulsewidths below 2 μs, and PRFs between 250 and 3,000 kHz.

- Earth exploration satellite and space research.
- Nonfederal fixed satellite services (FSS) (Earth-to-space).
- Nonfederal amateur radio service allocated on a secondary basis.

We complement this list with the current US U-NII allocation tables and rules illustrated in Figure 3.6. The radar-reliant services of ARNS, meteorological, radiolocation, and maritime radionavigation constitute the main users of this spectrum, being present over the entire swath of 5150–5850 MHz.

3.2.1.2 Policy history and context

Since 1996, national and international regulators including the European Communications Committee (ECC), the US FCC, the UK Office of Communications (OFCOM), and the International Telecommunication Union – Radiocommunication sector (ITU-R) have spent significant effort in the harmonization of the 5 GHz band for the introduction of unlicensed WASs [11,18]. These bands were envisioned to offer higher reliability and data transfer rates to systems traditionally operating in the industrial, scientific and medical (ISM) bands. The FCC placed these systems under its Part 15 rules in 1997 [16], designating the U-NII-1, U-NII-2A, and U-NII-3 frequencies for their operation. In 2003, the FCC extended Part 15 devices' authorization to operate in the U-NII-2C band. In Europe, the ECC published the ERC decision (99)23 in 1999, opening the U-NII-1, U-NII-2A, and U-NII-2C bands for the introduction of high performance radio area networks (HIPERLANs). The ECC stipulated that HIPERLAN devices should employ several interference mitigation techniques, namely, transmit power control (TPC) and the sensing-based dynamic frequency selection (DFS) mechanism to protect radar systems from interference. In light of the success of IEEE 802.11 technologies, Europe shifted its approach toward a technology-neutral standard in 2003 with the EN 301 893 report [19].

In 2003, the ITU-R at the World Radio Conference (WRC-03) agreed on the allocation of WASs on a noninterfering basis. To avoid potential cochannel interference to military, weather, navigation, and radiolocation radar systems, the ITU-R enforced the use of DFS in U-NII-2A and U-NII-2C bands. DFS performance criteria and validation procedures were standardized in the M.1652 recommendation [18].

The DFS functionality, as initially developed, was incapable of ensuring the protection of U.S. Terminal Weather Doppler Radar (TDWR) systems [12]. Field studies carried out by the NTIA and Federal Aviation Administration (FAA) identified several causes of interference, including some WASs not following FCC certification and enforcement requirements, users applying software changes to disable DFS functionality, the considerations made for radar protection from WASs' adjacent-channel emissions not being adequate, and some modes of operation of TDWR systems being overlooked during the establishment of DFS certification requirements. As a response, the FCC issued interim procedures in 2010 restricting outdoor operation in the U-NII-2C band to professionally installed devices, disallowing operations in the 5600–5650 MHz band, and stipulating a minimum distance of 35 km between TDWR and WASs whose frequency is separated by less than 30 MHz. To respect this distance, WAS operators could consult an online database that stores information on TDWR locations [12].

In 2014, the FCC issued their first Report and Order (R&O), modifying DFS technical requirements and test procedures to avoid interference to TDWRs [16]. In particular, the new rules required that WASs monitor for radar signals in the entirety of their emission bandwidth, increase their sensing threshold, and add new radar waveforms to the compliance measurement procedures. The use of the geolocation database (GL-DB) and the 5600–5650 MHz band operations proposed in 2010 were discarded, as they became redundant under the new rules. The R&O also extended U-NII bands to 5.85 GHz and removed the U-NII-1 band indoor-only restrictions.

In Europe, several instances of interference to weather radar systems were also reported. The main causes of interference observed by the ECC involved either the DFS mechanism being disabled or lack of compliance with the ETSI standards [20]. ETSI further revised EN 301 893 to include radar systems using staggered PRF and PWs below 1 μs and dictated DFS sensing times of 10 min in the 5600–5650 MHz band.

3.2.1.3 Incumbent protection

Several compatibility and interference studies have been performed to assess the coexistence between WASs and incumbents, such as radar and satellite systems. As a consequence of those studies, regulators have specified several operational rules and mandated the use of multiple interference mitigation techniques to enable sharing. Depending on the band and country, WAS devices' EIRP has been limited to either 200 mW or 1 W. Furthermore, in specific channels, WAS deployments were restricted to indoor use only, leveraging the additional through-wall attenuation to reduce interference. For the U-NII-2A and U-NII-2C bands, the use of TPC and DFS is currently mandated in both the United States and Europe.

Despite most of the reported DFS interference issues with TDWR systems having been addressed by regulators via additional recommendations, these instances served as examples of the potential risks that spectrum sharing can incur for radar services that concern safety-of-life if insufficient interference and coexistence assessments are performed. These cases also demonstrated the importance of making spectrum sharing technologies reconfigurable to permit upgrades for new capabilities, policies, standards, and protocols. The DFS mechanism still remains particularly sensitive to these issues. To check whether a device is DFS-compliant, it must be tested against multiple simulated signals that are representative of a variety of real radar systems. Since not all current and future radar signals in the 5 GHz band are considered in these tests, DFS compliance provides only limited confidence that incumbents will be protected from interference.

In addition to incumbents' interference reports, DFS functionality has also been subject to criticism regarding its implementation practicality, which has led to the current congestion of non-DFS channels in the 5 GHz band [21]. WLAN operators consider this technology too expensive to implement in cheaper WLAN devices, and its long and frequent channel availability check procedures cause battery and bandwidth inefficiencies. Furthermore, most of the current DFS implementations suffer from high false positive rates, which force devices to move to non-DFS bands and remain there for long periods of time. As a possible cause for the latter issue, operators point out the inability of WLAN devices to distinguish incumbent radar signals from other WLAN emissions.

3.2.2 3550–3700 MHz band in the United States

The FCC is currently finalizing rules regarding an initiative to open the 3550–3700 MHz band in the United States to lightly licensed small cells, which it refers to as Commercial Broadband Radio Service Devices (CBSDs) [22]. Several industry proponents including Google, Ericsson, and Qualcomm have applauded this endeavor and have expressed interest in the utilization of this band for LTE networks. The WLAN Working Group, on the other hand, has decided not to develop a standard for this band, expressing concern over its excessively large exclusion zones [23].

3.2.2.1 Radio environment

The 3550–3700 MHz band currently accommodates:

- Shipborne radar systems. The U.S. Navy uses the band for high-powered radars on guided missile cruisers [22].
- Ground-based firefinder radar systems owned by the U.S. Army and used in the detection of enemy projectiles.
- Airborne radar systems used by the U.S. Air Force to assist pilots in formation flying and to support drop-zone training.
- Nonfederal FSS (space-to-Earth) in the 3600–3650 MHz portion for which the FCC has licensed stations to operate in 35 American cities.
- Nonfederal terrestrial wireless broadband services in the 3650–3700 MHz band. Users are authorized to operate for a finite term through nonexclusive nationwide

licenses and are required to register all individual base stations (BSs). Broadband stations must employ a contention-based protocol and their EIRP is limited to either 25 W per 25 MHz or 1 W per 25 MHz depending on whether they are fixed or mobile stations, respectively.

3.2.2.2 Policy history and context

In March 2010 the FCC, as mandated by the U.S. Congress, released the "National Broadband Plan" that established a roadmap of initiatives and incentives to provide affordable broadband access to all American citizens [24]. Some of the recommendations of this plan include the following: (i) the FCC would make 500 MHz of government-owned spectrum available for broadband use by 2020, with 300 MHz suitable for mobile use by 2015; (ii) releasing additional spectrum for unlicensed use; (iii) creating new avenues for innovative spectrum access models, such as opportunistic use of spectrum; and (iv) initiating proceedings to enhance research on new spectrum-access techniques. As a continuation to the National Broadband Plan, President Obama released the Presidential Memorandum "Unleashing the Wireless Broadband Revolution" encouraging the NTIA and other independent agencies to coordinate with the FCC on the creation of methods for the measurement of on-going spectrum utilization, and the identification of new candidate bands for use by wireless broadband systems on an exclusive, shared, licensed, and/or unlicensed basis, while ensuring no loss of critical government capabilities [25].

In October 2010, NTIA released its "Fast Track Report," where it assessed the "Near-Term Viability of Accommodating Wireless Broadband Systems in the 1675–1710 MHz, 1755–1780 MHz, 3500–3650 MHz, 4200–4220 MHz, and 4380–4400 MHz Bands." Of these bands, NTIA ultimately recommended 3550–3650 MHz as the band that could be available by 2015 on a spectrum sharing basis with Department of Defense (DoD) radar and nonfederal FSS systems [26]. Some of the reasons that led to the selection of this band for fast tracking were as follows: (i) WiMAX equipment has already been deployed in the neighboring 3650–3700 MHz band; (ii) federal operations are geographically limited; and (iii) this band is already reserved for fixed services in other parts of the world. The NTIA's technical analysis showed that large exclusion zones, covering approximately 60% of the U.S. population, would be required to protect DoD systems from interference due to their high transmitter power levels and high-gain antennas.

In December 2012, the FCC, reflecting on the NTIA's Fast Track Report and recommendations of the President's Council of Advisors on Science and Technology (PCAST), put forth an NPRM proposing the creation of a new commercial broadband service (CBS) in the 3550–3650 MHz band [27]. In this report, the FCC envisioned small cells as the ideal type of technology for CBS use for the following reasons: (i) this band's high frequency leads to higher attenuation and penetration losses, making it less ideal for exclusive licensed commercial mobile use; (ii) larger attenuation facilitates dense deployments of small cells, with reduced risk of harmful interference; and (iii) the large incumbent exclusion zones make the band less attractive for macrocell deployments. Access to the spectrum by CBS users is set to be managed by a Spectrum Access System (SAS) following a three-tier priority-based access model. The three tiers of services are (i) incumbent access (IA); (ii) priority access (PA); and (iii) general authorized access (GAA). IA would include federal and FSS users currently operating in the 3.5 GHz band. For the PA tier, the NPRM proposed the authorization of critical QoS services, such as hospital and public safety entities, that would be assigned exclusive channels in specific regions with ensured interference protection through traditional long-term licenses. GAA users would operate in an opportunistic manner, not causing interference to, yet accepting interference from, other users. The FCC also sought advice in the report on whether to extend the proposed rules to the 3650–3700 MHz band.

In response to several reconsideration petitions and comments, the FCC in its licensing Public Notice (PN) in 2013, and Further Notice of Proposed Rulemaking (FNPRM) in 2014, revised and refined some of the initial licensing concepts and defined the set of rules for CBS in the 3.5 GHz band [28,29]. In particular, the revised framework expanded eligibility for access to the PA tier using mechanisms of competitive bidding for users to be assigned licenses. Moreover, the PN and FNPRM established the SAS's role in coordinating the space–time–frequency assignments of PA and GAA tier users and set guidelines for operation and certification of SAS implementations.

In March 2015, the NTIA submitted a letter to the FCC proposing a reduction of the incumbents' exclusion zones by as much as 77% [30]. NTIA's letter also suggested the addition of an environmental sensing capability (ESC) in the regulatory framework consisting of a sensor network that informs the SAS about the geographical position and band of operation of incumbent DoD systems.

In April 2015, the FCC released an R&O establishing a comprehensive regulatory scheme for the entirety of the 3550–3700 MHz spectrum, including licensing, technical rules, and operation [31]. In particular, it adopted a finite-term licensing mechanism that enables users to transition between GAA and PA tiers depending on local spectrum demand. The revised framework prohibits CBSDs from causing interference and from claiming protection from federal stations aboard vessels, ground-based radar sites, and nonfederal FSS stations. However, airborne radar receivers in the band must not claim protection from CBSDs. The R&O also confirmed the use of ESC as an enabler of commercial use inside coastal exclusion zones. The R&O was further revised and finalized in 2016 [22].

3.2.2.3 Spectrum access system

According to the proposed rules in the FCC's 2016 R&O, access to the 3550–3700 MHz band is to be performed under a three-tier framework. The first tier consists of IA users that include federal shipborne and ground-based radar systems, FSS Earth stations, and, for a limited time period, legacy terrestrial wireless systems in the 3650–3700 MHz portion of the band. The second and third tiers accommodate the CBS users, PA, and GAA, whose spectrum access is coordinated by the SAS on a space–time–frequency basis. To become a part of the PA tier, a CBSD has to request and be assigned a Priority Access License (PAL) through competitive bidding. A PAL consists of an exclusive nonrenewable authorization to use a 10 MHz channel in the 3550–3650 MHz portion in a single census tract for three years. The application for a PAL ensures these users' protection from other PA and GAA users' interference. Finally, the GAA tier accommodates licensed systems that operate on an opportunistic basis throughout the whole 3550–3700 MHz band.

The role of the SAS is two-fold: to coordinate the three-tier authorization framework and to manage CBS users' dynamic frequency allocations so that CBSDs only operate in regions and channels where they do not cause interference to IA or other licensed PA users. To that end, all spectrum assignment and PAL requests can only be made after the CBS users register and provide a report of their operational parameters to the SAS. These parameters include their location, antenna height, gain, beamwidth and azimuth, category, serial number, contact information, air interface technology, and deployment profile (e.g. indoor/outdoor). The CBS users also have to maintain an active connection to the SAS to facilitate reporting of their local interference levels, to inform the SAS of any operational parameter change (e.g. location), and to receive possible assignment and licensing updates. This framework allows the SAS to perform propagation analysis, interference mitigation procedures, and provides an OFF switch that disconnects or forces CBSDs to vacate the channel when their proper operation has been compromised or there is an expected risk of interference to IA users.

3.2.2.4 Incumbent protection

The FCC has adopted a two-phased approach for the protection of incumbents. During phase one, the IA users are protected primarily through static exclusion zones that reflect their ranges and possible geographic locations. These zones include 35 cities where FSS sites are located, ground-based radar sites, and all the U.S. coastlines. Nonfederal IA operators are required to register their parameters and frequency assignments in the SAS to claim protection from lower tiers. Phase two will commence once the ESC sensor network is developed and deployed, which will inform the SAS about the channels, range, and proximity of shipborne and ground-based radar systems on a dynamic basis.

DoD shipborne radar systems represent the main contributor, in terms of covered area, to the currently defined 3.5 GHz exclusion zones. It was originally estimated that, given static propagation model-based exclusion zones, only approximately 40% of the U.S. population could benefit from this band [27]. This low percentage was determined due to the use of conservative exclusion zones of up to 250 km from the U.S. shoreline to account for long transmit ranges and unknown/unplanned locations of Navy radar systems. Through a collaboration between the NTIA and DoD, a reevaluation of these protection distances was made using more sophisticated terrain-dependent propagation models, area specific clutter losses, building attenuation losses, and user density prediction models. The result, illustrated by the slightly darker line in Figure 3.7, was a reduction of the total coastal exclusion zone area by 77% [30]. Ground-based radar sites were also affected by this revision, as their exclusion radii were reduced from 40 to 3 km. Although significant geographically, the same report shows that this revision represents relatively little change in terms of the percentage of U.S. population affected, with still more than 45% unable to access the band.

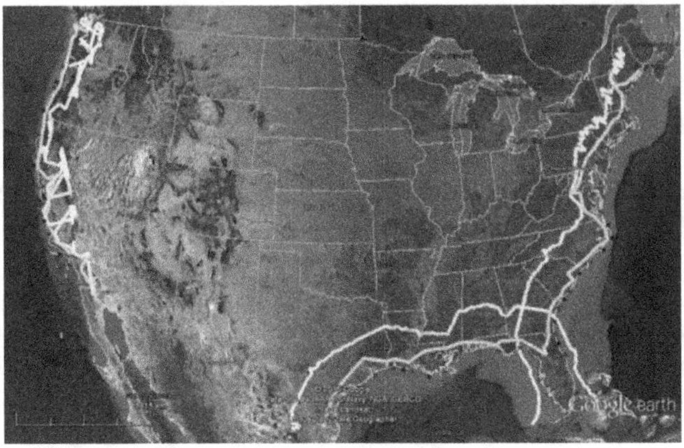

Figure 3.7 Exclusion zones defined for the 3.5 GHz band based on protection requirements of U.S. Navy radar systems. The previous and new exclusion zones, after NTIA's revision through updated propagation models, are represented by lighter and darker lines, respectively (taken from the FCC's R&O [31])

With the adoption of ESC during the second phase, access for CBSDs in the coastal areas will be increased. Taking into account the limited number of shipborne naval systems operating in this band, and the fact that these systems spend long periods of time away from the coast, it is expected that these protection zones will be available for commercial use a great majority of the time. Shipborne systems' long transmit ranges, geographically limited positions, and location confidentiality requirements also allow and mandate a sparse distribution of ESC devices solely located in the vicinity of protection zones, which will significantly reduce the deployment costs normally associated with sensor network technology.

3.3 Spatial sharing techniques

Spectrum sharing through geographical separation, which we denote as spatial sharing, has been the main approach proposed by regulators to enable coexistence between distinct radio technologies. This sharing model's operating principle consists of drawing static or dynamic exclusion zones around incumbent systems where SUs are not allowed to transmit and thereby cause harmful interference. The sizes of these zones are proportional to the antenna gain, interference sensitivity, path loss, shadowing, and any existing uncertainty regarding the location of incumbent receiver devices, and to the transmit power employed by SUs.

Exclusion zones can either be defined according to theoretical propagation models or based on RF measurements obtained at SUs or a dedicated sensor network.

The choice of which method to use strongly depends on the radio environmental factors of the band where spatial sharing is to be employed. For instance, while propagation models tend to be inaccurate at accounting for terrain elevations and wall penetration losses, RF measurements show stronger sensitivity to phenomena like the presence of hidden receivers. Choosing an inadequate spatial sharing method may make it challenging to protect incumbents from interference without the use of overly conservative exclusion zones or expensive SU hardware.

In this section, we consider the viability of different spectrum-access techniques proposed in the literature, namely, the use of a GL-DB, beacon signaling, spectrum sensing, cooperative sensing, and radio environment maps (REMs) within this context of spatial sharing in radar bands. In particular, we qualitatively assess how each of these techniques' specifications will be affected by the radio environmental factors discussed in Section 3.1.2.

3.3.1 Geolocation database (GL-DB)

In this spectrum-access technique, a centralized database stores information about the radio parameters and positions of PUs, which it then uses to define exclusion zones and channels where SU operation is not allowed. An SU seeking to access the spectrum first estimates its position using localization technology such as a global positioning system (GPS) and then reports the resulting coordinates to the database. The database then replies with a set of channels that are available for SU use based on the querying device's operating parameters and location [32–35].

The GL-DB calculates the interference between communication systems using theoretical propagation models rather than actual RF measurements. To that end, the GL-DB has to be aware of the PU's EIRP, center frequency, bandwidth, antenna height, location, and expected duration of channel usage. The GL-DB must likewise know the SU's bandwidth, transmit power and deployment density, and some band and terrain-specific path loss considerations. To avoid prohibitively high complexity, the database divides the terrain into multiple geographical zones/areas, for which spectrum availability is computed separately and is considered homogeneous. The state of each area remains cached to allow fast consultation by SUs until a change in a PU's operational parameters is reported. The techniques used to discern exclusion zones and other protection specifications may vary with the policy and regulations employed in each country [32,33]. In [35], the authors compared the FCC and ECC approaches for database technologies in TV White Space (TVWS), demonstrating that the former was less protection-oriented, leading to increased throughput capacity for SUs.

With initiatives such as TVWS, the SAS in the United States, and Licensed Shared Access (LSA) in Europe, the GL-DB concept has become the predominant and preferred technology by regulators for spectrum sharing. There are several reasons for this preference. First, a centralized database access provides a universal, extensible interface and a practical method to define policies, authorize and coordinate users' spectrum access, and control interference to incumbents. In contrast to DFS in the 5 GHz band, interference reported by incumbents can now be avoided by tuning the level of conservativeness of a GL-DB's propagation models or by removing SUs from specific channels. Moreover, geolocation solves the issue of hidden terminals that made spectrum sensing impractical in the TVWS. There are, however, some drawbacks associated with the adoption of this technology. First, it requires SUs to be equipped with some form of localization technology, making it inadequate for low-cost devices. Second, it requires that incumbents report their operational parameters and geographic location. For this reason, it cannot efficiently cope with systems that are placed on mobile platforms. Finally, since it relies on theoretical propagation models, its spectral efficiency may be lower than methods that rely on RF measurements.

GL-DB represents a reasonably attractive solution to spectrum sharing in radar bands. In fact, it was proposed by the FCC as a primary spectrum-access technique to enable the coexistence between maritime radar systems and small cells [27]. We highlight two key advantages for the use of this technology in radar spectrum. The first is that the registration of most incumbent systems and their respective parameters is fairly simple due to their low number and well-planned operation configurations. Many civilian weather, air terminal, shipborne and airborne radar systems already register their positions in databases that can be accessed in real time via the internet. Second, the great majority of surveillance and weather forecast radar stations are static. Hence, coexistence of SUs with these systems can be spectrally efficient and does not require high update rates and low latencies in the dissemination of the database's channel occupancy information.

From a propagation perspective, it is challenging to set exclusion zones efficiently in these bands due to the fact that radar systems' very long ranges and tilted antennas make the path between them and SUs likely to be obstructed by buildings or the Earth's curvature, as well as the dependence on terrain elevation. As shown in [11], a radar exclusion zone at 5 GHz is still far from being circular and is not easily predictable through theoretical propagation models if no considerations are made regarding the terrain elevation or other possible sources of attenuation.

There are some concerns regarding the applicability of GL-DB to radionavigation bands where mobility is involved. Drawing exclusion zones around airborne or shipborne systems' current location would either require very complex databases, with very fast update rates and low information dissemination latencies, or the use of large error regions. An error region, represented in Figure 3.8 by R_{err}, is an extra safety margin added to an incumbent system's exclusion zone to compensate for the

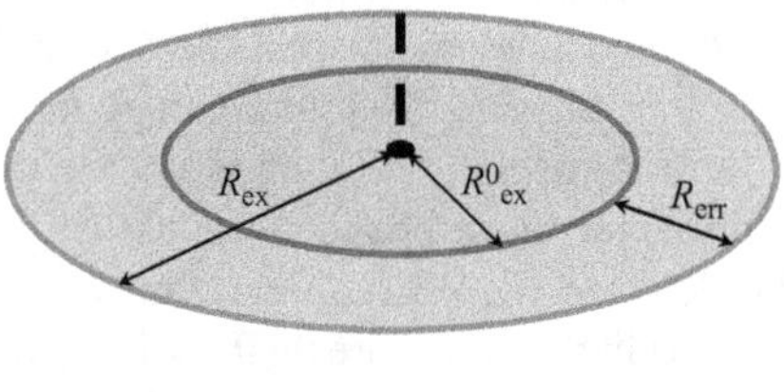

Figure 3.8 The effective radius of the incumbent's exclusion zone (R_{ex}) is equal to the radius of the exclusion zone in case the incumbent is static (R^0_{ex}) plus an error distance (R_{err})

uncertainty in its location or parameters. For example, for the case of a radar antenna aboard an airplane flying at 900 km/h and a GL-DB with an information dissemination delay of 5 min, the exclusion zone radius would increase by $R_{\text{err}} = 900\frac{5}{60} = 75$ km. The tracking of these positions would present further implementation challenges in the context of military systems where, due to security concerns, radar operators may oppose sharing or may obfuscate any information that could potentially disclose their system's activity or location. One possible way to circumvent this issue in maritime bands, proposed by the FCC, consists of drawing exclusion zones around all coastal areas [31]. However, since a large percentage of the world population lives in these same areas, this strategy is highly inefficient in both space and frequency and would prevent a significant percentage of users from accessing the spectrum.

Several theoretical interference studies for radar bands have already been conducted [3,15,36,37]. In [3] both cochannel and adjacent channel interference generated by a single secondary device were studied for the L, C, and S bands. The authors considered the C band to possess the best sharing conditions where, according to the tests, cochannel and adjacent channel coexistence is possible at a distance greater than or equal to 45 and 17 km, respectively. In [15], the authors studied the impact of aggregate interference from multiple, uniformly distributed devices on the meteorological radar band at 5.6 GHz. Their findings highlight the technical difficulties associated with the measurement of aggregate interference on the PU based on propagation models, since the degree of interference hugely depends on the propagation environment and, in particular, the path loss exponent. This work was further extended in [36] where nonuniform user distribution scenarios were considered.

3.3.2 *Spectrum sensing and DFS*

Spectrum sensing is the primary mechanism a cognitive radio (CR) employs to achieve awareness of its radio environment. Briefly, this mechanism comprises any signal processing procedures an SU may perform on its received samples to detect the presence of PUs in its channel or evaluate the risk of causing harmful interference to them. Spectrum sensing has received significant attention by regulators and the CR community for its simplicity, as it does not require changes to the incumbents' operation, SUs to be equipped with GPS, or any additional infrastructure [4].

Spectrum sensing as an enabling technique for coexistence with radar systems has already been proposed and implemented in the 5 GHz band as part of DFS [11,18]. Before initiating transmission, a system complying with DFS determines the channel availability by first checking for the presence of pulses with peak power above −64 dBm for a time interval, or channel availability check time (CACT), of 1 min. If a radar signal is detected, the WLAN device must vacate the channel in less than 10 s and then avoid this channel for more than 1 hour. Since a 1-min CACT may not be enough to detect some slowly rotating radars, SUs are also required to employ an in-service monitoring scheme in which they interleave their data transmissions with short in-band sensing procedures [18,38].

In contrast to TVWS, radar bands may present a favorable environment for spectrum sensing. The channel reciprocity conditions, the absence of hidden terminals, and radar systems' high transmit powers allow SUs to use very high detection thresholds. In light of the ITU-R interference studies for the 5 GHz band [18], it can generally be expected that these thresholds (e.g. −64 dBm or higher) will be well above the noise floor, thereby reducing the impact of RF impairments and noise uncertainty. Moreover, the radar signals' distinctive pulsed nature facilitates their discrimination from other communication-type emissions.

Although the detection of high power pulses is straightforward, a radar antenna's directivity and long rotation periods require that SUs employ very long sensing times. This issue is particularly acute for the case of weather radar systems, which employ elevation scanning and may take several minutes to perform a full antenna revolution. In Europe, for example, the DFS's CACT is 10 min for TDWR bands, which has caused several vendors to avoid these bands, as they would otherwise lead to significant power consumption and wasted spectrum resources. Techniques like DFS also lack the flexibility and context-awareness of modern DSA techniques. DFS's one-size-fits-all threshold and CACT, defined according to worst case scenarios, leads to overly conservative exclusion zones for some incumbent systems. For instance, some of the high-power, fast-rotating radar systems analyzed by ITU-R in [18] could be protected with sensing periods as low as 5 s and/or thresholds as high as −36.4 dBm.

3.3.2.1 Beaconing

In this approach, incumbents cooperate with SUs by sharing information regarding their spectrum utilization through signaling devices known as beacons. Although an attractive solution for efficient spectrum sharing, beacons raise several implementation issues, one of them being that deployment usually requires significant changes to legacy systems' infrastructure [4]. These changes are not only unattractive to PUs but may also be time-consuming or infeasible to implement when the technology is too widespread. The lack of a global consensus on what band the beacons should occupy represents another barrier to the deployment of this method in the near future.

To date, beacon signaling in radar bands has not received much attention in the literature, mainly due to the focus on GL-DB and DFS for these bands by regulatory entities [18,27]. We envision two possible types of beacon signaling mechanisms that could be employed in radar bands. The first would involve altering the radars' infrastructure to integrate the beaconing device or embedding data into the signals transmitted by the radar. The beacon data could contain, for instance, the radar's required detection threshold, incurred interference, or location, thereby avoiding some of the inefficiencies involved with spectrum sensing. If this data was transmitted at a power similar to the respective radars' pulses, it could be easily decoded outside these systems' exclusion zones. The main issue with this approach is that, despite the number of radar systems being low, changing their infrastructure would be very expensive and take several decades to complete. In contrast, the second approach would be to place separate beacons close to radar sites or in harbors and airports, which would inform SUs that they are approaching an exclusion zone and provide a list of clear channels.

In order not to affect incumbents' operation, beacons could employ several different interference mitigation techniques, including

- the use of a dedicated frequency channel;
- the transmission of Channel Allocation Frames inside the "inter-measurement gaps" [39] for weather radar that are the periods of time occurring after the radar's reception of expected scattering and before the start of the next pulse;
- placement of an additional antenna at radar sites with a nonintersecting pattern, where this antenna would request SUs to interrupt their communications before being swept by the radar antenna's mainbeam [2];
- in-band omnidirectional transmission using a low duty cycle waveform, which radar systems can easily suppress without additional changes to their infrastructure [12].

3.3.3 Cooperative spectrum sensing

Environment/propagation factors such as multipath fading, shadowing, and the presence of hidden terminals can significantly hinder an SU's ability to detect spectrum opportunities through conventional local sensing. Another limitation of local sensing is that in sharing scenarios where PUs transmit intermittently, it requires that SUs spend a significant portion of their time monitoring the channel, thereby wasting spectrum resources that could be used for transmission. The impact of these phenomena could, however, be mitigated if individual sensing results were shared between SUs at different positions, thus facilitating a combined decision regarding the availability of a specific channel. This mechanism, called cooperative spectrum sensing, has received significant attention in the CR literature, being proposed as a possible way to circumvent the limitations of local sensing in TVWS.

The main obstacle to the implementation of cooperative sensing in DSA has been the fact that it often implies the use of a global Common Control Channel (CCC). A CCC is not envisioned to be available to SUs in the near future as it would require regulators and industry bodies to agree on a separate frequency channel allocation, radio access technology (RAT), and protocol for exchanging sensing data. While other alternatives to a global CCC exist, namely, underlay and opportunistic channels, they are less reliable and increase the overhead of DSA networks [40]. Other challenges also exist related to the lack of performance guarantees provided by cooperative sensing. Its achievable detection levels depend on the number of nodes involved in the cooperating process, commonly referred to as data fusion, and the existence of spatially correlated shadowing across nodes [41].

Cooperative sensing is not envisaged to be as attractive for radar bands as it is for TVWS due to the absence of hidden node problems in these bands. In such scenarios, the use of this technique may actually lead to the occurrence of exposed terminal problems, and consequently to the waste of spectrum opportunities. Moreover, cooperative sensing does not address some of the limitations of local sensing in radar bands, namely, the inability to determine and apply a detection threshold on a user-specific basis according to each different radar system's protection requirements. There are, however, some benefits cooperative sensing can bring. The first is that it would address the issue of long sensing periods and periodicities that DFS, for instance, currently requires to protect weather-forecast systems. By coordinating the timing and combining sensing efforts, the total time spent by a group of nodes monitoring the spectrum would be significantly reduced. The second advantage is that cooperative sensing could provide an extra safety margin of protection to safety-critical systems such as ARNS.

3.3.4 Radio environment map (REM)

Instead of relying on quasi-static PU's parameter information and theoretical propagation models, a database could receive real-time RF measurement data from measurement capable devices (MCDs), which may include SUs or a dedicated sensor network. The database system would then aggregate the received data and, through interpolation methods, draw a REM showing channel occupancy or signal strength distributions across space. This mechanism has two clear advantages over traditional static GL-DBs. First, RF measurements consider effects such as terrain elevation and other sources of attenuation that are challenging to take into account through conventional theoretical propagation models. As a result, the use of REMs can lead to smaller exclusion zones and higher spatial sharing efficiency than that provided by GL-DB. Second, real-time RF information may allow a database to cope with incumbent systems' mobility, even when the latter do not disclose their location. The main obstacle to the use of REMs, which has limited its widespread adoption, derives from the costs associated with the deployment and maintenance of MCDs, and the power and bandwidth resources needed to communicate with the database.

The FCC has recently shown particular interest in the application of a rudimentary form of REM and sensor networks in the 3.5 GHz band, which it denotes as ESC. Under the proposal in [30], dedicated MCDs would be placed along the whole U.S. coast and would inform the SAS of the proximity and channels of operation of shipborne DoD systems. With the introduction of ESC, the 3.5 GHz band coastal exclusion zones would become dynamic, meaning that SUs could operate inside them but would need to be ready to vacate their channel when they are at risk of causing harmful interference to a PU. To prevent SAS users from pinpointing DoD systems' locations based on the reported RF measurements, MCDs would need to be placed sufficiently far apart.

In general, REM is an attractive spectrum-access technique for radar bands. We argued in [42] that the typically long transmit ranges of radar systems, which in many cases require exclusion zones with a radius over 50 km, relax the requirements for high resolution spatial data for spectrum sensing. Moreover, the geographically sparse deployment of radar stations, many of which are near coastal areas and airports, allows a more localized placement of MCDs. Due to these two reasons, we expect a generally low number of MCDs to be deployed in these bands, thereby driving down the cost of dedicated sensor networks. The MCDs' report and REM update rates are also expected to be low in most sharing scenarios, which can significantly reduce the amount of traffic circulating between MCDs and the REM, as well as the REM's complexity. Shipborne radar systems move relatively slowly when compared to their transmit ranges. Also, MCD may only send information when they are illuminated by a radar antenna mainbeam, which occurs less than 1% of the time.

3.4 Beyond traditional sharing schemes

From the radio environmental factors listed in Section 3.1.2 and the discussion presented in Section 3.3, it can be argued that spatial sharing methods such as DFS and GL-DB are fairly simple and effective approaches to spectrum sharing in radar bands. This attractive combination may explain why they are the currently proposed techniques by national regulators and ITU-R to populate the 3.5 and 5 GHz bands with low-power communication devices. On the other hand, multiple studies show that this form of sharing can still prevent a large percentage of the world's population from accessing the radar spectrum [27]. The reason for this limitation is that these approaches require very large exclusion zones, and radar systems are more densely deployed in coastal regions, and near harbors and airports, where population densities likewise tend to be higher.

Overcoming these issues requires that SUs employ more advanced interference mitigation techniques that allow them to transmit inside radar systems' exclusion zones without disrupting their services. As an example, consider primary radar systems with highly directional rotating antennas. From a temporal sharing perspective, a considerable amount of spectrum opportunities can be exploited in this scenario by allowing SUs to transmit when the radar antenna's mainbeam is pointing in a different direction. Another possible approach consists of employing multiantenna beamforming schemes as a way to reduce interference in the direction of radar antennas.

In this section, we provide a brief overview of the literature on more advanced sharing techniques and define the coexistence scenarios where they may be applicable in radar bands.

3.4.1 Temporal sharing

Temporal sharing comprises any channel access mechanism that leverages the time-domain cyclic patterns present in the traffic transmitted by an incumbent. Differently from spectrum sensing, an SU performing temporal sharing does not rely on its most recent RF measurement to infer a channel's current availability state, but rather on the long-term observation and identification of reoccurring time sequences or patterns in incumbents' channel access behavior. Examples of such patterns may include frequency hopping sequences, cyclic transmission time frames, or periodic antenna steering motions.

Radar bands present promising features for temporal sharing. Most surveillance and radionavigation systems utilize mechanically rotating antennas with constant scan periods to detect objects in their surroundings. An SU could leverage the signal strength oscillations created by such rotations and apply signal processing techniques to estimate the radars' ASPs. Once this information is obtained, the SU could then predict the exact instants when it will be illuminated by the radar and thus interrupt its own transmissions accordingly to avoid interference.

For the case of weather-forecast services, their radar systems generally display quasi-periodic slowly rotating scan patterns, transitioning between several rotation speeds throughout a day-period [7]. These features make them less suited to temporal sharing based solely on spectrum sensing and signal processing performed at the SU. On the other hand, the authors in [7] have shown that weather systems' rotation speed transitions happen very infrequently and are generally scheduled several hours in advance. Temporal sharing in this scenario could, therefore, still materialize through a close cooperation between SUs and incumbents [7]. Incumbent systems could inform a centralized database about their location and scheduled speed transitions, which the SUs would consult to acquire information about the radar in their environment, as well as the current rotation period.

There are, however, some exceptions where periodic or quasi-periodic scan patterns are not observed. One example is military missile control systems, which are used to track air objects' trajectories. Another example is ESA or proposed multiple input multiple output (MIMO) radar systems, whose electronically steered beams operate in an intentionally unpredictable manner from the perspective of a remote receiver [43]. These systems are, however, currently less predominant.

Several interference studies have been made concerning the use of temporal sharing in radar bands [36,37,44]. In [37], the authors quantitatively assess the achievable temporal opportunities if SUs were capable of switching off their transmissions when illuminated by a radar antenna main lobe. To account for the issue of aggregate interference, the authors considered that SUs would also refrain from using a channel if it is already being occupied by other nearby SUs. In [36], the authors evaluate the impact of aggregate interference on temporal sharing performance when the spatial distribution of SUs is not uniform. In this study, SUs perform a scheme called Temporal Dynamic Frequency Selection (DFS-T) that involves interrupting SU transmissions at the instants of time when the received signal strength from a radar exceeds a threshold. The results of this study showed the great improvement DFS-T provides in terms of reduced exclusion zones when compared to the conventional DFS scheme. In [44], the authors also evaluated the temporal sharing scenario between a rotating radar and SUs; however, differently from previous studies, their analysis was focused on the adequacy of this sharing model for different communication services including voice-over-IP (VoIP), video on demand, peer-to-peer file sharing, automatic meter reading, file transfer, and web browsing. The authors concluded that the data link interruptions caused by radar antenna mainbeam interference would make temporal sharing not suited for real-time transfer of small files and VoIP.

In [45], we provided a demonstration of the temporal sharing mechanism for software defined radio (SDR), the architecture of which is illustrated in Figure 3.9. In particular, it was shown that temporal sharing can be applied even in radio environments where more than one radar coexist in the same frequency band. This arrangement is a common occurrence in maritime bands where multiple ships may navigate at the same time in a harbor. To achieve this goal, the SU's sensing chain was separated into three main blocks: pulse detection, pulse sorting, and scan analysis.

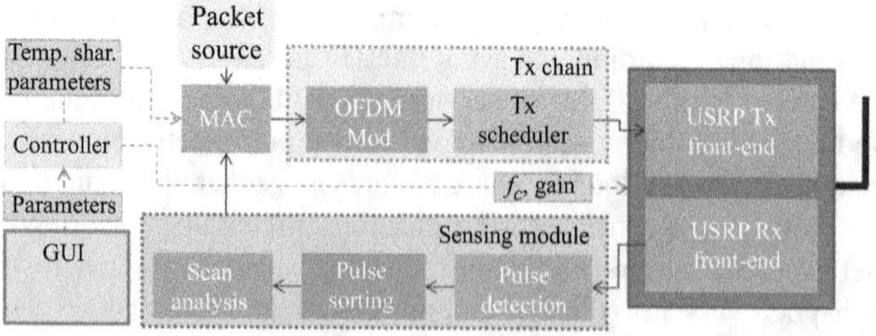

Figure 3.9 Block diagram of a temporal sharing-capable SU implementation for SDR

During pulse detection, the SU intercepts transmitted radar pulses and codes each pulse into a set of parameters or pulse description words (PDWs), namely, duration, center frequency, time of arrival, and amplitude. Pulse sorting then clusters pulses based on their PDWs into groups, with each group representing a different radar emitter. Through scan analysis, the SU then estimates the scan parameters of ASP and radiation pattern for each recognized radar emitter. The collected information is then passed to the SU's medium access control (MAC) module and used to predict future instances of interference between the SU and the multiple radars in the RF environment.

In [46], we proposed and analyzed the performance of several pulse sorting and scan analysis algorithms for the temporal sharing architecture in Figure 3.9. These techniques were tailored for low mobility radio environments where one or more radar systems are operating with constant antenna rotation speeds and do not employ low-probability-of-intercept (LPI) techniques to mask their presence. When such conditions are not met, we suggest that SUs should resort to more conservative techniques such as the current 3.5 GHz SAS.

The electronic intelligence literature suggests several pulse sorting techniques for the separation of superimposed radar signals. However, we have found that most of these techniques are ill-suited for temporal sharing as they are designed to store and perform batch processing over large sets of data without any real-time decision constraints. In [46], we developed new low complexity pulse sorting algorithms that operate in an online manner and yield superior performance than the alternatives at coping with the burstiness of radar signals caused by their antenna motion. Figure 3.10 illustrates one such pulse sorting algorithm that correctly deinterleaves the pulses of two maritime radar systems that were encountered during our measurement campaign at Cork Naval Base, Ireland.

Given the high detection thresholds needed to protect radar systems from harmful interference, thermal noise is not expected to significantly impact the capability of an SU to sense and estimate the radar system's parameters and to predict temporal opportunities. Instead, it can be argued that the phenomenon of inter-scan channel decoherence should not be ignored since it reduces the periodic features of the interference channel, potentially leading to erroneous transmit decisions. A considerable

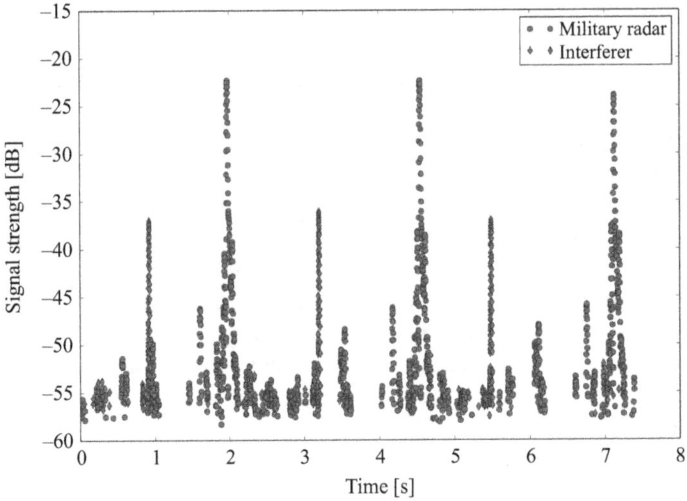

Figure 3.10 Detected pulses from two radar systems that were operating simultaneously near Cork Naval Base, Ireland. The circles correspond to pulses transmitted by the military ship Aisling, while the diamonds correspond to pulses from a cargo ship docked in the harbor

part of the work in [46] was focused on improving the robustness of proposed radar scan parameter estimation and tracking techniques to compensate for this effect. In particular, a new test statistic for parameter estimation was proposed, along with a stopping criterion for adaptive sensing, which were shown to provide interference probabilities similar to those achieved when SUs have clairvoyant knowledge of radar parameters and experience ideal synchronization.

3.4.2 Cognitive beamforming

Beamforming is an effective approach used in multi-antenna systems to improve the experienced signal to noise ratios (SNRs), rejection of interference, or, in the case of Multi-User MIMO (MU-MIMO), to enable the transmission of parallel streams to multiple devices. Its basic idea consists of adjusting the weights of different antenna elements to obtain a beam radiation pattern with the desired amplitude and direction signatures.

Beamforming can be employed both for signal reception and transmission. Receive beamforming allows users to increase their signal to interference plus noise ratio (SINR) by maximizing and nulling the array gain in the directions of the transmitter and interferers, respectively. It is generally simpler to implement receive than transmit beamforming since it can be applied by each user individually without affecting the remaining users in the network, and it relies on channel state information (CSI) obtained locally. In contrast, transmit beamforming may need to consider the performance of the entire system jointly, and obtaining CSI generally requires feedback from receivers or channel reciprocity that can only be assumed in time division duplexing (TDD) low-mobility systems.

With the introduction of MIMO in communication standards like WiMAX, LTE, and WLAN, the application of beamforming to spectrum sharing and CR scenarios, usually referred to as cognitive beamforming, has become an active area of research in recent years. In this setting, SUs would choose sets of transmit antenna weights or precoders that optimize their quality-of-service (QoS) while keeping interference at PUs below a certain stipulated limit.

Cognitive beamforming has the potential to remove incumbents' exclusion zones altogether. However, like conventional multiantenna schemes, its achievable performance is strictly dependent on the quality of the CSI available to SUs. Consequently, its applicability is limited to scenarios where either there is cooperation between PUs and SUs in the form of CSI or incurred interference feedback exchanges, or when channel reciprocity can be assumed.

Several beamforming research studies can already be found in the literature under the umbrella of CR or interference mitigation in cellular systems [47–60]. In [47–50], the authors considered the scenario where transmitters possess perfect channel estimates, either in the form of instantaneous CSI or covariance-based CSI. In [50], the authors propose a low-complexity algorithm for downlink SU user equipment (UE) selection that, when combined with adaptive transmit beamforming at the cognitive base station (BS), can jointly reduce interference on a single PU Tx–Rx pair and maximize the cognitive network throughput. Neither of these studies is deemed realistic for DSA considering the inevitable quantization errors and latencies of PU-SU feedback channels, or in the cases of channel reciprocity the issues of imperfect RF front-end calibration and thermal noise. In [52], the concept of an effective interference channel (EIC) is introduced as a way to account for a PU Tx–Rx pair equipped with several antennas, and the implications of nonideal blind channel estimation techniques on the leakage of interference at the PU are assessed. In [51], a cognitive beamforming mechanism was proposed and implemented in a real-time LTE-based CR on EURECOM's experimental OpenAirInterface platform. For PUs, the authors considered a single antenna TDD LTE BS and UE, whose specifications were known to the SU. A calibration technique was employed to restore the reciprocity characteristic of TDD systems lost due to RF front-ends mismatch.

Since methods that rely on perfect CSI are very sensitive to channel uncertainties, several authors have shifted their focus to a class of algorithms called robust beamformers [53–60]. One of the possible approaches to design these algorithms is to consider that the estimation errors of SUs are norm-bounded by ellipsoid contour regions. The role of the SU is then to compute a precoder constrained by these regions under worst case conditions [53–56]. Another approach for developing robust schemes is based on outage probabilities [57–60]. In these cases, the distribution of CSI errors is known, and worst-case error constraints are replaced by probabilistic constraints that the interference caused to the PU remain below a predefined value.

As discussed in Section 3.1.2, most radar systems present the key features of colocation of Tx and Rx antennas, and use of the same frequencies for Tx and Rx. This means that channel reciprocity between SUs and PUs is generally a fairly reasonable assumption to make after transceiver calibration, and cooperation between systems through feedback links can be avoided. Another important advantage of radar spectrum is the usual spatially sparse deployment of radar systems. A general rule of beamforming is that the number of nodes a transmitter can avoid causing interference to is limited by the number of antennas it possesses. Considering the lower densities of PUs per radio environment in radar bands, we envision that an SU with cognitive beamforming capabilities will, in general, only require from 2 to 8 antennas to operate without causing harmful interference.

On the other hand, a major obstacle to the feasibility of cognitive beamforming in radar bands is the rapid changes in the interference channel caused by the radar scan patterns, which we refer to as *intrascan channel decoherence*. A multipath environment composed of several reflectors with different physical features such as positions and cross-sections will cause an interference channel to decohere on the order of a few milliseconds.

The literature on the application of cognitive beamforming to radar bands is still limited. In [61,62], the authors propose the employment of multiple antenna techniques for spectrum sharing between radar and cellular systems. However, these papers focus on the shaping of the beamforming pattern at the radar side, a solution that would not be applicable for legacy radar systems. Moreover, the high density of LTE systems that the 3.5 GHz band may accommodate in the future, along with the authors' assumption of a real-time feedback link between military and civilian systems for CSI exchange, make this approach very challenging to implement.

3.4.2.1 Offline beamforming

In our own work [14,63], we focused on the use of cognitive beamforming by multiantenna LTE small cells to reduce their interference to radar systems. To overcome the issue of fast intrascan channel decoherence caused by the constant rotation of the radars' antennas, we proposed the use of offline beamforming techniques that do not rely on the recently estimated interference CSI, but rather on a set $\mathcal{S}$ of multiple CSIs obtained for different radar antenna orientations. Assuming that both the radar and SU have static or quasi-static positions, it is expected that the interference channel possesses cyclic characteristics across multiple radar antenna scans, and that previous CSIs obtained for a specific radar antenna orientation angle are still valid when those angles are revisited in subsequent antenna sweeps. This assumption was validated during the measurement campaign in Cork as shown in Figure 3.5. Contrary to traditional beamforming schemes or temporal sharing, the SU under this scheme would try to jointly cancel interference for all previously estimated CSIs. In this way, one can forgo the requirement of instantaneous channel estimation that is present in traditional beamforming, as well as the requirement that the SU be synchronized with the radar antenna rotation that arises for temporal sharing.

Figure 3.11 illustrates a sharing scenario that comprises a radar system whose antenna orientation follows a specific scan pattern $(\phi(t), \theta(t))$ and the downlink (DL) channel between a small cell evolved node base station (eNB) and five UEs. The DL channel center frequency coincides with the radar's frequency of operation. The eNB utilizes a half-wavelength spaced uniform linear antenna array comprised of four

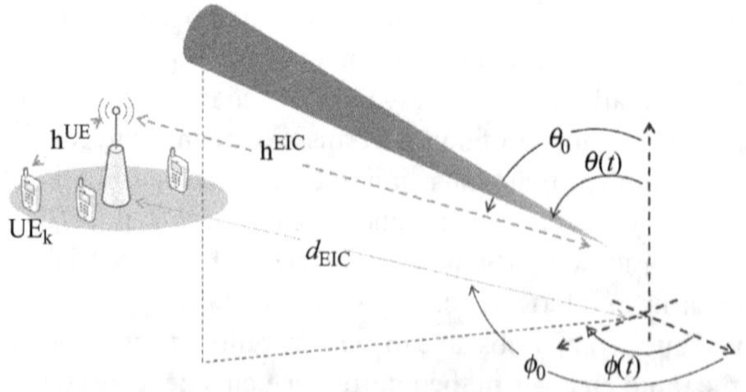

Figure 3.11 Spectrum sharing between a small cell downlink and a radar system

elements, with each element modeled as a dipole. The eNB transmits a single stream per UE and serves one UE per resource block (RB). The UEs, in turn, are equipped with one dipole antenna.

For a single OFDM subcarrier, the symbols $s \in \mathbb{C}$ transmitted by the eNB reach the radar after being subject to the transformation

$$\mathbf{y} = \mathbf{h}^{\mathrm{EIC}}\mathbf{w}s + \mathbf{z}. \tag{3.1}$$

Here $\mathbf{w} \in \mathbb{C}^{4\times1}$ is the eNB's applied precoder, $\mathbf{z} \in \mathbb{C}$ is AWGN at the radar's receive antenna, and $\mathbf{h}^{\mathrm{EIC}} \in \mathbb{C}^{1\times4}$ is the EIC, which encompasses the effects of the propagation channel, the hardware front-ends, antenna radiation patterns of the eNB and radar, and the current orientation $(\phi(t), \theta(t))$ of the radar's antenna.

The offline cognitive beamforming problem then consists of finding the RB allocation and precoders that maximize throughput for the UEs while maintaining the instantaneous interference at the radar below a predefined threshold I_{max} for all past CSI measurements. This problem is stated mathematically as

$$\sum_{f=1}^{N_{\mathrm{RB}}} \left\| \mathbf{h}^{\mathrm{EIC}}[k,f]\mathbf{w}[f] \right\|^2 \leq I_{max}, \forall k \in \mathcal{S}, \tag{3.2}$$

where N_{RB} is the number of LTE RBs, $\mathbf{h}^{\mathrm{EIC}}[k,f]$ is the EIC at RB frequency index f and time index k, $\mathbf{w}[f]$ is the selected precoder for the RB frequency index f, and $\mathcal{S}$ is the set of eNB's from past EIC measurements. Finding the appropriate precoder for each UE's RB at frequency f can be achieved by solving the multiobjective optimization problem

$$\begin{aligned} \underset{\mathbf{w}}{\text{minimize:}} \quad & \sum_{i=1}^{|\mathcal{S}|} \mu_i \left\| \widehat{\mathbf{h}}^{\mathrm{EIC}}[k]\mathbf{w} \right\|^2 + \mu_{|\mathcal{S}|+1} \left\| \widetilde{\mathbf{h}}^{\mathrm{UE}}\mathbf{w} - P_{\mathrm{eNB}} \right\|^2 \\ \text{s.t.} \quad & \|\mathbf{w}\|^2 = P_{\mathrm{eNB}}, \end{aligned} \tag{3.3}$$

where $\widehat{\mathbf{h}}^{\mathrm{EIC}}$ is the estimate of the EIC, $\widetilde{\mathbf{h}}^{\mathrm{UE}}$ is the normalized eNB-UE channel, and P_{eNB} is the eNB's transmit power. The weights μ_i, $i = 1, \ldots, |\mathcal{S}|$ scale each

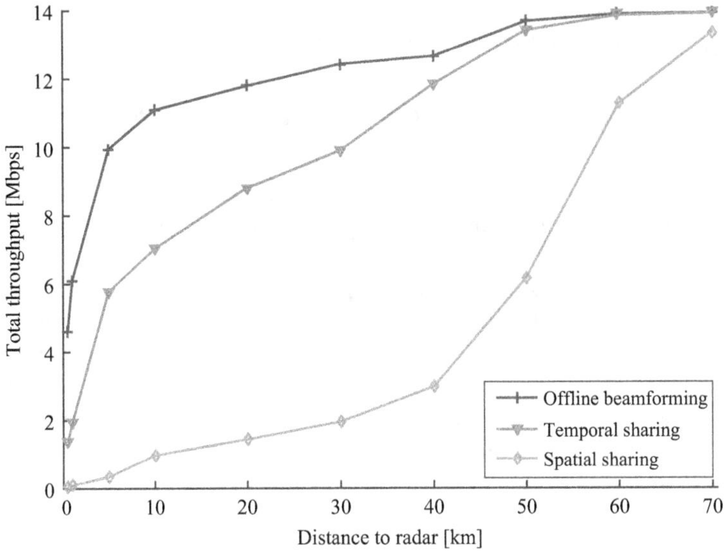

Figure 3.12 LTE small cell's total throughput for different sharing schemes and distances to an incumbent radar. The channel bandwidth was set to 5 MHz and the radar's maximum interference-to-noise ratio threshold to −6 dB

interference constraint and $\mu_{|\mathcal{S}|+1}$ does likewise for the UE's throughput term. The optimization problem (3.3) is a quadratically constrained least squares (QCLS) program that can be solved via the procedure in [64].

One aspect not addressed in this multiobjective optimization transformation is the selection of the weights μ_i for each of the $|\mathcal{S}| + 1$ constraints in (3.3). A simple approach we follow in this study is to set μ_i equal for all constraints.

Figure 3.12 provides a comparison between spatial sharing, temporal sharing, and the offline beamforming scheme in (3.3) for a small cell coexisting with a magnetron radar in a 5 MHz channel. The radar maximum INR threshold was set to −6 dB. RBs for which this limit could not be respected were left empty. Under both the spatial and temporal sharing schemes, the eNB used the maximum ratio transmission (MRT) precoder. The main difference between these two schemes is that, under the temporal sharing scheme, the eNB has knowledge of the current radar antenna rotation phase and can thus more confidently infer whether its caused interference will exceed the limit of −6 dB.

As can be seen in Figure 3.12, offline beamforming provides the highest spectral efficiency gains, allowing eNBs to coexist with radars at distances lower than 5 km. On the other hand, its requirement for static terminals to keep the inter-scan channel decoherence low makes it unsuitable for some scenarios, such as handheld devices' uplink. This approach also hinders the use of spatial multiplexing and MU-MIMO schemes to increase spectral efficiency since the extra degrees of freedom provided by multiple antennas are used to reduce interference instead.

3.4.3 Coexistence through cooperation and codesign

Spectrum sharing between communication and military systems is a particularly challenging problem due to the operational security requirements of military systems. For instance, it is not uncommon for military radars to use LPI techniques to avoid being jammed or detected by hostile forces.

In 2012, DARPA released its Shared Spectrum Access for Radar and Communications (SSPARC) program seeking the development of technology for coexistence between military radars and other military and commercial communication systems [43,61,65,66]. Contrary to prior work, the SSPARC program focused on cooperative spectrum sharing approaches operating under the assumption that primary and secondary systems could share information on their respective spectrum activity in near real time. This program specified three main tasks:

- *Coexistence* – Development of technical sharing approaches between legacy radar and communication systems.
- *Codesign* – Long-term development of new, more advanced radar and communication systems jointly designed with coexistence taken into account from their inception.
- *Fundamental limits* – Theoretical analysis to determine joint radar and communication performance bounds for spectrum sharing.

Within the *coexistence* task, several papers have been published assessing the impact of interference between current wireless broadband communication standards such as LTE, WLAN, and WiMAX, and radar systems [67–69]. This work has shed some light on how radar transmission parameters, particularly duty cycle and PRF, may disrupt traditional communications' physical layer (PHY) and MAC mechanisms, and on the overall impact of these parameters on communication systems' throughput and packet transmission delay. Specifically, [69] showed that optimization of communication parameters like packet length, number of retransmissions, and forward error correction schemes can be selected in such a way that interference from radar systems is minimized.

As part of the *codesign* task, several authors have focused on the design of future MIMO radar technology capable of avoiding interference to and from communication systems through precoding techniques [61,66,70,71]. The proposed techniques assume that the MIMO radar would obtain real-time CSI from small cells via a feedback link. In [43], the authors study the feasibility and spectral efficiency of frequency separation techniques, coordinated by a context-aware SAS.

Under the umbrella of *fundamental limits*, [72–76] has assessed coexistence from an information theoretical perspective, developing metrics to account for the impact of interference between radar and communication systems. In [67,72], the authors propose information-theoretic frameworks to evaluate trade-offs between radar systems' tracking performance and communication systems' throughput. By using a new metric to quantify radar performance called estimation rate, the authors in [72] also analyze performance bounds of diverse cooperative signaling schemes. In [74], codesign is posed as a joint optimization problem involving the maximization of the weighted sum of radar probability of detection and communication capacity. It was shown that, through water filling, communication systems would be able to allocate transmit power to the quieter segments of the spectrum. Finally, [75,76] focused on the characterization of mutual interference between radar and communication waveforms.

3.4.4 Open challenges

While the literature on spectrum sharing between radar and communication systems is expanding at a rapid pace, there are still multiple issues that remain unsolved or require further analysis. In this subsection, we summarize some of these issues and discuss how they may affect the practicality of spectrum sharing schemes that different research groups have proposed.

One major issue is the lack of radar-communication channel models that reliably account for the effects of multipath, antenna rotation, and the resulting channel decoherence over time and space. As pointed out in [14,61], the efficacy of spectrum sharing techniques such as precoding/beamforming and temporal sharing is directly correlated to these phenomena. To circumvent the lack of available models, some have modeled channel uncertainties via random distributions or by limiting them by error regions with predefined norms. However, there are not yet sufficient measurements to quantify the expected magnitude of these channel estimation uncertainties across the wide variety of scenarios an SU may encounter.

To avoid the inefficiencies of DFS and geolocation-based solutions while addressing the wide variety of existent radar systems, most research groups have been advocating spectrum sharing in radar bands as a cooperative problem [7,43,61,63]. Most of these proposed solutions require a context-aware SAS that coordinates spectrum access and selects the most appropriate techniques that SUs should employ to avoid causing interference to incumbents. However, we find there is a lack of work in the literature on possible architectures for this SAS and what burdens the use of the proposed sharing techniques place upon it.

Another topic that has not been comprehensively addressed is how proposed sharing techniques can be efficiently integrated with current small cell technology standards like Licensed Assisted Access-LTE (LAA-LTE) or LTE-Unlicensed (LTE-U). Temporal sharing, being an opportunistic access technique, does not provide guarantees regarding the percentage of time that spectrum will be available to SUs. Similarly, cognitive beamforming may cause strong variations in QoS when UEs move to or from a position that is closely aligned with the interference channel. This lack of strong assurances regarding spectrum availability may present some challenges to LTE networks that will have to manage the partitioning of users' data traffic between licensed and unlicensed bands based on defined latency and power-consumption requirements. For unlicensed spectrum standalone solutions like MulteFire or WLAN, these issues are exacerbated because UEs also have to react quickly to sudden changes in their experienced QoS and search for alternative eNBs with which to connect.

Temporal sharing and cognitive beamforming may also require that communication system operators follow different methodologies to position their eNBs and deploy their networks. For instance, small cells will be able to exploit more spectrum opportunities if they are placed in NLOS locations relative to the radar stations. In some cases, this arrangement may conflict with an operator's goal to maximize system coverage. Another aspect of cell deployment in this context is the impact of uneven range or QoS as a function of direction when employing beamforming techniques. Consider the example of an indoor eNB intended to serve a full room in an enterprise. If this cell displays a longer range in one particular direction, because it is facing away from a radar station, then the eNB should be positioned in such a way that said direction covers the whole room. It may, however, be challenging for an operator to know which are the most favorable positions without significant effort dedicated to testing.

3.5 Secondary radio access technologies

Regulatory and industry bodies envision small cells and WLAN as the main types of RATs to coexist with radar systems in the 3.5 and 5 GHz bands on a secondary basis. Several groups, including Ericsson, 3GPP, and Qualcomm, are currently expending significant effort in the deployment of LTE networks in the 5 GHz band where WLAN is already well established [17]. On the other hand, in the 3.5 GHz spectrum, only LTE seems to be currently under consideration. The WLAN Working Group has been hesitant to invest in this band due to excessive restrictions created by the current FSS and radar exclusion zones [23]. Its position may change, however, as more innovative sharing technologies like ESC are deployed.

Both the 3.5 and 5 GHz radar bands offer exceptional opportunities to small cell and WLAN technologies. First, these bands contain a large amount of underutilized spectrum that cellular operators can use to offload mobile data and WLANs can use to relieve current congestion problems in the 2.4 GHz ISM band. Second, the large attenuation and penetration losses characteristic of frequencies above 3 GHz combined with the radar systems' large exclusion zones make these bands less attractive for licensed commercial macrocell deployments. Third, higher penetration losses allow denser small cell deployments and greater frequency reuse for urban and indoor areas.

3.5.1 LTE in unlicensed spectrum

The term "small cell" encompasses any type of operator-controlled short-range BSs including femtocells, picocells, and microcells that are used to complement mobile phone service of larger macrocells. Femtocells were originally described as residential products connected using standard domestic internet broadband. Picocells and microcells were, in turn, used for enterprise/business premises and public outdoor spaces, respectively.

Recently, the 3GPP, the LTE-U Forum whose members include Verizon, Ericsson, Alcatel-Lucent, Qualcomm, and Samsung, and other groups have been investing

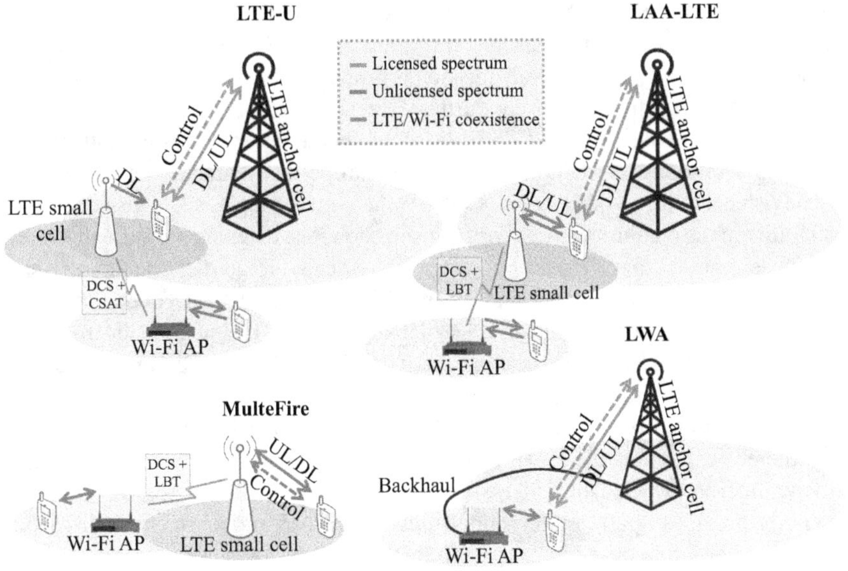

Figure 3.13 Different network architecture models to extend LTE and cellular services to unlicensed spectrum

considerable effort in the development of LTE small cell networks that operate in the unlicensed 5 GHz spectrum as a way to boost indoor coverage [17]. The FCC and LTE-U Forum also anticipate a widespread use of unlicensed LTE networks in the 3.5 GHz spectrum as GAA tier users [29,77].

There are currently three main types of LTE unlicensed access: LTE-U, LAA-LTE, and MulteFire. The deployment of each of these technologies will depend on the regulation restrictions imposed in each different country and the type of entities owning the LTE infrastructure. Figure 3.13 illustrates these three approaches along with LTE Wi-Fi Aggregation (LWA), which is described in Section 3.5.3.

LTE-U leverages 3GPP Release 10 [LTE-Advanced (LTE-A)] and will be commercialized and deployed in the United States, China, and Korea. LTE-U comprises only the downlink, relying on carrier aggregation with an "anchor" licensed spectrum network for most of the control channel, signaling, and uplink data exchanges [78]. To coexist with WLAN systems in the 5 GHz spectrum, LTE-U may employ dynamic channel selection (DCS) and Carrier-Sensing Adaptive Transmission (CSAT). DCS enables eNBs to frequently monitor and select the channel with the lowest LTE and WLAN activity for supplemental DL transmission. This procedure can be performed either through energy detection, which is technology-agnostic, or through WLAN preamble and LTE signaling identification mechanisms. The latter have the particular advantage of allowing an eNB to estimate the number of WLAN access points (APs) and LTE eNBs present in the channel. In the event that no clean channel is available, the eNB has to rely on the CSAT algorithm to coexist with other nodes. Using CSAT, an eNB defines a CSAT TDM cycle that lasts a few hundreds of milliseconds. This cycle is separated into two periods, one for LTE transmission and the other for sensing the medium. The duty cycle of transmission is proportional to the amount of time that measured channel activity is absent during the sensing period.

LAA-LTE is based on the LTE Release 13 specification and is compliant with regulatory requirements of India, Japan, and Europe that mandate the use of LBT for coexistence with WLAN devices [78]. When employing LBT, eNBs check the availability of the medium before every transmission. The listening period, called clear channel assessment, has a duration greater than or equal to 20 ms. The transmission period, in turn, lasts between 1 and 10 ms. Like LTE-U, LAA-LTE operates with an "anchor" licensed band and uses carrier aggregation to integrate licensed and unlicensed spectra. However, in some of its releases, LAA-LTE can utilize both uplink and downlink in a TDD fashion in the unlicensed spectrum.

Finally, MulteFire promotes a standalone version of LTE in unlicensed spectrum, thereby expanding the use of LTE to entities that do not own licensed spectrum, namely, internet service providers, enterprises, and small businesses. As such, MulteFire provides operators the opportunity to offload and augment their mobile networks when no licensed carrier is available [79]. It is a solution promoted and standardized by the MulteFire Alliance that was formed by Qualcomm and Nokia as founding members, with Ericsson and Intel as general members. This technology will be built on top of LAA-LTE and therefore relies on DCS and LBT mechanisms.

3.5.2 Licensed LTE

Not many predictions have yet been made regarding what type of RATs will make use of the 3.5 GHz spectrum under PA tier rules. However, considering the current licensing model that allows users to transit from GAA to PA through the acquisition of a PAL, we expect that both tiers will employ similar radio technologies, potentially TDD-LTE.

We envision that once an operator acquires a PAL for exclusive access to a 3.5 GHz channel within a certain census tract, its eNBs no longer needs to employ LBT or CSAT to coexist with other users and technologies. Under such conditions, operators may resort to more conventional licensed spectrum standards such as LTE or LTE-A to ensure a higher QoS for their users.

Several papers have been published assessing the robustness of the LTE standard and its constituent algorithms to the low duty cycle interference caused by a radar. In [67,80], the authors performed experimental studies, evaluating the impact of 3.5 GHz radar systems' interference on a TDD-LTE eNB. In particular, [67] showed that LTE systems can coexist with SPN-43 military radar systems in the same band at a distance of 5 km with an error rate of nearly 0. Using link simulations, [81] analyzed the impact of radar interference on LTE reference and data subcarriers, on different modulation and coding orders, and on the LTE soft-decoder. It was shown that, despite the use of coding mechanisms, LTE is very sensitive to bursty interference. The authors demonstrated that the main cause behind this phenomenon was the log-likelihood ratio errors in the LTE soft-decoder that assume Gaussian interference and then proposed a simple method to circumvent this problem.

3.5.3 WLAN

WLANs share many of the characteristics of femtocells. However, WLANs are not primarily designed to assist or be controlled by mobile operators and utilize a distinct RAT. The WLAN technology is already well established in the 5 GHz radar bands. To operate in this band, WLAN devices have to employ DSA techniques such as DFS to protect radar systems and LBT protocols to avoid interfering with other unlicensed devices.

An important exception to the conventional unplanned WLAN model is LWA, currently under study to be part of 3GPP Release 13 for use in the 5 GHz band [82]. LWA has emerged as an alternative to LTE over unlicensed spectrum with the aim to resolve existing disputes between 3GPP and the Wi-Fi alliance over the coexistence of LTE-U and LAA-LTE with WLAN systems. This approach uses WLANs APs to tunnel part of the LTE data in 802.11 MAC frames and transmit them over unlicensed spectrum. The UE will then use a split traffic data plane and aggregate the data flows coming from both an LTE anchor station and WLAN APs in different bands. Hence, through this method, no changes to the LTE channel access schemes are needed, and the already widespread Wi-Fi networks can be repurposed, thereby avoiding extra costs.

3.6 Conclusions

In this chapter, we have provided an introduction to spectrum sharing in radar bands from a communication perspective, including a review of pertinent literature. This review focuses on four different aspects: (i) sharing-relevant radio environmental features, (ii) current policy context, (iii) adequacy of existing spectrum-access techniques, and (iv) SU technologies that may be deployed in this spectrum.

Some of the radar systems found in the 225–6000 MHz frequency ranges were characterized in terms of operating principle, waveform properties, beam/scanning patterns, spectrum bands, and other deployment aspects. Based on these characteristics, a list of key radio environment factors was identified that are relevant to spectrum sharing. Key challenges associated with the exploitation of spectrum opportunities in these bands include the diversity of existent incumbent radars, their very long transmit ranges, heavier deployment in high population density areas, their crucial role in public safety services, and the exclusion zone and channel uncertainties that emerge from their antennas' directivity and fast rotation. On the other hand, these bands also offer some favorable features for sharing such as the absence of hidden terminals, low density of incumbents, and easy detectability of their signals through spectrum sensing.

Current regulator initiatives to make the 5150–5925 MHz and the 3550–3700 MHz radar bands available to broadband radio services was discussed. To address the technical issues arising from these initiatives, primarily the avoidance of undue interference to the incumbent radar systems, regulatory and industry bodies have proposed a variety of possible solutions with various trade-offs. For example, DFS and the SAS present some important design and technical limitations that restrict a large portion of broadband users from benefiting from these initiatives.

DSA was discussed in the context of spectrum sharing in radar bands. The pros and cons of several conventional spectrum-access techniques were analyzed, including GL-DB, spectrum sensing, cooperative sensing, REM, and beacon signaling, with the argument made that none of these solve the key issue of radar systems' very large exclusion zones that may cover a large portion of cities and coastal regions. In contrast, prospective solutions that do address this issue involve the use of temporal sharing and cognitive beamforming, which can allow SUs to operate at shorter distances from radars without causing influential interference. Although promising, a qualitative evaluation and literature review of these techniques show that their application to sharing in radar bands is still not well understood and requires further investigation. In particular, it is important to recognize that certain radio environmental factors like fast, and sometimes irregular, radar antenna rotation speeds may affect the performance of these techniques and prevent them from being universally applicable to all coexistence scenarios an SU may face.

Finally, several small-cell technologies were reviewed that industry is envisioning to deploy in the 3.5 and 5 GHz bands. The fundamental differences were discussed between the LTE-U, LAA-LTE, MulteFire, and LWA standards, which 3GPP and other groups are currently proposing to coexist with WLAN devices in the 5 GHz band.

3.7 Looking ahead

The 3.5 GHz CBRS initiative represents an important step toward the PCAST's general goal of opening up sparsely utilized government-owned radar bands for commercial communication applications. The NTIA initially proposed this band for its particularly attractive features for spectrum sharing. In particular, it was argued that the geographically limited operation of Navy ships and the limited number of ground-based radars and satellite systems would facilitate the deployment of a GL-DB spectrum access scheme like the current SAS. Taking into account the very low number of shipborne radar systems operating in this band and the fact that these systems spend a significant portion of their time at sea, ESC was later suggested as a way to dynamically unlock regions near the U.S. coast for secondary use.

While the combination of GL-DB technology with ESC allows for a significant reduction of the original radar exclusion zones in the 3.5 GHz band, some doubts have been raised on whether this sharing framework is fully extensible to other radar bands (military or nonmilitary). Other bands generally accommodate a higher density of systems, which limits the efficiency gains that stem from ESC. On the other hand, relying solely on GL-DB technology has been shown to limit spectrum usage in areas of higher population density where it is needed the most.

In addition to its spectrum being underutilized, at least from a communication perspective, most civilian radar systems have an area of coverage that is rather orthogonal to those of commercial communication systems. For instance, the antennas of these radars are generally directed to scan aerial and sea spaces, while communication systems focus on providing land coverage. This aspect suggests that the long-term solution to exploit radar spectrum lies in making these two types of radio systems better at coexisting with each other. In this chapter we summarized some of the coexistence solutions being explored in the literature, such as adaptive antennas, CR techniques, and radar–communication codesign methodologies. While the potential spectral efficiency gains brought by these techniques are encouraging, there are several challenges for their wide adoption.

The first main issue is that none of these advanced coexistence techniques is universally applicable to all types of radar and communication systems, policies, and radio environments. For example, cognitive beamforming and temporal sharing rely on relatively static interference channels. Furthermore, policy and security aspects may limit the amount of information SUs can obtain from incumbents in military bands. Interference mitigation using MIMO radar, on the other hand, is not applicable to legacy radar systems that are expected to remain in operation for decades. To circumvent these problems, we envision all these techniques as extensions of more conservative sharing frameworks to which devices revert in more challenging coexistence scenarios. However, the transitions between different techniques and the enforcement of interference protection requirements are still open problems.

Further research is also needed on the impact of the use of flexible coexistence techniques on radio hardware cost, design complexity, and the ability to meet different traffic requirements (e.g. energy consumption, latency, and robustness to mobility). The introduction of cognitive capabilities into commercial standards like LTE and WLAN without violating backwards compatibility and service guarantees has always been a challenging problem. Further, the implementation of one or multiple coexistence techniques in a standard may lead to more expensive radio equipment, thereby discouraging industry from investing in spectrum sharing. We predict that the growing market adoption of SDR technology and concepts such as slicing for 5G will help address some of these issues. In this context, radio devices will be able to tailor their RAT for specific traffic and coexistence requirements rather than trying to accommodate all types of services and possible spectrum neighbors.

References

[1] A. Martian, C. Vladeanu, I. Marcu, and I. Marghescu, "Evaluation of spectrum occupancy in an urban environment in a cognitive radio context," *International Journal On Advances in Telecommunications*, vol. 3, no. 3, pp. 172–181, 2010.

[2] Cave Audit Team, "Review of bandsharing solutions – Final report produced for: The cave independent audit of spectrum," ref. 72/05/R/281/R, no. 1, 2005.

[3] M. Hamid and N. Björsell, "Geo-location spectrum opportunities database in downlink radar bands for OFDM based cognitive radios," in *IEEE Conference on Communication, Science & Information Engineering*, 2011.

[4] F. Paisana, N. Marchetti, and L. A. DaSilva, "Radar, tv and cellular bands: Which spectrum access techniques for which bands?" *IEEE Communications Surveys & Tutorials*, vol. 16, no. 3, pp. 1193–1220, 2014.

[5] NTIA, "Developing forward thinking rules and processes to fully exploit spectrum resources: An evaluation of radar spectrum use and management," in *International Symposium on Advanced Radio Technologies (ISART), Special Publication SP-12-485*, 2011.

[6] R. G. Wiley, *ELINT: The Interception and Analysis of Radar Signals*. Norwood, MA: Artech House, 2006.

[7] Z. Khan, J. J. Lehtomaki, R. Vuohtoniemi, E. Hossain, and L. A. Dasilva, "On opportunistic spectrum access in radar bands: Lessons learned from measurement of weather radar signals," *IEEE Wireless Communications*, vol. 23, no. 3, pp. 40–48, 2016.

[8] ITU-R, "Efficient Use of the Radio Spectrum by Radar Stations in the Radiodetermination Service," Tech. Rep., 1998.

[9] ITU-R, "Characteristics of radiolocation radars, and characteristics and protection criteria for sharing studies for aeronautical radionavigation and meteorological radars in the radiodetermination service operating in the frequency band 2700–2900 MHz," Tech. Rep., 2003.

[10] F. H. Sanders, R. L. Sole, B. L. Bedford, and T. Pawlowitz, "Effects of RF interference on radar receivers," NTIA Report TR-06-444, 2006.

[11] ETSI, "Broadband radio access networks (BRAN); 5 GHz high performance RLAN; guide to the implementation of dynamic frequency selection (DFS)," Technical Report 102 651 v1.1.1, pp. 1–23, 2009.

[12] NTIA, "Evaluation of the 5350–5470 MHz and 5850–5925 MHz bands pursuant to section 6406(b) of the middle class tax relief and job creation act of 2012," Tech. Rep., 2013.

[13] QUASAR, "Initial report on the tolerance of legacy systems to transmissions of secondary users based on legacy specifications," *Deliverable ICT-248303/WP3/D3.1/100630*, pp. 1–45, 2010.

[14] F. Paisana, "Spectrum Sharing Techniques for Radar Bands," PhD Thesis, Trinity College, Dublin, 2017.

[15] M. Tercero, K. W. Sung, and J. Zander, "Impact of aggregate interference on meteorological radar from secondary users," in *IEEE Wireless Communications and Networking Conference (WCNC)*, Mar. 2011, pp. 2167–2172.

[16] FCC, "Revision of Part 15 of the commission's rules to permit Unlicensed National Information Infrastructure (U-NII) devices in the 5 GHz band," First Report and Order (R&O) FCC 14-30, ET Docket No. 13-49, 2014.

[17] Alcatel-Lucent, Ericsson, Qualcomm, Samsung, and Verizon, "LTE-U technical report: Coexistence study for LTE-U SDL v1.0," LTE-U forum, Tech. Rep., 2015.

[18] ITU-R Recommendations M.1652-1, "Dynamic frequency selection in wireless access systems including radio local area networks for the purpose of

protecting the radiodetermination service in the 5 GHz band," *Std.*, vol. 1, 2011.
[19] ETSI, "Broadband radio access networks (BRAN); 5 GHz high performance rlan; harmonized en covering the essential requirements of article 3.2 of the r&tte directive," 2003.
[20] ECC, "The current status of DFS (Dynamic Frequency Selection) in the 5 GHz frequency range," Tech. Rep. Report 192, 2014.
[21] T. Ngo, "Why Wi-Fi stinks - and how to fix it," *IEEE Spectrum*, vol. 53, no. 7, pp. 44–50, 2016.
[22] FCC, Order on Reconsideration and Second Report and Order, FCC 16-55, GN Docket No. 12-354, 2016.
[23] A. P. Stephens, "Comments of IEEE 802.11," IEEE 802.11 Working Group, Second FNPRM Comments, GN Docket No. 12-354, 2015.
[24] FCC, "Connecting America: National Broadband Plan," 2010.
[25] B. Obama, "Presidential Memorandum: Unleashing the Wireless Broadband Revolution," *Memorandum for the Heads of Executive Departments and Agencies*, 2010.
[26] NTIA, "An assessment of the near-term viability of accommodating wireless broadband systems in the 1675–1710 MHz, 1755–1780 MHz, 3500–3650 MHz, 4200–4220 MHz, and 4380–4400 MHz bands," Fast Track Report, 2010.
[27] FCC, "Enabling innovative small cell use in 3.5 GHz band," Washington, Notice of Proposed Rulemaking (NPRM) & Order Docket 12-148, 2012.
[28] FCC, "Commission seeks comment on licensing models and technical requirements in the 3550–3650 MHz band," Public Notice (PN) FCC 13-144, GN Docket No. 12-354, 2013.
[29] FCC, "Amendment of the commission's rules with regard to commercial operations in the 3550–3650 MHz band," Further Notice of Proposed Rulemaking (FNPRM), FCC 14-49, GN Docket No. 12-354, 2014.
[30] E. Drocella, J. Richards, R. Sole, F. Najmy, A. Lundy, and P. Mckenna, "3.5 GHz exclusion zone analyses and methodology," Tech. Rep., 2015.
[31] FCC, Report & Order and Second Further Notice of Proposed Rulemaking (FNPRM), FCC 15-47, GN Docket No. 12-354, 2015.
[32] ECC, "Technical and operational requirements for the operation of white space devices under geo-location approach," Report 186, no. January, 2013.
[33] FCC, "Second memorandum opinion and order," ET Docket No. 10-174, 2010.
[34] OFCOM, "Digital dividend: cognitive access," Tech. Rep., 2009.
[35] M. Denkovska, P. Latkoski, and L. Gavrilovska, "Geolocation database approach for secondary spectrum usage of TVWS," in *19th Telecommunications Forum (TELFOR)*, Nov. 2011, pp. 369–372.
[36] M. Tercero, K. W. Sung, and J. Zander, "Temporal secondary access opportunities for WLAN in radar bands," in *14th International Symposium on Wireless Personal Multimedia Communications (WPMC)*, 2011, pp. 1–5.
[37] P. Latkoski, J. Karamacoski, and L. Gavrilovska, "Indoor broadband use of 2.7–2.9 GHz radar spectrum: Case of Macedonia," in *15th International*

Symposium on Wireless Personal Multimedia Communications (WPMC), 2012, pp. 589–593.

[38] H.-A. Safavi-Naeini, S. Roy, and S. Ashrafi, "Spectrum sharing of radar and Wi-Fi networks: The sensing/throughput tradeoff," *IEEE Transactions on Cognitive Communications and Networking*, vol. 1, no. 4, pp. 372–382, 2015.

[39] Z. Horváth and D. Varga, "Channel allocation technique for eliminating interference caused by RLANs on meteorological radars in 5 GHz band," *Infocommunications*, vol. LXIV, pp. 24–34, 2009.

[40] B. F. Lo, "A survey of common control channel design in cognitive radio networks," *Physical Communication*, vol. 4, no. 1, pp. 26–39, 2011.

[41] I. F. Akyildiz, B. F. Lo, and R. Balakrishnan, "Cooperative spectrum sensing in cognitive radio networks: A survey," *Physical Communication*, vol. 4, no. 1, pp. 40–62, 2011.

[42] F. Paisana, Z. Khan, J. Lehtomaki, L. A. DaSilva, and R. Vuohtoniemi, "Exploring radio environment map architectures for spectrum sharing in the radar bands," in *23rd International Conference on Telecommunications (ICT)*, 2016.

[43] A. Lackpour, A. Rosenwinkel, J. R. Guerci, A. Mody, and D. Ryan, "Design and analysis of an information exchange-based radar/communications spectrum sharing system (RCS3)," in *IEEE Radar Conference (RadarConf)*, 2016, pp. 1–6.

[44] R. Saruthirathanaworakun, J. M. Peha, and L. M. Correia, "Opportunistic primary-secondary spectrum sharing with a rotating radar," in *International Conference on Computing, Networking and Communications (ICNC)*, Jan. 2012, pp. 1025–1030.

[45] F. Paisana, J. F. Santos, N. J. Kaminski, J. M. Marquez-Barja, N. Marchetti, and L. A. DaSilva, "Implementation of temporal spectrum sharing for radar bands," in *International Symposium on Dynamic Spectrum Access Networks (DySPAN)*, 2015, pp. 271–272.

[46] F. Paisana, N. Kaminski, N. Marchetti, and L. A. DaSilva, "Signal processing for temporal spectrum sharing in a multi-radar environment," *IEEE Transactions on Cognitive Communications and Networking (TCCN)*, vol. 3, no. 2, pp. 123–137, 2017.

[47] F. Rashid-Farrokhi, K. J. R. Liu, and L. Tassiulas, "Transmit beamforming and power control for cellular wireless systems," *IEEE Journal on Selected Areas in Communications*, vol. 16, pp. 1437–1450, 1998.

[48] M. Schubert and H. Boche, "Solution of the multiuser downlink beamforming problem with individual SINR constraints," *IEEE Transactions on Vehicular Technology*, vol. 53, no. 1, pp. 18–28, 2004.

[49] M. Bengtsson and B. Ottersten, "Optimal downlink beamforming using semidefinite optimization," *Proc. of 37th Annual Allerton Conference on Communication, Control, and Computing,*, pp. 987–996, 1999.

[50] K. Hamdi, W. Zhang, and K. B. Letaief, "Joint beamforming and scheduling in cognitive radio networks," *GLOBECOM – IEEE Global Telecommunications Conference*, pp. 2977–2981, 2007.

[51] B. Kouassi, L. Deneire, B. Zayen, *et al.*, "Design and implementation of spatial interweave LTE-TDD cognitive radio communication on an experimental platform," *IEEE Wireless Communications*, vol. 20, pp. 60–67, 2013.

[52] R. Zhang, F. Gao, and Y.-C. Liang, "Cognitive beamforming made practical: Effective interference channel and learning-throughput tradeoff," *IEEE Transactions on Communications*, vol. 58, no. 2, pp. 706–718, 2010.

[53] L. Zhang, Y.-C. Liang, Y. Xin, and H. V. Poor, "Robust cognitive beamforming with partial channel state information," *IEEE Transactions on Wireless Communications*, vol. 8, no. 8, pp. 4143–4153, 2009.

[54] G. Zheng, S. Ma, K. K. Wong, and T. S. Ng, "Robust beamforming in cognitive radio," *IEEE Transactions on Wireless Communications*, vol. 9, no. 2, pp. 570–576, 2010.

[55] I. Wajid, H. Nikolaeva, and M. Pesavento, "Iterative robust downlink beamforming in cognitive radio networks," in *Cognitive Radio Oriented Wireless Networks and Communications (CROWNCOM)*, 2011, pp. 375–379.

[56] I. Wajid, M. Pesavento, Y. C. Eldar, and D. Ciochina, "Robust downlink beamforming with partial channel state information for conventional and cognitive radio networks," *IEEE Transactions on Signal Processing*, vol. 61, no. 14, pp. 3656–3670, 2013.

[57] Y. Rong, S. A. Vorobyov, and A. B. Gershman, "A robust linear receiver for uplink multi-user mimo systems based on probability-constrained optimization and second-order cone programming," in *IEEE Sensor Array Multichannel Signal Processing Workshop*, 2004.

[58] B. K. Chalise and A. Czylwik, "Robust uplink beamforming based upon minimum outage probability criterion," in *Global Telecommunications Conference (GLOBECOM)*, pp. 3974–3978, 2004.

[59] Y. Rong, S. A. Vorobyov, and A. B. Gershman, "Robust adaptive beamforming using probability-constrained optimization," in *IEEE Workshop on Statistical Signal Processing*, pp. 934–939, 2005.

[60] B. K. Chalise and A. Czylwik, "Robust downlink beamforming based on outage probability specifications," *IEEE Transactions on Wireless Communications*, vol. 6, no. 10, pp. 3498–3503, 2007.

[61] A. Khawar, A. Abdel-Hadi, and T. C. Clancy, "Spectrum sharing between s-band radar and LTE cellular system: A spatial approach," in *IEEE International Symposium on Dynamic Spectrum Access Networks: SSPARC Workshop*, 2014, pp. 7–14.

[62] Z. Geng, H. Deng, and B. Himed, "Adaptive radar beamforming for interference mitigation in radar-wireless spectrum sharing," *IEEE Signal Processing Letters*, vol. 22, no. 4, pp. 484–488, 2015.

[63] F. Paisana, D. Finn, N. Marchetti, and L. DaSilva, "Reduction of radar band exclusion zones through database-aided beamforming," in *IEEE International Symposium on Dynamic Spectrum Access Networks (DySPAN)*, 2015, pp. 309–318.

[64] W. Gander, G. H. Golub, and U. von Matt, "A constrained eigenvalue problem," *Linear Algebra and its Applications*, vol. 114–115, pp. 815–839, 1989.

[65] J. Chapin, "Shared Spectrum Access for Radar and Communications (SSPARC)," *DARPA BAA-13-24*, 2012. [Online]. Available: www.darpa.mil

[66] H. Deng and B. Himed, "Interference mitigation processing for spectrum-sharing between radar and wireless communications systems," *IEEE Transactions on Aerospace and Electronic Systems*, vol. 49, no. 3, pp. 1911–1919, 2013.

[67] J. H. Reed, A. W. Clegg, A. V. Padaki, *et al.*, "On the co-existence of TD-LTE and radar over 3.5 GHz band: An experimental study," *IEEE Wireless Communications Letters*, vol. 5, no. 4, pp. 368–371, 2016.

[68] A. Lackpour, M. Luddy, and J. Winters, "Overview of interference mitigation techniques between WiMAX networks and ground based radar," in *20th Annual Wireless and Optical Communications Conference (WOCC)*, 2011.

[69] V. Ramaswamy and J. T. Correia, "Understanding the impact of RADAR interference on IEEE 802.11 MAC in spectrum sharing environments," in *IEEE Military Communications Conference (MILCOM)*. IEEE, 2015, pp. 73–78.

[70] A. Babaei, W. H. Tranter, and T. Bose, "A practical precoding approach for radar/communications spectrum sharing," in *International Conference on Cognitive Radio Oriented Wireless Networks (CROWNCOM)*, pp. 13–18, 2013.

[71] A. Khawar, A. Abdelhadi, and T. C. Clancy, "On the impact of time-varying interference-channel on the spatial approach of spectrum sharing between S-band radar and communication system," in *2014 IEEE Military Communications Conference*, pp. 807–812, 2014.

[72] G. M. Jacyna, B. Fell, and D. McLemore, "A high-level overview of fundamental limits studies for the DARPA SSPARC program," in *IEEE Radar Conference (RadarConf)*, 2016.

[73] A. R. Chiriyath, B. Paul, and D. W. Bliss, "Radar-Communications Convergence: Coexistence, cooperation, and co-design," *IEEE Transactions on Cognitive Communications and Networking (TCCN)*, vol. 3, no. 1, pp. 1–12, 2017.

[74] C. D. Richmond, P. Basu, R. E. Learned, J. Vian, A. Worthen, and M. Lockard, "Performance bounds on cooperative radar and communication systems operation," in *IEEE Radar Conference (RadarConf)*, 2016.

[75] M. P. Masarik and N. S. Subotic, "Cramer-Rao lower bounds for radar parameter estimation in noise plus structured interference," in *IEEE Radar Conference (RadarConf)*, 2016.

[76] J. R. Guerci, R. M. Guerci, A. Lackpour, and D. Moskowitz, "Joint design and operation of shared spectrum access for radar and communications," in *IEEE National Radar Conference*, pp. 761–766, 2015.

[77] R. E. Quirk, "Reader Forum: FCC spectrum rules create business opportunities," RCRWirelessNews, 2015. [Online]. Available: http://www.rcrwireless.com/20150922/opinion/reader-forum-fcc-spectrum-rules-create-business-opportunities-tag10

[78] Qualcomm Incorporated, "LTE in unlicensed spectrum: Harmonious coexistence with Wi-Fi," White Paper, no. June, pp. 1–19, 2014.

[79] D. Chambers, "MulteFire lights up the path for universal wireless service," ThinkSmallCell, White Paper, 2016.

[80] F. H. Sanders, J. E. Carroll, G. A. Sanders, and R. L. Sole, "Effects of radar interference on LTE base station receiver performance," U.S. Department of Commerce – National Telecommunications and Information Administration, Tech. Rep., 2014.

[81] H.-A. Safavi-Naeini, C. Ghosh, E. Visotsky, R. Ratasuk, and S. Roy, "Impact and mitigation of narrow-band radar interference in down-link LTE," in *Communications (ICC), 2015 IEEE International Conference on*. IEEE, 2015, pp. 2644–2649.

[82] 4G Americas, "LTE aggregation & unlicensed spectrum," White Paper, 2015.

Chapter 4

Radar spectrum sharing: history, lessons learned, and ways forward

Frank Sanders[1]

4.1 Introduction

Since the earliest practically useful radio transmissions began about a century ago, the most obvious method for the different radio services of mobile radio, commercial broadcasting, fixed point-to-point links, Earth-to-space and space-to-Earth communications, navigation, radar, and trunked and cellular communication systems to use spectrum without mutually interfering with one another has been to divide the spectrum into distinct service-dependent bands. The engineering thinking has historically been that it is easier for transmitters and receivers of a single given service to share a band than it is for different services. Thus was born the familiar lineage of colorful spectrum wall charts produced by governments around the world. For example, Figure 4.1 illustrates the U.S. government's National Telecommunications and Information Administration (NTIA) spectrum wall chart.

Spectrum charts like Figure 4.1 are constructed on a service-by-band basis, with each colored block representing a band reserved for one or more particular services. Even casual observers will immediately notice that some spectrum bands have two (and in some instances three or more) blocks stacked within a single band. These represent bands that are shared by multiple services.

Observers may also note that two of the radio services seem to have a remarkable status on these charts in that they mostly occupy their own exclusive bands, with little to no sharing with other services. These exclusive bands are reserved for commercial broadcasting (television and radio) and radiolocation, otherwise known as radar.

Of these two services, broadcasting does not concern us here. Suffice it to say that commercial broadcasting by definition covers large areas of the Earth's surface with high-power signals that are supposed to be received without interference by relatively inexpensive receiver sets, both fixed and mobile. Such technical factors are not conducive to sharing with other types of radio systems. To understand how and why radar achieved this special status, as well as the motivation behind the prospect of radar spectral sharing, it is useful to first consider some of the early radar history.

[1]National Telecommunications & Information Administration (NTIA), USA

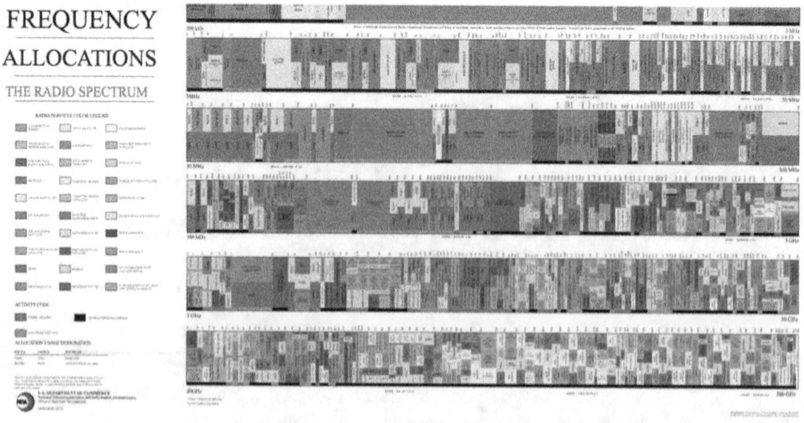

Figure 4.1 Details of the NTIA spectrum wall chart

4.2 Early radar development

As for the reasons that little spectrum sharing has historically been accomplished between radars and other systems, and that to this day the radar bands on the charts are mostly single-color, the story is not quite so simple as it is for broadcasting. However, the history needs to be understood in order to know what barriers and obstacles need to be addressed if radars are to share spectrum with other, non-radar systems from this point forward. The first reason is rooted in a confluence of world history and technological development in the 1930s and 1940s.

As air power developed and a renewal of European war loomed in the 1930s, Britain, in particular, became alarmed at its inability to detect aircraft at long distances. Douhet's strategic aerial bombing theory, enthusiastically endorsed in the 1920s and 1930s by Generals William Mitchell and Walter Wever in the United States and Germany, respectively, and by Sir Hugh Trenchard in the United Kingdom, called for massed-formation air strikes against population and manufacturing centers.

Effective technical counter-measures to that strategy were lacking in the 1920s and 1930s. This deficiency was a matter of much concern in the United Kingdom and the United States, two nations that had until then been effectively protected by naval surface fleets and that could now be simply overflown by incoming aircraft. Scientists and engineers in the United Kingdom and the United States responded with development efforts for high-power radio systems that could detect beam disturbances caused by aircraft at sufficiently large distances to allow for civilian evacuation to shelters and the direction of air defenses by their militaries. Some experimental *radio direction and ranging* systems ("radars") were built by the late 1930s—none too soon as later events would prove. From the beginning, the early radars operated at relatively high frequencies compared to other contemporary radio systems (mostly communication and navigation).[1] Even the earliest European, American, and Japanese radars operated at frequencies of several hundred megahertz or more when most communication and navigation (e.g. LORAN) radio systems were still using frequencies below 50 MHz. This early-on arrangement put radars into higher-frequency parts of the spectrum than were used by other radios.

4.2.1 The reason behind higher radar frequencies

There was an eminently practical reason for radars to operate in frequency bands higher than other radio systems. Built as military systems for use on far-flung and highly dynamic battlefields, radars more often than not needed to be mobile or at least transportable. They needed to be deployed on ships, on aircraft, and at ground stations that could be moved on short notice. As a practical matter, their antennas needed to be manageably compact and for airborne radars even downright small; fitting into spaces ranging from about ten meters wide on ships' masts to just a few centimeters across within pods underneath aircraft wings.

At the same time, however, these antennas had to produce high-gain directional beams to achieve the requisite high-resolution angular target[2] discrimination, an adequate concentration of transmitted pulse power on targets, and subsequently adequate reception of weak pulse-echoes returning from targets. As wave phenomena, radar signals can only be concentrated into tight beams by antenna aperture sizes that are significantly larger than the wavelengths they are radiating and receiving. The result is that, if an antenna is constrained to be relatively compact to meet operational requirements and yet at the same time must generate a tight, highly directional-gain beam, then its operational wavelength must be much smaller than its own aperture size. For feasible antenna dimensions of a meter down to about a centimeter or less, operational frequencies must therefore be correspondingly high, from around 300 MHz to 18 GHz or higher.[3]

Early ground-based transportable radars like the U.S. Army's SCR-270 and Germany's railroad-car Giant Wurzburg (Figure 4.2) operated at cutting-edge very high frequencies (VHF) and ultra-high frequencies (UHF) of about 100 and 600 MHz, respectively. The British breakthrough invention of the microwave (1 GHz and above) magnetron high-power oscillator tube[4] introduced a quantum leap in operational

[1] With the notable exception of the famous British Chain Home network, which used modified commercial high-frequency (HF) broadcast transmitters.

[2] Throughout this chapter, "target" refers to any discrete object in space that reflects pulse echoes back to radar receivers. The term is rooted in historical military parlance but is now used in the radar technical community for anything (e.g. buoys, fishing boats, and commercial air traffic) that radar operators are trying to locate or track.

[3] A specialty sub-class of radars operates at even lower frequencies between 5 and 30 MHz. These frequencies support over-the-horizon radio propagation for early warning of incoming long-range ballistic missiles and for detection and tracking of ships, periscope wakes, and aircraft at distances of thousands of miles. The antenna arrays for these radars have dimensions on the order of a linear kilometer, with two separate (bistatic) arrays usually being used for transmission and reception.

[4] Magnetron tubes produce such high power from such compact packages that they are still widely used.

Figure 4.2 A rare Giant Wurzburg radar parabolic dish lying in ruins prior to salvage by the Historical Electronics Museum. Designed to work at about 600 MHz, the reflector's size demonstrates a physical drawback of operating radars at frequencies below 1 GHz

frequencies. The first magnetrons resonated at a wavelength of 10 cm and the corresponding, then-astonishingly high, frequency of 3 GHz. Other magnetrons soon followed at frequencies of 9 GHz and higher. These magnetrons' short wavelengths and correspondingly small antennas were eminently practical for use in aircraft and on ships, and their frequencies also happened to have excellent atmospheric propagation characteristics. Early microwave long-range radars (e.g. the TPS-1D (Figure 4.3)) at ground stations and on ships used somewhat lower frequencies but still routinely operated at microwave frequencies around 1.3 GHz.

4.2.2 Radar spectrum band development

Thus in the 1940s, radar systems staked out numerous early claims in the VHF, UHF, and microwave bands while other radio systems were still operating at frequencies an order of magnitude lower. In the post-War years, those claims were stabilized in spectrum allocations that reserved large spectrum swaths for nearly exclusive use by radars: 5–28 MHz for over-the-horizon military early-warning and tracking applications; 200 MHz for long-range naval and space surveillance; 420–450 MHz for space surveillance and long-range air surveillance; 902–928 MHz for long-range air surveillance; 1215–1400 (now 1390) MHz for long-range air surveillance;

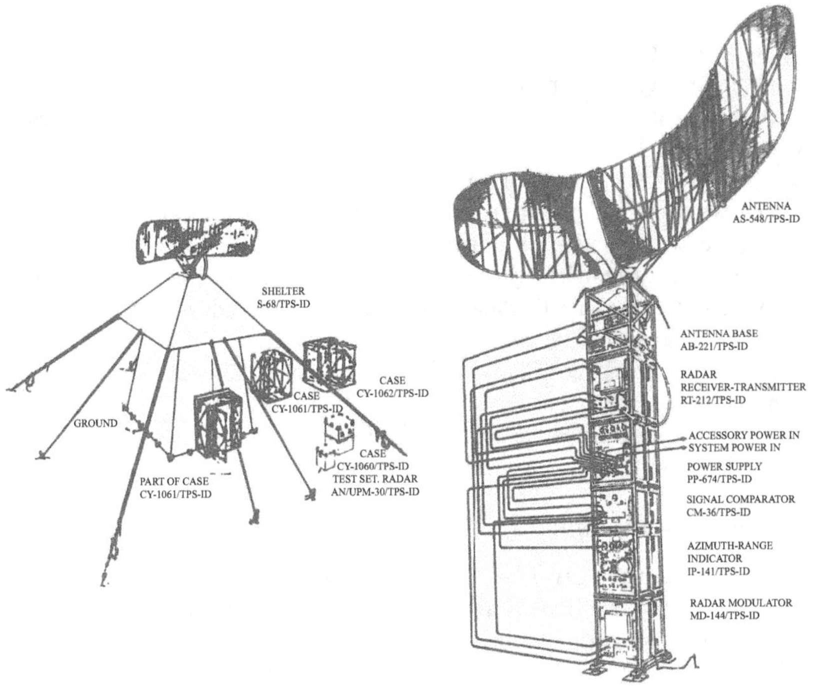

Figure 4.3 The TPS-1D radar as shown in the original Military Handbook 162B for radars. This early 1.3 GHz long-range system was the direct antecedent of Air Route Surveillance Radars that are still in use today. Propagation characteristics in this part of the spectrum are excellent for long-range applications and wavelengths are short enough to allow for relatively compact antennas

2700–3700 (now 3690) MHz for tactical surveillance, weather surveillance, navigation and air traffic control; 5250–5925 MHz for weather surveillance, severe storm warning, mapping, and navigation and air defense; 8500–10,500 MHz for navigation, weather surveillance, mapping and fire control; and then still more bands above 10 GHz, all the way up to 40 GHz.

The bands above 5–28 MHz ("HF") were given exotic letter designations that were originally meant to confuse the enemy but now just cause general confusion. From low to high, they are P, L, S, C, X, Ku, K and Ka.[5] Not only were band designations militarily coded, but radars were themselves considered to be "secret" by

[5] Radar spectrum bands are not to be confused with completely different NATO band designations that run sequentially from A through J and above. Observers must determine from context whether a given band designation (e.g. "C band") refers to radar versus NATO spectrum. Confusing the situation even further, NATO band edges are numerically round numbers that do not correspond to actual band edges used by operational radio systems.

nature: they were exotic, space-age inventions that reached out through distance, darkness, clouds, and fog to find and locate everything from tornadoes to supersonic aircraft to atomic-bomb-tipped intercontinental missiles. They guided manned spacecraft rendezvousing and docking in orbit for Moon missions. They had cachet, they were the future, and only a handful of people understood much about them (still mostly true today; some things never change).

There was a cultural bias from the 1940s right up to the present day not to inquire too much about either radars or the spectrum bands they use, nor to inquire too much about whether anyone might perhaps want to share the use of those bands.[6] To this day, it is the author's experience that many electronics engineers learn little about radars in the course of their formal education, with communication systems instead being the technology area that receives almost all attention in the field of radio engineering. This institutional blind spot in electronics engineering continues to stymie discussion about radar systems outside the small community of niche-specialist engineers who design and develop them, to the detriment of spectrum-sharing development.

4.2.3 Radars need quiet spectrum to work well

Not only were numerous, wide spectrum bands established for radars with no other native radio systems in proximity at the time, those bands were mostly radio-quiet (or at least radio not-so-noisy), since naturally occurring atmospheric noise decreases rapidly at frequencies above 50–100 MHz. This situation was ideal because radar receivers do have a kind-of secret little problem: they can barely pick out the echo returns from discrete targets and from weather (specifically hydrometeors: raindrops, snowflakes, and hail). Even though pulsed radar transmitters generate the highest peak effective radiated power (ERP) levels of any radio transmitters (typically more than a gigawatt and sometimes 100 GW or more), their target-echo returns are usually received at power levels that are so low that they are sometimes *weaker* than naturally occurring thermal electron noise in their receivers. Radar designers have to play sophisticated games to acquire their own pulse echoes effectively.

This problem brings us to another reason that radar operators did not then, and do not now, encourage others to share their spectrum: non-radar spectrum sharing tends to increase noise levels in radar bands, and therefore tends to degrade the effectiveness and reliability of radar systems. With noise-limited target detection being the Achilles Heel of many radar systems, the entities operating radars expressly do *not* want additional noise to be introduced into their operational bands, as can happen when sharing is implemented.

Electronic counter-countermeasures (ECCM) features of some radar systems might be expected to mitigate interference from non-radar systems and make spectrum sharing more practical. However, ECCM features are not always necessarily as effective as one might like when it comes to filtering out interference from non-radar transmitters, which is another reason that the radar bands on spectrum wall charts have tended to remain single colors.

[6]Then as of now, good advice for avoiding trouble includes not asking too much about local radar systems.

4.2.4 Why frequency bands allocated to radars require large bandwidths

Another question that arises from the inspection of spectrum charts is: why are radar bands so wide? The reason is because radars do not commonly use channelized frequencies like other radio systems. Radars mostly use pulsed radio energy because high power can be concentrated in relatively short pulses that are localizable in time and therefore space. This combination of high power (for short periods) and localizability makes pulses perfectly suited for radar applications.

However, pulsed emission spectra are inherently wideband. The Fourier-transform power spectrum of a pulse follows a $(\sin(x)/(x))^2$, or sinc-squared, envelope that spreads sidelobes over a theoretically infinite spectrum, which is obviously far greater than a neatly channelized radio signal. Furthermore, the spectral content of real radar emissions is actually not even bound by the sprawling sinc-squared envelope due to spectral regrowth effects caused by high-power oscillators and amplifiers. This expansive spectral content is yet another factor that has tended to discourage spectrum sharing between radars and other radio systems.

4.2.5 Why radars have tended to have their own spectrum allocations

Lastly, we cannot discount the fact that radar operational missions are considered critical to national safety, economies (as in weather surveillance), and defense postures, and that radar systems are inherently expensive, high-value assets that are difficult to redesign or replace. In an age in which many things are equipped with beacon systems that are linked to radio navigation (e.g. GPS) satellite signals, radars are still very much needed to locate and track both man-made objects and natural phenomena that are inherently noncooperative and thus cannot (or will not) be GPS-equipped.

Examples include ships, periscopes, aircraft, land vehicles, and people who are attempting to avoid detection in a variety of military and national security scenarios. Also included are rocks, buoys, and other obstacles in waterways, such as bridge supports, piers, and other vessels to avoid collisions, as well as aircraft that are either not beacon-equipped or whose beacons are broken or have been intentionally turned off, as happened during the September 11, 2001, attacks. This list further includes weather phenomena that produce active precipitation, such as thunderstorms and associated microbursts, hailstorms, rainstorms, snowstorms, tornadoes, and hurricanes, and additional diverse applications such as tracking of bird migrations, insect concentrations, airborne particulate contaminants, and ground mapping for commercial, military, and scientific (e.g. archaeological) purposes. Given the importance of these missions and applications, the inherent susceptibility of radar receivers to interference, and the high cost of radars and their dedicated platforms (e.g. airborne warning and control system aircraft), we clearly see yet another reason that spectrum sharing with radar has not historically been encouraged.

4.3 Regulation of radar spectrum in the United States and worldwide

In the United States and its territories, federal radar systems and how they use the spectrum allocated to them are regulated by the Radar Spectrum Engineering Criteria (RSEC). The RSEC is contained in Chapter 5.1 of the NTIA *Manual of Regulations & Procedures for Federal Radio Frequency Management*, more commonly known as the *Redbook*. The RSEC is used by NTIA to establish the amount of spectrum a radar system is permitted to use based on its operating frequency, function, waveform parameters, and power output. Five categories of radars (A, B, C, D, and E) are identified, and operating criteria are specified for each.

The RSEC defines an emission mask to limit allowable emissions based on the waveforms for each category of radar, the shape of which is determined by the aforementioned parameters. These masks were originally developed in the 1970s and are always under review so that they can be adapted as new types of radar systems are developed, recognizing that spectrum needs can likewise change. The Federal Communications Commission (FCC) defines criteria in the Code of Federal Regulations (CFR) to which manufacturers must adhere for nonfederal radar systems. Typically, these are small craft and commercial shipboard radar systems, weather radars used by academia and television stations, and airborne radars used on commercial or private aircraft. NTIA and the FCC work together to ensure any frequency bands allocated to or shared by radar systems are used efficiently.

Internationally, it is advantageous for radar regulations to be coherent across national boundaries. Most of the work of international standardization is performed under the aegis of the International Telecommunication Union (ITU) Radiocommunication Sector (ITU-R). The main reference for international radar emission regulations, for example, is ITU-R SM.1541 [1]. Annex 8 of Rec. SM.1541 closely parallels the RSEC.

4.4 The advent of radar band sharing

In the last forty years, and especially the last twenty, the non-radar bands on spectrum charts have become increasingly populated and crowded. Non-radar spectrum users have increasingly searched for available, unused spectrum. This scarcity has consequently made the radar bands increasingly conspicuous for their wide, single-color bandwidth markings on the charts.

This situation led to a few obvious questions. Could radar bands be shared with non-radar systems, at least at some times, in some places, and under some conditions? What advances in technology have potentially made such sharing feasible, where at one time it would not have been practical or safe? If such sharing might be entertained, what might be its possibilities and its limits?

The rest of this chapter will explore, briefly and at a high level, the most basic elements of radar spectrum engineering and requirements. Radar bands and their dominant applications are briefly reviewed along with radar emission mask limits. This discussion is necessary to inform the ensuing discussion that focuses on experience to date of radar spectrum sharing with non-radar systems using dynamic frequency selection (DFS) technology at 5.6 GHz. The lessons learned in this context ultimately inform the possibilities and limitations of all future radar spectrum-sharing endeavors.

4.5 Radar 101: essential knowledge for spectrum sharing

Radar designs are tightly constrained by the laws of physics. These constraints are at the heart of what is, and is not, physically possible for radar spectrum-sharing schemes. In this section, we examine these constraints by going through the exercise of designing a hypothetical radar. This radar is the one the author would build if he were trapped on a desert island with only coconuts, metaphorically speaking, available as construction material.

4.5.1 Specification of what the radar must do

We begin by specifying what task the radar will be expected to perform. This unit will be required to sweep a radio beam through space to obtain range and direction of aircraft at altitudes as high as 9 km (30,000 feet) above the ground with a range resolution of $^1/_{12}$ nautical mile (about 150 m). At that altitude, a radio beam emitted by a transmitter at a zero-degree elevation angle will reach a distance (due to the Earth's curvature) of a little more than 200 nautical miles (390 km). Thus the radar has a range requirement $r = 390$ km. For that distance, we shall specify that the radar be capable of detecting targets that have a reflective cross-section of $\sigma = 0.1\ \text{m}^2$.[7]

4.5.2 Radar receiver inherent internal thermal noise and other losses

Targets can only be sensed if pulse echoes returning from a distance of 390 km or less produce enough power in the radar receiver to be detected at a level sufficiently higher than the thermal electron noise in the receiver circuitry. The theoretical limit for thermal noise in a bandwidth B is kTB, where k is Boltzmann's constant (1.38×10^{-23} W/K/Hz) and the temperature T is taken to be room temperature at 290 K. Thus at room temperature $kT = 4 \times 10^{-21}$ W/Hz. In decibel terms, this inherent noise is -204 dBW/Hz $= -174$ dBm/Hz. The receiver circuitry noise level is therefore this value multiplied by an inherent, real-world noise factor (NF). A realistically achievable value for NF is 3 (or 5 dB).

The radar receiver will also have some inherent dissipative losses in the waveguides, connectors, and so forth that are placed in front of the first active element, a radio-frequency (RF) low-noise amplifier (LNA). These pre-LNA losses, collectively denoted as L_{Rx}, might typically be a factor of 2, for another 3 dB of overall

[7]The cross section of reflectivity σ is only a fraction of the physical cross section of a target. A target with a 0.1 m^2 reflectivity cross section might, for example, be the nose cone of a missile coming directly toward the radar, with the missile itself having a diameter of about 1.5 m and thus a nose-on physical cross section of about 1.8 m^2.

performance degradation. Thus, the radar receiver's overall internal noise must be adequately exceeded by a target-echo power that will be (−174 dBm/Hz + 5 dB + 3 dB) = −166 dBm/Hz.

At this point, we need to pivot from the consideration of spectral noise density (power per hertz) to the absolute power level of the noise in the receiver (just power). To accomplish this task, the receiver's internal spectral power density is multiplied by the receiver's bandwidth B. This bandwidth is determined by recognizing that the optimal signal-to-noise (SNR) ratio for a receiver that is trying to detect unmodulated pulse echoes (i.e. no modulating waveform) is achieved when B is equal to the inverse of the pulse width. Since it was specified that the radar should localize targets in range to within about $^1/_{12}$ of a nautical mile, and given that the range resolution of an unmodulated pulse is half the width of the pulse, this radar therefore needs to produce pulses that are $^1/_6$ of a nautical mile (equal to about 300 m = 1,000 ft) long in space. For speed of light propagation, this arrangement corresponds to a pulse width of 1 μs in time, so that B needs to be (1/(1 μs)) = 1 MHz = 10^6 Hz. In decibel terms, multiplication of the spectral noise density by this bandwidth equates to an increase of 10 log(10^6) = 60 dB. In other words, the thermal noise in the radar receiver is therefore N = (−166 dBm/Hz + 60 dB) = −106 dBm. If a target is to be (barely) detectable with a power equal to this noise level,[8] then the echo power in the receiver needs to be set at a design goal of −106 dBm.

4.5.3 The radar antenna

With the above detectable-power requirement in hand, we are nearly ready to determine the radar transmitter's output power requirement. The only remaining item to consider is the expected amount of antenna gain. At this radar's range requirement of 390 km, the spectrum between about 300 MHz to 1.6 GHz has good propagation characteristics. To keep the antenna size manageably small, we want to operate at the higher end of that range, and so select a frequency of 1.3 GHz,[9] which is where long-range air surveillance radars have operated since the Second World War. The physics and engineering of antenna design allow us to build a reasonably compact (about 8 m wide) antenna at this frequency having a directional gain g of 1,000 (or G = 30 dBi) relative to an isotropic radiator. Since the radar is to be monostatic, we shall use the same antenna for transmission as for reception.

4.5.4 Radar wave propagation to and from a target

Assume simple free-space radio propagation to and from any illuminated targets so that the energy transmitted by the radar spreads and dissipates on a spherical wavefront according to $1/d^2$, where d is the distance traveled. When some of this energy encounters a target, a fraction is absorbed, a fraction is scattered in various

[8]The sum of the target power (−106 dBm) and the noise power (−106 dBm) thus being −103 dBm.
[9]Radio navigation satellite systems (RNSS) such as GPS, Glonass, and Galileo use frequencies near 1.6 GHz for the same reason: excellent propagation is achieved at long ranges through the earth's atmosphere.

directions, and a fraction is reflected directly back toward the radar. A portion of this direct-reflected fraction is captured by the receiver since the reflected energy spreads out on another spherical wavefront as it propagates back towards the radar, again according to $1/d^2$ as a function of d.

4.5.5 *Peak power the radar must transmit*

Denote p_{Rx} as the power received by the radar that is reflected from a point target at a distance $d=r$, the radar's range requirement. In terms of peak transmit power p_{Tx}, this received power can be expressed as

$$p_{\mathrm{Rx}} = p_{\mathrm{Tx}} \frac{g^2 \sigma}{\left(4\pi r^2\right)^2}, \tag{4.1}$$

which can alternatively be solved in terms of the transmit power as

$$p_{\mathrm{Tx}} = p_{\mathrm{Rx}} \frac{16\pi^2 r^4}{g^2 \sigma}. \tag{4.2}$$

This relationship can also be expressed in decibel terms (using the convention $10\log_{10}(a)=A$) as

$$P_{\mathrm{Tx}} = P_{\mathrm{Rx}} + 10\log_{10}(16\pi^2) + 40\log_{10}(r) - 10\log_{10}(\sigma) - 2G. \tag{4.3}$$

The $40\log_{10}(r)$ term is what really drives the transmitter power requirement. For $r=390$ km and the other values specified above, the required peak transmitter power (in dBm) is therefore found to be

$$\begin{aligned} P_{\mathrm{Tx}} &= -106\,\mathrm{dBm} + 22\,\mathrm{dB} + 40\log_{10}(3.9\times 10^5) - 10\log_{10}(0.1) - 2(30\,\mathrm{dB}) \\ &= -106\,\mathrm{dBm} + 22\,\mathrm{dB} + 224\,\mathrm{dB} + 10\,\mathrm{dB} - 60\,\mathrm{dB} \\ &= +90\,\mathrm{dBm} = 1\ \mathrm{MW\ pulse\ peak\ power}. \end{aligned} \tag{4.4}$$

This peak power level of 1 MW driven by the transmitter into the antenna is quite typical for air-search and meteorological radars using conventional high-power tubes.[10]

4.5.6 *Radar pulse repetition rate*

The peak transmit power of 1 MW must be sustained for the entire microsecond pulsewidth, meaning that each pulse contains a total energy of 1 joule. A sequence of these pulses is transmitted with an adequate separation in time to allow for a complete round trip of each pulse before the next pulse is emitted. The round-trip light-speed

[10]Solid-state transmitter radars only generate 1/10 to 1/100 as much power as tube radars, so they have to use pulses that are proportionally 10 to 100 times longer, respectively, than the nominal 1 microsecond derived for this hypothetical radar. Such stretching of the pulse is needed to maintain the same amount of total energy in each pulse that is necessary for successful detection. Such long pulses are coded using phase or frequency modulation so that they can be compressed in receivers to recover range data with high resolution. Pulse compression, as this approach is called, is used in a variety of tube radars as well.

travel time for 2 × 390 km = 780 km is 2.6 ms. Thus, the pulse repetition rate will be (1 pulse)/(2.6 ms) = 385 pulses per second (pps or simply Hertz).

4.5.7 Radar pulse echo integration for effective detection

In reality, radars do not normally perform single-pulse detection because the probability of actually detecting a target clearly based on a single pulse echo is rather low. Instead, radar receivers integrate the echoes from multiple pulses to achieve a sufficiently high cumulative probability of target detection P_d. Theoretical–analytical studies have shown, and experiments have borne out, that P_d improves with an increasing number n_p of integrated pulse echoes until an asymptotic roll-off in detection probability occurs at around $n_p = 15$ to 20 pulses. For this hypothetical radar, we shall select $n_p = 15$.

4.5.8 Radar beam-scanning interval

With a 30-dBi antenna gain, the 3-dB beamwidth is about 1 angular degree. Since we must concentrate $n = 15$ pulses within each 1-degree-wide 3-dB beamwidth, each degree around the horizon needs to be scanned for an interval of 15 × 2.6 ms = 0.04 s/deg. Therefore, a complete 360-degree horizon scan for this radar requires 360° × 0.04 s/deg = 14 s. In other words, target positions will be updated once every 14 s, which is ordinarily acceptable for targets at long ranges.

In terms of parameterization, the radar design is now complete. The parameters we have derived here are very similar to those of actual long-range air surveillance radars and long-range weather radars. The major difference for weather radars is that, because of the weakness of weather echoes, they typically operate with $n_p = 30$ to 40 pulses within the search beam to provide more integration gain each scan (now roughly 15 to 16 dB). As such, their scanning intervals are concomitantly increased to about 30 to 40 s.

This exercise in the design of a radar is intended to demonstrate that radar transmitter and receiver characteristics are not arbitrary. These design choices are primarily driven by the exact values of required radar range, range resolution, and target cross-section. Once those parameters have been selected, radars to some extent design themselves. Conversely, it is not possible for radar designs to be arbitrarily altered to accommodate spectrum-sharing schemes, as most radar design parameters have been fixed in advance by the radar's operational requirements, the laws of physics, and engineering constraints on factors such as maximum antenna size and weight.

4.6 Radar receiver susceptibility to interference in spectrum-sharing scenarios

In the above radar design exercise, we set the radar pulse echo power to be equal to the receiver's own internal noise level. In reality, most radars are designed to operate with somewhat reliable target detection performance when the pulse echoes are at only about 1/4 the receiver's internal thermal noise power level; in decibel terms this

SNR is −6 dB. A pulse echo at this level is added with the 0 dB receiver noise via superposition to produce a blip in the receiver's integrator that is $10\log(1+{}^1/_4)=1$ dB above the thermal noise floor. This blip is not sufficient for single-pulse detection but, as noted above, with 15 or 20 integrated pulse echoes the desired targets can be reliably detected.

Similarly, however, any interference in a radar receiver that raises the overall internal noise by 1 dB or more (that is, any interference power I that realizes an interference-to-noise ratio (INR) of −6 dB or more) will begin to degrade the radar's P_d for desired targets. This often-used interference threshold value of $\mathrm{INR}_{\mathrm{thresh}} = -6$ dB is not only theoretically derived but has been borne out in many experimental and field studies with operational radars, as summarized for many radar types in [2]. This convergence of theory and observations is why a value of −6 dB is often specified as an interference threshold for spectrum-sharing scenarios.[11]

4.7 Detecting our hypothetical radar for DFS purposes

Now suppose that we are going to design a detect-and-avoid spectrum-sharing system that will work with our self-designed radar. The spectrum-sharing system is supposed to detect radar signal power when the sharing system's own transmissions are beginning to couple into the radar receiver at the critical threshold value of $\mathrm{INR}_{\mathrm{thresh}} = -6$ dB. An example of this type of spectrum-sharing system is called DFS and has been deployed in the United States with Unlicensed National Information Infrastructure (U-NII) systems at 5.6 GHz. For now, consider the fundamental physical underpinning of DFS, which is the threshold at which it must detect the radar in order to avoid exceeding $\mathrm{INR}_{\mathrm{thresh}} = -6$ dB in the radar receiver.

For our hypothetical radar, the power received from a U-NII transmitter is (in dB)

$$P_{\mathrm{Rx\_radar}} = P_{\mathrm{Tx\_UNII}} + G_{\mathrm{Tx\_UNII}} + L_{\mathrm{prop}} + G_{\mathrm{Rx\_radar}} \tag{4.5}$$

where $P_{\mathrm{Tx\_UNII}}$ is the U-NII emitted power, $G_{\mathrm{Tx\_UNII}}$ is the U-NII transmit antenna gain, and L_{prop} represents the propagation losses between the DFS transmitter and the radar receiver. For a single pulse, the radar's interference threshold is

$$\begin{aligned} P_{\mathrm{radar\_interf\_thresh}} &= 10\log(kTB_{\mathrm{radar}}) + \mathrm{NF}_{\mathrm{radar}} + L_{\mathrm{Rx\_radar}} + \mathrm{INR}_{\mathrm{thresh}} \\ &= N + \mathrm{INR}_{\mathrm{thresh}}, \end{aligned} \tag{4.6}$$

which is a function of bandwidth B_{radar} that was set to 1 Megahertz above.

At a U-NII access point (AP) equipped with a DFS-monitoring capability, it is determined if the radar's interference threshold is exceeded by ascertaining the power received at the U-NII AP, which is

$$P_{\mathrm{Rx\_UNII\_AP}} = P_{\mathrm{Tx\_radar}} + G_{\mathrm{Tx\_radar}} + L_{\mathrm{prop}} + G_{\mathrm{Rx\_UNII\_AP}} \tag{4.7}$$

[11]The value is not always −6 dB. Some agencies with especially sensitive radar systems (e.g. meteorological radars) may require even lower thresholds for harmful interference levels for spectrum-sharing arrangements.

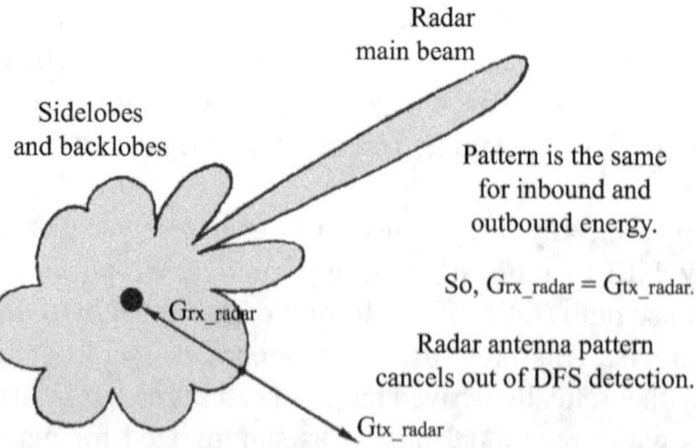

Figure 4.4 Radar antenna gain subtracts out of the DFS equations via reciprocity

for $G_{\text{Rx\_UNII\_AP}}$ the receive antenna gain at the U-NII AP. Since the propagation losses are invariant to the direction of propagation, we can solve both (4.5) and (4.7) in terms of L_{prop} and equate the result, yielding

$$\begin{aligned} &P_{\text{Rx\_radar}} - P_{\text{Tx\_UNII}} - G_{\text{Tx\_UNII}} - G_{\text{Rx\_radar}} \\ &\quad = P_{\text{Rx\_UNII\_AP}} - P_{\text{Tx\_radar}} - G_{\text{Tx\_radar}} - G_{\text{Rx\_UNII\_AP}}, \end{aligned} \tag{4.8}$$

noting that the U-NII AP and U-NII transmitter can be separate and use different antennas. The result in (4.8) can be rearranged and simplified as

$$P_{\text{Rx\_UNII\_AP}} = P_{\text{Tx\_radar}} + G_{\text{Rx\_UNII\_AP}} + P_{\text{Rx\_radar}} - P_{\text{Tx\_UNII}} - G_{\text{Tx\_UNII}}, \tag{4.9}$$

where we have invoked $G_{\text{Tx\_radar}} = G_{\text{Rx\_radar}}$ by reciprocity of the radar antenna pattern, which is illustrated graphically in Figure 4.4.

Setting $P_{\text{Rx\_radar}}$ in (4.9) equal to the radar interference threshold from (4.6), and using the previously determined values of $N = -106$ dBm for our hypothetical radar with 1 MHz bandwidth and $\text{INR}_{\text{thresh}} = -6$ dB, we obtain $P_{\text{Rx\_radar}} = -112$ dBm. Additional typical values for radar and U-NII parameters are $P_{\text{Tx\_radar}} = 1\text{ MW} = +90$ dBm as determined before, $P_{\text{Tx\_UNII}} = 4\text{ W} = +36$ dBm, and $G_{\text{Rx\_UNII\_AP}} = 0$ dBi and $G_{\text{Tx\_UNII}} = +6$ dBi for the U-NII AP and transmitter, respectively. Substituting these values into (4.9) results in a received power at the AP of

$$\begin{aligned} P_{\text{Rx\_UNII\_AP}} &= (+90\text{ dBm} + 0\text{ dBi} - 112\text{ dBm} - 36\text{ dBm} - 6\text{ dBi}) = -64\text{ dBm} \\ &= P_{\text{DFS\_thresh}}. \end{aligned} \tag{4.10}$$

This level, −64 dBm, is the DFS detection threshold specified by the FCC in the Code of Federal Regulations [3]. The calculation given above shows (in essence) the origin of that threshold value for DFS-based spectrum-sharing systems.

Reviewing the terms in (4.9) and (4.10) for the DFS detection threshold for U-NII AP monitors, we note the following:

- Radar antenna pattern gain values *do not matter* for DFS detection, and main-beam, sidelobe, and backlobe levels are not part of the equation;

- The DFS detection threshold depends on (and increases with) radar transmitter peak power;
- The DFS detection threshold decreases with the effective isotropic radiated power (EIRP) of U-NII transmitters.

These are the most important aspects to understand on the technical side of DFS-based spectrum sharing.

4.7.1 The DFS spectrum-sharing concept

The original technical ideas behind 5 GHz spectrum-sharing U-NII devices were:

- Sharing should work so long as U-NII receivers (associated with U-NII transmitters but not necessarily co-located with them) could detect local, incumbent radar signals and would not operate on those locally used radar frequencies.
- Radars, as pre-existing incumbents in the bands, would have primary status over the new spectrum-sharing systems. The new U-NII systems' status would be secondary. The U-NII systems would be operated on a not-to-interfere basis with radar systems.
- Radar receivers' noise-limited performance (as described above) would be the technical limiting factor for the new spectrum-sharing paradigm.
- Prior to commencing operation at any location, a U-NII system would verify that no radar signals were present on a candidate U-NII frequency. Subsequently, if a U-NII system using a frequency in the 5 GHz band detected a local radar signal on the same frequency, the U-NII transmitter(s) would be commanded to vacate the local radar frequency and the associated receiver(s) would re-tune accordingly. All of these sensing, detection, and avoidance operations were to occur in timeframes that would eventually be specified.

Upon the detection and avoidance of the frequencies used locally by a radar, the 5 GHz U-NII systems would by necessity need to have available, and make use of, alternative frequencies on a dynamic, *ad hoc* basis. This dynamic approach to detection and interference mitigation by switching to a new frequency when a protected system was sensed would become the newly developed DFS concept.

4.7.2 Uniqueness of DFS for real-world spectrum sharing

At the time DFS was proposed, the idea of identifying frequencies used locally by one system and then making use of unused frequencies in the same band, or perhaps even in an entirely different band, was not entirely new. Software-defined radio (SDR) system designers had been working with the same idea for some years. The SDR development community typically assumed that a single type of service (frequency-channelized communications) would share a band with the same type of service.

However, the idea of developing a fully automated detect-and-avoid functionality for sharing spectrum between communication systems and radars was entirely new. For the first time, automated sharing was being proposed for two completely different services in a single band. Moreover, one of those services (radar) would be allowed to function with no change in its operations (i.e. no modifications to the radar hardware or software) while the new service would literally work its way into the radar band without producing any harmful interference to the radar receivers.

4.7.3 Timeline of international (ITU-R) and U.S. national development of DFS

Development of DFS would be an inherently international undertaking. Administrations that had a particular interest in the concept and that actively participated in the idea's development included the United States, the United Kingdom, France, and Germany. This interest was likely due to these countries having significant pre-existing radar assets to protect in the 5 GHz band that had been proposed for DFS. A brief history of DFS development is provided in Section 4.7.4.

4.7.4 DFS introductory efforts, 1996

The 5 GHz DFS U-NII development began in the late 1990s with industry proposals to the U.S. FCC for the authorization of new types of radio devices that would be used for unlicensed wireless data networks. The FCC began a rulemaking process that led to the DFS-technology requirement for detection and avoidance of incumbent radar signals by unlicensed wireless devices in designated U-NII bands. In the ITU-R, the 2003 World Radio Conference (WRC-03) would eventually enable a corresponding allocation on a worldwide basis.

In 1996, the FCC released a Notice of Proposed Rulemaking (NPRM) for U-NII devices [4]. This document proposed to make available 350 MHz of spectrum at 5.15–5.25 GHz and 5.725–5.875 GHz for use by unlicensed NII/SUPERNet devices for the purpose of short-range wireless local area networks (WLANs). Among the proposed rules were (1) a proposed modification to Part 15 of the FCC rules to allow for unlicensed NII device operation in the proposed bands; (2) a proposed minimum of technical standards to allow technical flexibility in designs; (3) a proposed minimum attenuation of 50 dB by the unlicensed devices for all emissions outside the proposed bands or in accordance with CFR 15.209 (whichever is lower); (4) no proposals for a channeling plan; (5) proposed use of a "listen before talk" spectrum-sharing etiquette; and (6) proposed spectrum-sharing criteria for incumbent and proposed 5 GHz systems to prevent interference. This latter clause directly related to the later requirements and implementation of DFS.

4.7.5 Initial FCC R&O and MO&O policy statements, 1997–1998

On January 9, 1997, the FCC released a Report and Order (R&O) on U-NII devices [5] that set the allowable band at 300 MHz from 5.15 to 5.35 GHz and 5.725 to 5.825 GHz. This document stated the needs and advantages of U-NII devices and formally reiterated the findings of the NPRM. It also stated some potential conflicts with mobile satellite service devices and addressed public concerns from amateur radio and other vendors and agencies that had plans for use of the spectrum allocated for U-NII. There was no mention in this R&O of potential conflicts with radar systems or of a need for DFS.

On June 17, 1998, the FCC released a Memorandum Opinion and Order (MO&O) [6] that limited U-NII devices to a peak power of one watt and directional antennas of up to 23 dBi gain. Furthermore, it addressed the original industry petition, accepting some recommendations and rejecting others. This MO&O addressed for the first time the potential susceptibility of U-NII devices to high-power government radar systems. However, it did not address the possibility of interference to radars from U-NII devices and there was still no mention of DFS technology.

4.7.6 WRC-03 and Recommendation M.1652, Circa 2003

At WRC-03 a proposal was made for a new allocation in the 5250–5350 and 5475–5725 MHz bands for a new mobile unlicensed service. The U.S. administration actively promoted this allocation under the condition that the new service had to use a DFS approach to protect radar operations. The United States was successful in obtaining the new allocation under the DFS provision by the conclusion of WRC-03. ITU-R Recommendation M.1652, "Dynamic frequency selection in wireless access systems including radio local area networks for the purpose of protecting the radiodetermination service in the 5 GHz band" [7] was also developed to roughly outline the guidelines, limits, and operating procedures for the U-NII devices.

Specifically, ITU-R M.1652 provided the following:

- Representative 5 GHz radar system characteristics derived from the U.S. radar data sets;
- Detection thresholds for radar signals based on the U.S. data;
- Requirements for determining channel availability prior to data transmissions based on the U.S. data and studies;
- Speed requirements for vacating DFS channels when radar signals are detected.

At this point, no prototype U-NII devices had been tested nor had the DFS approach been proven in any laboratory to reliably detect radar signals and subsequently vacate channels.

4.7.7 Determination of protection criteria, late 1990s through mid-2000s

The rules developed for DFS were based on a multi-year fundamental research program by U.S. government spectrum engineers into the levels at which interference causes radars to degrade target detection. This program began in the late 1990s when the DFS concept was first proposed and continued into the mid-2000s. The results of the study were published in a comprehensive technical report [2], the significant results of which include:

- Radar receivers begin to lose desired targets when interference levels are 6 dB *below* the radar's own receiver noise (the INR = −6 dB threshold stated before);
- Substantial loss of radar performance occurs when an interference level is *equal* to the radar receiver noise level (i.e. when INR = 0 dB);

- Performance degradation occurs for both radar antenna mainbeam and sidelobe coupling, and it depends only on the amount of coupled interference power, not on the direction relative to the radar antenna orientation.

4.7.8 First DFS implementation steps in the United States, 2003–2004

With the frequency allocation in place and the domestic rulemaking process started, the implementation of domestic rules and procedures for permitting these devices to operate under the CFR was begun. The FCC initiated a rulemaking in 2003 under docket 03-122 for gathering information on the use of 5250–5350 and 5475–5725 MHz bands for U-NII devices.

By Spring 2004, three companies had developed prototype U-NII devices that they submitted to the US Institute for Telecommunications Sciences (ITS) laboratory (part of the NTIA) in Boulder, Colorado, for initial tests to determine if these devices could detect radar signals and perform the required actions to change their operating frequencies. The initial set of tests was based on a limited number of radar testing waveforms. ITS engineers spent considerable time and effort to develop the first system that would generate radar signals while also monitoring the U-NII responses to see if they detected the radar signals and vacated the operating channel within the required time intervals. Other test certification laboratories were also involved in the process and offered their comments and guidance on how to develop a DFS-capable U-NII test system.

4.7.9 DFS certification testbed development, 2005–2006

In 2005 and 2006, bench testing of prototype DFS devices from multiple manufacturers took place at the ITS Laboratories in Boulder, Colorado, followed by field tests against some actual radars. These tests included a fully operational air defense radar that tracked commercial air traffic at short and long ranges while interference was injected into its receiver, and further tests in which DFS systems were used to try to detect the presence of the defense radar while it was operating normally. Thus, testing began 9 years after the initial FCC NPRM for U-NII at 5 GHz and 2 years after Recommendation M.1652 actually described DFS.

These initial tests accomplished two goals: (1) they showed that DFS devices could actually detect representative radar waveforms, and (2) they validated a test-and-measurement setup that could be used by the FCC and independent certification labs for certification of DFS devices.[12] NTIA then passed on the DFS certification hardware design and associated custom-written test-and-measurement software that are needed for certification testing to the FCC's Columbia, Maryland laboratory.

[12]The ITS lab does not perform certification testing; it develops testing approaches and then transfers them to other agencies and to industry.

Table 4.1 DFS spectrum-sharing parameters and requirements

Parameter	DFS requirement
Radar signal detection threshold in DFS receivers	−62 or −64 dBm in 1 MHz bandwidth
Channel availability-check interval before any channel can be used	60 s
Channel non-occupancy period after radar detection	30 min
Maximum interval allowed for channel move after radar detection	10 s
Maximum intervals allowed for housekeeping transmissions during a channel move	200 ms + approx. 60 ms over the remaining 10 s period

4.7.10 DFS certification requirements developed, 2005–2006

The DFS rules for operation in the United States were codified in CFR 47, Part 15.407. These rules include:

- *Power-on test*: Verify that the device monitors the initial channel for 1 min after it completes the power-on cycle and that it does not transmit during that time.
- *Initial detection tests*: The device must be able to detect a radar signal within the first six seconds and final six seconds of the initial channel check.
- *In-service monitoring test*: This test is the most comprehensive. DFS U-NII devices must detect six different waveform categories ***representative*** of six **types** of radars operating in the 5 GHz bands. *Operational radar waveforms are not exactly replicated.* Hundreds of radar-detection trials are performed for the certification of each device.
- *30 Minute non-occupancy test*: When a previously used DFS channel has been occupied by a radar signal, the U-NII device must be verified as not attempting to use it again for at least the next 30 min.

The DFS protocol is summarized in Table 4.1.

4.7.11 DFS compliance testbed constructed and early testing, 2005–2006

The DFS compliance testbed that engineers from ITS and NTIA's Office of Spectrum Management (OSM) developed and constructed at the ITS laboratory in Boulder, CO, USA, between 2005 and 2006 included these features:

- A radar signal generator (itself a complicated undertaking) and RF signal synthesizer;
- Capability to produce bursts of unmodulated, frequency modulated ("chirped"), and frequency-hopping pulses in all 5 GHz DFS bands;
- Variable and user selectable capability for RF frequencies, number of pulses per test-burst, pulse width, pulse repetition interval, and chirp bandwidth;

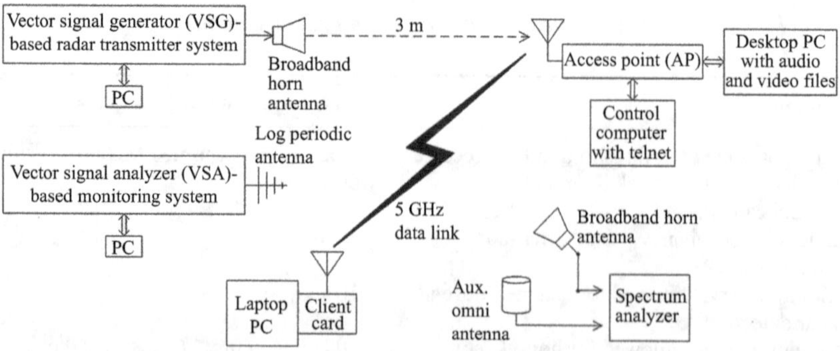

Figure 4.5 DFS performance-testing system developed by the NTIA and now used by the FCC

- Power control of RF radar pulses;
- Use of a programmable vector signal generator for pulse production;
- A radar pulse timing measurement system;
- Monitoring capability for RF activity on U-NII channels;
- Use of a vector signal analyzer and spectrum analyzer for fine and coarse measurement of the RF emissions of a U-NII AP and client transmissions over 12 s;
- Development of a customized MATLAB® routine for analysis of all collected data;
- Implementation of two systems that are synchronized so that the press of a button starts an in-service test and collects data for 12 or 24 s;
- For in-service tests, an MPEG file is streamed from one computer to another using a DFS AP and client to load the RF channel. The APs *must* have DFS functions built-in, while radar detection capabilities are optional for clients. European rules are different.

The DFS test-and-certification system that was developed is shown in Figure 4.5. Starting in August 2005, NTIA engineers began testing prototype DFS devices that were provided by industry. These tests were performed using an early version of the DFS testbed that subsequently helped NTIA to improve the testbed and helped industry by providing feedback on how well their early DFS devices functioned against clean-simulated radar pulses. The certification hardware and software were transferred from NTIA to the FCC in 2006.

4.7.12 Initial DFS deployment experience, 2006–2009

DFS-capable U-NII devices were available to consumers by July 2006, three years after WRC-03 and 10 years after the initial FCC R&O for U-NII. A wide variety of DFS-capable 5 GHz U-NII devices were soon certified by the FCC and sold by several manufacturers. Both frame-based and 802.11 devices were marketed; other types have also been certified including low data-rate and hand-held devices. The devices were marketed to wireless internet service providers (WISPs) for use as radio backhaul links between remote locations and central stations with wireline connections to the internet.

The NTIA later undertook random off-the-shelf spot-checks of commercially available products to verify their DFS performance. In 2008, a certified product was found by the NTIA and FCC to not be detecting radar signals, with post-certification changes to the device's firmware being the cause. The changes were nominally unrelated to DFS and were not supposed to disable DFS, but did. Ongoing NTIA and FCC spot-checks identified additional issues with some devices' FCC certification labels.

In early 2009, the Federal Aviation Administration (FAA) reported interference to its 5 GHz microburst-warning terminal Doppler weather radars (TDWRs). The interference was found to be caused by DFS-capable 5 GHz UNII transmitters. Significant interference areas included the San Juan area in Puerto Rico and the New York City area.

Subsequent studies conducted by the NTIA, FAA, and FCC, with assistance and cooperation from industry, identified cases in which some devices did not adequately detect radar signals in the field even though they had passed DFS type-certification tests. One cause was U-NII DFS software that allowed operators to change the country code from the United States to other countries (e.g. Brazil). Making this change disabled DFS functionality since most countries did not have DFS rules at that time. The devices, unhindered by DFS software operation, would then operate across local radar frequencies and cause interference to local radars, including safety-of-life-mission TDWRs.[13] Additional factors also played a role, including radar testing waveforms that were not sufficiently robust to ensure detection of radar signals at field locations by devices that had nevertheless performed adequately in lab testing.

4.7.13 Ongoing DFS deployment experience, 2010–present

Government and industry worked together to improve certification-testing parameters to resolve these problems. Notable results included the development and implementation of more robust DFS testing protocols and elimination of the user-accessible country code option in DFS software. On July 27, 2010, the FCC issued a Memorandum, "Elimination of Interference to Terminal Doppler Weather Radar (TDWR)." The memo summarized information about 5 GHz U-NII interference to TDWRs, with its contents based on multi-agency inputs including previous OSM-ITS troubleshooting results with these systems. The memo encouraged WISPs to use good practices in the deployment and use of DFS-equipped U-NII systems, including an on-line database developed by the government and industry to better coordinate the use of DFS-capable U-NIIs in areas near TDWR stations.

[13]TDWRs are one of a number of tools that are used to detect microbursts at airports and route air traffic around them. A major impetus for microburst-warning technology development was the August 1985 crash (with 137 fatalities) of Delta Flight 191, while attempting to land through a microburst at the Dallas/Fort Worth International Airport.

Some interference problems do continue to occur in the 5 GHz band used by DFS-equipped U-NIIs. However, most of these are caused *not* by DFS-capable devices, but rather by illegally imported and deployed transmitters that have nothing to do with DFS technology; they simply happen to be tuned to the DFS band.

An ongoing problem with DFS U-NIIs is that some firmware updates have caused unintentional failures in DFS functionality. Since DFS U-NIIs do not undergo re-testing after initial certification, it is possible for firmware updates that are sent to certificated devices to accidentally (that is, unintentionally) disable those devices' DFS operations. Devices that receive such updates may begin to cause interference even though they have had a long previous track record of interference-free operation.

Looking back on the history of DFS design, development, and deployment brings us to the first lesson learned from the DFS experience:

> **Lesson 1:** The development time for dynamic spectrum technologies, even when government and industry work closely together on the necessary technical and regulatory framework, can be something on the order of a decade because innovation requires considerable advance work in the absence of existing implementations. The more innovative the new sharing scheme, the longer the advance-work timeline can be expected to be.

4.8 Technical assumptions of DFS

DFS is based on several technical assumptions, including:

- Radar signals can be detected at given thresholds in the presence of background U-NII traffic;
- Detection of radar signals by a single AP in a U-NII network can provide adequate protection of radars from the transmissions of the entire network, including the transmissions from the AP's remotely located client units;
- APs can use a radar-detection threshold that is adequately sensitive to ensure that radar receivers will be protected when their associated radar transmitter emissions are detected by the APs at those thresholds;
- A sufficiently large and robust set of radar testing waveforms can be devised to ensure that if a U-NII system passes DFS testing with those waveforms in a laboratory, then the system's detection of operational radar waveforms in the field will be robust enough to adequately protect radar receivers from harmful interference;
- Firmware updates sent to DFS-equipped U-NIIs after the initial FCC DFS-certification testing will not accidentally corrupt or deactivate those systems' DFS functionality, despite the fact that DFS systems do not need to be re-certified after such firmware updates have been developed and distributed to the WISP customers using the now-deployed U-NII networks;
- DFS-equipped U-NIIs will always be properly installed and operated by users.

As with any system based on assumptions, there are limits to the conditions under which these assumptions hold true. These assumptions are examined further in the following sections.

4.8.1 Assumption: radars can be detected while U-NII message traffic is occurring

The fact that a U-NII AP can only monitor for radar signals during selected time intervals between regular message packets reduces the total probability per unit time that a radar signal will be detected. If the probability per unit time is greater than zero, however, then a radar signal should *eventually* be detected. The question then becomes how long that interval will be. This problem is addressed for 5 GHz U-NIIs by testing each new model hundreds of times for each radar waveform while message traffic is occurring. The U-NIIs must pass each test with certain scores for individual radar waveforms and another score for composite detection performance across all radar waveforms. These test results provide confidence that field-deployed DFS units can adequately detect ambient radar signals.

4.8.2 Assumption: detection of radar signals by APs protects radars from all network transmissions

This assumption is true to the extent that signal propagation paths of clients and APs are as good as, or at least nearly comparable to, those of local radars. This assumption is thought be mostly true but may not necessarily always completely hold due to variations in local geography and population density (i.e. mountains and large buildings) that can produce shadowing/occlusion effects.

4.8.3 Assumption: radar-detection thresholds are adequate to protect radars from harmful interference

The problem of determining adequate radar detection thresholds to ensure protection of radar receivers from U-NII transmissions is central to the entire DFS concept. The heart of the assumption is this: *a particular received power level from a radar transmitter received at a DFS AP monitoring location can be equated to a known (accurately predictable) power level in the radar's receiver circuitry.* To arrive at this detection threshold the following calculation must be performed. First, determine the interference power level in the radar receiver at which the radar target-detection probability begins to degrade from a nominal baseline condition. For radars that ITS and OSM engineers at NTIA examined [2], this interference power turned out to consistently be 6 dB *below* the radar receiver internal thermal noise level (i.e. INR = −6 dB).

The next step relies on reciprocity. Working under the assumption that signals propagate the same regardless of the direction between the radar and the U-NII, calculate the power level arriving *from the radar at the AP monitor* that equates to the condition that the *U-NII signals are arriving at the INR = −6 dB power level* in the radar receiver. For 5 GHz radars with characteristics of those published in M.1652, this power level has been shown to be about −64 dBm/MHz when received with an AP antenna (0 dBi gain).

Obviously the correctness of the detection-threshold condition is based on a number of individual assumptions, none of which is necessarily always true at all times and places, but all of which are probably true most of the time under most conditions. In other words, this condition is thought to be good, but not necessarily always perfect. It is supposed to be good enough to provide adequate protection of radar receivers from harmful interference.

4.8.4 *Assumption: radar waveform testing is sufficiently robust*

It might be thought that the key to adequately robust radar waveform testing of DFS systems would be to exactly replicate the radar signals that currently occur in the shared-spectrum band, and then to simply test against those. This approach has not been pursued, however, for a number of reasons. Firstly, DFS is supposed to be a *general* spectrum-sharing solution that will work with *any* radar signals that might occur in the shared band, which includes those from known radars at present, those that may occur in the future, and those that may occur from radars that were not thought to be in the band at the time of testing. Secondly, there are radar signals in the 5 GHz band for which governments cannot share exact emission parameters. With these factors in mind, radar test waveforms must *not* exactly replicate any one type of 5 GHz radar. Rather, the five test waveforms that have been developed are intended to have parameters of pulse width and repetition rate that *span the pulse-parameter range* of existing and expected future radars in the band. This approach ensures that DFS design engineers do not hard-code single radar pulse types into their detection algorithms, and thus that they keep DFS algorithms generally applicable to any radar signals that might occur in the band. And at the same time, the confidentiality of the government radars' pulse parameters is preserved.

For example, it was discovered after initial DFS-equipped 5 GHz U-NIIs were deployed that some U-NIIs were not detecting all radar signals in the shared 5 GHz band. Part of the reason turned out to be insufficiently designed radar test waveforms. That original set was eventually expanded to make the waveforms robust enough to ensure that tests conducted with them would ensure adequate DFS field performance.

This experience with radar test waveforms raises an important point: *any non-trivial spectrum-sharing scheme can be expected to be imperfect in some ways, and imperfections will tend to cause interference when the initial sharing units are field-deployed.* Thus spectrum managers and engineers should plan in advance of field deployments to expend time and budgets on sorting out and resolving interference problems that are likely to occur with initial deployments. The *causes* of the failures will not be known in advance (or else they would otherwise have been fixed in advance), but the *occurrence of failures due to unanticipated problems* with the sharing technology should be expected to be likely.

4.8.5 Assumption: firmware updates installed in DFS units after initial certification will not cause DFS to be impaired or disabled

As noted above, DFS-equipped 5 GHz U-NIIs are not recertified after a given model has passed DFS testing and has received an FCC approval for operation (an FCC sticker placed on each production unit and visible to purchasers is the proof of this certification). Subsequent to certification and approval of their units for sale, U-NII manufacturers often provide firmware upgrades for their units. These upgrades are developed by the manufacturer's design engineers and are then downloaded to units that have already been purchased and deployed at field locations. It has not been deemed practical or necessary to recertify every U-NII model for DFS performance after each new firmware upgrade throughout the subsequent operational history of the units. Note that it has been estimated informally by manufacturers that their DFS-equipped U-NIIs have about a 10-year half-life after they are purchased. The more expensive the units, the longer will be the likely half-life.

This approach makes good sense from the standpoint of cost and practicality, as DFS performance testing is costly and time-consuming. However, a problem with this approach is that it is possible for the DFS performance of certified U-NII models to be accidentally impaired or disabled by the firmware update. Since the engineers modifying the firmware do not normally have access to DFS testbeds of their own (such testbeds being costly and complicated, as described above), they may not be able to absolutely *ensure* that all code changes will definitely *not* accidentally impair or disable DFS functionality.

Some cases where this has in fact happened have been discovered and documented by the NTIA and FCC. Although the firmware has been fixed in every case, no solution has been identified so far for the root of this problem.

4.8.6 Assumption: DFS-equipped U-NIIs will be properly installed and operated

This assumption has been found to be imperfect. Installers are supposed to understand what DFS is, and that impairment or disablement of DFS functionality can have serious safety consequences. Actions that installers are *not* supposed to take include intentional disablement or circumvention of DFS functionality and use of any antennas for the U-NIIs other than the ones provided by the manufacturers. In reality, there have been cases in which installers have not adhered to DFS protocols. This situation was especially true during the earliest period of initial deployment, but still happens occasionally to this day.

One feature that a few installers and operators violated early on was the prohibition against disabling the DFS software and firmware functionality of their units. In the early days of deployment, manufacturers sold their units with software that included user-selectable "country codes." Operators were supposed to truthfully select the U-NII's country of operation from a drop-down menu. These individual country codes allowed manufacturers to tailor their software and firmware operations to the country in which the U-NII was used without needing to sell different versions of the devices on a country-by-country basis. For the United States, this arrangement of course meant that DFS functionality would be activated when "USA" was selected from the installation menu.

Nevertheless, word spread that DFS U-NIIs could "improve" their performance by *not* having "USA" selected as the country code and selecting another country instead. With DFS disabled, the 60-s initial channel-check disappeared, and the units would never try to switch from one channel to another since they would, of course, never detect any local radar signals.

Another problem was that some installers substituted their own antennas (typically high-gain parabolic dishes) for the lower-gain patch-style antennas that were supplied with the units. This modification likewise changed the U-NII antenna gain, thereby invalidating the DFS detection threshold that was developed based on the lower-gain antenna.

4.8.7 Ongoing need for enforcement in DFS bands

Finally, we include not an assumption but a practicality of the real world. It has turned out to be too much to expect that all operators of radio systems at 5 GHz, including supposedly professional DFS U-NII installers and operators, would always adhere to all rules. The solutions for combatting non-adherence have been multi-fold. First, the selectable country-code option has been eliminated; all units sold in the United States now have DFS functionality hard-coded into their firmware. Second, specialized RF antenna connectors are now provided with DFS units, making it more difficult for installers and operators to substitute their own antennas for the proper antennas that come with the units. In short, steps have been taken by U-NII manufacturers to make nonadherence to rules more difficult and therefore less likely to occur.

As for illegal retuning of otherwise legal transmitters as well as the illegal operation of wholly illegal transmitters, the only solution seems to be stepped-up surveillance and enforcement activities by the government when such transmitters cause interference to 5 GHz radar receivers. Ultimately, enforcement actions by the FCC have turned out to be a very important component of DFS implementation. The need for this robust enforcement activity was not anticipated by the developers of the DFS concept. Table 4.2 contains a summary list of these actions from January 2007 to July 2016.[14] Table 4.2 shows a total of 47 such actions over the nearly ten-year period. This result is an average of one DFS-related enforcement action every 2.5 months for almost a decade. Most of these actions have involved interference to safety-of-life TDWRs so there is clearly an ongoing need for enforcement to make DFS work without causing harmful interference. Given that even as of 2016 these enforcement actions continue to be numerous, it appears that the need for an ongoing enforcement component of DFS will probably be a *permanent feature* of this spectrum-sharing technology. This continuing enforcement need apparently represents an unavoidable, ongoing cost to taxpayers for this version of spectrum-sharing technology.

[14]Based on data extracted from https://www.fcc.gov/general/u-nii-and-tdwr-interference-enforcement.

Table 4.2 Summary of FCC DFS-related enforcement actions from 2007 to 2016

Date	Company affected by FCC enforcement action	FCC enforcement action
07-29-2016	Towerstream Corporation, Middletown, Rhode Island	ORDER & CONSENT DECREE
07-13-2016	Wide Cloud Communications, Hialeah, Florida	NOUO
06-13-2016	Aptech Networks Corp., Doral, Florida	NOUO
05-31-2016	Sooner Wireless, LLC, Norman, Oklahoma	NOUO
05-31-2016	Del Nero Communications Management, LLC, Norman, Oklahoma	NOUO
04-28-2016	Fastlane Internet, Francesville, Indiana	NOUO
03-11-2016	Satell SJNet, Corp., San Juan, Puerto Rico	NOUO
03-13-2015	Directlink, LLC, Parker, Colorado	PFR Order
03-10-2015	5 GHz Devices Causing Interference to Essential Safety Systems Must Cease Operations—Letter Issued by FCC Enforcement Bureau Tampa, Florida Office	LETTER
10-01-2014	CMARR, Inc., San Juan, Puerto Rico	FORFEITURE ORDER
06-24-2014	CMARR, Inc., San Juan, Puerto Rico	NAL
02-21-2014	Directlink, LLC, Parker, Colorado	FORFEITURE ORDER
02-07-2014	Skybeam Acquisition Corporation, Englewood, Colorado	FORFEITURE ORDER
02-07-2014	Rapidwave, LLC Saratoga Springs, Utah	FORFEITURE ORDER
11-21-2013	VPNet, Inc., San Juan, Puerto Rico	ORDER & CONSENT DECREE
11-18-2013	Ayustar Corporation, San Juan, Puerto Rico	ORDER & CONSENT DECREE
08-06-2013	Towerstream Corporation, Middletown, Rhode Island	NAL
02-12-2013	Argos Net, Inc., Caguas, Puerto Rico	FORFEITURE ORDER
01-16-2013	American Wireless Broadband, Chicago, Illinois	NOUO
01-04-2013	Directlink, LLC, Parker, Colorado	NAL
12-13-2012	HiPoint Technology Services Inc., Chicago, Illinois	NOUO
09-27-2012	Skybeam Acquisition Corporation, Englewood, Colorado	NAL
09-27-2012	AT&T, Inc., San Juan, Puerto Rico	FORFEITURE ORDER
03-23-2012	VPNet, Inc., San Juan, Puerto Rico	NAL
03-23-2012	Argos Net, Inc., Caguas, Puerto Rico	NAL
03-23-2012	Insight Consulting Group of Kansas City, LLC, Kansas City, Missouri	FORFEITURE ORDER
03-13-2012	The WIFI Store, Toa Baja, Puerto Rico	NOUO
03-13-2012	David Robles dba dmwireless, Manati, Puerto Rico	NOUO
12-20-2011	ExtraSpace Storage, Henderson, Nevada	NOUO
09-19-2011	Sling Broadband, LLC, Hollywood, Florida	FORFEITURE ORDER
07-29-2011	Sling Broadband, LLC, Hollywood, Florida	NAL
07-29-2011	Insight Consulting Group of Kansas City, LLC, Kansas City, Missouri	NAL
07-29-2011	Ayustar Corporation, San Juan, Puerto Rico	NAL
07-28-2011	Rapidwave, LLC, Saratoga Springs, Utah	NAL

(*Continues*)

Table 4.3 (Continued)

Date	Company affected by FCC enforcement action	FCC enforcement action
06-02-2011	Crucito Marrero dba CMARR, San Juan, Puerto Rico	NOUO
02-17-2011	AT&T, Inc., San Juan, Puerto Rico	NAL
02-11-2011	Utah Broadband, Sandy, Utah	NAL AND ORDER
11-24-2010	Ayustar Corporation, San Juan, Puerto Rico	PFR Order
08-13-2010	Wireless Metro LLC, Houston, Texas	NOUO
08-13-2010	NSN Wireless LP, Houston, Texas	NOUO
07-20-2010	Sling Broadband, LLC, Miami, Florida	NOUO
04-26-2010	Axxcelera Broadband Wireless, Inc.	ORDER & CONSENT DECREE
04-15-2010	Airspan Networks (Israel), Ltd.	ORDER & CONSENT DECREE
04-14-2010	Motorola, Inc.	ORDER & CONSENT DECREE
01-28-2010	Ayustar Corporation, San Juan, Puerto Rico	FORFEITURE ORDER
04-20-2009	Aeronet Wireless Broadband, San Juan, Pu	NOUO
01-23-2007	Neptuno Networks, Inc., San Juan, Puerto Rico	NAL

NOUO: Notice of Unauthorized Operation, PFR: Petition for Reconsideration, NAL: Notice of Apparent Liability

4.9 More lessons learned

Continuing from the first lesson learned described in Section 4.7.13, additional lessons learned are:

> **Lesson 2:** The use of dynamically based spectrum-sharing technologies requires permanent, ongoing government expenditures for testing facilities and trained engineering staff for permanent surveillance of such devices. This state must continue for as long as any given class of dynamic spectrum-sharing devices continues to be sold in markets. In other words, dynamic spectrum sharing is unavoidably associated with ongoing surveillance costs. This is not to say that the expenditures are not worth the advantages that society gains from more and better spectrum sharing; it is just the reality that such sharing is not free of ongoing costs.
>
> **Lesson 3:** Any nontrivial, innovative, dynamic spectrum-sharing technology introduction, no matter how carefully it is initially devised, will likely result in some harmful interference cases when it is initially deployed.
>
> **Corollary to Lesson 3:** When new, innovative dynamic spectrum-sharing technologies are introduced to a band, resources should be set aside in advance to accommodate the initial harmful interference cases that will almost inevitably occur.
>
> **Lesson 4:** Field surveillance will be necessary to identify spectrum-sharing devices that have been illegally marketed without proper regulatory certification.

Lesson 5: If installers and operators are able to modify spectrum-sharing devices to disable spectrum-sharing features, some may do so.
Corollary 1 to Lesson 5: Manufacturers need to restrict the ability to disable or modify their devices' spectrum-sharing functionality as much as possible.

4.10 Challenges for manufacturers and vendors

The development and introduction of DFS technology into what was essentially a radar-only band required a very close collaboration between industry, which designed and built actual DFS-equipped U-NIIs, government spectrum regulators, and government agencies that use multiple types of 5 GHz radars to accomplish their missions. This section focuses on some of the challenges that manufacturers faced in designing and developing DFS-capable systems at 5 GHz.

4.10.1 Challenges to the communications community in understanding radar systems

As noted in the introduction to this chapter, there is a deep technical divide between the radar and communication engineering communities. Radio communication system design and development is a vast worldwide industry that makes use of widely known and shared principles and approaches for research and development, prototype engineering, and manufacturing engineering. It sustains enormous markets that significantly support national economies.

Radar systems, in contrast, are designed and developed by a relatively small engineering community and are sold to niche customer markets. Radar design principles and requirements are not secret but are sometimes obscure. For the most part, only a small community of technical specialists has much understanding of radar system design and operation. The technical characteristics and requirements of radar systems arguably tend to be somewhat of a mystery to the radio communication design and development industry.

One of the most significant aspects of ITU-R M.1652 was that it represented an attempt by a number of countries, not least of which the U.S. administration, to shed more light on a number of these points for 5 GHz radars where possible. Regarding the exact technical characteristics of individual types of radars, it has been the author's experience that spectrum-sharing development does not require that exact radar characteristics be disclosed in order for sharing to move forward. In fact, it is arguably better for spectrum sharing to *not* be based on any one set of exact radar characteristics, as discussed above.

However, even with M.1652 in hand, industry was faced with the need to begin to design radar-detection algorithms for communication systems while not necessarily having much knowledge about the radar signals that they were going to be expected to detect. Virtually their only official source for this information has been M.1652, and even it only describes in fairly general terms the characteristics of the types of radar signals that AP receivers would have to identify and avoid.

Designing radar-detection algorithms (which must function in the midst of U-NII data traffic) was not a simple undertaking. The DFS goal of yes-no threshold detection of radar signals was technically challenging and furthermore had never before been implemented in commercially sold devices intended for use by consumers.[15]

Thus, it was that industry U-NII manufacturers worked closely with the U.S. government agencies, chief among them the NTIA and FCC, at the engineering level for several years from the late 1990s to the early 2000s to devise working prototypes of DFS receivers that would effectively monitor for radar signals. Much of this collaborative effort occurred during industry visits to the ITS laboratory in Boulder, Colorado. During these visits, the government provided helpful information and feedback to industry for DFS development, as described further below.

4.10.2 Difficulty of detecting general, not specific, radar waveforms

There was a persistent challenge between industry and government during DFS development concerning the extent to which DFS monitors would or would not search for a specific set of radar signals. The government insisted that radar-model-specific algorithms should *not* be used for DFS. This difficulty was exacerbated by the fact that DFS rules and protocols were developed by people who were not themselves necessarily the engineers who would be responsible for actually writing and coding the DFS detection algorithms. After the rules had been agreed upon and the responsibility for actually making DFS systems fell to manufacturer's engineers, those engineers sometimes had misunderstandings about exactly what they were supposed to be looking for in between their U-NII data packets. Chief among these misunderstandings was the idea that there was "a" single type of radar in the 5 GHz band, and that it alone needed to be detected. Thus there was a presupposition by some design engineers that they only a needed an algorithm to find this one waveform, and not have an algorithm that would detect radar pulses in general, across a wide range of waveform values for pulse width and pulse repetition rate.

4.10.3 Lack of industry testbeds for DFS

A further challenge related to the above-described design-information disconnect was that industry did not have any testbeds of its own for testing DFS radar monitors. This lack was entirely reasonable on both technical and financial grounds. The knowledge of how to even begin to design a testbed such as the one shown in Figure 4.4 was initially unavailable among U-NII manufacturers. Such testbeds are typically only built by niche entities that specialize in electronic intelligence and electronic warfare systems.

The financial aspect was also a challenge. Even if a system such as shown in Figure 4.4 had been completely sketched out by the government for industry, the cost

[15] A government requirement for 5 GHz DFS U-NIIs in the United States is that they are prohibited from collecting or retaining in memory the characteristics of the radar pulses that they receive during their operation.

to build it might have been deemed prohibitive considering the relative smallness of the U-NII market that it would serve. It also would seem wasteful for several U-NII manufacturers to each build their own DFS testbed when a single such testbed at a single location such as Boulder would suffice to meet everyone's needs and could be used on a shared basis instead (albeit by only one manufacturer at a time).

4.10.4 Development of the NTIA testbed and its use by industry and FCC

The development of the DFS testbed (Figure 4.4) by ITS and OSM engineers represented a significant investment by the government to move 5 GHz spectrum sharing forward. In its totality, development of this testbed absorbed about two engineering years of government resources, including the procurement and programming of specialized test-and-measurement equipment and of specialized software for controlling the measurements, and recording and analyzing the resulting data sets.

In 2005, the U-NII manufacturing industry was invited by NTIA to make use of this testbed on a no-cost basis. Any U-NII manufacturer could send its prototype device(s) to the lab as long as it/they were accompanied by knowledgeable DFS design engineers who could operate the systems' software on an as-needed basis during testing.

After the initial testbed was built and used to verify the performance of both early DFS monitors and of the testbed itself, the NTIA transferred the testbed technology to the FCC and to industry test houses that would, in the long run, use the NTIA testbed design to perform routine ongoing testing of DFS U-NII systems. The ITS-built testbed was, however, retained at ITS and is used to this day to perform some spot-checks of off-the-shelf DFS U-NIIs and to troubleshoot individual types of DFS U-NIIs that have been flagged by the FCC or other government agencies as having potential problems with effective detection of radar signals. This issue has arisen mainly with regard to the examination of firmware upgrade effects on DFS devices.

Starting in 2005, three U-NII manufacturers with DFS-capable prototypes accepted the NTIA offer to use the new testbed at ITS in Boulder. Each of these companies was scheduled to have up to two weeks of time at the lab to test their devices, although none actually ended up needing more than a week. Each manufacturer worked at the lab alone, with only ITS and OSM engineers to assist; there was no overlap between manufacturers using the testbed.

The fundamental information challenge (described above) about specific versus generalized radar waveform detection was only finally resolved when the three manufacturers sent DFS design staff to perform early testing on prototype DFS monitoring devices. It was at that point, when the government system illustrated in Figure 4.4 was used to emit radar pulses at the DFS receivers, that some of the industry engineers finally were able to see first-hand exactly what it was that their systems were supposed to detect. To their great credit, once they finally became aware what the 5 GHz radar test pulses looked like, they were all able to get their prototype DFS monitors running reasonably well in fairly short order.

It is difficult to overstate the need for eventual face-to-face technical exchanges between spectrum engineers on the government and industry sides when the nuts-and-bolts engineering of spectrum-sharing systems need to be accomplished. Design goals and specifications for sharing systems are necessary, but there is no substitute for human interchanges such as occurred when the government and industry people began to work on testing in Boulder. The exchanges do not need to be long or complicated, but without them huge misunderstandings can develop about what systems are supposed to do and how they are supposed to operate. This observations leads to the next lesson learned.

> **Lesson 6:** There is no substitute for face-to-face interchanges of critically important design information about technical requirements when spectrum-sharing systems and protocols are being developed.

4.11 Challenges for development of DFS test-and-certification protocols

Spectrum-sharing development offers the advantage of knowing exactly which systems are already incumbent in a proposed sharing band, and what most of the technical characteristics of those systems are. However, this process also suffers from the disadvantage of not necessarily knowing exactly what the technical characteristics will be of future systems that may be introduced into the band. Something is typically known of the likely general technology that will be used in the new sharing systems, but specifics may be unavailable because no one, not even industry itself, may know exactly what may be deployed in the future.

This uncertainty may be caused by the desire to remain technology-neutral regarding what is going to be newly introduced to a band. Thus only the broadest guidance outlines may be provided, such as "communication systems will be introduced…." In the case of the 5 GHz band, it was agreed at WRC-03 that the new systems would be "wide area networks and radio local area networks (WANs and RLANs)," which is fairly broad. Only later was the U.S. implementation at 5 GHz limited to U-NIIs, and even that is a broad technology area. To cite just one factor, there are two general types of U-NIIs: those based on 802.11 technology for their data transmissions and those that use frame-based transmissions. It can be an open question as to whether one type of spectrum-sharing technology can or will work across all available types of communication systems within a single service such as U-NII.

4.11.1 Advantages: working from a blank slate

Specifications for the exact requirements of spectrum-sharing protocols may have to be somewhat vague and generalized. Some sharing requirements may depend on the characteristics of the new systems that are to be introduced to the band, and as noted above it is possible that no one will know what those characteristics will be.

A positive aspect of this situation is that it may be regarded as an opportunity for some party, either a government agency, industry, or both, to break the deadlock by considering that the new spectrum sharing is going to be a blank slate for new development. This condition can then become an opportunity for innovation. Having said that, some initiatives may be required on both sides to take advantage of this situation; large organizations do not necessarily encourage such moves by individual engineers.

In the case of 5 GHz DFS development, the deadlock was broken over time by numerous technical interchanges in which certain aspects of sharing were negotiated. These included:

- Only AP units would monitor for radar signals;
- The government would provide industry with a monitoring threshold for radar signals;
- The government would not ask industry how their radar detection algorithms worked;
- Industry would submit their DFS devices for general-purpose radar waveform detection testing rather than radar-specific testing;
- Industry could use any basis for U-NII data packet protocols (i.e. both 802.11-based and frame-based systems were valid). They would all be submitted to a single set of tests and would be subject to a uniform set of pass-fail criteria.

4.11.2 Disadvantages: areas of developmental doubt and uncertainty

Neither the government nor the industry may necessarily be able to specify other important aspects of the spectrum-sharing protocols in advance of actual sharing implementation. This situation arises because some protocols may depend on design characteristics of the new sharing devices that cannot be known until the devices (or at least good prototypes) have finally been built.

In the case of 5 GHz DFS, two such problems had to be solved. The first was to determine the power levels in 5 GHz radar receivers that would cause harmful interference. This parameter had not been defined before for radar receivers. Thus government engineers (ITS and OSM) undertook a multi-year, roughly million-dollar effort to determine harmful interference thresholds for a number of different radar system receivers. This work involved numerous other agencies including the Department of Transportation (specifically the FAA), the Department of Defense (the Army and Navy), and the Department of Homeland Security (the Coast Guard). Industry was also involved as they were manufacturers of the radar systems. Some of the work was even performed by NTIA engineers in cooperation with the UK administration at Southampton and Funtington sites in that country.

In the end, the government was able to provide an INR (−6 dB) that would be considered a threshold for harmful interference, and which could be converted into the detection threshold that would be required of DFS radar monitors. Thus, the problem was solvable in advance of DFS deployment.

The other main problem that occurred in this area was the question of how much frequency change would be required of the DFS units when they had detected a radar. This problem was not so easily solved in advance of DFS-system construction because it depends on two factors, only one of which could be accommodated in advance: the width and shape of radar receiver intermediate frequency (IF) stages and the roll-off rate of U-NII spectrum emissions. The radar IF question was dealt with by measuring radar IF stage responses for every radar type that was checked for an INR response. However, the U-NII roll-off rates could not be known until the U-NIIs were built, and the U-NIIs were not going to be built until the rules for sharing were in place.

In the end, the original DFS rules simply called for U-NIIs to vacate their operational channels when they sensed a radar signal. Thus, they could move as little as one channel over in their band when they sensed a radar.

This one-channel minimum jump requirement turned out to not always be adequate in practice, but it could not be modified until DFS U-NIIs were built and a government agency (NTIA) had an opportunity to measure the emissions of an entire family of such devices. These measurements were eventually published in an NTIA Technical Report [8] and were used as the basis for refined estimates of how much frequency (channel) change would be needed to preclude interference to radars when DFS U-NIIs sensed them.[16] From these observations arose the next lesson learned.

> **Lesson 7:** Some technical spectrum-sharing parameters (such as out-of-band spectrum roll-off) may not be known until after spectrum-sharing devices have been built and possibly deployed.

4.12 Interference cases after initial DFS deployment

In late 2008, the FAA began to experience a puzzling new type of interference at multiple TDWR stations in the continental U.S. and the Possession of Puerto Rico. The worst of the interference occurred at the San Juan, Puerto Rico TDWR and the second-worst interference at New York City area TDWRs. FAA teams met with some initial success in examining the interference. In 2009, with some indications that the interference might be caused by 5 GHz DFS-equipped U-NIIIs in these areas, FAA contacted NTIA and requested a joint effort to examine and resolve the problem. Because the interference sources were suspected to be non-government systems, the FCC was also contacted for support and assistance.

4.12.1 Identification of interfering DFS-equipped devices at San Juan

It was agreed that, since the interference characteristics were similar at all TDWR locations, the detailed study of a single such location would be sufficient as its results would be applicable to all locations. The place where the interference problem was worst was selected as the study location: San Juan.

[16] In practice, many U-NIIs ended up not moving a single channel at a time upon radar detection, but instead simply changed to the ISM band at 2.4 GHz to preclude interference.

The approach for identifying the interference sources was methodical:

- First, examine and document the interference characteristics on the radar data display, including observation of how many radial interference lines (strobes) that appear.[17]
- Second, examine and document the appearance of interference in the time domain in the radar receiver IF stage. This examination showed that the interfering signals were coming from multiple sources that used both 802.11 and frame-based data packet protocols, which of course meant that the problem originated from U-NII transmitters.
- Third, partially disassemble the radar's RF front end and examine the responses to interference and to input test signals. This action was used to determine whether any front-end overload was occurring. As part of this stage, the radar's RF LNA was tested to determine its overload point and to compare that power level to the interference power level.
- Fourth, check and document the exact correlation between every azimuth where interference was occurring and the type of signal causing interference at each of those azimuths.
- Finally, perform field-work with the FCC to locate interfering transmitters and determine their technical characteristics, including whether they were adhering to DFS rules.

The field-work effort took two weeks and was a nontrivial operation for the government, but it was ultimately highly successful. Several interfering U-NIIs were identified and located. In the end it was determined that one U-NII had been illegally imported and operated in the TDWR band without DFS functionality. Other units were nominally legal as purchased but had been illegally modified to operate with after-market antennas. Further, some units seemed to have been bought and set up legally, but nevertheless were causing interference to the TDWR, which was not supposed to happen. The results were documented in an NTIA Technical Report [9].

The problem with the San Juan DFS U-NIIs that had passed certification testing and had been properly installed, but had nevertheless failed to detect a TDWR in the field had to be resolved. To do so, NTIA and FAA set up a series of tests and measurements with the FAA TDWR in Oklahoma City, which has an engineering unit at the Aeronautical Center used for training air traffic controllers that could be used for spectrum experimentation without interrupting service in any airport operations.

The resulting analysis, as documented in another NTIA Technical Report [10], showed that the protocols that were being used for radar test waveforms for DFS certification could be passed by DFS monitors that still would not reliably detect actual TDWR stations. The problem was rectified by improving the radar test waveforms.

It is important to note that this problem did not occur for all DFS U-NIIs that had been certified. Some devices certified with the original radar waveforms did detect

[17]Multiple strobes could indicate either multiple sources of interference, or a single interference source that was hitting the radar hard enough to couple into both the mainbeam and the radar antenna pattern sidelobes (producing what is referred to as a "rising-sun" display pattern).

TDWRs reliably. In other words, the problem of insufficiently designed radar-testing waveforms was only discovered after several different models of DFS-equipped U-NIIs had been put through the testing and deployed in the field.

Another outcome of the work at San Juan and Oklahoma City was that the FCC developed a great deal of in-house proficiency at searching for and finding U-NII transmitters that cause interference to TDWRs. This capability has been used many times over the last ten years, as the reader may glean from the 47 entries for DFS-related enforcement actions in Table 4.2.

> **Lesson 8:** Some flaws in spectrum-sharing device testing may only be discovered after many spectrum-sharing models have been tested, certified, and deployed.
> **Lesson 9:** Certification testing requirements for spectrum-sharing devices need to be as robust as possible.

4.13 Ongoing DFS spectrum sharing maintenance

It should be clear at this point that DFS spectrum-sharing technology is not a fire-and-forget method, but requires ongoing maintenance. This situation is likely to be true for future, more complex spectrum-sharing technologies as well.

4.13.1 Continuing monitoring of DFS devices

As noted above, DFS-equipped systems can fail after they have been certified and field-deployed. The primary reason for such failure of properly installed and operated systems seems at present to be firmware updates that can unintentionally impair or disable DFS. Other than waiting for interference reports, this problem calls for ongoing spot-check surveillance of DFS-equipped U-NIIs that are for sale to the public.

At the ITS laboratory in Boulder, NTIA periodically selects, on a random basis, some 5 GHz DFS U-NIIs for such testing. The units are purchased from commercial outlets without interaction with manufacturers. A version of the DFS testbed system in Figure 4.4 is used for the testing.

The NTIA also undertakes DFS performance checks on U-NIIs that have been reported to be causing interference to TDWRs even though those units have been DFS-certified. As already noted, the problem has usually turned out to be new firmware updates that have unintentionally affected DFS functionality. All of this work requires some ongoing support from NTIA (for the Boulder lab work) and from government field offices (to identify interfering U-NII transmitters).

4.13.2 Consideration of more complex future radar waveforms

Future radars may be built at 5 GHz in the DFS sharing bands that could result in an eventual need for modifications or additions to the DFS radar test waveforms. Keeping up with new 5 GHz radar development requires continuing, ongoing support in this area.

Lesson 10: Continuous surveillance of implemented sharing technologies is needed for the lifetime of a complex spectrum-sharing technology.
Lesson 11: Some modifications or additions may need to be made to spectrum-sharing technologies subsequent to field introduction.
Lesson 12: Finally, this lesson is presented as a look back on the totality of the entire DFS experience. Testing in a real-world testbed prior to widespread introduction of a new spectrum-sharing technology might be a good idea. Of course, such testbeds take time and resources to build, and taking the time for testing may conflict with the desire to deploy new technologies as quickly as possible.

4.14 Looking forward to future spectrum sharing

The development and deployment of DFS-based 5 GHz spectrum-sharing technology has arguably been largely successful, though achieving this success has required considerable use of government resources on an ongoing basis. As spectrum sharing between radars and non-radar systems are considered in other bands, the lessons described above ought to be heeded. Fundamentally, substantial resources will likely be needed to support sharing in new bands. These resources will be needed on both the front end (in design and development phases of sharing technologies) and at the back end, on an ongoing basis after such technologies have been deployed. It is unlikely that complex spectrum-sharing schemes between radars and non-radar systems will ever be fire-and-forget in the sense that they can be deployed without the need for ongoing work by engineers and spectrum managers to keep them functional.

References

[1] International Telecommunications Union, Radiocommunication Sector (ITU-R), “Unwanted Emissions in the Out-of-Band Domain,” ITU-R Recommendation M.1541-6, August 2015. https://www.itu.int/rec/R-REC-SM.1541/en
[2] F. Sanders, R. Sole, B. Bedford, D. Franc, and T. Pawlowitz, “Effects of Interference on Radar Receivers, NTIA Report TR-06-444, U.S. Department of Commerce, September 2006. http://www.its.bldrdoc.gov/2481.aspx
[3] Code of Federal Regulations, General Technical Requirements for DFS, Title 47, Sub-Part 15.407.
[4] Federal Communications Commission, Notice of Proposed Rulemaking, “NII/SUPERNet at 5 GHz, ET Docket No. 96-102, FCC 96-193. May 6, 1996. https://transition.fcc.gov/Bureaus/Engineering_Technology/Notices/1996/fcc96193.txt
[5] Federal Communications Commission, Report and Order, “Amendment of the Commission’s Rules to Provide for Operation of Unlicensed NII Devices in the 5 GHz Frequency Range,” Amendment to ET Docket No. 96-102, January 9, 1997 https://transition.fcc.gov/Bureaus/Engineering_Technology/Orders/1997/fcc97005.txt

[6] Federal Communications Commission, Memorandum Opinion and Order, "Amendment of the Commission's Rules to Provide for Operation of Unlicensed NII Devices in the 5 GHz Frequency Range," June 24, 1998. https://transition.fcc.gov/Bureaus/Engineering_Technology/Orders/1998/fcc98121.txt

[7] ITU-R, "Dynamic Frequency Selection in Wireless Access Systems Including Radio Local Area Networks for the Purpose of Protecting the Radiodetermination Service in the 5 GHz Band," Recommendation M.1652, May 2011. https://www.itu.int/rec/R-REC-M.1652/en

[8] J.E. Carroll, G.A. Sanders, F.H. Sanders, and R.L. Sole, "Case Study: Investigation of Interference into 5 GHz Weather Radars from Unlicensed National Information Infrastructure Devices, Part III," NTIA Technical Report TR-12-486, U.S. Dept. of Commerce, July 2012. https://www.its.bldrdoc.gov/publications/2677.aspx

[9] J.E. Carroll, F.H. Sanders, R.L. Sole, and G.A. Sanders, "Case Study: Investigation of Interference into 5 GHz Weather Radars from Unlicensed National Information Infrastructure Devices, Part I," NTIA Technical Report TR-11-473, U.S. Department of Commerce, November 2010. https://www.its.bldrdoc.gov/publications/2548.aspx

[10] J.E. Carroll, G.A. Sanders, F.H. Sanders, and R.L. Sole, "Case Study: Investigation of Interference into 5 GHz Weather Radars from Unlicensed National Information Infrastructure Devices, Part II," NTIA Technical Report TR-11-479, U.S. Department of Commerce, July 2011. https://www.its.bldrdoc.gov/publications/2554.aspx

Chapter 5
Spectrum use, congestion, issues, and research areas at radio-frequencies

Eric L. Mokole[1], *Tapan K. Sarkar*[2], *Miguel Angel Lagunas*[3], *and Magdalena Salazar Palma*[4]

5.1 Introduction

The objective of this chapter is to provide a high-level perspective on the growing conflict over use of the radio-frequency (RF) spectrum, a precious and highly sought resource extending from below 1 MHz to above 100 GHz, caused by the accelerating demand for consumer use via 4G and soon-to-be 5G wireless communications. The world at large now faces serious spectrum-compatibility problems that require new and innovative solutions—increased spectral congestion and crowding are especially challenging. However, anticipated improvements in electromagnetic (EM) systems up to 300 GHz are beginning to be realized. Less restrictive constraints on communication systems, inherent in one-way propagation paths and much less expensive components, have allowed that community to design and develop more diverse waveforms and systems. Consequently, commercial cellular systems are proliferating at incredible rates, resulting in extremely spectrally dense environments and fierce competition for spectrum that traditionally has been the almost exclusive province of radars as primary legal users. For radar applications, however, the promise is being realized much more slowly, and the inundation of communication devices from the commercial sector has caused significant radar–communication interference problems. In addition, radar and communication systems are important components of military operations, and advances in waveform-diversity signal and data-processing techniques that are likewise relevant to spectrum sharing offer the promise of significantly improved performance.

Although the ensuing discussion is strongly influenced by the lead author's experience with the U.S. Department of Defense, it is in many respects a microcosm of the world at large and provides a referential context. After giving a brief

[1]The MITRE Corporation, Programs & Technology, USA
[2]Electrical Engineering & Computer Science Department, Syracuse University, USA
[3]Centre Tecnologic de Telecomunicacions de Catalunya, Spain
[4]Department of Signal Theory and Communications, Universidad Carlos III de Madrid, Spain

background, the impacts of spectral interactions between communication and radar systems are discussed along with some essential EM theory for radar and communications, selected radar spectrum-mitigation efforts, and suggested research areas for spectral harmony. Identified research areas are aimed at mitigating existing and expected interference problems and consequently leading to harmonious co-existence among the various competing users, specifically between radar and communication systems. The increasing need for this finite resource is driving innovation into new methods that enable different systems to flourish. Examples include adaptive spectrum avoidance/notching, opportunistic use of other emissions like passive multistatic radar and various forms of radar-embedded communication, spectrum-efficient waveform design accounting for transmitter and antenna effects, and radar/communication co-existence and co-design strategies. These endeavors necessarily require consideration of all the different facets of the problem, especially the relevant EM theory, systems engineering, signal processing, and regulatory constraints.

The RF spectrum is used for a wide range of purposes such as communications, radio navigation, radio and television broadcasting, radar, and countermeasures. In particular, vitally important radar applications include air traffic control, geophysical monitoring of Earth resources from space, automotive safety, search-and-rescue operations, humanitarian demining, severe weather tracking, and surveillance/tracking/identification for public safety and national defense. Despite the importance of such radars, spectrum that traditionally has been the almost exclusive province of radars as the primary legal users is being fiercely contested worldwide by very lucrative and widely used communication systems (especially commercial cellular systems). Insufficient numbers of radars, with the possible exception of air-traffic-control systems, are manufactured to justify a business case for their production. In contrast, commercial cellular systems are exceedingly profitable, provide a tremendous number of services to all strata of society, and are proliferating at incredible rates, resulting in extremely spectrally dense environments and fierce competition for spectrum where radar has had primacy. For example, IEEE activities in this vein include the Internet of Things, the Internet of Space, Smart Village, the Special Interest Group on Humanitarian Technology (SIGHT), and 5G [1–3]. For radar applications, however, the promise of new and innovative technology is being realized much more slowly, and the inundation of communication devices from the commercial sector has caused significant radar–communication interference problems.

The growing worldwide contention over radar spectrum usage and management is fundamentally a contest between the commercial benefits of wireless expansion and the need for scientific research and national-defense/public-safety systems (Figure 5.1). Nearly all radar and communication services need greater bandwidth, which only exacerbates the competition for spectrum.

In the following, some efforts over the past twenty years to mitigate radar–communication interference are highlighted, and suggestions for requisite research and possible ways to achieve spectral harmony are postulated. For example, to mitigate power-amplifier-induced adjacent-band interference, spectrally confined transmit waveforms and hardware for power-amplifier circuits to implement such waveforms were developed in a laboratory setting at the U.S. Naval Research Laboratory, Baylor

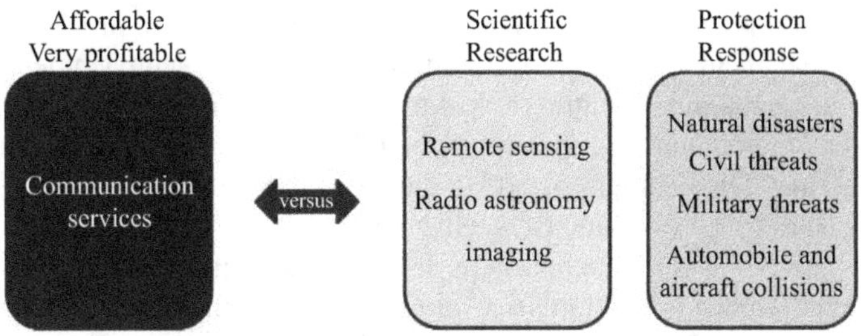

Figure 5.1 EM spectrum contention between communication services and radar applications such as scientific research and protection responders

University, and the University of Kansas. Finally, performance issues and capability gaps associated with these efforts are discussed. In particular, the impact of not properly incorporating EM theory into system design, especially the antenna, signal representations, and proper characterizations of propagation media are presented.

The physics of EM propagation and the vast operational differences between the various spectrum users clearly make the commoditization of spectrum usage extremely problematic. Different frequency bands experience different amounts of atmospheric attenuation, reflection, refraction, and penetration within the various levels of the troposphere and ionosphere, both of which significantly impact propagation depending on the operating frequency. All spectrum users share a common fundamental payload: information—be it for dissemination (communications), for self-reference via comparison of different emissions (navigation), or for extraction from and exploitation of the environment itself (radar). However, the nature and source of the information are radically different for each modality, thereby necessitating significantly different operational paradigms.

Modalities such as communications involve one-way propagation to convey information from one location to another, with the transmit power and receiver sensitivity being largely driven by the separation distance and the amount of atmospheric attenuation. In contrast, sensing modalities like radar involve two-way propagation such that, for the same separation distance as the communication link, a radar would require orders of magnitude greater transmit power and receiver sensitivity just to contend with EM attenuation by the medium and system losses. Furthermore, an extremely important goal of many radars is to exploit aspects of the illuminated environment to discriminate the signatures of potential targets of interest from among an overwhelming collection of reflected echoes in its field of view (FoV)—akin to listening for a whisper in a hurricane. Finally, while different measures may be taken to ensure the quality of information obtained at a communication receiver (interleaving, forward error correction coding, channel equalization), the information the radar seeks may be non-cooperative (stealthy or combative) targets.

Despite the rather incompatible aspects of these modalities, economic forces dictate that radar systems must become more interoperable with commercial communications. To do so requires that radar (1) drastically curtail the spectral regrowth of high-power emissions into adjacent bands through improved technology and operational methods and (2) contend with the steady encroachment of interference from commercial emitters into radar bands. While the latter is governed by the Federal Communications Commission (FCC) within the United States and by the International Telecommunication Union (ITU) internationally, there is no guarantee that commercial spectrum users will actually adhere to emission specifications at all times (litigation is not an option during a combat mission). That said, publicly available communication standards may aid in identifying the structure of an interference source so as to effect its cancellation or at least partial suppression. With regard to curtailing adjacent-band spectral regrowth of radar emissions, for obvious reasons of national security, the precise form of military radar emissions cannot be made available to the non-military spectrum users with which a military radar could potentially interfere. Consequently, the impetus is largely on the radar community to remediate the interference it may cause to other spectral occupants. This solution could potentially take the form of spectral avoidance (frequency hopping), spectral/spatial nulling on transmit, modifications to transmit hardware, or some combination thereof.

5.2 Impacts of EM spectral interactions between communication and radar systems

The tremendous expansion and proliferation of commercial cellular devices over the last two decades have had a significant effect on the other users of the spectrum. Many traditional users are being supplanted through national legislations that reduce their status in favor of the commercial cellular community. For example, several radar bands are being impacted.

- At S-band (2.7–2.9 GHz and 3.4–3.7 GHz), 4G and fixed wireless access (FWA) are being considered for sharing and primacy.
- At C-Band (5.25–5.725 GHz), wireless LANs and unlicensed (license-exempt) users may get primary status.
- At X-band (10–10.45 GHz and 10.5–10.68 GHz), fixed and mobile users may be given co-primary status, and FWA deployments are expected soon.
- Ku-band (13.75–14 GHz) has been shared with the frequency selection service (FSS) as determined at the WARC-92 (World Administrative Radio Conference). Initially, only a few big FSS terminals were deployed; however, small ubiquitous terminals are now being considered.

For example, in the United States, the National Telecommunications and Information Administration (NTIA) and the FCC identified 589 MHz in federal/shared (335 MHz) and nonfederal (254 MHz) bands for potential reallocation [4] based on the 2010 National Broadband Plan [5].

- U.S. federal/shared bands
  - 1695–1710 MHz and 1755–1780 MHz;
  - 3550–3650 MHz;
  - 5350–5470 MHz and 5850–5925 MHz.

- U.S. nonfederal bands
  - 2305–2320 MHz and 2345–2360 MHz;
  - 1915–1920 MHz and 1995–2000 MHz;
  - 2000–2020 MHz and 2180–2200 MHz;
  - 2155–2180 MHz;
  - 2020–2025 MHz;
  - 144 MHz from 512–698 MHz.

5.2.1 Impacts of radars on communication systems

Of the documentation on radar interference to communications systems [6], two examples are selected for illustrative purposes. In the early 2000s, the U.S. Navy realized that it had a nontrivial electromagnetic compatibility (EMC) problem with WiMAX (Worldwide Interoperability for Microwave Access) systems. WiMAX is a wireless industry coalition dedicated to the advancement of IEEE 802.16 standards for broadband wireless access (BWA) networks. Consequently, field tests (2007) and subsequent bench tests (2010) were devised to assess the impact that navy radars might have on WiMAX systems.

The field tests were conducted at Wallops Island, Virginia, USA. The test configuration consisted of the AN/SPY-1 B/D radar (circled in Figure 5.2), a WiMAX base station (BS), and two single-subscriber devices (SS#1 and SS#2) in a realistic geographic scenario over a 50 square-mile area. The BS was located on the Assateague Lighthouse, with the antenna 154 feet above the ground in the search FoV of the SPY-1. The SS#1 was located at the NASA Visitors Center just outside the SPY-1 search sector, and SS#2 was located at Wishart Point well outside the search FoV, but still exposed to significant energy via the radar's sidelobe (Figures 5.2 and 5.3).

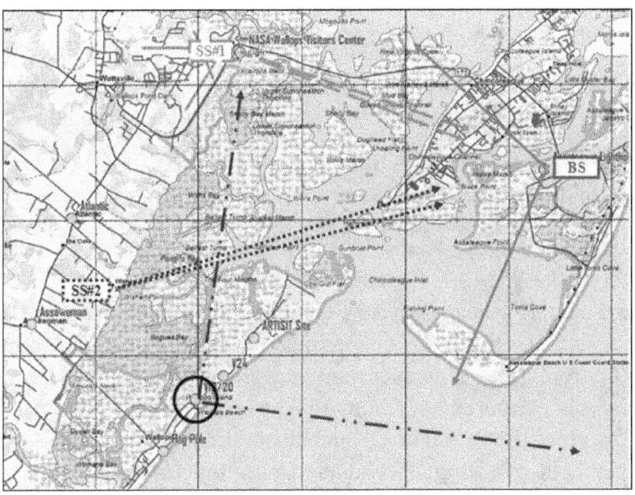

Figure 5.2 Distributed WiMAX test locations around AN/SPY-1 B/D radar on Wallops Island, Virginia, USA

Figure 5.3 AN/SPY-1 B/D radar and WiMAX BS antenna on Assateague Lighthouse on Wallops Island

To assess the impact of the SPY-1 on WiMAX, the packet loss was measured for various communication links. Although no WiMAX equipment burn-out occurred during the tests, significant interference was observed. When the radar operated in a fixed single band that included the WiMAX center frequency, the packet loss was often severe:

- 30–45 percent for the low-power mode on the SS-to-BS links;
- >50 percent for the high-power mode on the SS-to-BS links;
- 14–17 percent for the low-power mode on the BS-to-SS links;
- 27–42 percent for the high-power mode on the BS-to-SS links.

When the radar operated in a fixed single band that was 40 MHz from the WiMAX center frequency, the packet loss was less severe but still significant:

- 10–30 percent on the SS-to-BS links;
- 3–15 percent on the BS-to-SS links.

Since the tests at Wallops Island could not assess the impact of a radar at the constellation level of the WiMAX system, a coexistence experiment was conducted to emulate a radar operating near a wireless communication system using 64-quadrature amplitude modulation (QAM). This experiment was performed in a laboratory bench test using a Textronix real-time spectrum analyzer serving as the BS receiver [7]. Figure 5.4 displays measurement results for the received constellation. The constellation diagram plots the in-phase (I) and quadrature (Q) diagram of all possible transmitted symbols in a digital 6-bit configuration (64 possible symbols). The received constellation looks normal (64 evenly spaced symbols) when the communication system operates in the absence of the radar (left frame). However, when the

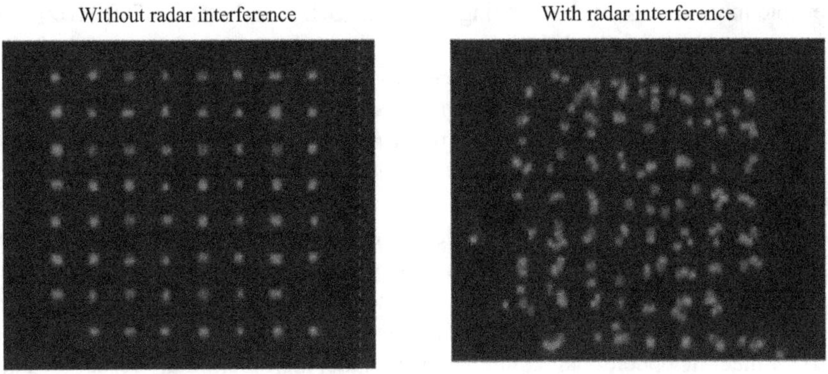

Figure 5.4 I-Q symbol diagram of a received 64-QAM constellation for an experimental bench test involving wireless and radar systems without radar interference (left frame) and during the operation of a nearby radar system in a neighboring frequency band (right frame)

radar is operating in a neighboring frequency band, severe degradation of the QAM constellation is manifested as a smearing of the individual symbol representations (right frame). The radar transmitter causes this degradation by leaking into the frequency band of the wireless communication system, which in turn drives the receiver amplifier into its nonlinear regime, causing the constellation symbols to be detected inaccurately [8]. In other words, many of the transmitted symbols are received and detected near or in regions where they would be incorrectly detected as different symbols.

5.2.2 Impacts of communication systems on radars

The impacts of communication systems on radars include loss of primacy in a band; operational restrictions; desensitization (range reduction), design constraints (emissions restrictions), and the unavailability of wide spectral swaths needed to achieve high-range resolution. In particular, the spectrum from 2 to 4 GHz, known as S-band in the U.S. radar community, is being highly contested as the commercial communication industry continues to apply intense pressure on the U.S. government to allow 4G systems based on WiMAX and the increasingly more prominent long-term evolution (LTE), along with impending 5G systems, to operate in bands allocated for radar. In 2010, the FCC promulgated President Obama's National Broadband Plan [5], which delineated the process by which any reallocation of this spectrum may be determined. On July 20, 2012, the U.S. President's Council of Advisors on Science and Technology (PCAST) released a set of recommendations to the President on how to realize the full potential of government-held spectrum to spur economic growth by facilitating spectrum sharing as a mainline approach to spectrum management [9]. Then on April 28, 2016, the FCC put final rules in place for a new Citizens Broadband Radio Service to enable innovative spectrum sharing in the 3.5 GHz Band (3550–3700 MHz) [10]. This sequence of events completes the demotion of radar as primary user in the 3550–3700 MHz band that began in 2004 when the FCC initiated a proceeding to foster the introduction of wireless broadband operations in the 3650–3700 MHz band [11].

The S-band regime is attractive from a radar standpoint because it provides a trade-off between long-distance search/track capabilities (usable propagation distance improves as frequency decreases due to less attenuation) and reasonable antenna size (antenna aperture size generally decreases as frequency increases). While this band is also attractive to the wireless industry for many of the same reasons, most communication systems are cell-based, using relatively shorter distances (radii of 25–50 km), and do not require the significant long-range capability that radar systems need for meeting operational requirements. In 2006, the U.S. military recognized that conflicts would arise between radar and these commercial wireless systems and began discussing the need for R&D to address potential solutions to the growing problem of spectral congestion.

One casualty of the demotion of radar in the 3500–3700 MHz band is the ship-based air surveillance radar, otherwise known as the AN/SPN-43C, which was used for carrier air traffic control within a 50 nautical-mile radius up to an altitude of 30,000 feet. The AN/SPN-43C ensured safe and expeditious shipboard approach, departure, and marshal control to support fleet aircraft operations for aircraft carriers and amphibious assault ships. After 15–20 years of seeking an alternative frequency of operation, the U.S. Navy finally awarded Saab Defense and Security USA an initial contract in 2016 to provide the AN/SPN-50 (V)1 as a partial replacement for the currently deployed AN/SPN-43C air traffic radars aboard the U.S. Navy's aircraft carrier (CVN) and amphibious assault (LH) class ships [12]. The order includes an initial procurement of three radars over a period of four years. However, it remains unclear what the ultimate solution to this problem will be.

In addition to affecting military radars, communication systems can cause serious problems to civil radars. At least three different sites (Grand Rapids, Michigan; Jacksonville, Florida; and St. Louis, Missouri) for the NEXRAD WSR-88D weather radar belonging to the U.S. National Weather Service (NWS) have experienced interference and performance degradation from transmitters of 4G WiMAX BSs [13]. For example, four WiMAX BSs west of Lake Michigan in Wisconsin degraded the service of the NEXRAD system to the east of the lake in Grand Rapids (Figure 5.5), indicated by the three radial strobes piercing the radar image (cloud-like formation) in Figure 5.6 [14]. As a baseline condition, Figure 5.6(a) shows the Grand Rapids NEXRAD display when no interference was present. The cloud-like response near the radar is a normal, clear-air, interference-free baseline condition. In contrast, these strobes contaminated NEXRAD's data images of reflectivity, velocity, and spectral width as evidenced by the diminished cloud and intensity in Figure 5.6(b). Once the interference was recognized, efforts were undertaken to mitigate the problem with a filter on WiMAX transmitters. However, not all fixes to interference are so straightforward.

NEXRAD WSR-88D Radar
Grand Rapids Michigan, USA

Figure 5.5 NEXRAD WSR-88D weather radar in Grand Rapids Michigan, USA. Photographs from the NTIA Report 13-490; https://www.ntia.doc.gov/files/ntia/publications/13-490_1_.pdf

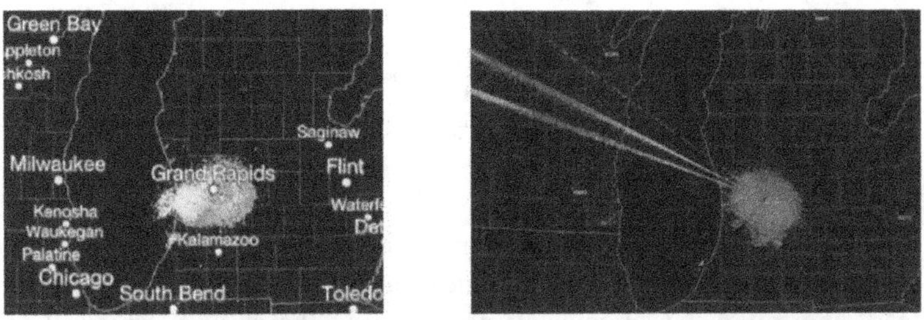

(a) Baseline clear-air reflectivity display with no interference for NEXRAD in Grand Rapids Michigan, USA

(b) Interference Strobes from 4 WiMAX 4G Base Stations in Wisconsin on NEXRAD display in Grand Rapids Michigan, USA

Figure 5.6 Radar displays of NEXRAD WSR-88D weather radar in Grand Rapids Michigan, USA (a) without interference and (b) with interference. Photographs from the NTIA Report 13-490; https://www.ntia.doc.gov/files/ntia/publications/13-490_1_.pdf

5.3 Radar spectrum-mitigation efforts

5.3.1 Selected efforts since 1998

Several low-level R&D efforts to address RF-spectrum compatibility issues began in the U.S. radar community around 1998. Two areas of interest involved RF propagation in confined ship spaces [15,16] and mitigation of interference by high-power radars to other users in adjacent bands.

Adjacent-band interference generated by high-power radars is caused by spectral regrowth arising from nonlinearities in their power amplifiers (PAs) that generate third-order and higher intermodulation products. Because these PAs operate in saturation, nonlinear effects often create significant interference. These adjacent-band emissions are further exacerbated by the fact that the power in the spectral regrowth regions is directly proportional to the peak power of the radar, which can be as high as ~10 MW. To overcome this problem, spectrally confined transmit waveforms [17] and hardware for PA circuits to implement these waveforms [18] were investigated with some success in a laboratory setting. To realize better interference reduction, these early efforts were extended to study joint optimization of the physical transmitter and driving waveform [18–20], with such investigations still ongoing.

Understanding the propagation of RF energy in confined shipboard environments is extremely important to the operability and effectiveness of communication systems on naval vessels. For example, single-frequency communication systems are known to suffer from multipath interference that can render them ineffective in such spaces. In contrast, spread-spectrum and frequency-hopping systems can operate effectively in these same environments in the absence of other RF signals. Unfortunately, the EM interactions of external and internally generated RF signals within ship structures are not well understood. For example, the transfer characteristics of these signals relative to the very complicated environment of a ship (steel bulkheads, watertight doors, ducts, mast cable transits, compartment penetrations and obstructions, and steel deck plates) are not well characterized. Moreover, the complex interactions of communication signals with other RF signals like radar are not well understood for confined ship spaces.

In an initial attempt to obtain some insight into RF radiation and propagation in this environment, a broadband bistatic measurement system (800–3000 MHz) was designed, and initial measurements were taken on the ex-USS *Shadwell* in 1998 [15,16]. An underlying purpose of this study was to determine the limitations of shipboard emergency communications, especially during fires when watertight compartment doors are closed. To quantify the distribution of energy from a source in ship compartments, data were sampled and received power levels were collected for various open–closed door configurations, for selected locations of the transmitter and receiver on three levels of the ship, and for different signal polarizations. Although no major conclusions were reached on what EM modes were actually propagated, a few inferences were drawn: (1) received power levels appeared to be sufficient to ensure communication over reasonably long propagation distances for inter-deck and intra-deck communication when watertight doors were closed;

(2) several possible mechanisms for enabling such communication were identified; and (3) significantly more testing in a controlled environment was necessary to achieve even the most rudimentary explanations of shipboard EM radiation and propagation. Accurate modeling of the RF EM interactions associated with naval vessels remains an open research area.

In 2002, two activities were initiated to understand and mitigate RF interference, the NATO Task Group of the Sensors and Electronics Technology (SET) Panel of the Collaboration Support Office (http://www.cso.nato.int/), SET-066 "Frequency Sharing between Communication and Radar Systems," and the Waveform Diversity (WD) working group. The overall objectives for the SET-066 program of work (active until 2006) were:

- to quantify the performance of military systems under conditions of interference from proposed re-allocation of the EM spectrum;
- to assess the effects of system performance changes on military capability;
- to propose methods to evaluate likely levels of interference between civil and military EM systems;
- to evaluate possible technology solutions to allow systems to work in the same part of the EM spectrum;
- to consider implications of new COTS (commercial off-the-shelf) wireless communication schemes;
- to review possible impacts of the allocation of frequency masks for ultra-wideband (UWB) devices.

Although a final report for SET-066 was not completed, its findings are summarized in Annex C of the final report for SET-182 "Radar Spectrum Engineering and Management" [21].

The Waveform Diversity working group, which began in the United States, had a much broader scope, with a major focus on radar spectrum issues and solutions [22]. This working group, which conducted numerous international workshops and conferences, eventually became a committee of the Radar Systems Panel of the IEEE Aerospace and Electronic Systems Society (AESS) and is still active. The SET-066 Task Group marked the beginning of several RF spectrum activities by the SET Panel that continue to this day, beginning with the lecture series SET-119 "Waveform Diversity for Advanced Radar Systems." In particular, the recently completed four-year SET-182 Task Group developed experiments and models to exploit transmitter, receiver, and waveform designs toward more optimal spectrum use. For example, PA linearization techniques and more spectrally cooperative waveforms that support radar transmitter emissions with lower spectral sidelobes were investigated. In addition, the SET Panel recently completed Task Group SET-179 on "Dynamic Waveform Diversity and Design" [23], Specialists' Meetings SET-204 on "Waveform Diversity" [24], SET-SCI-230 on "Reconfigurable and Scalable Multi-Function RF Systems in a Congested EM Spectrum" [25], and the SET-216 Lecture Series on "Cognition and Radar Sensing" [26].

Another U.S.-sponsored spectrum activity is the International Symposium on Advanced Radio Technologies (ISART), a government-sponsored conference that brings together government, academia, and industry leaders for collaboration on groundbreaking developments in and applications of advanced radio technologies. The ISART is sponsored by the Institute for Telecommunication Sciences (ITS), the research and engineering laboratory of the NTIA, which performs cutting-edge telecommunications research and engineering with other federal entities and the private sector. ITS supports the NTIA by addressing the telecommunications concerns of federal/state/local governmental agencies, private corporations/associations, and international organizations and by performing research and engineering that enables the U.S. government, national/international standards organizations, and private industry to manage the radio spectrum and to ensure that innovative new technologies are recognized and effective.

The ISART conference has been held 15 times since 1998 to discuss sensitive spectrum issues through oral presentations, workshops, and panel discussions. The previous symposium (ISART 2016) featured spectrum forensics—spectrum measurements that support interference monitoring, investigation, and enforcement. The most recent conference (ISART 2017) explored applications and the associated technical challenges at millimeter-wave frequencies from five perspectives: regulation, industry, standards, measurement and modeling, and systems. More information can be found at the ISART website [27].

5.3.2 *Current efforts*

Currently, many efforts addressing spectrum issues are ongoing throughout the world. Internationally, the NATO SET-227 Task Group on "Cognitive Radar" started in 2015 and will run through 2018. In 2012, the Radar Systems Panel of the IEEE's AESS formed its Waveform Diversity Committee with the goals of informing and educating scientists, engineers, and policy makers on the challenges to radars operating in congested and diminishing spectrum environments, describing differences in requirements between radar and communication systems operating in said environments, and identifying key R&D areas like cognitive radar and spectrum sharing. The committee's efforts recently yielded two special issues [28,29] and a tutorial [30] in the *AESS Magazine* on waveform diversity that featured radar spectrum.

In the United States, efforts at jointly optimizing the physical transmitter and driving waveforms and at embedding communication symbols in physical radar waveforms have been investigated in the last five years by Baylor University [8], the University of Kansas [31], and the U.S. Air Force Research Laboratory [32] to realize even better interference reduction through spectrally cleaner waveform designs. In addition, earlier efforts by the U.S. Naval Research Laboratory on shared-spectrum multistatic radar [33–35] and space-time adaptive nulled radar waveforms [36] are continuing. More recently, the U.S. Army Research Laboratory and the U.S. Army's Intelligence and Information Warfare Directorate (I2WD) of the Communications-Electronics Research, Development and Engineering Center (CERDEC) have become heavily involved in cognitive radar, a spectrum-sensing technique to enhance the performance of radars in the presence of harmful RF interference by adapting the bandwidth of the transmitted waveform [37,38] and to determine whether LTE communications and radars can coexist [39]. Since 2013, a DARPA program [40] on Shared Spectrum Access for Radar and Communications (SSPARC) has been attempting to achieve spectrum sharing between these modalities.

5.4 Suggested research areas for radar–communications spectral harmony

To address the spectrum problem, much research is needed to fill the gaps in the science-and-technology (S&T) knowledge bases of the radar and communication communities. Suggested research areas include, but are not limited to: (1) the design of efficient PAs (solid-state and tube) and transmitter configurations to provide improved spectral purity; (2) adaptable transmit filters and antenna technology for active arrays that more completely and accurately integrate EM theory and signal processing; (3) adaptive/cognitive waveform design to avoid spatial/spectral interference on transmit; (4) optimization of radar emissions accounting for non-ideal nonlinear aspects of the transmitter; (5) development of radar emission structures that induce minimal interference to commercial users in adjacent spectral bands; (6) innovative receiver designs supporting digital signal processing for in-band reception and adjacent-band interference rejection; and (7) coding at RF instead of baseband.

5.4.1 Adjacent-band interference mitigation for radar emissions

The many occurrences of EM interference (EMI) from high-power radars to communication systems like WiMAX for BWA and LTE for high-speed wireless communication have demonstrated the urgency for reducing adjacent-band interference caused by out-of-band noise and spectral regrowth. Out-of-band noise, which is particularly problematic for cross-field amplifier (CFA) tubes, is caused by the random fluctuations in electron-beam density of the electron cloud that spirals out from the cathode to the slow-wave structure of the tube. Spectral regrowth on the other hand is a result of nonlinearities in the transmitter that produce spurious signals at frequencies outside the operational bandwidth. Collectively, these spectrally expanded emission effects often extend into adjacent bands, causing interference to other spectral occupants. Because the amplifier of a high-power radar operates in saturation, these nonlinear effects can be quite pronounced. Moreover, these adjacent-band emissions are further exacerbated by the fact that the power in regions of spectral regrowth is directly proportional to the peak power of the radar, which can be as high as 10s of megawatts.

To address these sources of interference, the May 2013 edition of the NTIA Radar Spectrum Engineering Criteria (RSEC) [41] dictates considerably more stringent spectral roll-off requirements in the United States than has previously been the case (some modes increase from the previous 20 dB/decade to 30 or even 40 dB/decade). Of course, meeting these requirements is driving research into more spectrally clean high-power transmitter technologies, which has three general aspects: development of spectrally cognizant device technology; transmitter topology design; and hardware-in-the-loop (HiLo) waveform design.

While the nonlinearity of the PA is the predominant source of spectral regrowth, a holistic treatment of the entire transmitter and the waveform-modulated pulse is necessary to compensate for this effect. Given a desired waveform to be modulated onto a high-power pulse, the transmitter components preceding the PA (power supply, exciter, modulator) induce some amplitude modulation (AM) to AM distortion and AM to phase modulation (PM) distortion of the waveform, thus compounding the subsequent spectral regrowth distortion emitted from the PA. Furthermore, the rapid switching that occurs in the rise and fall of the pulse produces significant contributions to the frequency extent of the spectral spreading. A "slowing down" of the rise and fall times, a significant technical challenge by itself for high-power operation, produces further distortion of the intended waveform, which translates into lost sensitivity. Thus, it is necessary that all physical components and the intended waveform be made to operate in a coordinated manner to mitigate spectral regrowth while maintaining acceptable radar performance. This coordination will require new transmitter components and architectures and a much greater interaction between transmitter design and waveform design, so as ultimately to enable the requisite precise tailoring of physical radar emissions. Some recent research demonstrated that it is possible to maximize power-added efficiency (PAE) in a gallium-nitride (GaN) PA while simultaneously reducing interference to adjoining channels [8]. Clearly, additional research along these lines is warranted. Other PA research areas include amplitude tracking in PAs to avoid saturation (nonlinearity), developing a means to assess PA spectral characteristics, and developing other PA architectures.

Microwave-tube PAs in high-power radar transmitters will be a reality for at least the next 25 years in military radars, which predominantly employ CFAs and klystrons. The disadvantage of CFAs is that they generate high levels of out-of-band noise induced by AM-to-AM and AM-to-PM effects. Although klystrons are linear devices that generate less noise, they are also bulky and require lead shields to protect people from the high-intensity X-rays that they emit. Currently, two competing technologies (the multibeam klystron (MBK) and GaN-based solid-state PAs (SSPAs)) are attempting to overcome the deficiencies of current CFAs and klystrons.

Research and development on MBKs has reduced the AM-to-AM and AM-to-PM effects, subsequently realizing a 30 dB reduction in spurious emissions compared to the CFAs in the existing transmitters. Also to the positive, an MBK is much smaller than a klystron, does not have a klystron's shielding requirements for protection from X-rays, and fits within the smaller space constraints of a CFA (see the example of an 8-beam MBK in Figure 5.7). Another smaller-length MBK has 18 beams, generates 500 kW of peak power with a 1.0 dB bandwidth of 350 MHz, and enables a 5 percent duty cycle that could double the average power of existing transmitters. Although MBKs may provide part of the solution to mitigating adjacent-band interference, the technology requires better longer-life cathodes to support the larger instantaneous bandwidths needed for many radar applications. Further R&D is necessary if the potential of MBK technology is to be realized.

Figure 5.7 Eight-beam four-cavity 600-kW MBK of the U.S. Naval Research Laboratory [42]

In GaN-based SSPAs, GaN provides a wide-bandgap semiconductor material for the component transistors. An SSPA sums the powers from individual GaN transistors while maintaining isolation between adjacent ones with an isolated in-phase combining architecture. This innovative technology provides high-gain, high-efficiency, and excellent pulse fidelity that produces decreased AM–PM effects, phase-noise, and spectral regrowth while realizing increased power density and efficiency [43]. While microwave tubes are used in most high-power radars, GaN SSPAs (e.g. Model VSS3607 by Communications & Power Industries—see http://www.cpii.com/product.cfm/8/61/278) are now appearing in some military air traffic control (ATC) and search radars. In the ATC systems (2.7–2.9 GHz), GaN transistors are combined into 1.3 kW air-cooled bricks that are power summed with radial and waveguide combiners for high-power transmission [43]. This architecture reduces the second- and third-order harmonics of the transmitted signal by more than 30 dB. From a practical standpoint, this technological approach has some weaknesses that require additional R&D. First, the cost to generate the amount of average power to obtain a peak power of 1 MW or higher is significant and has at least two contributors: (1) the high cost of the hardware associated with the requisite large number of transistors and power converters to produce such levels of power; and (2) the substantial costs of the cooling and corresponding detailed thermal analyses needed to dissipate the enormous heat that the system would generate. Moreover, SSPAs will undoubtedly have EMC issues that tubes do not. Since it is very difficult programmatically and cost-wise to retrofit legacy tube-based high-power radars into solid-state radars, SSPAs are probably only applicable to new high-power systems.

5.4.1.1 Adaptive RF filter technology

At the component level, the use of adaptive RF filters may provide a means to attenuate the spurious emissions generated by the PA [43–45]. With the infusion of advanced active-array radar technology, adaptive filters could prove critical in attenuating out-of-band emissions, particularly in radars that frequency hop and for emerging concepts in cognitive sensing. For radar, a major technological hurdle to be overcome for a post-PA filter approach is minimizing the insertion loss of the filter, as this loss of transmit power translates directly into a sensitivity loss. Chapter 10 discusses this topic in detail.

5.4.1.2 Transmitter topology

Beyond the impact of individual components, the overall transmitter topology can also be exploited to reduce spurious emissions. Such methods to diminish adjacent-band interference have their genesis in the 1930s, when they were directed at contending with the effects from vacuum-tube PAs in high-power AM broadcast transmitters. The Chireix [46] and Doherty [47] PA configurations are classical techniques developed during this time. This general class of transmitter topologies is broadly categorized as linear amplification with nonlinear components (LINC) [48,49]. In addition, research efforts should explore the efficacy of more recent PA topologies to suppress the spurious emissions induced by nonlinearities in high-power PAs.

Recently, the growth of 4G wireless technology that relies on amplitude and phase modulations to encode information (64 QAM is quite common) has generated a resurgence of interest and research into mitigation techniques for spectral regrowth in PAs, for both communications and radar. The primary driver of this research has been the need to contend with the large peak-to-average power ratio (PAPR) that occurs for orthogonal frequency division multiple access (OFDMA) waveforms that serve as the basis for WiMAX and LTE 4G communication systems [50]. While these efforts in the wireless community may be leveraged to some degree, the distinct requirements of radar (pulsed operation, wide instantaneous bandwidth, very high transmit power, constant envelope or nearly so) necessitate investigations into radar-specific PA topologies. Moreover, the impact of the imminent 5G protocol needs to be studied, because it may require different research solutions as well. Also to be explored are methods to parameterize the nonlinearities of transmitter topologies so as to enable optimization, potentially in concert with desired waveforms. Chapters 11 and 12 provide an in-depth discussion of this topic.

5.4.1.3 Linearizing radar transmitters

As noted earlier, exploratory research was conducted on developing spectrally confined transmit waveforms and using the Chireix technique to implement such waveforms for radar [17,18]. Now if the constant-envelope requirement (a staple of radars) is relaxed, thus allowing AM effects, there consequently would be sufficient freedom in waveform design to ensure physical emissions with excellent spectral containment. By converting amplitude excursions into two separate phase excursions, a pair of high-efficiency PAs can be employed in parallel, thus maintaining high-power operation while achieving some cancelation of out-of-band spurious products.

The linearity of a PA is established by supplying just enough power-supply voltage (proportional to the input voltage level) to prevent the PA from saturation [51,52]. To acquire information about the spectral characteristics at the output of the PA to use linearization methods (e.g. envelope tracking) effectively, a recent innovation has been developed using a technique from Wirtinger Calculus [19,53]. This technique assumes an operating point of a transistor's large-signal voltage/current curve and computes the Fourier coefficients for that operating point, which can then provide the magnitude and phase information ascribed to those signal components generated at that operating point. The importance of this formulation is that by appropriately parameterizing PA operation, the performance can be optimized, thus streamlining the PA design process.

In addition, it is possible to maximize the PAE of a PA and minimize the adjacent channel power ratio (ACPR) in a combined manner that determines the best trade-off between PAE and ACPR. Further, while earlier research that attempted to optimize transmitted waveforms and PA circuitry independently yielded reduced out-of-band spectral spreading [17,18], it is believed that joint optimization of the physical transmitter and driving waveform may be realized to achieve even greater performance in terms of spectral containment while minimizing amplitude variation (loss). See Chapter 12 for further details.

5.4.2 Radar waveforms

The final aspect to mitigating the adjacent-band interference induced by radar emissions is the determination of appropriate radar waveforms. It is well known that the nonlinear aspects of a transmitter distort radar waveforms, resulting in sensitivity loss from increased range sidelobes. As different component technologies and transmitter topologies are developed, the resulting impact on radar emissions must also be investigated. While the primary focus may be on reducing out-of-band emissions, great care must be taken not to do so at the cost of sensing performance. Thus, it is imperative that waveform design be performed jointly with transmitter and antenna design. To optimize system performance while reducing various types of interference, both to and from a radar, a holistic system design approach must be employed. Any good approach must keep these objectives in mind when the system is designed, starting from signal generation and radiation to signal reception.

Recently, new forms of waveform implementation have been explored that enable optimization of a waveform while accounting for the impact of the transmitter (via HiLo optimization [20,54]). Preliminary results demonstrate how an increase in range-sidelobe level can be partially mitigated by designing waveforms that are essentially "tuned" to the specific distortion profile of the given transmitter hardware. Such approaches become particularly important if the amplitude envelope of a pulse is modified to minimize the spectral spreading that is otherwise induced by the rapid rise and fall times of a high power pulse. Recent advances in HiLo emission design using a form of the continuous phase modulation (CPM) implementation, otherwise denoted as polyphase-coded frequency modulation, can be leveraged to optimize waveforms specific to various transmitter configurations. In conjunction with research on components and hardware topology, this approach can be used to determine emitted radar signals so that they meet prescribed spectral masks while achieving the required sensitivity. This topic is further discussed in [30] and Chapter 6.

Other related areas that merit further investigation include proactive/dynamic/adaptive waveform design for spatial/spectral interference avoidance on transmit, waveform implementation strategies (CPM, 64 QAM, binary phase shift keying, etc.), error-correction coding geared toward high-power radar interference, the development of spectrally cognizant device technology, means for identifying other spectral users, new sampling hardware and processing modules for analog-to-digital conversion (ADC) at RF, and filters and post-PA filters with minimal insertion loss.

5.4.3 Innovative antenna elements and arrays

For antenna/array design and selection, research is needed that includes remediation of near-field coupling effects from the system's mounting platform and other proximate assets, and that addresses the implications of antenna mutual coupling and element patterns to radar emission design [15,16], including the spectral filtering effect that has been observed as a function of spatial transmit angle [55]. For adaptable antenna technologies of active arrays, it is necessary to integrate EM theory and signal processing. Equally importantly, the antenna should be accurately integrated into the subsystems behind the aperture to increase control of parameters that affect the performance of the radiating system. More specifically, research is needed on the incorporation of antenna characteristics into the transmitter as design parameters such that the combination of PA, antenna feed, and radiating structure yield optimal performance.

5.4.4 Innovative radar receivers

New analog receiver designs and adaptive RF filter technology are required for interference mitigation and receiver optimization, respectively. In particular, receiver and filter designs must support digital signal processing for in-band reception and adjacent-band interference rejection, while also providing analog interference rejection prior to digitization so that the usable dynamic range can be maximized. The DARPA Signal Processing at RF (SPAR) program is presently focused on this challenge.

5.4.4.1 Radar receiver design using digital beamforming

Recent advancements in ADC technology permit the use of direct RF sampling architectures in receiver design. The most flexible and therefore costliest solution is to put an ADC at every element in the array, thus realizing a digital beamforming system. In contrast, the simplest solution is to combine all elements in the analog domain using phase shifters and/or time-delay units and rely on only one ADC, thus realizing an analog beamforming system. The trade-off here lies in the fact that a system that utilizes an ADC at every element will likely be more resistant to jammer interference than the analog counterpart. Analog beamforming systems rely on interferers lying outside the main beam where the phased array pattern intrinsically nulls spatially separated interference. The analog beamforming system is essentially a very high-gain antenna that allows interferers in the main beam to be amplified. Alternatively, a digital beamforming system acts as a large array of low-gain antennas where digital signal processing can be applied prior to signal summation and amplification, which allows frequency filtering, spatial filtering (nulling), and other interference mitigation to be applied while the received radar signal is relatively small. Since each element has only a low-gain antenna in front of it, the input compression point of a digital beamforming system is much higher than that of an equivalent analog beamforming system. Often hybrid solutions are designed to reap some of the benefits of digital beamforming while still using some degree of analog beamforming, because the cost of the digitizing hardware is so high. However, as technology advances and ADC prices fall, digital beamforming becomes a more attractive solution. The DARPA Arrays at Commercial Timescales (ACT) program is presently focused on this endeavor.

Pertinent research should investigate the degree of digital beamforming at the subarray level that is required to mitigate interference issues and to enable more efficient use of the available spectrum within an allocated band. In particular, a model that compares analog and digital beamforming for large arrays (shipboard-sized or larger) needs to be developed to assess the capability for interference rejection.

5.4.4.2 Receiver interference rejection

Establishing interoperability requirements between radars and communication systems in a congested spectrum environment necessitates modeling and characterizing the interference between them. In addition, methods of mitigation and metrics of receiver design need to be studied. Research must include the impact of geographic location, temporal behavior, frequency, polarization, and waveform coding to assess spectrum sharing between radar and communication systems. The topology of radar receivers, both analog and digital, must be explored to determine their limitations in processing signals of interest in a conflicted spectrum environment so that innovative and cost-effective analog receivers and digital signal processing can be developed. Research topics include (1) using adaptive amplifier linearization techniques for low-noise amplifiers (LNAs) in a receiver front end to prevent compression and the subsequent generation of debilitating in-band third-order intermodulation products; (2) using fast-sampling ADCs and adaptive digital filters to enhance performance and mitigate interference; and (3) reduction of high-power interference at the analog front end of the receiver prior to the ADC.

While increasing spectral congestion drives the imperative for better spectral confinement of radar emissions, it likewise necessitates enhancements to radar interference rejection on receive due to the increased likelihood of being the recipient of adjacent-band interference and intentional/unintentional encroachments into radar bands. Though interference rejection approaches may take numerous different forms, they may be coarsely divided into two general categories: sample-data oriented (unstructured) and structure oriented. The former relies heavily on the received data to provide the structure (e.g. via an interference covariance matrix) so that the interference can "cancel itself," while the latter operates via presumed a priori knowledge of the interference structure (e.g. standardized communication waveforms and protocols). Speaking generically, trade-offs exist between requirements on sample data (sufficient sample support, homogeneity, etc.) and limitations on presumed knowledge from calibration errors, such that practical interference cancellation is typically some hybridization of both.

The most common forms of interference cancellation for radar address clutter, jamming, or both. For radar, clutter is pervasive and not necessarily coherent, with the radar emission thus requiring adaptivity in the spatial and Doppler domains (STAP) [56], where the clutter often is assumed to be of sufficiently low rank. Noncoherent jamming can generally be suppressed via spatial nulling, while coherent (deceptive) jamming remains a topic of on-going research.

The potential incursion of interference from numerous spatially dispersed wireless communication users may greatly exacerbate the already difficult task of interference cancellation. The spatial dispersion of cellular telephony greatly limits the effectiveness of spatial nulling. Likewise, since some users will be traveling at low speeds (walking or standing still) while others may be in a moving vehicle, the effectiveness of Doppler nulling is also limited. Knowledge of these interfering emissions that might be leveraged are the published communication standards [57] that dictate the structures these signals may take. Thus, a major component of research should be investigating how to incorporate communication standards into the interference cancellation algorithms of a radar.

Recent work also suggests that the radar range domain may be an untapped resource for interference cancellation by exploiting the structure of the radar emission [58]. Both simulations and experimental results have shown that the echoes from two radars occupying the same spectrum can be separated at a common receiver via this form of adaptive processing. In fact, interference from distributed sources was demonstrated to be reduced by more than 25 dB [59]. This new research domain within which to perform adaptive processing raises the possibility of achieving greater selectivity between interfering radars (e.g. from ionospheric ducting), as well as has the potential for new ways to combat neutral interference from commercial spectrum occupants.

Another possible research area is to exploit multidimensional modeling, like the well-known STAP formulation that relies on the coupling of space and time (Doppler), to increase multiplicatively the adaptive degrees of freedom for interference cancellation. It is this increase in coupling that makes the low-rank clutter assumption feasible (relative to the now high degrees of freedom). By incorporating the new forms of fast-time (range) domain interference cancellation into this multidimensional framework, there is the potential to acquire enhancements necessary to contend with the growing spectrum congestion problem. Of course, doing so also incurs the well-known "curse of dimensionality" for high-dimensional adaptivity that must likewise be addressed.

5.4.5 Propagation

Many researchers and practitioners routinely model the propagation channels of broadband wireless systems as free space and the antennas as isotropic point sources. Consequently, essential fundamental physics is excluded from their analyses. This simple approach does not include the effects of the transmitting and receiving antennas, integral parts of any wireless system, because point sources are not physically realizable. The smallest physically realizable radiating current element is a short dipole, which can only produce an omnidirectional pattern. Further, the platform on which the antenna is mounted and the surrounding environment are also often ignored, which are very bad assumptions as discussed in Section 5.5.2. It is therefore extremely important to revisit propagation research to remediate the near-field coupling effects between proximate assets on the same platform, the implications of antenna mutual coupling and element patterns, and the RF fratricide that can occur from atmospheric ducting. Also, special emphasis should be placed on accurately modeling EMI and the unintended emissions or coupling that can occur. Likewise, more accurate theoretical predictions and modifications to the transmit/receive hardware and signal processing are needed to account for these neglected sources of interference, which would make anticipated results significantly better. Additionally, research should be conducted on improved co-site analysis methods, which permit better prediction and adaptive mitigation of co-site interference, resulting in more robust systems that will resist intra-system interference and external interferers.

5.4.6 Adaptive and cognitive emission control

One may think of adaptive emission control as being reactive to a situation, while cognitive (also known as fully adaptive) emission control is proactive. As an example of the latter, an emission could be modified to suppress a portion of its energy spatially or spectrally. To provide adaptive (reactive) or cognitive (proactive) emission control, a transmitter must have adaptable transmit filters, and the designs of a transmitter and waveforms should be unified and jointly optimized.

While mitigation of adjacent-band interference will ensure radar compliance to standards for spectral masks, and thus substantially reduce the interference that a radar may cause to other spectral occupants, there may still be occurrences of interference if a radar is forced to share spectrum with other users. Because these occurrences are situationally dependent, the impetus is on the radar to be "smart" about how energy is transmitted through adaptive/cognitive emission control. Both types of control are based on learning paradigms and really only differ in terms of the timing (*a posteriori* for adaptive versus *a priori* for cognitive) of transmitter interference suppression.

The details of the many different learning paradigms notwithstanding, the key steps to transmitter-oriented interference suppression are (1) identification of another spectral occupant for which interference is occurring (or will occur); and (2) modification of one's own RF emission to suppress the energy in a given spatial direction and/or frequency band. Of course, it is necessary to accomplish suppression of transmitter-induced interference without degrading the sensing performance of a radar or the received power at an intended communication receiver—or at least by maintaining the degradation at an acceptable level. See Chapters 15 and 16 for further discussion on spectrum sensing.

Recent preliminary work has explored the capability of achieving space–time adaptive nulling (STAN) of radar emissions that allow interference suppression on transmit simultaneously in frequency and spatial angle [59] (see Chapter 18). The benefit of this approach over simply performing spatial nulling or frequency nulling separately is that it minimizes where the radar cannot look in frequency and space jointly, thereby reducing the impact on sensing effectiveness. Open research questions remain, however, including how to determine the proper space–time locations of nulls, how best to predict null drift from relative motion, and determining the impact of array mutual coupling and calibration on the accuracy of spatial nulling.

At the other extreme from STAN are spectrally fragmented emissions wherein a radar transmitter is only allowed to operate in multiple disjoint bands that would not otherwise be sufficient in terms of radar performance if taken individually [60]. It is interesting to note that the 4G wireless standard LTE employs what is referred to as carrier aggregation [61], whereby carriers in disjoint frequency bands can be simultaneously emitted from the same transmitter. The fundamental requirement of this approach is that inaccessible bands, particularly those in between bands being used by the transmitter, do not encounter increased interference. The difficulty of this requirement lies in the fact that nulls are intrinsically narrow (such as employed by STAN), while the need is to suppress potentially broad swaths of spectrum during transmission. Given the resources to be applied by commercial interests to solve this problem, it bears investigation on how radar may be able to leverage the resulting technology as well. In fact, it may be that innovative topology configurations (Section 5.4.1.2) could facilitate this type of emission.

With regard to STAN and spectrally disjoint operation for radar, emission schemes need to be devised that focus on affecting the physical emission that contains nulls (narrow or broad) while maintaining effective sensing performance. In particular, the impact on a saturated PA must be considered and treated, either through an appropriate topology configuration that enables some AM effects or through a constrained form of waveform optimization. A related aspect of this research is conceiving appropriate receive processing to compensate for performance limitations of such an emission.

5.4.7 Radar–communications co-design

In the opinion of the authors, the co-design of a communication and radar system minimally requires innovative refinement of the concept of an antenna, coding the transmitted signals at RF, joint optimization of the spatiotemporal energy distribution (radar) and channel capacity (communications), and definition of RF functionality through software. In addition, accurate measurements and characterizations of radar interference on communications and vice versa are essential. For example, accurate knowledge of intermodulation products caused by LNA compression and limiter-induced noise is required for both functionalities. In addition, the deleterious effects of mutual interference on ADC and subsequent digital signal processing in a receiver need to be quantified. In communications, the intermodulation products and limiter-induced noise are manifested as an inability to discern correct symbols (recall Figure 5.4), whereas the impact on radar performance is the interpretation of wireless interference as jamming.

5.4.7.1 Code-transmitted signal at RF

An alternate approach for operating radar and communication systems over the same band is to code at RF, which was initially envisioned by Shannon [62,63] and subsequently investigated by Viterbi [64]. Coding at RF is similar to the systems used in GPS and satellite communication and can also be carried out for radar.

In the existing communication systems, signal coding is performed at baseband, followed by an up-conversion to RF. However, if a second layer of coding is introduced at RF, as is performed in satellite communications [64], subsequent matching to the transmit coding in the receiver can facilitate a significant relative reduction in RF interference via the associated coding gain. This approach is also used for GPS-type signals to avoid multipath and other interference. In radar, to enhance the signal level, sometimes a Barker code is applied at RF. However, a possible shortcoming of this method for radar is that the processing delay may be too long. Consequently, an open research area is to develop new codes that reduce processing time for real-time radar operation.

5.5 Some essential EM theory for communications and radar

Much research and development on wireless RF systems treat the associated signal processing and EM analyses independently and make unphysical simplifying assumptions to obtain more tractable mathematics, which results in the improper inclusion of the underlying Maxwellian physics—thereby yielding inaccurate performance predictions for these systems. Examples include analyzing and optimizing ideal waveforms without enforcing physically relevant differentiability conditions; using channel capacity without incorporating the Maxwell–Poynting theory; treating antennas in free space instead of including the mounting structure and scattering from proximate objects; and achieving maximum power transfer of an antenna instead of good radiation efficiency. Selected examples are discussed in the next several subsections. The main point of this section is that, to effectuate realistic and accurate characterizations of real-world systems, one needs to use the underlying physics of each problem and to merge EM theory and signal processing intelligently. The following highlights several pertinent concepts and examples that are not always well appreciated but can nonetheless have an important bearing on spectrum sharing research.

5.5.1 Physically realizable waveforms

As just noted, many RF system analyses use idealized analytical waveforms like sinusoids, polyphase codes, etc. that may lead to erroneous conclusions. So one may reasonably ask: what constitutes a physically realizable waveform, and how might using an ideal waveform impact predictions? Since answering the second question requires comparisons between simulations and measurements, which requires a much more extensive discussion, it is not considered here (though the reader is directed to [54] for an example). The first question is inextricably intertwined with the underlying Maxwellian physics. Specifically, a real voltage source $V(t)$ that is driving an antenna must have finite duration and must satisfy Maxwell's equations and/or the associated transmission-line equations. In either case, $V(t)$ must be *at least* twice continuously differentiable and must satisfy the end conditions,

$$V^{(n)}(0) = 0 = V^{(n)}(T),$$

for $n = 0, 1, 2$, and possibly greater, where T is the duration of the waveform that starts at $t = 0$ and $V^{(n)}$ is the nth derivative [65]. Although it complicates the mathematics, any optimization of waveforms subject to some criteria should include these constraints. These waveform conditions seem to match measured signals for real systems [66]. Otherwise, unphysical artifacts are introduced, possibly leading to incorrect performance predictions.

As an example, consider the UWB waveform denoted as the T-pulse that is time-limited (finite duration), has its energy concentrated primarily in a finite band, and its spectral envelope conforms to an arbitrarily shaped spectral mask [67,68]. Such pulses also enjoy the advantage of having a linear phase over the frequency band of interest and are orthogonal to their shifted versions of one or more baud times, where a baud time is the time interval between successive pulses.[1]

The design of this manner of UWB signal is outlined in [69,70] and is performed via steepest descent optimization. Figure 5.8(a) and (b) depict time and frequency domain representations of a T-pulse having a duration of 1.53 ns that conforms to the FCC UWB spectral mask, which is denoted by the straight lines in Figure 5.8(b).

It is also important to note that for a UWB pulse to be transmitted and received without distortion, different antennas must be chosen for transmission and reception, because the impulse response of a transmitting antenna is time differentiated when the same antenna is used on receive [71]. Therefore, it is not possible to use similar antennas for transmit and receive without distorting broadband signals [71]. In metrology experiments, Dr James R. Andrews [72, pp. 102–109] showed that input waveforms to a biconical antenna do not distort the radiated waveforms and that a transverse electromagnetic (TEM) horn does not distort the received incident EM field. Further, his experiments revealed that when a bicone/TEM horn transmit and receive configuration is used for pulse propagation, this system will not distort broadband waveforms.

5.5.2 Notions of far field and antenna pattern for communication cell sizes

To treat BS and mobile antennas as independent objects in free space, with the mobile unit being in the far field of the BS, is physically inaccurate. To illustrate the fallacy of these two assumptions, simply model a BS by a 15-cm half-wave dipole antenna operating at 1 GHz on top of a 20-m high tower, which is typical in mobile communications. For the simple case that the ground is perfectly conducting, EM theory implies that the far field begins at a distance of $2D^2/\lambda$ from the source of the EM fields,

[1]To put the term in historical context, "baud" originates from the French engineer Jean-Maurice-Émile Baudot, who invented the 5-bit teletype code, and baud rate refers to the number of signal or symbol changes that occur per second, where a symbol is one of several voltage, frequency, or phase changes.

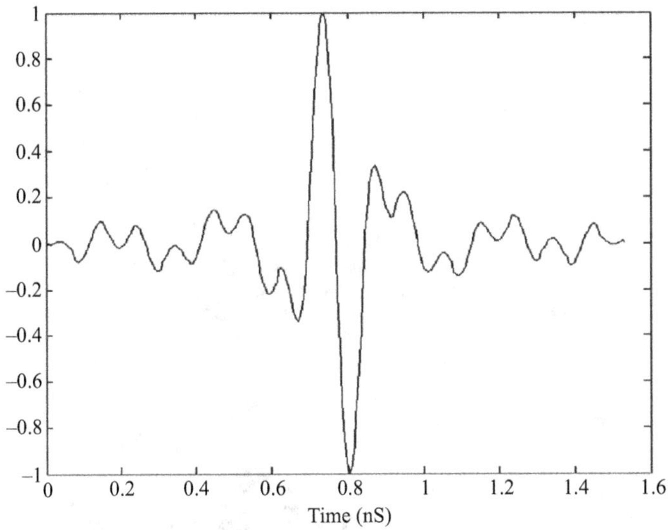

(a) Designed T-pulse in the time domain

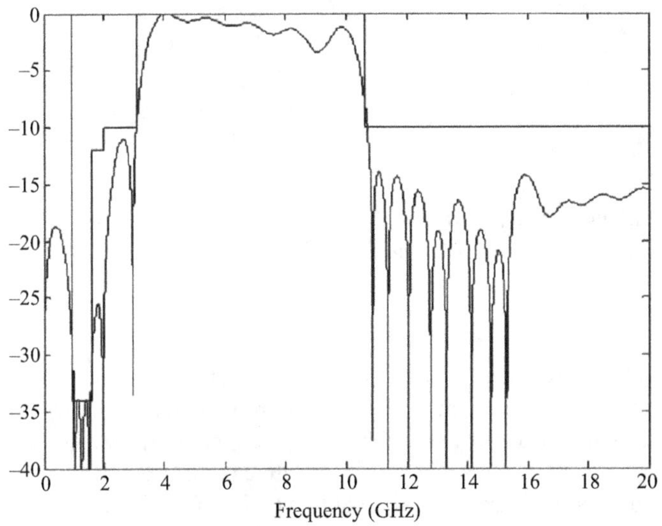

(b) Spectral magnitude of the T-pulse in the frequency domain fitting the FCC mask

Figure 5.8 Designed UWB T-pulse in the time and frequency domains

where λ is the transmission wavelength and D is the size of the radiating structure. Note that D is NOT the maximal dimension of the physical size of the antenna; rather it is a measure of the distance spanned by the antenna and its entire support structure (the total radiating structure), including the passive and the active components of the

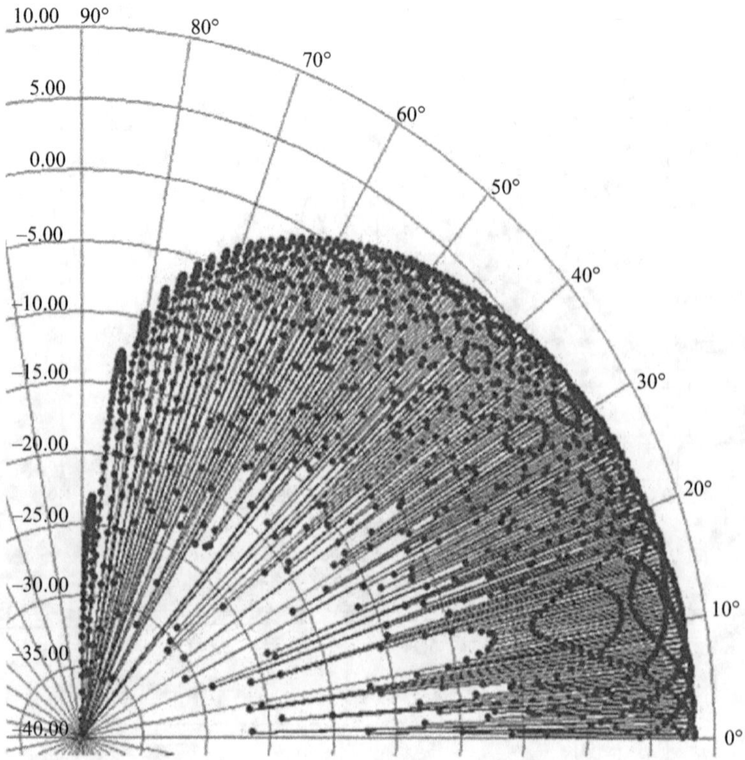

Figure 5.9 Simulated field pattern of a 15-cm half-wave dipole located 20 m above a perfectly conducting ground [74]

emanating field. Consequently, the far field of the radiating structure does NOT start at 0.15 m (=2 × 0.15 × 0.15/0.3 m) as determined from the size of the 15-cm half-wave dipole (this calculation appears far too often in the literature). Rather, the far field is defined by the effective *image* of the dipole relative to the ground, which implies that the far field of this structure starts at 10,666 m (=2 ×(2 × 20)(2 × 20)/0.3 m) from the dipole. Since typical cell sizes are less than 3 km, mobile subscribers are usually in the near field of the BS. Consequently, talking about antenna patterns and beam forming at BSs in typical communication cells makes no physical sense.

Applying an accurate computational EM code [73] yields the simulated far field depicted in Figure 5.9. As the reader can observe, the far field for this BS structure is quite different from the usual *figure-of-eight* pattern of a dipole in free space that is often assumed, as evidenced by the vertical lobing. While the envelope does resemble the dipole pattern, clearly one should expect elevation nulls at various angles.

The relevant quantity for characterizing the radiation in the near field of an antenna is the electric field intensity, which varies significantly as the figure clearly indicates. Furthermore, antenna patterns are not definable in the near field, and channel modeling may be neither relevant nor pertinent for many cell sizes. Therefore, in mobile-communication environments where antennas are routinely mounted on top of towers and buildings, the far field often is several kilometers away—hence may be outside the boundaries of the cell for which the BS antenna was designed to service. Consequently, beamforming in such situations may be meaningless since the mobile receivers are always in the near field. How to apply the rules of basic physics to address these issues is discussed in Chapter 5 of Reference [72].

5.5.3 Maxwellian-based use of capacity

The point of this subsection is to make readers aware of the different models of channel capacity (Shannon, Gabor, Hartley–Nyquist–Tuller) for wireless systems [62,63,75–78], the implications of their use, and the need to incorporate the physical properties of each real system accurately in analyses to ensure that it performs according to design predictions [79,80]. Current research on wireless RF systems often makes unphysical simplifying assumptions and treats the associated signal processing and EM analyses independently, resulting in improper inclusion of the underlying physics (Maxwellian theory) of such systems and consequently leading to inaccurate and erroneous performance predictions. To achieve accurate modeling and predictions, EM theory and signal processing must be intelligently merged, and the relevant physics must be connected to any abstract mathematical system analyses.

The integration of Maxwellian physics in system analyses and designs provides an engineering flavor that the unrealistic "models" commonly used in the signal processing and communication literatures do not. For example, it is common to read that a wireless channel is considered from baseband to baseband, forgetting that within this channel there are processing stages that can be controlled and designed (antenna, front-end, I–Q, analog–digital, etc.) and that the propagation medium necessarily imposes difficulties for proper communication through the principles of physics (noise figure and path loss). Under the baseband-to-baseband premise, the Shannon capacity is taken as the major performance figure when claiming novelty or improvement in proposed processing. Yet amazingly, one cannot find any mention of capacity when buying or designing a front-end or antenna. Equally amazingly, in the signal processing literature, plots of the number of bit errors per unit time (or bit error rate (BER)), and symbol error rate with respect to signal-to-noise ratio (SNR), and the ratio of the energy per bit to the noise spectral density (E_b/N_0), respectively, have been replaced by the Shannon capacity formula. Shannon himself cautioned against inappropriately using his formulation [81]. In fact, the physical layer is often trivialized to the extent that individuals claim experience on physical-layer technologies without knowledge of how an antenna and the front-end impact the performance of the wireless communication link, which is akin to claiming expertise in "discrete" signal processing without knowing ADC technology. For these reasons, among others, it is important to heed the arguments of Ivrlac and Nossek [82,83] who articulate the importance of EM and circuit theory in signal processing and communications.

As previously noted, some research has been conducted on simultaneously optimizing power transfer and the rate of information transfer. To the authors of this chapter, it is unclear how implementing such optimization procedures with a real antenna system will achieve the desired goal, since the response of a system over its bandwidth is not uniform [84,85]. Consequently, it may be necessary to send signals over some portions of the band that are not necessarily contiguous. Thus, in such optimization procedures, it is relevant to use the Tuller channel capacity [76] instead of the Shannon and Gabor capacities. In Shannon's channel capacity theorem, the communication channel is optimized for white background noise, which is not really the case in this situation, and is the very reason Gabor modified the Shannon's formulation to deal with interference. Although the Gabor capacity does treat the interference-limited case, the sensitivity of a system in practice is determined by the sensitivity of the receiver—hence is related to the minimum strength of the electric field that can be handled by that receiver. This practical consideration led Tuller [77] to create his capacity theorem, which subsequently led to his development of the first operating system for ground-to-air communication. More detailed discussions of the differences and relevance of these three differing definitions of channel capacity can be found in [79,80].

Finally, for a wireless system where a transmitting antenna is involved, it is necessary to consider the radiation efficiency of the antenna, in addition to coding, because to compare the channel capacity of the various systems, the input power to the transmitting antenna must be fixed and the received power is induced by the transmitted power density received by a receiving antenna. Recall from basic EM theory that an antenna does NOT radiate power, it radiates a power density through the radiating fields, and when the received power density is integrated by the receiving antenna, it provides the received power. Therefore, without specifying the transmitter and the receiving antennas and without keeping the input power to the transmitting antenna fixed, it is difficult to characterize the received power.

5.5.4 Antenna pattern and placement for communications

By simulating BS antennas for cellular communications at different heights in various environments, Sarkar *et al.* [72] argue that the electric field strength radiated by an antenna closer to the ground may be larger than when it is deployed on top of a high tower (say at heights of 10 m and 20 m). Also, simulations in [74] indicate that tilting an antenna slightly towards the sky minimizes the interference between the direct ray and the reflected ray from the Earth, yielding a further enhancement of the electric field intensity.

Therefore, deploying a simple antenna near the ground without the tower may make the BS significantly less expensive to construct and maintain, provided the conclusions of the simulations are validated. Consequently, how well predictions from typical approximations match observed data from fielded systems needs to be considered, which effectively means one has to consider whether the underlying physics admits such approximations. To that end, Prof. Monai Krairiksh of King Mongkut's Institute of Technology in Ladkrabang, Thailand currently is measuring the radiated

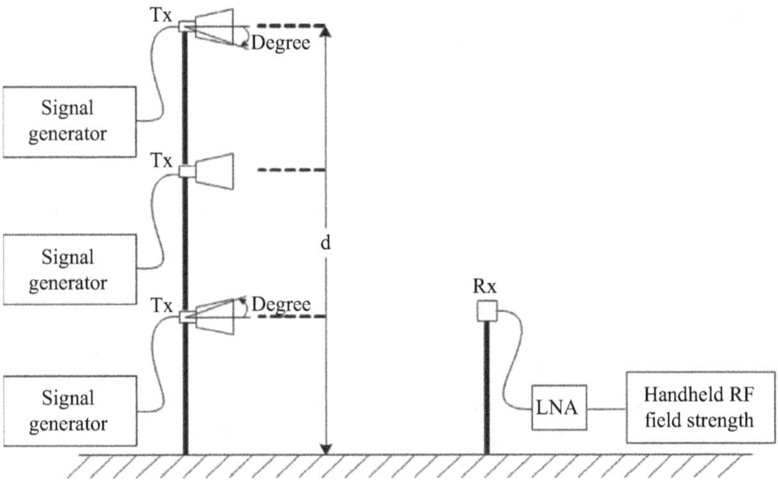

Figure 5.10 Experimental setup for IEEE AP-S SIGHT-funded program to test the efficacy of lowering the height and changing the boresight angle of BS antennas

field for an antenna in various environments (urban, suburban, rural, within a tree canopy) to determine to what extent the conjectures articulated in Reference [72] are valid in practice. This program is sponsored by the IEEE Antennas and Propagation Society (AP-S) under its standing committee on the SIGHT. Although initial data has been taken, the data analyses are not yet complete. Figure 5.10 shows the experimental setup. The end goal is to develop a cheap, low-cost, high-efficiency antenna for BSs in any environment. If viable, the proposed approach is inexpensive because the most expensive component for a BS, the tower, becomes unnecessary and the electronics and antenna can be deployed close to the ground. These results may likewise impact the exploitation of such communication emissions for passive radar (see Chapter 7).

5.5.5 Essential antenna properties for system design

Perhaps the most important component in designing radar and communication systems is the antenna. Essential properties to consider in antenna design are (1) an antenna is a spatial filter and a temporal filter simultaneously; (2) an antenna radiates power density (NOT power); and (3) improved radiation efficiency is the goal instead of maximum power transfer [72,80]. However, current antenna designs are usually based on the power transfer between the transmitter and the antenna and NOT on radiation efficiency. Unfortunately, this approach dissipates half of the generated power. Consequently, to ensure much better performance, antennas need to be designed to achieve maximum radiation efficiency. With this objective, one might consider radiators like the dielectric resonator antenna, which has a radiation efficiency exceeding

95 percent. New system concepts developed according to these properties may go a long way towards addressing the complex problem of spectrum sharing.

5.6 Closing observations

The increased commercialization of spectrum usage, particularly within the RF and microwave bands, is severely inhibiting the use of radar by the military to secure the homeland and operate abroad and by the civil sector to maintain public safety, safe air travel, situational awareness during severe weather, and remote sensing of naturally occurring physical phenomena. The tremendous increase in spectrum utilization by the wireless telecommunication industry is expected to continue as a result of the growing demand for spectrally voracious applications such as wireless video. While the economic benefit cannot be denied, this continued demand for more commercial access to spectrum comes at the detriment of national security and public safety from the loss and degradation of radar resources. It is therefore imperative to determine the degree to which spectrum sharing feasibly can be achieved.

5.6.1 Radar–communications spectral harmony

As was stated earlier, radar and communication systems have grave EMC issues with immense consequences. Basically, mitigating the interference between high-power radars and communication systems has at least two approaches: (1) independently modifying current systems to achieve some semblance of coexistence; and (2) co-designing future communication–radar systems, as was very recently attempted in Phase I of the DARPA SSPARC Program (2013–2015) [86].

Much work is being undertaken on the coexistence approach with some success, but these efforts are hampered by the complexity of existing high-power radars, which are expensive and have been operational for decades. These longstanding legacy systems are based on the technologies at the time of their initial operational capability. Consequently, the hardware of high-power radars built before 1995 (and there are many) limits the number and flexibility of transmitted waveforms. Even the modular structure and flexibility of the hardware in high-power radars built after 1995 may be insufficient to reduce the significant cost associated with adding or modifying waveforms and hardware in those systems to mitigate their impact on other RF users, especially on communication systems.

To date, co-design approaches have been dominated either by radar or by communications and consequently have not been adequately co-designed, good intentions notwithstanding. These activities suffer from entrenched ideologies in both communities. Unfortunately, the communications and radar contingents are not sufficiently knowledgeable of the other community's needs, requirements, and hardware limitations, which can lead to flawed designs. For example, a joint radar–communication system should NOT be limited to a small number of waveforms, because the system must be able to adapt dynamically and to anticipate new technologies and techniques. In the authors' opinion, the program lead for a co-designed system must be an expert in designing and building both modalities. Consequently, technical approaches to solving this problem must have an excellent approach and strong leadership that can treat communications and radar equally and be forward looking and innovative, taking advantage of current and future research. Finally, insufficient joint radar–communication system research has been conducted, and this S&T gap needs to be closed to support radar–communication co-design.

For a co-designed system to perform well, perhaps the following three statements should be guiding principles: (1) the system should be low-power for short-range applications and high-power for long-range applications; (2) the basic principles for radar and communication operations and good performance metrics for jointly optimizing both modalities need to be delineated; and (3) the energy returned from objects (radar) and the channel capacity (communication) should be simultaneously maximized.

For the sake of discussion, consider the interference between communication and radar systems by partitioning the problem into four classes:

1. military radar versus military communication;
2. commercial radar versus commercial communication;
3. military communication versus commercial communication;
4. military radar versus commercial communication.

The co-design of a joint military radar–communication system (type I) is the most amenable to development, because both modalities can be more easily mandated within a military structure to share system details, information, hardware, procedures, and resource allocation. In addition, such developments would probably be subsidized by a government. An early example of R&D on operating simultaneous RF functionalities was the Advanced Multifunction Radio Frequency Concept (AMRFC) Program, funded by the U.S. Office of Naval Research and led by the Radar Division of the U.S. Naval Research Laboratory. The AMRFC Program demonstrated the integration of many shipboard RF functions (radar, communications, electronic warfare, calibration) through a common set of broadband array antennas, signal and data processing, signal generation, and display hardware [87].

A commercial communication–radar system (type II) may be more difficult to design and develop, because it must satisfy more stringent economic and ecologic constraints. Co-designing a military–commercial communication system (type III) may not be as difficult to design, because military communications often use commercial developments with more stringent constraints and may be able to work within the architecture of the commercial communications while maintaining the requisite security. By far, the most difficult, possibly impossible, development is the co-design of a military radar and commercial communication system since a country's military radar is severely constrained by national-security considerations. For example, while a military radar may know a communication system's specifications and infrastructure, security requirements preclude the converse, thus creating a substantial informational asymmetry.

Because high-power radars require enormous transmitted powers and extremely sensitive receivers, their designs involve a complex set of trade-offs. High power is critical for projecting surveillance and other functionalities to great distances, and the extremely sensitive receivers are essential for detecting a transmitted signal that has diminished by as much as 10^{18} at 10 nautical miles along the two-way propagation path of a monostatic radar (transmit and receive antennas are co-located). A monostatic radar typically searches a volume of space by scanning through a set of range–azimuth–elevation cells determined by the range resolution (proportional to the duration τ of one pulse in the absence of waveform modulation), the azimuth beamwidth, and the elevation beamwidth. Within each of the many range cells that comprise a search volume, a radar transmits a series of pulses that are separated by some predetermined inter-pulse time T (here $\tau + T$ is the pulse repetition interval), during which the radar is performing the critical function of receiving (listening) instead of transmitting. The duty cycle D_C $(= \tau/(\tau + T))$ is an important parameter in design trades for a radar. Often D_C is less than 10 percent – hence the radar *appears* to operate only 10 percent of the time along a particular bearing. In addition, when a radar is looking at bearings that are significantly far from a given bearing, the radar appears to be operating significantly less than 10 percent of the time, which has led to two *misconceptions* in the communications community: (1) the radar is using spectrum inefficiently; and (2) the radar frequency is available to other users because it is NOT transmitting. The listening periods are critical to the successful operation of a radar, and communication transmissions will seriously degrade radar performance if the radar cannot keep communication transmissions out of its operating band. Further, while a distributed radar system may significantly improve radar performance and EMC with other users in some eventualities, it is not a panacea for all applications.

Clearly, significantly more R&D over an extended period of time is needed to address RF spectral harmony. Regardless of the situation, any technical efforts must account for the physical EM properties of the RF system (antennas, amplifiers, tuners, filters, transmitters, receivers, etc.), if one wishes to achieve realistic and accurate characterizations of the system and its impact on other RF users.

Equally evident is the need for much better collaboration among the RF user communities and regulators. For example, radar regulators must play a stronger role in helping the wireless communications community develop systems that are more robust to radar emissions. Further, it is essential to have better specifications and conformance to specifications by all users. Since they have the largest market share, it is incumbent upon industry and policy makers in wireless communications to do their homework on attaining a better understanding of the requirements and operation of other users of the RF spectrum and on how their transmissions affect the performance of another system.

It is extremely important to have the requisite resources to obtain experimental assessments of the theory and systems of prospective solutions. Test facilities for experiments and functioning test beds with built-in prototype engineering design models are essential for the various users of the RF spectrum. Essential components of an appropriately flexible test bed are an arbitrary waveform generator, a transmitter, and a software-defined receiver. A test bed should permit independent manipulation of all components and should support a variety of transmitters and PAs in the transmission chain. Moreover, a flexible test bed should accommodate reconfiguration via software to allow different research projects to utilize a common resource without the downtime associated with reconfiguring hardware.

In closing, we believe that a holistic integration of the underlying EM theory, signal processing, and system engineering is needed if a system is to be designed and constructed so that it can perform its functions successfully while being a good spectral neighbor. Minimally, improvements are expected in the following areas: designs of better innovative arrays that jointly optimize spatiotemporal energy distribution (radar) and channel capacity (communications); jointly optimized power-amplifier circuitry and waveform design in the transmitter chain; more realistic EM modeling of propagation, propagation channels, radiation, and scattering; innovative receiver designs supporting digital signal processing for in-band reception and adjacent-band interference rejection; and much better spectral containment.

References

[1] https://www.ieee.org/special_interest_group_on_humanitarian_technology.html. [Accessed 3 April 2017]

[2] http://www.ieee.org/about/technologies/index.html. [Accessed 3 April 2017]

[3] http://iot.ieee.org/. [Accessed 3 April 2017]

[4] *Fifth Interim Progress Report on the Ten-Year Plan and Timetable.* U.S. Department of Commerce: April 2015. Available from https://www.ntia.doc.gov/files/ntia/publications/ntia_5th_interim_progress_report_on_ten-year_timetable_april_2015.pdf. [Accessed 28 March 2017]

[5] *Connecting America: The National Broadband Plan.* U.S. Federal Communications Commission: 17 March 2010. Available from https://transition.fcc.gov/national-broadband-plan/national-broadband-plan.pdf [Accessed 28 March 2017]

[6] *Compatibility between the mobile service in the band 2500–2690 MHz and the radiodetermination service in the band 2700–2900 MHz*, European Conference of Postal and Telecommunications Administrations: Electronic Communications Committee, ECC Report 174, March 2012. Available from http://www.erodocdb.dk/docs/doc98/official/pdf/ECCRep174.pdf [Accessed 2 March 2017]

[7] Baylis, C., Cohen, L., and Jost, R., 'Introduction to radar transmitters'. Presented at Short Course at IEEE Radar Conference; Kansas City, MO, USA, May 2011.

[8] Baylis, C., Fellows, M., Cohen, L., and Marks II, R.J., 'Solving the spectrum crisis'. *IEEE Microwave Magazine*. 2014; July/August, pp. 94–107.

[9] *Realizing the full potential of government-held spectrum to spur economic growth*. United States: President's Council of Advisors on Science and Technology (PCAST); 20 July 2012. Available from https://obamawhitehouse.archives.gov/sites/default/files/microsites/ostp/gorenberg_ppt.pdf [Accessed 28 February 2017]

[10] *FCC puts final rules in place for new citizens broadband radio service*. United States: Federal Communications Commission; FCC News; 28 April 2016. Available from https://apps.fcc.gov/edocs_public/attachmatch/DOC-339104A1.pdf [Accessed 15 March 2017]

[11] *FCC begins rulemaking proposing to allow wireless broadband operations in the 3650–3700 MHz band*. United States: Federal Communications Commission; FCC News; 15 April 2004. Available from https://apps.fcc.gov/edocs_public/attachmatch/DOC-246146A1.pdf [Accessed 15 March 2017]

[12] http://fbodaily.com/archive/2016/10-October/01-Oct-2016/FBO-04293934.htm [Accessed 15 March 2017]

[13] Allmon, L., *Spectrum issues*. 2011. Available from https://www.roc.noaa.gov/WSR88D/PublicDocs/TAC/2011/SpecIssues_TAC_Presentation_March_2011.pdf [Accessed 15 March 2017]

[14] Sanders, F.H., Sole, R.L., Carroll, J.E., Secrest, G.S., and Allmon, T.L., *Analysis and Resolution of RF Interference to Radars Operating in the Band 2700–2900 MHz from Broadband Communication Transmitters*. NTIA Report 13-490, U.S. Department of Commerce, National Telecommunications and Information Administration, October 2012.

[15] Mokole, E.L., Parent, M., Valenzi, J., *et al*. 'Initial bistatic measurements of electromagnetic propagation in an enclosed ship environment'. *Proceedings of the 1999 Antenna Applications Symposium Volume I*. Monticello, IL, USA, 15–17 September 1999. Dayton: Air Force Research Laboratory, Sensors Directorate; 1999. pp. 1–29.

[16] Mokole, E.L., Parent, M., Samaddar, S.N., Valenzi, J., and Tomas, E. *Radiofrequency Propagation Measurements in Confined Ship Spaces aboard the Ex-USS Shadwell*. NRL Formal Report NRL/FR-5340-00-9951, August 18, 2000.

[17] Faust, H.H., Connolly, B., Firestone, T.M., Chen, R.C., Cantrell, B.H., and Mokole, E.L., 'A spectrally clean transmitting system for solid-state phased-array radars'. *Proceedings of the IEEE 2004 Radar Conference*; Philadelphia, USA, 26–29 April 2004. Piscataway, NJ, USA: IEEE; 2004. pp. 140–144.

[18] de Graaf, J., Faust, H., Alatishe, J., and Talapatra, S., 'Generation of spectrally confined transmitted radar waveforms: experimental results'. *Proceedings of the IEEE 2006 Radar Conference*; Verona, NY, USA, 24–27 April 2006. Piscataway, NJ, USA: IEEE; 2006. pp. 1–8.

[19] Baylis, C., Wang, L., Moldovan, M., *et al.*, 'Designing transmitters for spectral conformity: power amplifier design issues and strategies'. *IET Radar, Sonar & Navigation*. 2011; **5**(6): 681–5.

[20] Jakabosky, J., Blunt, S.D., Cook, M.R., Stiles, J., and Seguin, S.A. 'Transmitter-in-the-loop optimization of physical radar emissions'. *Proceedings of the IEEE 2012 Radar Conference*; Atlanta, GA, USA, 7–11 May 2012. Piscataway, NJ, USA: IEEE; 2012. pp. 874–79.

[21] Collaboration Support Office of NATO Science and Technology Organization (France). *Radar Spectrum Engineering and Management*.

Paris: Technical Report STO-TR-SET-182; April 2017. Accessible from https://www.sto.nato.int/ [Accessed 24 December 2017]

[22] Wicks, M., Mokole, E., Blunt, S., Amuso, V., and Schneible, R. (eds.). *Principles of Waveform Diversity & Design*. Raleigh, NC: SciTech Publishing, Inc.; 2010.

[23] Collaboration Support Office of NATO Science and Technology Organization (France). *Dynamic Waveform Diversity and Design*. Paris: Technical Report STO-TR-SET-179; May 2016. Accessible from https://www.sto.nato.int/ [Accessed 24 December 2017]

[24] Collaboration Support Office of NATO Science and Technology Organization (France). *Waveform Diversity*. Paris: Specialists' Meeting STO-MP-SET-204, Berlin, Germany; 29–30 September 2014. Program accessible from https://www.cso.nato.int/Detail.asp?ID=7452 [Accessed 14 April 2017]

[25] Collaboration Support Office of NATO Science and Technology Organization (France). *Reconfigurable and Scalable Multi-function RF Systems in a Congested EM Spectrum*. Paris: Specialists' Meeting STO-MP-SET-SCI-230, The Hague, The Netherlands; 18–19 April 2016. Abstract accessible from https://www.cso.nato.int/activity_meta.asp?act=7501 [Accessed 14 April 2017]

[26] Collaboration Support Office of NATO Science and Technology Organization (France). *Cognition and Radar Sensing*. Paris: Lecture Series SET-216; September–October 2015. Program accessible from https://www.cso.nato.int/Prog/SET-216-7743.pdf [Accessed 14 April 2017]

[27] Home page International Symposium on Advanced Radio Technologies (ISART). Available from https://www.its.bldrdoc.gov/isart/isart-home.aspx.

[28] Mokole, E.L., Adve, R.S., and Goldstein, S. (eds.). *Waveform Diversity: Part I*. Special Issue of *IEEE AESS Magazine*. 2016; **31**(10).

[29] Mokole, E.L., Adve, R.S., and Goldstein, S. (eds.). *Waveform Diversity: Part II*. Special Issue of *IEEE AESS Magazine*. 2016; **31**(12).

[30] Blunt, S.D. and Mokole, E.L. 'Overview of Radar Waveform Diversity'. *IEEE AESS Magazine*, 2016; **31**(11): 2–42.

[31] McCormick, P.M., Blunt, S.D., and Metcalf, J.G. 'Simultaneous radar and communications emissions from a common aperture, part i: theory'. *Proceedings of the IEEE 2017 Radar Conference*; Seattle, WA, USA, 1–6 May 2017. Piscataway, NJ, USA: IEEE; 2017. pp. 1–6.

[32] Sahin, C., Jakabosky, J., McCormick, P.M., Metcalf, J.G., and Blunt, S.D. 'A novel approach for embedding communication symbols into physical radar waveforms'. *Proceedings of the IEEE 2017 Radar Conference*; Seattle, WA, USA, 1–6 May 2017. Piscataway, NJ, USA: IEEE; 2017. pp. 1–6.

[33] Gerlach, K., Shackelford, A.K., and Blunt, S.D. 'Combined multistatic adaptive pulse compression and adaptive beamforming for shared-spectrum radar'. *IEEE Journal of Selected Topics in Signal Processing*. 2007; **1**(1): 137–46.

[34] Higgins, T., Webster, T., and Mokole, E.L. 'Passive multistatic radar experiment using wimax signals of opportunity. part 1: signal processing'. *IET Radar, Sonar & Navigation*. 2016; **10**(2): 238–47.

[35] Webster, T., Higgins, T., and Mokole, E.L. 'Passive multistatic radar experiment using wimax signals of opportunity. part 2: multistatic velocity backprojection'. *IET Radar, Sonar & Navigation*. 2016; **10**(2): 248–55.

[36] Shackelford, A.K., de Graaf, J., Talapatra, S., Gerlach, K., and Blunt, S.D. 'Shared-spectrum multistatic radar: preliminary experimental results'. *Proceedings of 2007 International Waveform Diversity and Design Conference*; Pisa Italy, 4–8 June 2007. Piscataway, NJ, USA: IEEE; 2007. pp. 443–7.

[37] Martone, A.F., Gallagher, K.A., Sherbondy, K.D., *et al.*. 'Adaptable bandwidth for harmonic step-frequency radar'. *International Journal of Antennas and Propagation*. 2015; **2015**:1–15.

[38] Martone, A., Dietlein, C., Govoni, M., Sherbondy, K., and Pulskamp, J. 'Tuning technology for adaptable radar bandwidth'. *Proceedings of 2016 IEEE MTT-S International Microwave Symposium (IMS)*; San Francisco, CA, USA, 22–27 May 2016. Piscataway, NJ, USA: IEEE; 2016. pp. 1–3.

[39] Labib, M., Marojevic, V., Martone, A.F., Reed, J.H., and Zaghloul A.I. 'Coexistence between communication and radar systems—a survey'. *arXiv:1709.08573 [eess.SP]*. 2017; 1–15.

[40] United States: Federal Business Opportunities (FedBizOpps) home page. Available from https://www.fbo.gov/ [Accessed 14 April 2017]

[41] https://www.ntia.doc.gov/files/ntia/publications/redbook/2013/5_13.pdf [Accessed 15 April 2017]

[42] Abe, D.K, Nguyen, K.T., Pershing, D.E., *et al.* 'Demonstration of a 600 kW multiple-beam klystron amplifier: a first-pass design success'. *2006 NRL Review*. Washington, DC: Naval Research Laboratory; 2006. pp. 1–3.

[43] Rauscher, C., Kirchoefer, S.W., Pond, J.M., Guyette, A.C., and Jachowski, D.R. 'A compact ridge-waveguide contiguous-channel frequency multiplexer'. *IEEE Transactions on Microwave Theory & Techniques*. 2009; **57**(3): 647–56.

[44] Jachowski, D.R., Rauscher, C., Kirchoefer, S.W., Kruppa, W., Guyette, A.C., and Pond, J.M. 'Tunable filter and multiplexer for improved transmitter electromagnetic compatibility'. *2009 NRL Review*. Washington, DC: Naval Research Laboratory; 2009. pp. 177–180.

[45] Jachowski, D.R. and Rauscher, C. 'Frequency-agile bandstop filter with tunable attenuation'. *Proceedings of IEEE International Microwave Symposium*; Boston, MA, USA, June 2009. pp. 649–52.

[46] Chireix, H. 'High power outphasing modulation'. *Proceedings of the IRE*. 1935; **23**(11): 1370–92.

[47] Doherty, W.H. 'A new high efficiency power amplifier for modulated waves'. *Proceedings of the IRE*. 1936; **24**(9): 1163–82.

[48] Cox, D. 'Linear amplification with nonlinear components'. *IEEE Transactions on Communications*. 1974; **22**(12): 1942–5.

[49] Birafane, A., El-Asmar, M., Kouki, A.B., Helaoui, M., and Ghannouchi, F.M. 'Analyzing linc systems'. *IEEE Microwave Magazine*. 2010; **11**(5): 59–71.
[50] Han, S.H. and Lee J.H. 'An overview of peak-to-average power ratio reduction techniques for multicarrier transmission'. *IEEE Wireless Communications*. 2005; **12**(2): 56–65.
[51] Hoversten, J., Schafer, S., Roberg, M., Norris, M., Maksimovic, D., and Popovic, Z. 'Codesign of pa, supply, and signal processing for linear supply-modulated rf transmitters'. *IEEE Transactions on Microwave Theory & Techniques*. 2012; **60**(6): 2010–20.
[52] Raab, F.H., Asbeck, P., Cripps, S., *et al.* 'Power amplifiers and transmitters for rf and microwave'. *IEEE Transactions on Microwave Theory & Techniques*. 2002; **50**(3): 814–26.
[53] Baylis, C., Marks, R.J., Martin, J., Miller, H., and Moldovan, M. 'Going nonlinear'. *IEEE Microwave Magazine*. 2011; **12**(2): 55–64.
[54] Blunt, S.D., Cook, M., Jakabosky, J., de Graaf, J., and Perrins, E. 'Polyphase-coded fm (pcfm) radar waveforms, part i: implementation'. *IEEE Transactions on Aerospace & Electronic Systems*. 2014; **50**(3): 2218–29.
[55] Cordill, B.D, Seguin, S.A., Cohen, L.S., de Graaf, J., and Parent M. 'Spectrum attributes of frequency-steerable phased array antennas'. *Proceedings of 2012 IEEE International Symposium on Electromagnetic Compatibility*; Pittsburgh, PA, USA, August 2012. Piscataway, NJ, USA: IEEE; 2012. p. 571.
[56] Ward, J. *Space–time Adaptive Processing for Airborne Radar*. MIT Lincoln Lab Technical Report #1015, Lexington, MA, USA, December 1994.
[57] Zhang, J., Ariyavisitakul, S.L., and Tao, M. (eds.). 'Special section on 'LTE-Advanced and 4G Wireless Communications'. *IEEE Wireless Communications*. 2012; **50**(12): 102–86.
[58] Blunt, S.D. and Gerlach, K. 'Multistatic adaptive pulse compression'. *IEEE Transactions on Aerospace and Electronic Systems*. 2006; **42**(2): 572–84.
[59] Gerlach, K., Frey, M.R., Steiner, M.J., and Shackelford, A. 'Spectral nulling on transmit via nonlinear fm radar waveforms'. *IEEE Transactions on Aerospace & Electronics Systems*. 2011; **47**(2): 1507–15.
[60] Lindenfeld, M.J. 'Sparse frequency transmit-and-receive waveform design'. *IEEE Transactions on Aerospace & Electronics Systems*. 2004; **40**(3): 851–61.
[61] Yuan, G., Zhang, X., Wang, W., and Yang, Y. 'Carrier aggregation for lte-advanced mobile communication systems'. *IEEE Communications Magazine*. 2010; **48**(2): 88–93.
[62] Shannon, C.E. 'A mathematical theory of communication'. *Bell System Technical Journal*. 1948; **27**: 379–423 (July) and 623–56 (October).
[63] Shannon, C.E. 'Communication in the presence of noise'. *Proceedings of the IRE*. 1949; **37**(1): 10–21.
[64] Viterbi, A.J. 'Wireless digital communication: a view based on three lessons learned'. *IEEE Communications Magazine*. 1991; **29**(9): 33–36.
[65] Samaddar, S.N. and Mokole, E.L. 'Some basic properties of antennas associated with ultrawideband radiation' in Baum, C.E., Carin, L., and

Stone, A.P. (eds.) *Ultra-Wideband, Short-Pulse Electromagnetics 3*. New York: Plenum Press; 1997. pp. 147–64.

[66] Ross, G.F. 'A time domain criterion for the design of wideband radiating elements'. *IEEE Transactions on Antennas Propagation*. 1968; **16**(3): 355–6.

[67] Hua, Y. and Sarkar, T.K. 'Design of optimum discrete finite duration orthogonal Nyquist signals'. *IEEE Transactions on Acoustics, Speech & Signal Processing*. 1988; **36**(4): 606–8.

[68] Sarkar, T.K., Salazar-Palma, M., and Wicks, M.C. *Wavelet Applications in Engineering Electromagnetics*. Norwood, MA, USA: Artech House; 2002.

[69] Mei, Z., Sarkar, T.K., and Salazar Palma, M. 'The design of an ultrawideband t-pulse with a linear phase fitting the fcc mask'. *IEEE Transactions on Antennas and Propagation*. 2011; **59**(4): 1432–6.

[70] Yeung, S.H., Mei, Z., Sarkar, T.K., and Salazar Palma, M. 'An ultrawideband t-pulse fitting the fcc mask using a multiobjective genetic algorithm'. *IEEE Microwave and Wireless Components Letters*. 2012; **22**(12): 615–7.

[71] Dyab, W.M.G., Sarkar, T.K., García-Lampérez, A., Salazar-Palma, M., and Lagunas, M.A. 'A critical look at the principles of electromagnetic time reversal and its consequences'. *IEEE Antennas and Propagation Magazine*. 2013; **55**(5): 28–62.

[72] Sarkar, T.K., Salazar-Palma, M., and Mokole, E.L. *Physics of Multiantenna Systems and Broadband Processing*. Hoboken, NJ: John Wiley & Sons, Inc.; 2008.

[73] Zhang, Y., Sarkar, T.K., Zhao, X., *et al. Higher Order Basis Based Integral Equation Solver (HOBBIES)*. New York: John Wiley & Sons, Inc.; 2012.

[74] Hwang, S., Medouri, A., and Sarkar, T.K. 'Signal enhancement in a near-field mimo environment through adaptivity on transmit'. *IEEE Transactions on Antennas Propagation*. 2005; **53**(2): 685–93.

[75] Gabor, D. 'Theory of communication'. *Journal of the IEE*. 1946; **93**(3): 429–57.

[76] Tuller, W.G. 'Theoretical limitations on the rate of transmission of information'. *Proceedings of the IRE*. 1949; **37**(5): 468–78.

[77] Tuller, W.G. 'Information theory applied to system design'. *Transactions AIEE*. 1950; **69**: 1612–14.

[78] Gabor, D. 'Communication theory and physics', *IEEE Transactions on Information Theory*. 1953; **1**(1): 48–59.

[79] Sarkar, T.K., Burintramart, S., Yilmazer, N., *et al.* 'A look at the concept of channel capacity from a Maxwellian viewpoint'. *IEEE Antennas and Propagation Magazine*. 2008; **50**(3): 21–50.

[80] Sarkar, T.K., Mokole, E.L., and Salazar Palma, M. 'Relevance of electromagnetics in wireless systems design'. *IEEE Aerospace and Electronic Systems Magazine*. 2016; **31**(10): 8–19.

[81] Shannon, C.E. 'General treatment of the problem of coding'. *Transactions IRE on Information Theory*. 1953; **1**(1): 102–4.

[82] Ivrlac, M.T. and Nossek, J.A. 'Towards a circuit theory for communication'. *IEEE Transactions on Circuits and Systems I*. 2010; **57**(7): 1663–83.

[83] Ivrlac, M.T. and Nossek, J.A. 'The multiport communication theory'. *IEEE Circuits and Systems Magazine*. 2014; **14**(3): 27–44.

[84] Caspers, E.P., Yeung, S.H., Sarkar, T.K., *et al.* 'Analysis of information and power transfer in wireless communications'. *IEEE Antennas and Propagation Magazine*. 2013; **55**(3): 82–95.

[85] Sarkar, T.K., Yeung, S.-H., Salazar-Palma, M., Lagunas, M.A., and Perez-Neira, A.I. 'The effect of broadband matching in simultaneous information and power transfer'. *IEEE Antennas and Propagation Magazine*. 2015; 57(1): 192–203.

[86] http://www.darpa.mil/program/shared-spectrum-access-for-radar-and-communications [Accessed April 2017]

[87] Tavik, G.C., Hilterbrick, C.L., Evins, J.B., *et al.* 'The advanced multifunction rf concept'. *IEEE Transactions on Microwave Theory & Techniques*. 2005; **53**(3): 1009–20.

Part II

Systems engineering perspectives

Chapter 6
Spectrally efficient communications and radar

Erik S. Perrins[1], Shannon D. Blunt[1], Patrick M. McCormick[1], and Brandon Ravenscroft[1]

6.1 Introduction

The concept of bandwidth is both generally well understood and overly vague. It is well understood to be a measure of the amount of spectral content. However, the precise nature of that measure is all too often poorly defined. The invaluable utility of bandwidth as a single descriptive number that one can use to assess spectral usage, to perform theoretical analysis or simply to provide a rough sense of scale, is rather obvious. However, as we continue to move toward an ever more interference-limited spectrum, particularly in light of this push for cohabitation by completely different modes (e.g. communications, radar, navigation), it is clear that more expansive notions of spectral content and spectral containment become necessary. One can likewise consider the spectral efficiency of some mode in terms of the degree of performance (e.g. data rate, range resolution, location accuracy) that can be achieved relative to the given spectral content/containment.

The traditional approach to spectrum allocation is based on spectral-containment requirements through the imposition of masks that are applied according to the particular mode of operation. While the fact that this rigid approach greatly underutilizes spectrum is driving us toward the far greater flexibility that sharing will entail, it is also important that we continue to explore new ways in which greater spectral efficiency can be achieved as well. Indeed, the various notions of spectrum sharing and the efficient use of spectrum are essentially two sides of the same coin. As such, with an eye toward ultimately enabling new spectrum sharing capabilities, in this chapter we focus on the attributes of spectral efficiency for the communication and radar modes.

6.2 Communication spectral efficiency

We begin by examining the fundamental attributes of a communication system to provide the appropriate context for spectral efficiency. A basic linearly modulated

[1]Electrical Engineering & Computer Science Department, University of Kansas, USA

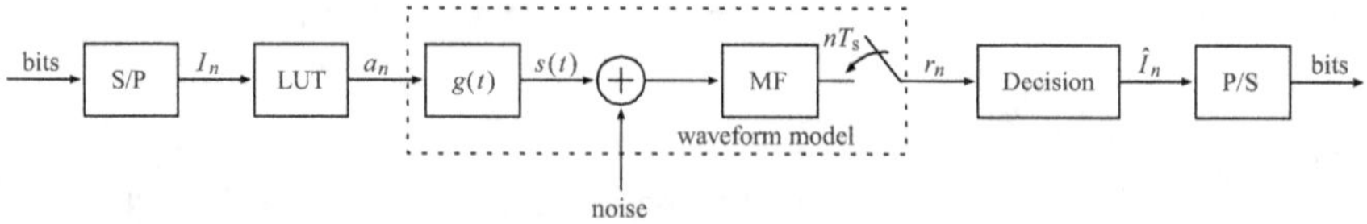

Figure 6.1 Block diagram of a basic linearly modulated communication system

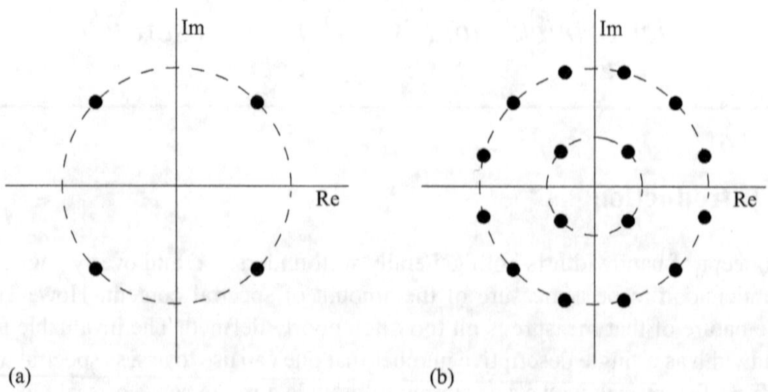

Figure 6.2 Signal constellation for (a) quadrature (4-ary) phase shift keying (QPSK) and (b) 16-ary amplitude phase shift keying (APSK)

communication system is shown in block-diagram form in Figure 6.1 and is described in greater detail below.

6.2.1 Basic linear modulation schemes

The information-bearing bits undergo a serial-to-parallel (S/P) operation where they are grouped into "symbols," each consisting of m bits. The symbol at time index n is represented by a decimal index I_n in the range $0 \leq I_n \leq M - 1$, where $M = 2^m$. The bits arrive at a rate of R_b bits/s, and symbols are produced at a rate of $R_s = R_b/m$ symbols/s. The symbol rate is less than the bit rate when $m > 1$. These relationships also mean that the bits are spaced apart by $T_b = 1/R_b$ seconds, and the symbols are spaced by $T_s = 1/R_s = mT_b$ seconds.

The symbol indexes are fed to a look-up table (LUT) that returns a complex-valued point a_n from the desired *signal constellation*. Figure 6.2 illustrates two example constellations. The first is quadrature (4-ary) phase shift keying (QPSK) and the second is 16-ary amplitude phase shift keying (APSK). Table 6.1 shows the contents of the LUT for the QPSK example. The orthogonality between the real (inphase or I) and the imaginary (quadrature or Q) axes allows for greater information throughput. The energy of a given constellation point is $|a_n|^2$, and the average energy of the signal constellation in the LUT is denoted as E_{avg}. For the QPSK example, each constellation

Table 6.1 Look-up table (LUT) for QPSK example

Bits	I_n	a_n
00	0	$(A/\sqrt{2})(-1-j)$
01	1	$(A/\sqrt{2})(-1+j)$
10	2	$(A/\sqrt{2})(+1-j)$
11	3	$(A/\sqrt{2})(+1+j)$

point has an energy of A^2 and $E_{\text{avg}} = A^2$. Therefore, the peak-to-average power ratio (PAPR) of the QPSK constellation is unity. The average symbol and bit energies are related by the identity $E_{\text{avg}} = mE_{\text{b}}$.

6.2.2 Detection in additive white Gaussian noise

Turning attention to the right-hand side of Figure 6.1, after passing through an additive white Gaussian noise (AWGN) channel, the complex-valued constellation point arrives at the receiver in the form

$$r_n = a_n + w_n, \tag{6.1}$$

where $\{w_n\}$ is a sequence of independent zero-mean Gaussian random variables each with variance $N_0/2$. The maximum-likelihood decision rule is to select the constellation point from the LUT that is closest to r_n in terms of Euclidean distance, i.e.

$$\hat{I}_n = \underset{0 \le I_n \le M-1}{\arg\min} |r_n - \text{LUT}[I_n]|^2. \tag{6.2}$$

The detected symbol index $\hat{I}_n$ is then converted to m bits that subsequently undergo a parallel-to-serial (P/S) operation to form the final output.

Equation (6.1) is known as the *discrete model* of a communication system. It is a valid abstraction for modeling the two ends of the communication system, but all too often it is used in the literature without ever invoking the spectrum-occupying aspects that must be considered in a practical system.

6.2.3 The waveform model for linear modulations

The middle portion of Figure 6.1 contains the items that comprise the *waveform model*. The principal features of this model are the modulation pulse $g(t)$ and its corresponding matched filter (MF), whose impulse response is $g^*(-t)$ where $(\cdot)^*$ denotes the complex conjugate. The cascade of the modulation pulse and the MF response is the autocorrelation function of $g(t)$, which we refer to as $p(t)$.

As Figure 6.1 shows, the constellation points are fed to the modulation pulse in the form of an impulse train spaced apart by T_s seconds, which results in the final signal that is to be transmitted,

$$s(t) = \sum_k a_k g(t - kT_s). \tag{6.3}$$

This signal is said to be *linearly modulated* because the transmitted signal is the superposition of data symbols, each scaling (modulating) the amplitude of a pulse.

Because $s(t)$ is the transmitted signal, its spectral properties are extremely important. Before moving on to this spectral characterization, we point out two practical considerations that must not be overlooked. The first issue is that the transmitted/received signals must be up/down converted between complex-baseband and RF passband. The techniques for doing so are straightforward and are common to both radar and communications. The second issue is that $s(t)$ must pass through a linear power amplifier (PA) at the transmitter; because of this requirement, the PAPR of $s(t)$ is a key concern. As noted earlier, one component of PAPR is the signal constellation itself, i.e. the contents of the LUT. The other component is the modulation pulse $g(t)$ and its interactions with scaled and shifted versions of itself. As such, even a constant-envelope constellation such as QPSK in Figure 6.2(a) tends to have a PAPR greater than unity.

PA nonlinearities result in two kinds of signal distortion: (a) a distorted signal constellation, which translates to a loss in detection efficiency at the receiver, and (b) a distortion commonly referred to as "spectral regrowth," where spurious sidelobes are introduced to the carefully constructed transmitted spectrum (to be discussed shortly). The former is merely a penalty that translates into a small power loss, while the latter can be catastrophic in today's spectrum-constrained environment. To avoid this problem, the PA cannot be operated in its most power-efficient manner, which is in full nonlinear saturation. Instead, the input level must be "backed off" from the saturation point by an amount at least equal to the PAPR of $s(t)$ to allow for enough "headroom" for the peak signal levels to pass through without distortion. The result is a loss in transmitter power efficiency that must be accounted for when calculating end-to-end link budgets.

Figure 6.1 shows that $s(t)$ passes through an AWGN channel and is then fed to the MF, which results in the signal

$$r(t) = \sum_k a_k p(t - kT_s) + w(t) \tag{6.4}$$

for $w(t)$ the AWGN filtered by the MF. The properties of this signal are also extremely important because the information it contains must be detected. The requirements of both (6.3) and (6.4) can be jointly satisfied by careful design of the modulation pulse $g(t)$.

The power spectral density (PSD) of the transmitted signal $s(t)$ in (6.3) is almost always expressed as [1, Chapter 3]

$$P_s(f) = \frac{1}{T_s}|G(f)|^2 P_a(f), \tag{6.5}$$

where $G(f)$ is the Fourier transform of $g(t)$ and $P_a(f)$ is the PSD of the discrete-time random process $\{a_n\}$. If we make the typical assumption of independent and equiprobable data symbols, then $P_a(f) = E_{\text{avg}}$.

However, (6.5) does a poor job of relating the impact symbol rate R_s has on the PSD. To address this limitation, we define a *unit modulation pulse* $g_1(t)$ that has unit energy and a symbol time of $T_s = 1$. From $g_1(t)$ a unit-energy modulation pulse can be obtained for any desired value of T_s by simple use of time scaling (stretching), i.e.

$$g(t) = \frac{1}{\sqrt{T_s}} g_1\left(\frac{t}{T_s}\right) \overset{\mathscr{F}}{\longleftrightarrow} G(f) = \sqrt{T_s}G_1(fT_s) = \sqrt{T_s}G_1\left(\frac{f}{R_s}\right). \tag{6.6}$$

Applying the time scaling property of the Fourier transform from (6.6) to the PSD expression in (6.5) yields

$$P_s(f) = E_{\text{avg}}\left|G_1\left(\frac{f}{R_s}\right)\right|^2. \tag{6.7}$$

As such, it is easy to see that the PSD of the communication signal is dominated by the following two factors:

1. The basic shape of the spectrum is determined by the magnitude-squared of the Fourier transform of the unit modulation pulse $|G_1(f)|^2$. This relationship has important implications because $g_1(t)$ cannot have infinite duration in practice, and therefore, while $|G_1(f)|^2$ may fall off and become very small for some frequencies, it cannot achieve a final value of zero. In other words, in terms of *absolute bandwidth*, all practical communication signals have an infinite bandwidth. To work around this technicality, we must use other bandwidth definitions instead of absolute bandwidth, examples of which are 3-dB bandwidth, 60-dB bandwidth (used in aeronautical telemetry [2]), 99.9% bandwidth, 99.99% bandwidth, etc. As discussed at the beginning of this chapter, because it has no single definition, the notion of "bandwidth" is prone to ambiguity when taken from one context and used in another.
2. The basic shape of the spectrum is scaled (stretched) by the symbol rate, $f \to f/R_s$. Therefore, regardless of the particular definition that is employed, the bandwidth of a communication signal scales linearly with the symbol rate. The bandwidth also scales (stretches) with the *bit* rate, $f \to mf/R_b$, except that with higher order modulation schemes (i.e. larger values of m), the amount of scaling is tempered by the increased value of m. Thus, when the bit rate is held constant, higher order modulation schemes have an inherent spectrum efficiency advantage over lower order modulation schemes.

Having established the PSD of (6.3), we now turn to the design requirements of (6.4). As Figure 6.1 suggests, (6.4) is sampled at the instant $t = nT_s$, thereby yielding

$$r(nT_s) = \sum_k a_k p(nT_s - kT_s) + w(nT_s) = a_n + w_n \tag{6.8}$$

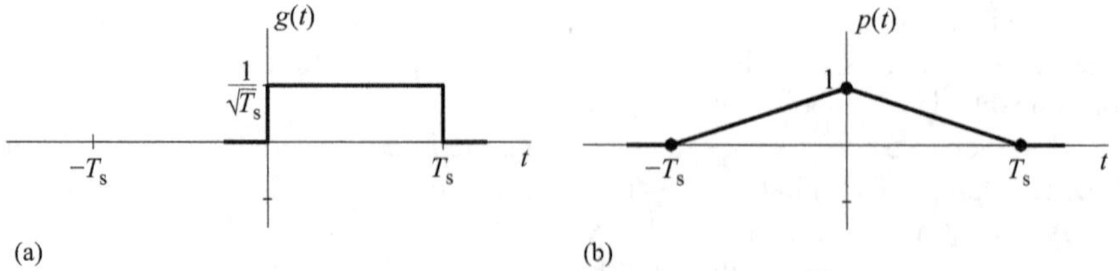

Figure 6.3 (a) Nonreturn-to-zero (NRZ) pulse g(t); (b) corresponding autocorrelation function p(t), including symbol-spaced samples that satisfy the condition in (6.9)

where the right-most expression matches the discrete model in (6.1) when the following condition on $p(t)$ is satisfied:

$$p(kT_s) = \begin{cases} 1 & k = 0 \\ 0 & k \neq 0. \end{cases} \tag{6.9}$$

When this condition is met, there is no intersymbol interference (ISI) and the desired symbol can be detected using (6.2). Another viewpoint is that (6.9) allows for a pulse train that is orthogonal to symbol-spaced time shifts.

Figure 6.3 shows the most basic and widespread example of the modulation pulse, which is a rectangle shape of duration T_s that is called the "nonreturn-to-zero" (NRZ) pulse in communication literature (cf. [1]). Its autocorrelation function $p(t)$, including symbol-spaced samples, is also shown in Figure 6.3. Note that $p(t)$ is nonzero only during the time interval $[-T_s, T_s]$. As such, it is clear that the NRZ pulse (and any other length-T_s pulse for that matter) satisfies the condition in (6.9). The drawback of using the NRZ pulse in practice, however, is that its Fourier transform is the sinc function, which has sidelobes that decay rather slowly, thus yielding poor spectral efficiency for just about any bandwidth definition.

What is needed is a way of satisfying the condition in (6.9) for band-limited pulses, i.e. pulses of infinite duration. The Nyquist no-ISI theorem is the solution to this problem. It states that a pulse $p(t) \overset{\mathcal{F}}{\longleftrightarrow} P(f)$ that satisfies the frequency-domain condition [1, Chapter 9]

$$\sum_k P(f - kR_s) = \text{a constant} \quad \forall f \tag{6.10}$$

likewise satisfies the time-domain no-ISI condition in (6.9). Figure 6.4 shows the Fourier transform of the spectral raised-cosine (SRC) pulse, which has the symmetry needed to satisfy (6.10). The SRC has a parameter called the *roll-off factor*, $0 \leq \alpha \leq 1$ and has an *absolute bandwidth* of $R_s(1 + \alpha)$. The roll-off factor is a measure of the *excess bandwidth* of the pulse. Figure 6.4 shows the SRC pulse for three values of the roll-off factor:

1. $\alpha = 0$. In this case, the absolute bandwidth of the communication signal is at its minimum and is equal to the symbol rate R_s. There is no excess bandwidth. The

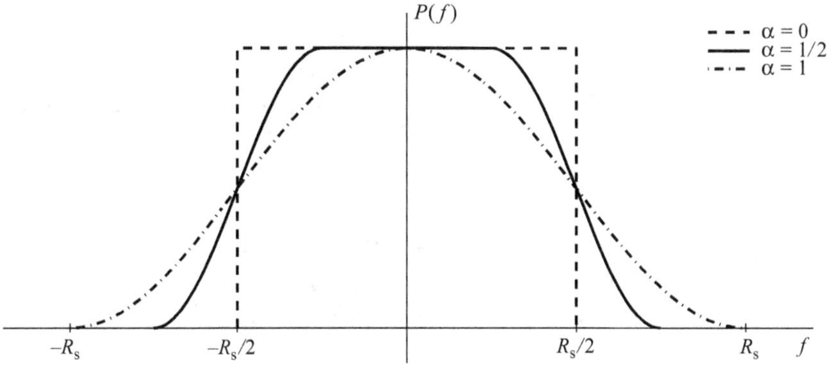

Figure 6.4 Fourier transform of the spectral raised-cosine (SRC) pulse with various values of the roll-off factor α

autocorrelation function $p(t)$ is the sinc function. There are numerous roadblocks that prohibit this case from being used in practice, including (a) the sinc function decays slowly at the rate of $1/t$ and therefore requires a very long pulse length before being windowed (truncated), and (b) a minuscule sampling offset (timing error) in (6.8) results in an uncontrolled amount of ISI.

2. $\alpha = 1/2$. This case is representative of those "in the middle." The absolute bandwidth of the communication signal is $1.5R_s$ (50% excess bandwidth). The pulse decays rapidly in the time domain, and when windowing the pulse a reasonable balance can be struck to mitigate spectral distortion and ISI, as discussed in [3] for example. When α goes closer to zero, the windowed duration of the pulse must increase, and the system becomes more sensitive to synchronization (timing) errors.
3. $\alpha = 1$. This case is the other extreme of satisfying the frequency-domain condition (6.10). The absolute bandwidth of the communication signal is $2R_s$ (100% excess bandwidth).

One final detail must be mentioned. The Nyquist no-ISI theorem results in the design of the autocorrelation function $p(t)$, but it is the *modulation pulse* $g(t)$ that is needed. The Fourier transform relationship between these two is $P(f) = |G(f)|^2$. Therefore, the square-root of the SRC pulse is taken in the frequency domain, i.e. $G(f) = \sqrt{P(f)}$ (conveniently, $P(f)$ is real and positive), before transforming back to the time domain. The result of these steps is the square-root raised-cosine (SRRC) modulation pulse $g(t)$ that is shown in Figure 6.5 for $\alpha = 1/2$, along with its autocorrelation function $p(t)$ and symbol-spaced samples. Note that $g(t)$ does not have zero-crossings that satisfy (6.9), nor does it need them, as this condition is a constraint that is placed on (and met by) its autocorrelation function $p(t)$.

We have summarized the waveform model that is at the heart of the basic linearly modulated communication system in Figure 6.1. The central feature of the waveform model is the modulation pulse $g(t)$. It is this pulse along with the symbol rate R_s that

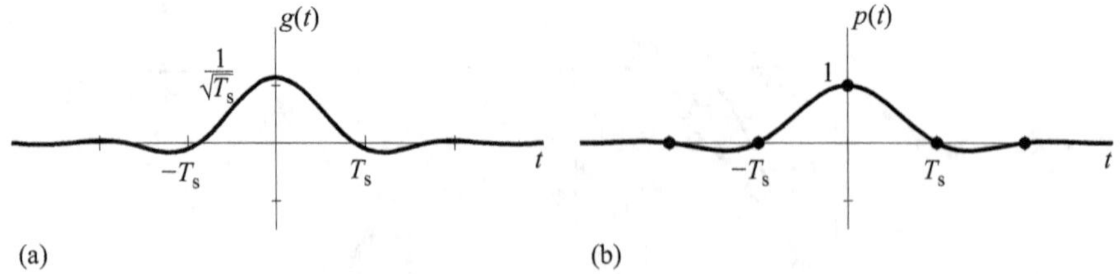

Figure 6.5 (a) Square-root raised-cosine (SRRC) pulse g(t) with roll-off factor $\alpha = 1/2$, (b) corresponding autocorrelation function p(t) including symbol-spaced samples that satisfy the condition in (6.9)

determine the spectral characteristics of the communication signal. This pulse must also be designed with ISI and synchronization requirements in mind. The result of this careful design effort is a bandlimited waveform model that exploits the orthogonality of $p(t)$ to time shifts of T_s; as such, when the MF output is properly sampled, the waveform model disappears and the remainder of the receiver processing follows the discrete detection model.

6.2.4 Orthogonal frequency division multiplexing

A signaling format that is very widespread in wireless communications is orthogonal frequency division multiplexing (OFDM). Consider a (potentially large) block of bits. In Figure 6.1, the bits were formatted into a single symbol stream $\{a_n\}$, where n is a *time* index. With OFDM, the bits are also formatted into complex-valued symbols $\{B_k\}$, but the viewpoint is slightly different because k is viewed as an index for harmonically related subcarriers (sinusoids). In other words, k is an index in the *frequency* domain. No transform was taken to arrive in the frequency domain. We simply assign each data symbol to its own subcarrier, rather than assigning them sequentially in time. When all N subcarriers have a data symbol B_k to transmit, then they are all transmitted together in parallel. The description provided here is for a single such transmission, with the understanding that a system in practice would have one parallel transmission followed by another, and so on.

The modulation pulse for the kth subcarrier is

$$p_k(t) = e^{j2\pi kt/T}\text{rect}_{[0,T]}(t) \qquad 0 \le k \le N-1, \tag{6.11}$$

where T is the duration of the (relatively long) parallel transmission, and $\text{rect}_{[0,T]}(t)$ is defined as unity over the interval $[0, T]$ and is zero otherwise. It is straightforward to verify that these pulses form an orthonormal set of functions. One parallel transmission of OFDM is then given by

$$s(t) = \sum_{k=0}^{N-1} B_k p_k(t) = \sum_{k=0}^{N-1} B_k e^{j2\pi kt/T}\text{rect}_{[0,T]}(t). \tag{6.12}$$

Therefore, instead of transmitting the single symbol stream $\{a_n\}$ in rapid sequential fashion using a relatively brief T_s, one large block of symbols $\{B_k\}$ is transmitted in parallel using a relatively long T.

What is gained by this arrangement? Figure 6.1 considered the AWGN-only channel, where the goal was to design a waveform that was spectrally efficient and ISI-free in such an environment. However, if we now consider that the signal passes through a linear time invariant channel over which we have no control, plus additive noise, then the sampled MF outputs are no longer ISI free. This unfortunate situation motivates the use of techniques such as channel equalizers.

But now a different question arises: can we design modulation pulses that remain orthogonal when they pass through such a channel? The pulses in (6.11) and the resulting OFDM signal in (6.12) are the answer to this question. Also, if T is sufficiently long, then the subcarriers are spaced closely together. When such a signal passes through a frequency-selective channel, the attenuation experienced by any one subcarrier can be approximated as being flat, which requires only trivial equalization.

Furthermore, the form of (6.12) bears a striking resemblance to the inverse discrete Fourier transform. As such, with a few minor tweaks (such as adding the so-called cyclic prefix), the transmitter and receiver processing can be implemented efficiently using the fast Fourier transform (FFT), cf. e.g. [1, Chapter 11].

These attributes are some of the motivations for the widespread use of OFDM. However, OFDM does have one important drawback: the PAPR of $s(t)$ in (6.12) can be quite large, often in excess of 10 dB. A simple numerical example illustrates this point. In an application such as aeronautical telemetry, a transmitter output of 10 W is typically needed to close the link over a large test range. Another aspect of aeronautical telemetry is the severe size, weight, and power (SWaP) constraints that are placed on the telemetry devices installed on test articles. A nonlinear PA operating in full saturation may have an efficiency that approaches 70%. A constant-envelope signal such as continuous phase modulation (CPM), which is discussed below, can be used with such a PA. Even so, the miniaturized, high-power transmitter has the difficult task of dissipating around 4 W in the form of waste heat (10 W/70% = 14.3 W). In contrast, to generate an OFDM signal for this arrangement, the PA input must be backed off considerably so that an efficiency of around 10% is all that can be expected (based on a PAPR of 10 dB). Thus, a transmitter output of 10 W would result in 90 W of waste heat, which cannot be accomplished in the face of SWaP constraints.

6.2.5 Continuous phase modulation

As discussed above, CPM is a type of constant-envelope modulation, and this attribute has motivated its use wherever miniaturized, high-power, battery-operated transmitters are required. Examples of CPM's use include the Global System for Mobile Communications, Bluetooth, Digital Enhanced Cordless Telecommunications (cordless home phones), sensor networks, and aeronautical telemetry as previously mentioned.

The aeronautical telemetry community has a long history with CPM going back to the 1970s. Their original CPM waveform is known as PCM/FM (pulse code

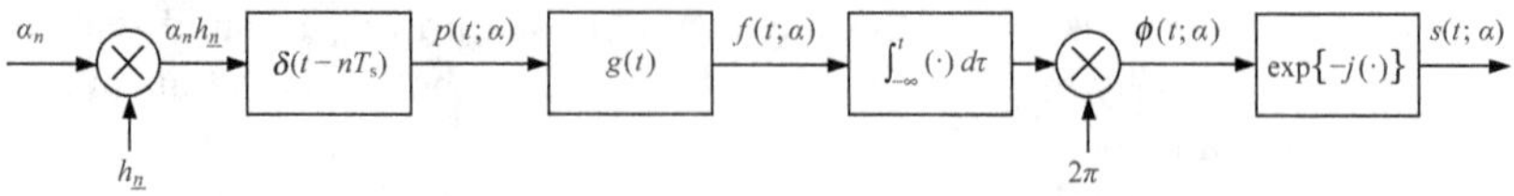

Figure 6.6 Block diagram of a generic CPM modulator

modulation/frequency modulation) and has poor spectral efficiency. Spectrum reallocations of frequency bands in 1997 exposed the need for more bandwidth efficient alternatives to PCM/FM. This need gave rise to the Advanced Range Telemetry (ARTM) program [4] that resulted in two "tiers" of bandwidth efficient waveforms. The waveform search was limited to CPMs. The results of the ARTM program are found in the IRIG 106 standard [2].

The main ARTM Tier I waveform is a type of shaped offset QPSK designed for the telemetry group (SOQPSK-TG). The Tier I waveforms were chosen to have twice the spectral efficiency of PCM/FM while maintaining the same detection efficiency of PCM/FM with limiter-discriminator detection (due to its chronological position as the "status quo," PCM/FM is sometimes referred to as "ARTM Tier 0").

For ARTM Tier II, a partial response, multi-h CPM, denoted ARTM CPM, was adopted as part of the IRIG-106 standard. This waveform was intended to have three times the spectral efficiency of PCM/FM while promising the same detection efficiency as PCM/FM with limiter-discriminator detection. Due to its sensitivity to adjacent channel interference (ACI), ARTM CPM transmissions cannot be spaced as closely together as originally hoped, and so its spectral efficiency relative to PCM/FM is on the order of 2.5 [2].

A classic reference on the subject of CPM is [5], and the interested reader is referred there for additional background information on the topics covered in this section. A block diagram of a generic CPM modulator is shown in Figure 6.6. The modulation process begins at the far left of the figure with the *current symbol* α_n. Each symbol has a duration of T_s seconds, is drawn from an M-ary pulse amplitude modulation (AM) alphabet, i.e. $\alpha_n \in \{\pm 1, \pm 3, \ldots, \pm(M-1)\}$, and carries $m = \log_2(M)$ bits of information.

The next step in Figure 6.6 is to scale the current symbol by the (current) modulation index $h_{\underline{n}}$. In general, there may be multiple, say N_h, modulation indexes. The most common case is when $N_h = 1$, otherwise known as *single-h* CPM. The more exotic case is when $N_h > 1$, which is known as *multi-h* CPM. The notation in Figure 6.6 is general and allows for multi-h CPM. Here the underlined subscript notation is defined as "modulo-N_h," such that $\underline{n} \triangleq n \bmod N_h$. A typical assumption is that the modulation indexes can be represented as rational numbers, with a common denominator across the modulation indexes of the form

$$h_{\underline{n}} = \frac{2k_{\underline{n}}}{p}. \tag{6.13}$$

We determine p by expressing each modulation index as a fraction and taking p as the least common denominator.

Next in Figure 6.6, the discrete-time symbol sequence $\alpha \triangleq \{\alpha_i\}$, with the accompanying modulation indexes, is converted to a continuous-time impulse train

$$p(t;\alpha) \triangleq \sum_i \alpha_i h_{\underline{i}} \delta(t - iT_s), \tag{6.14}$$

where $\delta(t)$ is the continuous-time *delta function*. The impulse train in (6.14) is used as the input to a filter with impulse response $g(t)$, where in this context $g(t)$ is commonly referred to as the *frequency pulse*. The frequency pulse is subject to the following constraints:

- *Finite duration*. The frequency pulse is zero outside the time interval $(0, LT_s)$, where L is a positive integer. When $L = 1$, the pulse is said to be *full-response*, and when $L > 1$ the pulse is said to be *partial-response*.
- *Finite area*. The frequency pulse is normalized to have area 1/2.

The signal at the output of the frequency pulse filter is the *frequency signal*

$$f(t;\alpha) = \sum_i \alpha_i h_{\underline{i}} g(t - iT_s). \tag{6.15}$$

The next operation in Figure 6.6 is an integrator, whose input is the frequency signal $f(t;\alpha)$ and whose output, after being scaled by 2π, is the *phase signal*

$$\phi(t;\alpha) \triangleq 2\pi \int_{-\infty}^{t} f(\tau;\alpha)\, d\tau = 2\pi \sum_i \alpha_i h_{\underline{i}} q(t - iT_s). \tag{6.16}$$

In the case of single-h CPM, the modulation index h is no longer time-varying and can be pulled outside the summation in the right-most expression in (6.16) to join the factor 2π (this simplification can also be performed in Figure 6.6). The *phase response* $q(t)$ in (6.16) is defined as

$$q(t) \triangleq \begin{cases} 0 & t < 0 \\ \int_0^t g(\tau)\, d\tau & 0 \le t < LT_s \\ 1/2 & t \ge LT_s. \end{cases} \tag{6.17}$$

The final operation in Figure 6.6 is to form the complex baseband CPM waveform, which is given by

$$s(t;\alpha) \triangleq \exp\{j\phi(t;\alpha)\}. \tag{6.18}$$

In summary, a particular CPM waveform is parameterized by three quantities:

1. the symbol alphabet size $M = 2^m$;
2. the frequency pulse, which includes the pulse shape $g(t)$ and its duration L; and
3. the set of modulation indexes $\{h_{\underline{i}}\}_{i=0}^{N_h-1}$.

Broadly speaking, CPM schemes are made more bandwidth efficient by adjusting these parameters in certain ways. Because CPM is a nonlinear modulation, the exact impact of these adjustments is difficult to control in a direct manner and is more of

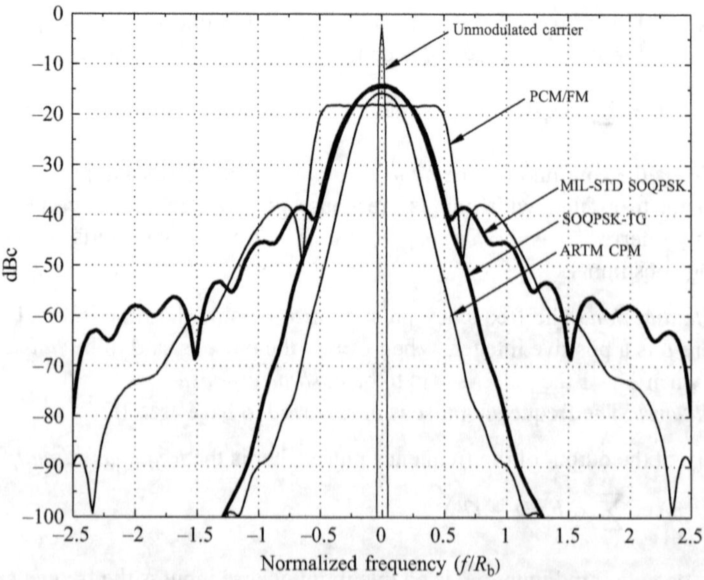

Figure 6.7 The power spectral densities (PSDs) of PCM/FM, MIL-STD SOQPSK, SOQPSK-TG, and ARTM CPM. The frequency scale is normalized to the bit rate, R_b. An unmodulated carrier is shown for reference

an *ad hoc* exercise. In general, a longer, smoother frequency pulse $g(t)$ results in a PSD whose sidelobes fall off more smoothly; a smaller modulation index h generally results in a PSD with a narrower main lobe. A larger symbol alphabet $M = 2^m$ might increase the bandwidth when it is normalized by the *symbol rate*, $f \to f/R_s$, but typically results in greater spectral efficiency when viewed in terms of the *bit rate*, $f \to mf/R_b$. Multi-h CPMs generally allow for a "coding gain" at the expense of greater receiver complexity with minimal bandwidth expansion.

The PSDs of four CPM variants are shown in Figure 6.7. They are

- PCM/FM, which has parameters $M = 2$, $h = 7/10$, a frequency pulse with $L = 2$, and the shape of a time-domain raised-cosine (2RC) [2];
- Military standard SOQPSK (MIL-STD SOQPSK), which has parameters $M = 2$, $h = 1/2$, a frequency pulse with $L = 1$, and the shape of a time-domain rectangle (1REC) [6];
- Telemetry-group SOQPSK (SOQPSK-TG), which has parameters $M = 2$, $h = 1/2$, a frequency pulse with $L = 8$, and a custom shape [2];
- ARTM CPM, which has parameters $M = 4$, $h = \{4/16, 5/16\}$, and a 3RC frequency pulse [2].

The aeronautical telemetry community typically defines bandwidth as the point where the PSD crosses −60 dB relative to an unmodulated carrier (also shown in Figure 6.7). Using that definition, SOQPSK-TG has twice the spectral efficiency of PCM/FM, and ARTM CPM has three times the spectral efficiency of PCM/FM. The differences between the PSDs for MIL-STD SOQPSK and SOQPSK-TG illustrate the effect of using longer, smoother frequency pulses, which come at the expense of larger signal memory and therefore greater receiver complexity.

6.2.6 *Channel capacity and the fundamental limits on spectrum efficiency*

It is useful to consider this discussion of spectral efficiency in the context of the amount of data that can be transmitted. The capacity of a bandlimited AWGN channel is given by the famous equation [1, Chapter 6]

$$\frac{C}{W} = \log_2\left(1 + \frac{P}{N}\right) = \log_2\left(1 + \frac{C}{W}\frac{E_\mathrm{b}}{N_0}\right) \quad \text{bits/s/Hz,} \tag{6.19}$$

where C is measured in bits/second (the same as R_b), W is the bandwidth of the channel in Hz, and P and N are the signal and noise power, respectively. These last two quantities can be subdivided as $P = E_\mathrm{b}R_\mathrm{b}$ and $N = WN_0$. Then, by setting the bit rate equal to the ultimate limit, i.e. $R_\mathrm{b} = C$, we have the right-most expression in (6.19). We treat the term C/W as the limit on *bandwidth efficiency*, and its units are bits/s/Hz.[1] The region below the capacity limit is where error free transmission can occur.

Figure 6.8 plots C/W vs. E_b/N_0 and labels this curve as the "unconstrained AWGN capacity." In order to achieve this limit, the distribution of the data stream must be continuous and Gaussian, which is nonsensical in a digital modulation context. Therefore, it is more instructive to use the capacity constrained by a particular modulation scheme. Figure 6.8 shows two such constrained capacity curves:

- 16-ary APSK from Figure 6.2. This curve reaches a maximum value of 4 bits/s/Hz for large values of E_b/N_0 and is based on the assumption of a modulation pulse with zero excess bandwidth (i.e. a sinc function). Figure 6.8 also shows the performance of two error control coding schemes for such a system. The first uses a rate-3/4 low-density parity check (LDPC) code (2 bits/s/Hz), and the second uses a rate-1/2 LDPC code (4/3 bits/s/Hz) [7]. Both example schemes use an SRRC modulation pulse with roll-off factor $\alpha = 1/2$. For these systems, "error free" means "E_b/N_0 needed to achieve a bit error rate (BER) of 10^{-10}." These LDPC schemes have the same performance when paired with 16-QAM; however, 16-APSK has a more attractive PAPR of 1.06 dB compared with the 2.55 dB for 16-QAM.
- SOQPSK-TG from Figure 6.7. This curve reaches a maximum value of 2/3 bits/s/Hz based on a 60 dB bandwidth of approximately $1.5R_\mathrm{b}$ from Figure 6.7. Figure 6.8 also shows the performance of two SOQPSK-TG schemes. The first is for uncoded SOQPSK-TG (2/3 bits/s/Hz), and the second is for the same rate-1/2 LDPC code as used with APSK (1/3 bits/s/Hz). We also use the same BER threshold for "error free."

[1] One of the most significant means of achieving bandwidth efficiency in modern wireless communications is geographical reuse of spectrum. A smaller footprint for each cell translates directly into greater reuse of the available spectrum. This is a major topic in its own right and is outside the scope of this discussion.

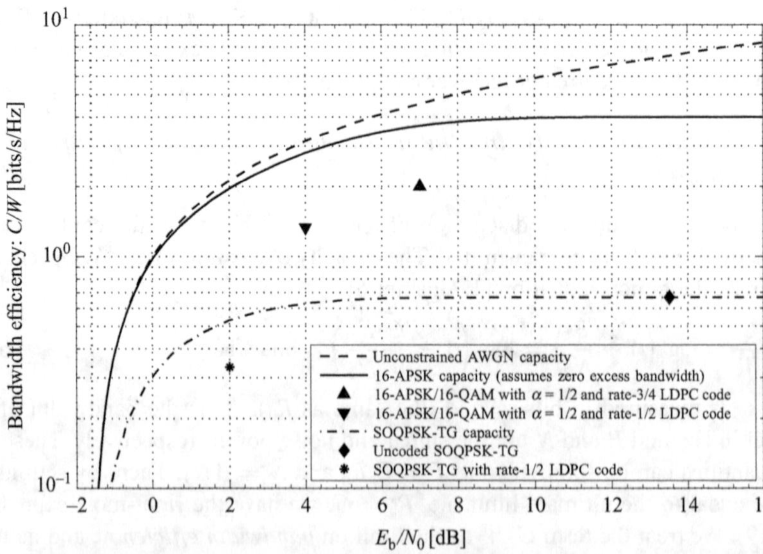

Figure 6.8 Information theoretic limits on bandwidth efficiency and the bandwidth efficiency of certain modulation and coding schemes

These example schemes show that coded APSK can achieve double or triple the bandwidth efficiency of uncoded SOQPSK-TG at much lower values of E_b/N_0. In other words, the rate-3/4 APSK system has a 6.5 dB power margin over SOQPSK-TG, while the rate-1/2 APSK system has a 9.5 dB power margin over SOQPSK-TG. It is possible that the bandwidth efficiency of APSK may be greater still if it is more tolerant of ACI than is SOQPSK-TG. Motivated by this result, the aeronautical telemetry community is currently investigating the feasibility of coded APSK in future telemetry systems. A key unknown in this research is the amount of PA “back-off” needed in order to generate a spectrally clean APSK signal at high power. Thus, the power margin of APSK over SOQPSK-TG in Figure 6.8 will decrease significantly when this important system-design aspect is included in the analysis.

6.3 Radar spectral efficiency

Although the amount of spectrum (or loosely speaking, bandwidth) employed by communications translates into notions of data rate and capacity, for radar the occupied bandwidth corresponds to the achievable range resolution. Further, while communication spectral demands tend to be driven by the need for ever higher data rates, radar requirements tend to focus more on improving sensitivity and robustness to interference, both external and self-induced (i.e. clutter), though enhanced resolution is also often desired. Of course, both sets of demands are largely just the application-specific attributes of the necessity for more efficient use of spectrum. Here, we consider this notion of spectral efficiency from the radar perspective.

6.3.1 Radar spectral content

It is well known that radar bandwidth B is inversely proportional to range resolution δR according to the simple relationship

$$\delta R = \frac{c}{2B}, \tag{6.20}$$

for c the speed of light. For example, a bandwidth of 150 MHz corresponds to a 1 m range resolution. Of course, as mentioned above, there are many ways in which to define bandwidth, with typical measures for radar including root-mean-square (RMS) [8,9], an X dB power level relative to the peak (−3 dB, or half-power, is quite common) [10], and percent energy (e.g. 99%) [11]. Likewise, radar range resolution is often specified according to the mainlobe of the receiver MF response, typically either −3 dB down from the mainlobe peak or the separation from the peak to the first null (also referred to as the Rayleigh resolution). See Sections 20.3 and 20.7 of [12] for an in-depth discussion on these relationships.

Within the context of spectrum sharing, it is more useful to talk in broader terms by addressing the overall power spectrum of the radar emission and the corresponding radar waveform autocorrelation, which is also the MF response when Doppler is negligible. Denote $s(t)$ as the complex baseband representation of the waveform that is modulated onto a pulse of pulsewidth T [easily extensible to continuous wave (CW) by allowing the pulse duty cycle to become 100%]. The power spectrum is thus $|S(f)|^2$, for $S(f)$ the Fourier transform of $s(t)$, and the autocorrelation is

$$r(\tau) = \int_{-T}^{T} s(t)s^*(t+\tau)dt \tag{6.21}$$

in terms of delay τ. The autocorrelation is also the zero Doppler component of the more general delay–Doppler ambiguity function

$$\chi(\tau, f_D) = \int_{-T}^{T} e^{j2\pi f_D t} s(t)s^*(t+\tau)dt \tag{6.22}$$

for f_D the Doppler frequency. Likewise, the power spectrum $|S(f)|^2$ is the Fourier transform of $r(\tau)$.

As exemplified in Figure 6.9 using a standard linear FM (LFM) waveform, the goal of radar waveform design essentially involves the realization of an MF response (autocorrelation) that achieves specified requirements on mainlobe width (i.e. range resolution) and sufficiently low sidelobe levels. The most common metrics for evaluation, and by extension optimization, of sidelobes are the peak sidelobe level (PSL) and the integrated sidelobe level (ISL), both of which assess aspects of the ambiguity function. The PSL metric is typically defined as

$$\Phi_{\mathrm{PSL}}[\chi(\tau, f_D)|_{f_D=0}] = \max_{\tau}\left|\frac{\chi(\tau,0)}{\chi(0,0)}\right| \text{ for } \tau \in [\tau_{\mathrm{main}}, T] \tag{6.23}$$

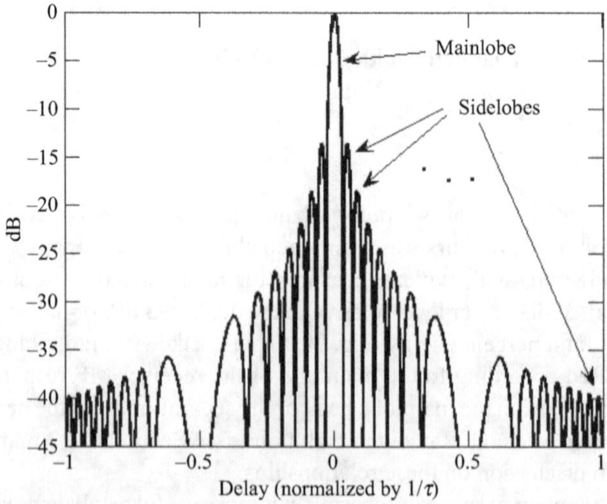

Figure 6.9 Autocorrelation of an LFM radar waveform

where, accounting for symmetry of $|\chi(\tau, f_D)|$ about the peak at $\tau = 0$, the interval $[0, \tau_{\text{main}}]$ corresponds to the mainlobe while the interval $[\tau_{\text{main}}, T]$ contains sidelobes. Likewise, ISL is defined as

$$\Phi_{\text{ISL}}[\chi(\tau, f_D)|_{f_D=0}] = \frac{\int_{\tau_{\text{main}}}^{T} |\chi(\tau, 0)|^2 d\tau}{\int_0^{\tau_{\text{main}}} |\chi(\tau, 0)|^2 d\tau}. \tag{6.24}$$

Both of these metrics could also readily be extended to include a region around zero Doppler as well.

An interesting performance benchmark in terms of PSL is the bound associated with hyperbolic FM waveforms (also known as linear period modulation) that are associated with some species of bats and are widely used in sonar applications due to their robustness to Doppler. This bound is stated as [13]

$$\text{PSL}_{\text{HFM bound}} = [-20\log_{10}(BT) - 3] \text{ dB} \tag{6.25}$$

for BT the time–bandwidth product of the waveform, which itself is a very useful measure of waveform dimensionality. As discussed above, there are a variety of ways in which to define bandwidth B. In [13] and subsequently in [14], the definition of B used in the context of (6.25) is an analytical bandwidth corresponding to the total amount of swept frequency in a functional description of an FM waveform. More recently, [15] used 3 dB bandwidth for B in (6.25) to evaluate PSL performance relative to this benchmark for polyphase-coded FM (PCFM) and related waveforms (which will be discussed shortly) where an analytical notion of swept bandwidth cannot be clearly delineated. The particular approach is really not important as long as it is applied consistently to compare a given set of waveforms.

Given the Fourier relationship between the autocorrelation and power spectrum, it is clear that waveform design can alternatively be viewed from the perspective of designing the emission spectral content, which naturally permits the inclusion of specifications regarding spectral containment and even avoidance regions that are becoming increasingly important due to growing congestion. Such a perspective takes on particular importance when one considers that some radar modes (typically for imaging) require bandwidths on the order of several 100s of megahertz up to several gigahertz.

Given the prospect of using the frequency domain to design radar emissions, it might at first appear that the design of $s(t)$ could be accomplished by simply establishing the appropriate $S(f)$ that adheres to the spectral emission requirements of the radar. In fact, it is known that a power spectrum $|S(f)|^2$ that tapers off towards the band edges corresponds to a waveform autocorrelation with low sidelobes [16]. For example, a Gaussian-shaped power spectrum is associated with a Gaussian-shaped autocorrelation that theoretically possesses no sidelobes at all. Thus, one could employ a frequency domain performance metric such as the frequency template error (FTE) [15]

$$\Phi_{\mathrm{FTE}}\left[S(f), W(f)\right] = \left(\frac{1}{f_{\mathrm{H}} - f_{\mathrm{L}}}\right) \int_{f_{\mathrm{L}}}^{f_{\mathrm{H}}} \left| |S(f)|^p - |W(f)|^p \right|^q df, \tag{6.26}$$

where f_{L} and f_{H} demarcate the edges of the frequency interval of interest (which should include enough of the spectral roll-off region to provide sufficient fidelity) and $|W(f)|^2$ is the desired power spectrum. For $p = 1$ and $q = 2$, the FTE metric defines a form of frequency-domain mean-square error (MSE). As a quick aside, note that a time-limited pulse obviously cannot be bandlimited and so the inclusion of some of the spectral roll-off greatly reduces the unavoidable error that arises from considering only a portion of the theoretically infinite spectral content.

In addition to spectral containment, practical radar waveforms also contend with the rigors of a radar transmitter that must provide sufficient energy-on-target to accommodate for the losses induced by scattering and two-way signal propagation, along with the presence of noise and interference. Without getting into the myriad complexities of clutter/interference characteristics, target scintillation, atmospheric effects, and RF system effects (see Chapter 11 for a discussion of the latter), it should generally be understood that for a radar to detect a nominal target (10 dB SINR after various gains and interference cancellation is reasonable) at some adequate distance (application specific) may necessitate the emission of up to megawatts of peak power (see example derivation in Chapter 4). Achieving this amount of transmitted power without irreparably distorting the intended waveform therefore involves the imposition of additional requirements on the waveform signal structure. The most notable requirement is that the waveform should possess a constant amplitude so that the transmitter PA, itself the most prominent source of possible distortion, can be operated in saturation so that power efficiency is maximized, just like in the telemetry example provided earlier in this chapter.

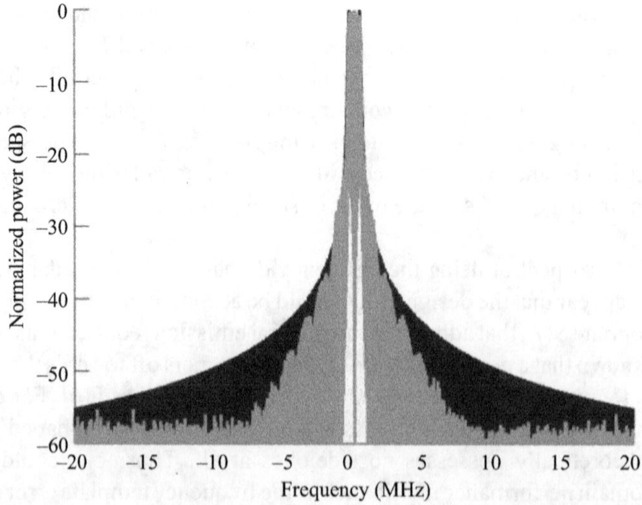

Figure 6.10 Measured power spectrum of PCFM-implemented LFM waveform before the transmitter (black) and after transmitter distortion (gray) [17]

It is also necessary to consider the radar emission spectral containment, albeit for the purpose of minimizing distortion caused by transmitter spectral shaping (i.e. approximate bandlimiting) as opposed to the Fourier relation to autocorrelation sidelobes or the impact on other spectral users. As an example, Figure 6.10 depicts the measured power spectrum of a PCFM-implemented LFM waveform before and after it is distorted by the transmitter of an S-band radar testbed containing a class AB solid-state GaN PA, with the transmitted version captured in a loopback configuration. The two versions are very similar above 35 dB, while below that point, the transmitter distortion is observed to suppress much of the broader spectral roll-off. It is important to note that despite an LFM waveform having constant amplitude and good spectral containment, it is still modestly distorted. Further, while it would appear that the transmitter is actually improving the spectral containment in this case, the distortion will impose a signal-to-noise ratio (SNR) mismatch loss on receive and high-power radars tend to use class C or higher tube-based PAs that realize spectral regrowth which is not observed here.

In contrast to the well-behaved LFM waveform, now consider the measured power spectrum of a P4-coded waveform that is illustrated in Figure 6.11. A P4 code [14,18] is a discretized version of an LFM waveform and is implemented here using the theoretical polyphase code structure of

$$s_{\mathrm{PC}}(t) = \frac{1}{\sqrt{T}} \sum_{n=1}^{N} \exp(j\theta_n)\mathrm{rect}\left[\frac{t-(n-1)T_{\mathrm{C}}}{T_{\mathrm{C}}}\right] \text{ for } 0 \leq t \leq T, \tag{6.27}$$

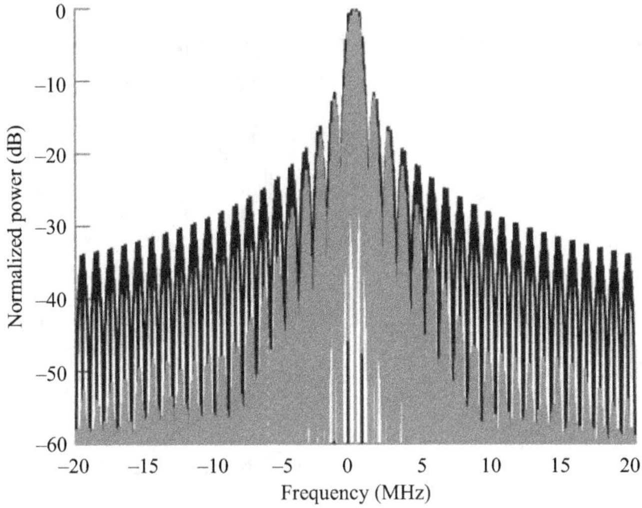

Figure 6.11 Measured power spectrum of P4-coded waveform before the transmitter (black) and after transmitter distortion (gray) [17]

where θ_n is the nth of N chips in the code and T_C is the temporal extent of each chip. The abrupt transitions between chips translate into an extended spectral skirt having a $\sin(x)/x$ roll-off that in Figure 6.11 is shown to be significantly distorted by the inherent spectral shaping of the transmitter. As above, while this aspect might seem beneficial from a spectral containment perspective, the resulting distortion can produce significant losses. For this coded structure, an additional form of loss arises in the form of range straddling in which the receiver-sampled version of such a waveform greatly differs from the assumed version that was sampled with a different delay offset. This theoretical implementation of coded waveforms is particularly susceptible to straddling effects due to the lack of good spectral containment (see [17]).

The impact of transmitter spectral shaping of the extended spectral skirt in Figure 6.11 is also evident when examining the amplitude envelope of the pulse after experiencing transmitter distortion (see Figure 6.12). Where the LFM waveform on the left side of Figure 6.12 exhibits an expected small amount of AM ripple, the P4-coded waveform on the right side of Figure 6.12 experiences multiple deep amplitude nulls that correspond to abrupt phase changes due to the structure of (6.27). The depth of each null is directly related to the amount of phase change that is occurring, with the maximum phase change of $\pm\pi$ involving a transition through the center of the phase circle where the amplitude is zero. If constant power is being supplied, these amplitude nulls likewise correspond to power that is converted to heat instead of being radiated, thus meaning less energy on target (by a small amount overall), increased phase noise (due to heat), and possibly even damage to the transmitter. In short, it is better to use waveforms that are continuous, like FM, and possess good spectral containment. In the next section, a coded form of FM is described that borrows from the

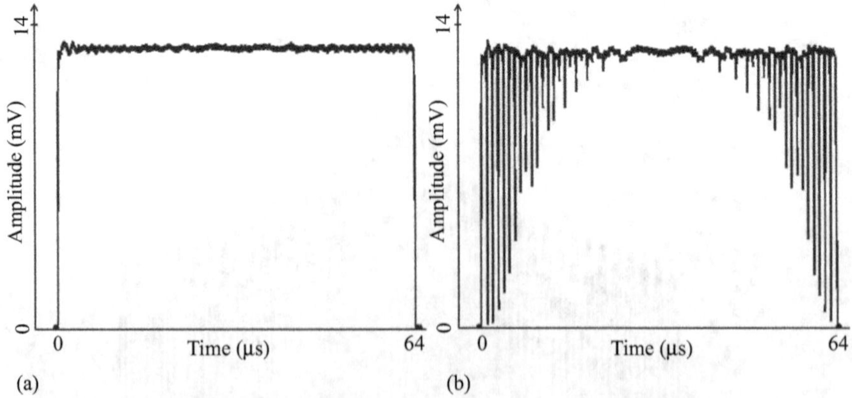

Figure 6.12 Pulse amplitude envelope after loopback capture of (a) measured LFM waveform and (b) measured P4-coded waveform [17]

CPM framework described in the previous section to achieve both constant amplitude and good spectral containment.

6.3.2 Designing for radar spectral containment: holistic system perspective

The CPM signaling scheme [5] discussed earlier in this chapter enables communication applications to achieve both high power efficiency (constant amplitude permits operation of the transmitter PA in saturation) and good spectral containment by virtue of being a digital form of FM. It was recently shown [15,17] that a modified form of CPM denoted as PCFM can likewise be used to implement and optimize arbitrary polyphase radar codes as physically realizable FM waveforms that are likewise inherently amenable to the stresses of a high-power radar transmitter. Here we consider the utility of such waveforms in an holistic system perspective to address the demands of an increasingly congested spectral environment.

Leveraging the CPM communication scheme depicted in Figure 6.6, the PCFM implementation can be expressed as [17]

$$s_{\text{PCFM}}(t;\mathbf{x}) = \exp\left\{j\left(\int_0^t g(\tau) * \left[\sum_{n=1}^{N} \alpha_n \delta(\tau-(n-1)T_{\text{p}})\right]d\tau + \theta_0\right)\right\}. \tag{6.28}$$

where the sequence of delay-shifted $\delta(t)$ terms comprise an impulse train with separation T_{p}, the shaping filter $g(t)$ must integrate to unity over the interval $[0, T_{\text{p}}]$, the operation $*$ denotes convolution, and τ is used here as the integration variable. The value θ_0 is the initial waveform phase (generally arbitrary so could be set to 0) and the vector $\mathbf{x} = [\alpha_1\ \alpha_2\ \cdots\ \alpha_N]^T$, for $(\bullet)^T$ the transpose operation, is the length N phase-change code, the values of which can be viewed as instantaneous frequencies due to their subsequent integration in (6.28). In Chapter 23, it is also shown how the

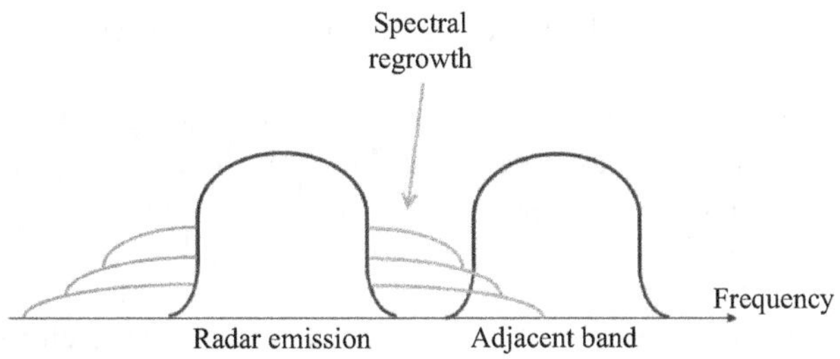

Figure 6.13 Spectral regrowth due to nonlinear harmonics in the radar emission

PCFM radar implementation can be combined with the original CPM communication framework to realize a multifunction radar/communication emission structure.

The implementation in (6.28) could be used to generate FM radar waveforms from existing length $N + 1$ polyphase codes using the conversion

$$\tilde{\alpha}_n = \theta_n - \theta_{n-1} \text{ for } n = 1, \ldots, N \tag{6.29}$$

followed by

$$\alpha_n = \begin{cases} \tilde{\alpha}_n & \text{if } |\tilde{\alpha}_n| \leq \pi \\ \tilde{\alpha}_n - 2\pi \,\text{sgn}(\tilde{\alpha}_n) & \text{if } |\tilde{\alpha}_n| > \pi \end{cases}, \tag{6.30}$$

for sgn(•) the sign operation. However, since polyphase codes tend to be designed according to implementation via (6.27), which was just shown to possess practical limitations, it is generally better to optimize the phase-change values in $\mathbf{x}$ directly by assessing performance metrics applied to the physical PCFM waveform from (6.28). In fact, if the shaping filter $g(t)$ is rectangular, (6.27) and (6.28) can, respectively, be viewed as zeroth-order and first-order code implementations, where the latter realizes continuous, piece-wise linear phase changes. Moreover, second-order and third-order PCFM implementations have also been defined and optimized [19,20] to realize greater design freedom. The notion of "overcoding" has likewise been proposed as a means to more fully exploit the innumerable continuous phase functions that are possible [21].

When designing waveforms to achieve greater sidelobe performance and/or spectral containment (which are themselves related prospects) the distortion of the waveform by the transmitter, as exemplified in Figures 6.10 and 6.11, becomes an important consideration. This distortion comprises both linear and nonlinear attributes, with the former realizing a filtering effect that serves to curtail excess spectral content outside of the passband (as prominently demonstrated in Figure 6.11) while the latter introduces intermodulation effects that alternatively lead to spectral regrowth (see Figure 6.13). Many different approaches have been developed to combat transmitter nonlinearity, typically either through the use of linearizing transmit architectures such as the Kahn and Doherty techniques, envelope tracking, outphasing methods, etc. [22] or via predistortion techniques [23] that rely on parameterized transmitter models whereby the nonlinear behavior is estimated and compensated prior to the PA.

More recently, the notion of transmitter/waveform co-design was suggested [24] that takes inspiration from the astounding capabilities of echo-locating mammals [25]. Examples of this broader design perspective include joint optimization for power-added efficiency and linearization capability [26,27] and the development of the Smith Tube [28,29], a generalization of the well-known Smith Chart used in RF engineering, to permit inclusion of other parameters such as waveform bandwidth into the design metric. Chapter 12 discusses this joint approach in detail.

It was shown in [15] that transmitter distortion can also be addressed by direct extension of the PCFM optimization scheme. In so doing, the optimization metric (e.g. PSL, ISL, FTE, etc.) is applied to the transmitter-distorted version of the PCFM-implemented waveform, where this resulting physical emission inherently captures all of the salient aspects of the transmit chain. It was demonstrated in [15] that both model-in-the-loop (MiLo) and hardware-in-the-loop (HiLo) optimization formulations are possible, with the former using a high-fidelity model of the transmitter (e.g. [30]) and the latter using the actual transmitter hardware. Of the two, the MiLo approach is significantly faster to converge because the emulation of the physical emission and its optimization are completely contained within a simulated environment. In contrast, the HiLo is much slower due to the latencies involved with loading the waveform onto hardware (e.g. an arbitrary waveform generator (AWG)) and then subsequently capturing the resulting emission either in loopback or free-space (likely within an anechoic chamber to prevent signal contamination). However, where the MiLo model cannot perfectly represent all the intricacies of the transmitter, HiLo has the advantage of designing the emission to the specific transmitter from which it will be launched. As such, it is clear that hybridization of these two could be rather beneficial.

With regard to spectral containment, these MiLo and HiLo emission optimization schemes become particularly important when one considers the impact of pulse rise/fall-time. While some radars do employ CW waveforms, these tend to be for shorter range applications in large part due to the limited transmit peak power that can be achieved for CW. In contrast, pulsed radars can realize much higher peak powers. Consequently, most radars tend to be of the pulsed variety. However, the rapid switching involved with the ultrashort rise/fall-times of the pulse introduces spectral content that, theoretically, rolls off with a $\sin(x)/x$ structure. For example, Figure 6.14 illustrates the spectral content of an optimized PCFM waveform with and without the inclusion of the pulse rise/fall, noting that the latter is useful to assess the spectral content of the waveform alone, but the former represents the actual physical emission of the pulsed radar. Further, it should be understood that the spectral skirt induced by the pulse rise/fall is still much less severe than that observed in Figure 6.11 for a coded waveform based on (6.27) because the code structure involves rapid switching of the phase values over the entire waveform, instead of just at the ends of the pulse.

To continue to achieve even better spectral containment, it becomes necessary to address the rise/fall-time of the pulse. On the one hand, a rather obvious solution is simply to slow down or amplitude taper the pulse shape. However, this seemingly obvious solution can be far more difficult to actually accomplish on real radar systems

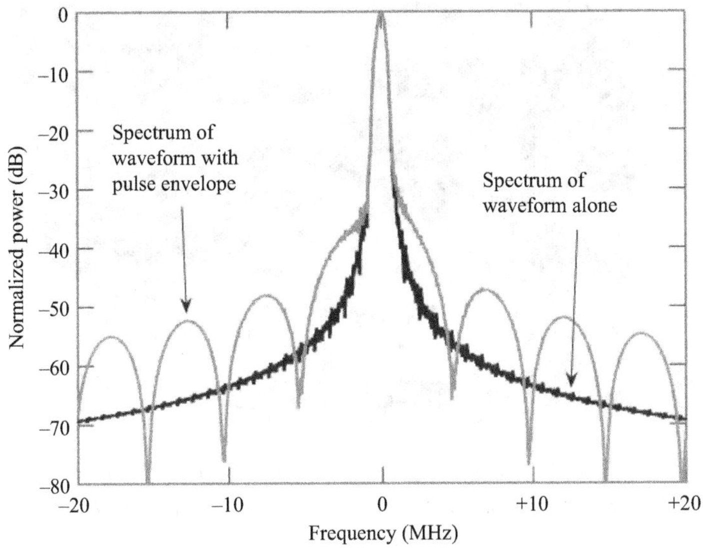

Figure 6.14 Power spectrum of optimized FM waveform with and without inclusion of the pulse envelope

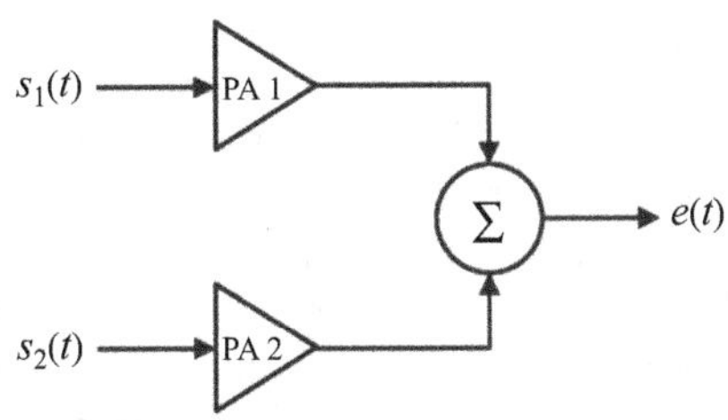

Figure 6.15 Transmitter out-phasing architecture

that radiate high power and thus require high power efficiency. In such case, the PA (typically some form of vacuum tube [31]) behaves like a high-speed switch. In recent years, linearized transmitter architectures like those discussed above (see [22]) have been examined as a means to improve the spectral containment of high-power radar emissions, which inherently necessitates some degree of AM. For example, in [32,33] the Chireix outphasing scheme was experimentally demonstrated to achieve excellent spectral containment for a waveform convolved with a Gaussian-weighted sinc kernel. The general outphasing concept, depicted in Figure 6.15, involves the use of two constant amplitude waveforms that are individually amplified using separate high-power/high-efficiency PAs and then subsequently combined such that their relative phase offset as a function of time realizes the desired AM.

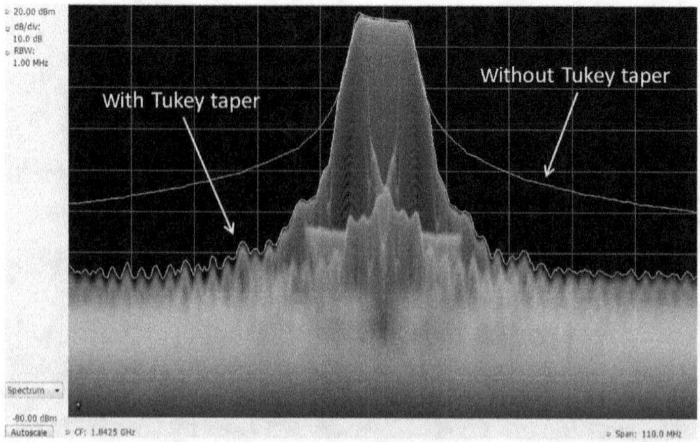

Figure 6.16 Loopback measured power spectrum of 64 μs LFM waveform with (lower trace) and without (higher trace) a Tukey taper on the first and last 4 μs of the pulse

This outphasing approach, which belongs to the broader class of linear amplification using nonlinear components (LINC) architectures, was also recently used to control the spectral spreading caused by the abrupt rise/fall of a high-power radar pulse [34]. Figure 6.16 illustrates the power spectrum of a LINC-implemented 64 microsecond LFM waveform with and without the use of a Tukey taper that shapes the first and last 4 μs of the pulse. This result was implemented in hardware using a dual-channel AWG to produce constant amplitude waveforms $s_1(t)$ and $s_2(t)$, with their resulting combination possessing the desired tapering captured in loopback using a real-time spectrum analyzer. With the vertical divisions in Figure 6.16 demarking 10 dB increments, it is observed that this shaping of the pulse rise/fall is able to improve the spectral containment by nearly 20 dB. However, this pulse amplitude shaping can also incur significant degradation of waveform range sidelobes, particularly for optimized waveforms, due to deviation from the optimal waveform structure. That said, the notion of HiLo design can be employed to optimize the emission inclusive of the LINC architecture. Figure 6.17 depicts the measured power spectrum for PCFM emissions that are HiLo optimized either using one transmit channel without tapering or using the LINC approach with a Tukey taper. Note that, while not a formal condition imposed during optimization, the spectral mainlobe now exhibits the roll-off towards the band edges that is associated with waveforms possessing lower range sidelobes [16].

While the result in Figure 6.17 demonstrates that it is possible to physically realize optimized emissions that control the pulse rise/fall to improve spectral containment, a couple of important comments are in order. First, any power that is generated but not emitted by the antenna must be appropriately dealt with to avoid the buildup of excess heat in the transmitter. This "waste" power can be a nontrivial issue for high-power radar systems. Second, it should be noted that it was necessary to have picosecond

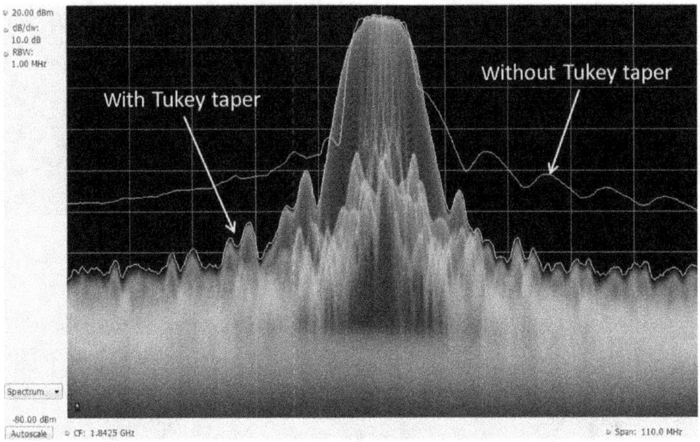

Figure 6.17 Loopback measured power spectrum of 64 μs hardware-in-the-loop optimized FM waveform with (lower trace) and without (higher trace) a Tukey taper on the first and last 4 μs of the pulse

timing granularity in the dual-channel AWG to maintain sufficient synchronization between the two waveforms to ensure convergence of the resulting emission during optimization. In an earlier instantiation of this experiment, this level of timing control was not available due to the use of separate single-channel AWGs, and consequently, convergence could not be achieved due to timing drift during the HiLo optimization process. In short, good spectral containment can be achieved for high-power radar systems in principle, but it necessitates the consideration of alternative transmitter architectures, a high degree of timing synchronization, the means to dispose of waste power/heat, and the means to generate very high fidelity waveforms. The latter is likewise true for the realization of advanced waveform diversity capabilities [35].

6.3.3 Some practical aspects of sharing radar spectrum

In light of the above discussion on the requirements and restrictions of the radar transmitter, consider some of the effects that can be expected when a radar is obliged to share spectrum. In particular, we shall examine (a) the degree to which a radar can use coherent knowledge of another signal (in this case another radar) to extract the interference-limited response of the desired radar, and (b) some of the practical implications of notching the radar transmit spectrum for the purpose of avoiding other spectrum users.

6.3.3.1 Coherent rejection of shared-spectrum interference

Adaptive interference rejection typically involves estimation of second-order statistical properties in the form of an interference covariance matrix $\mathbf{R}$ (e.g. [36]), from which an adaptive filter is formulated as $\mathbf{w} = \mathbf{R}^{-1}\mathbf{s}$, for $\mathbf{s}$ a steering vector that is matched to the desired signal structure (e.g. a spatial beamformer). However, the very

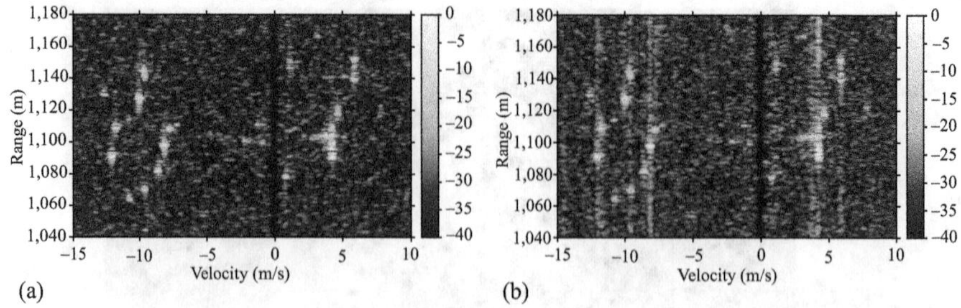

Figure 6.18 Free-space measured range-Doppler response of moving targets for (a) only a down-chirp LFM waveform transmitting and (b) two radars simultaneously emitting an up-chirp LFM and down-chirp LFM, respectively [37]

idea of spectrum sharing implies some degree of simpatico relationship in which each user has some expectation that the other exists and therefore can anticipate having at least some knowledge of the other signal structures. For example, Chapter 4 discusses the use of dynamic frequency selection by communication devices that have already been deployed in the 5 GHz band that are required to detect the presence of radar signals using generic knowledge of the radar waveforms that may be employed. It was likewise suggested in [24] that the standardization of commercial communications leads to signal structures (e.g. 4G and beyond are expected to use OFDM) that may potentially be leveraged by the radar to improve interference cancellation at the waveform level.

An approach that may ultimately enable this manner of coherent communication signal cancellation in the radar receiver was recently experimentally demonstrated [37] for the related application of radar/radar spectrum sharing. Specifically, the previously developed multistatic adaptive pulse compression (MAPC) algorithm [38], originally derived for theoretical polyphase codes, was modified to facilitate the separation of received echoes from two or more radars operating concurrently in the same band and using practical FM waveforms.

As an example, Figure 6.18 illustrates free-space measurements collected from the rooftop of Nichols Hall on the University of Kansas (KU) campus. The left panel of this figure shows the usual case of only a single radar emission in the band, with multiple moving targets clearly visible (these are vehicles traversing the intersection of 23rd and Iowa streets in Lawrence, KS). In contrast, on the right side of the figure, prominent streaks are observed that are caused by the presence of a second radar emission occurring simultaneously in the same band. These streaks would degrade detection performance in subsequent downstream processing, and this particular example is far from a worst-case scenario since both transmitters are colocated and illuminated the same scene, and thus the echoes each produce is of commensurate amplitude.

Now consider Figure 6.19 in which waveform-domain adaptive cancellation is performed via the MAPC algorithm [37,38]. Normally, radar receive processing is performed by first applying pulse compression (matched filtering or similar) of the

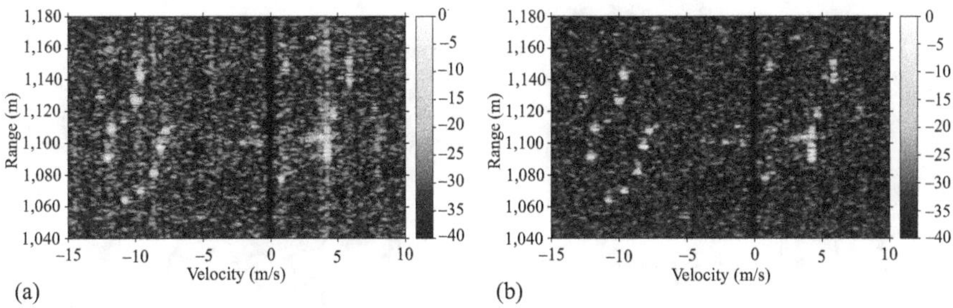

Figure 6.19 Free-space measured range-Doppler response of moving targets for two simultaneous radar emissions in which (a) MAPC is applied before Doppler processing and (b) MAPC is applied after Doppler processing and range-walk compensation [37]

transmitted waveform in "fast-time" followed by Doppler processing across the set of pulses in "slow-time." The left panel of Figure 6.19 shows that replacing normal matched filtering with MAPC does provide some reduction of the streaking observed in the right panel of Figure 6.18, though clearly some of the interference still remains. The reason for this effect is that the adaptive processing is able to suppress the interference to the level of the noise floor, but subsequent Doppler processing realizes an additional coherent integration gain on the order of the number of pulses in the coherent processing interval (CPI) that likewise integrates the residual interference.

This limitation on interference suppression can be addressed by switching the order of the pulse compression and Doppler processing stages. The right panel of Figure 6.19 reveals that doing so exploits the, now prior, integration gain of Doppler processing so that MAPC can suppress the interference to a relatively much lower noise floor. By accomplishing waveform-level suppression of in-band interference, this result illustrates that, given sufficient knowledge of communication signal structures, it may be possible to suppress communication interference in the radar receiver via a similar approach.

However, it should be noted that reversing the order of pulse compression and Doppler processing necessitates range-walk compensation in the radar receiver to ensure adequate fidelity. Further, the Doppler domain is often used in conjunction with clutter cancellation and, in the case of airborne radar, likewise combined with spatial processing to obtain sufficient adaptive degrees of freedom to account for angle-Doppler coupling of the clutter. In short, waveform-domain adaptive processing may facilitate some spectrum sharing capabilities if these (and potentially other) practical effects can be addressed.

6.3.3.2 Radar spectrum nulling on transmit

Notching the radar transmit spectrum, also referred to as transmit nulling, is a type of adaptive waveform design in which knowledge of the current spectral environment is used to avoid portions of the in-band spectrum that are occupied by other users

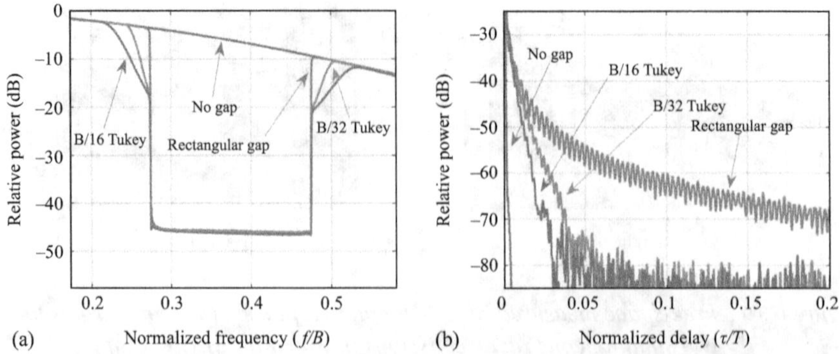

Figure 6.20 Loopback measured stationary spectral notch (or gap) and tapering in terms of (a) the power spectrum and (b) the associated integrated autocorrelation for an FMCW noise waveform with $BT = 11$ million [50]

[39–51]. This adapt-on-transmit approach to interference avoidance, which also makes it a particular form of cognitive radar, reduces the interference caused by the radar to other spectrum users and likewise reduces the interference to the radar by virtue of the notch also residing in the waveform MF. Chapters 18 and 23 discuss some particular methods for the design of notched waveforms, the former for the case of space-time transmit nulling and the latter as a means to facilitate a multifunction mode involving tandem-hopped radar and communications. Here we consider some of the system-level issues that arise in practice for transmit notching schemes.

It is generally understood, and was analytically derived in [46], that notching the transmit spectrum of a radar waveform incurs a penalty in terms of the range sidelobe level that can be achieved. Intuitively, this effect can be appreciated by considering that the spectral notch in the passband of a radar emission represents a reduction in design degrees of freedom commensurate with the fraction of the passband that is removed. Since the overall bandwidth is not reduced, and thus the range resolution is not degraded, this loss in degrees of freedom manifests as an increase in the range sidelobes.

It was recently shown [49–51] that a newly developed form of FM noise waveform that can realize both spectral notches and sufficiently low sidelobes by virtue of achieving exceedingly high dimensionality (BT exceeding 10 million has been demonstrated). In so doing, further practical effects begin to emerge. For example, Figure 6.20 illustrates the impact of the shape of the spectral notch (or “gap” in the figure) on the range sidelobe response for a notched FMCW waveform having $BT = 11 \times 10^6$ that was experimentally measured in a loopback configuration. When the notch has a rectangular shape, regardless of the overall shape of the waveform power spectrum (which is presumably designed to be at least moderately good), the range sidelobe response is dominated by a $\sin(x)/x$ roll-off, which makes sense when considering that this spectrum can be expressed as the linear combination of the

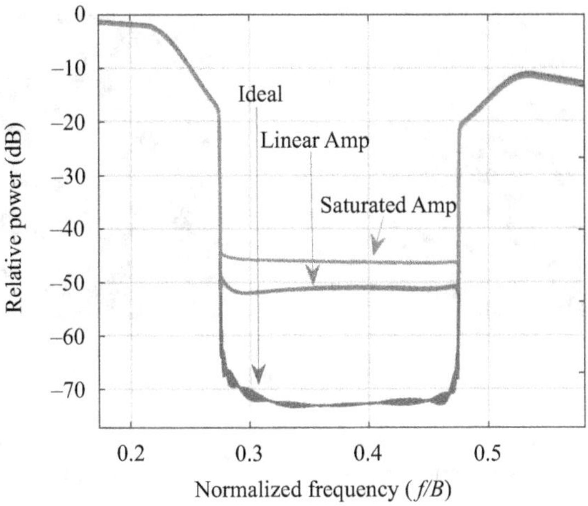

Figure 6.21 Power spectrum of notch depth after amplification [50]

unnotched spectrum and the negative of the spectral content removed by the notch. However, in the same way that tapering in time or frequency can reduce sidelobes in the other domain, so too can tapering of the notch edges as depicted in Figure 6.20. Note that the achievable relative sidelobe floor for this type of waveform is better than $-10\log_{10}(BT)$, which in this case amounts to roughly -70 dB.

The practical achievement of transmit spectral notches is also greatly affected by the fidelity of the radar emission, which is the waveform (likely designed under idealistic conditions of very high fidelity) distorted by the radar transmitter. Figure 6.21 shows how the measured notch from Figure 6.20 (the $B/16$ Tukey tapered case) was actually designed to be 30 dB deeper than was ultimately realized in loopback due to distortion from the PA when operated in saturation. Slightly less degradation of the notch depth (about 5 dB) is incurred when the PA (here a modest solid-state Class A) is operated in a linear mode. When one considers that the radar may emit a megawatt or more of peak power, it is clear that such a degradation in realizable notch depth could be a serious practical deterrent to deploying spectral transmit nulling. That said, as discussed in the previous section and further detailed in Chapter 12, transmitter-in-the-loop waveform design and the prospect of joint transmitter/waveform optimization may hold the key to operationally produce these signals in the future.

Finally, assuming that waveforms with adequate notch depth can be practically achieved for high-power radars, there still remains a potential practical issue if the interference spectral location(s) is/are nonstationary during the radar CPI. Such an occurrence could certainly arise as communication systems themselves become more spectrally nimble combined with the "internet of things" vision for the connectivity of a massive number of devices. The first important factor that arises in this situation is the latency involved with the radar (a) sensing the new interference spectral location(s) followed by (b) adequately adjusting the notch location(s) within the radar

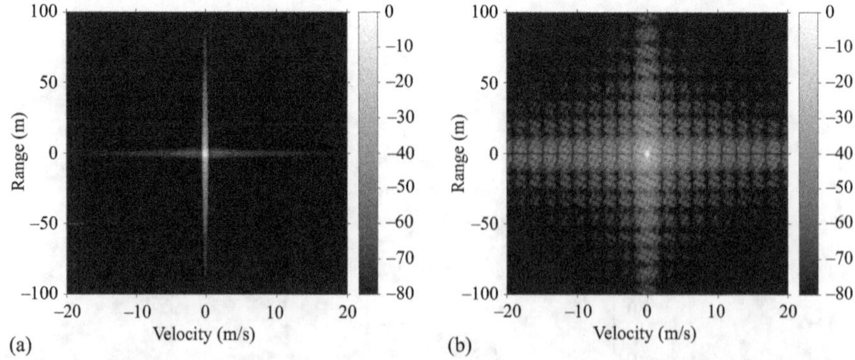

Figure 6.22 Range-Doppler point spread function of an FMCW noise waveform for (a) a static notch and (b) a notch location that changes ten times during the CPI [50]

emission structure. The former is determined by propagation delays and the speed of interference detection signal processing (e.g. see [52]), while the latter is dictated by how quickly new waveforms can be designed, which involves an optimization process that can be achieved using repeated FFTs. Continued advances in processing speeds and parallelization provides some optimism that minimal latencies can be achieved.

The more difficult attribute of nonstationary in-band interference is the impact it can cause on radar receive processing. Figure 6.22 shows the delay-Doppler point spread function for the case of a stationary notch during the CPI as well as the case in which the notch changes to a different location (here ten times) during the CPI. While the former provides a more or less standard response despite the presence of the notch, the latter case of notch movement exhibits substantial smearing across delay/Doppler that translates into residual clutter that is not adequately suppressed. In other words, even if the notch(es) could clairvoyantly keep pace with the changing in-band interference, this manner of changing the radar emission structure produces its own penalty in terms of radar performance degradation. Of course, ongoing work is likewise exploring how this effect could be compensated.

6.4 Conclusions and looking ahead

In this chapter, we have examined the nature of spectral efficiency (i.e. how well spectrum is utilized) and its role in radar/communication spectrum sharing. With the demand for finite spectrum continuing to increase for all modes, it stands to reason that "better use of spectrum" goes hand-in-hand with any and all attempts at sharing. A related factor that must also be a part of the discussion on spectrum sharing/efficiency is the ubiquitous need for power efficiency to meet severe SWaP constraints (e.g. aeronautical telemetry), to achieve high peak power for radar transmitters, and to extend battery life on myriad wireless devices. Finally, to these juxtaposed demands on spectrum and power efficiency, one overlays the application requirements such as high data-rates and capacity for communications or useful waveform designs for radar, which themselves are clearly significant oversimplifications of the actual system specifications. In short, these systems are already extremely complex before we even begin to ask questions about how to make them share spectrum.

Within the particular perspective of spectral efficiency, there remain some open topics to be explored. For example, despite decades of strict adherence to CPM, the aeronautical telemetry community is investigating the system engineering trade-space involved with using APSK due to the prospect of realizing even greater bandwidth efficiency. Conversely, new radar waveform implementations based on CPM have emerged, which provide greater freedom for range sidelobe reduction while providing good spectral containment. Finally, with the continued development of new interference cancellation and avoidance techniques, it is important to consider the practical aspects as well, such as the actual fidelity one can achieve with a distortion-inducing high-power transmitter and the operational trade-offs that could be required to implement a new capability.

References

[1] J. Proakis and M. Salehi, *Digital Communications*. New York, NY: McGraw-Hill, 2008.

[2] Range Commanders Council Telemetry Group, Range Commanders Council, White Sands Missile Range, New Mexico, *IRIG Standard 106-15: Telemetry Standards*, 2015. (Available on-line at http://www.irig106.org).

[3] M. Rice, C. Josephson, and E. Perrins, "Optimizing coded 16-APSK for aeronautical mobile telemetry," in *Proc. Int. Telemetering Conf.* (Las Vegas, NV), Oct. 2017.

[4] T. Chalfant and C. Irving, "Range telemetry improvement and modernization," in *Proc. Int. Telemetering Conf.* (Las Vegas, NV), pp. 294–303, Oct. 1997.

[5] J. B. Anderson, T. Aulin, and C.-E. Sundberg, *Digital Phase Modulation*. New York, NY: Plenum Press, 1986.

[6] D. I. S. Agency, "Department of defense interface standard, interoperability standard for single-access 5-kHz and 25-kHz UHF satellite communications channels." Tech. Rep. MIL-STD-188-181B, Department of Defense, Mar. 1999.

[7] S. Pathak and E. Perrins, "LDPC-coded APSK for aeronautical telemetry," in *Proc. Int. Telemetering Conf.* (Las Vegas, NV), Oct. 2017.

[8] "Standard radar definitions." IEEE Standard 686-2008, May 2008.

[9] A. W. Rihaczek, *Principles of High-Resolution Radar*. New York, NY: McGraw-Hill, 1969.

[10] R. A. Scholtz, "How do you define bandwidth?," in *Proc. Int. Telemetering Conf.* (Los Angeles, CA), Oct. 1972.

[11] F. Sabath, E. L. Mokole, and S. N. Samaddar, "Definition and classification of ultra-wideband signals and devices," *URSI Radio Science Bulletin*, vol. 2005, no. 313, pp. 12–26, Jun. 2005.

[12] M. A. Richards, J. A. Scheer, and W. A. Holm, *Principles of Modern Radar, vol. I: Basic Principles*. Edison, NJ: SciTech, 2010.

[13] T. Collins and P. Atkins, "Nonlinear frequency modulation chirps for active sonar," *IEE Proceedings – Radar, Sonar and Navigation*, vol. 146, pp. 312–316, Dec. 1999.

[14] N. Levanon and E. Mozeson, *Radar Signals*. Hoboken, NJ: Wiley-IEEE Press, 2004.

[15] S. D. Blunt, J. Jakabosky, M. Cook, J. Stiles, S. Seguin, and E. L. Mokole, "Polyphase-coded FM (PCFM) radar waveforms, part II: optimization," *IEEE Transactions on Aerospace and Electronic Systems*, vol. 50, pp. 2230–2241, Jul. 2014.

[16] A. Johnston, "Improvements to a pulse compression radar matched filter," *Radio and Electronic Engineer*, vol. 53, pp. 138–140, Apr. 1983.

[17] S. D. Blunt, M. Cook, J. Jakabosky, J. D. Graaf, and E. Perrins, "Polyphase-coded FM (PCFM) radar waveforms, part I: implementation," *IEEE Transactions on Aerospace and Electronic Systems*, vol. 50, pp. 2218–2229, Jul. 2014.

[18] B. L. Lewis and F. F. Kretschmer, "A new class of polyphase pulse compression codes and techniques," *IEEE Transactions on Aerospace and Electronic Systems*, vol. AES-17, pp. 364–372, May 1981.

[19] P. S. Tan, J. Jakabosky, J. M. Stiles, and S. D. Blunt, "On higher-order representations of polyphase-coded FM radar waveforms," in *IEEE Intl. Radar Conf.* (Washington, DC), May 2015.

[20] P. S. Tan, J. Jakabosky, J. M. Stiles, and S. D. Blunt, "Higher-order implementations of polyphase-coded FM radar waveforms," *to appear in IEEE Transactions on Aerospace and Electronic Systems*.

[21] J. Jakabosky, S. D. Blunt, and B. Himed, "Optimization of over-coded radar waveforms," in *IEEE Radar Conf.* (Cincinnati, OH), May 2014.

[22] F. H. Raab, P. Aspeck, S. Cripps, *et al.*, "Power amplifiers and transmitters for RF and microwave," *IEEE Transactions on Microwave Theory and Techniques*, vol. 50, pp. 814–826, Mar. 2002.

[23] F. M. Ghannouchi and O. Hammil, "Behavioral modeling and predistortion," *IEEE Microwave Magazine*, vol. 10, no. 7, pp. 52–64, Dec. 2009.

[24] H. Griffiths, S. Blunt, L. Cohen, and L. Savy, "Challenge problems in spectrum engineering and waveform diversity," in *IEEE Radar Conf.* (Ottawa, Canada), Apr./May 2013.

[25] W. W. L. Au and S. W. Martin, "Why dolphin biosonar performs so well in spite of mediocre 'equipment'," *IET Radar, Sonar and Navigation*, vol. 6, pp. 566–575, Jul. 2012.

[26] C. Baylis, L. Wang, M. Moldovan, *et al.*, "Designing transmitters for spectral conformity: power amplifier design issues and strategies," *IET Radar, Sonar and Navigation*, vol. 5, pp. 681–685, Jul. 2011.

[27] J. Hoversten, S. Schafer, M. Roberg, M. Norris, D. Maksimovic, and Z. Popovic, "Codesign of PA, supply, and signal processing for linear supply-modulated RF transmitters," *IEEE Transactions on Microwave Theory and Techniques*, vol. 60, pp. 2010–2020, Jun. 2012.

[28] M. Fellows, M. Flachsbart, J. Barlow, *et al.*, "Optimization of power-amplifier load impedance and waveform bandwidth for real-time reconfigurable radar," *IEEE Transactions on Aerospace and Electronic Systems*, vol. 51, pp. 1961–1971, Jul. 2015.

[29] J. Barkate, M. Flachsbart, Z. Hays, *et al.*, "Fast, simultaneous optimization of power amplifier input power and load impedance for power-added efficiency and adjacent-channel power ratio using the power Smith tube," *IEEE Transactions on Aerospace and Electronic Systems*, vol. 52, pp. 928–937, Apr. 2016.

[30] E. Costa, M. Midrio, and S. Pupolin, "Impact of amplifier nonlinearities on OFDM transmission system performance," *IEEE Communication Letters*, vol. 3, pp. 37–39, Feb. 1999.

[31] B. Manz, "Advancing TWTs: the traveling wave tube lives on … and on," *AOC Journal of Electronic Defense*, vol. 32, pp. 26–30, Jul. 2009.

[32] R. Chen and B. Cantrell, "Highly bandlimited radar signals," in *IEEE Radar Conf.* (Long Beach, CA), Apr. 2002.

[33] H. H. Faust, B. Connolly, T. M. Firestone, R. C. Chen, B. H. Cantrell, and E. Mokole, "A spectrally clean transmitting system for solid-state phased-array radars," in *IEEE Radar Conf.* (Philadelphia, PA), Apr. 2004.

[34] L. Ryan, J. Jakabosky, S. D. Blunt, C. Allen, and L. Cohen, "Optimizing polyphase-coded FM waveforms within a LINC transmit architecture," in *IEEE Radar Conf.* (Cincinnati, OH), May 2014.

[35] S. D. Blunt and E. L. Mokole, "An overview of radar waveform diversity," *IEEE Aerospace and Electronic Systems Magazine*, vol. 31, pp. 2–42, Nov. 2016.

[36] J. Ward, "Space-time adaptive processing for airborne radar." MIT Lincoln Lab Technical Report ESC-TR-94-109, Dec. 1994.

[37] P. McCormick and S. Blunt, "Shared-spectrum multistatic radar: experimental demonstration using FM waveforms," in *IEEE Radar Conf.* (Oklahoma City, OK), Apr. 2018.

[38] S. D. Blunt and K. Gerlach, "Multistatic adaptive pulse compression," *IEEE Transactions on Aerospace and Electronic Systems*, vol. 42, pp. 1043–1057, Jul. 2006.

[39] M. Lindenfeld, "Sparse frequency transmit-and-receive waveform design," *IEEE Transactions on Aerospace and Electronic Systems*, vol. 40, pp. 851–861, Jul. 2004.

[40] M. Cook, T. Higgins, and A. Shackelford, "Thinned spectrum radar waveforms," in *Intl. Waveform Diversity and Design Conf.* (Niagara Falls, Canada), Aug. 2010.

[41] K. Gerlach, M. Frey, M. Steiner, and A. Shackelford, "Spectral nulling on transmit via FM radar waveforms," *IEEE Transactions on Aerospace and Electronic Systems*, vol. 47, pp. 1507–1515, Apr. 2011.

[42] I. Selesnick and S. Pillai, "Chirp-like transmit waveforms with multiple frequency-notches," in *IEEE Radar Conf.* (Kansas City, MO), May 2011.

[43] C. Nunn and L. Moyer, "Spectrally-compliant waveforms for wideband radar," *IEEE Aerospace and Electronic Systems Magazine*, vol. 27, pp. 11–15, Aug. 2012.

[44] L. Patton and B. Rigling, "Phase retrieval for radar waveform optimization," *IEEE Transactions on Aerospace and Electronic Systems*, vol. 48, pp. 3287–3302, Oct. 2012.

[45] L. Patton, C. Bryant, and B. Himed, "Radar-centric design of waveforms with disjoint spectral support," in *IEEE Radar Conf.* (Atlanta, GA), May 2012.

[46] S. Frost and B. Rigling, "Sidelobe predictions for spectrally-disjoint radar waveforms," in *IEEE Radar Conf.* (Atlanta, GA), May 2012.

[47] T. Higgins, T. Webster, and A. Shackelford, "Mitigating interference via spatial and spectral nulling," *IET Radar, Sonar and Navigation*, vol. 8, pp. 84–93, Feb. 2014.

[48] A. Aubry, A. D. Maio, Y. Huang, M. Piezzo, and A. Farina, "A new radar waveform design algorithm with improved feasibility for spectral coexistence," *IEEE Transactions on Aerospace and Electronic Systems*, vol. 51, pp. 1029–1038, Apr. 2015.

[49] J. Jakabosky, S. Blunt, and A. Martone, "Incorporating hopped spectral gaps into nonrecurrent nonlinear FMCW radar emissions," in *IEEE CAMSAP Workshop* (Cancun, Mexico), Dec. 2015.

[50] J. Jakabosky, B. Ravenscroft, S. Blunt, and A. Martone, "Gapped spectrum shaping for tandem-hopped radar/communications and cognitive sensing," in *IEEE Radar Conf.* (Philadelphia, PA), May 2016.

[51] J. Owen, B. Ravenscroft, B. Kirk, *et al.*, "Experimental demonstration of cognitive spectrum sensing and notching for radar," in *IEEE Radar Conf.* (Oklahoma City, OK), Apr. 2018.

[52] A. Martone, K. Ranney, K. Sherbondy, K. Gallagher, and S. Blunt, "Spectrum allocation for non-cooperative radar coexistence," *to appear in IEEE Trans. Aerospace and Electronic Systems*.

Chapter 7
Passive bistatic radar for spectrum sharing

Hugh Griffiths and Chris Baker***

7.1 Introduction

Passive bistatic radar (PBR) may be defined as a set of radar techniques that exploit existing signals such as broadcast, communications, or radionavigation emissions as their transmitting source. This arrangement contrasts with conventional monostatic radars that use their own dedicated transmitter and a single antenna for both transmitting and receiving, and where the form of the signal (usually pulsed) is optimised solely for the radar function. Thus, PBR represents a prospective way with which to address the spectrum congestion problem.

The term *PBR* is widely used and emphasises the bistatic configuration involved with the physical separation of transmitter and receiver, which in practice dictates many of the properties of this type of radar. Sometimes this name is shortened to *passive radar*. Other terms have also been suggested and used. For example, *passive coherent location* or *passive covert radar* have been employed, particularly by military users. The term *hitchhiking* is invoked when the transmitter source is an existing monostatic radar. Yet further terms that have been used for this sensing framework include *broadcast radar*, *non-cooperative radar*, *parasitic radar* and *symbiotic radar* [1].

The term *commensal radar* (literally 'from the same table') has been adopted to describe a PBR in which the illuminating source not only fulfils its primary purpose, but is also in some sense optimised as a radar illuminator – in terms of waveform and coverage [1], which is clearly of significant interest in the context of this chapter.[1]

The transmit sources used in PBR are often referred to as *illuminators of opportunity*, which means that their signals are optimised for some purpose other than radar. A further distinction has been made between cooperative sources and non-cooperative sources. In practice, these definitions are rather crude and there are a range of situations that actually exist. At one extreme, the waveform and coverage of the illuminator might be completely cooperative and under control of the PBR designer, and could

*Department of Electronic and Electrical Engineering, University College London, UK
**Department of Electronic, Electrical and Systems Engineering, University of Birmingham, UK

[1]The term *Commensal Radar* has alternatively been used by researchers in South Africa as a synonym for Passive Bistatic Radar, especially with non-cooperative illuminators. See Chapter 8.

even be dynamically varied in real time. At the other, an illuminator might be designed so as to be as difficult as possible to exploit for use as a radar signal by a PBR receiver. Clearly, there are a number of cases between these extremes.

The potential advantages of PBR are listed in [1]:

- Broadcast and communication transmitters tend to be sited on high locations, and hence achieve broad coverage.
- Since the system makes use of existing transmitters, the cost of a passive radar is likely to be much lower than a conventional radar.
- Similarly, there are no licensing issues due to the receive-only nature.
- It allows the use of frequency bands (particularly VHF and UHF) that are not normally available for radar purposes. Such frequencies may be beneficial in detecting stealthy targets, since the wavelength is on the same order as the physical dimensions of the target, and forward scatter provides a relatively broad angular scatter.
- Since the receiver emits no signal of its own, and as long as the receive antenna is inconspicuous, the PBR receiver may be undetectable and hence completely covert.
- It is difficult to deploy countermeasures against PBR. Any jamming will have to be spread over a range of directions, diluting its effectiveness.
- PBR does not require any additional spectrum. For this reason it has been termed 'green radar'.
- There is an enormous range of transmissions that may be used. In practice, almost any emission can be used as the basis of PBR.

However, there are also some significant disadvantages:

- The waveforms of opportunistic transmissions are not optimised for radar purposes, so care has to be taken when selecting these waveforms for passive use and in how to process them in the optimum manner.
- The transmit source is typically not under the control of the passive radar.
- For analogue signals, the ambiguity function (dictating resolution in range and in Doppler) depends on the instantaneous modulation, and some kinds of modulation are better than others. Digital modulation does not suffer from these problems, so is likely to be preferred.
- The waveforms are usually continuous (i.e. a duty cycle of 100%), so significant processing has to be performed to suppress the direct signal and multipath in order to detect weak target echoes.
- In common with all bistatic radars, the resolution in range and Doppler is poor for targets that are on or close to the baseline between the transmitter and receiver.

Chapter 1 of reference [1] provides an account of the history of PBR. In fact, some of the very earliest radar experiments in the 1920s and 1930s made use of broadcast transmitters [2,3]. In World War II, German engineers developed an ingenious bistatic radar system known as *Klein Heidelberg* that hitchhiked off transmissions from the

British *Chain Home* radars [4–6]. Interest in the subject of PBR then waned until the 1980s, when experiments were undertaken using broadcast (analogue television and FM radio) transmissions [7–11]. Over the next two decades research continued, largely by university groups for which the relative simplicity and low cost were attractive. However, the early years of the twenty-first century marked a change in the maturity of the subject, as a number of aerospace companies built and demonstrated PBR systems with substantially better performance. A defence business website estimated in 2016 that the commercial market for PBR would be worth more than US$10 billion over the next decade [12].

The objectives of this chapter are therefore to provide an introduction to the relevant properties of PBR, and then to consider how PBR may be able to help in addressing the spectrum congestion problem. The structure of the remainder of the chapter is as follows: Section 7.2 discusses the principles of bistatic radar, many of which result from the bistatic geometry, and introduces the bistatic radar equation. Section 7.3 considers the properties of PBR illuminator sources, in terms of the power density at the target, coverage and waveform. Section 7.4 describes a number of processing techniques that are specific to PBR, including the suppression of the direct signal at the PBR receiver. Section 7.5 provides some examples of practical PBR systems and their performance. Finally, Section 7.6 considers the role of PBR as a potential solution to the spectrum problem.

7.2 Bistatic radar principles

7.2.1 Bistatic radar geometry

Figure 7.1 depicts the two-dimensional geometry of a bistatic radar. Many of the properties of bistatic radar are a consequence of this geometry, and excellent treatments are provided by references [13] and [14]. Usually the transmitter and receiver will be fixed, though in the general case either or even both may be carried by moving platforms. The distance between transmitter and receiver is known as the bistatic baseline, denoted by L. The extended baseline is depicted as a dashed line in Figure 7.1, extending beyond both the transmitter and receiver. The range from the transmitter to a target is R_T, and the range from the same target to the receiver is R_R. In a north-referenced coordinate system, the directions of the target from the transmitter and from the receiver are denoted by θ_T and θ_R, respectively.

The angle at the target made by the transmitter and the receiver is the bistatic angle, denoted by β, the bisector of which is shown as a dashed line in Figure 7.1. The velocity of the target is denoted by v, having angle δ with the bisector of β. Straightforward geometry shows that

$$\beta = \theta_T - \theta_R. \tag{7.1}$$

Usually the bistatic range sum $(R_T + R_R)$ is measured by the difference in time of arrival between the direct signal along the baseline and the target echo, which

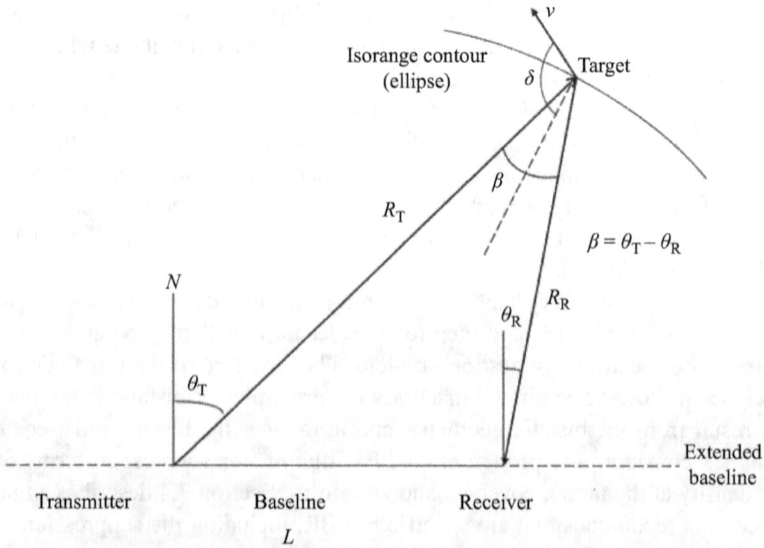

Figure 7.1 Bistatic radar geometry. The target velocity is v, producing an angle δ with the bisector of the bistatic angle β

assumes that the direct signal is detectable at the receiver and that L is known. The range of the target from the receiver can be deduced from

$$R_R = \frac{(R_T + R_R)^2 - L^2}{2\,(R_T + R_R + L \sin\theta_R)} \tag{7.2}$$

and

$$R_T = \frac{(R_T + R_R)^2 - L^2}{2\,(R_T + R_R - L \sin\theta_T)}. \tag{7.3}$$

Contours of constant bistatic range sum $(R_T + R_R)$ are ellipses, with the transmitter and receiver as the two focal points.

The Doppler shift f_D of the target echo is given by the rate of change of the bistatic range sum via

$$f_D = \frac{1}{\lambda} \frac{d}{dt} (R_T + R_R)\,. \tag{7.4}$$

In the simplest case where the transmitter and receiver are both stationary and only the target is moving we find that

$$\frac{dR_T}{dt} = v \cos\left(\delta - \frac{\beta}{2}\right) \text{ and } \frac{dR_R}{dt} = v \cos\left(\delta + \frac{\beta}{2}\right), \tag{7.5}$$

so that

$$f_D = \frac{2v}{\lambda}\left(\cos\delta\cos\frac{\beta}{2}\right). \tag{7.6}$$

It is thus observed that if the target lies on the baseline between transmitter and receiver, the target echo arrives at the receiver at the same instant as the direct signal, irrespective of where on the baseline the target is located. There is therefore no range information provided for such a target. Furthermore, there will be no Doppler information, since in equation (7.6) we see that $\beta = 180°$, so that $\cos(\beta/2) = 0$. This result can also be appreciated because at the instant the target crosses the baseline, the transmitter-to-target range is changing in an equal and opposite manner to the target-to-receiver range. This geometry is known as *forward scatter*, and although it provides little to no range or Doppler information, it can provide substantial enhancement of target signature. This point is discussed further in Section 7.2.3.

7.2.2 Bistatic radar equation

In its simplest form, the radar equation for a bistatic radar is derived in essentially the same way as for a monostatic radar:

$$\frac{P_R}{P_n} = \frac{P_T G_T}{4\pi R_T^2}\sigma_B \frac{1}{4\pi R_R^2}\frac{G_R\lambda^2}{4\pi}\frac{1}{kT_0BF} \tag{7.7}$$

in which

- P_R is the received target echo power,
- P_n is the receiver noise power,
- P_T is the transmit power,
- G_T is the transmit antenna gain,
- R_T is the transmitter-to-target range,
- σ_B is the target bistatic radar cross-section (RCS),
- R_R is the target-to-receiver range,
- G_R is the receive antenna gain,
- λ is the signal wavelength,
- k is Boltzmann's constant $= 1.38 \times 10^{-23}$ J/K,
- T_0 is the noise reference temperature $= 290$ K,
- B is the receiver effective bandwidth and
- F is the receiver effective noise figure.

This basic equation can be modified to include the effects of losses, the pattern propagation factors associated with the transmitter-to-target and target-to-receiver paths, and integration gain in the receiver.

Note that the numerator in (7.7) includes the transmit and receive antenna gains individually, and that the denominator includes the factor $1/(R_T R_R)^2$. This latter factor means that, for omnidirectional transmit and receive antenna patterns, contours of constant signal-to-noise ratio (SNR) are defined by $(R_T R_R) =$ constant, which are the *Ovals of Cassini*. However, when directional antennas are employed the contours are weighted according to the associated antenna patterns, and thus no longer have the same shape. The $1/(R_T R_R)^2$ factor also means that the SNR has a minimum value when $R_T = R_R$ (i.e. when the target is equidistant from the transmitter and receiver) and is maximum when the target is either close to the transmitter or close to the receiver. The factor $P_T G_T/4\pi R_T^2$ represents the power density Φ of the transmit signal at the target, and is an important property in determining the utility of a given transmission for PBR purposes.

7.2.3 Target signatures

The quantity σ_B in (7.7) is the target bistatic RCS. One of the advantages sometimes ascribed to bistatic radar is that it may offer a counter-stealth capability, since targets that are shaped and/or surface-treated to minimise their monostatic RCS may nevertheless have higher bistatic RCS. Of course, this property is not easy to verify, as there has been little reported on the subject in the open literature and measurements are much more challenging than for monostatic radar. For monostatic RCS only the target geometry needs to be considered, whereas for bistatic radar both the target geometry and the radar geometry have to be considered, thereby greatly adding to the number of measurements needed for a comprehensive description of bistatic RCS. Further, it will often be the case, especially for military targets, that such measurements are not published in the open literature.

For non-stealthy targets, the bistatic RCS is comparable to the monostatic RCS [15]. Work in the 1960s led to the bistatic equivalence theorem [16], which states that the bistatic RCS of a target is equal to the monostatic RCS measured at the bisector of the bistatic angle β and reduced in frequency by the factor $\cos(\beta/2)$, provided that (i) the target is sufficiently smooth, (ii) there is no shadowing, and (iii) retro-reflectors persist. Clearly these conditions are not always met, especially at large bistatic angles, so the theorem should be used with some care.

Another factor is that of resonance scatter that occurs when the spacing of scattering centres on the target is comparable to the radar wavelength. This effect may enhance target RCS at the VHF and UHF frequencies that are often used for PBR.

The forward scatter geometry, mentioned in Section 7.2.1, also offers the possibility of considerable enhancement of RCS. This effect can be understood by referencing Babinet's Principle [17] from physical optics (Figure 7.2). Suppose that there is an infinite screen placed between the transmitter and receiver, blocking the signal

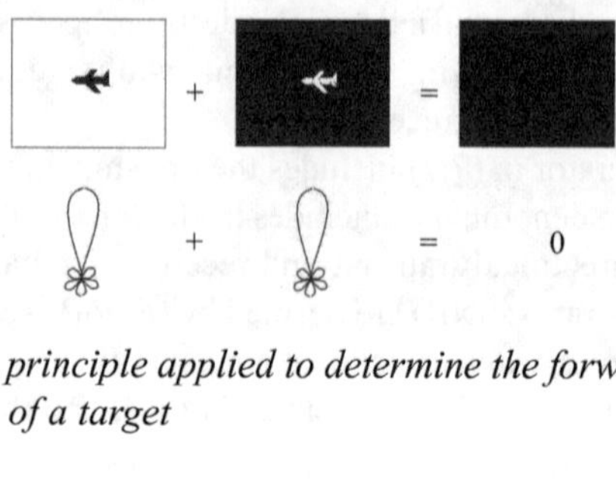

Figure 7.2 Babinet's principle applied to determine the forward scatter radar signature of a target

completely, so the received signal is zero. Now suppose that a target-shaped hole is cut in the centre of the screen. Babinet's Principle says that the signal diffracted through the target-shaped hole must be equal and opposite to the forward-scattered signal from the target – since the two contributions must add to zero. Calculation of the signal diffracted through an aperture of a given size and shape is a standard problem in electromagnetics and is readily evaluated, subsequently providing the forward scatter signature of the target directly.

Specifically, for a target of silhouette area A, the forward scatter RCS is approximately

$$\sigma_{FS} = \frac{4\pi A^2}{\lambda^2} \tag{7.8}$$

and the angular width of the scatter is

$$\theta_B = \frac{\lambda}{a} \text{ (rad)}, \tag{7.9}$$

where a is the linear dimension of the target in the appropriate plane (azimuth or elevation). Figure 7.3 plots these values as a function of frequency for an idealised small aircraft target with $A = 10\,\text{m}^2$ and $a = 10$ m, for which the monostatic RCS might be 0 dBm^2. It is observed that σ_{FS} is many tens of dB greater than the monostatic RCS, though at high frequencies the forward-scattered energy is concentrated in a rather narrow beam. In practice, lower frequencies (i.e. VHF or UHF) are likely to be preferred, where σ_{FS} is still substantial but the energy is spread over a broader swath of angles. This arrangement is quite compatible with PBR, where most of the broadcast and communication illuminators lie in the VHF and UHF bands.

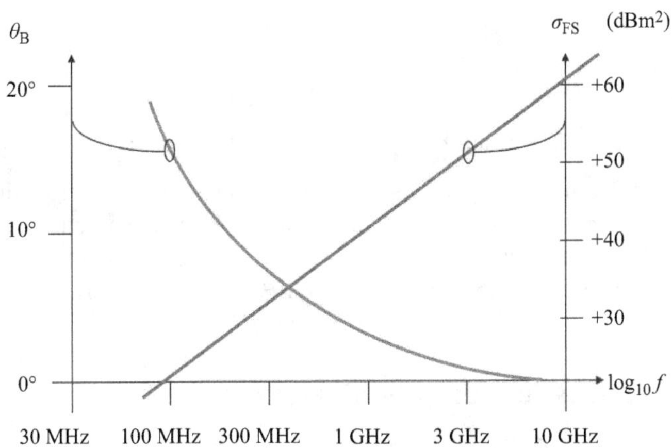

Figure 7.3 Forward scatter RCS σ_{FS} and angular scatter width θ_B for an idealised small aircraft target with $A = 10$ m^2 and $a = 10$ m

7.2.4 The ambiguity function in bistatic radar

The classical tool used to evaluate the performance of a radar waveform is the ambiguity function, devised by the British mathematician Philip Woodward in the early 1950s. The ambiguity function can be expressed as [18]:

$$|\psi(T_R, f_D)|^2 = \left| \int_{-\infty}^{\infty} s_t(t)\, s_t^*(t + T_R) \exp[j2\pi f_D t] dt \right|^2, \tag{7.10}$$

for $s_t(t)$ the radar waveform and T_R the delay time. The ambiguity function is the square magnitude of the output from a filter matched to waveform $s_t(t)$, and represents the point target response of the radar as a function of delay T_R and Doppler shift f_D. A plot of this (the *ambiguity diagram*) shows the resolution, sidelobe pattern, and presence of ambiguities in range and Doppler in an elegant and easily visualised manner.

The situation for bistatic radar is more complicated, however, because the ambiguity function depends not only on the waveform but also on the target location with respect to the transmitter and receiver. Indeed, at noted above, if the target is on or close to the bistatic baseline, the resolution in both range and Doppler is severely degraded.

This arrangement was analysed by Tsao *et al.* [19], who also explained the effect by pointing out that the simple linear relationships between target range and echo delay, and between target velocity and Doppler shift, that exist for a monostatic radar no longer hold for a bistatic radar. Hence the bistatic ambiguity function has to be expressed as a function of more variables than just time delay and Doppler shift. In [19], the ambiguity function for a bistatic radar was defined as

$$\begin{aligned} &\left| \psi(R_{RH}, R_{Ra}, V_H, V_a, \theta_R, L) \right|^2 \\ &= \left| \int_{-\infty}^{\infty} s_t(t - \tau_a(R_{Ra}, \theta_R, L))\; s_t^*(t + \tau_R(R_{RH}, \theta_R, L)) \right. \\ &\quad \left. \times \exp\left[j2\pi f_{DH}(R_{RH}, V_H, \theta_R, L) \; - \; 2\pi f_{Da}(R_{Ra}, V_a, \theta_R, L)\, t\right] dt \right|^2 \end{aligned} \tag{7.11}$$

where R_{RH} and R_{Ra} are the hypothesised and actual ranges (delays) from the receiver to the target, respectively, V_H and V_a are the hypothesised and actual target radial velocities with respect to the receiver, respectively, f_{DH} and f_{Da} are the hypothesised and actual Doppler frequencies, respectively, and θ_R and L are as defined in Figure 7.1.

7.3 Passive bistatic radar illuminators

The three most important factors in assessing the usefulness of PBR illuminators are (i) the power density at the target, (ii) coverage, and (iii) the waveform. In the following sections, we discuss each of these in turn.

7.3.1 *Power density incident upon a target*

The power density of a given illuminator at a target, under the assumption of free-space line-of-sight propagation, is $P_T G_T / 4\pi R_T^2$. In practice, this value may be modified by the pattern propagation factor, which takes into account effects such as tropospheric propagation, multipath and shadowing. From the standpoint of target detection in the PBR receiver, we wish this value to be as large as possible, though for a fixed amount of radiated power from a single transmitter, there is clearly a trade-off with coverage.

7.3.2 *Coverage*

Closely related to power density is the coverage, which is impacted by the number and disposition of available transmitters that may be on fixed or moving platforms. In the case of a moving platform, the revisit time also becomes a factor. The coverage of broadcast and communication transmitters is optimised according to the services they provide, with these transmitters frequently sited on hilltops or on tall buildings. The horizontal-plane coverage is often omnidirectional, though for some cell phone base stations it may be arranged in 120° sectors. The vertical-plane coverage is usually optimised so as to avoid wasting power above the horizontal, and thus in some cases the beams are tilted downwards. An assessment by Lockheed Martin of the coverage of the USA by FM radio transmitters showed that an aircraft at an altitude of 1,000 feet would be within line-of-sight of at least one transmitter over virtually the entire country, apart from a few very small and isolated regions in the Rocky Mountains.

Measured vertical-plane field strength patterns have been published for typical PBR transmitters such as VHF FM and digital video broadcast – terrestrial (DVB-T) [20,21]. These results can be replotted in a more meaningful form (Figure 7.4) to show the reduction in power density illuminating a target as a function of elevation angle. It can be seen that the VHF FM transmit antennas have relatively high sidelobes, but the array of the DVB-T transmitter provides greater control of the radiation pattern and hence lower sidelobes. Both, however, focus their coverage near the horizon (near sin(0°)) as one would expect.

The effect of this horizon-focused coverage on PBR performance is significant, as indicated by the radar equation from (7.7) in the alternative form

$$R_{\text{R-max}} = \sqrt{\frac{P_T G_T G_R \lambda^2 \sigma_B G_p}{(4\pi)^3 R_T^2 (S/N)_{\min} k T_0 B F Q_{\text{loss}}}}, \tag{7.12}$$

in which G_p accounts for the processing gain and Q_{loss} the losses. Inspection of (7.12) shows that for every 10 dB reduction in $P_T G_T$, the maximum detection range $R_{\text{R-max}}$ for a given target is reduced by a factor of 3.3. Thus, at the peaks of the elevation-plane lobes the effect is significant (Figure 7.5), while in the nulls between the lobes it is even more so [22].

Therefore, there is an attendant loss in PBR maximum detection range for aircraft at cruising altitudes around 30,000 feet. Hence, for the current disposition of transmitters there may be airspace that is not comprehensively covered. Addressing this

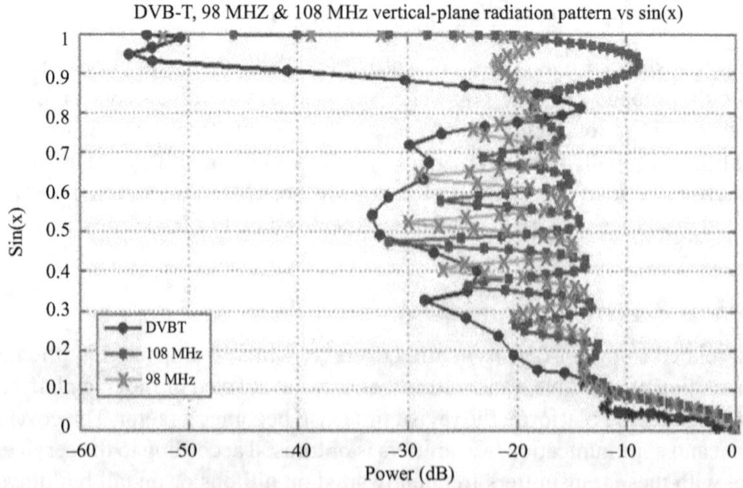

Figure 7.4 Measured vertical-plane radiation patterns of BBC VHF FM radio transmitters at 98 MHz, 108 MHz and an 8-bay DVB-T transmitter. The vertical scale is the sine of the elevation angle at the transmitter (i.e. sin(0°) is directed at the horizon and sin(90°) points straight up)

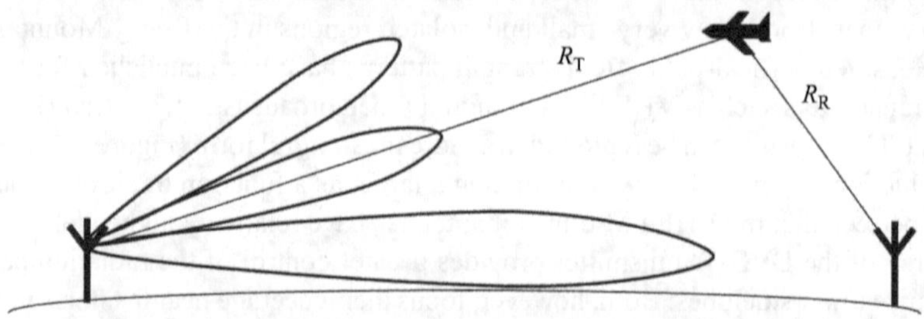

Figure 7.5 Reduced coverage in the elevation plane by illuminators of opportunity impairs PBR detection performance against air targets

deficiency would require a careful and detailed system assessment, which is likely to be high on the agenda of future work in this subject.

7.3.3 Waveforms

The waveforms produced by illuminators of opportunity are a major factor in determining the performance of the PBR system. As with any radar, the range resolution depends on the waveform bandwidth. As noted in the discussion regarding the bistatic ambiguity function in Section 7.2.4, the Doppler properties of a waveform are also rather important.

The waveforms used in early PBR experiments, namely FM radio and analogue TV, were far from ideal. The ambiguity function of an FM radio signal depends strongly on whether the modulating signal is speech or music, and even on the type of music. With speech, during pauses between words the bandwidth of the modulating signal is quite narrow, and hence the range resolution can be poor. With music that has a high spectral content, the bandwidth of the modulating signal is broad and hence the range resolution is rather better. For instance, a cacophony of rock music is likely to be better for PBR purposes than speech or music from a solo instrument.

The ambiguity function of an analogue TV signal also depends on the programme content, with the additional inclusion of pronounced range ambiguities associated with the 64 μs line repetition rate that correspond to multiples of 9.6 km in bistatic range. Thus, in general, PBR signals with analogue modulation have an ambiguity performance that is time-varying. Nevertheless, FM radio and analogue TV transmissions usually have high power, and so provided quite good target detection performance in the early experiments [7,9,11].

Digital modulation formats are now widely used, and are likely to be employed for any future broadcast or communication applications. From a PBR perspective they have the advantage of being more noise-like, and hence possess more favourable ambiguity function properties. Further, the ambiguity function does not depend on the nature of the programme content [23–27], thus providing greater consistency and predictability of PBR performance. It should therefore come as no surprise that digital modulations have become the waveform of choice for passive radar, especially at VHF and UHF frequencies where transmit powers are higher and coverage is wide.

These points are illustrated in Figure 7.6, which shows the ambiguity functions of a variety of PBR signal types that were captured in open air and digitised. Figure 7.6(a) shows the ambiguity function of a VHF FM radio station with speech modulation (BBC Radio 4). The peak and sidelobe structure are well defined, though the peak is relatively broad as a consequence of the low spectral content of the modulation. Figure 7.6(b) shows the equivalent result for a station with fast-tempo jazz music (Jazz FM). The peak and the sidelobe structure are correspondingly sharper because of the higher spectral content of the modulation. Figure 7.6(c)–(e) shows typical ambiguity functions for digital transmissions (DAB, DBV-TV, and GSM, respectively). These functions are more favourable for PBR purposes than signals with analogue modulation, such as in Figure 6(a) and (b), since the peak of the ambiguity function is narrower and the sidelobes are lower. Also, the general structures of these latter functions are time-invariant as they do not depend on the programme content.

The range of digitally modulated signals considered for PBR includes GSM, 3G and 4G cell phone signals, DAB, DVB-T and DRM broadcast signals, and Wi-Fi (802.11) and WiMAX (802.16). Several of these are based on orthogonal frequency-division multiplexing (OFDM), implemented as long-term evolution (LTE). The next sections therefore describe the basis of OFDM and LTE, and the potential utility of these signals as PBR sources.

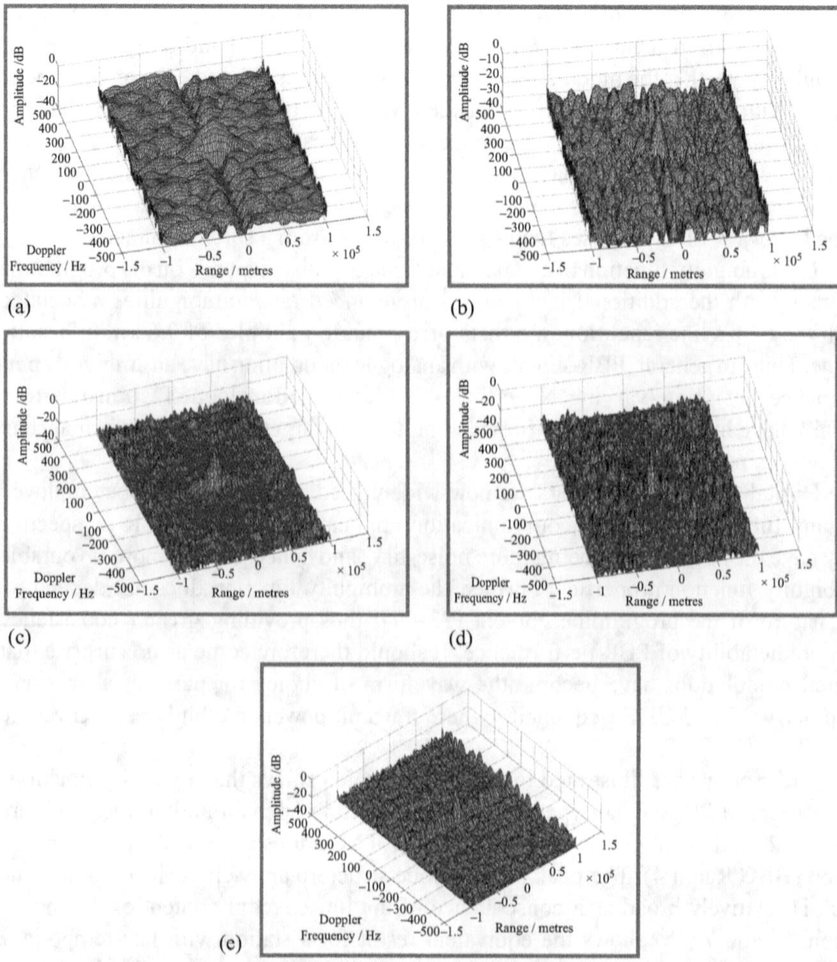

Figure 7.6 Measured ambiguity functions from (a) VHF FM BBC Radio 4 (speech), (b) VHF FM Jazz FM (fast tempo jazz music), (c) Digital Audio Broadcast (DAB) at 222.4 MHz, (d) Digital Video Broadcasting – Terrestrial (DVB-T) at 505 MHz and (e) GSM900 at 944.6 MHz. Reproduced, with permission, from [26]

7.3.4 Orthogonal frequency division multiplexing

In most cases, the communication or broadcast channel will consist of a direct path plus a number of multipath components, each with its own amplitude and delay. The direct path and multipath components may also be time-varying and Doppler-shifted. Multipath components may add destructively at the receiver, causing fading, and since the phase corresponding to a given delay is a function of frequency, the fading is likewise a function of frequency.

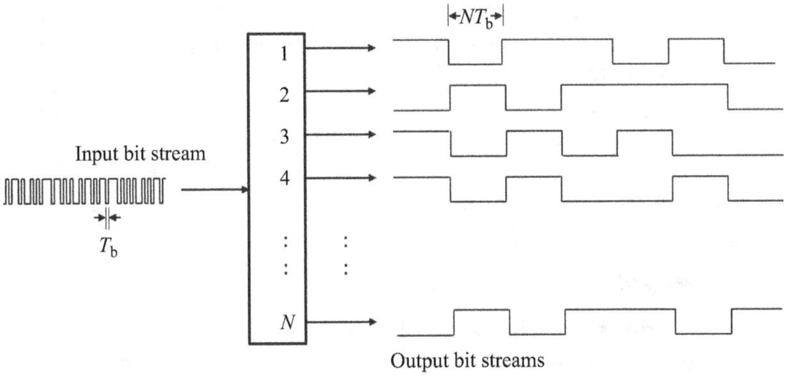

Figure 7.7 OFDM multiplexing of the digital serial bit stream into multiple parallel streams. The bit length of the output bit streams is now much greater than the maximum delay spread of the multipath

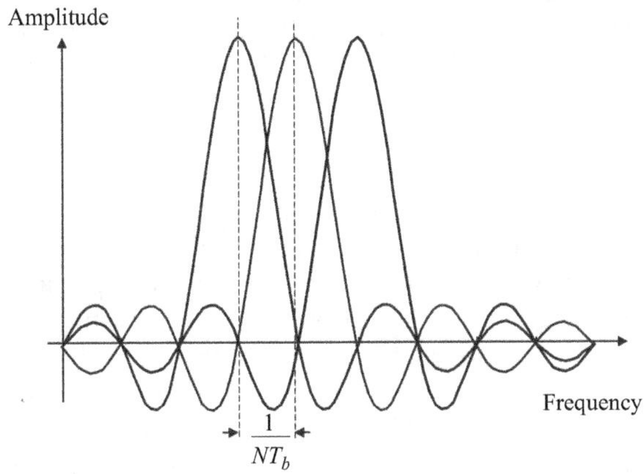

Figure 7.8 Orthogonal sub-carriers of an OFDM signal

If the delay spread associated with the multipath is comparable with the bit length, *intersymbol interference* occurs and the received signal is corrupted. This condition sets a limit on the data rate that can be transmitted through the channel.

To overcome this effect, the input bit stream (with bit length T_b) can be multiplexed into N parallel channels, each with bit length NT_b that is much greater than the maximum delay spread in the multipath (Figure 7.7). Consequently, the presence of multipath has relatively little effect. The parallel data streams are then modulated onto a set of sub-carriers spaced in frequency such that the nulls of the $(\sin x)/x$ modulated spectrum of one sub-carrier are coincident with the centre frequencies of all of the other sub-carriers (Figure 7.8). In other words, the signals are orthogonal to one another.

The sub-carrier spacing in frequency required to achieve this arrangement is $1/\tau$, where $\tau = NT_b$ is the bit length of the expanded bit stream. The overall signal, consisting of the modulated simultaneous sub-carriers, is then transmitted over the channel. In the receiver, each sub-carrier is individually demodulated, with the original data stream reconstituted by demultiplexing.

This form of modulation also allows the use of single-frequency networks, in which all of the transmitters for a given station share the same frequency and are synchronised together. OFDM has been shown to be an effective means of combatting the effects of multipath and forms the basis of many of the modulation formats used in modern communications and broadcast.

7.3.5 Long-term evolution

The LTE waveform is carrier-modulated digital data transmitted on OFDM sub-carriers. LTE offers data and voice services with downlink data rates up to 100 Mbit/s. The LTE base channel is defined according to channel bandwidths ranging from 1.4 to 20 MHz, divided across a number of OFDM sub-carriers ranging from 72 to 1,320. LTE may operate across several frequency bands; 32 bands are defined by the standardisation body 3GPP [28] with frequencies ranging from 729 MHz to 3.8 GHz. Operators are allocated specific spectrum bands depending on licensing and bandwidth requirements, with widths that are multiples of 5 MHz. For example, in the UK operators are allocated bands of 5, 10, 15, 20, 25 and 35 MHz [29]. LTE uses various modulation formats depending on the type of information transmitted and the quality of the wireless channel, with QPSK, 16-QAM and 64-QAM specifically defined in the standard [30].

The multiple access scheme of LTE is based on OFDM and is termed orthogonal frequency-division multiple access. Users are allocated bandwidth according to demand and traffic loading in a given cell. The basic unit of allocation in LTE is termed the LTE resource block, which is based on a 1 ms repeating sub-frame divided into two 0.5 ms slots. Each slot contains 12 sub-carriers with 15 kHz fixed spacing and either 6 or 7 OFDM symbols (depending on the length of the cyclic prefix used). A single sub-carrier and one OFDM symbol define what is termed a resource element, which is the smallest information unit of LTE [28,31]. Therefore, an LTE time domain signal is based on the aggregation of sub-frames into 10 ms frames.

Figure 7.9 shows the ambiguity function of the LTE signal, which is observed to approximate the ideal 'thumbtack' form with a sharp peak and a more or less uniform sidelobe level, though there are also periodic range sidelobes that are undesirable from a radar point of view. This result illustrates an important general point about digital modulation formats – the ambiguity functions are favourable, but the arrangement of the data into blocks, with synchronisation information in the form of preambles, cyclic prefixes, and pilots, introduces undesirable range sidelobes.

However, since the features of the signal that cause these unwanted sidelobe peaks are known *a priori*, it is possible to modify the reference signal within the PBR receiver to realise a mismatched filter that improves the cross-ambiguity function. This approach was first proposed by Saini and Cherniakov [32], and has been applied with some success for several of the waveforms considered in this section. A further issue is the use of so-called single frequency networks, as a means to combat multipath. Here, all of the transmitters for a given station share the same frequency and are synchronised together. Poullin [33] discusses the implications of this arrangement for

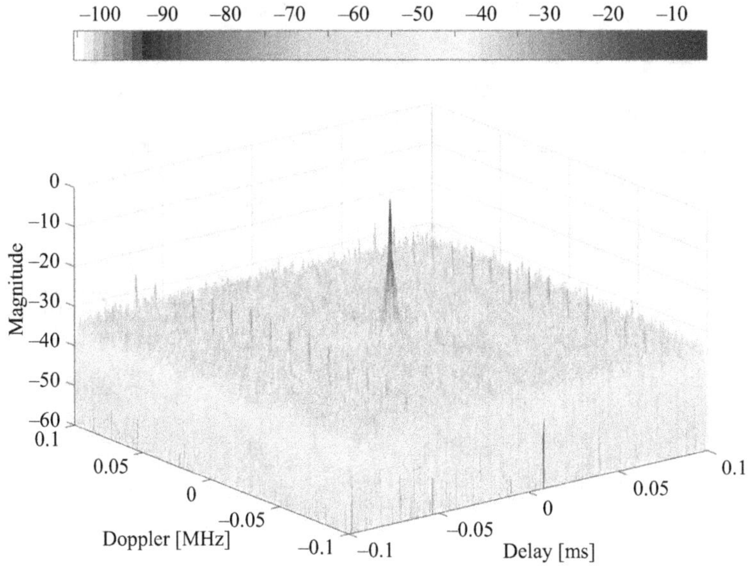

Figure 7.9 Ambiguity function of LTE signal, which approximates the 'thumbtack' ideal that possess a uniform, low sidelobe level. However, note there are also periodic range sidelobes which are undesirable from a radar perspective [34]

PBR purposes. Table 7.1 summarises the properties of a variety of PBR illuminator sources – including common broadcast and communication signals, as well as a space-based synthetic aperture radar (SAR) illuminator – and their associated type of modulation and power density at a target at a representative range.

7.4 Passive bistatic radar techniques

7.4.1 Direct signal suppression

Almost all of the signals that may be used for PBR have a 100% duty cycle – in other words, they radiate continuously, rather than in a pulsed manner like a conventional radar signal. This arrangement means that at the PBR receiver it is necessary to suppress the direct path signal so that it does not otherwise mask the desired weak target echoes [36]. At the same time, it is necessary to obtain a clean version of the direct signal at the PBR receiver to use in the detection processing. Consequently, most PBR approaches employ two separate receiver channels – one to capture the direct signal and one to collect the target echoes – and additionally employ various processing methods to suppress the direct signal and multipath.

This situation is complicated by the fact that the direct signal may be very strong, especially if the PBR receiver is in the line-of-sight of the transmitter. The dynamic range requirements on the PBR receiver may therefore be quite stringent, possibly 100 dB or more.

Table 7.1 Summary of typical parameters of PBR illuminators of opportunity

Transmission	Frequency	Modulation, bandwidth	$P_T G_T$	Power density(note 1) $\Phi = \frac{P_T G_T}{4\pi R_T^2}$
HF broadcast	10–30 MHz	DSB AM, 9 kHz	50 MW	−67 to −53 dBW/m^2 at $R_T = 1{,}000$ km
VHF FM	88–108 MHz	FM, 200 kHz	250 kW	−57 dBW/m^2 at $R_T = 100$ km
Analogue TV	~550 MHz	vestigial sideband AM, 5.5 MHz	1 MW	−51 dBW/m^2 at $R_T = 100$ km
DAB	~220 MHz	digital, OFDM 220 kHz	10 kW	−71 dBW/m^2 at $R_T = 100$ km
DVB-T	~750 MHz	digital, 6 MHz 220 kHz	8 kW	−72 dBW/m^2 at $R_T = 100$ km
Cell phone basestation (GSM)	900 MHz	GMSK, FDMA/TDMA/ FDD, 200 kHz	100 W	−71 dBW/m^2 at $R_T = 10$ km
Cell phone basestation (3G)	2 GHz	CDMA, 5 MHz	100 W	−71 dBW/m^2 at $R_T = 10$ km
Wi-Fi 802.11	2.4 GHz	DSSS/OFDM, 5 MHz	100 mW	−41 dBW/m^2 (note 2) at $R_T = 10$ m
WiMAX 802.16	2.4 GHz	QAM 1.25–20 MHz	20 W	−88 dBW/m^2 at $R_T = 10$
GNSS	L-band	CDMA, FDMA 1–10 MHz	200 W	−134 dBW/m^2 at Earth's surface
DBS TV	Ku-band 11–12 GHz	analogue and digital	300 kW	−107 dBW/m^2 at Earth's surface
Satellite SAR(note 3)	9.6 GHz	chirp pulse 400 MHz (max)	28 MW	−54 dBW/m^2 at Earth's surface

Notes: [1]assuming free space line-of-sight propagation
[2]would be subject to additional attenuation due to propagation through walls
[3]parameters from SARs carried by the COSMO-SkyMed series of satellites [35]

Reference [1] describes four techniques that may be used to suppress the direct signal and multipath. These are (i) physical shielding, (ii) Fourier processing, (iii) adaptive beamforming, and (iv) adaptive filtering. In practice, a combination of more than one of these is likely to be necessary.

Physical shielding has the benefit of being simple, but it is unlikely to achieve more than about 10 dB of suppression. It does, however, reduce the dynamic range requirement of subsequent receiver processing.

Adaptive beamforming can be used to steer a spatial null in the direction of the direct signal. This approach likewise reduces the dynamic range requirement of subsequent receiver processing.

A widely used adaptive filtering algorithm has been described by Colone *et al.* [37,38]. This particular technique estimates the delay and amplitude of the direct signal and then subtracts it with the appropriate weight from the receive signal path through the use of a transversal filter architecture. The process is repeated for the next-strongest multipath component, and so on successively to derive the weight coefficients of the filter. Palmer *et al.* [39] describe further work on this subject and present a performance comparison of different approaches.

7.4.2 Processing gain and performance prediction

When evaluating the performance of a practical PBR system, the radar equation in (7.7) should be modified to include the effect of processing gain, i.e. the integration gain. Since the PBR antenna is fixed, the processing gain is limited by the decorrelation time of the target, the maximum value of which is set by the maximum time T_{max} that the target echo remains coherent, which in turn depends on how much the target is manoeuvring and/or accelerating. A rule of thumb value for this time, obtained by setting the change in one-way path length due to the radial component of target acceleration A_R to be equal to half a wavelength as

$$\frac{A_R T_{max}^2}{2} = \frac{\lambda}{2}, \tag{7.13}$$

results in

$$T_{max} = \left(\frac{\lambda}{A_R}\right)^{1/2}. \tag{7.14}$$

Thus, the realisable processing gain is $G_p = BT_{max}$, for receiver effective bandwidth B. As an example, a VHF FM radio waveform of bandwidth 50 kHz and an integration time of 1 s, yields $G_p = 47$ dB, which is substantial. This result can be inserted into the radar equation of (7.12), along with appropriate values of Φ from Table 7.1, expected bistatic RCS σ_B, and a suitable value for receiver effective noise figure F (perhaps 25 dB) to account for limited suppression of the direct signal, multipath, and other signals. In so doing, contours of constant detection range, i.e. Ovals of Cassini [40], can be plotted for a given illuminator.

7.4.3 Target detection, localisation and tracking

The process of detection, localisation, and tracking [41] is slightly different to that of conventional radars because most PBR sources are omnidirectional in azimuth coverage. Further, at the VHF and UHF frequencies that are commonly used, the receive antenna beamwidths are quite broad, often on the order of 90°.

A PBR receiver may measure bistatic range, echo direction of arrival, and Doppler shift of the echo from a given target. The target range measurement from a single bistatic transmit-receive pair establishes an iso-range ellipse on which the target lies. The addition of a second transmitter provides range measurement information that defines a second iso-range ellipse that intersects the first at several locations. These intersections introduce ambiguities known as 'ghosts' that need to be resolved in the detection and tracking processing. Of course, this arrangement becomes more complicated when there are multiple targets present.

Various approaches to target tracking in PBR have been investigated and reported, including Kalman filters [11] and probability hypothesis density filters [42].

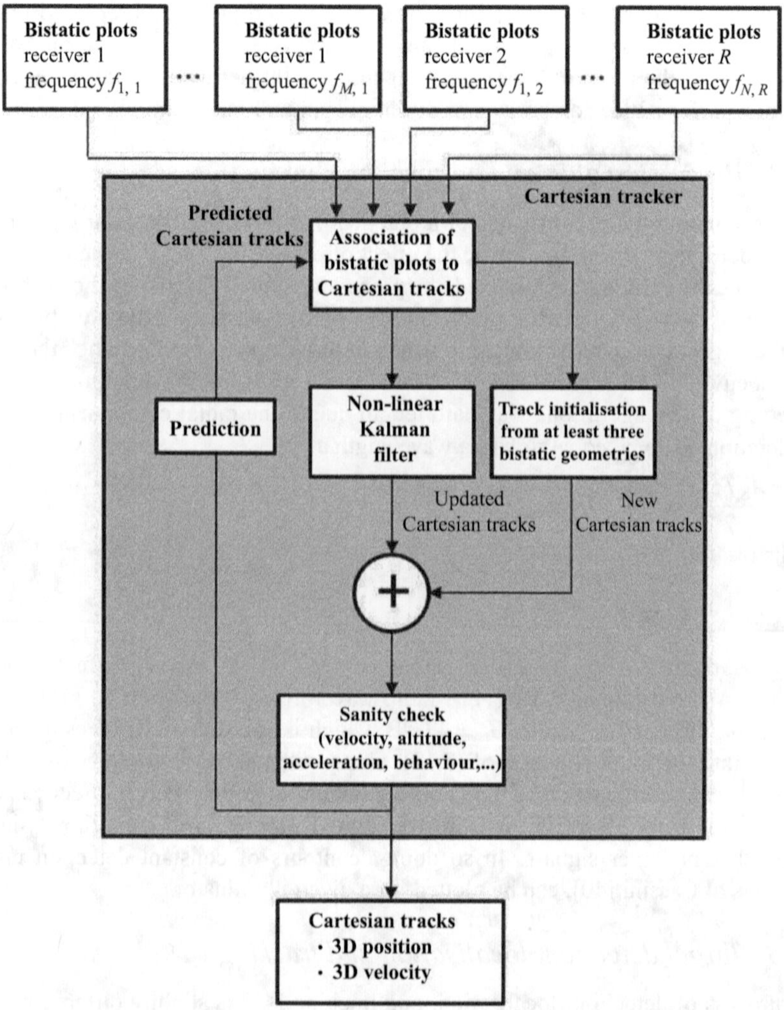

Figure 7.10 Tracking architecture used in the Thales HA-100 system. The tracker uses the plots from each bistatic transmit-receiver pair (adapted from [43])

Figure 7.10 shows the approach taken in the Thales' HA-100 system [43] that uses information from both FM radio and DVB-T illuminators, exploiting the relative advantages of both, and is typical of the new generation of PBR systems.

Figure 7.11 also shows the antenna of the PBR system developed by Airbus Defence and Space (formerly CASSIDIAN) [44,45]. Here it can be seen that the antenna comprises two eight-element circular arrays, both vertically polarised, covering the two bands used by this sensor.

Figure 7.11 *Antenna of the PBR system developed by the Airbus Defence and Space*

7.5 Passive bistatic radar and the spectrum problem

We now turn to the most important question: How can PBR techniques help address the spectrum congestion problem? Or, more specifically:

- To what extent might conventional radar applications be replaced by PBR approaches?
- If so, which illuminators, what system configurations, and what geometries are appropriate?
- What performance is achievable, and with what degree of reliability?

Since the modern form of PBR is still an emerging technology, we do not attempt to offer definitive answers. Nevertheless, the large number of research publications in journals and at major radar conferences on the subject, along with the results reported by several major aerospace companies, gives confidence that the subject will continue to grow and that it may have a significant part to play in addressing spectrum congestion.

7.5.1 PBR in air traffic management

Two examples where PBR may supplement or replace conventional radar operation arise in air traffic management (ATM). The first of these is aimed at developing countries where an advanced ATM radar infrastructure does not presently exist, but a network of FM broadcast radio with excellent coverage does. In South Africa, Inggs *et al.* demonstrated that a PBR system can be constructed very simply, at low cost, and achieve good performance. This result is described in detail in Chapter 8.

The second example is in the United Kingdom, where an advanced ATM radar network does exist, but where there is substantial pressure on the current spectrum allocation. Thales UK has been awarded a substantial contract worth £1.5 billion over 22 years to provide a safer, more efficient, and sustainable ATM service for the UK's armed forces and which would use both active and passive radar technology [46]. Another description of the work of Thales UK in this area [47] states that:

> The civil aviation industry is examining the potential for adopting this technology [passive bistatic radar] as a less expensive alternative to the conventional radars used for approach and landing. Here again, given the exacting safety and dependability requirements of the civil air transport sector, signals of opportunity are not deemed sufficiently reliable. However, studies are underway as part of the SESAR programme to evaluate the potential of passive radar systems using dedicated radio transmitters. Depending on the findings of these studies, the future role of passive radar in civil airspace surveillance should be clearer by about 2020.

In the United States, the Federal Aviation Authority (FAA) have an intent to vacate the upper part of L-band that is currently occupied by Air Route Surveillance Radars. The value of this part of the spectrum is very high as indicated in Chapter 2 and there are many suitors that covet this band. One answer is to migrate to S-band, but it is not so well suited to very long-range operation. Another is to use just the lower part of L-band, but this too is likely to come under pressure for re-allocation and may only be a short-term measure. A third possibility is to use passive radar, perhaps in combination with S-band systems. Options are being examined through an FAA sponsored programme but the best answers are, as yet, unclear.

Clearly, PBR plays a large part in the thinking aimed at future ATM systems, including the use of radio transmitters dedicated also to the radar function, and similar thinking and developments may be expected in other countries.

7.6 Summary and conclusions

This chapter has provided a brief introduction to the properties of PBR, and considered how PBR may be able to help address spectral congestion. The problem itself is already severe, and is likely to get worse. But there are some grounds for optimism, and PBR certainly has a part to play in providing solutions.

Some specific conclusions:

1. The spectral congestion problem is far better addressed collaboratively rather than by the various interested parties taking retrenched positions.

2. The change in maturity of PBR that has occurred in the second decade of the twenty-first century means that PBR can now be seriously considered for some applications that have previously only been served by conventional radar.
3. The widespread adoption of digital modulation schemes in communications and broadcasting provides favourable ambiguity functions for radar purposes; though, synchronisation information in these signals in the form of preambles, cyclic prefixes, and pilots, introduces undesirable range sidelobes. These effects can, to some extent, be suppressed by mismatched filtering in the PBR receiver.
4. The largest gains are likely to be not so much in devising waveforms that fulfill communication/broadcast functions as well as facilitating radar, but in optimising coverage of illuminator sources to include regions where radar coverage is desired. A good example of this is in elevation, where currently the coverage towards air targets may be poor, thereby yielding substantially degraded performance.

Acknowledgements

We are most grateful to the many people who have helped with the work described in this chapter, in particular, Shannon Blunt, Izzat Darwazeh, and Mike Inggs. We are also grateful to the organisations who have funded work on the subject over many years, including THALES UK, the UK Ministry of Defence, the Royal Academy of Engineering, the UK Engineering and Physical Sciences Research Council (EPSRC), the IET A.F. Harvey Research Prize and the Office of Naval Research (Global).

References

[1] Griffiths, H.D. and Baker, C.J., *An Introduction to Passive Radar*, Artech House, Norwood, MA, March 2017.

[2] Griffiths, H.D., ‘Early history of bistatic radar,’ *EuRAD Conference 2016*, London, 6/7 October 2016.

[3] Blanchard, Y., ‘A French pre-WWII attempt of Air Warning Radar: Pierre David’s “Electromagnetic Barriers”,’ *URSI Radio Science Bulletin*, No. 358, pp. 18–34, September 2016.

[4] *Air Scientific Intelligence Interim Report, Heidelberg*, A.D.I. (Science), IIE/79/22, 24 November 1944, Public Records Office, Kew, London (AIR 40/3036).

[5] Griffiths, H.D. and Willis, N.J., ‘Klein Heidelberg—the first modern bistatic radar system,’ *IEEE Transactions on Aerospace and Electronic Systems*, Vol. 46, No. 4, pp. 1571–1588, October 2010.

[6] Griffiths, H.D., ‘Klein Heidelberg: new information and further insight,’ *IET Radar, Sonar and Navigation*, Vol. 11, No. 6, pp. 903–908, June 2017.

[7] Griffiths, H.D. and Long, N.R.W., ‘Television-based bistatic radar,’ *IEE Proceedings*, Vol. 133, Pt. F, No. 7, pp. 649–657, December 1986.

[8] Ogrodnik, R.F., 'Bistatic laptop radar: an affordable, silent radar alternative,' *IEEE Radar Conference*, Ann Arbor, MI, pp. 369–373, 13–16 May 1996.

[9] Sahr, J.D. and Lind, F.D., 'The Manastash Ridge radar: a passive bistatic radar for upper atmospheric radio science,' *Radio Science*, Vol. 32, No. 6, pp. 2345–2358, 1997.

[10] Howland, P.E., 'Target tracking using television-based bistatic radar,' *IEE Proceedings of Radar, Sonar and Navigation*, Vol. 146, No. 3, pp. 166–174, June 1999.

[11] Howland, P.E., Maksimiuk, D., and Reitsma, G., 'FM radio based bistatic radar,' *IEE Proceedings of Radar, Sonar and Navigation*, Vol. 152, No. 3, pp. 107–115, June 2005.

[12] https://www.asdreports.com/news.asp?pr_id=1701 accessed 8 September 2016.

[13] Jackson, M.C., 'The geometry of bistatic radar systems,' *IEE Proceedings*, Vol. 133, Part F., No. 7, pp. 604–612, December 1986.

[14] Willis, N.J., *Bistatic Radar*, SciTech Publishing, Raleigh, NC, 2004.

[15] Ewell, G.W., Bistatic radar cross section measurements, Chapter 7 in *Techniques of Radar Reflectivity Measurement*, (N.C. Currie ed.), Artech House, Norwood, MA, 1989.

[16] Kell, R.E., 'On the derivation of bistatic RCS from monostatic measurements,' *Proceedings of the IEEE*, Vol. 53, pp. 983–988, 1965.

[17] Born, M. and Wolf, E., *Principles of Optics* (sixth edition), Pergamon Press, London, p. 559, 1980.

[18] Woodward, P.M., *Probability and Information Theory, with Applications to Radar*, Pergamon Press, 1953; reprinted by Artech House, Norwood, MA, 1980.

[19] Tsao, T., Slamani, P., Varshney, P., Weiner, D., Schwarzlander, H., and Borek, S., 'Ambiguity function for a bistatic radar,' *IEEE Transactions on Aerospace and Electronic Systems*, Vol. 33, No. 3, pp. 1041–1051, July 1997.

[20] Millard, G.H., *The Introduction of Mixed-Polarization for VHF Sound Broadcasting: the Wrotham Installation*, Research Department Engineering Division, British Broadcasting Corporation, BBC RD 1982/17, September 1982.

[21] O'Hagan, D.W., Kuschel, H. , Ummenhofer, M., Heckenbach, J., and Schell, J., 'A multi-frequency hybrid passive radar concept for medium range air surveillance,' *IEEE AES Magazine*, Vol. 27, No. 10, pp. 6–15, October 2012.

[22] O'Hagan, D.W., Griffiths, H.D., Ummenhofer, S.M., and Paine, S.T., 'Elevation pattern analysis of common passive radar illuminators of opportunity,' *IEEE Transactions on Aerospace and Electronic Systems*, Vol. 53, No. 6, pp. 3008–3019, December 2017.

[23] Ringer, M.A. and Frazer, G.J., 'Waveform analysis of transmissions of opportunity for passive radar,' *Proceedings of the ISSPA'99*, Brisbane, pp. 511–514, 22–25 August 1999.

[24] Griffiths, H.D., Baker, C.J., Ghaleb, H., Ramakrishnan, R., and Willman, E., 'Measurement and analysis of ambiguity functions of off-air signals for passive coherent location,' *Electronics Letters*, Vol. 39, No. 13, pp. 1005–1007, 26 June 2003.

[25] Griffiths, H.D. and Baker, C.J., 'Measurement and analysis of ambiguity functions of passive radar transmissions,' *IEEE International Radar Conference RADAR 2005*, Arlington, VA, IEEE Publ. No. 05CH37628, pp. 321–325, 9–12 May 2005.

[26] Baker, C.J., Griffiths, H.D., and Papoutsis, I., 'Passive coherent radar systems—Part II: waveform properties,' *IEE Proceedings of Radar, Sonar and Navigation*, Vol. 152, No. 3, pp. 160–168, June 2005.

[27] Thomas, J.M., Griffiths, H.D., and Baker, C.J., 'Ambiguity function analysis of digital radio Mondiale signals for HF passive bistatic radar,' *Electronics Letters*, Vol. 42, No. 25, pp. 1482–1483, 7 December 2006.

[28] *3GPP standard, LTE; Evolved Universal Terrestrial Radio Access (E-UTRA);* Base Station (BS) radio transmission and reception (3GPP TS 36.104 version 10.2.0 Release 10).

[29] http://media.ofcom.org.uk/news/2013/winners-of-the-4g-mobile-auction/ accessed 5 September 2016.

[30] LTE: Evolved Universal Terrestrial Radio Access (E-UTRA), 3GPP standard document, ETSI TS – B6.211 V10.0.0 (2011).

[31] Dahlman, E., Parkvall, S., and Sköld, J., *4G-LTE/LTE-Advanced for Mobile Broadband*, Elsevier, Cambridge, MA, 2013.

[32] Saini, R. and Chemiakov, M., 'DTV signal ambiguity function analysis for radar application,' *IEE Proceedings of Radar, Sonar and Navigation*, Vol. 152, No. 3, pp. 133–142, 2005.

[33] Poullin, D., 'Passive detection using digital broadcasters (DAB, DVB) with COFDM modulation,' *IET Proceedings of Radar, Sonar and Navigation*, Vol. 152, No. 3, pp. 143–152, June 2005.

[34] Griffiths, H.D., Darwazeh, I., and Inggs, M.R., 'Waveform design for commensal radar,' *IEEE International Conference RADAR 2015*, Arlington, VA, pp. 1456–1460, 11–14 May 2015.

[35] Torre, A. and Capece, P., 'COSMO-SkyMed: the advanced SAR instrument,' *5th International Conference on Recent Advances in Space Technologies (RAST)*, Istanbul, 9–11 June 2011.

[36] Griffiths, H.D. and Baker, C.J., 'Direct signal suppression,' Chapter 4 in *An Introduction to Passive Radar*, Artech House, Norwood, MA, March 2017.

[37] Colone, F., Cardinali, R., and Lombardo, P., Cancellation of clutter and multipath in passive radar using a sequential approach, *IEEE Radar Conference*, Verona (NY), USA, 393–399, 24–27 April 2006.

[38] Colone, F., O'Hagan, D.W., Lombardo, P., and Baker, C.J., 'A multistage processing algorithm for disturbance removal and target detection in passive bistatic radar,' *IEEE Transactions on Aerospace and Electronic Systems*, Vol. 45, No. 2, pp. 698–722, April 2009.

[39] Palmer, J.E. and Searle, S.J., 'Evaluation of adaptive filter algorithms for clutter cancellation in passive bistatic radar,' *2012 IEEE Radar Conference*, Atlanta, GA, pp. 493–498, 7–11 May 2012.

[40] Griffiths, H.D. and Baker, C.J., 'Passive coherent radar systems—Part I: performance prediction,' *IEE Proceedings of Radar, Sonar and Navigation*, Vol. 152, No. 3, pp. 153–159, June 2005.

[41] Griffiths, H.D. and Baker, C.J., 'Passive radar performance prediction,' Chapter 5 in *An Introduction to Passive Radar*, Artech House, Norwood, MA, March 2017.

[42] Tobias, M. and Lanterman, A.D., 'Probability hypothesis density-based multi-target tracking with bistatic range and Doppler observations,' *IEE Proceedings of Radar, Sonar and Navigation*, Vol. 152, No. 3, pp. 195–205, June 2005.

[43] Klein, M. and Millet, N., 'Multireceiver passive radar tracking,' *IEEE AES Magazine*, Vol. 27, No. 10, pp. 26–36, October 2012.

[44] Edrich, M. and Schroeder, A., 'Multiband multistatic passive radar system for airspace surveillance: a step towards mature PCL implementations,' *Proceedings of the International Radar Conference RADAR 2013*, Adelaide, Australia, pp. 218–223, 10–12 September 2013.

[45] Edrich, M., Schroeder, A. and Meyer, F., 'Design and performance evaluation of a mature FM/DAB/DVB-T multi-illuminator passive radar system,' *IET Radar, Sonar and Navigation*, Vol. 8, No. 2, pp. 114–122, February 2014.

[46] https://www.thalesgroup.com/en/countries/europe/united-kingdom/aerospace/air-traffic-management accessed 31 December 2017.

[47] https://www.thalesgroup.com/en/worldwide/aerospace/case-study/passive-radar-activists accessed 31 December 2017.

Chapter 8

Symbiosis for communications, broadcasting and sensor systems in the white space TV band

Michael R. Inggs[1]

8.1 Introduction

This book, as a whole, highlights the congestion of the lower microwave bands, the competition for spectrum allocation, and fiscally enforced regulation being imposed by governments in the form of spectrum licensing. Due to this congestion, systems that provide a combination of functions (communications and sensing, for example) will find favour from legislators and can share the spectrum rental costs. In this chapter, we discuss the potential for systems that are the combination of a telecommunications/broadcast application and a sensing function. Specifically, the white space wireless regional area network (WRAN) that is carefully governed by specification IEEE802.22 [1] offers enormous potential. Besides the use for communications, the detection and tracking of moving targets (air, land, and sea vehicles, humans, and animals) is also possible, as well as geoscience monitoring (e.g. soil moisture, biomass, rainfall rate).

In the next section, we discuss how the rapid replacement of analogue TV systems with spectrally efficient digital broadcast TV (DBTV) has suddenly released for replanning a large percentage of the 50–800 MHz bands, otherwise known as 'white spaces' (WS). Fortunately, a negotiation concerning present and future use of this valuable spectrum has emerged, centred on the WRAN specification.

Coexistence between sensor systems has also been discussed in the recent literature. A key element is the detection of other sensor emissions combined with techniques to intersperse the new sensor's waveforms among the other detected waveforms to alleviate mutual performance degradation (see Chapter 15 and [2]). We propose in Section 8.7.2 that the sensor community can learn from the 802.22 standardisation in this regard to facilitate sensor coordination across all bands, especially those allocated for such use (e.g. the L, X, and Ku bands).

Section 8.2 discusses how the WS concept came about and its frequencies of operation. In this arrangement, users collaborate using a combination of databases listing licenced operators and a carefully constructed intelligent radio system that senses

[1]Department of Electrical Engineering, University of Cape Town, South Africa

band occupancy. Unlicenced users are restricted to short time occupancy of channels that are measured to be clear. It should also be noted that the frequencies involved (50–800 MHz) have excellent propagation through materials, including foliage, which is useful for a variety of sensing modes.

The terminology used by both the communications and sensor communities for different types of collaboration is highly fragmented. To that end, in this chapter, we follow de Bary's [3] biological definitions of *symbiosis* to clarify a somewhat arbitrary nomenclature used in the present radar literature [4,5]. This new taxonomy for electromagnetic (EM) sensors is explained in Section 8.3.2, with Section 8.3 as a whole introducing a new biologically-inspired terminology that provides an accurate description of this manner of collaboration. The biological hierarchy of symbiosis includes:

- *commensalism*, where the systems use one another's resources without impacting either;
- *mutualism*, where systems are totally dependent on one another, and even co-design is implied;
- *parasitism*, which does not seem to describe a meaningful relationship between sensors and other spectrum users.

This taxonomy is yet another example of where engineering has learned from the study of biological systems, e.g. bio-inspired systems and neural networks. We speculate, further, that all EM sensors may eventually be forced to evolve into a symbiosis, either with other sensors or other users. This section then continues with a discussion of sensor networks, which fit nicely into this symbiotic relationship and can include communications/radar sensor collaboration. Important concepts in this context are how target parameters can be extracted, with some examples provided.

Wilson-Langman *et al.* [6] proposed that WS spectrum might well be utilised in a wider context to include sensing, and we go on to discuss more specifically how sensors and communications can collaborate in the WS frequency band. It can be argued that a symbiotic relationship is needed to enable coexistence between dissimilar users, with many different possible modes of collaboration being possible.

Based on the WS standard, a large number of applications for this new, symbiotic sensor system are then examined. The performance of these systems, based on the relatively low power requirement in the WS standard, relies heavily on the network configuration, i.e. many low power sensors distributed spatially and in frequency. This requirement also places this arrangement well into the growing collection of sensor network concepts. As mentioned above, the collaboration defined in the IEEE 802.22 standard could be used much more widely by the sensor community within their own licenced bands, thereby increasing efficiency.

This chapter is not about providing complete mathematical analysis of some systems but rather to show a new generation of sensor engineers that there are an enormous number of exciting possibilities, provided we are prepared to break from the somewhat rigid suite of sensors of the past. These ideas are summarised at the end of the chapter.

8.2 The white space standard and its evolution

The WS standard is an outstanding example of pragmatism by pivoting to a new paradigm for shared communications in a relatively short period of time. EM sensor developers can learn from this example and should take note of how the WS standard came about to facilitate a revolution in their own field. We first look at the historical aspects, and then on to the technical components.

The scope of this chapter is, by necessity, somewhat limited due to the sheer complexity of the array of standards (IEEE) that have arisen to govern WS evolution. There are a large number of articles addressing technical aspects of the standard (e.g. see [1,7,8]), with the key units of a WS communication system discussed in Section 8.2.2. We begin with a very light, historical overview of this evolution that will hopefully be subsequently repeated in the development of future sensor systems.

8.2.1 A historical perspective

Broadcast systems have a large influence on daily life, and by extension, commercial and governmental interests. To this day, television remains one of the most important news and entertainment mediums. However, because of its long history and the requirement of maintaining backwards compatibility, TV was tied to old, analogue standards that were difficult to change due to the necessary investment in new transmitter and consumer receivers. As a result of their early implementation prior to World War II, TV and radio sprawled across the bands from 50 to 800 MHz, interspersed with audio broadcast and two-way radio. In this chapter, we refer to 'WS frequencies' as an imperfect summary of these bands. Interestingly, it is worth noting that even the modern form of DBTV is moving towards obsolescence due to the growth in streaming via fibre and similar technology. This shift will subsequently free up more valuable spectrum.

It should be noted that radar systems have been just as guilty of 'sprawl' due to their early implementation, and this book hopefully starts us on the process of modernisation of thinking concerning sensor systems in general as well. In fact, the question should be asked whether the WS movement is just the start of a much more cooperative relationship between myriad different spectrum users.

Once introduced, the broadcast technology for digital satellite TV was rapidly adopted since both the transmitter and consumer-end receiver were new installations, thus avoiding the requirement of backwards compatibility aside from the display. The digital format offered clear advantages in terms of efficient spectrum use that translate into the provision of a large number of channels. As a result, this new format put pressure on the existing, sprawling analogue systems. A plan evolved to close down the analogue TV channels and replace them with a smaller number of broadband digital TV (DTV) broadcasts. Industry and government worked together to make this transition in a careful manner.

The changeover was viewed as an evolutionary process, and thus it was understood that new standards for governing the use of the band should take into account the existing legacy users while providing a far more flexible framework for new, possibly

Table 8.1 Attenuation as a function of frequency through various construction materials illustrating the advantage of the WS frequency bands (i.e. below 1 GHz) [10]

Material	**0.57 GHz (dB)**	**1 GHz (dB)**	**2 GHz (dB)**	**5.7 GHz (dB)**	**0.57 to 5.7 (ΔdB)**
Brick 89 mm	1.5	3.5	5.4	15	13.5
Brick 267 mm	4.8	7	10.5	38	33.2
Composite brick 90 mm/ Concrete wall 102 mm	12	14	18	42	30
Composite brick 90 mm/ Concrete wall 203 mm	21.5	25	33	71.5	50
Masonry 203 mm	9.5	11.5	11	12.75	3.25
Masonry 610 mm	26.5	27.5	30	46.5	20
Glass 6 mm	0.4	0.8	1.4	1.1	0.7
Glass 19 mm	2.5	3.1	3.9	0.4	2.1
Plywood (dry) 6 mm	0.15	0.49	0.9	0.1	0.05
Plywood (dry) 32 mm	0.85	1.4	2	0.9	0.05
Reinforced concrete 203 mm 1% steel	23.5	27.5	31	56.5	33
Reinforced concrete 203 mm/ 2% steel	27.5	30	36.5	60	32.5

unlicenced, users. These negotiations took place entirely in the context of new data network users, as well as some existing broadcast systems.

Governments are now focused on a future in which spectrum can be bought or leased. The income from these ‘incentive auctions’ [9] could be used to motivate inefficient spectrum users to develop technology that enables sharing with others, especially in the unlicenced mode. For example, many governments agree that the internet should be made broadly available to rural inhabitants, thereby providing greater education and commerce opportunities to a segment of society that is otherwise often excluded. The WS frequencies are attractive for this purpose.

The EM sensor community is well behind in this type of collaborative thinking. However, the success of the WS transition may increase the governmental and commercial pressure on the sensor community to give up sole ownership of some bands and collaborate, such as has already occurred in the 5 GHz band (see Chapter 4).

8.2.2 White space overview

The WS frequencies are ideal for propagation in precipitation and provide significant penetration (i.e. less attenuation) through foliage and buildings (see Table 8.1). These attributes indicate that the WS bands would be excellent for various sensor applications such as vehicle and aircraft tracking, detection of people/animal motion, and surface penetration for soil moisture and vegetation change monitoring. Based on a recent

Table 8.2 Useful standards governing the operation of systems in the WS (and more generally in the area of wireless communications)

IEEE standard	Applicability/Domain
IEEE802.1	The IEEE 802.1 working group (WG) is chartered to develop standards and recommend practices in the following areas: 802 LAN/MAN architectures; inter-networking among 802 LANs, MANs and other wide area networks; 802 Security; 802 overall network management; and protocol layers above the MAC and LLC layers. The 802.1 WG has four active task groups: time-sensitive networking; security; data centre bridging; and OmniRAN
IEEE802.2	The 802.2 WG developed the logical link control (LLC). This WG is now disbanded, and readers should refer to ISO/IEC 8802-2, which is 'stabilised' (active but not updated) instead
IEEE802.11af	The WG for WLAN standards
IEEE802.19	The wireless coexistence WG that develops standards for coexistence between unlicenced devices
IEEE802.22	The WG for WRANs, enabling broadband wireless access using cognitive radio technology and spectrum sharing in WS bands. This group was the recipient of the IEEE Standards Emerging Technology Award

See http://www.ieee802/x, where x is the group and appropriate standard.

patent [6], the possible inclusion of sensor networks into the WS standard is the primary subject of this chapter.

The teams working on the proposed WS transition realised that an accommodation had to be made to allow existing, licenced systems to continue operation until ultimate technical obsolescence, while at the same time providing a host of other users with varying bandwidth and time of occupancy. This process requires careful management and, to make things even more difficult, is different from place to place.

The solution came in the form of 'cognitive radio', i.e. radios that intelligently address spectral congestion by sensing available channels for operation. The application that emerged from these considerations is captured in the IEEE 802.22 specification for a white space WRAN [1]. The discussions and agreements on this topic have produced a useful set of standards, some in third and even fourth levels of revision, as shown in Table 8.2. An early paper by Peha [2] shows enormous insight as to how 'cognitive' radio technology would ease the task of cooperation between users. Comprehensive discussions on the operation of the WRAN, on topics ranging from EM radiation standards to the physical layer for data transfer, can be found in [8]. Unsurprisingly, the provision of wide area internet access to rural and peri-urban regions has arisen as a prime use.

Users of the WS bands must be aware of geographically distributed, licenced users and avoid their frequencies. In addition, WS users must be able to sense other ad hoc users and seek available frequencies within which to operate. This arrangement clearly has a strong resonance with radar/communication spectrum sharing.

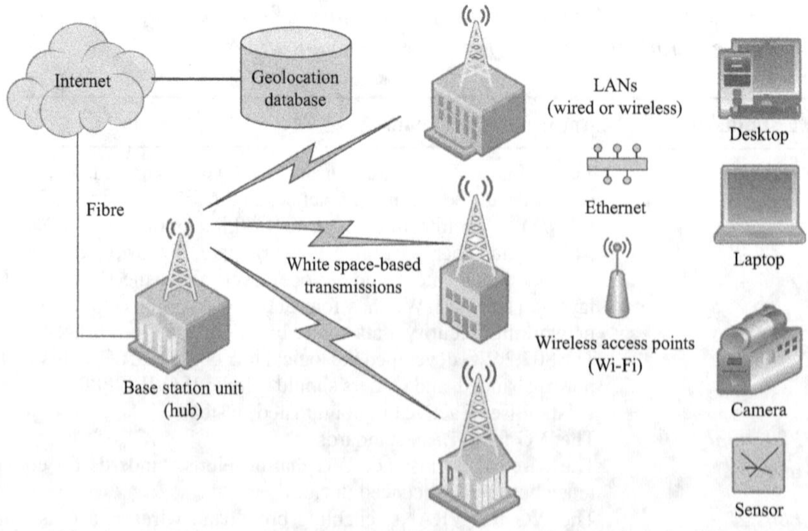

Figure 8.1 Cartoon view of a WS TV band network (adapted from [11]). Base stations (BSs) and customer furnished equipment (CFE) are connected wirelessly. The BSs also have a link to the geolocation database via the internet, which is critical for self-management of the WRAN. In addition, unlicenced mobile users can be connected to CFEs using various networking technologies

Figure 8.1 is a cartoon view of a WRAN and illustrates important components such as base stations (BSs) and customer furnished equipment (CFE). The BS has a direct internet link to a geolocation database (GDB). The GDB has information about licenced users (e.g. broadcasters), their positions and frequency of operation, and bandwidth. CFEs connect to local infrastructure. In addition, a provision is made for unlicenced, low bandwidth mobile users. The BS, CFEs, and mobiles must observe strict control of frequency and band occupancy time using cognitive radio technology and various spectrum-sensing techniques.

The topology is important if WS infrastructure is to be utilised as a sensor as well. Generally, the WRAN BS and CFE are positioned for lowest path loss. As discussed later, this arrangement is not ideal for a sensor since the direct signal may overload sensitive receivers, thereby reducing dynamic range.

8.2.3 Key WRAN radiator parameters

Before assessing the use of the WRAN as a sensor, it is important to first consider some of the pertinent operating parameters, namely the frequency range, transmit effective isotropic radiated power (EIRP), bandwidth, and typical antenna height. These parameters are listed in Table 8.3, though it is important to note that these values may vary by region.

Table 8.3 Important RF characteristics of the WRAN [8], which may vary by region

Parameter	Value and units	Remarks
Frequency range	54–862 MHz	Depending on region
Transmit EIRP	4 W	Maximum for BSs
	4 W	Maximum for CFEs
	20 W	Allowed for BSs in some regions
Bandwidth	6, 7, or 8 MHz	Depending on region
Antenna height	30 m	Above local terrain

In Section 8.3, it is shown that there are number of ways in which the nodes of the WRAN can also be utilised as a sensor network, varying from opportunistic use of the WRAN transmitters (commensal use) to full collaboration, i.e. choice of modulation to suit both network and sensor needs (mutualistic). We define a biologically inspired taxonomy to make this clear in Section 8.3.2.

8.3 Networks of sensors and a taxonomy

Consider the prospective migration from a single, monolithic sensor such as a radar to the concept of a network of sensors using EM waves to detect, classify, and identify objects within a volume of interest. How this objective can be achieved is, of course, dependent upon the architecture, operating frequency band, bandwidth, and measurement method. Based on use of the WRAN as a symbiotic partner, it is clear that a sensing system operating in this context would possess many low-power transmitting nodes. Using the taxonomy introduced in this section, we refer to this arrangement as a commensal system, though it could become mutualistic if the communication and sensing functions are co-designed with the WRAN architecture.

8.3.1 Early networks of sensors: radar

The term 'radar', shorthand for radio detection and ranging, evolved in the United States during the Second World War and represents a class of sensor that emits an EM wave to measure the time delay of the resulting echoes, thereby inferring distance to the objects from which the EM wave is reflected. In Britain, it was initially referred to as radio direction finding, but the U.S. term was eventually adopted. Rapidly, the sensor evolved to also measure Doppler (radial velocity) and, using directional antenna technology, azimuth and elevation angles as well [12].

Much earlier than this wartime surge in technology, Christian Hülsmeyer in Germany patented a sensor (the Telemobiloscope [13]) modelled on Marconi's communication technology that could determine the presence of a reflecting object on the water for the purpose of avoiding marine collisions. This device, however, did not actually have the capability of range measurement.

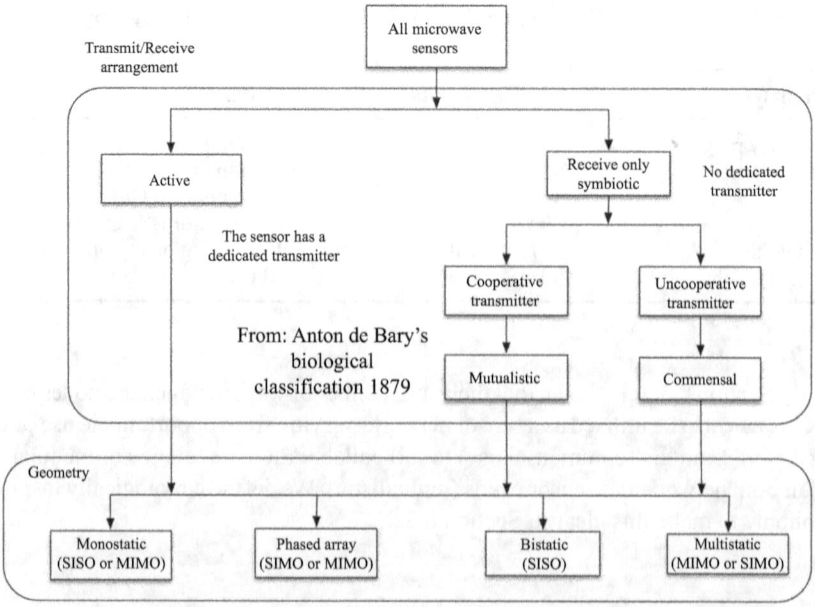

Figure 8.2 Proposed taxonomy for a network of sensors involving collaboration with other users of the EM spectrum. This taxonomy absorbs what is currently referred to as 'radar' as a special case of a sensor that provides its own illumination. Note the important distinction of a commensal relationship and a mutualistic. In the former case, the sensor may simply use a transmitter of opportunity. In the latter case, there is a more intimate sharing, including co-design of the communication and radar waveforms

8.3.2 A proposed taxonomy for networks of sensors

The term 'sensor' is used here to collectively represent the means with which information is extracted from the environment via the use of EM waves. Specifically, we consider the sensing function in *symbiosis* with other uses of the same EM waves. The term symbiosis, in particular, is adopted to leverage the classification taxonomy from the field of biology, where it has been observed that certain organisms (plants and animals) have evolved a state of coexistence, and in some cases it is even mutually beneficial. As discussed above, Anton de Bary's [3] symbiosis definition of the interaction between organisms contains the three subdivisions of commensalism, mutualism, and parasitism. Now consider how these concepts could be applied within the context of EM spectrum sharing (see Figure 8.2).

8.3.2.1 Commensal systems

Commensalism between organisms implies that neither system degrades the other. In a sensor/communication symbiosis, though, the sensor might not exist if it were not for the commensal partner providing EM illumination. The simplest example is the case of a sensor system that utilises an FM radio broadcast to track aircraft (Inggs *et al.* [14] describe such systems and how they are implemented in some detail). This topic has been, and continues to be, heavily investigated by many different groups. Given that existing broadcasters such as analogue TV, digital video broadcast (DVB), and digital audio broadcast (DAB) will continue providing high power illumination signals, there will likewise continue to be excellent commensal sensor opportunities.

In the current literature, a large number of terms are used for this type of system. For example, passive coherent location (PCL) [15] was invented for the Lockheed Martin *Silent Sentry* system. In general for this manner of sensing, signals produced by transmitters of opportunity are correlated with signals from the volume of interest, which contain delayed and Doppler-shifted target echoes. Although this naming is accurate in terms of the description of the signal model, it does not describe the relationship between the sensor and other systems (in this case, broadcast transmitters).

Other popular terms are 'passive radar' [5], 'passive bistatic radar', and 'passive multistatic radar' [4, Chapter 1]. However, it can be argued that the sensor is not 'passive', and the exact relationship between transmitter(s) (of opportunity) and receiver(s) is undefined. We believe the biological taxonomy solves many of these issues.

8.3.2.2 Mutualistic systems

Mutualism means that two organisms collaborate to mutual benefit. In the spectrum sharing context, we can consider the co-design of communication and sensing functions, where compromises are necessary to ensure efficient operation of both. The WS sensor network symbiosis that is the focus of this chapter fits into this mutualistic classification and is described in the patent [6].

Specifically, the communication system, in this case a WRAN, is modified to pass on sensor signals to a separate processing chain for correlation processing. Due to this modification at the design level of the WRAN, it is clearly a mutualistic usage.

8.3.2.3 Parasitic systems

In nature, parasitism leads to a degradation of one organism due to the parasite harvesting resources, often leading to the death of the host. There does not seem to be an analogue of this arrangement in the EM sensor network context.

8.3.3 Parameter extraction with sensor networks

The objective of a sensor can be changed, and may be as simple as detecting 'presence' as in the case of the telemobiloscope [13]. Our view is that the objective of a sensor (or network of sensors) is first to estimate the position of an object of interest, and then possibly move on to classification and perhaps even identification. The standard view of this process is shown in Figure 8.3. An early paper by Llinas and Hall [16] provides an excellent review of sensor networks.

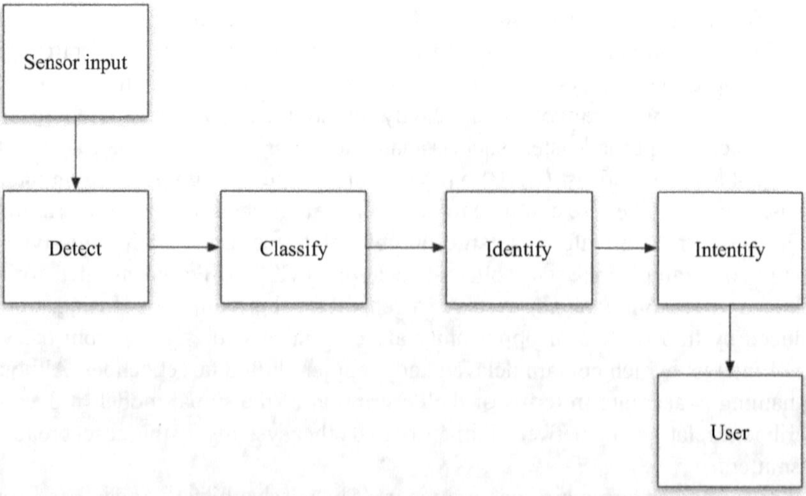

Figure 8.3 Sensor processing flow: detect, classify, identify, and 'intentify'. The latter term has been contrived to signify processing used to determine the intent of the target, i.e. maybe a potential attacker. Many sensors cannot go beyond the first step but can provide a probability of the target being real

Sometimes overlooked in discussions is that the objective of a sensor (network) is not the parameters measured internally, but the meaningful information that can be gleaned. In this process, we extract the position of the object in one or two dimensions. Using a number of snapshots, we might be able to track the object, which subsequently informs the processes of classification, identification, and even determining the intent of the object. The latter process can be referred to as 'intentify'. It should be apparent that the configuration of the sensor has a large bearing on the parameters that it might extract, and that a spatially distributed set of sensors can add enormously to the information that can be extracted.

Most are familiar with how a monostatic radar uses time delay to estimate range from the transceiver to a target and back, and how Doppler frequency is used to estimate relative radial velocity. Using a narrow beam antenna, and based on the assumption that the target is within the beam, the azimuth and elevation angles can likewise be estimated.

In the case of multistatic EM sensors, though, the techniques available for parameter estimation increase considerably, and it is actually beyond the scope of a single chapter to describe them all. A few techniques are explained in this section that are relevant to a WS sensor network. The designer of a WS-based sensor system will also have to determine whether the communication and sensor components will be mutualistic or commensal. A detailed example is provided later in the chapter (Sections 8.4 and 8.5).

Note that the relatively low operating frequency of a WS system implies larger antenna structures and also some difficulty in forming narrow beams to achieve good angular resolution, which is typically important for sensing applications. However, we show that the spatial diversity inherent in the IEEE 802.22 standard combined with intercommunication between sensors allows for spatial/angular resolution to be achieved in other ways.

8.3.3.1 Monostatic sensing

There are copious introductory texts and research papers discussing the performance of what are known as monostatic radar systems [12,17]. Simple expressions for the signal-to-noise ratio (SNR) under different operating conditions are available, and from these, we can readily derive the probability of detection p_d for an arbitrary probability of false alarm p_fa.

Within this new WS sensor concept, we could consider a mutualistic monostatic sensor in which the WS BS (see Figure 8.1) transmits pulsed blocks of data that have a frequency spectrum that is sufficient for the range resolution requirement of the proposed system. The pulse must be carefully timed, but could still carry data in the modulation of the pulse. Depending on the type of radar sensor to be implemented (pulse, pulse-Doppler, CW, FMCW [12]), different expressions for p_d and p_fa are obtained. The monostatic sensor is able to measure range according to the resolution dictated by its bandwidth. Note the requirement for precise time and carrier frequency knowledge to determine position and Doppler, which is discussed in further detail in Section 8.6.

To measure angle requires a narrow beam antenna, which of course is counter to the needs of a WS communication system to achieve wide area coverage. It would therefore be necessary to provide a narrow beam receiver antenna that is steerable over the search space. Monostatic sensors can also use combinations of sub-beams to form sum and difference patterns that, when combined, provided greater angle estimation accuracy [18].

If the communication stream is continuous, then a commensal relationship would be appropriate. Segments of the continuous transmit waveform would be used as sequential matched filters at the receiver to provide range estimates of targets in the beam. The continuous transmit signal could limit the achievable dynamic range of the sensor receiver, but since it would have a separate receive chain, some cancellation techniques could be deployed [19].

8.3.3.2 Bistatic sensing

We use the geometry proposed by Jackson [20] (see Figure 8.4) to define a transmitter illuminating a target, with the echo from the target received not only by a possible monostatic receiver but also by a bistatic receiver placed along the horizontal axis. Some simple analysis [4] shows that if targets move anywhere on a plane, the corresponding range and Doppler shifts are mapped onto ellipses. The Doppler ellipses are zero along the line joining transmitter and receiver.

The three-dimensional situation is clearly much more complex and is not discussed further here, though the important concept of what a bistatic radar measures can be inferred from Figure 8.4. In the two-dimensional case, the iso-range ellipses are symmetric around the baseline, with a similar set of ellipses for bistatic Doppler

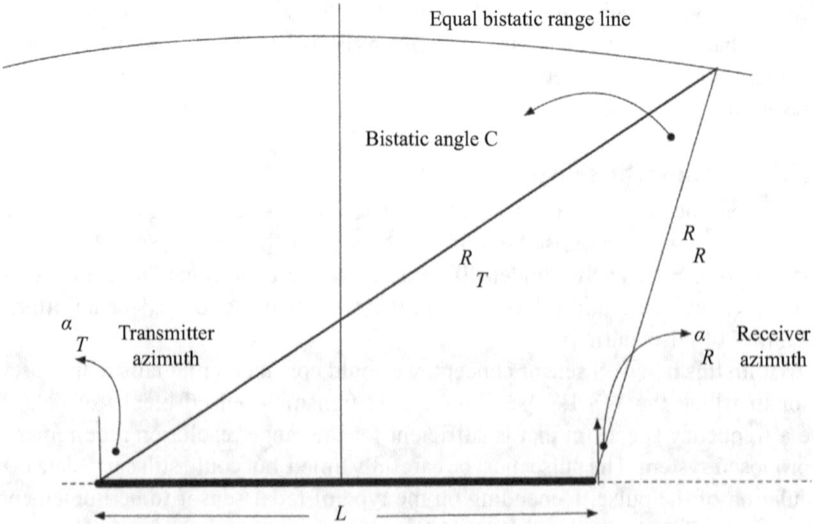

Figure 8.4 The bistatic geometry defined by Jackson [20]. The bistatic illumination angle is $\beta = \alpha_T - \alpha_R$. The baseline is shown by distance L, which is bisected by a line pointing towards true North that is also the line of zero Doppler shift. An iso-range ellipse is also shown. The ranges R_T and R_R are from the transmitter to the target and from target to the receiver, respectively

shift at right angles to this arrangement. A target will thus appear ambiguously on two of these curves, and a multistatic geometry is required to resolve this ambiguity.

8.3.3.3 Multistatic sensing

A multistatic sensor is shown in Figure 8.5, which is simply a superposition of multiple bistatic radars. We see that the system can utilise wide beam antennas and can recover spatial resolution using the high delay resolution provided by the bandwidth of the illumination. This capability is accomplished because the different bistatic range ellipses intersect, thereby improving spatial resolution as well as significantly reducing the illuminated clutter area, which improves the signal-to-clutter ratio. Clearly, combining two bistatic pairs whose baselines are not orthogonal will result in an increase in range uncertainty and increase the illuminated clutter. We extend these ideas to using Doppler resolution in Section 8.3.3.4.

Work at the University of Pisa [21–24] has shown how the spatially separated set of received signals for a multistatic sensor can be combined and the resulting tracking accuracy that can be achieved over the area of coverage. This work is highly relevant to the planning of a WRAN-based sensor network.

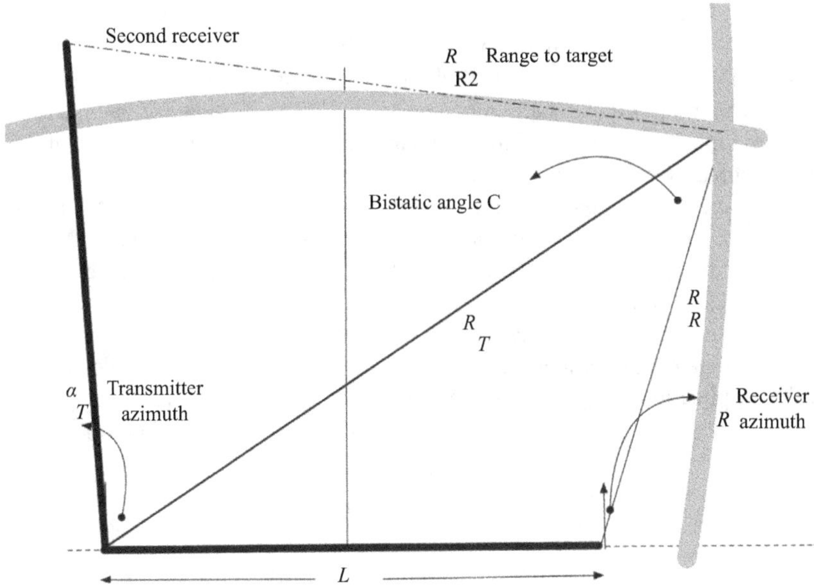

Figure 8.5 A 'tri-static' sensor consisting of two bistatic pairs (one transmitter and two receivers). The overlap of the bistatic ellipses (the grey arcs) is small, providing good localisation. The two receivers can also have narrow beams, which would further assist with ambiguity resolution. The Doppler arcs (not shown) yield additional information

8.3.3.4 Doppler-only sensing

Doppler-only tracking by multistatic sensor networks has been explored extensively in the sonar community via the use of large arrays of acoustic sensors deployed on the ocean bottom to track submarines. Maasdorp subsequently extended this work to the EM sensor field, specifically for FM band commensal systems [25–27].

Applying logic similar to Section 8.3.3.3, a target with a specific bistatic velocity only appears at the intersection of the Doppler arcs, thus providing enhanced SNR. In addition, if the illumination time is reasonable (a few seconds might be possible), the Doppler extraction process further improves the SNR. Clutter values should be low in the Doppler space except where processing artefacts [28] are large. Maasdorp extended this work further to identify the best configuration of receivers for a single transmitter [27].

8.3.3.5 Multilateration

Multilateration [29] has gained popularity for navigation and air traffic management (ATM). Multilateration relies on the time of flight of a wave from a target to a source. With just two measurements, the position cannot be resolved, but if a number of time difference measurements can be combined at different receivers, the ambiguities can be resolved. One example of multilateration is the British 'Gee' system used during World War II as an early form of navigation for bombing.

Recently, commercial offerings relying on beacon responses from the aircraft itself have been installed and are in operation [30]. The advantage is that the receive stations are much simpler than full secondary surveillance radar (SSR) [31] installations, though they suffer from the same disadvantage of requiring cooperative targets and a network of interrogators. Given that the transponder in the aircraft is part of a secondary radar system, this multilateration system can be classified as being commensal with the SSR, i.e. the multilateration system does not need to be coherent or know the content of the secondary radar signal.

Note that because multilateration relies on instrumented aircraft and ground-based interrogators, it cannot be viewed as completely dependable. For example, security applications requiring aircraft positions cannot completely rely on the availability of this data because the airborne beacon could be switched off.

8.3.4 Applications of EM sensor networks

In this section, we review a number of potential sensor applications that could be added to the communication functionality of WRAN. The signal and data processing is then covered in the sections that follow. In Sections 8.4 and 8.5, two specific examples are examined in detail. To be clear, we only consider applications where the transmitters and receivers of the WRAN are used to implement the sensor. In contrast, we do not consider using the WRAN nodes to implement a triangulation system that communicates to a base site via the WRAN network, or the monitoring of self-emissions of a target.

8.3.4.1 Aircraft detection and tracking

The sensor network for ATM is heavily dependent on radar sensors (en route and terminal) that use highly inefficient waveforms, with little attempt at spectrum sharing through coordination. However, the desirability of the L and S bands for communication uses is putting more and more pressure on these sensors, and unrealised leasing costs for this spectrum will make air-traffic control (ATC) sensors very expensive to operate. We believe the current generation of radars might be replaced with symbiotic sensors, with FM band commensal systems a highly attractive option [14].

ATC sensors tend to monitor only certain air corridors designated for licenced air traffic. Of course, a general air picture is important for security applications. Included in this concept would be border security, and in the case of conflict, areas of interest where own and enemy forces are operating.

8.3.4.2 Vehicle detection and tracking

The tracking of vehicular traffic, as for aircraft monitoring, has become an essential tool for efficient use of highways and the controlled airspace. Like en route and terminal ATC, the particular sensing application depends on the traffic density.

Urban highways

The flow of commuter vehicles in and out of cities is a worldwide problem. It is interesting to see that a sensor network comprised of GPS-enabled cell phones and a data link to a central server (Google and Waze) has evolved very rapidly to provide extremely accurate traffic density and delay information to users. This arrangement is an example of a collaborative sensor network.

In the maritime world, the automatic identification system (AIS) [32] has become mandatory in most economic zones (coastal areas). A ship that does not wish to be tracked switches off its AIS and regulatory authorities then have to depend on the skin echo produced by traditional radar. Similarly, an individual motorist not part of a collaborative network is essentially invisible.

It does not seem practical to track large numbers of individual vehicles using a WS sensor due to resolution issues and data density, but certainly exceptional speeds (high Doppler) could be noted along with an approximate corresponding location. This information could be used to cue other enforcement resources to investigate further.

Rural security

Peri-urban small holdings are subject to deprivations from nearby cities. Crimes range from crop and stock theft, to more serious armed robberies of household goods [33]. To be able to remove the often bulky stolen goods (such as livestock, flat screen TVs, etc.), a vehicle is driven along roads that generally carry very little traffic.

A sensor system that could locate the presence of unusual traffic along such roadways, especially at night, would be extremely useful to rural dwellers and police forces. A WS commensal or mutualistic system would be useful in this regard. The WS communication services would also be critical for warning the relevant persons based on centrally produced signal and data processing.

8.3.4.3 Maritime/waterway traffic detection and tracking

Clearly, a commensal WS-based system would not illuminate in a seawards direction due to the lack of communication users in that direction. However, there is spillover energy, and in the case of rivers, canals, narrow sea channels, and other inland waterways, good cover is still possible. Furthermore, the clutter from the water regions tends to be low, thereby aiding detection and tracking. This topic has been investigated through the implementation of a commensal sensor based on DTV emissions [34].

The main user for this manner of sensor is likely to be law enforcement as a means to curtail illegal smuggling and human trafficking, as well as illegal fishing. As mentioned above, the AIS relies on the good behaviour of waterborne vessels, but such criminal activity often involves the perpetrators switching off the AIS. In this case, only sensors relying on 'skin' reflections such as a WS sensor or a conventional radar are useful. Switching off a transponder is, of course, already a sign of bad intent.

8.3.4.4 Human detection and tracking

Detecting people moving about an area is an important security need, both for government/commercial entities and private individuals. Currently, a host of sensor modalities are available, including acoustic, optical, and RF/microwave.

Through-wall monitoring

The use of EM sensors to detect and even image persons moving behind a wall has been of great interest for many years, with applications in emergency services (earthquake building collapse, terrorism bomb damage) as well as security services (asymmetric warfare, hostage situations, etc.). Many such systems are deployed by security services and their actual performance is, of course, not publicised.

The WS frequency bands are good for wall penetration (see Table 8.1) and there appears to be scope for exploring the use of a commensal sensor system based on the WS network in this context. In urban areas, high power legacy TV transmitters might be available, but in rural areas the installation of WS communication systems may be more feasible. The success of through-wall experiments using 2.4 GHz Wi-Fi [35,36] indicates this could be promising application. Using our proposed classification, these systems are commensal sensors since they utilise the Wi-Fi signal without requiring direct collaboration with the Wi-Fi system.

Care of elderly/infirm

The care of the elderly and infirm can be very expensive. Enabling a person to live in their own home instead of a dedicated care facility has clear psychological and cost benefits, as long as their physical well-being can be monitored remotely. For example, recent work has demonstrated how in-home radar could be used to detect if a person has fallen [37]. Wi-Fi-based approaches have likewise been investigated [38,39] and show promising results.

Similar to the previous application for through-wall monitoring, the WS frequency bands are attractive in this regard. If research finds that more powerful illumination of the premises to be monitored is required, then the cognitive radio aspects of the WS standard could be utilised to install a bespoke transmitter in the building of interest. This illuminator would adaptively adjust in time and frequency according to the WS standard.

8.3.4.5 Earth observation applications

Earth observation is predominantly performed by EM sensors, particularly in the optical and near-optical bands. WS sensors may also play a role in this context.

Precipitation monitoring

Researchers have shown that by monitoring the received amplitude of cell phone carrier signals or the inter-cell microwave links, it is possible to obtain fairly accurate information on rainfall rate based on the calibrated amount of attenuation [40]. A similar application is possible with a WS sensor system, as illustrated in Figure 8.6.

Biomass monitoring

The WS frequencies are good for foliage penetration [41], which means that the signal strength between nodes of a WS sensor system will depend on the foliage density. Differential measures of attenuation over time could therefore provide biomass indicators. Radical changes to the environment, such as harvesting of woody biomass, clearing for construction, etc., would be detected.

Figure 8.6 Conceptual precipitation measurement using a multistatic WS-based system. Given that there are enough links sampling widespread rain, a rough spatial distribution could be obtained

Soil moisture
The measurement of soil moisture using EM waves is well established [42]. Wave propagation between nodes of a WS sensor will be influenced by hydrometeors, but the reflectivity also varies with moisture content. Therefore, through complicated multipath interactions, the collective signals between nodes will likewise vary with soil moisture. This interaction is very similar to the use of global navigation satellite system (GNSS) signals for soil moisture measurements [43]. Of course, the accurate calibration of such a means of measurement is rather complex.

River flow measurements
There are a number of applications of EM sensors to measure river flow and currents [44,45]. Much of this work has been accomplished using HF surface-wave radar, though more recently radars operating in the VHF and UHF bands have been considered. The WS frequencies are clearly suitable for this task and a commensal sensor would be possible. Spatial resolution in two dimensions is required, so a convenient transmitter and a small network of receivers would need to be deployed.

8.4 Commensal implementation of a WS sensor

Consider the implementation of a commensal WS sensor network. We assume that there is a WS communication system operating in an area. Ideally, the number of stations is large to provide good coverage with some redundancy. This arrangement permits probably the easiest design of a sensor system since the commensal relationship means that no attempt is made to modify what is being transmitted or to use the existing WS system receivers.

The sensor network will employ its own network of receivers, relying on the WS system to provide transmissions of opportunity. The reason for recommending a separate network of receivers is that the sensor receivers need to be adequately

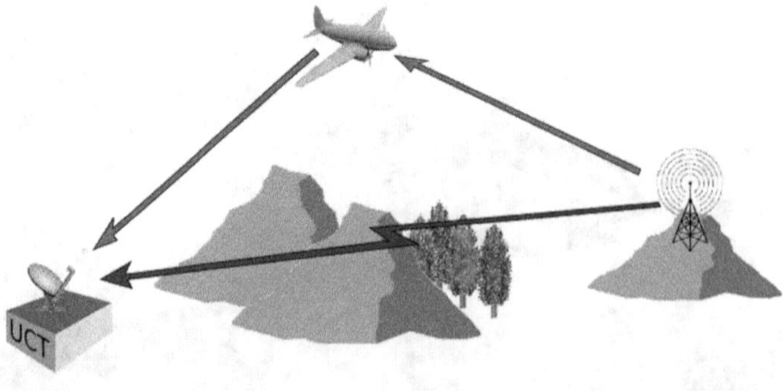

Figure 8.7 Conceptual diagram of an FM band commensal system [14]

isolated from the direct path signals to facilitate the capture of reflected echoes with finite dynamic range while avoiding saturation of the receiver digitiser. A recent paper outlines the complex process of siting a commensal radar system [14]. This type of system is also referred to as passive coherent location (PCL), and a previous paper discusses the sensitivity of such a system [46].

At this time, we are not aware of the deployment of any such WRAN-based system since WRAN itself is still in early trials in a number of parts of the world. However, it is instructive in this context to consider what can be achieved using the FM band broadcast transmitters in an area. FM broadcast operates in the WS spectrum, but as a licenced user similar to DVB and DAB, which can also be used in the commensal sense. A mutualistic version would require modification of the existing broadcast systems.

8.4.1 FM band commensal aircraft sensing

The FM broadcast band is likely to remain popular for quite some time. Many of the issues relating to multipath and corrupted demodulation of DAB do not exist in the simple analogue modulation of the 88–108 MHz band. The sensor we describe is completely commensal; as long as the FM transmitter is operating, the associated sensor can successfully detect aircraft out to bistatic ranges exceeding 600 km and even small targets such as microlites out to 30 km [14,47]. The operation of such a commensal sensor has been investigated using many different sources of illumination [15,34]. In Figure 8.7, the transmitter signal is cross correlated with the Doppler shifted, time-delayed echoes from the target.

The design of such systems was outlined in [14]. There are some early commercial implementations, but as yet no widespread deployment. The South African Council for Scientific and Industrial Research [48] is implementing an FM band network for this purpose and intends to operate it over a 3-year period to assemble a hard set of detection and tracking statistics to facilitate comparison with conventional ATC radars.

Although this example is not really embedded into the WS network approach, it does use the reserved FM broadcast band. We have shown [47] that the reference

Figure 8.8 *Positions of the transmitter and receivers for the FM band commensal sensor [27]*

signal can be sent to a common processing centre or to each node, where Doppler and delay can then be extracted. These backhaul links could be implemented on a nearby WRAN. Thus, the commensal sensor also becomes a user of the WS infrastructure, though it does not manage the emissions on which it depends.

The FM band commensal sensor tested out the concept of Doppler-only tracking using the Gauss–Newton filter [27,49,50] (see Figure 8.8), which performs much better than Kalman filters; in fact, the invariant-extended Kalman filter is a simplified version of the Gauss–Newton filter [50]. As illustrated in Figure 8.9, the accuracy achieved is excellent.

8.5 Mutualistic implementation of a WS sensor

The previous section discussed a commensal sensor implementation based on the WRAN. In this section, we outline what might be achieved if the design became mutualistic. The design concept presented here is similar to the consideration of the commensal system above. However, the communication and the sensor functions are now designed jointly from the ground up to provide a good compromise for both.

The transmitters and receivers of the mutualistic partners both contribute to the sensing function. Note that doing so necessitates a reduction of data bandwidth due to the interchange of sensor data as well as communication payloads. The network configuration is the same as for the WRAN shown in Figure 8.1. However, the BS

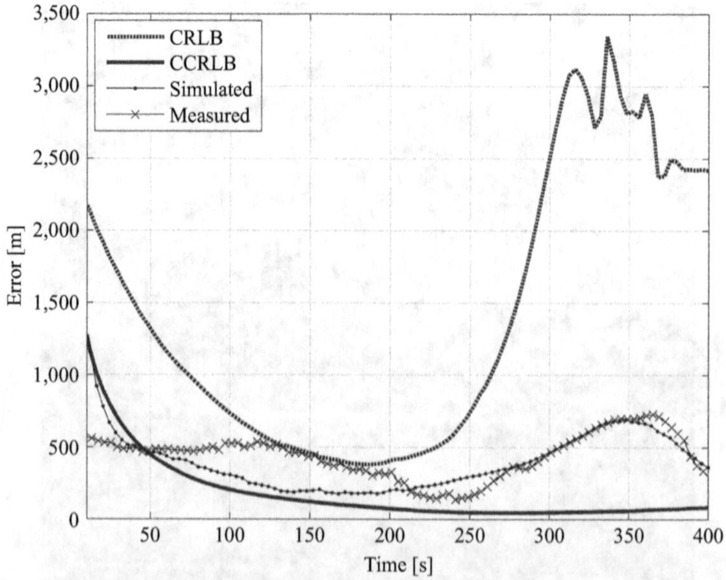

Figure 8.9 Tracking accuracy of a Gauss–Newton filter based on Doppler-only measurements for four stations in an FM band commensal sensor [27]. The Cràmer–Rao Lower Bound (CLRB), and a cumulative version thereof (CCLRB) are introduced by Maasdorp [27] to estimate the best tracking that can be achieved

and CFE units now also perform a sensing function by measuring delay and Doppler shift. Notice also that the sensor system becomes a combination of monostatic sensors (BS to target, reflected back to BS, measuring delay and Doppler shift), as well as multiple bistatic pairs (BS to CFEs measuring bistatic delay and Doppler), per the parameter estimation discussion in Sections 8.3.3.1–8.3.3.3.

At present, we are not aware of such systems in planning or test. A limitation of this approach is the rather modest EIRP generated by BS transmitters, which can vary according to geographical region. However, when viewed as a sensor network, each of the relatively short links does provide target information. The well-known adage that 'a system's performance far exceeds the sum of its parts' applies in this case. In other words, a monostatic radar is easy to categorise using a receiver operating curve [12] that predicts detection probability as a function of SNR, with the probability of false alarm as a parameter. In contrast, when one considers a sensor network, the statistical analysis becomes far more complex and has not been widely addressed in the literature. The approaches of Maasdorp and Stinco [16,21,27] are a good start.

The importance of accurate time and frequency alignment emerges as a key enabler of this arrangement. For example, bistatic time delay (and thus range) can only be measured with as much accuracy as the degree of timing alignment between a given geographically separated transmitter and a receiver pair. The same applies to

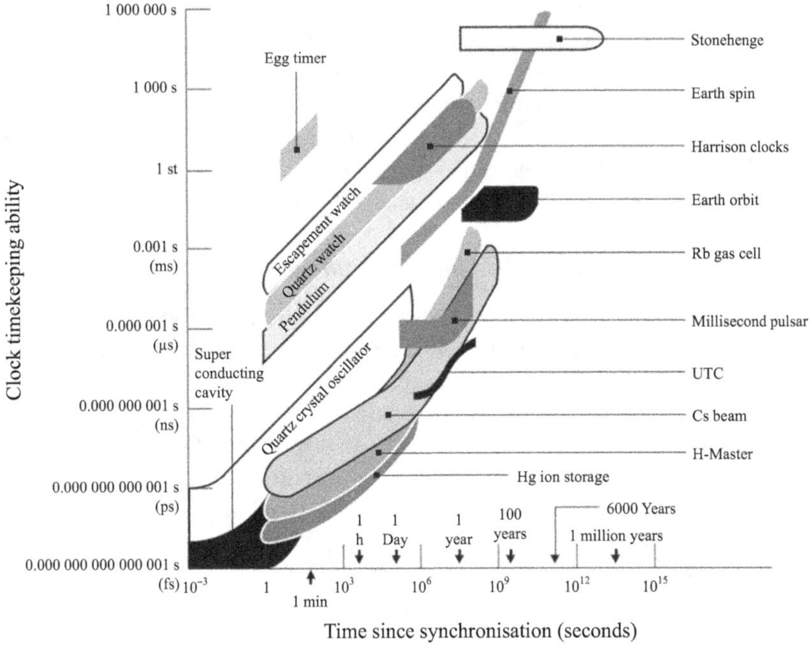

Figure 8.10 The timekeeping ability of various clock sources [52]

the carrier frequency alignment for Doppler accuracy at the bistatic nodes. This topic is discussed further in the following section.

The resolution of the sensor (monostatic or bistatic) depends, in the usual way, on the bandwidth of the communication data packet that also serves as a sensor waveform. Similarly, the Doppler resolution depends on the amount of time over which the target is illuminated. It was shown in [51] that for aircraft targets illumination times of up to 4 s are quite common. However, that result is for a situation involving only 100 kHz bandwidth of an FM broadcast, with associated range resolution of 150 m. In contrast, for a system using 20 MHz bandwidth (7.5 m resolution) and a target velocity of 300 m/s, range walking (the target moves between range cells) occurs within 7.5/300 = 25 ms.

8.6 Time and frequency alignment

The mutualistic system proposed in the previous section requires very precise alignment of data in time and frequency to achieve acceptable range and Doppler accuracy. Communication networks generally achieve this alignment using either clock recovery or through the distribution of timing signals. Time-keeping technology [52] has evolved significantly, and many options for primary and secondary sources exist (Figure 8.10).

In this section, we consider the timing needs of a sensor relative to the standard network timing protocols that WRAN may employ. GNSS disciplined oscillators and the White Rabbit (WR) [53] network standard are also examined. A particular danger of the GNSS method of maintaining time/frequency alignment is the very real threat of denial of service by jamming GNSS. A system such as WR, however, requires a large and possibly expensive network installation (physical cables), though such a network can also carry data and provides a useful link between BSs (see Figure 8.1).

8.6.1 Necessity of time/frequency alignment

Since the detailed operation of a mutualistic system based on WRAN is still at the conceptual stage, one cannot define all the requirements for timing and frequency accuracy, though it is possible to sketch out some rough expectations. For example, for a monostatic radar, a root-mean-squared (RMS) timing accuracy of 1 ns yields a range accuracy of about 0.15 m. On the other hand, resolution of bistatic measurements depends on the bandwidth of the joint symbiotic system and is independent of timing accuracy. For the FM band commensal system described in Section 8.8, only about 100 kHz of bandwidth is available, thus providing a range resolution of 1.5 km. In contrast, a 20 MHz DTV signal can achieve more like 7.5 m.

Doppler is another way of tracking targets [27] and is determined based on a succession of phase measurements. Thus, the random phase error between the transmitter and receiver must be known to a sufficient degree, with performance established by the phase noise spectrum of the transceiver hardware (up- and down-conversion oscillators). Recall as well, that for multistatic radar systems, the phase noise is uncorrelated between receivers.

8.6.2 Hard wired timing

In this arrangement, components of the sensor network are synchronised (time and frequency) via cables. The NetRAD [54] was originally synchronised this way. However, GPS disciplined oscillators (GPSDOs) [55] were later added due to the physical limitation of Canbus [54] constraining node spacings to hundreds of metres at best. Transmission of timing over long distances via cable, be it through fibre or copper, can be expensive. The WR [53] solution developed by CERN (the European Organisation for Nuclear Research) is a compromise that uses standard fibre with special WR switches. WR over a radio link is still being investigated.

8.6.3 SONET and other network technologies

Synchronous optical networks (SONETs) [56] provide fairly precise timing signals over digital networks. Since SONET was designed for synchronising frames and packets of data, the jitter requirements are not onerous and are generally on the order of a few microseconds between hops of the network.

Cell phone technology (such as LTE) [57] has tight timing requirements since the scheduling of the transmission of data between BSs and clients, along with synchronising between other cells, has to be carefully controlled to achieve optimum filling

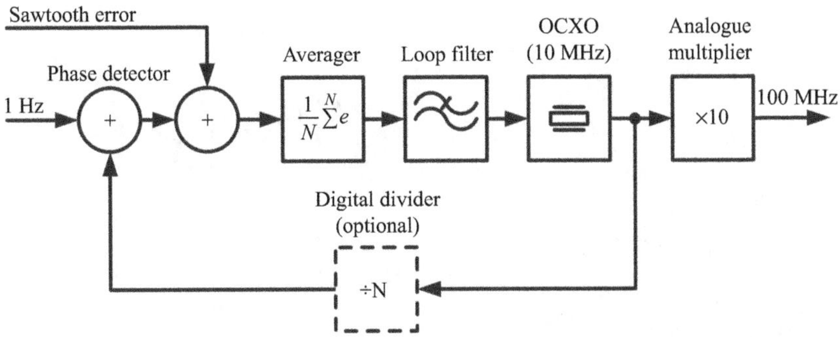

Figure 8.11 Functional overview of the UCT GPSDO. The measured phase discrepancy between the 1 and 10 Hz signals is sawtooth corrected, averaged, and fed to a second-order infinite impulse response (IIR) loop filter. The filter output is then used to compensate for the measured error by steering the ovenised crystal oscillator (OCXO). Finally, the 10 MHz OCXO is up-converted to 100 MHz using an analogue multiplier

of the time-domain slots. However, the achievable timing jitter of LTE still does not suffice for sensors requiring nanosecond accuracy to achieve a given bistatic range accuracy.

The largest source of jitter occurs when timing signals pass through data multiplexers. New standards and equipment have been developed that allow timing packets to bypass these switches (with unpredictable backplane delays), with a goal of achieving better timing accuracy than SONET. This enhanced time resolution has enabled LTE and new standards to be implemented.

8.6.4 GPS disciplined oscillators

Figure 8.11 shows the functionality of GPSDO[1] developed at the University of Cape Town (UCT) using long-term GPS-derived time (based on caesium clocks) to discipline the internal crystal oscillator to achieve good short-term and poor long-term stabilities (Figure 8.12). This approach realises about 4 ns RMS timing accuracy, resulting in a monostatic range uncertainty of about 60 cm.

The practical implementation of this approach is particularly difficult, even including the correct choice of amplifiers and filters. Performance measurement is clearly a science in its own right. The UCT GPSDO implementation is shown in Figure 8.13.

Frequency synthesis by phase-locked loop (PLL) is a widely understood technique. However, peculiar to the GPSDO is its low frequency 1 Hz reference and

[1]This section was contributed by Stephan Sandenbergh.

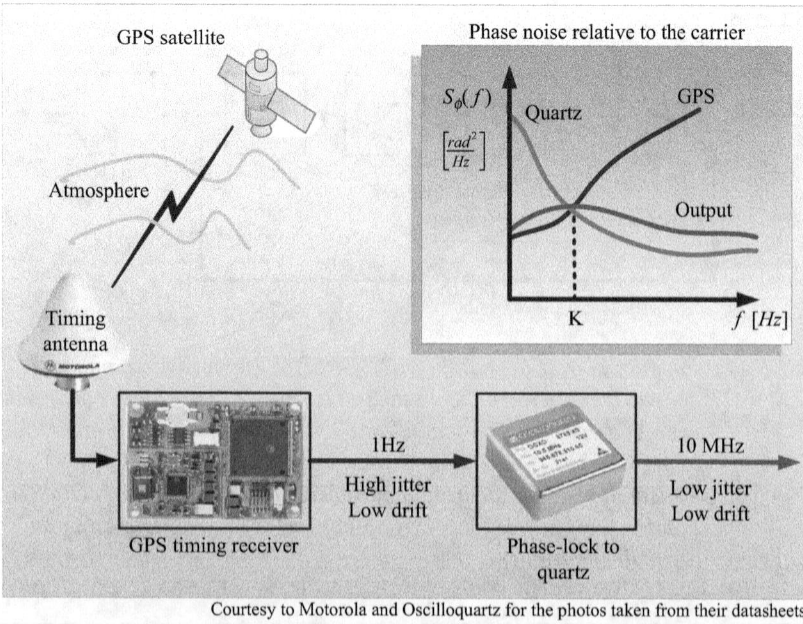

Figure 8.12 Overall GPSDO operational cartoon. The optimum PLL filter transition frequency is set where the phase noise curves of the GPS system and the quartz oscillator cross, providing the output compromise phase noise

narrow-loop bandwidth. In the UCT GPSDO [58], the 1 Hz GPS timing reference is locked to a 10 MHz OCXO. Then, analogue multiplication is used to acquire a pure 100 MHz frequency reference.

The 1 Hz GPS time reference is either phase compared to a 10 MHz quartz oscillator directly or after digitally dividing the 10 MHz signal. Direct phase comparison circumvents jitter being added by the digital division process. However, this approach limits the maximum measured phase error to only one period of the quartz oscillator.

The GPS receiver introduces a quantisation error (also referred to as sawtooth error) onto the 1 Hz GPS time reference. This error is a function of the drift rate and frequency of the local oscillator that is internal to the GPS receiver [59]. Conveniently, the GPS receiver supplies an estimate of the sawtooth error prior to each 1 Hz pulse. Hence, each sawtooth error is removed by applying a corresponding offset to the phase detector (PD). Other calibrations could also be achieved in a similar fashion. In an attempt to partially remove some of the remaining zero-mean jitter, the sawtooth corrected phase errors are averaged for tens to thousands of seconds. The maximum number of samples that can be averaged is dictated by the frequency at which the OCXO frequency is adjusted.

These averaged error values are then fed to a second-order infinite impulse response (IIR) loop filter. The output of this filter drives a digital-to-analogue

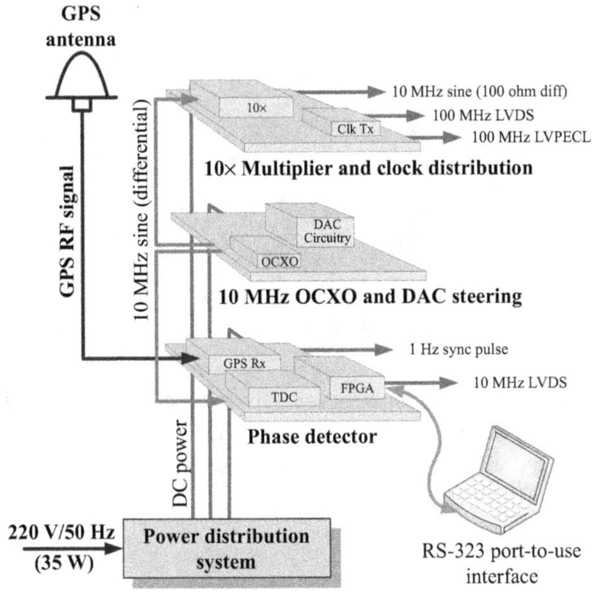

Figure 8.13 The UCT GPSDO, which has a modular design-separating functionality across different printed circuit boards (PCBs). The bottom board contains the phase detector, GPS receiver, and other digital circuitry including an FPGA and RS-232 communications. The FPGA is programmed with a parallel picoBlaze architecture that manages user communication and other digital tasks such as phase detection, oscillator steering, and programming of the clock distribution. The middle board contains the OCXO and the digital-to-analogue converter (DAC) steering circuitry. The top board contains the 10 × analogue multiplier, up-converting the 10 MHz OCXO output to 100 MHz, and the clock distribution

converter (DAC) that steers the 10 MHz oscillator to compensate accurately for the measured phase discrepancy. The exact number of averages and the specific bandwidth of the loop filter depend on the stability of both the 1 Hz reference and the OCXO. Finally, the sinusoidal 10 MHz OCXO output is up-converted to 100 MHz using an analogue multiplier.

The UCT GPSDO was used to synchronise NetRAD [60–63], an experimental network radar jointly developed by University College London and UCT. Some synchronisation results are shown [64] in Figures 8.14–8.16. In conclusion, it seems that with GPSDO technology very carefully implemented, it should be possible to achieve WS sensor time accuracy below 5 ns. Of course, when combining the inputs from a number of sensors, their geometry will play a major part.

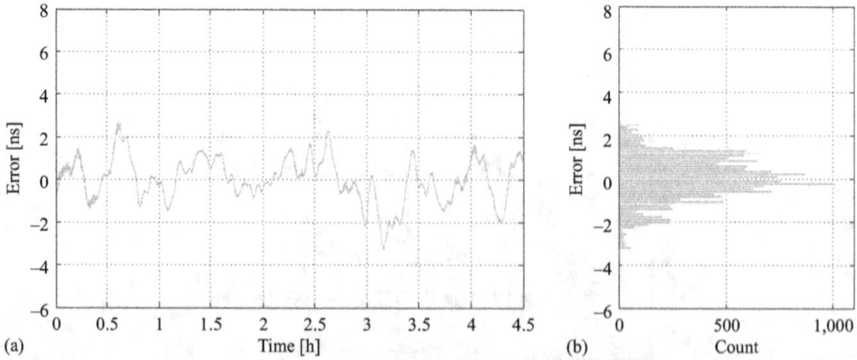

Figure 8.14 Relative phase error between two co-located UCT GPSDOs, with (a) showing PLL filter averaging over 1,000 s and (b) illustrating the histogram after removal of a 5 ns offset

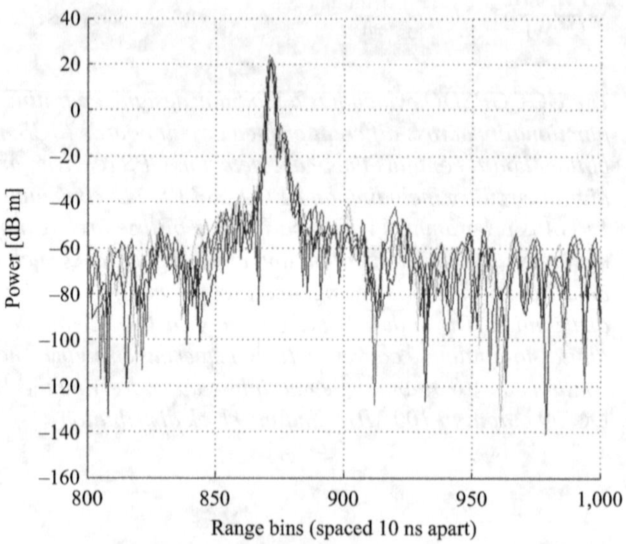

Figure 8.15 The direct breakthrough, which when received from the transmitter's sidelobe, is a good indicator of how synchronously the nodes of a bistatic radar are triggered. Without a precise measurement of distance, the synchronisation accuracy cannot be determined. However, the direct breakthrough does provide a good measure of synchronisation precision. In this case, the bistatic baseline was approximately 2.3 km. The figure shows the matched filtered direct breakthrough for four separate synchronisation events performed over a period of 20 min. The peaks are within 1 range bin (10 ns) of one another, which is consistent with the timing errors measured between two co-located GPSDOs in the laboratory

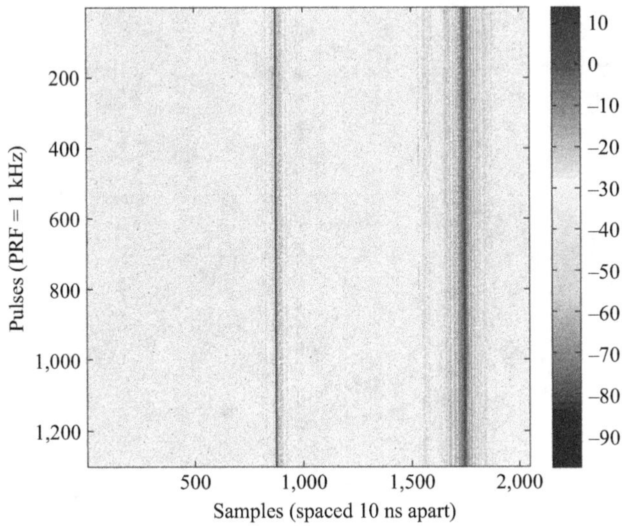

Figure 8.16 A bistatic time range plot of a lighthouse (Roman Rock, Simon's Town, South Africa). The first peak at roughly bin 870 is the direct breakthrough received from the transmitter's sidelobes. The second peak at roughly bin 1,700 is that of the lighthouse. The bistatic baseline was approximately 2.3 km

8.6.5 White Rabbit

The WR[2] project undertaken by CERN, GSI, and industry partners is an ethernet implementation of an accurate synchronisation network. Figure 8.17 depicts a basic WR network (WRN) that follows a clock hierarchy topology. A stable atomic or GPS clock is propagated from the first switch, known as the *Grandmaster* switch, to further switches and nodes down the hierarchy. A best master clock algorithm selects the highest quality clock available, thereby allowing redundant timing sources to be installed for failure prevention. A single WRN can synchronise thousands of nodes with fibre links at distances up to 10 km [65]. Using only WR equipment, clock accuracies below a nanosecond and with low picosecond level jitter are achievable for any node with respect to the Grandmaster clock.

The WR protocol is an extension of the precision time protocol (PTP or IEEE-1588 v2) with the addition of synchronous ethernet (SyncE). PTP calculates the clock offset of a slave device relative to its master using time-of-flight calculations based on time-stamped ethernet frames. Ethernet frames are hardware time-stamped at the PHY level, allowing software latencies found in the network time protocol to be avoided. However, PTP itself is limited by the granularity of the transmission clock,

[2]This section was contributed by Simon Lewis and Stephan Sandenbergh.

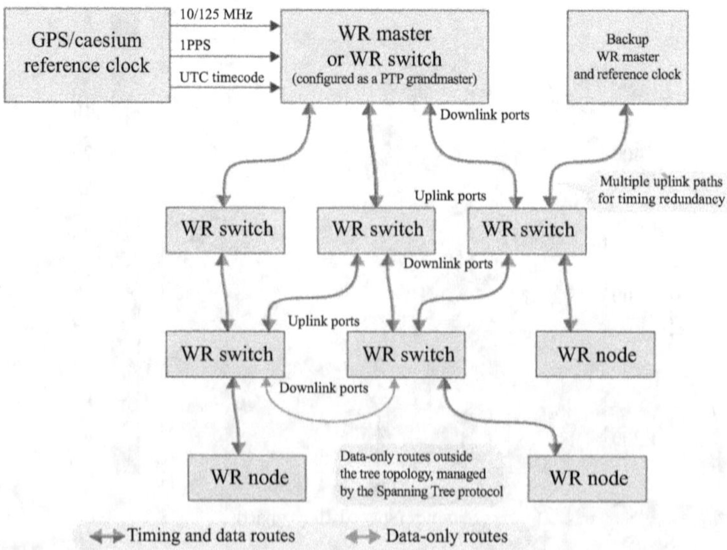

Figure 8.17 Infrastructure-based precision time distribution using the White Rabbit standard [65]

which is 8 ns in the case of Gigabit ethernet with a 125 MHz clock rate, as used by a standard WRN.

With the addition of SyncE to PTP, a common network frequency allows data to be synchronously transmitted based on a time-division multiplexing approach similar to SONET. Slave PLLs allow the clock skew between master and slaves to be adjusted and provides a fine resolution measurement of the phase difference using a digital variant of the dual mixer time difference (DMTD) technique.

8.6.5.1 Performance of White Rabbit in radar

The standard implementation of WR delivers clock accuracies below 1 ns, primarily focusing on the accuracy of a clock with respect to an absolute time reference. The system is therefore not concerned with providing a low-phase-noise frequency reference, which is important to high performance radar. The short-term phase stability of a standard WR card is thus a concern, though the long-term stability of the system exhibits atomic level performance.

In [66,67], Inggs, Sandenbergh, and Lewis proposed a WR disciplined oscillator (WRDO) to meet accuracy concerns for range and Doppler. A WRDO can be developed using a similar rationale to the GPSDO. Instead of using the pulse-per-second (PPS) signal from GNSS, a WR PPS from a WR node or switch can be used to discipline a low-phase-noise phase-locked oscillator. A more accurate trigger signal from WR will improve the range accuracy of the bistatic pair, whilst the low-phase-noise oscillator will determine the Doppler accuracy.

8.7 Conclusions and looking ahead

8.7.1 Summary

This chapter has shown that the migration of TV broadcast to DVB has opened up a frequency band that is extremely favourable for a number of sensor applications. The remarkable flexibility of the IEEE 802.22 standard was discussed and how a number of sensor functions can be integrated into the underlying communication network.

The key to the WS standard has been the spirit of cooperation between existing users and new users. The U.S. government has encouraged this sharing of spectrum. Making revenue streams from spectrum leasing available to existing users [9] is a great incentive. The other clear dividend to both existing and new users is the sharing of very expensive bandwidth lease costs.

The WS WRAN is a form of cognitive radio, operating with access to geographic databases of licenced users, as well as a spectrum sensing function to ensure that clashes are avoided. The ability to handle unlicenced users is a big advantage to rapid adoption. The relatively high power produced by BSs makes this technology suitable for wide area broadcast coverage in rural and peri-urban regions. This same wide coverage likewise makes WRAN transmitters suitable for symbiotic use in sensing applications.

We have also considered how a new generation of sensors in the WS can be defined as a 'symbiosis' by using a simple but elegant terminology taken from biology. This taxonomy is then introduced via a few examples of possible systems.

Finally, several WS sensing applications were described, varying from *commensal* (no requirements from the source of radiation) to *mutualistic* (full cooperation with the WS geographic database is required). Sensors require careful timing and low-phase-noise for the transceivers, and we showed some practical ways in which this can be achieved, from hard-wired connections to the use of GNSS signals.

8.7.2 Looking ahead

It is clear that there is a great deal to do in terms of establishing WS sensor technology. We have outlined the successes of commensal applications such as FM broadcast band sensors, DTV, and so on which are the 'low hanging fruit', while the more difficult issues of mutualistic designs remain open problems.

We believe that the breakthrough in cooperation between users of the TV WS bands, as negotiated and specified by IEEE 802.22, is a landmark that should become a template to what may become known as the 'EM sensor community'. This degree of cooperation may well lead to a future where the sensor bands (L, S, and X in radar terminology [68]) and many other bands used by Earth observation and radio astronomy could be shared cooperatively.

No one will escape the financial squeeze that will be imposed by governments [69] and their determination to open the EM spectrum to their citizens using financial incentives. A cooperative EM sensor community will be very important. For example, the geographic database of the IEEE 802.22 standard could be adapted to deal with passing satellite radiometers in L band, thus making the band available for other users between passes. ATM radars could likewise be replaced by networks of sensors.

These new sensors will possess a spectrum sensing function and be able to share data more widely, thus allowing for regulation of the bands in question and removal of transgressors. At present, this process is performed in an ad hoc fashion [70] with a great deal of on-site intervention. Hopefully, this sharing movement will allow us to claw back inefficient use of spectrum by 'Going Green', as we are doing with pollutants and energy usage.

References

[1] IEEE 802.22 Working Group. Available from http://www.ieee802.org/22/

[2] J. M. Peha, "Sharing spectrum through spectrum policy reform and cognitive radio," *Proceedings of the IEEE*, vol. 97, pp. 708–719, Apr 2009.

[3] A. De Bary, "The phenomenon of symbiosis," *Karl J. Trubner, Strasbourg, Germany*, 1879.

[4] N. Willis, *Bistatic Radar*. Raleigh, NC: Scitech, 2004.

[5] H.D. Griffiths, C.J. Baker, *Passive Radar*. Norwood, MA: Artech House, 2017.

[6] A. Wilson-Langman, M. R. Inggs, and A. K. Mishra, "Symbiotic radar and communication system," Feb 17 2014. 14/768,382.

[7] A. B. Flores, R. E. Guerra, E. W. Knightly, P. Ecclesine, and S. Pandey, "IEEE 802.11 AF: a standard for TV white space spectrum sharing," *IEEE Communications Magazine*, vol. 51, no. 10, pp. 92–100, 2013.

[8] A. K. Mishra and D. L. Johnson, *White Space Communication*. New York, NY: Springer, 2015.

[9] President's Council of Advisors on Science and Technology (PCAST), "Realizing the full potential of government-held spectrum to spur economic growth," Jul 2012.

[10] W. C. Stone, "NIST construction automation program report no. 3: Electromagnetic signal attenuation in construction materials," *Building Fire Res. Lab., Nat. Inst. Standards Technol., Gaithersburg, MD, Tech. Rep. NISTIR*, vol. 6055, 1997.

[11] W. Edwards, J. Gebauer, and B. Reinicke, "White space networks: architecture, application, and opportunity," *Computer*, vol. 47, no. 12, pp. 64–71, 2014.

[12] S. Kingsley and S. Quegan, *Understanding Radar Systems*, vol. 2. Raleigh, NC: SciTech Publishing, 1999.

[13] F. Kingsley, *The Applications of Radar and Other Electronic Systems in the Royal Navy in World War 2*. London: Palgrave Macmillan, 1995, p. 267.

[14] M. Inggs, C. Tong, R. Nadjiasngar, G. Lange, A. Mishra, and F. Maasdorp, "Planning and design phases of a commensal radar system in the FM broadcast band," *IEEE Aerospace and Electronic Systems Magazine*, vol. 29, no. 7, pp. 50–63, 2014.

[15] Unknown, "Silent sentry: innovative technology for passive, persistent surveillance." Available from http://quoteinvestigator.com/2014/05/22/solve/, May 2014.

[16] J. Llinas and D. L. Hall, "An introduction to multi-sensor data fusion," in *Circuits and Systems, 1998. ISCAS '98. Proceedings of the 1998 IEEE International Symposium on*, vol. 6, pp. 537–540, May 1998.

[17] M. I. Skolnik, *Radar Handbook*. New York, NY: McGraw-Hill, 2008.

[18] S. M. Sherman and D. K. Barton, *Monopulse Principles and Techniques*. Norwood, MA: Artech House, 2011.

[19] M. H. Kabutz, A. Langman, and M. R. Inggs, "Hardware cancellation of the direct coupling in a stepped CW ground penetrating radar," in *Geoscience and Remote Sensing Symposium, 1994. IGARSS '94. Surface and Atmospheric Remote Sensing: Technologies, Data Analysis and Interpretation., International*, vol. 4, pp. 2505–2507, Aug 1994.

[20] M. C. Jackson, "The geometry of bistatic radar systems," *IEE Proceedings F – Communications, Radar and Signal Processing*, vol. 133, pp. 604–612, Dec 1986.

[21] P. Stinco, M. S. Greco, F. Gini, and B. Himed, "Cognitive tracking in IEEE 802.22 symbiotic radars," in *2016 24th European Signal Processing Conference (EUSIPCO)*, pp. 2211–2214, Aug 2016.

[22] M. Greco, P. Stinco, F. Gini, A. Farina, and M. Rangaswamy, "Cramer–Rao bounds and Tx–Rx selection in a multistatic radar scenario," in *2010 IEEE Radar Conference*, pp. 1371–1376, May 2010.

[23] P. Stinco, M. S. Greco, F. Gini, and B. Himed, "IEEE 802.22 passive radars: multistatic detection and velocity profiler," *IEEE Transactions on Aerospace and Electronic Systems*, vol. 52, pp. 2298–2313, Oct 2016.

[24] P. Stinco, M. S. Greco, F. Gini, and M. L. Manna, "Non-cooperative target recognition in multistatic radar systems," *IET Radar, Sonar Navigation*, vol. 8, no. 4, pp. 396–405, 2014.

[25] F. Maasdorp, R. Nadjiasngar, and M. Inggs, "A Cramer Rao analysis on receiver placement in a FM band commensal radar system based on Doppler only measurements," in *Radar Conference (Radar), 2014 International*, pp. 1–6, IEEE, 2014.

[26] F. Maasdorp, M. Inggs, and R. Nadjiasngar, "Target tracking using Doppler-only measurements in FM broadcast band commensal radar," *Electronics Letters*, vol. 51, no. 19, pp. 1528–1530, 2015.

[27] F. Maasdorp, *Doppler-only target tracking for a multistatic radar exploiting FM band illuminators of opportunity*. PhD thesis, University of Cape Town, Dec 2015.

[28] C. A. Tong, *A scalable real-time processing chain for radar exploiting illuminators of opportunity*. PhD thesis, University of Cape Town, 2014.

[29] C. James, *Multilateration: Radar's Replacement?* 2007. Available from www.aviationtoday.com/2007/04/01/multilateration-radars-replacement/

[30] C. James, *Multilateration Consortium (with ADS-B)*. Available from www.multilateration.com/

[31] R. M. Trim, "Mode S: an introduction and overview (secondary surveillance radar)," *Electronics Communication Engineering Journal*, vol. 2, pp. 53–59, Apr 1990.

[32] International Maritime Organisation, *Automatic Identification System (AIS)*. Available from www.navcen.uscg.gov/?pageName=aismain

[33] Afriforum, *South African Farm Murders*. Available from https://africacheck.org/factsheets/factsheet-statistics-farm-attacks-murders-sa/

[34] M. Ummenhofer, J. Schell, J. Heckenbach, H. Kuschel, and D. W. D. Hagan, "Doppler estimation for DVB-T based passive radar systems on moving maritime platforms," in *2015 IEEE Radar Conference (RadarCon)*, pp. 1687–1691, May 2015.

[35] K. Chetty, G. E. Smith, and K. Woodbridge, "Through-the-wall sensing of personnel using passive bistatic Wi-Fi radar at standoff distances," *IEEE Transactions on Geoscience and Remote Sensing*, vol. 50, pp. 1218–1226, Apr 2012.

[36] B. Tan, K. Woodbridge, and K. Chetty, "A wireless passive radar system for real-time through-wall movement detection," *IEEE Transactions on Aerospace and Electronic Systems*, vol. 52, pp. 2596–2603, Oct 2016.

[37] M. Amin, Y. Zhang, F. Ahmad, and K. Ho, "Radar signal processing for elderly fall detection: the future of in-home monitoring," *IEEE Signal Processing Magazine*, vol. 33, pp. 71–80, Mar 2016.

[38] B. Tan, A. Burrows, R. Piechocki, *et al.*, "Wi-Fi based passive human motion sensing for in-home healthcare applications," in *2015 IEEE 2nd World Forum on Internet of Things (WF-IoT)*, pp. 609–614, Dec 2015.

[39] Q. Chen, K. Chetty, K. Woodbridge, and B. Tan, "Signs of life detection using wireless passive radar," in *2016 IEEE Radar Conference (RadarConf)*, pp. 1–5, May 2016.

[40] H. Messer, A. Zinevich, and P. Alpert, "Environmental monitoring by wireless communication networks," *Science*, vol. 312, no. 5774, pp. 713–713, 2006.

[41] M. E. Davis, *Foliage Penetration Radar*. Raleigh, NC: SciTech, 2011.

[42] F. T. Ulaby, P. C. Dubois, and J. van Zyl, "Radar mapping of surface soil moisture," *Journal of Hydrology*, vol. 184, no. 1, pp. 57–84, 1996.

[43] N. Rodriguez-Alvarez, X. Bosch-Lluis, A. Camps, *et al.*, "Review of crop growth and soil moisture monitoring from a ground-based instrument implementing the interference pattern GNSS-R technique," *Radio Science*, vol. 46, pp. 1–11, Dec 2011.

[44] D. Barrick, M. Evans, and B. Weber, "Ocean surface currents mapped by radar," in *Proceedings of the 1978 IEEE First Working Conference on Current Measurement*, vol. 1, pp. 59–65, Jan 1978.

[45] C. C. Teague, D. E. Barrick, P. M. Lilleboe, R. T. Cheng, and C. A. Ruhl, "Long-term UHF RiverSonde river velocity observations at Castle Rock, Washington and Threemile Slough, California," in *Proceedings of the IEEE/OES Eighth Working Conference on Current Measurement Technology, 2005*, pp. 85–89, Jun 2005.

[46] H. D. Griffiths and C. J. Baker, "Passive coherent location radar systems. Part 1: performance prediction," *IEE Proceedings – Radar, Sonar and Navigation*, vol. 152, pp. 153–159, Jun 2005.

[47] M. Inggs and C. Tong, "Commensal radar using separated reference and surveillance channel configuration," *Electronics Letters*, vol. 48, no. 18, pp. 1158–1160, 2012.

[48] F. D. V. Maasdorp, C. A. Tong, A. Lysko, M. R. Inggs, and D. W. O'Hagan, "The design and development of a FM band passive radar test-bed for long term qualification testing," in *2017 IEEE Radar Conference (RadarConf)*, pp. 1515–1520, May 2017.

[49] N. Morrison, *Tracking Filter Engineering*. London: The Institution of Engineering and Technology, 2012.

[50] R. Nadjiasngar, *On improving the performance of the Gauss-Newton filter*. PhD thesis, University of Cape Town, Nov. 2013.

[51] M. R. Inggs, C. Tong, A. K. Mishra, and F. Maasdorp, "Modelling and simulation in commensal radar system design," in *Radar Systems (Radar 2012), IET International Conference on*, pp. 1–5, IET, 2012.

[52] D. W. Allan, N. Ashby, and C. C. Hodge, *The Science of Timekeeping*. Agilent application note 1289, 2000.

[53] *White Rabbit Project*. Available from www.ohwr.org/projects/white-rabbit/wiki/synchronization

[54] International Standards Organisation, *ISO 11898-1:2015 Preview Road Vehicles – Controller Area Network (CAN)*. Available from www.iso.org/standard/63648.html

[55] J. S. Sandenbergh and M. R. Inggs, "A common view GPSDO to synchronize netted radar," in *2007 IET International Conference on Radar Systems*, pp. 1–5, Oct 2007.

[56] K. Iniewski, C. McCrosky, D. Minoli, *Network Infrastructure and Architecture: Designing High-Availability Networks*. New York: Wiley Telecom, 2008.

[57] *IEEE 802.16 Working Group on Broadband Wireless Access Standards*. Available from www.ieee802.org/16/

[58] J. S. Sandenbergh and M. R. Inggs, "A common view GPSDO to synchronize netted radar," in *Radar Systems, 2007 IET International Conference on*, pp. 1–5, IEE, 2007.

[59] J. B. Bullock, T. M. King, H. L. Kennedy, E. D. Berry, and G. Zanfino, "Test results and analysis of a low cost core GPS receiver for time transfer applications," in *Frequency Control Symposium, 1997, Proceedings of the 1997 IEEE International*, pp. 314–322, IEEE, 1997.

[60] T. Derham, K. Woodbridge, H. Griffiths, and C. Baker, "The design and development of an experimental netted radar system," in *Radar Conference, 2003. Proceedings of the International*, pp. 293–298, IEEE, 2003.

[61] T. Derham, S. Doughty, K. Woodbridge, and C. Baker, "Design and evaluation of a low-cost multistatic netted radar system," *IET Radar, Sonar & Navigation*, vol. 1, no. 5, pp. 362–368, 2007.

[62] M. Ritchie, H. Griffiths, M. Inggs, F. Fioranelli, and S. Sandenbergh, "Analysis of polarimetric bistatic sea clutter using the NetRAD radar system," *IET Radar, Sonar & Navigation*, vol. 10, no. 8, pp. 1356–1366, Apr 2016.

[63] M. R. Inggs, H. D. Griffiths, K. Woodbridge, W. Al-Ashwal, and G. E. Inggs, "Experimental networked radar using GPS-disciplined reference oscillators," in *IET Cognitive Systems with Interactive Sensors (GOGIS)*, 2010.

[64] J. S. Sandenbergh and M. R. Inggs, "Synchronizing network radar using all-in-view GPS-disciplined oscillators," in *IEEE Radar Conference on*, IEE, 2017.

[65] T. Wlotowski, *Precise time and frequency transfer in a white rabbit network.* Master's thesis, Warsaw University of Technology, 2011.

[66] M. Inggs, S. Sandenbergh, and S. Lewis, "Investigation of white rabbit for synchronization and timing of netted radar," in *IEEE Radar Conference 2015*, (Johannesburg, South Africa), IEEE, 2015.

[67] M. Inggs, S. Sandenbergh, and S. Lewis, "Evaluating an off-the-shelf white rabbit system to synchronize network radar via optic fibre," in *IEEE Radar Conference 2017* (Seattle, Washington), IEEE, 2017.

[68] U.S. Department of Commerce, *United States Frequency Allocations*. Available from www.ntia.doc.gov/files/ntia/publications/spectrum_wall_chart_aug2011.pdf

[69] E. Noam, "Spectrum auctions: yesterday's heresy, today's orthodoxy, tomorrow's anachronism. Taking the next step to open spectrum access," *The Journal of Law and Economics*, vol. 41, no. S2, pp. 765–790, 1998.

[70] R. Oliva, E. Daganzo, Y. H. Kerr, *et al.*, "SMOS radio frequency interference scenario: status and actions taken to improve the RFI environment in the 1400–1427-MHz passive band," *IEEE Transactions on Geoscience and Remote Sensing*, vol. 50, no. 5, pp. 1427–1439, 2012.

Chapter 9

Fusion of radar sensing, data communications, and GPS interoperability via software-defined OFDM architecture

Dmitriy Garmatyuk[1]*, Yu Tong (Jade) Morton*[2]*, and Saba Mudaliar*[3]

9.1 Overview of OFDM in radar and communications

This chapter considers a variety of functionalities that are typically assigned to distinctly different devices, each designed separately and each requiring its own power supply, analog front end (AFE), physical space, spectral allocation, and computational processing power. Specifically, these devices comprise a radar-based sensor, a data communication transceiver, and a radio frequency (RF) position/navigation system. The rationale behind this "division of labor" is well understood. Certain types of waveforms are simply better suited to particular tasks. For example, linear frequency modulation (LFM) chirp signals are a veritable radar staple, while a variety of keying modulation formats are used for communications. There is an obvious redundancy in this approach; a redundancy that often leads to a critical increase in power consumption, spectral overcrowding, instrument weight and size, and demands upon computing resources. When combined, these increases may represent a formidable challenge especially when these systems are designed for autonomous, long-endurance platforms such as unmanned (aerial, ground, underwater) vehicles. Consequently, instead of choosing system parameters that are best suited to each function, it is worth investigating a prospective compromise solution that allows for a combination of these functions via judicious signal design. A uniting factor throughout this chapter is thus the signal and system architecture format that enables the fusion of the aforementioned functionalities. We explore this potential via the orthogonal frequency division multiplexing (OFDM) software-defined radar system (SDRS) concept.

OFDM as a method of digital modulation is not a new technique—its concept was first explored in the 1960s by several researchers (see [1]). However, it was not until

[1]Department of Electrical & Computer Engineering, Miami University, USA
[2]Department of Aerospace Engineering Sciences, University of Colorado, USA
[3]Sensors Directorate, Air Force Research Laboratory, USA

the beginning of the 2000s that this technique and corresponding system architecture started being considered for wideband applications. At the time there were two major reasons precluding ultra-wideband (UWB) implementation of OFDM: (1) the U.S. Federal Communications Commission (FCC) ban on commercial use of wide swaths of spectrum, which was lifted for extremely low-power signals in 2002 (discussed further in Section 9.2.2); and (2) the unavailability of fast, inexpensive analog-to-digital and digital-to-analog converters [2], which are significant components of an OFDM system that effectively determine its feasible bandwidth. Once these two hurdles were overcome, UWB OFDM quickly became a focus of research efforts in industry and academia with an emphasis on commercial broadband communications [3–5]. Implementation of OFDM signals and system architecture in radar, on the other hand, has not been as widely explored. Notable early research concerning OFDM-coded radar signals was initiated at Tel Aviv University by Levanon in the 2000s via analyses of multicarrier phase coded radar signal performance [6,7]. In the United States, a 2001 Air Force Research Laboratory (AFRL) "in-house" report proposed using OFDM for an interferometer implementation [8]. Levanon's work was further expanded to develop a family of OFDM-like multi-carrier bi-phase signals in 2006 [9], around which time OFDM for radar was starting to receive wider attention. Doppler tolerance and the wideband ambiguity function (WAF) of OFDM radar signals were discussed in [10] and [11], respectively. The potential for combining radar and communications functionality in multiple-input/multiple output (MIMO) systems using OFDM signals was proposed in [12]. In [13], a simulation study of synthetic aperture radar (SAR) imaging using UWB OFDM radar waveforms was described; and constant-envelope OFDM (CE-OFDM) for radar and communications was suggested in [14]. In 2009, we described the first approach to the design of an UWB OFDM SDRS [15,16] that was based on the general SDRS concept first introduced by Wiesbeck in 2001 [17]. In SDRS and more specifically in software-defined SAR (SD-SAR) systems, a concept was envisioned to facilitate fully software-configurable transmitters and modular digital receiver data processors with downloadable software for a variety of scenarios. To maximize the performance of a multiple-use SDRS system, it may be desirable—or even necessary—to employ UWB signals and components, which allow for best flexibility in functionality. It should be noted (cf. Scholtz and Lee [18]) that obtaining analytical, closed-form solutions for environments in UWB swaths of spectrum may often lead to significant inaccuracies, or such solutions may not even exist. This limitation led to the development of the first experimental UWB OFDM SDRS by our group at Miami University [19], which was then tested in a variety of scenarios including ranging, SAR imaging, data communications, and subsequently radar-assisted positioning and localization [20,21].

Recently, an exhaustive theoretical assessment of OFDM radar sensor performance in certain scenarios was performed by Sen and Nehorai [22–28]. Specifically, an adaptive OFDM radar signal for moving target detection in urban settings was proposed and analyzed in [22]. Adaptive design of a WAF of OFDM signals aimed at reducing range sidelobes was developed in [23]. In [24], OFDM radar was introduced for potential use in a MIMO configuration for low-grazing angle tracking due to the richer information about the target afforded by OFDM waveforms' frequency diversity. Multi-objective optimization (MOO) of OFDM radar signals—which included exploitation of the frequency and spatial diversity afforded by usage of wideband OFDM waveforms—was investigated [25] in the context of target detection reformulated as sparse-signal spectral estimation. Subsequent work included analysis of OFDM radar signals for multi-target tracking under sparsity conditions [26], space-time adaptive processing of OFDM radar signals [27], and OFDM signal design for micro-Doppler estimation [28].

Thus, with OFDM being recognized as a viable radar waveform since the early 2000s, the last five years of investigations by various research groups have brought about more than 50 publications on topics ranging from implementation of OFDM for MIMO-SAR, to its use in passive radar, to fusion of sensing and communications via OFDM. The latter is the focus of this chapter, along with the general concept of achieving interoperability among a variety of RF services via an OFDM SDRS architecture.

The chapter begins with design considerations for UWB OFDM SDRS, in which we convey our own experience in developing such a system, emphasizing the software-defined architecture as well as transmit power limitations and requirements. We then discuss radar–communication fusion using the SDRS and give an example of practical implementation. The subsequent section focuses on the covertness of radar-communication waveforms, which is proposed to be achieved via randomization of OFDM subcarrier coefficients and demonstrated using two examples, along with a method of cross-range compression tailored to such signals for imaging radar. The coexistence of UWB OFDM and global navigation satellite system (GNSS) receivers through use of the same AFE is then discussed. Finally, we conclude with remarks about future perspectives in multifunctional UWB OFDM sensor systems and their implementations.

9.2 Design of an UWB software-defined system based on OFDM

It is useful to first review some of the general practices of designing an UWB system for in-lab use as a multi-functional, software-defined RF system. The particular focus here is upon the challenges and possible solutions for the assembly and operation of such a system. Specifically, we address the transmit power considerations that stem from the FCC requirements on UWB emission limits, and the corresponding system design adjustments that ensue. A method is also suggested for joint OFDM frequency-domain signal processing of sensing and communications based on I/Q (in-phase/quadrature-phase) reception and subcarrier magnitude extraction from non-synchronized received signals. This method may be used, for example, in cases involving MIMO implementations or backscatter communications.

9.2.1 General considerations for an UWB SDRS design

As noted above, OFDM lends itself to the design of a software-defined architecture quite well; the signal can be designed in software and sent to/from the AFE via a

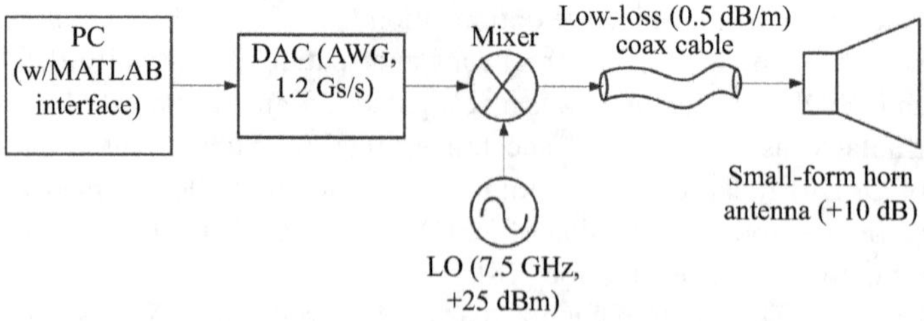

Figure 9.1 Transmitter block diagram

digital-to-analog converter and analog-to-digital converter (DAC/ADC) pair. Unlike most other radar waveforms, OFDM is generated by assigning weights to its frequency components (or subcarriers) and then translating the result into time-domain samples. Thus the first step in signal design involves populating the vector of subcarrier coefficients **S** with coefficients according to

$$\mathbf{S} = [0,\ S(1),\ \ldots,\ S(N),\ S(N),\ \ldots,\ S(1)], \tag{9.1}$$

where N is the total number of subcarriers in the baseband OFDM symbol and the structure of **S** reflects the desired discrete spectrum of the transmit signal. The case shown here has a symmetric spectrum which corresponds to an all-real time-domain signal. The zero-frequency component is nulled, resulting in a suppressed carrier for the up-converted transmit waveform. This vector is then sent to a fast Fourier transform (FFT) processor, followed by a DAC operating at sampling frequency f_s. The output from the DAC is an analog signal that, after up-conversion to carrier frequency f_c and amplification by a factor K, can be expressed as

$$s_{\text{transmit}}(t) = K \cdot \left\{ \sum_{n=1}^{2N+1} s_n \cdot \prod_{\tau=1/f_s} \left(t - \frac{n-1}{f_s} \right) \right\} \cos(2\pi f_c t),$$
$$t \in \left[0, \frac{2N}{f_s} \right]. \tag{9.2}$$

Here, $\prod_{\tau=1/f_s}$ represents a square pulse of duration $1/f_s$, and the coefficients s_n are time-domain amplitude values obtained from the FFT of the values $S(k)$ in **S** as

$$s_n = \frac{1}{2N+1} \sum_{k=1}^{2N+1} S(k) \cdot e^{j2\pi \frac{(k-1)(n-1)}{2N+1}} \quad \text{for } n = 1, 2, \ldots, 2N+1. \tag{9.3}$$

This signal-forming procedure is further illustrated by the transmitter block diagram in Figure 9.1, which represents the current implementation of Miami University's (MU) UWB OFDM SDRS. In the diagram, "LO" represents a solid-state local oscillator that is used simultaneously as a source of transmit power and for up-conversion of the baseband OFDM signals to 7.5 GHz carrier frequency. Likewise, "AWG" represents an arbitrary waveform generator that is an off-the-shelf DAC board [29].

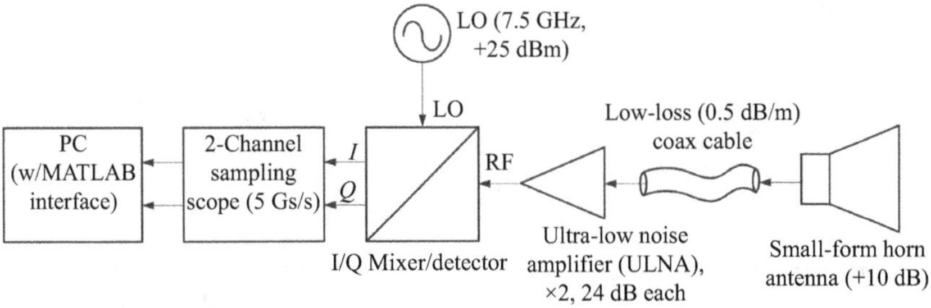

Figure 9.2 Receiver block diagram

The reception and processing of the received signals can be performed in approximately the reverse order of the transmit signal formation—i.e. amplification, down-conversion, and digitization via an ADC. One interesting aspect of OFDM is that, owing to its frequency domain modulation arrangement, it is possible to perform signal reconstruction even when the transmitter and the receiver are not synchronized, and this attribute holds for both communication and radar sensing functions. This operation can be accomplished via the receive architecture/procedure depicted in Figure 9.2.

The output from the I/Q detector in Figure 9.2 is

$$
\begin{aligned}
s_{\mathrm{RX\_I}}(t) &= \frac{A_{\mathrm{RX}}}{2} s_{\mathrm{TX}}(t - t_{\mathrm{m}}) \cos(\underbrace{2\pi f_c t_{\mathrm{m}} - \Delta\phi}_{\Theta_{\mathrm{m}}}) + w_{\mathrm{I}}(t) \\
s_{\mathrm{RX\_Q}}(t) &= \frac{A_{\mathrm{RX}}}{2} s_{\mathrm{TX}}(t - t_{\mathrm{m}}) \sin(\underbrace{2\pi f_c t_{\mathrm{m}} - \Delta\phi}_{\Theta_{\mathrm{m}}}) + w_{\mathrm{Q}}(t)
\end{aligned} \tag{9.4}
$$

where A_{RX} is the amplitude coefficient associated with channel and target scene effects, t_{m} is time delay, $\Delta\phi$ is the phase mismatch between the carrier frequency signal at up-conversion and down-conversion (during I/Q detection), and w_{I} and w_{Q} are cumulative representations of external disturbance and own system noise in the I and Q channels, respectively. After these signals are sampled and converted to the frequency domain, the essential information is still preserved when the I- and Q-channel results are combined and represented in the same format as the subcarrier coefficients from (9.1) as

$$
\begin{aligned}
\hat{\mathbf{S}}_{\mathrm{IQ}} &= \left[\hat{S}_{\mathrm{IQ}}(N),\ \ldots,\ \hat{S}_{\mathrm{IQ}}(1),\ 0,\ \hat{S}_{\mathrm{IQ}}(1),\ \ldots,\ \hat{S}_{\mathrm{IQ}}(N)\right] \\
\hat{S}_{\mathrm{IQ}}(k) &= \sqrt{\hat{S}_{\mathrm{I}}^2(k) + \hat{S}_{\mathrm{Q}}^2(k)}
\end{aligned}, \tag{9.5}
$$

where $\hat{S}_{\mathrm{I}}(k)$ and $\hat{S}_{\mathrm{Q}}(k)$ are

$$
\begin{aligned}
\hat{S}_{\mathrm{I}}(k) &= S(k) A_{\mathrm{RX}}\, e^{-j2\pi \frac{k \cdot f_s}{2N+1} \cdot t_{\mathrm{m}}} \cos\Theta_{\mathrm{m}} + w_{\mathrm{I},k} \\
\hat{S}_{\mathrm{Q}}(k) &= S(k) A_{\mathrm{RX}}\, e^{-j2\pi \frac{k \cdot f_s}{2N+1} \cdot t_{\mathrm{m}}} \sin\Theta_{\mathrm{m}} + w_{\mathrm{Q},k}
\end{aligned}. \tag{9.6}
$$

Averaging over time approximately produces

$$\overline{\hat{S}}_{\mathrm{IQ}}(k) \approx \left\{ \left(S(k) A_{\mathrm{RX}} \, e^{-j2\pi \frac{k \cdot f_s}{2N+1} \cdot t_{\mathrm{m}}} \right)^2 \cdot \left(\cos^2 \Theta_{\mathrm{m}} + \sin^2 \Theta_{\mathrm{m}} \right) \right\}^{\frac{1}{2}}$$
$$= S(k) A_{\mathrm{RX}} \, e^{-j2\pi \frac{k \cdot f_s}{2N+1} \cdot t_{\mathrm{m}}}. \tag{9.7}$$

Finally, note that no power amplifier is used in this transmitter chain and thus transmit power is supplied by the LO only. This implementation is aimed at preventing non-compliance with the UWB emission rules established by the FCC [30]. The question of just how much transmit power is allowed for the case of indoor UWB OFDM sensor system is not a trivial one and is considered next.

9.2.2 Transmit power considerations for indoor UWB OFDM SDRS

When the FCC rules regarding UWB emission limits were first put forth in its February 14, 2002 "Report and Order" (along with formal authorization for unlicensed use of the entire frequency band from 3.1 to 10.6 GHz by UWB devices), it stipulated that those systems that employ gating should be tested for peak power emissions with only the active transmission intervals gated on (recorded), essentially requiring quasi-continuous wave (CW) operation for testing purposes (see §15.521(d) in [30]). This condition was seen as unfair and too conservative by the Multi-Band OFDM Alliance Special Interest Group (MBOA-SIG) that represents "approximately 160 domestic and foreign companies seeking adoption by the IEEE 802.15 Task Group 3a of a UWB standard specifying the use of its MB-OFDM modulation" [31]. On August 26, 2004, the MBOA-SIG submitted a Petition for Waiver to the FCC requesting that the requirement of testing gated systems with emission gating always on (the virtual CW mode) be waived for indoor and handheld devices. This waiver was granted by the FCC on March 10, 2005: "…we are waiving that portion of 47 C.F.R. §15.521(d) that requires the emissions from UWB devices that employ gating to be measured with the emission gated on. The emissions from such systems shall be measured in their normal operating mode" [31]. The waiver also mentions that it applies only to such indoor and/or handheld UWB systems that operate in the 3.1–5.03 GHz and/or 5.65–10.6 GHz frequency ranges. According to the original "Report and Order" these systems, when tested in normal operation, must emit no more power than −41.3 dBm/MHz. Therefore, the expression [32]

$$P_{\mathrm{ave}} = \frac{1}{T_{\mathrm{PRI}}} \int_0^{T_{\mathrm{PRI}}} x_{\mathrm{T}}^2(t) dt, \quad \text{for} \quad x_{\mathrm{T}}(t) = \begin{cases} s_{\mathrm{transmit}}(t), & 0 \le t < \dfrac{2N}{f_s} \\ 0, & \dfrac{2N}{f_s} \le t < T_{\mathrm{PRI}} \end{cases} \tag{9.8}$$

is subject to this limit for any indoor UWB SDRS with pulse gating [32], where T_{PRI} is the pulse repetition interval (PRI) and $x_{\mathrm{T}}(t)$ is the gated transmit signal.

If we define the ungated maximum transmit power due to the LO as $P_{\mathrm{CW,max}}$ and take into account cable losses P_{Loss} and transmit antenna gain D_{ANT}, then the expression for this power is

$$P_{\text{CW,max}} = P_{\text{LO}} - P_{\text{Loss}} + D_{\text{ANT}}\,(\text{dBm}). \tag{9.9}$$

Applying the permitted power limit of −41.3 dBm/MHz and accounting for the actual transmit signal bandwidth B_{GHz} expressed in multiples of GHz, we obtain the maximum allowable power

$$P_{\max}(\text{dBm}) = -41.3 + 30 + 10\log_{10}(B_{\text{GHz}}), \tag{9.10}$$

where the factor of 30 accounts for the conversion from MHz to GHz. Applying signal gating, we arrive at the final expression for the maximum transmit power (in linear units)

$$P_{\max}(\text{linear}) = P_{\text{CW,max}}(\text{linear})\frac{2N}{f_s\,T_{\text{PRI}}}, \tag{9.11}$$

which then leads to a formula for the desirable pulse repetition frequency (PRF):

$$f_{\text{PRF}} \leq \frac{P_{\max}(\text{linear})f_s}{2N\,P_{\text{CW, max}}(\text{linear})}. \tag{9.12}$$

Example (PRF): Using the MU UWB OFDM SDRS parameters, we estimate the maximum allowable PRF as follows. First, the maximum allowable power in dBm is determined using (9.10) to be $-41.3 + 30 + 10\log_{10}(1.2) \approx -10.41$, or approximately 91 μW. Using (9.9), the maximum transmit CW power, accounting for the loss in a 3-m cable, is estimated as $25 - 1.5 + 10 = +33.5$dBm, or approximately 2.24 W. Using $N = 32$ subcarriers and $f_s = 1.2$ GHz, we arrive at the maximum PRF value of approximately 761 Hz, which is easily realized by zero-padding signal vectors from the MATLAB interface to the transmitter DAC.

Example (Maximum range): To verify that this system can operate in indoor conditions for the purpose of experimenting with target detection and imaging, the maximum free-space (ideal) achievable range of the SDRS can be estimated using the radar equation [33]

$$R_{\max} = \left\{\frac{P_{\text{LO}}G^2\lambda_c^2\sigma_{\text{TCR}}}{(4\pi)^3\cdot(kTBFL)\cdot SNR_{\min}}\right\}^{\frac{1}{4}} \tag{9.13}$$

and assuming a trihedral corner reflector (TCR) as a target. In (9.13), λ_c is the wavelength of the signal at the carrier frequency, σ_{TCR} is the radar cross section (RCS) of the TCR, and k is Boltzmann's constant. We set the antenna gain to $G = 10$ dB, the minimum received signal-to-noise ratio to $SNR_{\min} = 6$ dB, noise temperature $T = 293$ K, bandwidth $B = 1.2$ GHz, receiver noise figure $F = 2$ dB, and total losses in the antenna cable assembly to $L = 3$ dB. A square TCR with side length of $a = 20$ cm at carrier frequency $f_s = 7.5$ GHz realizes an RCS value of [34]

$$\sigma_{\text{TCR}} = \frac{12\pi a^4}{\lambda_c^2} \cong 37.7\text{m}^2. \tag{9.14}$$

Thus the free-space R_{max} estimate is approximately 63 meters, which even if reduced by half due to indoor conditions (e.g. multi-path losses) would still be deemed sufficient for most indoor testing.

9.3 Dual use of system bandwidth and transmit power via OFDM radar/communication signals

This section provides a brief overview of multifunctional RF systems that possess radar sensing and data communication options "fused" for the purpose of attaining reduction of component count, physical dimensions, and to allow for reuse of transmit power and bandwidth. Following the overview, we present an example of how OFDM-based systems and signals can be used efficiently to achieve the same goals while reducing complexity and providing for significant flexibility in waveform design.

9.3.1 Radar-embedded communications and radar/communication signals

The era of multifunctional RF sensor systems began around the end of the 1970s when a dual-use radar and communication sub-system was fielded in the Space Shuttle Orbiter by the U.S. National Aeronautics and Space Administration (NASA) [35]. While the subsystem had different signal processing components for each functionality, the radar and communication implementations did share a wideband transmitter, a two-channel receiver, and an antenna, thus providing significant savings on weight, physical dimensions, power requirements, and overall system complexity. Later, in the beginning of the 2000s a multifunction RF test-bed developed by the U.S. Naval Research Laboratory (NRL) under the Advanced Multifunction Radio Frequency Concept (AMRFC) program [36] sought to integrate such functionalities as radar sensing, electronic warfare, and communications into a single system.

Another example of fusing radar and communications into a single architecture is described in [37], where UWB LFM signals having opposite chirp slopes were employed to achieve orthogonality between the radar and communication channels. The authors of [37] reported an experimental range resolution of 63 cm and a bit error rate (BER) of 0.2% for simultaneous data communication at 1 Mb/s when the transmitter output was connected directly to the receiver input in a loopback configuration through a 60-dB attenuator. Both carrier and data clock signals were hardwired from the transmitter to the receiver as well.

The topic of radar–communication functionality "fusion" subsequently received further attention in both academia and industry, particularly from the perspective of automotive-based design; see [38] for an excellent overview. In scenarios requiring covertness of communications, radar signals may be used to conceal the transmitted data; for example, in [39], the authors propose to utilize coherent random noise signaling transmitted via orthogonally polarized channels such that the resultant waveform appears noise-like. OFDM can also be duplexed with random noise to create a radar-communications waveform [40] in which communication is carried out by the OFDM-modulated signal embedded in noise radar transmissions.

An interesting recent development in joint radar/communications is the backscatter communications approach [41], which does not rely on having line-of-sight (LOS) between transmitter and receiver. Rather, it uses reflections from a target scene common to two or more friendly platforms to convey data encoded onto the sensor waveforms. In such a scenario, the locations of the specific reflecting targets are unknown, as well as even the existence of a friendly platform in the same scene. Below, we consider an example of using the OFDM system architecture to construct and process radar/communication signals for the purpose of reusing pulse bandwidth and transmit power between these functions.

9.3.2 *Example: OOK OFDM signal performance in radar and communications*

While a radar signal designer can have significant freedom in choosing the composition of an OFDM pulse—its duration, number of subcarriers, and subcarrier amplitude and phase coefficients are all changeable on a pulse-to-pulse basis in software-defined systems—it is illustrative to consider a simple scenarios in which the subcarrier amplitude assignment is based on a slight variation of on/off-keying (OOK). That is, let us select values of vector $\boldsymbol{S}$ in (9.1) from the set $\{0, +1, -1\}$ on a random basis and evaluate the performance of the resultant radar and communication signals according to their ambiguity function (AF), peak sidelobe level (PSL), and BER. Signals generated in this way do not embed communication data into the radar waveforms. Rather, the exact same pulse can serve either as a radar signal (if its reflection off a target scene is received by a platform in sensing mode) or as a data signal (if received by a friendly platform in listening mode).

9.3.2.1 AF and PSL of UWB OOK-like OFDM radar signals

Conventionally, the narrowband form of the AF is used and the closed-form integral solution for analog cases is obtained before plotting the function [42]. Such an approach was taken in [10] to plot the AF of an OFDM signal consisting of 8 sub-bands spread over a 5 MHz bandwidth. However, a UWB OFDM signal should be treated differently due to its wide bandwidth. The error that results from using the narrowband approximation to compute a wideband signal's AF is derived and discussed in [11], which uses the conventional integral form of the AF definition. In [43], a similar approach is used to derive and optimize a narrowband AF of OFDM radar signals with up to 7 orthogonal sub-bands. In [44], we derived the discrete form of a UWB OFDM radar signal AF, since information extraction in OFDM systems is performed on a digital baseband waveform, which is

$$\chi\left(\tau_{\mathrm{m}}, \nu\right) = \left| \sum_{k=1}^{2N+1} s_{\mathrm{RX}}(t_k, \nu)\ \hat{s}_{\mathrm{RX}}(\tau_{\mathrm{m}} - t_k) \right|^2 \tag{9.15}$$

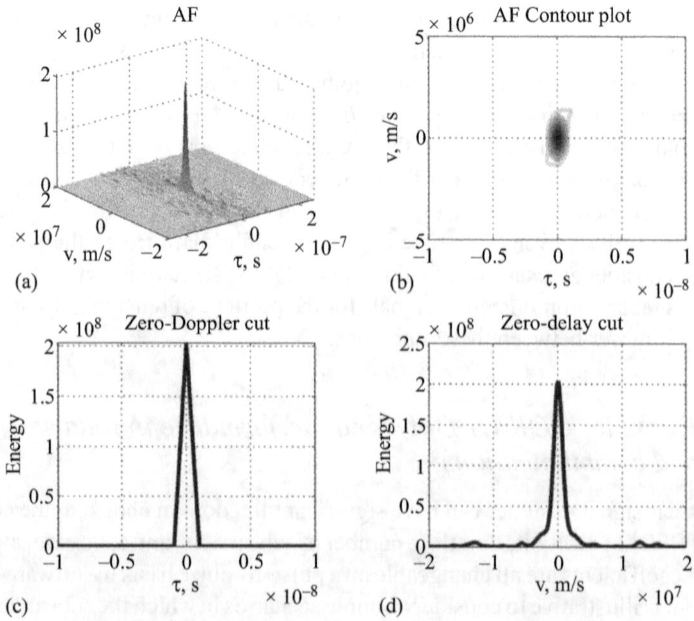

Figure 9.3 Example of an UWB OOK-like OFDM radar waveform with random subcarrier amplitudes selected from {0, +1, −1} for (a) a 3-D plot of the AF, (b) a contour plot of the center region, (c) the zero-Doppler cut, and (d) the zero-delay cut. (Reprinted, with permission, from [44])

for $s_{\mathrm{RX}}(t_k, v)$ the received signal reflected by a point target expressed as a function of discrete sampling time points $t_k = (k-1)\Delta t$ as

$$s_{\mathrm{RX}}(t_k, v) = \sum_{n=1}^{N} s_n \cos\left(\frac{2\pi n}{2N+1}\left[\frac{\left(\frac{c-v}{c+v}\right)t_k - \left(\frac{2R_0}{c+v}\right)}{\Delta t} - N\right]\right). \quad (9.16)$$

Here, the sampling period is $\Delta t = 1/f_s$, the radial velocity of the radar platform v is with respect to the illuminated target scene (assumed constant during the observation interval), c is the speed of light, R_0 is the point target's initial range, and $\hat{s}_{\mathrm{RX}}$ is the reference signal function obtained from (9.16) by setting $v = 0$. An example simulated AF is shown in Figure 9.3 for the case of a 500 MHz bandwidth OFDM pulse comprised of 128 subcarriers with amplitudes randomly selected from the ternary alphabet $\{0, +1, -1\}$, which in this instance resulted in 17 subcarriers with a zero amplitude weight. As reported in [23], the AF for OFDM radar signals is devoid of the prominent range-Doppler (RD) coupling that is commonly associated with LFM waveforms.

To illustrate the dependence of PSL on the number of subcarriers and the "fill ratio" of the OOK OFDM subcarriers (i.e. the number of subcarriers with non-zero weights divided by the total number of subcarriers N), Monte Carlo simulations were

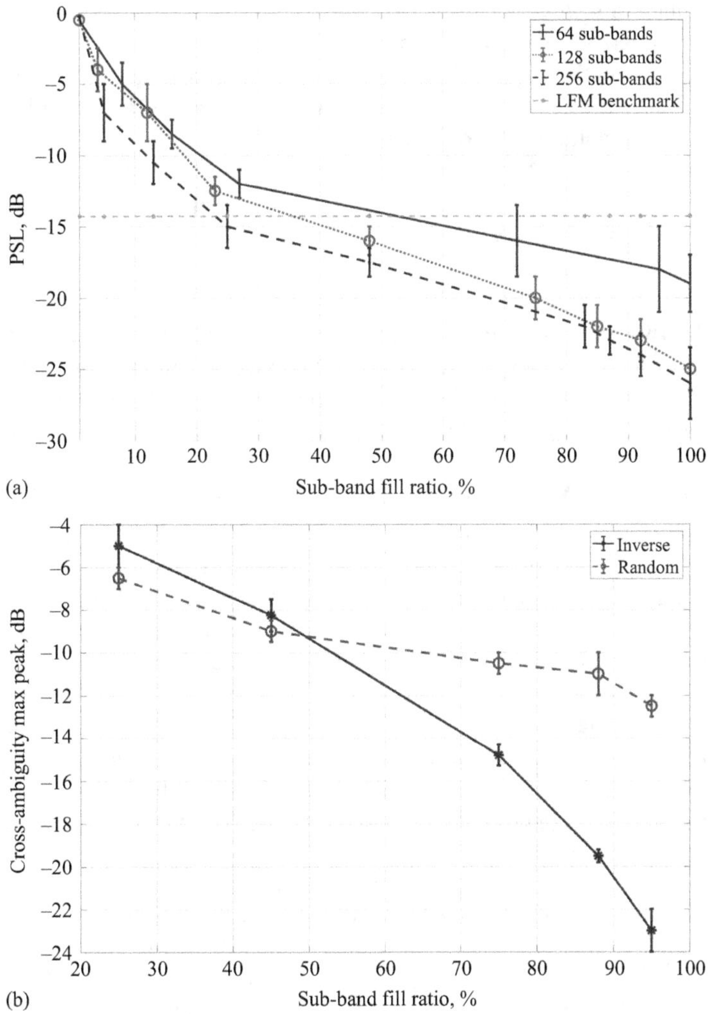

Figure 9.4 Monte Carlo simulation results for 500 MHz bandwidth OFDM radar signals for (a) the PSL performance and (b) the peak XAF performance for the 128 subcarrier case. (Reprinted, with permission, from [44])

performed for an ensemble of realizations. The PSL is defined as the ratio of the peak sidelobe value to the peak mainlobe value when s_{RX} and $\hat{s}_{RX}$ in (9.15) have identical subcarrier amplitude weights; thus PSL can be obtained from the resultant zero-Doppler cut of the AF to evaluate range reconstruction performance. Conversely, when s_{RX} and $\hat{s}_{RX}$ have different subcarrier amplitude weights, the cross-ambiguity function (XAF) likewise results from (9.15), which in this case is a measure of orthogonality between two randomly generated OOK-like OFDM pulses. The Monte Carlo PSL performance of a 500 MHz bandwidth OFDM radar signal with random subcarrier weights is illustrated in Figure 9.4(a) for various numbers of subcarriers and as a function of the subcarrier fill ratio. Figure 9.4(b) likewise plots the maxima of XAFs versus subcarrier fill ratio for the case of 500 MHz bandwidth and $N = 128$ subcarriers. In Figure 9.4(b), "inverse" refers to a simple method for cross-correlation minimization that involves the generation of two OOK-like OFDM pulses with subcarrier amplitudes distributed according to

$$\mathbf{S}_1 = \begin{cases} S_{k1}[n] = \pm 1, & \text{for } n \text{ odd} \\ S_{k1}[n] = 0, & \text{for } n \text{ even} \end{cases} \quad \text{and} \quad \mathbf{S}_2 = \begin{cases} S_{k2}[n] = 0, & \text{for } n \text{ odd} \\ S_{k2}[n] = \pm 1, & \text{for } n \text{ even} \end{cases} \tag{9.17}$$

so that the two signals have no nonzero weighted subcarriers in common. It is observed that for the particular cases in which subcarrier fill ratios exceed 50% and the total number of subcarriers $N > 64$, the OOK-like OFDM signals exhibit better PSL performance than the benchmark LFM chirp with same bandwidth and energy.

9.3.2.2 BER of UWB OOK-like OFDM communication signals

We can further evaluate the performance of UWB OOK-like OFDM signals by considering them in the context of data communications. Assuming the presence of a static flat-fading channel with additive white Gaussian noise (AWGN), the free-space loss model established by the IEEE 802.15.3a channel modeling committee [4], and inter-carrier interference (ICI) [19], we can derive an analytic expression for BER following the general procedure from [45,46].

In OOK modulation, the subcarrier coefficients S_k are drawn with equal probability from $X_k \in \{1, 0\}$. Following [46], we proceed by assigning a coefficient of 1 to the first subcarrier and formulating the BER as the probability of error $P_{\text{BER}} = P(r_1 < 0.5)$, where r_1 is the amplitude coefficient of the first subcarrier as reconstructed from the received signal. The formula for the characteristic function of r_1 is then identical to the expression derived for the case of binary phase shift keying (BPSK), which is equation (9.5) in [46]. This relationship leads to the result [19][1]

$$P_{\text{OOK}} = \frac{1}{2^{N-1}} \cdot \sum_{k=1}^{2^{N-2}} \frac{1}{\sqrt{2\pi\sigma^2}} \left\{ \begin{array}{l} \displaystyle\int_{-\infty}^{1/2} \exp\left(-\frac{(x-\theta_k)^2}{2\sigma^2}\right) dx \\ \displaystyle + \int_{-\infty}^{1/2} \exp\left(-\frac{(x-\beta_k)^2}{2\sigma^2}\right) dx \end{array} \right\} \tag{9.18}$$

where N is again the total number of subcarriers, σ is the standard deviation of AWGN, and θ_k and β_k are special coefficients that quantify the effect of ICI according to [46]

$$\begin{Bmatrix} \theta \\ \beta \end{Bmatrix}_k = \Re\left(1 \begin{Bmatrix} + \\ - \end{Bmatrix} \mathrm{Z}^T \mathbf{e}_\mathbf{k}\right). \tag{9.19}$$

[1]©2011 IEEE. Reprinted, with permission, from *IEEE Sensors Journal*, 2011, vol. 11(1), pp. 99–106.

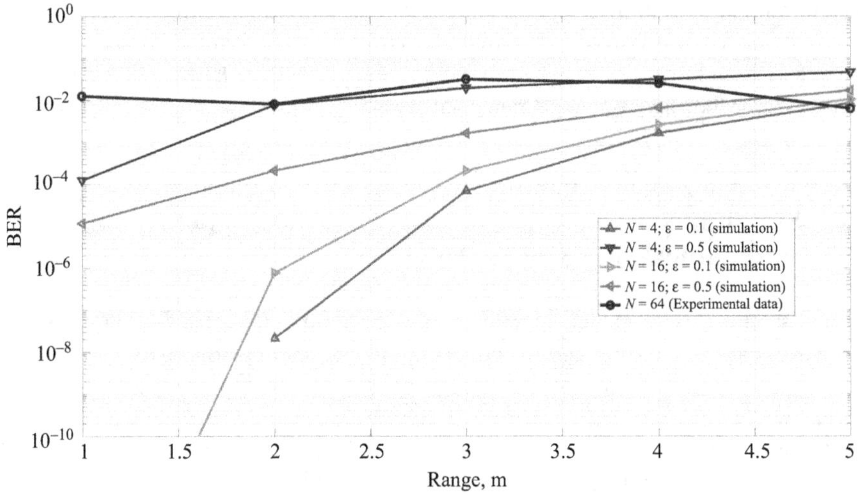

Figure 9.5 BER as a function of communication range (simulated and experimental data)

Here, $\Re(\cdot)$extracts the real part of the argument and the vector $\mathrm{Z} = (\mathrm{Z}_2\ \mathrm{Z}_3\ \ldots\ \mathrm{Z}_N)^T$ was defined in [46] to be

$$\mathrm{Z}_k = \frac{\sin(\pi[k-1+\varepsilon])}{N\sin\left(\dfrac{\pi[k-1+\varepsilon]}{N}\right)}\ \exp\left(j\pi\left(1-\frac{1}{N}\right)[k-1+\varepsilon]\right) \tag{9.20}$$

for ε the frequency offset normalized by subcarrier spacing Δf, and $\mathbf{e}_k$ a column vector obtained by representing the number $(2^{N-1}-k)$ in binary format, albeit with each zero replaced by (-1). Evaluation of the integrals in (9.18) leads to the final BER expression for OOK-modulated OFDM pulses [19] which takes the form

$$P_{\mathrm{OOK}} = \frac{1}{2^{N-1}}\sum_{k=1}^{2^{N-2}}\left\{Q\left(\sqrt{2\gamma}\left[\theta_k-\frac{1}{2}\right]\right)+Q\left(\sqrt{2\gamma}\left[\beta_k-\frac{1}{2}\right]\right)\right\}, \tag{9.21}$$

where $Q(\cdot)$is the Gaussian Q-function, $\gamma = E_b/N_0$ is the SNR (where $E_b = 1$) and $N_0 = 2\sigma^2$ is the AWGN power spectral density. Note that this result may have limited utility when N is sufficiently large because the number of inner products $\mathrm{Z}^T\mathbf{e}_\mathbf{k}$ that need to be computed via (9.19) for use in (9.18) is 2^{N-1}. For instance, $N=128$ requires in excess of 3.4×10^{38} inner product operations, thus requiring considerable computational power. A single inner product computation with $N=128$, performed on a PC with Intel® Core™ i7-6700 3.40 GHz processor and using MATLAB® R2016a, takes approximately 5 ms runtime, which renders implementation of (9.18) with $N=32$ or greater quite problematic. Figure 9.5 illustrates the simulation results for several values of N below 32, along with an experimentally obtained estimate of BER using the UWB OFDM SDRS for $N=64$. Here, we used the distance between transmit and receive antennas as an inverse function of SNR (in experiment) and converted SNR into communications range for the simulated cases. Due to a limited number of trials, the experimental result does not exhibit monotonic growth of BER with distance (as expected) for the chosen span of range values.

9.4 Simultaneous sensing and covert, ad-hoc asynchronous communications with OFDM

In this section, we consider approaches for the randomization of OFDM signals to realize covert sensing and communications. A method of cross-range compression (used in imaging radar signal processing) is also described that utilizes signals with randomized subcarrier coefficients to extract the phase history of target reflections. Collectively, these components describe a framework for designing and processing signals with low-probability of intercept/detection (LPI/LPD) potential to perform joint radar/communications functions in ad-hoc environments requiring covertness. Thus the environment under consideration may contain interceptors (i.e. "eavesdropping" platforms), while friendly platforms may lack synchronization.

9.4.1 Randomization of radar/communication signals for communications

Covert RF operations can be of great importance for some applications and there are multiple ways in which to implement LPI/LPD systems and waveforms. In [47], for example, methods of signal design for RF tags embedded in areas illuminated by a high-power pulsed radar are examined. The purpose of the design is to create waveforms that carry communication data from the tags to intended receivers, but also appear similar to the radar backscatter, thus providing the LPI/LPD attribute. Other approaches that achieve LPI/LPD performance include wideband noise radar [48,49], which is noted to provide a significant measure of protection against digital RF memory (DRFM)-based deception jamming [50], has a good potential for coexistence with other RF services, and realizes relatively low PSL characteristics. While true noise radar will use amplified samples of naturally occurring noise—e.g. a band-limited signal taken from a reverse-biased microwave diode [51]—other ways to approximate it with pseudo-random and stochastic-based signaling have been considered as well (cf. [52]). Randomization of UWB OFDM subcarrier amplitude weights can be regarded as part of the latter category. Below, we present an example of using this method to penalize an interceptor ("eavesdropper") in a scenario involving covert sensing as well as ad-hoc, non-LOS (line of sight), asynchronous communications. This scenario envisions a radar that performs a nondirectional, widebeam illumination of a scene containing friendly communication receiver platforms that may be positioned so as to be capable of receiving the transmitted radar signals through either direct or reflected paths. Due to the requirement of high transmit power for the radar/communication pulses, and the likely need to retransmit the same message multiple times to establish

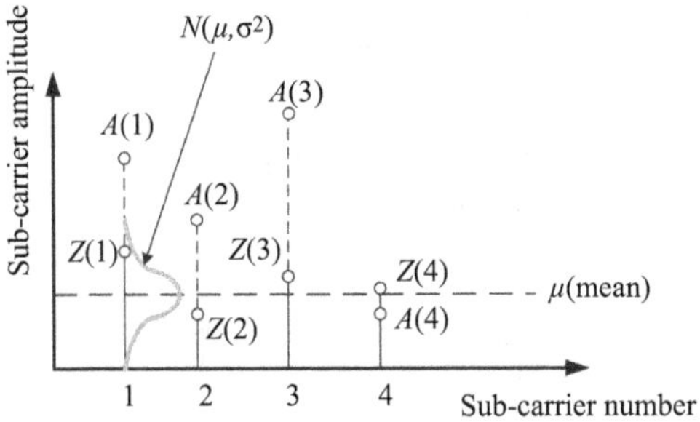

Figure 9.6 Graphical representation of subcarrier amplitude coefficient randomization

the initial links to the friendly platforms, such interceptor penalization may be highly desirable.

Consider that we are to retransmit the data vector $\boldsymbol{A}$ of length N (the number of OFDM subcarriers per each pulse), where $A(k)$ are real-valued and positive. The $A(k)$ coefficients are directly encoded onto the $S(k)$ amplitude coefficients of the OFDM subcarriers discussed in the previous sections. Next, a sample from the random process $Z(k)$ is added to each subcarrier as illustrated in Figure 9.6. The $Z(k)$ values are drawn from a normal distribution $N(\mu, \sigma^2)$ whose parameters are not known to the intended receivers [53].

Thus, the samples of the randomized OFDM radar-communication signal are constructed as

$$s_n^{\text{rand}} = \frac{1}{2N+1} \sum_{k=1}^{2N+1} (A(k) + Z(k))\ e^{j2\pi \frac{(k-1)(n-1)}{2N+1}}, \tag{9.22}$$

where one of the values in vector $\boldsymbol{A}$ is set to zero for all transmissions, i.e. $A(m) = 0$ and the value of index m is known to both the transmitting platform and all friendly receivers. This arrangement translates to the mth subcarrier carrying samples of the random process only, while the remaining $(N - 1)$ subcarriers carry both data and random samples.

Upon reception, the sampled signal is converted into the frequency domain using an FFT and the subcarrier coefficient amplitudes are obtained, per (9.5) and (9.6), as

$$\hat{S}_{\text{RX,IQ}}(k) \approx \begin{cases} (A(k) + Z(k))\ \Omega_{\text{A}} + w_k & k \neq m \\ (Z(k))\ \Omega_{\text{A}} + w_k & k = m \end{cases} \tag{9.23}$$

in which Ω_{A} is a cumulative amplitude coefficient that combines antenna gain, propagation losses, and target scene reflection. Thus, following several re-transmissions, the mean of the mth subcarrier amplitude coefficient becomes an effective estimator of μ via

$$\begin{aligned} E\left(\hat{S}_{\text{RX,IQ}}(k);\ k \neq m\right) &= E(\Omega_{\text{A}} A(k) + \Omega_{\text{A}} Z(k) + w_k) \approx \Omega_{\text{A}} A(k) + \Omega_{\text{A}} \mu, \\ E\left(\hat{S}_{\text{RX,IQ}}(k);\ k = m\right) &= E(\Omega_{\text{A}} Z(k) + w_k) \approx \Omega_{\text{A}} \mu \end{aligned} \tag{9.24}$$

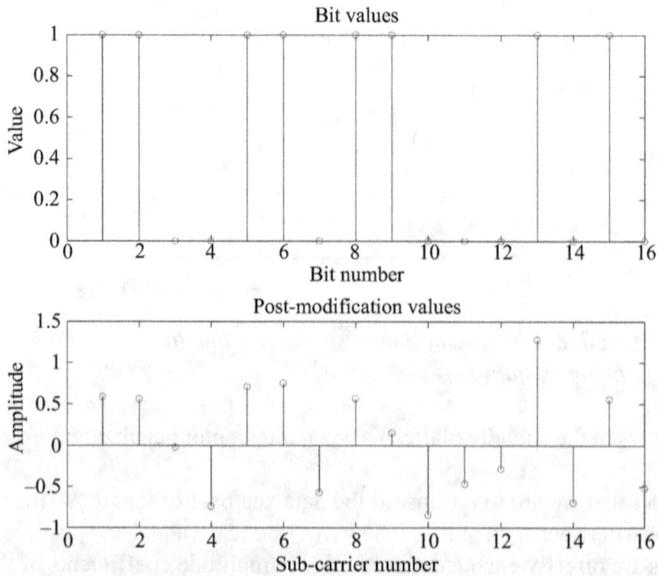

Figure 9.7 Original OOK data (top), and transmit signal with randomized subcarriers (bottom)

where $E(\cdot)$ denotes expectation. This result is then subtracted from each subcarrier coefficient to obtain an estimate of the data bit values

$$\hat{A}(k) = E\left(\hat{S}_{\mathrm{RX,IQ}}(k);\ k \neq m\right) - E\left(\hat{S}_{\mathrm{RX,IQ}}(k);\ k = m\right). \tag{9.25}$$

If the same signal is intercepted by an unauthorized platform, with possible knowledge of the method of encoding and the number of re-transmissions, but not the index m, the result of data reconstruction is

$$\hat{A}_k^{\mathrm{intercept}} \approx \Omega_{\mathrm{A}} A(k) + \Omega_{\mathrm{A}} \mu, \tag{9.26}$$

assuming the interceptor is at an equal distance and experiences the same propagation path for the sake of simplicity. Therefore, the interceptor is left with the modifier value and data but is unable to extract the data itself. An example below illustrates this method's performance under the described assumptions.

For this simple OOK encoding scheme, set $N = 16$ and draw the values of $Z(k)$ from a normal distribution with $\mu = -0.7$ and $\sigma^2 = 0.3$ with the assumption that SNR = +10 dB (for both the friendly and interceptor platforms). The original data (in the frequency domain) and one realization of its randomization are illustrated in Figure 9.7. Statistical performance of interceptor penalization was estimated using 10^4 trials. The optimal μ value [53] and OOK threshold of 0.5 (after normalization) were used. A set of variances was used for each retransmission value to allow for

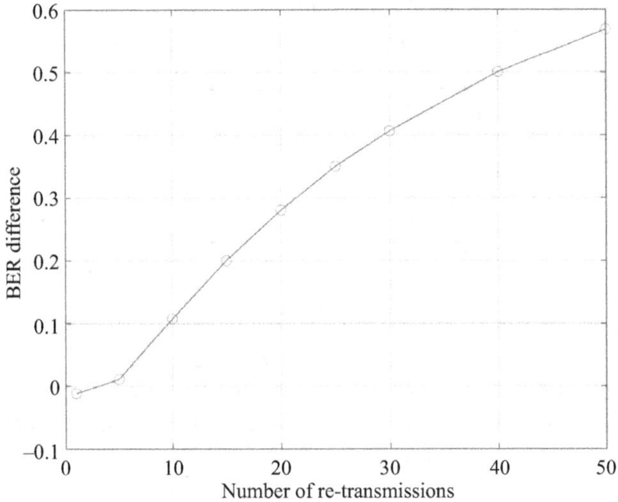

Figure 9.8 Illustration of BER improvement between the friendly receiver and interceptor

variations in normal distribution parameters. Figure 9.8 shows the result of the simulation, where the horizontal axis represents the number of signal re-transmissions and the vertical axis is the difference between a friendly platform and the interceptor platform's resultant BER in data reconstruction. It can be seen that a re-transmission value greater than 5 allows the friendly platform to outperform the interceptor.

We also performed a laboratory experiment with a UWB OFDM SDRS to test the backscatter communication efficiency like in Section 9.2.2. A trihedral corner reflector was placed 20 cm away from the transmit and receive antennas to serve as a target within the background of a typical indoor clutter environment. The resulting SNR was empirically estimated as being between 0 and +3 dB. The SDRS ran through 250 unique 16-bit data vectors, and tested retransmission values of 2, 5, and 10, yielding the results shown in Table 9.1. It is observed that the BER decreases significantly from 2 to 5 retransmissions, with diminishing improvement thereafter.

9.4.2 Randomization of radar/communication signals using stochastic sequences

Another approach to randomize OFDM radar/communication signals is to assign samples of a random process related to an underlying data symbol directly to the OFDM subcarrier amplitude coefficients. An example below illustrates this method and demonstrates its efficiency against an interceptor.

Suppose that we again wish to transmit useful data with radar pulses, thereby reusing the power and bandwidth, while penalizing a possible "eavesdropper." The data we wish to transmit is to be encoded using a certain parameter of a chosen

Table 9.1 Experimental BER for backscatter communications in an indoor setting with UWB OFDM SDRS

Number of retransmissions	BER, %
2	9.3
5	4
10	3.8

distribution. The transmit signal is designed so that the OFDM subcarrier coefficients $S(k)$ are samples of a stochastic sequence that represents a given distribution with the data-driven parameter. Upon reception, these coefficients are reconstructed, from which the useful parameter estimation is also made, thus leading to data recovery. The data reconstruction process can be summarized in the following three steps:

1. Recover the samples of a random process that are recorded onto $S(k)$ – i.e. the values of $S(k)$ are the samples of the random process itself;
2. Estimate the parameter of the distribution that was used to generate the random sequence;
3. Map the parameter estimate onto the preselected data scale to produce an estimate of the underlying data symbol.

It is clear from (9.23) that the two factors that must be considered before the $S(k)$ values are estimated are (1) the influence of Ω_A, which could also be frequency dependent as $\Omega_A(k)$, upon the parameters of the random sequence, and (2) the effect of external additive noise w_k that will corrupt the random sequence statistics. Let us assume that the random process' variance is used as the parameter from which the data is to be estimated. Then, the following applies:

1. If random process X is scaled in amplitude by a constant factor α the variance becomes

$$\sigma_{\alpha X}^2 = \mathrm{VAR}(\alpha X) = \alpha^2 \mathrm{VAR}(X), \tag{9.27}$$

where in this case α is the free-space signal attenuation between the transmitter and receiver, plus any contribution from reflections (i.e. α is $\Omega_A(k)$). To correct for the attenuation, channel estimation needs to be performed via a training procedure.

2. If there are two additive random processes X and Y, samples of which are received simultaneously, the relationship between the variances is

$$\mathrm{VAR}(\underbrace{\alpha X}_{\substack{\text{Original}\\ \text{process } X\\ \text{scaled by } \alpha}} + Y) = \mathrm{VAR}(X) + 2\alpha\,\mathrm{COV}(X, Y) + \mathrm{VAR}(Y), \tag{9.28}$$

where COV(X, Y) is the covariance of the two processes. Therefore, to extract just the variance of X, which contains the data, the following two steps are required: (1) estimate the variance of the external noise prior to reception; and (2) evaluate the covariance between the random process used to encode the data and the external noise. The latter can be either minimized by selecting an appropriate distribution of our own random sequence, or even ignored if we determine that such covariance is negligible.

Although it can be readily estimated from a random sequence, the variance is not a good candidate for encoding the data for applications requiring covert sensing and communications because it is distribution-independent and can be easily determined by any observer. Distributions involving multiple parameters are more useful for this purpose, where one of the parameters is used to encode data while other parameter(s) are known to both the transmitter and friendly receivers. This formulation was used to perform an illustrative simulation study, the results of which are described below.

As above, this simulation study focuses on the amount of interceptor penalization, expressed in terms of BER improvement of the friendly platform's data recovery relative to that of the interceptor. Here, we select a Weibull distribution, defined as

$$f(x;\lambda,k) = \begin{cases} \dfrac{k}{\lambda}\left(\dfrac{x}{\lambda}\right)^{k-1} e^{-(x/\lambda)^k} & x \geq 0 \\ 0 & x < 0 \end{cases} \tag{9.29}$$

which is dependent on scale parameter λ and shape parameter k. If the data are encoded onto the scale parameter (with the shape parameter known to both the transmitter and friendly receivers), then the data can be reconstructed at the receivers by estimating λ via

$$\hat{\lambda}_{\mathrm{W}} = \sqrt{\frac{\hat{\sigma}_{\mathrm{W}}^2}{\left[\Gamma\left(1+\dfrac{2}{k}\right)+\left(\Gamma\left(1+\dfrac{1}{k}\right)\right)^2\right]}} \tag{9.30}$$

where $\hat{\sigma}_{\mathrm{W}}^2$ is the receiver's estimate of variance based on the received samples of the Weibull random sequence, and Γ is the gamma function. The following assumptions regarding a realistic worst-case scenario with regard to an interceptor are used in the simulation study:

1. The interceptor knows that the data are encoded onto a parameter of a Weibull distribution and it knows which parameter is being used in this manner;
2. There is a maximum transmit variance value that can be achieved (e.g. $\sigma_{\mathrm{W}}^2 \leq 100$) and the interceptor knows this maximum value;
3. The interceptor does not know the shape parameter k, or when/if its value is changed from one transmission to another;
4. The interceptor knows the number of re-transmissions used, the total number of subcarriers N, and the particular subcarriers in which UWB OFDM signal transmissions occur.

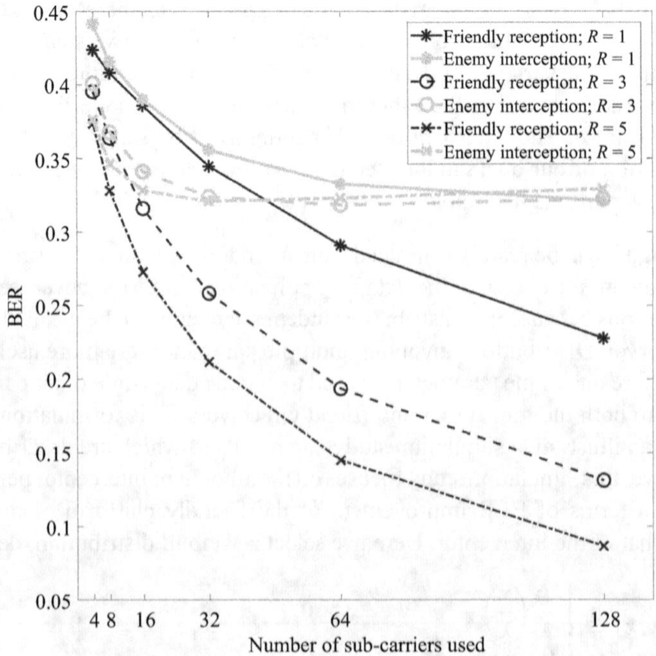

Figure 9.9 Interceptor penalization using a Weibull distribution and scale parameter estimation

Finally, it is also assumed that the external noise is Gaussian distributed. Under these conditions, the BER was simulated for various numbers of re-transmissions R and subcarriers N, with the result shown in Figure 9.9. For example, if $N = 64$ and $R = 5$, a random sequence of length $N \times R = 320$ is transmitted. It can be readily observed that after N is increased beyond about 32, the interceptor realizes little to no improvement in BER, yet the friendly platform BER does continue to improve. For $N = 128$ and $R = 5$, the gap between the BER for the friendly platform and the interceptor reaches approximately 0.25. To put it another way, the enemy will still reconstruct about every third bit incorrectly while a friendly platform will only reconstruct about every thirteenth bit incorrectly, leading to much better reception rate under the worst-case scenario described above.

Figure 9.10 also illustrates this method's performance under various levels of external Gaussian noise, i.e. BER versus SNR. It is observed that the method still provides an advantage over an interceptor even when SNR is below 0 dB.

It should also be noted that the BER values obtained in this scenario exceed those typically produced in conventional UWB communications between fully synchronized platforms. This deficiency is due to a much greater emphasis being placed on covertness in this case, in which the data are not transmitted or received directly but rather are inferred from the estimation of stochastic parameters from finite-length

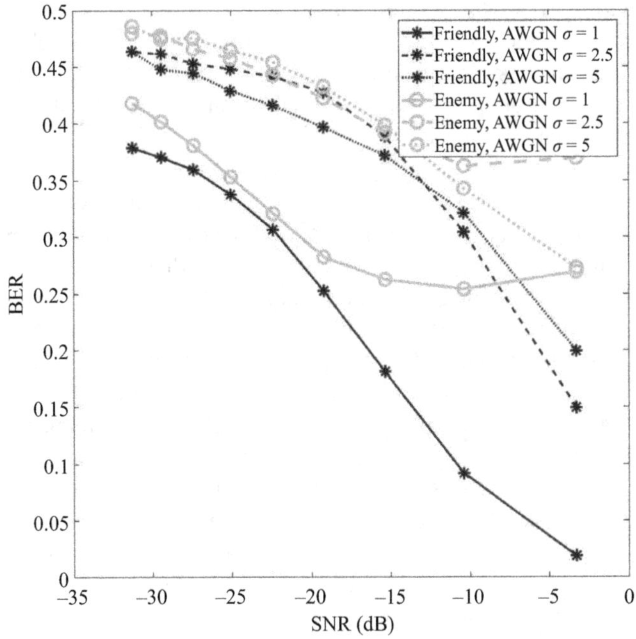

Figure 9.10 BER vs SNR using a Weibull distribution and scale parameter estimation

sequences. The BER therefore becomes a function of parameter estimation accuracy as opposed to relying on the traditional recovery of known, deterministic symbol patterns in additive noise.

The two approaches have illustrated a measure of flexibility with which randomization of OFDM signals can be constructed for the purpose of achieving LPI/LPD signaling. These randomized waveforms can be used as radar signals without any additional adjustment. In a similar vein, the next section outlines a method of cross-range compression for SAR applications based on phase history estimation that is applicable to UWB OFDM signals with random subcarrier amplitude coefficients.

9.4.3 Cross-range compression for SAR with randomized OFDM signals

SAR processing involves two major components: range compression and cross-range compression. The former is performed via matched filtering of the transmitted waveform while the latter can be addressed in a variety of ways that often depend on the particular type of waveform being used. One of the most common SAR waveforms is an LFM chirp, the structure of which allows for the implementation of computationally efficient frequency-domain methods such as range-Doppler (RD) processing that continue to receive interest. For example, several modified and generalized RD algorithms for wideband FMCW SAR were proposed in [54–56], exploiting the fact that the LFM waveform structure permits the signal phase to be expressed analytical, though an approximation of the analytical expression is widely used to achieve even higher computational efficiency.

In contrast, when arbitrary non-LFM waveforms are used for SAR to achieve adaptivity, exploit frequency diversity of target returns, mimic random noise to achieve LPI/LPD [50,57], and the like, different cross-range compression methods are required. In this section, we describe a method that numerically separates the azimuthal components of randomized OFDM SAR signals and estimates Doppler/phase histories of these components. Cross-range reconstruction can then be performed simply via matched filtering of the estimated phase histories. In applications where cross-range compression is performed in a similar manner, the phase history is typically extracted using a Doppler centroid, thereby yielding a quasi-single-frequency solution. However, if the location of the Doppler centroid (which often at the signal's carrier frequency) is subject to narrowband jamming and/or interference, cross-range compression becomes problematic. The method described below and in [58][2] does not suffer from this limitation because narrowband jammers can simply be rejected from a received OFDM signal by placing nulls in the transmitted signal's appropriate subcarrier locations.

In the following discussion, the stripmap SAR geometry shown in Figure 9.11 is assumed. The distance from the SAR platform at the mth position to a point target, and the radial component of the velocity between the platform and the target, are shown as R_{m0} and v_{rm}, respectively. For the simplicity of analysis, we designate the target to be a single static point-like reflector. We also assume a clutter model represented by zero-mean AWGN, which is the same model used in the derivation of the phase gradient autofocus method for point targets in [59].

After defining the OFDM subcarrier amplitude coefficients as a $(2N+1)$-length vector $\boldsymbol{S}$, we can form the SAR transmit signal

$$s_{\text{tx}}(t) = \frac{1}{2N+1}\sum_{k=1}^{2N+1} S(k)\, e^{j\frac{2\pi t}{2N+1}(k-1)f_s} \tag{9.31}$$

that is defined over the interval $t \in [0, N/f_s]$. The noiseless/clutter-less reflected signal at the mth SAR platform position from the point-like target is thus

$$s_{\text{rx}}(t) = A_m\, s_{\text{tx}}(t - D_m), \tag{9.32}$$

for A_m the complex target reflectivity coefficient and

$$D_m = \frac{2R_{m0}}{c - v_{rm}} \tag{9.33}$$

[2]©2012 IEEE. Reprinted, with permission, from *IEEE Geoscience and Remote Sensing Letters,* 2012, vol. 9(5), pp. 808–812.

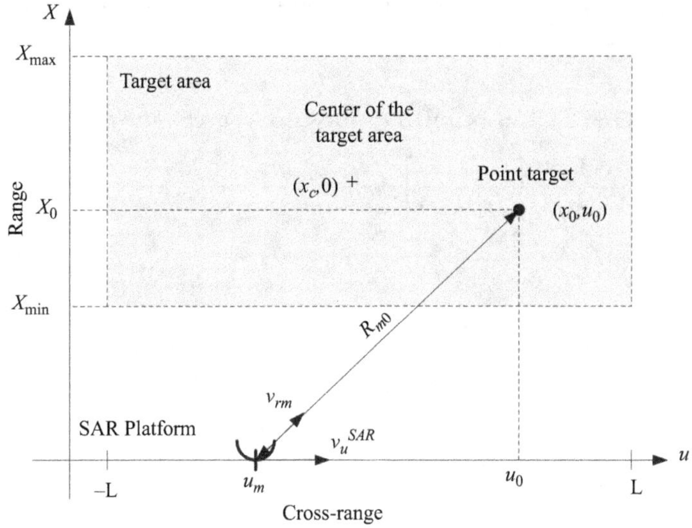

Figure 9.11 Stripmap SAR geometry (©2012 IEEE. Reprinted, with permission, from IEEE Geoscience and Remote Sensing Letters, 2012, vol. 9(5)).

the associated time delay. Using the OFDM representation of (9.31), sampling of (9.32) yields the sequence

$$s_{rx}(n) = \frac{A_m}{2N+1} \sum_{k=1}^{2N+1} S(k)\, e^{j\frac{2\pi\left(\frac{n-1}{f_s} - (D_m - \tau)\right)}{2N+1}(k-1)f_s} \tag{9.34}$$

for $n = 1, \ldots, N+1$, where τ is the delay associated with the given range bin. Upon reception, this signal is converted back to the frequency domain via

$$S_{rx}(k) = \sum_{n=1}^{N} s_{rx}(n+1)\, e^{-j2\pi\frac{kn}{2N+1}} + \sum_{n=N+1}^{2N+1} s_{rx}(2(N+1)-n)\, e^{-j2\pi\frac{kn}{2N+1}}$$
$$\text{for } k = 1, \ldots, N \tag{9.35}$$

to obtain the original N-length positive-frequency vector of subcarrier coefficients. In matrix form [60] the operation in (9.35) can be represented as

$$(2N+1)\ \mathbf{S}_{rx} = A_m \begin{bmatrix} \gamma_{1,1} & \cdots & \gamma_{1,2N} \\ \vdots & \ddots & \vdots \\ \gamma_{N,1} & \cdots & \gamma_{N,2N} \end{bmatrix} \begin{bmatrix} \psi^{1-2N} \\ \vdots \\ \psi^{2N-1} \end{bmatrix} = A_m \boldsymbol{\Gamma}\boldsymbol{\Psi} \text{ for } \psi = e^{-j\frac{\pi\beta}{2N+1}}, \tag{9.36}$$

where $\beta = (D_m - \tau) f_s$ specifies the phase history, and the term $\gamma_{k,r} = C_{k,r}\, S(r+1)$ for

$$C_{k,r} = \begin{cases} -\dfrac{j}{2}[\Theta_1(k,r) + \Theta_2(k,r)] & \text{for } k+r = \text{even},\ k \neq r \\ +\dfrac{j}{2}\left[\dfrac{1}{\Theta_1(k,r)} + \dfrac{1}{\Theta_2(k,r)}\right] & \text{for } k+r = \text{odd},\ k+r \neq 2N+1 \\ -\dfrac{j}{2}\Theta_1(k,r) + \dfrac{2N+1}{2} & \text{for } k = r \\ +\dfrac{j}{2}\dfrac{1}{\Theta_2(k,r)} + \dfrac{2N+1}{2} & \text{for } k+r = 2N+1 \end{cases}$$

$$\Theta_1(k,r) = \tan\left(\frac{\pi}{2}\frac{k+r}{2N+1}\right)$$

$$\Theta_2(k,r) = \tan\left(-\frac{\pi}{2}\frac{k-r}{2N+1}\right). \tag{9.37}$$

Including a clutter model as vector $\mathbf{N}_c$ comprised of independent and identically distributed (iid) AWGN samples and an interference/noise model as vector $\mathbf{N}_i$, the true received OFDM SAR signal vector is obtained as

$$\hat{\mathbf{S}}_{rx} = \mathbf{S}_{rx} + \mathbf{N}_c + \mathbf{N}_i. \tag{9.38}$$

The L^2-norm of the error between the received signal and the signal model from (9.36) can be expressed as

$$J(\beta) = \left\| \hat{\mathbf{S}}_{rx} - A_m \mathbf{\Gamma}\Psi \right\|^2, \tag{9.39}$$

the minimization of which provides an estimate of β. Equation (9.39) can be used directly for the case of known target reflectivity A_m. In the case of unknown A_m, however, it first needs to be estimated for subsequent use in (9.39) (see full derivation in [60]). The resulting expression is the least-squares estimate (LSE)

$$\hat{\beta} = \arg\max_{\beta} \left\{ \left\| \hat{\mathbf{S}}_{rx}^H \frac{\mathbf{\Gamma}\Psi}{|\mathbf{\Gamma}\Psi|} \right\|^2 \right\}. \tag{9.40}$$

Once this estimate of the phase history at the SAR's mth position is obtained (e.g. via grid search method), it can be used for cross-range profile reconstruction via matched filtering with the reference function. The latter is the Green's function (spherical phase function) corresponding to the center of the imaged target area [61]. Note that since all subcarriers of an UWB signal are equally spaced by a known amount, each subcarrier supplies information about the multiples of fundamental phase β. Thus, once β is estimated using all subcarriers, it can be implemented as if it was the phase of a single complex sinusoid via

$$\phi_m = e^{-j2\pi \frac{\hat{\beta}_m}{2N+1}}, \tag{9.41}$$

where $\hat{\beta}_m$ is the estimate of β at the SAR's mth position. The reference function for matched filtering is readily obtained as

$$\phi_m^{\text{ref}} = e^{-j2\pi \frac{(\hat{D}_m - \tau)}{2N+1} f_s} \tag{9.42}$$

for $\hat{D}_m$ an estimate of the signal's time delay for a static unit point-like target at the center of the scene determined by

$$\hat{D}_m = \frac{2\sqrt{u_m^2 + x_c^2}}{c - v_u^{\text{SAR}} \cos(\arctan\{x_c/u_m\},)} \tag{9.43}$$

with the geometric variables u_m, x_c, and v_u^{SAR} described in Figure 9.11.

Cross-range profile reconstruction is then performed by matched filtering via ϕ_m and ϕ_m^{ref}, yielding the cross-range profile function $\Phi(u)$. This process can achieve the accuracy and efficiency of a single-sinusoid signal's cross-range reconstruction, but can be implemented for randomized UWB OFDM SAR signals without any need for approximation or interpolation.

This method can be applied to scenes containing multiple targets. The spherical phase function for N_{T} point targets is generated as

$$\phi_m^{N_{\text{T}}} = \sum_{q=1}^{N_{\text{T}}} A_q e^{-j2\pi \frac{\beta_q}{2N+1}} \tag{9.44}$$

where A_q is the amplitude of the reflection of the qth point-like target and β_q is the associated fundamental phase (delay). The approach for cross-range profile reconstruction follows the same procedure as outlined above, i.e. we seek to obtain the LSE of (9.44) in the format of (9.41), and thus the total number of point-like targets does not need to be known.

A necessary step prior to the estimation of $\hat{\beta}_m$ in this case is normalization of the received signal's vector to eliminate amplitude variation due to summation of responses from several point-like targets. We simulated an extended target by forming a collection of 20 equally weighted equidistant point-like targets distributed along a line in cross-range from 40 to 90 m. Per the arrangement in Figure 9.11, the following parameters are specified for the simulation: $L = 150$ m, $x_c = 550$ m, $x_o = 520$ m, $u_o = 70$ m, $v_u^{\text{SAR}} = 35$ m/s, $f_{\text{PRF}} = 35$ Hz, and $f_s = 1$ GHz. The grid-search space is $[-N, N]$ to obtain the estimate $\hat{\beta}$, using a step of $\Delta\hat{\beta} = 0.75$ and 601 total steps in u. No noise or interference was included.

An example of phase history reconstruction using LSE for $N = 16$ subcarriers is shown in Figure 9.12(a). Here, "LS" denotes the result from least squares estimation and "SF" denotes the true equivalent single-frequency result, which represents an ideal scenario. It is observed that the real part of the estimated phase history closely follows the ideal (true) case. Examples of cross-range reconstruction performed for the cases of $N = 16$ and 32 subcarriers, contrasted with the corresponding single-frequency (ideally reconstructed Doppler centroid) cases, are shown in Figure 9.12(b). The reconstruction is performed successfully and improves with more subcarriers. With randomized subcarrier coefficients and the possibility

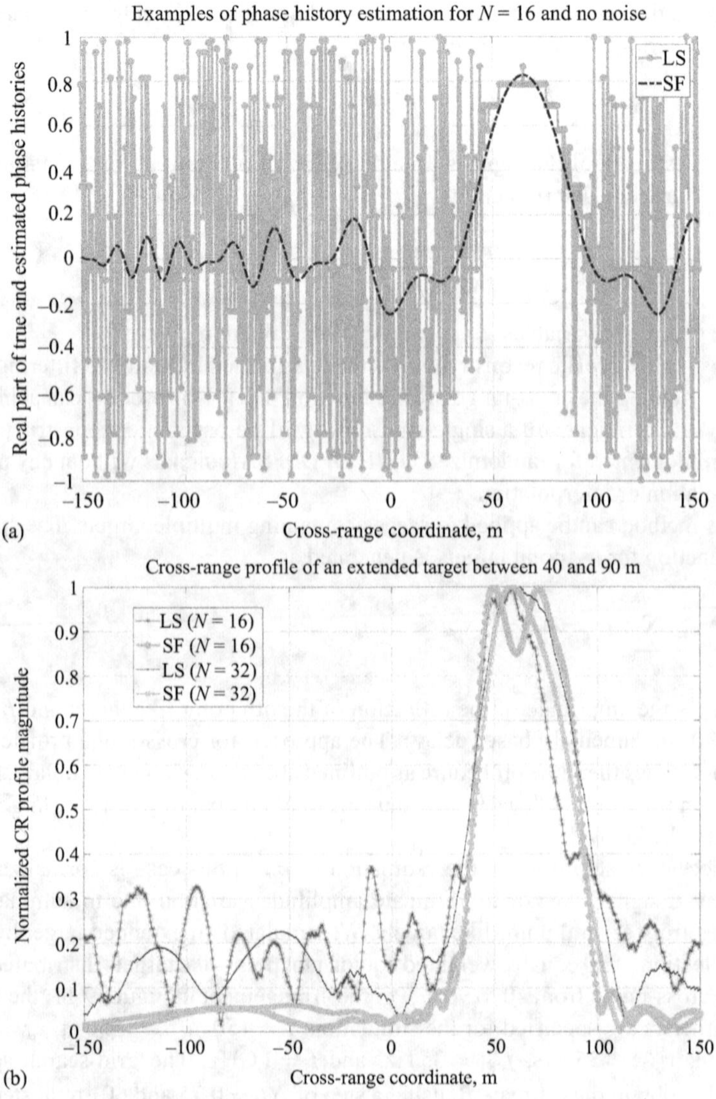

Figure 9.12 (a) Example of phase history estimation for extended target; (b) Cross-range profile reconstruction example (©2012 IEEE. Reprinted, with permission, from IEEE Geoscience and Remote Sensing Letters, 2012, vol. 9(5)).

of nulling certain subcarrier locations that are subjected to narrowband jamming or interference, this method of SAR signal processing offers unique benefits with respect to LPI/LPD performance (for more on narrowband jammer penalization, see [58]).

9.5 In-band co-existence of UWB OFDM radar signals and navigation satellite signals

Airborne reconnaissance, surveillance, location, and navigation operations are performed via a number of techniques, the most promising of which include the use of GNSS signals embedded with radar imagery of underlying terrain and objects. The relatively recent inclusion of unmanned aerial vehicles (UAVs) in battlefield operations has also brought about new challenges and opportunities, particularly in light of the increasing importance of military operations in urban terrain where the gap between short-range and long-range radar imaging scenarios is narrowing, with possible target ranges of less than 100 meters. Numerous sensor design challenges abound in these cases, though the requirement to preserve or even reduce device size, weight, power, and cost (also referred to as SWaP-C) is almost always present. One possible way of addressing this requirement is by using the same device for multiple purposes. We considered the combination of sensing and communications in the preceding sections. This section addresses the feasibility of in-band coexistence of UWB radar transceivers and GNSS receivers by utilizing the UWB OFDM system architecture, which also facilitates the simultaneous reception and analysis of GNSS signals, thus avoiding the need for a separate GNSS receive antenna and front end.

GNSS employs narrowband spread spectrum signals at L band carrier frequencies. The U.S. Global Positioning System (GPS) is the original GNSS with transmissions on two carrier frequencies: L1 at 1.57542 GHz and L2 at 1.2276 GHz. Legacy GPS satellites only transmit an open signal, denoted L1 C/A, at the L1 frequency for use in civil applications. The L1 C/A signal uses BPSK modulation with a bandwidth of 1.023 MHz. Since 2005, GPS modernization has led to four additional civil signals: two time-multiplexed signals denoted as L2 CL and L2 CM residing at L2, and two QPSK modulation signals named L5I and L5Q that operate at L5 (1.17645 GHz). Each of these new signals employs basic BPSK to modulate ranging codes onto the carrier with bandwidths from 1.023 MHz (L2) to 10.23 MHz (L5). The L2C signal is especially attractive for spectrum sharing in indoor and urban navigation settings because it has a low bandwidth, a pilot channel without data modulation, and a relatively long code period that provides long integration time for the processing of weak signals.

During the past decade, several other global and regional satellite navigation systems have also emerged or matured [62], including Russia's GLONASS, the European Union's Galileo, China's Beidou, Japan's QZSS, and India's IRNSS. All these systems transmit open signals on multiple L band carriers and have at least one 1.023 MHz narrow bandwidth signal. Together with modernized GPS, these multiple constellations of navigation satellites provide an average coverage of 25 satellites for most habitable places on the surface of the Earth. The availability of this large number of navigation signals and the increased spectral and modulation diversity offers more possibilities for coexistence.

Earlier studies on radar/navigation spectrum sharing focused on a common UWB waveform feature that may be considered a disadvantage in certain scenarios; that is, the instantaneous spectrum is predetermined by the radar system design and cannot be adjusted during operation. This constraint results in a medium-to-low opportunity for coexistence between these types of UWB radar signals and received GNSS waveforms [63], which may lead to an increased chance of GNSS receiver interruptions. Such situations create the need to establish elaborate isolation schemes and protocols, particularly if same-platform functionality of UWB radar and GNSS is desired. There are several approaches to interference cancellation that exist for GNSS (e.g. [64,65]). A comprehensive study of the UWB signals used for sensing and communications was performed by the U.S. National Telecommunications and Information Administration (NTIA) in 2001, including analysis of UWB interference effects on GPS receivers [66,67]. As explained in these reports, the definition of UWB signals used in the NTIA study was based on the traditional short-pulse concept, i.e. "UWB systems make use of narrow pulses and time-domain signal processing" [66]. Thus, the approach taken by the NTIA researchers was to survey the various ways that short-pulse UWB transmitters operate as means to "pollute" segments of the spectrum used by GPS. All the UWB signals and systems analyzed in those reports had fixed, wide instantaneous bandwidths that did not allow for spectrum management on a pulse-to-pulse basis.

Multicarrier systems such as OFDM, which are capable of producing time-varying spectra, were not included in the study (in fact, the term "OFDM" is not mentioned once in all three reports), because they were not considered appropriate for UWB implementation at the time of the study. As opposed to the traditional definition of UWB signals used by the NTIA study, OFDM coding technique may use fairly long pulses and perform signal processing in the frequency domain, thereby achieving UWB characteristics not via generation of ultra-short pulses but by populating the fixed bandwidth with user-defined spectral content on a pulse-to-pulse basis. Thus, OFDM provides a significant potential for coexistence between it and any narrowband system functioning on the same platform. The two systems could be used separately or they could complement one another in various types of environments. Despite these advantages, UWB OFDM architectures have not been considered for co-design with GPS. This section, based on [68],[3] introduces the signal model of such a system to demonstrate the potential for in-band coexistence of OFDM waveforms and GPS signals. A simulation example is used to illustrate GPS signal recovery and tracking within the context of the UWB OFDM radar model.

9.5.1 System and simulation setup for UWB radar and GPS receiver co-design

To evaluate the performance of simultaneous in-band reception of GNSS signals via a UWB-OFDM radar receiver, we use the most commonly recognized legacy GPS L1 C/A signal as an example. Without loss of generality, the receiver model considered is shown in Figure 9.13, where the received signal $s_{rx}(t)$ is a sum of UWB-OFDM and GPS signals, along with AWGN denoted as $n(t)$. This signal was modeled as

[3]©2011 IEEE. Reprinted, with permission, from *IEEE Transactions on Aerospace and Electronic Systems*, 2011, vol. 47(1), pp. 265–274

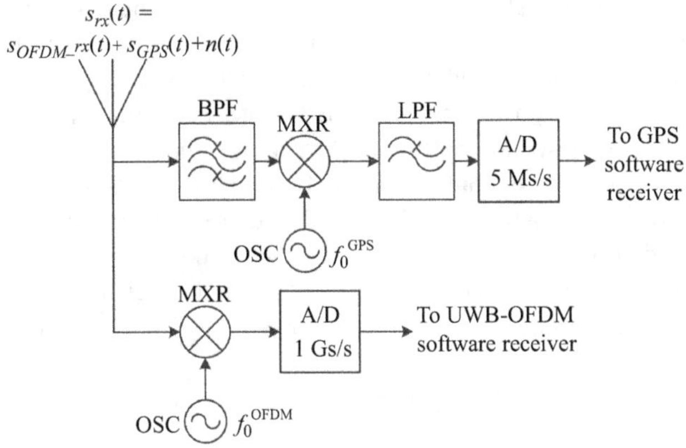

Figure 9.13 Block diagram of dual-use OFDM/GPS receiver frontend model (without amplifiers) (©2011 IEEE. Reprinted, with permission, from IEEE Transactions on Aerospace and Electronic Systems, 2011, vol. 47(1)).

a 20 GHz waveform to approximate a realistic analog form. The carrier frequency of the UWB-OFDM signal is 1.8 GHz and the GPS signal is the L1 band civilian instantiation that has a 1.57542 GHz carrier frequency.

The bandpass filter (BPF) was an elliptic filter designed in MATLAB to have the following parameters: center frequency = 1.57542 GHz; transition band = 10 MHz; maximum ripple in passband = 0.1 dB; and maximum ripple in stop-band = −60 dB. The sub-bands within the OFDM signal that need to be turned off to mitigate GPS interference were determined according to

$$f_0^{GPS} + \Delta f^{GPS} \cong f_0^{OFDM} + \frac{k_{max} f_s}{2N} \quad \text{and} \quad f_0^{GPS} - \Delta f^{GPS} \cong f_0^{OFDM} + \frac{k_{min} f_s}{2N},$$

$$\text{or} \quad k_{max} = \frac{2N}{f_s}\left(\left|f_0^{GPS} - f_0^{OFDM}\right| + \Delta f^{GPS}\right) \quad \text{and}$$

$$k_{min} = \frac{2N}{f_s}\left(\left|f_0^{GPS} - f_0^{OFDM}\right| - \Delta f^{GPS}\right), \tag{9.45}$$

where f_0^{GPS} and Δf^{GPS} are the center frequency and bandwidth of the GPS signal, respectively, f_0^{OFDM} is the up-converted OFDM signal's center frequency, and k_{max} and k_{min} are the maximum and minimum indices, respectively, of the sub-bands to be turned off (along with all those in between). For the system operating values listed above and rounding up/down for k_{max} and k_{min}, respectively, we find k_{min} to be 13 and k_{max} to be 16.

It should be pointed out that the extent of the GPS spectrum for which the UWB-OFDM radar signal's spectrum needs to be nulled is assumed to be relatively narrow and an OFDM signal by definition is not going to occupy the entire allocated spectrum. Similar to the approach discussed earlier in this chapter, the ternary alphabet $\{0, +1, -1\}$ is used for signal coding, with the expectation that some sub-bands will always be nulled at certain times. To UWB-OFDM radar imaging scenarios [13], we assume random sub-band coefficient distribution that necessarily results in an "uneven" radar spectrum on a pulse-to-pulse basis. This UWB-OFDM radar configuration does not suffer from the nulling of particular sub-bands any more than it is affected by random nulling of subcarriers for the purpose of pulse diversity. Thus, OFDM spectrum nulling to accommodate GPS is not a detriment to the radar or communication functions.

9.5.2 GPS software receiver model

A software GPS receiver operates on direct sampled RF or intermediate frequency signals. The ADC output shown in Figure 9.13 is used as an input to the GPS software receiver to perform signal acquisition, tracking, navigation data decoding, bit synchronization, and ultimately navigation processing. In previous studies, GPS signal acquisition performance was considered [69]. The acquisition process provides coarse estimation of a GPS signal's Doppler frequency and code phase. This coarse estimate can be used to initialize the code and carrier tracking loops (TLs) to obtain refined code and carrier phase, Doppler frequencies, and the navigation bit transition edge. Here, we focus on signal tracking performance.

A high-level GPS signal-TL is shown in Figure 9.14 that contains a carrier-TL and a code-TL. The digitized input data signal contains multiple baseband GPS signal and receiver noise samples. At the initial stage, the coarse Doppler frequency and code phase of a satellite obtained from the acquisition process are used to initialize the tracking. Once the TL is locked, the code-TL provides a so-called "prompt code" whose code phase is driven to align with that of the incoming signal. The carrier-TL provides either the carrier Doppler and carrier phase (if the carrier-TL is implemented as a phase-locked loop) or just the carrier Doppler frequency alone (if the carrier-TL is implemented as a frequency-locked loop). The code phase, carrier Doppler, and carrier phase are then fed back to the correlators to perform code and carrier "wipe off" for the next block of data. The carrier-TL is the weak link in the entire tracking process and its performance is a good indicator of that of the overall GPS receiver. Figure 9.15 shows a more detailed block diagram of the carrier-TL.

The first component of a carrier-TL is a correlator that removes (or "wipes off") the code and carrier from the input and then integrates the resultant signal. This correlator is the same prompt correlator as in the code-TL. With a software receiver, the real digitized input stream from Figure 9.13 can be immediately down-converted to obtain the complex baseband signal

$$s_{\mathrm{GPS}}(t) = a\, C_{\mathrm{a}}(t - \tau) D(t)\, e^{j(\omega_{\mathrm{d}} t + \theta_0)} + \varepsilon \tag{9.46}$$

where a is the amplitude, C_{a} is the CA code, τ is the CA code phase, D is the navigation data bit value, ω_{d} is Doppler frequency, and θ_0 is the initial carrier phase of a satellite signal. The receiver noise, signals from other satellites, and UWB leakage from

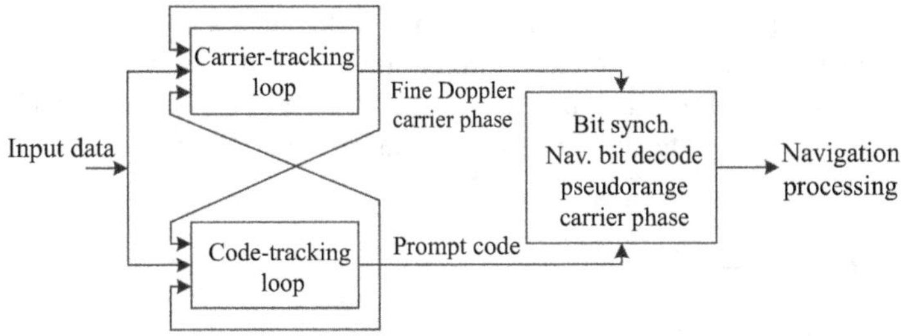

Figure 9.14 A high-level block diagram of the GPS signal-TL

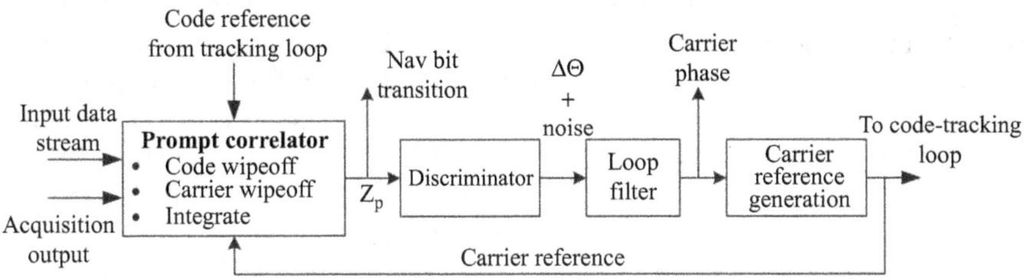

Figure 9.15 Block diagram of a carrier phase-TL

neighboring sub-bands are collectively represented by ε. Denoting the code reference generated by the code-TL as $C_a(t-\hat{\tau})$, and the carrier Doppler and phase generated by the carrier phase loop output as $\hat{\omega}_d$ and $\hat{\theta}_0$, respectively, the correlator output is

$$\begin{aligned} z_p &= \int_{T_I} s_{GPS}(t) C_a(t-\hat{\tau})\, e^{-j(\hat{\omega}_d t+\hat{\theta}_0)}\, dt \\ &= a\int_{T_I} D(t)C_a(t-\tau)C_a(t-\hat{\tau})\, e^{j[\Delta\omega_d t+\Delta\theta_0]}\, dt + \tilde{\varepsilon}, \end{aligned} \tag{9.47}$$

for $\tilde{\varepsilon}$ the correlator response to ε and T_I the integration interval of the carrier-TL correlator.

When the code delay-locked loop is locked, and if the integration does not occur across a navigation data bit transition, then (9.47) becomes

$$z_p = aD\,\mathrm{sinc}\left(\frac{\Delta\omega_d T_I}{2}\right) e^{j\left[\frac{\Delta\omega_d T_I}{2}+\Delta\theta_0\right]} + \tilde{\varepsilon}, \tag{9.48}$$

and simplifies further to

$$z_p \approx aD + \tilde{\varepsilon} \tag{9.49}$$

when the carrier-TL is locked. The navigation data bit transition can be extracted from the prompt correlator output.

The correlator output is used as input to a discriminator to generate a measure of the carrier phase error between the input signal and the carrier-TL output. The existence of the navigation data bit introduces phase ambiguities to the TL. Costas discriminators can tolerate the navigation data bit sign. Here, we use the Costas discriminator

$$\Theta = \tan^{-1}\left(\frac{Im(z_p)}{Re(z_p)}\right) \tag{9.50}$$

in the carrier-TL. Substituting (9.48) into (9.50) yields $\Theta = \Delta\theta$ for $\Delta\theta < |\pi/2|$ if there is no noise in the correlator output.

To reduce the impact of noise, a loop filter is needed. The carrier phase measurement is extracted from the output of this loop filter. The loop filter order determines the dynamic response performance of the carrier-TL. A second-order loop filter is sufficient to track a slow platform, while a third-order loop filter is needed to track platform acceleration. In the presence of strong leakage from insufficiently nulled OFDM sub-bands when the UWB signal power is high, additional signal processing needs to be incorporated into the TL to maintain lock. With the software receiver architecture, batch-based processing can be used to enhance GPS signal observability and to provide improved performance under reduced GPS SNR conditions. Other on-board sensors such as an inertial navigation sensor can also provide dynamic correction information to the receiver to reduce the TL performance threshold [70]. In this analysis, however, we evaluate the UWB-GPS dual-use platform under the hypothesis that the receiver is in a stand-alone mode and that only a conventional tracking procedure is used.

9.5.3 UWB OFDM and GPS coexistence modeling results

The received signal modeled in Figure 9.13,

$$s_{rx}(t) = s_{OFDM,rx}(t) + s_{GPS}(t) + n(t), \tag{9.51}$$

was generated at 20 GHz sample rate to approximate analog behavior. Both GPS and UWB-OFDM signals were produced by up-converting their baseband versions to respective carrier frequencies. The baseband signal representations are used to calculate the signal-to-signal ratio (SSR) defined as

$$\mathrm{SSR} = 10\log_{10}\left(\frac{\langle s_{GPS_{BB}}(t), s_{GPS_{BB}}(t)\rangle}{\langle s_{OFDM_{BB}}(t), s_{OFDM_{BB}}(t)\rangle}\right), \tag{9.52}$$

where $s_{GPS_{BB}}$ and $s_{OFDM_{BB}}$ represent the baseband GPS signal and the baseband OFDM pulse train, both with 10 μs total duration and sampled at 1 GHz. For the OFDM signal, all 32 subcarrier amplitude coefficients are assigned randomly on a pulse-to-pulse basis, except for subcarriers 13 through 16, which are nulled for the complete set of pulses to avoid interfering with the GPS signal. An illustration of the resultant spectra of the baseband GPS signal segment and OFDM pulse train for SSR ≈ −40 dB is shown in Figure 9.16.

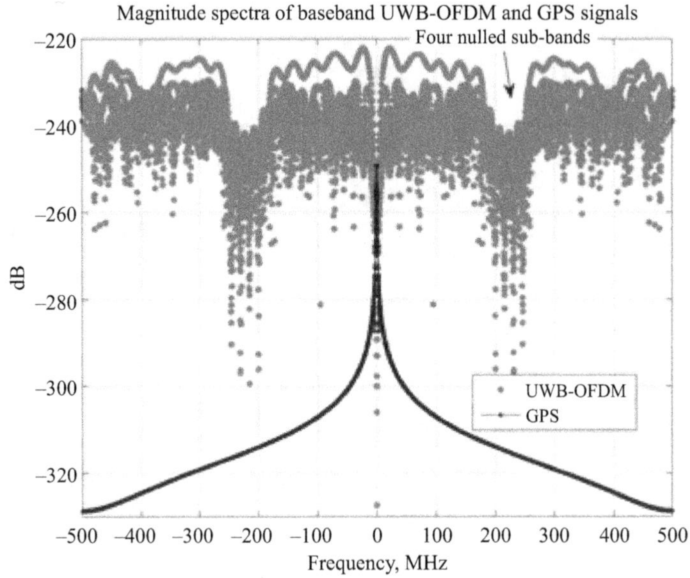

Figure 9.16 Magnitude spectra of 10 μs segments of baseband GPS and OFDM train pulse at SSR ≈ −40 dB. (©2011 IEEE. Reprinted, with permission, from IEEE Transactions on Aerospace and Electronic Systems, 2011, vol. 47(1))

The first test case involves the recovery of a generic signal located in the GPS band under various SSR conditions. Several 10 μs signal segments were generated and the recovered signal from s_{rx} was contrasted with the original signal copy to evaluate the error according to

$$\text{Error} = \sqrt{\frac{1}{N_s}\sum_{k=1}^{N_w}\left\{s_{\text{GPS - B}}(k) - \hat{s}_{\text{GPS - B}}(k)\right\}^2}, \tag{9.53}$$

where N_s is the total number of samples in the 10 μs interval, $s_{\text{GPS-B}}$ is the true signal located in a GPS band, and $\hat{s}_{\text{GPS-B}}$ is a GPS-band signal recovered from the received signal after BPF. The main objective of this exercise is to qualitatively assess GPS-band signal reproduction fidelity using the UWB-OFDM signal model and sub-band nulling approach of (9.45). Since the signal recovery is a frequency-domain operation at this stage, a long time-domain simulation was not necessary and we analyzed several 10 μs segments of time-domain data at various SSRs. Figure 9.17 shows the resultant error obtained from the simulations using (9.53). It is observed that even at a fairly low SSR of −25 dB, the recovered signal maintains high fidelity relative to the original waveform, thus enabling us to move forward with tracking performance characterization.

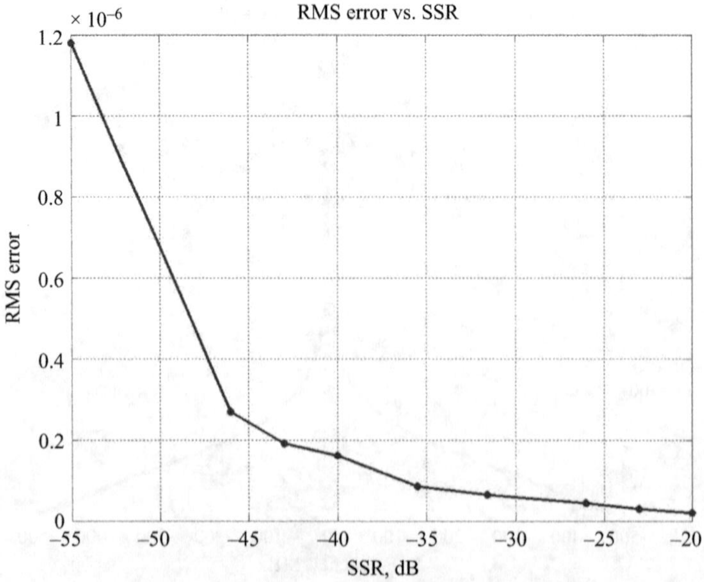

Figure 9.17 GPS-band signal recovery at various SSR: RMS error calculated for 10 μs segment. (©2011 IEEE. Reprinted, with permission, from IEEE Transactions on Aerospace and Electronic Systems, 2011, vol. 47(1)).

The tracking performance was evaluated for civil GPS signals located at the L1 band with 46 dB-Hz carrier-to-noise ratio (C/N_0), which is a nominal value. The correlations were performed using 1 ms and 10 ms data blocks. The Costas discriminator from (9.50) was applied to the correlator output. A second-order loop filter was implemented to reduce the noise contribution to carrier phase estimation at the discriminator output. The analysis was performed using two TL bandwidths: 10 Hz and 20 Hz. Due to the high sampling rate used to generate the input data, significant computational time was required to generate an adequate amount of UWB and GPS signals. As a result, we limit our evaluations to tracking 1 second of data. Figure 9.18 shows the recovered carrier phase for two scenarios: 10 Hz TL bandwidth with 10 ms integration time; and 20 Hz TL bandwidth with 1 ms integration time. As expected, a wider bandwidth loop output with shorter integration time converges to the true phase (30°) more quickly, but the error remains larger than the case having narrower bandwidth loop output and a longer integration time.

We also performed tracking performance evaluations for UWB OFDM with embedded GPS sequences for mean SSR values ranging from approximately −45 dB to −20 dB. Each sequence data length was 1 s. The standard deviation of the carrier phase error was computed for each sequence and for each of the TL bandwidths. The 20 Hz TL bandwidth case was again performed with 1 ms of integration time, and the 10 Hz TL bandwidth case with 10 ms of integration time. Figure 9.19 shows

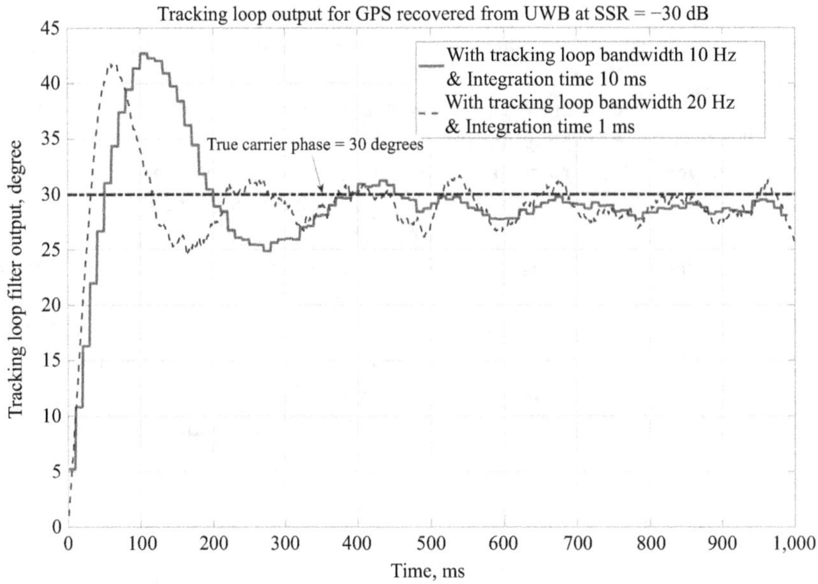

Figure 9.18 GPS signal recovery from UWB-OFDM: carrier phase-tracking output for two cases. (©2011 IEEE. Reprinted, with permission, from IEEE Transactions on Aerospace and Electronic Systems, 2011, vol. 47(1)).

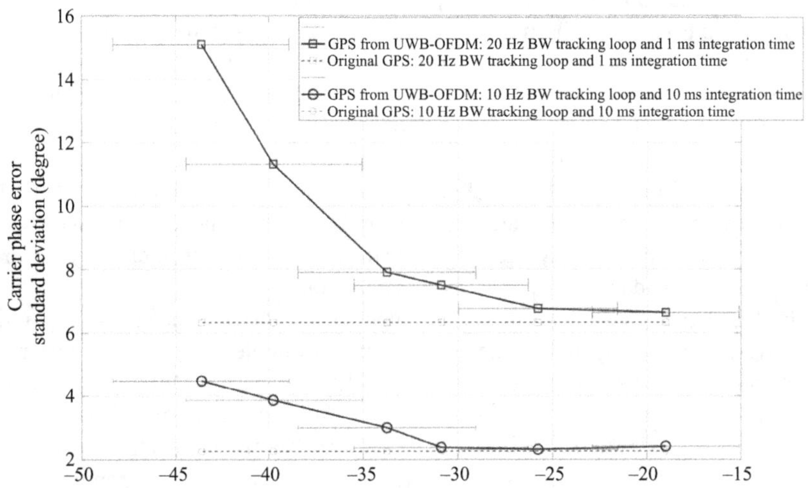

Figure 9.19 GPS receiver carrier tracking error standard deviation as a function of SSR. (©2011 IEEE. Reprinted, with permission, from IEEE Transactions on Aerospace and Electronic Systems, 2011, vol. 47(1)).

the standard deviation of carrier phase error as a function of the GPS to UWB SSR. Note that the original GPS signal with a nominal signal power maintains a carrier phase-tracking error at approximately 2.25° for the 10 Hz case and 6.3° for the 20 Hz case. When GPS is coexisting with UWB signals, if the SSR is above approximately −30 dB, the leakage from neighboring UWB sub-bands has less than a 1° impact on the carrier phase-TL performance. The impact becomes more noticeable at SSR below −35 dB with over 2° of increase in the carrier phase error standard deviation for the 20 Hz case and approximately 1° for the 10 Hz case. At SSR = −43 dB, the carrier phase error is approximately 15° for the 20 Hz case, more than double the amount from when the UWB OFDM signal was absent. The same level of SSR causes only about 2° of additional error for the 10 Hz case. At SSR = −50 dB, the carrier TL loses lock for the 20 Hz case. Thus, this study shows that there is strong potential for in-band coexistence of UWB OFDM and GPS systems, making a software GPS receiver an ideal platform to be integrated with a UWB OFDM SDRS to obtain position solutions.

9.6 Looking ahead: conclusions and perspectives

For next-generation radar, communication, and navigation systems there is a requirement for efficient spectrum utilization and lifecycle cost reduction. Spectrum congestion is a growing problem that seriously limits operational capabilities due to the increasing bandwidth of wireless communication networks, the use of net-centric unmanned systems, the need for flexibility in radar and communication spectrum to improve performance, the increasing demand on autonomous systems in challenging environments, and to overcome sophisticated countermeasures. Thus, there is a pressing need to develop sharing technology that enables spectrum access for radar, communication, and navigation systems to accomplish their respective evolving missions. One promising capability that can help satisfy these requirements is adaptive multi-mode RF systems. The realization of such capabilities necessitates research and development of smart systems that maintain performance in adverse conditions and mobile environments. To develop an efficient and robust multimode RF system, a combination of several technologies and techniques need to be investigated including waveform diversity, spread spectrum signals, software-defined apertures, knowledge-based adaptive sensing, and cognitive radar.

In this chapter, we have explored the scope of OFDM as a means to accomplish the above-described capabilities using illustrative examples. Backscatter communication concepts are useful for operation in non-LOS scenarios. OFDM is well suited for this type of problem especially when privacy and non-interference are important. Backscatter may be produced by a single target, or by several targets that work collectively. This ensemble of targets can be deterministic or stochastic depending on the model selected for the backscatter scene. Further, it is unrealistic in most situations to assume that the target scene is frequency-independent. In the future, these details can be readily added to the framework outlined here. Indeed, performance of OFDM schemes may not be the same in all situations and would need to be considered on a case-by-case basis.

Success with joint UWB OFDM radar and communications has led us to explore the feasibility of adding navigation and geolocation into the arrangement. One obvious concern that arises is the interference between navigation and radar signals. A simple way to avoid this interference is to null the OFDM radar subcarriers that fall within the GPS band. A simulation example was used to illustrate that such fusion does not degrade the performance of either the radar or navigation.

Future directions in radar sensing are heading toward small platforms and UAVs operating in either GPS-degraded or GPS-denied environments. In such situations, one needs to employ the radar for geolocation. Furthermore, "size, weight and power" (SWaP) considerations prohibit separate systems for radar, communication, and navigation. Therefore, new concepts in multifunction systems will continue to be explored. OFDM is definitely a viable option for such endeavors.

In summary, we presented several possible ways to reduce the complexity of RF systems, along with realizing savings in transmit power, AFE weight and size, computational requirements, and spectral allocation, by considering an OFDM-based signal and architecture design. Most of the described approaches also afford a measure of LPI/LPD capability for scenarios where covertness is desirable. Another benefit that can be derived from using OFDM is the potential to exploit frequency diversity of complex target scenes for the purpose of target identification. Multicarrier UWB waveforms are ideally suited to extracting unique frequency/angle profiles of various scenes, while OFDM signal processing allows for efficient analysis of individual subcarrier components (initial feasibility of this topic was explored in [71]).

Instead of separate modalities, one could also consider the functionalities above as a unified approach to location-aware information collection, processing, and dissemination. In so doing, it becomes clear that UWB OFDM SDRS could well become a candidate for implementation of a cognitive sensor system architecture via an extension of the cognitive radar concept [72,73]. Such a system would autonomously and adaptively interrogate and identify a target scene, establish its own location with respect to targets of interest, position itself appropriately, and exchange the resulting data with other friendly platforms in a covert manner. Several research questions toward that end were outlined above and it is likely that many more will arise as the topic is further developed.

References

[1] Chang, R. and Gibby, R. 'A theoretical study of performance of an orthogonal multiplexing data transmission scheme'. *IEEE Trans. Commun.* 1968, vol. 16(4), pp. 529–540.

[2] Le, B., Rondeau, T. W., Reed, J. H., and Bostian, C. W. 'Analog-to-digital converters'. *IEEE Signal Process. Mag.* 2005, vol. 22(6), pp. 69–77.

[3] Batra, A., Balakrishnan, J., Dabak, A., *et al. TI Physical Layer Proposal: Time-Frequency Interleaved OFDM* [online]. 2003. Available from http://www.ti.com/lit/wp/slla138/slla138.pdf [Accessed 11 Mar 2017]

[4] Batra, A., Balakrishnan, J., Aiello, G. R., Foerster, J. R., and Dabak, A. 'Design of a multiband OFDM system for realistic UWB channel environments'. *IEEE Trans. Microwave Theory and Tech.* 2004, vol. 52(9), pp. 2123–2138.

[5] Snow, C., Lampe, L., and Schober, R. 'Performance analysis of multiband OFDM for UWB communication'. *Proc. 2005 IEEE Int. Conf. Commun.*, Seoul, Korea, 2005. Vol. 4, pp. 2573–2578.

[6] Levanon, N. 'Multifrequency complementary phase-coded radar signal'. *IEE Proc. Radar, Sonar Navig.* 2000, vol. 147(6), pp. 276–284.

[7] Mozeson, E. and Levanon, N. 'Multicarrier radar signals with low peak-to-mean envelope power ratio'. *IEE Proc. Radar, Sonar Navig.* 2003, vol. 150(2), pp. 71–77.

[8] Antonik, P. A., Griffiths, H., Weiner, D. D., and Wicks, M. C. *Novel diverse waveforms*. Air Force Research Laboratory In-House Report (AFRL-SN-RS-TR-2001-52), 2001.

[9] Sverdlik, M. B. and Levanon, N. 'Family of multicarrier bi-phase radar signals represented by ternary arrays'. *IEEE Trans. Aerospace Electron. Sys.* 2006, vol. 42(3), pp. 933–953.

[10] Franken, G. E. A., Nikookar, H., and van Genderen, P. 'Doppler tolerance of OFDM-coded radar signals'. *Proc. 2006 European Radar Conf.*; Manchester, UK, 2006, pp. 108–111.

[11] Ruggiano, M. and van Genderen, P. 'Wideband ambiguity function and optimized coded radar signals'. *Proc. 2007 Eur. Radar Conf.*; Munich, Germany, 2007, pp. 142–145.

[12] Donnet, B. J. and Longstaff, I. D. 'Combining MIMO radar with OFDM communications'. *Proc. 2006 Eur. Radar Conf.*; Manchester, UK, 2006, pp. 37–40.

[13] Garmatyuk, D. S. 'Simulated imaging performance of UWB SAR based on OFDM'. *Proc. 2006 IEEE Int. Conf. on Ultra-Wideband*; Waltham, MA, 2006, pp. 237–242.

[14] Thompson, S. C. and Stralka, J. P. 'Constant-envelope OFDM for power-efficient radar and data communications'. *Proc. 2009 Intl. Waveform Diversity and Design Conf.*; Kissimmee, FL, 2009, pp. 291–295.

[15] Garmatyuk, D., Kauffman, K., Schuerger, J., and Spalding, S. 'Wideband OFDM system for radar and communications'. *Proc. 2009 IEEE Radar Conf.*; Pasadena, CA, 2009, pp. 1–6.

[16] Garmatyuk, D. and Kauffman, K. 'Radar and data communication fusion with UWB-OFDM software-defined system'. *Proc. 2009 IEEE Int. Conf. on Ultra-Wideband*; Vancouver, Canada, 2009, pp. 454–458.

[17] Wiesbeck, W. 'SDRS: Software-defined radar sensors'. *Proc. 2001 IEEE Int. Geosc. Rem. Sens. Symp. (IGARSS'01)*; Sydney, Australia, 2001, pp. 3259–3261.

[18] Scholtz, R. A. and Lee, J.-Y. 'Problems in modeling UWB channels'. *Proc. 2002 36th Asilomar Conf. Signals, Syst. Comput.*; Asilomar, CA, 2002, pp. 706–711.

[19] Garmatyuk, D., Schuerger, J., and Kauffman, K. 'Multifunctional software-defined radar sensor and data communication system'. *IEEE Sensors Journ.* 2011, vol. 11(1), pp. 99–106.

[20] Jameson, B., Garmatyuk, D., Morton, and Y. T. Jade. 'Cognitive radar for indoor positioning with software-defined UWB OFDM system'. *Proc. 2012 IEEE International Radar Conference*; Atlanta, GA, 2012, pp. 465–470.

[21] Cole, R., Jameson, B., Garmatyuk, D., and Morton, Y. T. J. 'Simultaneous indoor localization and detection with multi-carrier radar'. *Proc. 2014 IEEE Radar Conference*; Cincinnati, OH, 2014, pp. 881–886.

[22] Sen, S. and Nehorai, A. 'Target detection in clutter using adaptive OFDM radar'. *IEEE Signal Proc. Lett.* 2009, vol. 16(7), pp. 532–535.

[23] Sen, S. and Nehorai, A. 'Adaptive design of OFDM radar signal with improved wideband ambiguity function'. *IEEE Trans. Signal Proc.* 2010, vol. 58(2), pp. 928–933.

[24] Sen, S. and Nehorai, A. 'OFDM MIMO radar with mutual-information waveform design for low-grazing angle tracking'. *IEEE Trans. Signal Proc.* 2010, vol. 58(6), pp. 3152–3162.

[25] Sen, S., Tang, G., and Nehorai, A. 'Multiobjective optimization of OFDM radar waveform for target detection'. *IEEE Trans. Signal Proc.* 2011, vol. 59(2), pp. 639–652.

[26] Sen, S. and Nehorai, A. 'Sparsity-based multi-target tracking using OFDM radar'. *IEEE Trans. Signal Proc.* 2011, vol. 59(4), pp. 1902–1906.

[27] Sen, S. 'OFDM radar space-time adaptive processing by exploiting spatio-temporal sparsity'. *IEEE Trans. Signal Proc.* 2013, vol. 61(1), pp. 118–130.

[28] Sen, S. 'Adaptive OFDM radar waveform design for improved micro-Doppler estimation'. *IEEE Sensors Journ.* 2014, vol. 14(10), pp. 3548–3556.

[29] *Signatec PXDAC4800* [online]. Available from http://www.gage-applied.com/arbitrary-waveform-generators/14-bit/pcie/pxdac4800.htm [Accessed 12 March 2017]

[30] Federal Communications Commission. *Code of Federal Regulations, Title 47* [online]. 2010. Available from https://www.gpo.gov/fdsys/pkg/CFR-2010-title47-vol1/pdf/CFR-2010-title47-vol1-part15-subpartF.pdf [Accessed 12 March 2017]

[31] Federal Communications Commission. *FCC Order 05-58* [online]. 2005. Available from https://apps.fcc.gov/edocs_public/attachmatch/FCC-05-58A1.pdf [Accessed 12 March 2017]

[32] Corral, C. A., Emami, S., and Rasor, G. 'Ultra-wideband peak and average power limits'. *Proc. 2006 3rd IEEE Consumer Comm. and Networking Conference*; Las Vegas, NV, 2006, pp. 478–481.

[33] Skolnik, M. *Radar Handbook, Third Ed.* New York, NY: McGraw-Hill Education; 2008. pp. 1.10–1.12.

[34] Naval Air Warfare Center Weapons Division. *Electronic Warfare and Radar Systems Engineering Handbook* [online]. 2013, p. 4-11.2. Available from http://www.navair.navy.mil/nawcwd/ewssa/downloads/NAWCWD%20TP%208347.pdf [Accessed 12 March 2017]

[35] Cager, Jr. R. H., LaFlame, D. T., and Parode, L. C. 'Orbiter Ku-band integrated radar and communications subsystem'. *IEEE Trans. Commun.* 1978, vol. COM-26, pp. 1604–1619.
[36] Tavik, G. C., Hilterbrick, C. L., Evins, J. B., *et al.* 'The advanced multifunction RF concept'. *IEEE Trans. Microwave Theory Techn.* 2005, vol. 53, pp. 1009–1020.
[37] Saddik, G. N., Singh, R. S., and Brown, E. R. 'Ultra-wideband multifunctional communications/radar system'. *IEEE Trans. Microwave Theory Techn.*2007, vol. 55, pp. 1431–1437.
[38] Sturm, C. and Wiesbeck, W. 'Waveform design and signal processing aspects for fusion of wireless communications and radar sensing'. *Proc. IEEE.* 2011, vol. 99(7), pp. 1236–1259.
[39] Chuang, J., DeMay, M. W., and Narayanan, R. M. 'Secure spread spectrum communication using ultrawideband random noise signals'. *Proc. 2006 Military Commun. Conf. (MILCOM)*; Washington, DC, 2006, pp. 1–7.
[40] Surender, S. C. and Narayanan, R. M. 'UWB noise-OFDM netted radar: physical layer design and analysis'. *IEEE Trans. Aerospace and Electronic Sys.* 2011, vol. 47(2), pp. 1380–1400.
[41] Cnaan-On, I., Thomas, S. J., and Krolik, J. 'Multichannel backscatter communication and ranging for distributed sensing with an FMCW radar'. *IEEE Trans. Microwave Theory and Techn.* 2015, vol. 63(7), pp. 2375–2383.
[42] Levanon, N. and Mozeson, E. *Radar Signals*. New Jersey: John Wiley & Sons; 2004. pp. 34–35.
[43] Sebt, M. A., Sheikhi, A., and Nayebi, M. M. 'Orthogonal frequency-division multiplexing radar signal design with optimised ambiguity function and low peak-to-average power ratio'. *IET Radar, Sonar and Navig.* 2009, vol. 3, pp. 122–132.
[44] Schuerger, J. and Garmatyuk, D. 'Multifrequency OFDM SAR in presence of deception jamming'. *EURASIP J Adv Signal Proc*. 2010, vol. 2010, pp. 1–13.
[45] Santhananthan, K. and Tellambura, C. 'Probability of error calculation of OFDM systems with frequency offset'. *IEEE Trans. Comm.* 2001, vol. 49(11), pp. 1884–1888.
[46] Dharmawansa, P., Rajatheva, N., and Minn, H. 'An exact error probability analysis of OFDM systems with frequency offset'. *IEEE Trans. Comm.* 2009, vol. 57(1), pp. 26–31.
[47] Metcalf, J. G., Sahin, C., Blunt, S. D., and Rangaswamy, M. 'Analysis of symbol-design strategies for intrapulse radar-embedded communications'. *IEEE Trans. Aerospace and Electronic Sys.* 2015, vol. 51(4), pp. 2914–2931.
[48] Narayanan, R. M., Xu, Y., Hoffmeyer, P. D., and Curtis, J. O. 'Design performance and applications of a coherent ultra-wideband random noise radar'. *Opt. Eng.* 1998, vol. 37(6), pp. 1855–1869.
[49] Kulpa, K., Lukin, K. A., Miceli, W., and Thayaparan, T. 'Signal processing in noise radar technology'. *IET Radar, Sonar and Navig.* 2008, vol. 2(4), pp. 229–232.

[50] Garmatyuk, D. S. and Narayanan R. M. 'ECCM capabilities of an ultra-wideband bandlimited random noise radar'. *IEEE Trans. Aerospace and Electronic Sys.* 2002, vol. 38(4), pp. 1243–1255.

[51] Garmatyuk, D. S. and Narayanan, R. M. 'Ultra-wideband continuous-wave random noise Arc-SAR'. *IEEE Trans. on Geoscience and Remote Sensing.* 2002, vol. 40(12), pp. 2543–2552.

[52] Ahmad, F. and Amin, M. 'Stochastic model based radar waveform design for weapon detection'. *IEEE Trans. Aerospace and Electronic Sys.* 2012, vol. 48(2), pp. 1815–1826.

[53] Giza, P. J. 'Fusion of sensing and communication via OFDM'. M.S. thesis, *Miami University*, 2016.

[54] de Wit, J. J. M., Meta, A., and Hoogeboom, P. 'Modified range-Doppler processing for FM-CW synthetic aperture radar'. *IEEE Geoscience Rem. Sens. Lett.* 2006, vol. 3(1), pp. 83–87.

[55] Wang, W.-Q., Peng, Q., and Cai, J. 'Waveform-diversity-based millimeter-wave UAV SAR remote sensing'. *IEEE Trans. Geoscience Rem. Sen.* 2009, vol. 47(3), pp. 691–700.

[56] Zaugg, E. C. and Long, D. G. 'Generalized frequency-domain SAR processing'. *IEEE Trans. Geoscience Rem. Sen.* 2009, vol. 47(11), pp. 3761–3773.

[57] Axelsson, S. R. J. 'Noise radar using random phase and frequency modulation'. *IEEE Trans. Geoscience Rem. Sen.* 2004, vol. 42(11), pp. 2370–2384.

[58] Garmatyuk, D. 'Cross-range SAR reconstruction with multicarrier OFDM signals'. *IEEE Geoscience Rem. Sens. Lett.* 2012, vol. 9(5), pp. 808–812.

[59] Jakowatz, C.V. and Wahl, D. E. 'Eigenvector method for maximum-likelihood estimation of phase errors in synthetic-aperture-radar imagery'. *J. Opt. Soc. Am.* 1993, vol. 10(12), pp. 2539–2546.

[60] Garmatyuk, D. and Brenneman, M. 'Adaptive multi-carrier OFDM SAR signal processing'. *IEEE Trans. Geoscience Rem. Sen.* 2011, vol. 49(10), pp. 3780–3790.

[61] Soumekh, M. *Synthetic Aperture Radar Signal Processing with MATLAB Algorithms.* New York: Wiley; 1999.

[62] Betz, J. W. *Engineering Satellite-Based Navigation and Timing: Global Navigation Satellite Systems, Signals, and Receivers*. Hoboken, NJ: Wiley-IEEE Press; 2016.

[63] Hämäläinen, M., Hovinen, V., Tesi, R., Iinatti, J. H. J., and Latva-aho, M. 'On the UWB system coexistence with GSM900, UMTS/WCDMA and GPS'. *IEEE Journ. Sel. Areas Communications*. 2002, vol. 20(9), pp. 1712–1721.

[64] Fante, R. L. and Vaccaro, J. J. 'Wideband cancellation of interference in a GPS receive array'. *IEEE Trans. Aerospace and Electr. Sys.* 2000, vol. 36(2), pp. 549–564.

[65] Hwang, S-S. and Shynk, J. J. 'Multicomponent receiver architectures for GPS interference suppression'. *IEEE Trans. Aerospace and Electr. Sys.* 2006, vol. 42(2), pp. 489–502.

[66] Kissick, W. A. (ed.) *The Temporal and Spectral Characteristics of Ultra-wideband Signals (NTIA Report TR-01-383)* [online]. 2001. Available

from http://www.its.bldrdoc.gov/pub/ntia-rpt/01-383/01-383.pdf [Accessed 12 March 2017]

[67] Hoffman, J. R., Cotton, M. G., Achatz, R. J., Statz, R. N., and Dalke, R. A. *Measurements to Determine Potential Interference to GPS Receivers from Ultrawideband Transmission Systems (NTIA Report TR-01-384)* [online]. 2001. Available from https://www.its.bldrdoc.gov/publications/01-384.aspx [Accessed 12 March 2017]

[68] Garmatyuk, D., Jade, M. Y., and Mao, X. 'Radar and GPS system interoperability with UWB-OFDM signals'. *IEEE Trans. Aerospace and Electronic Sys.* 2011, vol. 47(1), pp. 265–274.

[69] Garmatyuk, D. and Morton, Y.T., 'On coexistence of in-band UWB-OFDM and GPS signals'. *Proc. 2007 Institute of Navigation (ION) National Technical Meeting*; San Diego, CA, 2007.

[70] Soloviev, A., Gunawardena, S., and van Graas, F. 'Deeply integrated GPS/low cost IMU for low CNR signal processing: flight test results and real time implementation'. *Proc. 2004 Institute of Navigation (ION) Global Navigation Satellite Systems (GNSS) Conference*; Long Beach, CA, 2004, pp. 1621–1631.

[71] Garmatyuk, D., Jameson, B., Cole, R., Morton, Y., and Mudaliar, S. 'Target scene frequency diversity exploitation with ultra-wideband orthogonal frequency division multiplexing radar'. *IET Radar, Sonar and Navig.* 2014, vol. 8(9), pp. 1247–1254.

[72] Haykin, S. 'Cognitive radar: A way of the future'. *IEEE Signal Proc. Mag.* 2006, vol. 23(1), pp. 30–40.

[73] Guerci, J. R. *Cognitive Radar: The Knowledge-Aided Fully Adaptive Approach*. Norwood, MA: Artech House, 2010.

Chapter 10

Adaptive RF multi-interference suppression for wideband radar/communication receivers

Roberto Gómez-García[1], Dimitra Psychogiou[2], Zhengyu Peng[3], José-María Muñoz-Ferreras[1], Changzhi Li[3], and Dimitrios Peroulis[4]

10.1 Introduction

With continued growth in spectrum demand, wideband high-frequency receivers are attracting considerable attention from the RF/microwave and wireless communities (e.g. [1]). The desired receiver modules should be capable of supporting multiple services that are allocated at different spectral bands—i.e. with distinct center frequencies, bandwidths, and power levels—thus leading to the need for spectrally agile and multiband operation [2]. In addition, they must coexist in a scarce electromagnetic environment that is becoming ever more crowded, where the probability of exposure to unintentional interference is dramatically increasing [3]. This scenario will become more common in the near future with the upcoming deployment of next-generation high-data-rate telecommunication services such as 5G mobile-communication systems, as well as the expansion of short-range radar-based high-resolution remote-sensing applications [1,4].

Wideband RF receivers in communication systems will have to make very efficient use of the spectrum in order to provide high quality-of-service (QoS) access to a large number of customers at high data rates, thus ensuring fast download of large files and streaming of high-quality video [5]. In a more evolved scenario, primary and secondary users can cohabitate through the use of temporally and spatially unused "white space" bands [5]. In this context, smart receivers must continuously sense/detect these transmission opportunities, with the ultimate purpose of efficiently providing more bandwidth-consuming services by maximally exploiting the finite spectrum [5].

On the other hand, short-range portable radar sensors are currently experiencing a growing widespread use in civilian applications including automotive, healthcare,

[1]Department of Signal Theory and Communications, University of Alcalá, Spain
[2]Department of Electrical, Computer, and Energy Engineering, University of Colorado Boulder, USA
[3]Department of Electrical and Computer Engineering, Texas Tech University, USA
[4]School of Electrical and Computer Engineering, Purdue University, USA

infrastructure monitoring, industry/manufacturing, indoor and outdoor scenarios, gesture identification, gaming, life detection after natural distances, and so forth [4,6]. Nowadays, radars can be very cost-efficient and portable, leading to the prospect of microwave sensor systems with advantageous features in terms of sensitive vibration detection, through-the-wall imaging/detection capabilities, and precise angle, range, and speed determination [6]. It is important to emphasize that radar systems require a large bandwidth (100s of Megahertz to a few Gigahertz) if high-range resolution capabilities are required [7,8]. Consequently, broadband receivers are envisaged as key modules in these novel wireless sensors for short-range civil applications [1].

To realize the flexibility and adaptability needed for these myriad tasks, it becomes clear that software is essential in future wideband receivers (and transmitters as well). This programmable capability is vital to effectively sense white spaces, to enable the cohabitation of primary and secondary users, to detect possible interference sources, to manage different available services, to adapt the QoS depending on the specific circumstances, to change the data rate, transmitted power, operating frequency, or even to electronically steer the transmit/receive antenna beams of smart devices [9–11]. All these functions depend on innovative software modules integrated into software-defined-radio (SDR) and cognitive-radio (CR) paradigms [9–11].

Considerable effort is also being dedicated to the development of advanced RF/microwave hardware, which is an essential part of wideband high-frequency receivers (and transmitters) [3]. Note that many of the promised features provided by modern SDR and CR approaches are actually characteristics enabled by properly engineered analog RF hardware [3]. As an example, to process multiple signal bands simultaneously, the associated RF electronics (e.g. antennas and signal preselection filters) must exhibit this multi-band capability [12–14]. The key concept under these innovative RF/microwave circuits is reconfigurability/adaptability [3], in which spectral agility plays a primary role [15].

The main components of fully reconfigurable RF circuits are microwave filtering devices that can change their spectral characteristics in terms of the number of transmission and rejection bands, their bandwidths and center frequencies, the out-of-band attenuation profile through dynamic transmission-zero positioning, and even the type of filtering profile (e.g. Chebyshev, Butterworth, Cauer, or Bessel/Gaussian) [15–17]. These high-frequency components are fundamental to all classes of transmit and receive chains to facilitate the implementation of SDR and CR paradigms with an efficient and dynamic use of the spectrum. In addition, these tunable filtering circuits can provide a reduction in terms of physical size and DC-power consumption of the associated equipment because many bulky frequency-static filters that exist in currently deployed front-end architectures can be replaced by a single spectrally agile device with comparable functionality [15–17].

In this chapter, a comprehensive review of recent contributions in the field of fully reconfigurable RF/microwave multi-notch filters is presented. These circuits are understood to be key components in future RF front-ends for wideband receivers to efficiently mitigate intentional and unintentional narrowband interference [18]. Specifically, tunable implementations for planar technologies are described in which the adaptive multi-notch filtering action is co-integrated with the wideband signal

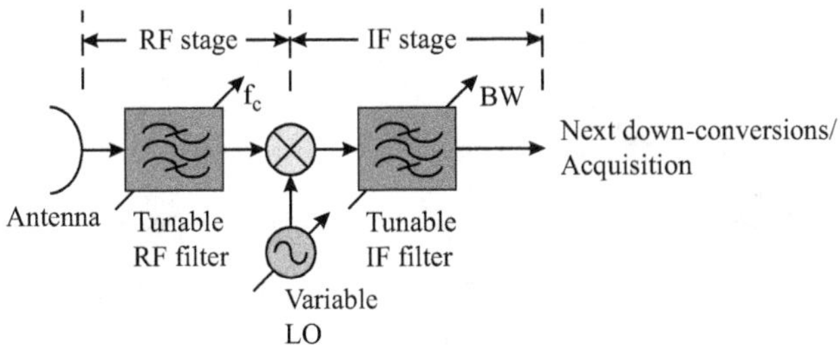

Figure 10.1 High-level schematic for the architecture of an adaptive superheterodyne wideband receiver

preselection process in a single dual-function RF filter. These filtering devices exploit different operating principles such as frequency-controllable circuit networks and coupled-resonator arrangements, whose advantages and disadvantages are discussed. Further, state-of-the-art mobile form factor high-Q filters for ultra-narrowband interference suppression based on mixed-technology acoustic-wave lumped-element resonators are also reported. A brief summary of future trends in this area is also provided at the end of the chapter.

For the specific case of coherent short-range frequency-modulated continuous wave (FMCW) high-resolution radars, this chapter also addresses the effects of narrowband interference on imaging performance through theoretical analysis and simulations. This example further justifies the necessity of integrating tunable notch filters in the receiver front-end. First, however, it is instructive to consider some of the conceptual architectures for wideband RF receivers and the effects interference has upon them. Spectrally adaptive multi-notch filters are key elements for these wideband receiver schemes that are applicable to dense and dynamic spectral environments containing a variety of interference sources.

10.1.1 Wideband receiver architectures for radar and communication systems

The superheterodyne architecture is a traditional scheme that can be extended for the SDR/CR paradigm [2,11]. A high-level view of this architecture for wideband receivers is shown in Figure 10.1, in which the exploitation of reconfigurable filters in the radio-frequency (RF) and intermediate-frequency (IF) stages is proposed to attain advantageous features.

The standard superheterodyne scheme is known to be suitable for processing signals with high dynamic range and can be designed so that the image band is attenuated by a fixed-frequency RF filter. One important disadvantage is that this RF filter must possess a passband that is broad enough to enable the reception of all the service channels since it is the IF filter that provides fine channel acquisition via tuning of the local oscillator (LO) frequency. If the fixed-frequency RF filter is very broad, the RF signals in other channels may experience interference at the IF stage due to spurious artifacts from the mixer.

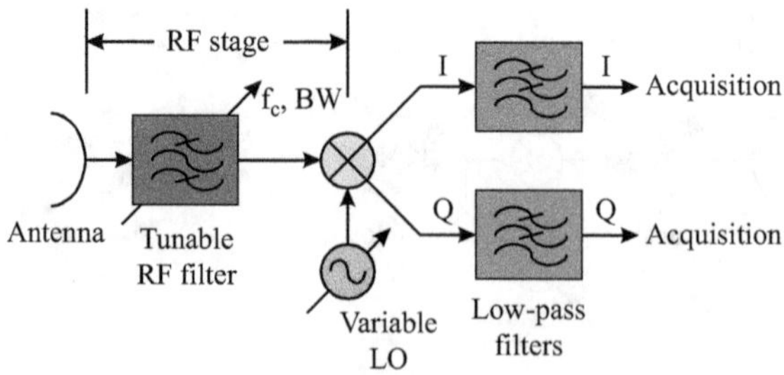

Figure 10.2 High-level schematic for the architecture of an adaptive zero-IF wideband receiver

In Figure 10.1, it is alternatively proposed to use a frequency-tunable RF filter with a reduced bandwidth to facilitate control of the desired channel along with attendant avoidance of out-of-band interference from spurious mixer artifacts that no longer coincide with the IF band. Note that the analog filter in the IF stage is also suggested to be tunable, but only in terms of instantaneous bandwidth of the passband, which can be determined adaptively. In any case, the superheterodyne architecture does not permit the simultaneous acquisition of several bands, though such could be feasible through the use of reconfigurable multi-passband filters.

Another well-known architecture for wideband receivers is the zero-IF scheme [2,19], the conceptual block diagram of which is detailed in Figure 10.2. This scheme considers the incorporation of a tune-all bandpass filter before the conversion stage, is very compact, and permits direct access to the in-phase (I) and quadrature (Q) components of the selected channel. Note that the use of an RF/microwave filter that can be tuned in center frequency and bandwidth together with a variable LO is proposed here, thus allowing for proper selection of the desired channel and simultaneous mitigation of out-of-band noise and interference. The main disadvantages of the zero-IF architecture include DC-offset issues and I/Q mismatches that may arise, especially in six-port circuit-based implementations for the mixing process. To circumvent these drawbacks, the so-called low-IF architecture has been devised, whose implementation is identical to the one depicted in Figure 10.1 that employs a single conversion. In this approach, the use of a low IF provides a solution to the DC-offset and I/Q-mismatch problems, though the issue with spurious artifacts in the image band reemerges [2,19].

Finally, the direct-sampling architecture in Figure 10.3 does not make use of frequency conversions [9,13]. Instead, it just proceeds with the acquisition of the filtered RF signals without utilizing mixers. The main disadvantage of this scheme is that the analog-to-digital converter (ADC) required to acquire the wideband signals is usually complex, even though subsampling rates could be used [13]. Note that the architecture in Figure 10.3 could employ a reconfigurable multi-band bandpass filter to enable the simultaneous and flexible acquisition of several bands at sub-Nyquist rates, thereby relaxing the operational demands on the ADC.

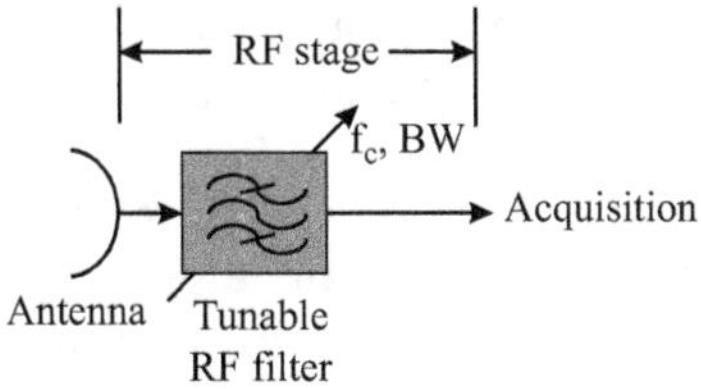

Figure 10.3 High-level schematic for the architecture of a direct-sampling wideband receiver

Upon consideration of the architectures in Figures 10.1, 10.2, and 10.3, it is clear that RF/microwave filtering devices possessing reconfigurable/adaptive attributes will likely play a critical role in future wideband RF receivers. When compared to conventional approaches that employ multiple front-ends to acquire several different services/bands, the use of tunable circuits permits a minimization of the required hardware with obvious associated advantages in terms of size, weight, DC power consumption, and cost (otherwise known as SWaP-C). Further, spectrally agile multi-notch filters should also be situated at the beginning of these wideband RF receiver architectures to protect them from intentional or unintentional narrowband interference to assure robust operation [18].

10.1.2 Narrowband interference in wideband receivers

In future wireless scenarios where the electromagnetic environment is highly congested and the exploitation of white space is permitted, the appearance of unintentional in-band interference could become a critical problem [5,18]. The effect of such interference on the performance of wideband RF receivers depends on multiple aspects, including the signal-to-interference ratio, the relative bandwidth of the interference, the particular operating mode (e.g. radar, communications, etc.), the modulation schemes of the desired and interfering signals, and their specific spectral attributes. Temporal and spatial aspects also play an important role in the probability of experiencing interference, this being more likely to occur during rush hour of a large city for example [5].

In the worst-case scenario where the power level of a captured in-band interference signals is much higher than that of the desired signal, and where the bandwidth of the former is comparable to the latter, the wideband receiver would suffer significant performance degradation. To some degree, this situation could still occur even if bandwidth-expanding techniques such as code-division multiple access (CDMA) were used to improve robustness [5]. Having said that, such an arrangement tends to be limited to cases involving intentional interference imposed on noncivilian systems.

A more common scenario involves the existence of unintentional narrowband interference signals that are effectively competing with the desired signal in the wideband receiver. If the relative interference power is modest, the accurate capture of the desired signal may undergo little to no degradation depending on its specific characteristics. For instance, the encoder/decoder block of a digital communication system employs error correction codes (e.g. Reed-Solomon or convolutional codes) to substantially reduce symbol classification errors, thus leading to a reasonable QoS [20].

However, if the narrowband interference signals are powerful enough, the performance of a wideband RF receiver can be seriously damaged. Possible damage notwithstanding, such interference to a radar can still produce significant errors in the localization, monitoring, and tracking of targets, as well as the presence of ghost images in the illuminated scene (e.g. a range-Doppler map or a synthetic aperture radar image). Further, the first active stages of the RF front-end, which contain low-noise amplifiers and down-conversion mixers, could become saturated, resulting in impaired sensitivity and possibly even a complete blinding of the wideband receiver [18].

In the case of front-end receiver saturation, attempts to perform adaptive interference cancellation, such as through the use of a digital phased-array antenna, would still fail because the saturation effect violates the linear superposition condition required for such methods. Thus, the rejection of high-power narrowband interference requires an analog approach. This solution can be achieved by the inclusion of a frequency-adaptive multi-notch filtering device at the beginning of the wideband receive chain that can efficiently suppress these dynamic interference sources before they desensitize the receiver. Since these adaptive notches reside within the operational bandwidth of the receiver, a modest performance degradation may be incurred as a trade-off for avoiding a complete blinding of the receiver. A case study on this topic is presented in the next section for an imaging application involving a high-resolution FMCW radar.

10.2 Case study: impact of narrowband interference on a wideband radar receiver

Portable short-range radars continue to become more miniaturized for a variety of applications (e.g. UAV sense-and-avoid, automotive blind-spot detection) and are being widely deployed [6]. With the expectation of progressively denser spectral environments, these systems will suffer from interference [18]. Here, a brief description of coherent high-resolution FMCW radar is provided and its unique advantages are detailed. The effects of narrowband interference on this type of system are then assessed through theoretical analysis and simulation. The impact of tunable notch filters in the receiver front-end to minimize these effects is also addressed [21].

10.2.1 Coherent FMCW radar

A high-level block diagram of a conventional FMCW high-resolution radar is illustrated in Figure 10.4. This architecture is based on the analog deramping (or dechirping) technique that consists of mixing the reflected echoes with a replica of the transmitted signal [22]. This mixing process gives rise to the so-called baseband "beat signal." In Figure 10.4, the signal reflected from a point scatterer and a continuous-wave (CW) interference signal are conceptually represented.

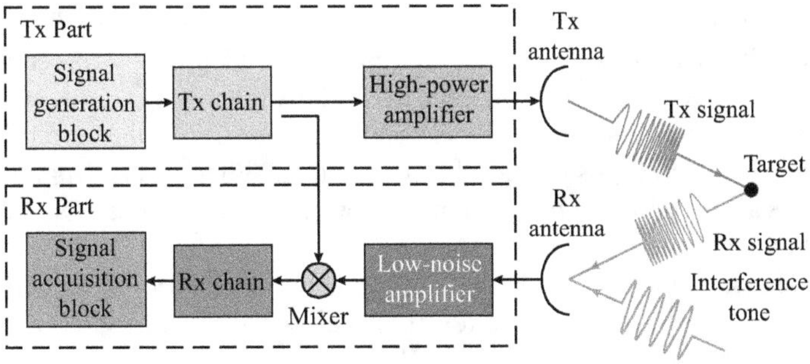

Figure 10.4 Architecture for a deramping-based coherent FMCW high-resolution radar. An interference tone incident at the receive antenna is also depicted

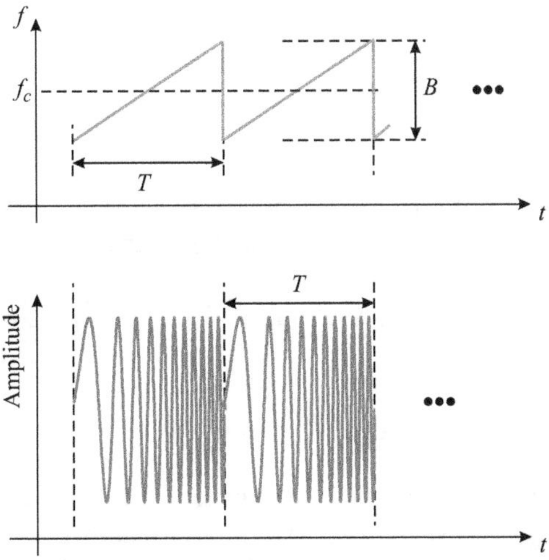

Figure 10.5 Instantaneous frequency vs. time (top) and time-domain representation of the transmitted waveform (bottom) for a coherent FMCW high-resolution radar

The FMCW "chirp" waveform transmitted by the radar is shown in Figure 10.5. This waveform consists of a periodic sequence of frequency ramps, with [22]

$$s_{\mathrm{Tx}}(t) = \exp\big(j(2\pi f_c t + \pi\,\gamma t^2)\big) \tag{10.1}$$

a complex analytic representation for one sawtooth interval of period T, where t represents the "fast-time" domain and is defined on $[-T/2, +T/2]$. Likewise, f_c is the center frequency, B is the bandwidth, and $\gamma(=B/T)$ is the chirp rate, which is the slope of the instantaneous-frequency ramp.

Consider that the point target in Figure 10.4 is located at a range $R(\tau)$ from the radar, where the inclusion of τ denotes an explicit dependence on "slow-time" that is sampled once per period T. The slow-time definition of $R(\tau)$ relies on the stop-and-go assumption that considers the target location to be fixed during the current period T and then hops to its next position for the next period. Although this hypothesis is not realistic, it is a very good approximation for slow targets and it greatly simplifies the subsequent math.

After algebraic manipulation, the beat signal over one period T that arises from mixing the reference waveform from (10.1) with the signal reflected by the point target can be approximately written as [22]

$$s_{\mathrm{b}}(t) = \sigma \exp\left(j\left(\left[\frac{4\pi\gamma R(\tau)}{c}\right]t + \frac{4\pi f_{\mathrm{c}} R(\tau)}{c}\right)\right), \tag{10.2}$$

where σ is the received target amplitude and c is the speed of light (3×10^8 m/s). From (10.2), it is clear that the baseband mixer response corresponding to a point target is a sinusoid whose frequency is proportional to the range $R(\tau)$. The Fourier transform of (10.2) followed by range axis scaling yields [22]

$$S_{\mathrm{b}}(f) = \sigma T \exp\left(j\left[\frac{4\pi f_{\mathrm{c}} R(\tau)}{c}\right]\right) \operatorname{sinc}\left(T\left[f - \frac{2\gamma R(\tau)}{c}\right]\right), \tag{10.3}$$

where f is the fast-time frequency variable associated with range and $\operatorname{sinc}(x) = \sin(\pi x)/(\pi x)$. The location of the sinc function peak in frequency is therefore proportional to the range $R(\tau)$.

It can be shown that the range resolution for a deramping-based FMCW radar is [22]

$$\Delta R = \frac{c}{2\gamma}\Delta f = \frac{c}{2(B/T)} \times \frac{1}{T} = \frac{c}{2B}, \tag{10.4}$$

which clearly simplifies to the well-known general expression for radar range resolution. From (10.4) we observe that if the transmitted instantaneous bandwidth is increased, a finer range resolution is obtained, which is useful for determining precise target locations in short-range applications.

The key advantage of the deramping-based architecture in Figure 10.4 is that the beat signal bandwidth B_{b} is much smaller than the transmitted bandwidth B. Specifically, these terms are related via [22]

$$B_{\mathrm{b}} = \frac{2\gamma R_{\max}}{c} \ll B, \tag{10.5}$$

where $R_{\max}$ is the maximum range from which echoes are expected to arrive. This relationship allows the ADC in the acquisition block to operate at a much lower rate, even to the point of permitting a low-end inexpensive unit. This aspect simplifies the entire architecture for short-range radar without compromising range resolution, which is still determined by transmit bandwidth B according to (10.4).

The constant phase term on the right-side of (10.2), which subsequently translates to the exponential factor in (10.3), corresponds to the phase history of the target as a function of slow-time τ. Specifically, this phase history is provided by the well-known expression [23]

$$\varphi(\tau) = \frac{4\pi f_c}{c} R(\tau). \tag{10.6}$$

It can be shown that a precise phase-based estimation of target range can be accomplished using the unwrapped phase history of (10.6). Further, it has been demonstrated that this phase-based range monitoring can realize sub-millimeter accuracy, which is rather convenient for the detection of small vibrations [22,23]. Moreover, if the range profiles collected over multiple periods are stacked to form a matrix and the Fourier transform subsequently applied in the slow-time dimension, Doppler information can be recovered, thus permitting the formation of range-Doppler maps [22,23].

To ensure that target phase history can be recovered, it is necessary that the radar system be coherent. Perfect coherence requires the use of complex circuitry based on a direct digital synthesizer (DDS) [22]. Fortunately, even with a free-running voltage-controlled oscillator and a reference signal, it is possible to properly synchronize the generation and acquisition blocks from Figure 10.4 to make the system coherent [23]. Therefore, despite the proliferation of incoherent FMCW radars, especially for automotive applications [24], it is feasible to construct portable and inexpensive coherent FMCW radar systems that are very well suited for short-range applications. Now consider the impact of in-band CW interference on the imaging performance of a coherent high-resolution FMCW radar system [18].

10.2.2 Impact and mitigation of CW interference on FMCW radar

Along with the block diagram of a coherent FMCW radar, Figure 10.4 also depicts an unintentional interference signal that is captured by the receive antenna. The complex analytic expression for this CW-interference tone is [18]

$$s_I(t) = \sigma_I \exp(j2\pi f_I t) \tag{10.7}$$

where f_I and σ_I are the associated frequency and received amplitude of the interference, respectively. It is assumed that the interference resides within the operational bandwidth of the receiver, and thus $f_I \in [f_c - B/2, f_c + B/2]$.

As discussed in Section 10.1.2, the unintentional interference in (10.7) may desensitize the radar receiver or even saturate the LNA in the radar receiver chain, thereby greatly degrading target detection capability. To avoid this situation, a notch-type filter needs to be incorporated into the wideband radar receiver prior to the LNA stage. Regardless, an attenuated version of the interference tone may still appear at the input to the deramping mixer, in which case it can be shown that the baseband beat signal corresponding to this CW-interference tone during one sawtooth period T is the chirping signal [18]

$$s_{bI}(t) = \sigma_{bI} \exp(j(2\pi(f_c - f_I)t + \pi\gamma t^2)), \tag{10.8}$$

with associated amplitude σ_{bI}.

Table 10.1 Parameters of the simulated coherent FMCW radar example

Center frequency (f_c)	10 MHz
Transmitted bandwidth (B)	1 MHz
Relative bandwidth (B/f_c)	10%
Ramp period (T)	10 ms
Chirp rate ($\gamma = B/T$)	0.1 MHz/ms
Coherent processing interval (CPI)	0.2 s
Number of range profiles (CPI/T)	20

The expression in (10.8) does not depend on slow-time τ, thus leading to the fact that the phase history for this unwanted signal has a zero value after formatting and processing the raw data. This ghost echo therefore appears at zero Doppler in a range-Doppler map. Further, the bandwidth of (10.8) is B, which means that the zero-Doppler ghost is spread in range (i.e. not focused like (10.8)), possibly with some aliasing since the reduced sampling rate of the ADC is much lower than B.

To verify the analysis, a coherent FMCW radar is simulated to illuminate a moving point-target while experiencing interference from a CW tone. Note that an RF-signal-level simulation is required to incorporate the effects of the additive interference and the notch filter. This requirement implies that an exceedingly high sampling frequency is necessary so that the radar and interference signals are parameterized realistically. To avoid this excessive computational burden, the simulations here make use of unrealistic parameters since the objective is to adequately emulate the effects of the interference and receiver notch on the imaging performance of the radar. Having said that, it is important to emphasize that the results derived from this academic example can be readily extrapolated to what would be obtained in practice.

The parameters for the simulated coherent FMCW radar are listed in Table 10.1. The range resolution is calculated to be $\Delta R = 150$ m in accordance with (10.4), whereas the unambiguous Doppler bandwidth is $1/T = 100$ Hz, which equivalently means the maximum unambiguous radial velocity is $c/(4Tf_c) = 750$ m/s. The coherent processing interval (CPI) is the total acquisition time that is used for the generation of the range-Doppler map.

In the simulation, it is assumed that the point-target is moving away from the radar in the radial direction. Table 10.2 lists the motion parameters for the scatterer. In particular, the unrealistic velocity $v = 1{,}900$ m/s has been chosen to illustrate target range migration. Also note that this velocity is ambiguous for the radar.

Finally, Table 10.3 details the parameters of the CW-interference tone. From Table 10.2, it can be deduced that a signal-to-interference ratio (SIR) of −20 dB is considered. To circumvent undesired aliasing, a simulation sampling frequency of 50 Mega-samples/s is used, which does provide some oversampling since the maximum frequency corresponds to a high-frequency replica at $2f_c = 20$ MHz. Further, no additive noise is included and conventional uniform windowing is employed in the fast- and slow-time dimensions.

Table 10.2 Parameters of the simulated point-target

Initial range (R_0)	1,000 m
Radial velocity (v)	1,900 m/s
Signal amplitude (σ)	1

Table 10.3 Parameters of the simulated CW-interference tone

Interference frequency (f_{I})	10 MHz
Signal amplitude (σ_{bI})	10

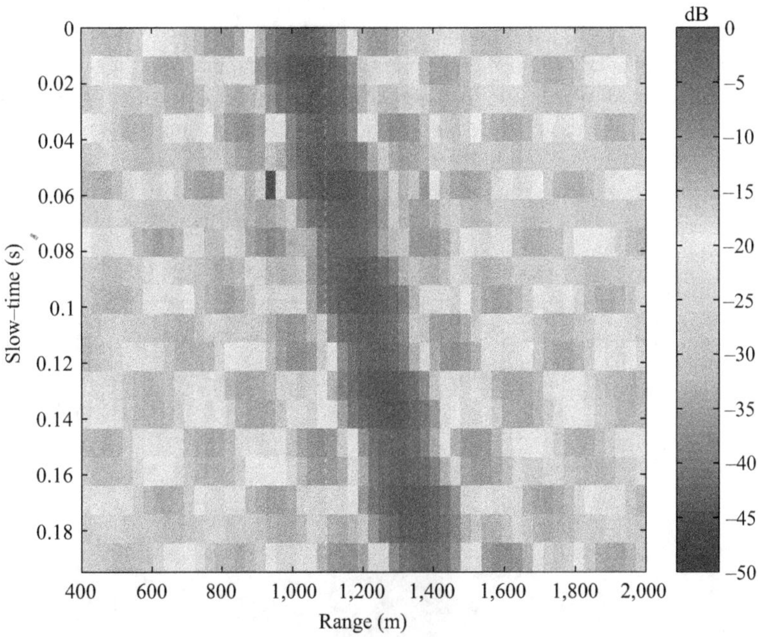

Figure 10.6 Range/slow-time data matrix simulating a moving target being illuminated by a coherent FMCW radar and combined with a narrowband interference tone

Figure 10.6 shows the 20 range profiles for this simulation, where the horizontal axis corresponds to range and the vertical axis to slow-time. Per Section 10.2.1, the result in Figure 10.6 has been obtained by applying an FFT to each fast-time interval of raw data followed by range scaling. The range history of the target is clearly identifiable in Figure 10.6. Indeed, it is verified that the target moves from

$R_0 = 1{,}000$ m to a final range of $R_0 + v$ (CPI) $= 1{,}380$ m, moving through several range cells in the process. The impact of interference in this range/slow-time map is evidenced by the relatively significant values not associated with the moving target.

After applying another FFT, this one in slow time (i.e. each column of the range-profile matrix in Figure 10.6), the corresponding range-Doppler map in Figure 10.7 is obtained. It is observed that the interference tone manifests as an increased response across range at zero-Doppler. On the other hand, the echo corresponding to the target is the typical defocused response one expects for a moving target whose translational motion has not been compensated.

The Doppler frequency for this target is $2vf_c/c = 126.7$ Hz, which ambiguously appears at 26.7 Hz in the first Nyquist interval in Doppler since the slow-time sampling frequency is $1/T = 100$ Hz. Further, the broadening in range corresponds to the range excursion of the target during the CPI (i.e. $v \times$ CPI $= 380$ m) and the observed Doppler spreading is the inverse of the time the target takes to move through a single range cell (i.e. $v/\Delta R = 1{,}900/150 = 12.7$ Hz).

For this specific simulation, the CW interference across range at zero-Doppler in Figure 10.7 would have little influence on radar performance since the target response is clearly isolated. However, in situations where the target is stationary or vibrating, such would not be the case. Moreover, front-end saturation (not shown here) due to higher interference power would degrade the target/interference separation shown in Figure 10.7, not to mention the possibility of completely blinding the radar receiver.

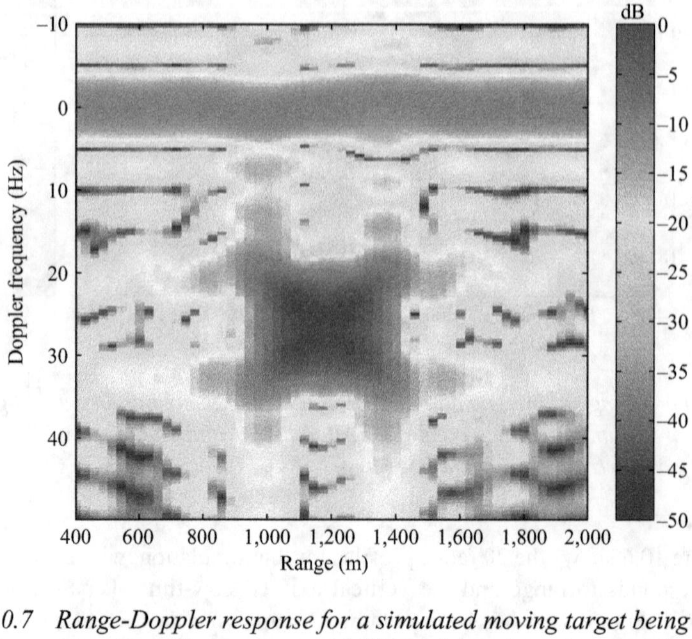

Figure 10.7 Range-Doppler response for a simulated moving target being illuminated by a coherent FMCW radar and combined with a narrowband interference tone

If the wideband RF receive filter has a notch tuned to the same frequency as the CW interference tone, the range-Doppler response from Figure 10.7 is replaced by that in Figure 10.8. Here, the 3-dB bandwidth of the notch is 10 kHz. It is observed that the large zero-Doppler signal from Figure 10.7 has now been adequately suppressed, while any distortion to the target response is not noticeable.

If the 3-dB bandwidth of the notch is alternatively increased to 1 MHz, the interference suppression is still quite effective, as shown in Figure 10.9. However, the target echo is now clearly distorted, thus indicating that notches should be made as narrow as possible while still achieving the required degree of interference rejection. Consequently, this result implies the need for ultra-narrowband, high quality-factor (Q) notch-type filters that can tune their transfer function to address interference at arbitrary locations within the receiver bandwidth.

10.2.3 Integration of tunable notch filters in the receiver front-end

The simulation results above illustrate the need for controllable filter notches within the analog RF front end of a wideband receiver to mitigate narrowband interference. Figure 10.10 illustrates the block diagram of a laboratory setup to integrate a notch filter into the receiver. Note that this notch-type filtering device must be frequency-tunable, since the center frequency of realistic narrowband interference sources are

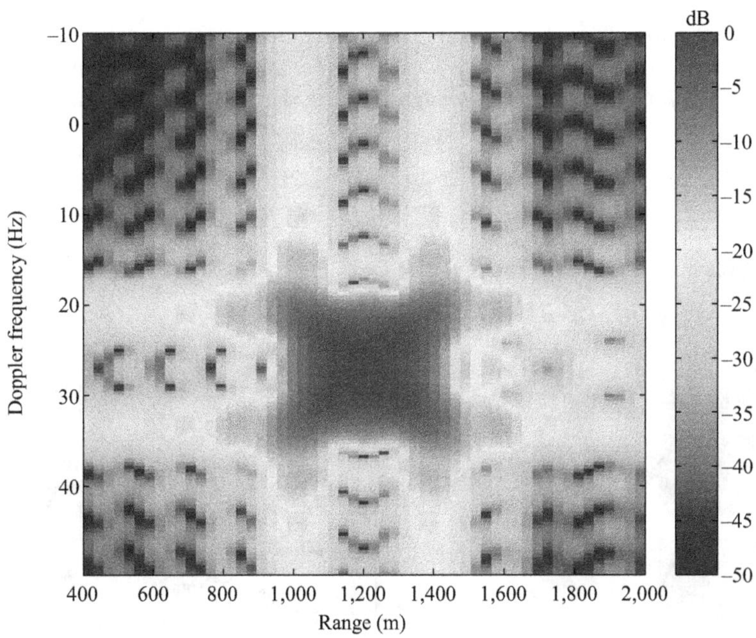

Figure 10.8 Range-Doppler response for a simulated moving target illuminated by a coherent FMCW radar after mitigation of the narrowband interference tone using a notch filter having a 3-dB bandwidth of 10 kHz

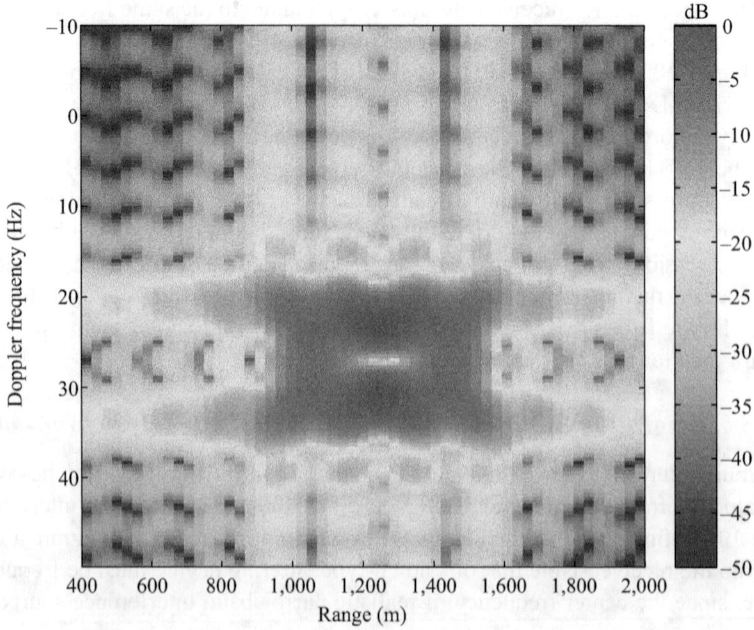

Figure 10.9 Range-Doppler response for a simulated moving target illuminated by a coherent FMCW radar after mitigation of the narrowband interference tone using a notch filter having a 3-dB bandwidth of 1 MHz

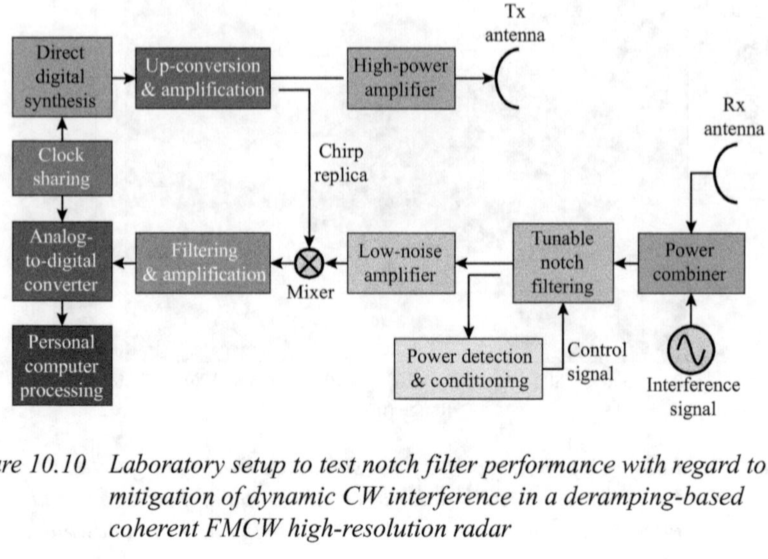

Figure 10.10 Laboratory setup to test notch filter performance with regard to mitigation of dynamic CW interference in a deramping-based coherent FMCW high-resolution radar

not known *a priori*. The laboratory setup shows that chirp generation can be achieved using a DDS combined with several mixers and/or frequency multipliers to up-convert the transmit signal. Further, the phase-coherence property can be easily obtained by sharing the clocks between the generation and acquisition blocks.

For the interference-suppression procedure, a power-detection closed-loop feedback control mechanism is used to track possible changes in interference frequency. The arrangement in Figure 10.10 is likewise extensible to cases requiring multiple tunable notches to address multiple simultaneous narrowband interference sources. The next section provides examples of state-of-the-art devices to achieve spectrally agile single-/multi-notch filters.

10.3 Spectrally agile RF/microwave multi-notch filters

Continued improvements in hardware miniaturization also supports the demand for co-integration of multiple RF functions in the same device. As such, a preferred design approach to realize the interference rejection capabilities discussed above may involve the co-synthesis of the reconfigurable multi-notch filter with the preselection wideband bandpass filter to realize a unique RF receiver component. The advantages of this multifunction concept include a more efficient implementation through simultaneous co-design of RF analog components, and reduced insertion loss due to the reduction of interconnect stages.

Despite their clear relevance, only a few high-frequency, wideband bandpass filters containing embedded notches have been proposed [25–31], and some of those [25–27] are frequency-static and do not permit frequency control of the notches. Of those with tunable notches, [28] employs a coplanar waveguide with spectrally agile notches, [29] involves a channelized double-multiplexer arrangement with controllable rejection bands, and [30,31] exploit variable-reactance terminated cross-shaped resonators. However, the disadvantages of these approaches include relatively narrow tuning ranges for the notches, poor notch attenuation due to their reconfigurable nature, and poor in-band power-matching. Further, these devices do not allow for simultaneous tuning of the center frequency of the wide passband in which the notches are embedded.

In the following, state-of-the-art planar implementations of wideband RF filters containing embedded adaptive notches for the purpose of multi-interference suppression are reported. These formulations exploit two different operational approaches, namely the signal-interference philosophy and tune-all coupled-resonator realizations that attain unprecedented levels of spectral controllability for both the in-band notches and the wide reception band. Further, emerging resonator-based high-Q filter implementations involving acoustic-wave lumped elements are discussed as a means to realize ultra-narrowband tunable and switchable notches.

10.3.1 Filter design for agile interference rejection

Based on a high-frequency extrapolation of transversal digital filters (see [32]), analog RF/microwave interference rejection filters employ circuit configurations whereby the input signal is divided into multiple different components that are individually filtered [33,34]. The overall desired filtering operation is then achieved by the frequency-dependent constructive and destructive interactions between these different components when they are subsequently recombined in the filter output node. Specifically, passband regions are obtained by forcing constructive combinations of signal components, while sharp-rejection stopbands are realized through destructive combinations that equivalently produce out-of-band transmission zeros (i.e. nulls).

The potential for this manner of interference-rejecting circuit network has been demonstrated for a wide variety of high-performance filtering applications and technologies [35–46]. For the case of frequency-static passive filters based on planar implementations, relevant examples include highly selective ultra-wideband bandpass and bandstop filters [35–38], multi-band filters that exhibit a large number of quasi-elliptic passbands [12], sharp-rejection filters with an ultra-broad stopband [39], and even multi-function passive RF/microwave components that simultaneously perform several analog signal processing operations within the same device (e.g. power distribution and impedance transformation combined with a single-/multi-band interference filtering capability) [40–42]. Likewise, examples of reconfigurable planar filters include bandwidth-switchable bandpass filters having an ultra-broad passband-width tuning ratio [43] and controllable notch-type filters with extended upper passband range [44]. All of these implementations exploit nonconventional filter structures to reject interference, such as through the use of pairs of dissimilar transmission line segments connected in parallel—which was the first published demonstration of a passive planar filter for interference cancellation [35]—or directional power couplers arranged in transversal schemes. In the following we consider a recent class of multifunction reconfigurable filter that provides highly efficient wideband signal acquisition while simultaneously suppressing multiple spectrally agile interferers in the band.

The circuit scheme and basic operating principles of this concept for a wideband RF bandpass filter containing embedded frequency-controllable notches are illustrated in Figure 10.11 [12]. This approach enables the dynamic mitigation of several interference sources that are simultaneously captured by the wideband receiver and is comprised of various transversal filter sections (TFSs) cascaded in series, while each TFS is shaped by three different electrical paths that are arranged in parallel. The organization of a TFS involves (i) transmission line paths 1 and 2 that realize the wideband bandpass filtering operation along with sharp-rejection stopbands that are produced by out-of-band transmission zeros, and (ii) a resonant-type path 3 consisting of inter-coupled lossy resonators that interacts with the path 1 transmission line to produce a quasi-absorptive notch embedded in the wide passband of the TFS. Thus, path 1 is used to perform both filter functions, which leads to benefits in terms of smaller circuit size and reduced insertion loss relative to more conventional solutions in which wideband filtering and notching are performed separately (e.g. by combining [35] and [47] in series through an additional transmission line segment).

The transfer function of each TFS is defined by the aggregation of the wideband filtering and notching operations. Further, by making the notches frequency-controllable via the resonators in path 3, multiple spectrally agile notches can

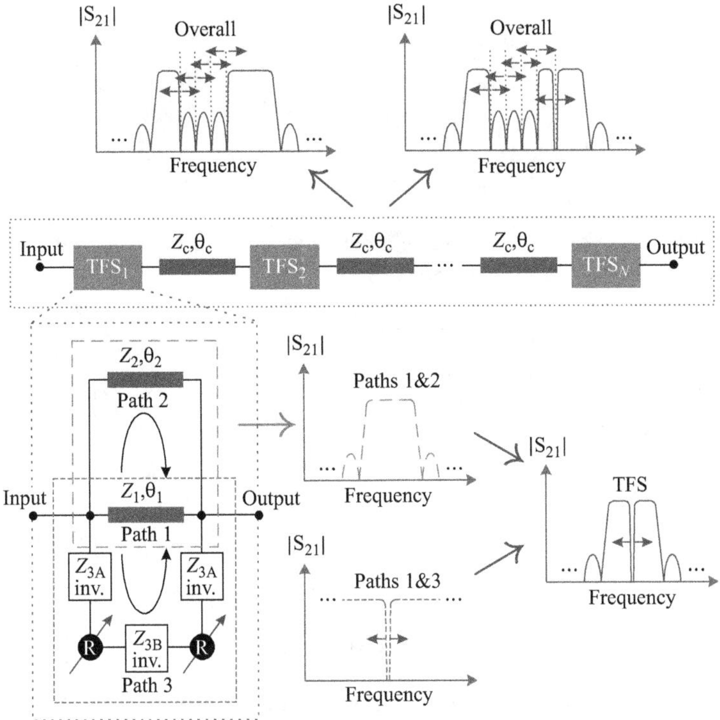

Figure 10.11 Circuit scheme and operating principles of wideband bandpass filter design involving frequency-tunable notches to reject multiple simultaneous interference sources as well as realize multi-band filtering profiles with controllable passband/stopband characteristics. Here, "TFS" denotes a transversal filtering section, "R" is a resonator, Z is the characteristic impedance, and θ is electrical length

be realized within the wide passband as shown in Figure 10.11. The controllable notches can likewise be exploited to reconfigure the bandwidths of multiple disjoint passbands.

As a proof-of-concept for the approach described above, a 50Ω-referred two-stage microstrip prototype was constructed that features two first-order rejection bands [48] (see Figure 10.12). This circuit has a center frequency of 2 GHz and 3-dB absolute bandwidth of 800 MHz for the passband, thus realizing a fractional bandwidth of 40%. It was fabricated on a Rogers 4003C microstrip substrate with a relative dielectric permittivity of 3.55, dielectric thickness of 1.524 mm, metal thickness of 35 μm, and dielectric loss tangent of 0.0027.

Per Figure 10.12, the transmission line paths of the TFSs (paths 1 and 2 in Figure 10.11) were realized as microstrip-line segments whose electrical lengths and

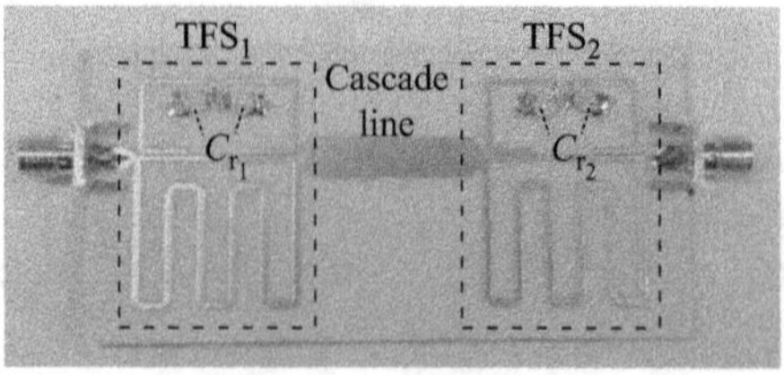

Figure 10.12 Microstrip prototype of a two-stage 1.6–2.4 GHz bandpass filter with embedded tunable notches

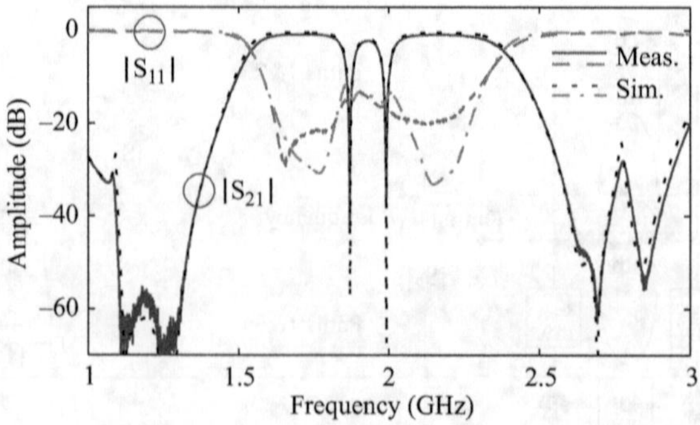

Figure 10.13 Simulated and measured power transmission (|S_{21}|) and reflection (|S_{11}|) parameters of the filter prototype in Figure 10.12 for one example state

characteristic impedances were selected to meet the specifications for the desired bandwidth according to the design procedure in [35]. The lossy resonant path 3 of each TFS was implemented by means of an inter-coupled capacitive-loaded resonator that interacts with the associated path 1. Mechanically adjustable capacitors (0.5–2.5 picofarads (pF)) from Johanson Manufacturing (denoted C_{r_1} and C_{r_2} in Figure 10.12) were inserted into these resonators to make them frequency-controllable, with their ends grounded by means of 1 mm diameter metallic via holes. The overall result is a filter containing a pair of quasi-absorptive first-order rejection bands that can be independently reconfigured in terms of their center frequency within the passband range of 1.6–2.4 GHz.

Using the commercial software tool Advanced Design System (ADS) from Keysight to simulate and an Agilent E8361A network analyzer to measure, the power transmission and reflection parameters of the filter circuit in Figure 10.12 are evaluated in Figure 10.13 for one example state. It is observed that the in-band insertion loss and input-power matching levels are better than 0.95 dB and 20 dB, respectively, across the 1.6–2.4 GHz passband, while the two notches reach depths of 43.5 dB (lower) and 38 dB (upper).

Independent center frequency tuning of the two notches is illustrated in Figure 10.14. Specifically, Figure 10.14(a) demonstrates how the lower frequency

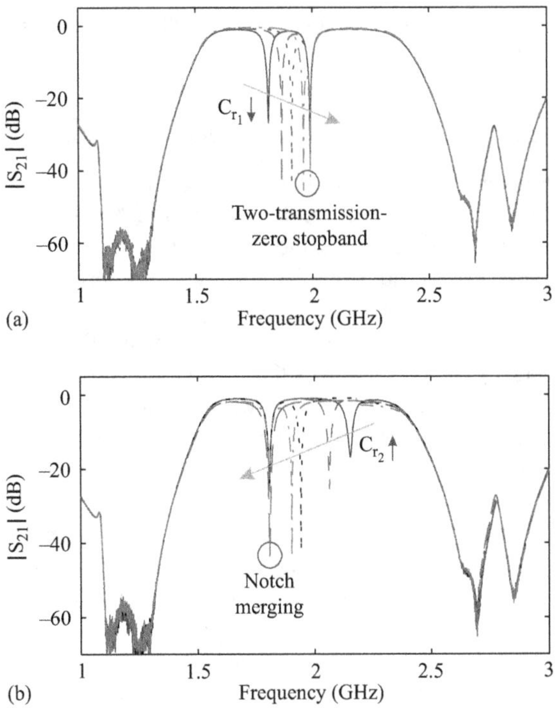

Figure 10.14 Measured demonstration of reconfigurable filter notches for the prototype in Figure 10.12 in terms of the power transmission parameter (|S_{21}|), with (a) independent frequency control of the lower notch for a fixed upper notch, and (b) independent control of the upper notch for a fixed lower notch

notch is spectrally controlled (denoted by $C_{r_1}\downarrow$), yielding a notch depth of 27 to 46 dB over the frequency range of 1.81 to 1.97 GHz while the upper notch remains fixed, and producing a two transmission-zero stopband when the notches coincide. Figure 10.14(b) likewise shows that tuning of the upper notch over the frequency range of 1.8 to 2.16 GHz (denoted by $C_{r_2}\uparrow$) realizes depths of 44 to 17 dB when the lower notch is fixed. Note that the notch merging condition implies that the combination of notches into a single stopband permits further control of the associated bandwidth and rejection level as well as a means to dictate the number of required notches without the need for RF switches.

As a final remark, when compared to other wideband bandpass filtering devices with embedded reconfigurable stopbands such as those in [28–31], this circuit shows much greater notch rejection depths and a broader tuning range. Further, this formulation can be practically scaled via the inclusion of addition TFS stages to realize an increased number of notches and selectivity levels. It should be noted, however, that the main shortcoming of this design is that the center frequency of the broad passband cannot be tuned. This limitation is the motivation for the alternative approach presented in the next section.

10.3.2 Filter design based on a coupled-resonator arrangement

Wideband RF receivers for emerging radar and communication capabilities under the spectrum-sharing paradigm may require much greater flexibility in how they select and subsequently utilize their band of operation. Consequently, in addition to facilitating reconfigurable notching for narrowband interference rejection, such receivers should also provide the means with which to change the center frequency of the passband. This section presents an RF filter architecture that addresses this need.

In recent years, it has been shown that coupled-resonator configurations are appropriate for the development of RF/microwave bandpass filters with fully tunable characteristics. Because the center frequency of such circuits is mainly determined by the natural frequencies of the constituent resonators, a reconfigurable capability can be easily realized by making these resonators spectrally controllable. This condition can be achieved by inserting variable-reactance elements into the filter resonators. Moreover, control of the overall passband bandwidth can be obtained by using inter-resonator coupling control techniques, thus permitting the bandwidth to be dynamically modified by varying the strengths of the various electromagnetic couplings. Fully tunable coupled-resonator bandpass filters has received considerable attention in the literature based on these principles as well as others including the dynamic allocation of transmission zeros to simultaneously reconfigure center frequency, bandwidth, and the out-of-band attenuation profile for quasi-elliptic transfer functions [49–52]. More advanced approaches even permit switching between different types of filtering functions—such as bandpass/bandstop commutation—to enable selectable transmission/suppression characteristics for desired signals and interference [53–56].

Here, we consider how a coupled-resonator approach could be employed to produce a wideband bandpass filter with embedded reconfigurable notches [57]. The associated concept and operating principles are illustrated in Figure 10.15, where it is observed that two different types of frequency-controlled resonators are used to achieve different characteristics of the overall transfer function. First, the resonators included in the direct bandpass input-to-output path facilitate synthesis of the main passband. For the filter network in Figure 10.15 comprised of K resonators, a Kth-order transfer function is realized, with control of the associated resonant frequencies enabling tuning of the center frequency of this passband component. Attached to each of these resonators via separate impedance inverters or couplings is a set of N additional resonators with different natural frequencies (the N-band bandstop filtering cells in Figure 10.15) that permits the creation of N in-band notches at different spectral locations. Collectively, this architecture thereby enables the generation of N, Kth-order stopbands throughout the passband. Synchronous tuning of the kth resonators across the N sets provides control of the center frequency of the kth stopband, while reconfigurable control of the bandwidth for the main passband and the notches can be attained by making all the inter-resonator coupling variable. This filter

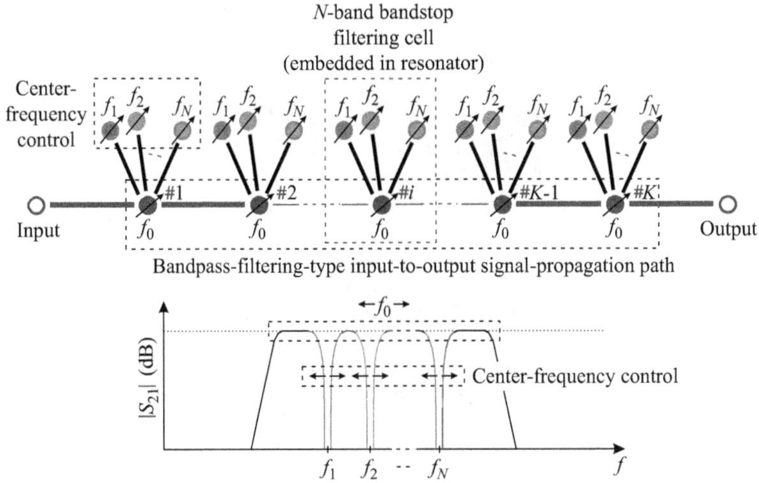

Figure 10.15 Concept and operating principles of a coupled-resonator wideband bandpass filter with tunable embedded notches, where circles indicate the resonators, lines the impedance inverters or couplings, f_0 is the resonant frequency of the resonators in the input-to-output path that determines the center frequency of the overall passband, and $f_1, f_2, \ldots, f_N$ are the natural frequencies of the resonators that define the center frequencies of the N, Kth-order stopbands

is scalable to supply any number of adaptable notches embedded in an arbitrary-order tunable passband with theoretically unlimited selectivity.

As a practical demonstration of the RF filter concept in Figure 10.15, a 50 Ω-referred mechanically tunable microstrip prototype is depicted in Figure 10.16. This circuit exhibits two third-order notches and was designed so that these notches and the overall wide passband can be frequency tuned over the 0.8 to 1.2 GHz range by means of the corresponding frequency-controllable resonators. For this implementation the same substrate used for the previous design was again employed. The tunable resonators that produce the wide passband and in-band notches were realized as uniform-impedance transmission line segments terminated in shorted, mechanically adjustable capacitors capable of variation over 1–5 pF. In particular, capacitor C_f provides center frequency control for the overall passband, while C_{r1} and C_{r2} permit tuning of the center frequencies of the lower and upper notches, respectively. The inter-resonator couplings were designed to be static through the use of 0.5 pF fixed capacitors, which determine the bandwidth of the notches, and quarter-wavelength transmission line segments that establish the bandwidth of the main passband.

Simulated and measured results for one example state of this filter prototype are shown in Figure 10.17. The measured insertion-loss and input-matching levels are 0.85 dB and 13.9 dB, respectively, over the entire passband. The measured center

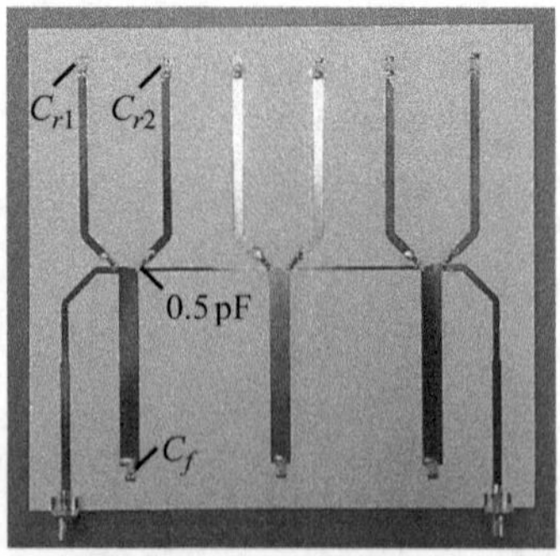

Figure 10.16 Microstrip prototype of a three-stage coupled-resonator bandpass filter with embedded tunable notches

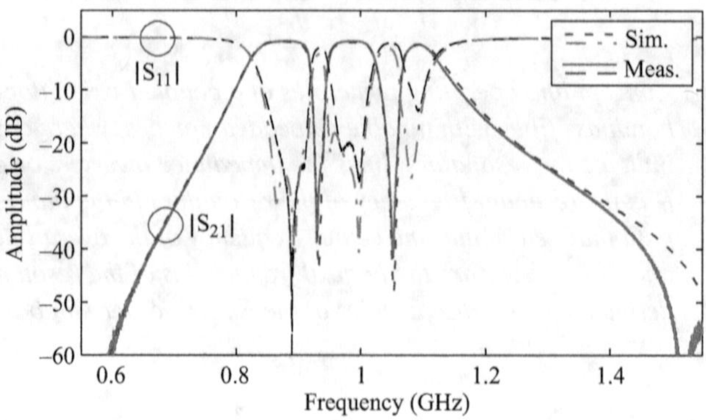

Figure 10.17 Simulated and measured power transmission ($|S_{21}|$) and reflection ($|S_{11}|$) parameters of the filter prototype from Figure 10.16 for one example state

frequency is 0.97 GHz and the 3-dB absolute bandwidth is 293 MHz, for a fractional bandwidth of 30.2%. The maximum attenuation levels for the lower and upper notches positioned at 0.93 GHz and 1.05 GHz were measured to be 38.5 dB and 41.5 dB, respectively.

The frequency reconfigurability of this filter design is corroborated in Figure 10.18. First, the ability to control the center frequency of the overall passband containing two embedded third-order stopbands is verified in Figure 10.18(a) for static notches and in Figure 10.18(b) for similarly tuned notches. Figure 10.18(c) and 10.17(d) shows how, for a static passband, the lower and upper notches can respectively be tuned in terms of center frequency while the other one remains fixed. Finally, the flexibility attained by merging the lower and upper notches is depicted in Figure 10.18(e).

These results demonstrate that a coupled-resonator wideband filter design can be readily tuned in terms of passband center frequency and the number/location of

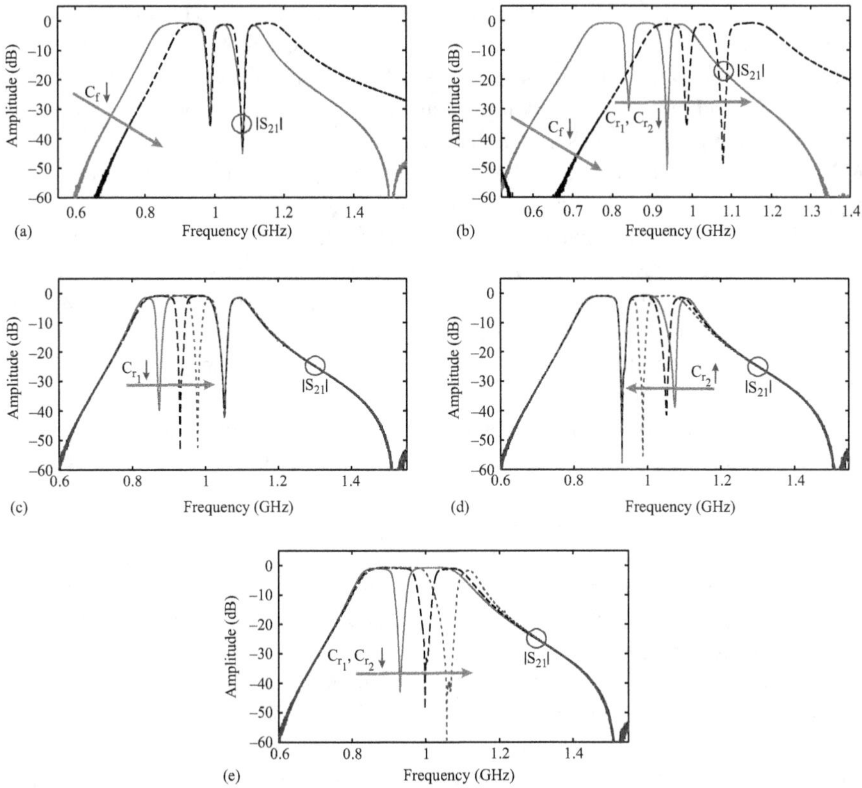

Figure 10.18 Measured demonstration of reconfigurable filter in Figure 10.16 in terms of the power transmission parameter ($|S_{21}|$), with (a) center frequency tuning of the overall passband for static stopbands, (b) passband center frequency tuning with similarly tuned stopbands, (c) center frequency tuning of the lower stopband, (d) center frequency tuning of the upper stopband, and (e) spectral merging of the two stopbands into a single wider one whose center-frequency is also tunable

spectral notches without the use of RF switches. This approach therefore exhibits superior spectral adaptivity to the design methodology in the previous section, albeit with band transitions that are less sharp and narrower overall passband. Consequently, we have one further design approach to consider.

10.3.3 Filter design based on hybrid acoustic-wave/lumped-element resonators

As illustrated in the previous example involving a chirped radar encountering narrowband interference, the width of the spectral notch used to reject this interference can have a deleterious impact on the radar if it is wider than necessary. However, the multi-notch filters discussed above involve planar topologies, which are known to suffer from low to moderate Q factors at microwave frequencies, thus limiting the possible notch depths and minimum bandwidths that can be achieved. In contrast, 3-D implementations using waveguides and dielectric resonators can realize ultra-narrowband high-Q bandstop filters [58–60], though they tend to possess a large size/weight due to the well-known inverse relationship between a resonator's Q-factor and its volume. Moreover, these approaches cannot be easily tuned in an electronic way while maintaining high-Q properties. For example, piezoelectric-actuated cavity resonators achieve our desired goal reasonably well, but at the expense of high DC biasing voltages [61]. Likewise, optically switched superconductor technologies have also been applied to design commutable narrowband notch filters, but they require external cryogenic cooling systems [62].

Acoustic-wave resonator circuits are among the most popular options for high-performance filters that have high Q (on the order of 1,000) and compact size (1–4 mm^3) [63]. Nevertheless, they also possess some major shortcomings, including (i) high insertion-loss for high-order filter realizations, (ii) the presence of spurious modes (e.g. Rayleigh and transverse) inherent to acoustic-wave resonators that can severely affect the filter performance, especially if operating near the fundamental mode, and (iii) spectral asymmetry of their transfer function that hinders tuning due to the lack of reconfigurable attributes.

In [64], acoustic-wave resonators were hybridized with lumped elements to circumvent some of the previous drawbacks in high-Q, mobile form factor bandpass filter designs. The idea proposed in [64] was to connect an additional inductor in parallel with the acoustic-wave resonator so that the series resonance of the composite hybrid cell (i.e. the desired high-Q resonance) is effectively decoupled from its resulting parallel resonance (i.e. the unwanted low-Q resonance). Based on this RF design technique, compact high-Q bandpass filters featuring nearly symmetrical transfer functions were implemented. This design has the added benefit of minimizing the anti-resonance intrinsic to acoustic-wave elements that otherwise negatively affects the entire filter.

This hybrid acoustic-wave lumped-element resonator was subsequently applied in [65–67] to design frequency-static, frequency-switchable, and bandwidth-tunable ultra-narrowband notch filters with reflective and absorptive behavior in their stopband(s). In some cases, these filters have exhibited an effective Q of up to 10,000. For this purpose, the hybrid lumped-element acoustic-wave resonator was exploited as a notch-type building block (see Figure 10.19) to facilitate different coupled-resonator bandstop filter configurations with or without additional controllable discrete components. Further details about these filters specific to spectrally agile RF interference suppression are as follows:

- Figure 10.20 shows the block diagram for a general N-channel architecture, a hardware prototype, and subsequent experimental results of a notch filter that can switch its ultra-narrow stopband between two discrete states [65]. This implementation is formed from a filter array shaped by two parallel-cascaded second-order

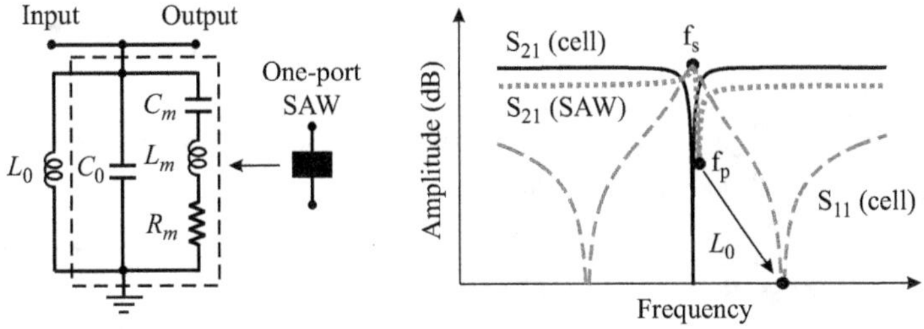

Figure 10.19 Circuit layout and operating principle of the notch-type cell based on a hybrid acoustic-wave lumped-element resonator

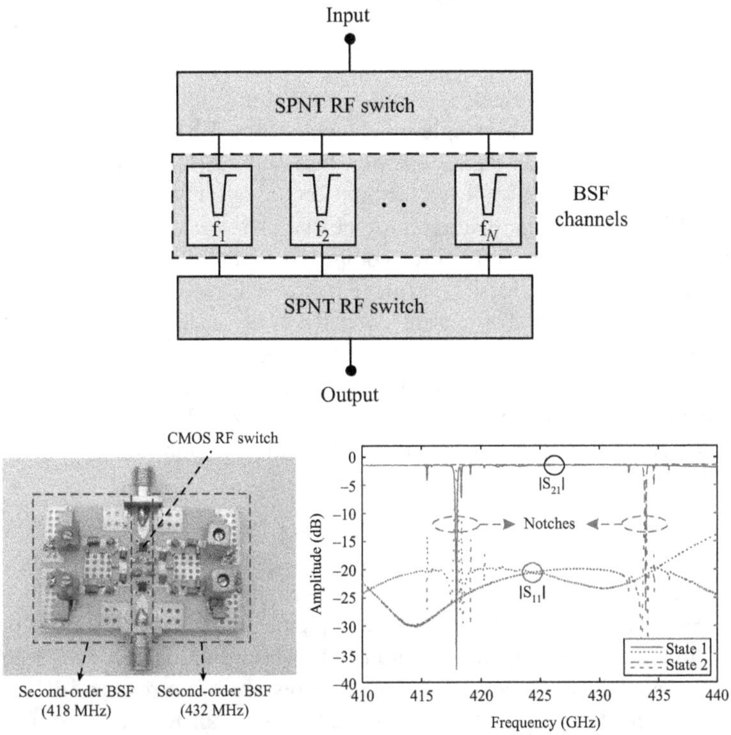

Figure 10.20 Block diagram of a general N-channel architecture, hardware prototype, and experimental results of a frequency-switchable/two-state second-order bandstop filter based on hybrid acoustic-wave lumped-element resonators. Here, BSF is a bandstop filter and SPNT refers to single-pole N-throw

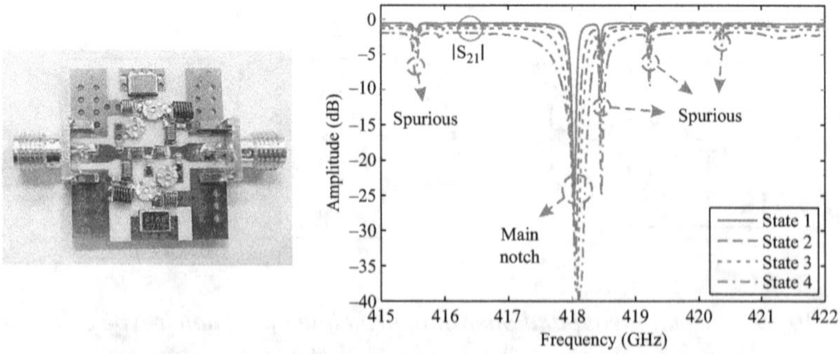

Figure 10.21 Prototype and experimental results of a bandwidth-tunable third-order bandstop filter based on hybrid acoustic-wave lumped-element resonators

acoustic-wave lumped-element resonator bandstop-filtering channels that can be arbitrarily selected via two single-pole double-throw RF switches from *Skyworks Solutions*. Over the passband, insertion loss and notch depths of 0.35 dB and 45 dB, respectively, are achieved for both commutation states.

- Figure 10.21 shows a hardware prototype and measured results of a bandwidth-tunable third-order notching filter comprised of three identical acoustic-wave lumped-element resonators. To achieve bandwidth reconfigurability, two additional lumped-element inverters are inserted between each of the hybrid resonators and the direct input-to-output electrical path of the filter so that the stopband bandwidth can be continuously tuned by varying one of the two impedance inverters. An analog bandwidth variation between 0.18 MHz and 0.71 MHz is measured, thus achieving a 4:1 bandwidth-tuning ratio with insertion loss and notch depths of 0.55–2 dB and 25–40 dB, respectively.

In summary, the combination of high-Q acoustic-wave resonators and low/moderate-Q lumped elements provides a means to preserve the merits of both technologies so that state-of-the-art ultra-narrowband bandstop filters with compact size and deep notch depths can be readily implemented. Although further research is needed to make these filters continuously tunable in terms of center frequency, they are envisioned as key devices for the suppression of narrowband RF interference in wideband RF receivers that may well become all too common for future communication and radar systems.

10.4 Looking ahead: future trends

In this chapter, microwave bandstop filters have been considered that can be reconfigured in terms of center frequency, bandwidth, and stopband attenuation as a possible hardware-oriented solution for the mitigation of dynamic RF interference in wideband receivers. Of course, additional power-detection control mechanisms are needed to establish where in frequency the passband and subsequent interference-rejecting notches should be located. Complex digital signal processing may also be required for interference detection/location, causing accumulated time delay to become a critical factor. In scenarios in which the spectral locations of interfering signals are known (e.g. co-located base stations), a more efficient approach could be the use of frequency-static bandstop filters centered at the interference positions that can react by themselves to the RF signal power level. Thus, the filter transfer function would change depending on the level of the input signal, so that high-power interference is attenuated at the analog level (bandstop filter state) while low-power interference is allowed through (all-pass filter state) and mitigated through signal processing.

Although different alternatives such as frequency-selective limiters with a static center frequency have been considered (e.g. [68,69]), innovative frequency-tunable power-dependent RF filters are still being investigated. One recent example is the circuit architecture reported in [70] that employs tunable, resonant all-pass networks with self-reconfigurable capabilities to provide bandstop filtering of high-power interference. This approach is based on the detuning of one of its resonators as a function of the RF signal power that is sensed in another part of the circuit, resulting in a power-activated all-pass-to-bandstop commutation when this level surpasses a threshold. Similar concepts were applied in [71] for the development of an evolved family of RF frequency-selective limiters that incorporate switchable multiplexing capabilities to achieve a more dynamic selection/power-limitation of desired signals along with multi-interference suppression. On the other hand, in parallel with [70] and [71] that addressed this topic at the circuit level, new physics phenomena are being investigated with a view to establishing new RF component classes for even higher power RF environments. One such example is plasma-based high-power widely tunable limiters [72]. All these efforts could open the way towards the invention of highly advanced multi-interference suppression modules that could find broad usage in complex threat/jamming-suppression scenarios envisioned for future electronic warfare and tactical communication systems.

Beyond the mitigation of dynamic RF interference in future crowded spectral environments, frequency domain reconfigurability of future transmit and receive chains for the purpose of sharing spectrum between radar and communications will be highly appealing. To that purpose, investigations on high-power tunable components will be of paramount importance [73], especially for long-range radar transmitters where algorithms to concurrently change the impedance matching and transfer function according to spectral requirements are considered to be vital [74]. In addition, integrating spatial reconfigurability via smart phased-array systems is also considered an import notion to which much research effort will be devoted [75].

10.5 Conclusions

RF interference is a critical problem for next-generation wireless communications and radar systems operating in increasingly congested electromagnetic environments. As their receiver modules must operate over large portions of the RF spectrum, such systems could be easily damaged by intentional or unintentional narrowband interference appearing in their operating band. Thus, highly robust hardware mechanisms must be developed to protect these RF systems from this undesired phenomenon and guarantee their co-existence. Within this context, this chapter has reviewed recent research findings in the field of spectrally agile RF filters for dynamic multi-interference mitigation in wideband receivers. Followed by a simple case study in which the effect of this problem was illustrated based on the imaging performance of a high-resolution FMCW radar affected by narrowband CW interference, different types of state-of-the-art tunable multi-notch filters were presented. These filters include planar realizations based on transversal filter and coupled-resonator arrangements, in which the reconfigurable notches are integrated with the signal-preselection wide passband filtering process to achieve compact devices with low insertion loss. Also, mobile form factor high-Q notch filter components have been described based on emerging hybrid acoustic-wave lumped-element resonators for ultra-narrowband multi-interference suppression. These filters exhibit bandwidth-tunable and frequency-switchable notch-filter realizations. Finally, emerging trends in frequency-adaptive multi-notch filtering devices were discussed in the context of their effective deployment in real systems, as well as prospects to enable them to operate under high-power conditions and make their behavior sensitive to the input RF signal level.

References

[1] R. Gómez-García, J. P. Magalhaes, J. M. Muñoz-Ferreras, J. M. N. Vieira, N. B. Carvalho, and J. Pawlan, "Filling the spectral holes: novel/future wireless communications and radar receiver architectures," *IEEE Microwave Magazine*, vol. 15, no. 2, pp. 45–56, Mar.–Apr. 2014.

[2] J. M. Muñoz-Ferreras, R. Gómez-García, and F. Pérez-Martínez, "RF front-end concept and implementation for direct sampling of multiband signals," *IEEE Transactions on Circuits and Systems II: Express Briefs*, vol. 58, no. 3, pp. 129–133, Mar. 2011.

[3] W. J. Chappell, E. J. Naglich, C. Maxey, and A. C. Guyette, "Putting the radio in "software-defined radio": Hardware developments for adaptable RF systems," *Proceedings of the IEEE*, vol. 102, no. 3, pp. 307–320, Mar. 2014.

[4] J. M. Muñoz-Ferreras, Z. Peng, R. Gómez-García, G. Wang, C. Gu, and C. Li, "Isolate the clutter: pure and hybrid linear-frequency-modulated continuous-wave (LFMCW) radars for indoor applications," *IEEE Microwave Magazine*, vol. 16, no. 4, pp. 40–54, May 2015.

[5] N. B. Carvalho, A. Cidronali, and R. Gómez-García, *White Space Communication Technologies*. Cambridge: Cambridge University Press, 2014.

[6] C. Li, Z. Peng, T.-Y. Huang, *et al.*, "A review on recent progress of portable short-range non-contact microwave radar systems," *IEEE Transactions on Microwave Theory and Techniques*, vol. 65, no. 5, pp. 1692–1706, May 2017.

[7] M. A. Richards, J. A. Scheer, and W. A. Holm, *Principles of Modern Radar: Basic Principles*, 2nd ed. Edison, NJ: SciTech Publishing, 2010.
[8] M. I. Skolnik, *Introduction to Radar Systems*, 3rd ed. New York, NY: McGraw-Hill, 2002.
[9] J. Mitola, "The software radio architecture," *IEEE Communications Magazine*, vol. 33, no. 5, pp. 26–38, May 1995.
[10] R. Bagheri, A. Mirzaei, M. E. Heidari, *et al.*, "Software-defined radio receiver: dream to reality," *IEEE Communications Magazine*, vol. 44, no. 8, pp. 111–118, Aug. 2006.
[11] P. Cruz, N. B. Carvalho, and K. A. Remley, "Designing and testing software-defined radios," *IEEE Microwave Magazine*, vol. 11, no. 4, pp. 83–94, Jun. 2010.
[12] R. Gómez-García, J. M. Muñoz-Ferreras, and M. Sánchez-Renedo, "Microwave transversal six-band bandpass planar filter for multi-standard wireless applications," in *2011 IEEE Radio Wireless Symposium*, Phoenix, AZ, USA, Jan. 2011, pp. 166–169.
[13] J. M. Muñoz-Ferreras, R. Gómez-García, and F. Pérez-Martínez, "Multiband radar receiver design approach for minimum bandpass sampling," *IEEE Transactions on Aerospace and Electronic Systems*, vol. 49, no. 2, pp. 774–785, Apr. 2013.
[14] J. M. Muñoz-Ferreras and R. Gómez-García, "A deramping-based multi-band radar sensor concept with enhanced ISAR capabilities," *IEEE Sensors Journal*, vol. 13, no. 9, pp. 3361–3368, Sep. 2013.
[15] R. Gómez-García and A. C. Guyette, "Reconfigurable multi-band microwave filters," *IEEE Transactions on Microwave Theory and Techniques*, vol. 63, no. 4, pp. 1294–1307, Apr. 2015.
[16] D. Psychogiou, R. Gómez-García, and D. Peroulis, "Fully-reconfigurable bandpass/bandstop filters and their coupling-matrix representation," *IEEE Microwave and Wireless Components Letters*, vol. 26, no. 1, pp. 22–24, Jan. 2016.
[17] R. Gómez-García, A. C. Guyette, D. Psychogiou, E. J. Naglich, and D. Peroulis, "Quasi-elliptic multi-band filters with center-frequency and bandwidth tunability," *IEEE Microwave and Wireless Components Letters*, vol. 26, no. 3, pp. 192–194, Mar. 2016.
[18] J. M. Muñoz-Ferreras, Z. Peng, C. Li, and R. Gómez-García, "Effects and mitigation of interference tones on coherent FMCW short-range radars," in *2016 IEEE MTT-S International Wireless Symposium*, Shanghai, China, Mar. 14–16, 2016, pp. 1–4.
[19] M. A. T. Sanduleanu, M. Vidojkovic, V. Vidojkovic, A. H. M. van Roermund, and A. Tasic, "Receiver front-end circuits for future generations of wireless communications," *IEEE Transactions on Circuits and Systems II: Express Briefs*, vol. 55, no. 4, pp. 299–303, Apr. 2008.
[20] B. Sklar, *Digital Communications: Fundamentals and Applications*, 2nd ed. New Delhi: Pearson New International, 2013.

[21] D. Psychogiou, R. Gómez-García, and D. Peroulis, "Signal-interference bandpass filters with dynamic in-band interference suppression," in *2016 IEEE Radio Wireless Symposium*, Austin, TX, USA, Jan. 2016, pp. 1–4.

[22] G. Wang, J. M. Muñoz-Ferreras, C. Gu, C. Li, and R. Gómez-García, "Application of linear-frequency-modulated continuous-wave (LFMCW) radars for tracking of vital signs," *IEEE Transactions on Microwave Theory and Techniques*, vol. 62, no. 6, pp. 1387–1399, Jun. 2014.

[23] Z. Peng, J. M. Muñoz-Ferreras, Y. Tang, C. Liu, R. Gómez-García, L. Ran, and C. Li, "A portable FMCW interferometry radar with programmable low-IF architecture for localization, ISAR imaging, and vital sign tracking," *IEEE Transactions on Microwave Theory and Techniques*, vol. 65, no. 4, pp. 1334–1344, Apr. 2017.

[24] J. Hasch, E. Topak, R. Schnabel, T. Zwick, R. Weigel, and C. Waldschmidt, "Millimeter-wave technology for automotive radar sensors in the 77 GHz frequency band," *IEEE Transactions on Microwave Theory and Techniques*, vol. 60, no. 3, pp. 845–860, Mar. 2012.

[25] F. Wei, Z. D. Wang, F. Yang, and X. W. Shi, "Compact UWB BPF with triple-notched bands based on stub loaded resonator," *IET Electronics Letters*, vol. 49, no. 2, pp. 124–126, Jan. 2013.

[26] J. Wang, W. Hong, H. Tang, Y. Zhang, J. Chen, and J. Zhou, "Ultra-wide band bandpass filter with multiple frequency notched bands based on SIW and SIR technology," in *39th European Microwave Conference*, Rome, Italy, Sep. 28–Oct. 2, 2009, pp. 268–271.

[27] S. Kumar, R. D. Gupta, and M. S. Parihar, "Multiple band notched filter using C-shaped and E-shaped resonator for UWB applications," *IEEE Microwave and Wireless Components Letters*, vol. 26, no. 5, pp. 340–342, May 2016.

[28] M. Nosrati, N. Vahabisani, and M. Daneshmand, "Compact MEMs based ultrawide-band CPW band-pass filters with single/double tunable notch-bands," *IEEE Transactions of Components, Packaging, and Manufacturing Technology*, vol. 4, no. 9, pp. 1451–1460, Sep. 2014.

[29] R. Gómez-García and A. C. Guyette, "Two-branch channelized passive filters for lowpass and bandpass applications," in *2016 IEEE MTT-S International Microwave Symposium*, San Francisco, CA, USA, May 22–27, 2016.

[30] W. Feng, Y. Shang, and W. Che, "Ultra-wideband bandpass filter with reconfigurable notch band using TCSRs," *IET Electronics Letters*, vol. 51, no. 23, pp. 1893–1894, Nov. 2015.

[31] H. Wang, K.-W. Tam, S.-K. Ho, W. Kang, and W. Wu, "Design of ultra-wideband bandpass filters with fixed and reconfigurable notch bands using terminated cross-shaped resonators," *IEEE Transactions on Microwave Theory and Techniques*, vol. 62, no. 2, pp. 252–265, Feb. 2014.

[32] A. V. Oppenheim and R. W. Schafer, *Discrete-Time Signal Processing*, ser. Signal Process., 2nd ed. New York, NY, USA: Prentice-Hall.

[33] C. Rauscher, "Microwave active filters based on transversal and recursive principles," *IEEE Transactions on Microwave Theory and Techniques*, vol. 33, no. 12, pp. 1350–1360, Dec. 1985.

[34] C. Rauscher, "Microwave channelized active filters—a new modular approach to achieving compactness and high selectivity," *IEEE Transactions on Microwave Theory and Techniques*, vol. 44, no. 1, pp. 122–132, Jan. 1996.

[35] R. Gómez-García and J. I. Alonso, "Design of sharp-rejection and low-loss wideband planar filters using signal-interference techniques," *IEEE Microwave and Wireless Components Letters*, vol. 15, no. 8, pp. 530–532, Aug. 2005.

[36] M.-A. Sánchez-Soriano, E. Bronchalo, and G. Torregrosa-Penalva, "Compact UWB bandpass filter based on signal interference techniques," *IEEE Microwave and Wireless Components Letters*, vol. 19, no. 11, pp. 692–694, Nov. 2009.

[37] M.-A. Sánchez-Soriano, G. Torregrosa-Penalva, and E. Bronchalo, "Compact wideband bandstop filter with four transmission zeros," *IEEE Microwave and Wireless Components Letters*, vol. 20, no. 6, pp. 313–315, Jun. 2010.

[38] W. J. Feng and W. Q. Che, "Ultra-wideband bandpass filter using broadband planar Marchand balun," *IET Electronics Letters*, vol. 47, no. 3, pp. 198–199, Feb. 2011.

[39] R. Gómez-García, M.-A. Sánchez-Soriano, M. Sánchez-Renedo, G. Torregrosa-Penalva, and E. Bronchalo, "Low-pass and bandpass filters with ultra-broad stopband bandwidth based on directional couplers," *IEEE Transactions on Microwave Theory and Techniques*, vol. 61, no. 12, pp. 4365–4375, Dec. 2013.

[40] R. Gómez-García, M. Sánchez-Renedo, and J.-M. Muñoz-Ferreras, "A type of planar array-antenna feeding network with single/multi-band filtering capability," *IEEE Antennas and Wireless Propagation Letters*, vol. 9, pp. 1271–1274, 2010.

[41] R. Gómez-García, R. Loeches-Sánchez, D. Psychogiou, and D. Peroulis, "Single/multi-band Wilkinson-type power dividers with embedded transversal filtering sections and application to channelized filters," *IEEE Transactions on Circuits and Systems I: Regular Papers*, vol. 62, no. 6, pp. 1518–1527, Jun. 2015.

[42] R. Loeches-Sánchez, D. Psychogiou, R. Gómez-García, and D. Peroulis, "Transformers with incorporated filtering capabilities exploiting signal-interference principles," in *2015 IEEE International Conference on Microwave, Communications, Antennas and Electronic Systems*, Tel Aviv, Israel, Nov. 2–4, 2015.

[43] M.-A. Sánchez-Soriano, R. Gómez-García, G. Torregrosa-Penalva, and E. Bronchalo, "Reconfigurable-bandwidth bandpass filter within 10–50%," *IET Microwave, Antennas and Propagation*, vol. 7, no. 7, pp. 502–509, May 2013.

[44] A. C. Guyette, "Design of fixed- and varactor-tuned bandstop filters with spurious suppression," in *40th European Microwave Conference*, Paris, France, Sep.–Oct. 26–1, 2010, pp. 288–291.

[45] J. Ni, I. Vendik, D. Kholodnyak, and J.-S. Hong, "A compact bandpass filter based on right- and left-handed transmission line sections," *IEEE Microwave and Wireless Components Letters*, vol. 23, no. 6, pp. 279–281, Jun. 2013.

[46] K. W. Tam, P. Vitor, and R. P. Martins, "MMIC active filter with tuned transversal element," *IEEE Transactions on Circuits and Systems II: Analog and Digital Signal Processing*, vol. 45, no. 5, pp. 632–634, May 1998.
[47] D. R. Jachowski, "Compact, frequency-agile, absorptive bandstop filters," in *2005 IEEE MTT-S International Microwave Symposium*, Long Beach, CA, USA, Jun. 12–17, 2005, pp. 513–516.
[48] D. Psychogiou, R. Gómez-García, and D. Peroulis, "Signal interference-based bandpass filters with frequency reconfigurable in-band rejection bands," in *2017 USNC-URSI National Radio Science Meeting*, Boulder, CO, USA, Jan. 4–6, 2017.
[49] I. C. Hunter and J. D. Rhodes, "Electronically tunable microwave bandpass filters," *IEEE Transactions on Microwave Theory and Techniques*, vol. MTT-30, no. 9, pp. 1353–1360, Sep. 1980.
[50] A. R. Brown and G. M. Rebeiz, "A varactor-tuned RF filter," *IEEE Transactions on Microwave Theory and Techniques*, vol. 48, no. 7, pp. 1157–1160, Jul. 2000.
[51] M. Sánchez-Renedo, R. Gómez-García, J.I. Alonso, and C. Briso-Rodríguez, "Tunable combline filter with continuous control of center frequency and bandwidth," *IEEE Transactions on Microwave Theory and Techniques*, vol. 53, no. 1, pp. 191–199, Jan. 2005.
[52] A. C. Guyette, "Intrinsically switched varactor-tuned filters and filter banks," *IEEE Transactions on Microwave Theory and Techniques*, vol. 60, no. 4, pp. 1044–1056, Apr. 2012.
[53] D. Psychogiou, R. Gómez-García, and D. Peroulis, "Fully-reconfigurable bandpass/bandstop filters and their coupling-matrix representation," *IEEE Microwave and Wireless Components Letters*, vol. 26, no. 1, pp. 22–24, Jan. 2016.
[54] E. J. Naglich, J. Lee, D. Peroulis, and W. J. Chappell, "A tunable bandpass-to-bandstop reconfigurable filter with independent bandwidths and tunable response shape," *IEEE Transactions on Microwave Theory and Techniques*, vol. 58, no. 12, pp. 3770–3779, Dec. 2010.
[55] J. Lee, E. Naglich, H. H. Sigmarsson, D. Peroulis, and W. J. Chappell, "New bandstop filter circuit topology and its application to design a bandstop-to-bandspass switchable filter," *IEEE Transactions on Microwave Theory and Techniques*, vol. 61, no. 3, pp. 1114–1123, Mar. 2013.
[56] Y.-H. Cho and G. Rebeiz, "0.7–1.0-GHz reconfigurable bandpass-to-bandstop filter with selectable 2- and 4-pole responses," *IEEE Transactions on Microwave Theory and Techniques*, vol. 62, no. 11, pp. 2626–2632, Nov. 2014.
[57] D. Psychogiou, R. Gómez-García, and D. Peroulis, "Fully-adaptive multi-band bandstop filtering sections and their application to multi-functional components," *IEEE Transactions on Microwave Theory and Techniques*, vol. 64, no. 12, pp. 4405–4418, Dec. 2016.
[58] J. R. Qian and W. Z. Zhuang, "New narrow-band dual-mode bandstop waveguide filters," *IEEE Transactions on Microwave Theory and Techniques*, vol. MTT-31, no. 12, pp. 1045–1050, Dec. 1983.

[59] R. S. Snyder, "Quasi-elliptic compact high-power notch filters with selectivity both below and above the notched passband," in *2014 IEEE MTT-S International Wireless Symposium*, X'ian, China, Mar. 24–26, 2014, pp. 1–4.

[60] A. Ananad and X. Liu, "Capacitively tuned electrical coupling for reconfigurable coaxial cavity bandstop filters," in *2015 IEEE MTT-S International Microwave Symposium*, Arizona, AZ, USA, May 17–22, 2015, pp. 1611–1614.

[61] D. Psychogiou and D. Peroulis, "Advances in high-*Q* tunable filter technologies," *International Journal of Advances in Engineering Science and Applied Mathematics*, vol. 7, no. 4, pp. 170–176, Dec. 2015.

[62] K. F. Raihn, N. O. Fenzi, G. L. Hey-Shipton, E. R. Saito, P. V. Loung, and D. L. Aidnik, "Adaptive high temperature superconducting filters for interference rejection," *IEEE Transactions on Microwave Theory and Techniques*, vol. 44, no. 7, pp. 1374–1381, Jul. 1996.

[63] P. Warder and A. Link, "Golden age for filter design: innovative and proven approaches for acoustic filter, duplexer, and multiplexer design," *IEEE Microwave Magazine*, vol. 16, no. 7, pp. 60–72, Aug. 2015.

[64] D. Psychogiou, R. Gómez-García, R. Loeches-Sánchez, and D. Peroulis, "Hybrid acoustic-wave-lumped-element resonators (AWLRs) for high-*Q* bandpass filters with quasi-elliptic frequency response," *IEEE Transactions on Microwave Theory and Techniques*, vol. 63, no. 7, pp. 2233–2244, Jul. 2015.

[65] D. Psychogiou, R. Gómez-García, and D. Peroulis, "High-*Q* bandstop filters exploiting acoustic-wave-lumped-element resonators (AWLRs)," *IEEE Transactions on Circuits and Systems II: Express Briefs*, vol. 63, no. 1, pp. 79–83, Jan. 2016.

[66] D. Psychogiou, R. Gómez-García, and D. Peroulis, "Continuously-tunable-bandwidth acoustic-wave resonator-based bandstop filters and their multi-mode modeling," in *46th European Microwave Conference, London*, United Kingdom, Oct. 3–7, 2016.

[67] D. Psychogiou, R. Gómez-García, and D. Peroulis, "Acoustic wave resonator-based absorptive bandstop filters with ultra-narrow bandwidth," *IEEE Microwave and Wireless Components Letters*, vol. 25, no. 9, pp. 570–572, Sep. 2015.

[68] P. Phudpong and I. C. Hunter, "Frequency-selective limiters using nonlinear bandstop filters," *IEEE Transactions on Microwave Theory and Techniques*, vol. 57, no. 1, pp. 157–164, Jan. 2009.

[69] C. Collado, A. Hueltes, E. Rocas, *et al.*, "Absorptive limiter for frequency-selective circuits," *IEEE Microwave and Wireless Components Letters*, vol. 24, no. 6, pp. 415–417, Jun. 2014.

[70] A. C. Guyette, E. J. Naglich, and S. Shin, "RF-power-activated and signal-tracking tunable bandstop filters," *IEEE Transactions on Microwave Theory and Techniques*, vol. 65, no. 5, pp. 1534–1544, May 2017.

[71] E. J. Naglich and A. C. Guyette, "Frequency-selective limiters utilizing contiguous-channel double multiplexer topology," *IEEE Transactions on Microwave Theory and Techniques*, vol. 64, no. 9, pp. 2871–2882, Sep. 2016.

[72] A. Semnani, S. O. Macheret, and D. Peroulis, "A high-power widely tunable limiter utilizing an evanescent-mode cavity resonator loaded with a gas discharge tube," *IEEE Transactions on Plasma Science*, vol. 44, no. 12, pp. 3217–3280, Dec. 2016.

[73] D. Peroulis, S. P. Pacheco, and L. P. B. Katehi, "RF MEMS switches with enhanced power-handling capabilities," *IEEE Transactions on Microwave Theory and Techniques*, vol. 52, no. 1, pp. 59–68, Jan. 2004.

[74] C. Baylis, M. Fellows, L. Cohen, and R. J. Marks II, "Solving the spectrum crisis: Intelligent, reconfigurable microwave transmitter amplifiers for cognitive radar," *IEEE Microwave Magazine*, vol. 15, no. 5, pp. 94–107, Jul. 2014.

[75] G. M. Rebeiz and K.-J. Koh, "Silicon RFICs for phased arrays," *IEEE Microwave Magazine*, vol. 10, no. 3, pp. 96–103, May 2009.

Chapter 11

Transmitter architectures for radar/communication spectral coexistence

Randy Jost[1] *and Larry Cohen*[2]

While considerable progress has been made using illuminators of opportunity for sensing purposes (see Chapters 7 and 8), there will continue to be a need for high-power radar transmitters. As such, this chapter considers transmitter architectures for the purpose of spectral coexistence among radar and communication systems. Specifically, it asks the question: what are the transmitter structures and designs that will best allow both radars and communication systems to operate concurrently in the same portion of the electromagnetic (EM) spectrum, with neither degrading the other to at least some predefined level of performance? The authors attempt to address this question by describing the approach that is used in most existing radar and communication systems and by proposing approaches for new and future systems.

Before proceeding, the more fundamental question of "what is a transmitter?" is addressed. A general answer might be a device for transferring energy in the form of a signal via some propagation path from one location to another, where an additional device (a receiver) detects the energy and converts it into a form from which information can be extracted. Under this general definition, many devices can be considered to be transmitters, including such disparate things as a power plant, the Sun, or a biological cell. For the purpose of this discussion, we can refine the definition of a transmitter to encompass the means to emit radio frequency (RF) signals that traverse a propagation path either for the purpose of extracting information from the environment based on the subsequent reflections, such as in the case of a radar, or to convey information to another location, as in the case of communications. This RF emission may be only a carrier wave, e.g. a continuous-wave (CW) radar that provides sensing of Doppler information based on relative radial motion, or it may consist of a carrier wave and an impressed signal that couples additional information to be reflected to or decoded at the receiver.

For the energy transfer to work efficiently, the transmitter and receiver should be "matched" to each other; that is, whatever signal the transmitter generates, the receiver should be designed to capture as much of the signal, with as much fidelity, as

[1]Ball Aerospace, USA
[2]U.S. Naval Research Laboratory, USA

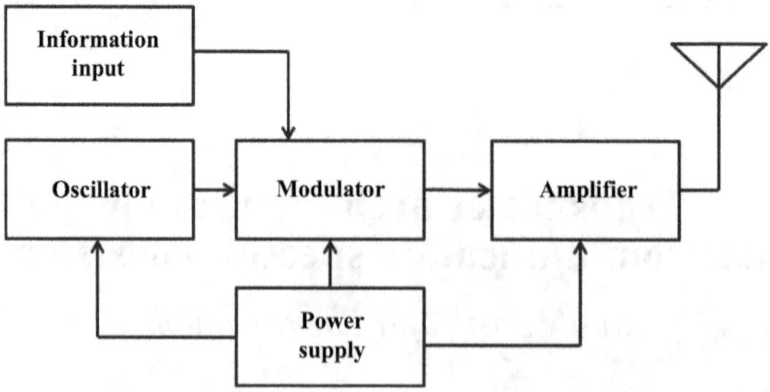

Figure 11.1 Block diagram of a basic transmitter

possible. Thus, in many important respects, the transmitter architecture must mirror or complement the receiver architecture as closely as possible, even in the case of a radar where transmitted pulses (coded or otherwise) are designed to interact advantageously with the targets they illuminate. The reader should note that this capability is especially important for transmitter architectures that rely on an associated receiver to provide feedback information that allows the transmitter to operate more effectively or efficiently in an environment with many other transmitter–receiver combinations.

Given the preceding discussion, the key components of a typical RF transmitter are:

- the information input (signal source);
- the oscillator (RF carrier);
- the modulator;
- the power supply;
- the high-power amplifier (HPA); and
- the output impedance-matching section (antenna).

A typical arrangement of a basic transmitter with these components and a low-level modulator is shown in Figure 11.1.

The scope of this chapter is limited to the transmitter portion of a generic radar or communication system, mostly focusing on descriptions of current approaches for transmitter design to provide a departure point from which to propose architectures for future implementations. The first section briefly discusses the EM and operational environments from a radar perspective. The EM environment includes the portion of the spectrum in which sensing is performed along with the related propagation effects, while the operational environment includes practical issues such as how the transmitter interacts with other proximate receivers and emitters, and the regulatory framework that defines and constrains what the transmitter is permitted to do.

After that, a brief discussion of various transmitter schemes is presented in Section 11.2, from the first RF implementations to current architectures, so as to describe what items and functions are common to all transmitters. This overview

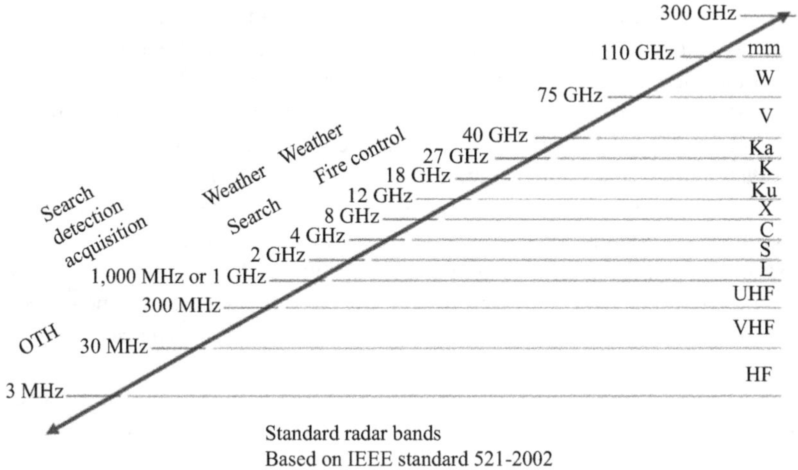

Figure 11.2 Radar applications versus frequency bands

provides a background from which to propose future transmitter arrangements since all future systems must carry out these same functions. Section 11.3 then reviews the components that realize particular transmitter functions and also provides descriptions of the types of components that will be required to implement future architectures. Following that is a section on the measurement of transmitter emissions, which is necessary both to be able to assess the success of a specific radar architecture in meeting its requirements and to understand the parameters that must be measured to provide the feedback needed to implement adaptive and so-called cognitive transmitter architectures. The chapter closes with a brief outline of prospective research directions to facilitate more advanced transmitter architectures, along with references and suggested reading for additional information on transmitters.

11.1 Spectral, propagation, and regulatory environment for transmitters

Radars typically operate from 3.0 MHz to 94 GHz, with some remote sensing conducted at medium frequencies (0.3–3 MHz) and extremely low frequencies (3–30 Hz), and newer pioneering radars operating as high as 300 GHz. Figure 11.2 [1] displays radar applications by typical frequency bands. For example, search radars often operate between 2 and 4 GHz; however, HF radars (3.0–30 MHz) are also used for over-the-horizon coverage, surface-wave systems for distances up to 250 nautical miles (nmi), and sky-wave systems for distances up to 2,000 nmi. Sky-wave HF radars use ionospheric refraction to achieve such long-distance coverage. Many of these radars are of the solid-state (SS) variety, employing class-C SS power amplifiers (PAs) to maximize efficiency. In the United States, however, HF radars have secondary status with respect to communications; that is, if this radar causes interference, it is required to change its operating frequencies. The other bands of the RF spectrum also contain radars for various applications. Communication systems operate in these very same bands (3 MHz to 100 GHz) and depend upon regulatory or technical means to prevent interference between radars and other communication systems.

Radars require spectrum to operate efficiently and effectively, with the particular amount of spectrum and radiated power dependent on their mission, expected distances to targets, and the physical sizes of targets (which relates to their radar cross sections). For example, long-range air-traffic-control (ATC) radars in the United States operate in the 1215–1390 MHz band transmitting megawatts of power so that they can detect and monitor small targets within the national airspace for hundreds of miles from any one radar site. Likewise, an airborne weather radar might operate at X-band with detection ranges much shorter than in L- and S- bands. Recently, radars in S-band have been experiencing challenges from the wireless communication community in the frequency band above 3500 MHz.

In the United States, the National Telecommunications Information Administration (NTIA) has frequency allocation authority over government-owned spectrum, which includes that used by a majority of radar modes. The infamous *Red Book*, aka the *Manual of Regulations and Procedures for Federal Radio Frequency Management* [2], contains the radar spectrum engineering criteria (RSEC). Topics addressed in the RSEC include (1) the emission bandwidth contained within a −40 dB emissions mask; (2) the roll-off and spurious emission masks beyond the −40 dB limit mask; (3) radar tenability; (4) antenna patterns; (5) frequency tolerance; (6) receiver selectivity; (7) image and spurious rejection; and (8) radiation from local oscillators. The most important of these eight topics is the emission bandwidth, which is depicted in Figure 11.3 for a nominal radar pulse in the time domain [1]. Note the pulse center and rising/falling (leading/trailing) edges in the figure. Care should be taken to ensure that the time required for the rising and falling edges is not arbitrarily chosen to be too fast so as to avoid spectral spreading, which often results in adjacent-band interference to other users of the spectrum. To illustrate the RSEC, Figure 11.4 shows the same pulse in the frequency domain. The in-band emissions are contained within the stovepipe that decreases from the maximum emission value at the zero (reference) point down to −40 dB. Beyond −40 dB, the emissions are designated as unwanted, and these unwanted emissions are further subcategorized as *out-of-band* (OOB) and *spurious*.

The RSEC provides computational formulas that set the limits of the −40 dB stovepipe and demarcation limits between the OOB and spurious emissions. Guidance for making emission measurements is included within the NTIA Technical Report TR-05-420 and can be summarized by:

1. measurements of radiated emissions are preferred, as opposed to measuring power through a directional coupler connected to a waveguide;
2. peak detection in the spectrum analyzer should be used;
3. the measurement bandwidth of a pulsed-CW emission should be as close as possible to t^{-1}, where t is the pulse width at the 50% voltage level;

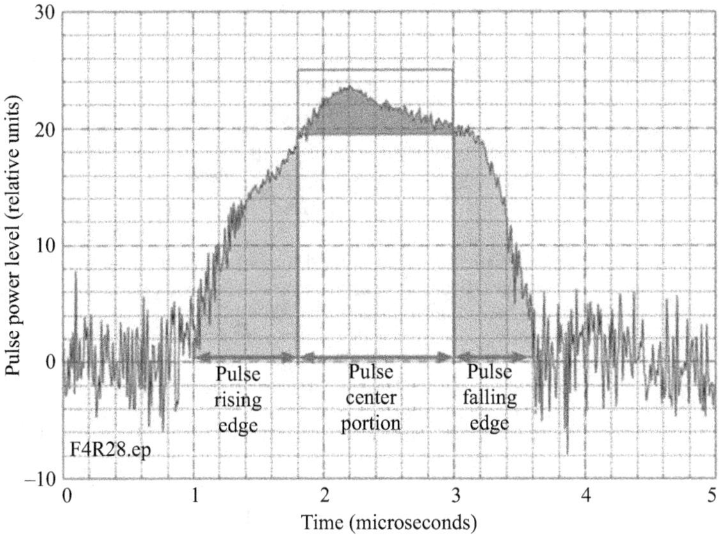

Figure 11.3 Radar pulse in the time domain

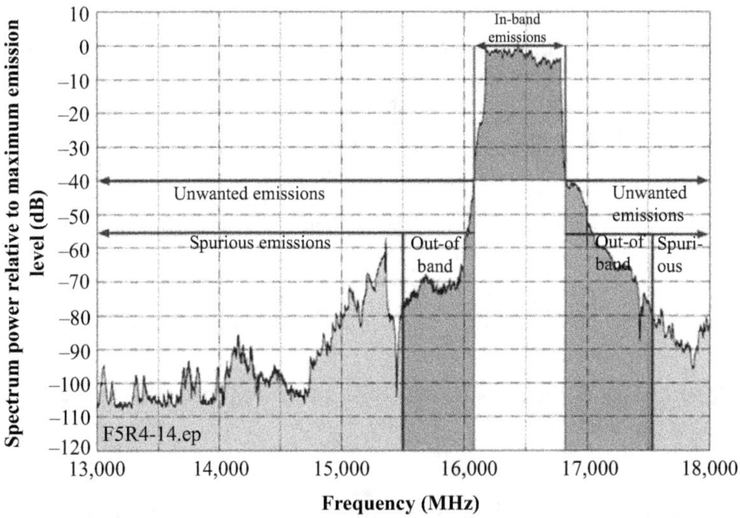

Figure 11.4 Radar pulse emissions criteria in the frequency domain

4. the measurement bandwidth of an FM-pulsed chirp should be $(B/t)^{1/2}$, where B is the frequency sweep range during each pulse; and
5. the measurement bandwidth for a phase-coded signal, such as a Barker code, is the chip width^{-1}.

Table 11.1 Determination of RSEC measurement bandwidth (B_m)

Radar modulation type	RSEC Measurement bandwidth (B_m)
Non-FM pulsed and phase-coded pulsed	$B_m \leq (1/t)$ where $t =$ emitted pulse duration (50% voltage) or phase-chirp (sub-pulse) duration (50% voltage). **Example for non-FM pulsed**: If emitted pulse duration is 1 μs, then $B_m \leq 1$ MHz. **Example for phase-coded pulsed**: If radar transmits 26 ms duration pulses, each pulse consisting of 13 phase-coded chips that are each 2 μs in duration, then $B_m \leq 500$ kHz.
FM-pulsed (chirped)	$B_m \leq (B_c/t)^{1/2}$ where $B_c =$ frequency sweep range during each pulse and $t =$ emitted pulse duration (50% voltage). **Example**: If radar sweeps (chirps) across frequency range of 1.3 MHz during each pulse, and if the pulse duration is 55 μs, then $B_m \leq 154$ kHz.
CW	$B_m \leq 1$ kHz; see sub-paragraph 4.2 of RSEC Chapter 5 for RSEC Criteria B, C and D. **Example**: $B_m = 1$ kHz
FM/CW	$B_m \leq 1$ kHz; see sub-paragraph 4.2 of RSEC Chapter 5 for RSEC Criteria B, C and D. **Example**: $B_m = 1$ kHz
Phase-coded CW	$B_m \leq (1/t)$ where $t =$ emitted pulse duration (50% voltage). **Example for phase-coded pulsed**: If chip duration is 2 μs, then $B_m \leq 500$ kHz.
Multi-mode radars	Calculations should be made for each waveform type as described above, and the minimum resulting value of B_m should be used for the emission spectrum measurement. **Example**: A multimode radar produces a mixture of pulse modulations as used in the above examples for non-FM pulsed and FM-pulsed. These values are 1 MHz and 154 kHz, respectively. Then $B_m \leq 154$ kHz.

Table 11.1 summarizes the computation of measurement bandwidths B_m for various types of radar modulations.

11.2 Transmitter architectures: past, present, and future

Early transmitters were nothing more than energy storage devices that released their energy across a "spark gap," either due to a voltage breakdown in the gap or to some manual triggering mechanism like a switch or telegraph key. Figures 11.5–11.7 show examples of early radio transmitters that were capable of transmitting information to a receiver, and had the same general attributes as all transmitters despite the crudeness of some of their components. These early transmitters formed the basis for the development of the transmitters of today (Figures 11.8 and 11.9). Although current versions look very different, they possess the same basic attributes of the early ones.

Figure 11.5 *The first oscillator of Hertz. Two copper wires, each 1 m in length, supported on rods of sealing wax. The large spheres are of sheet zinc, and are 30 cm in diameter. The base is 260 × 7.5 cm, ca. 1886 [3]*

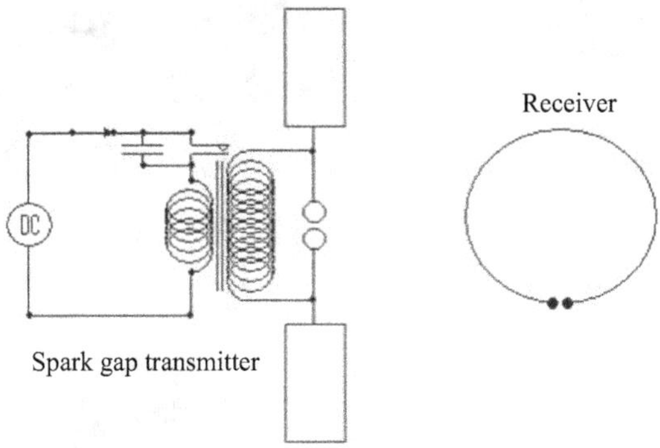

Figure 11.6 *1887 experimental setup of transmitter Hertz's apparatus [3]*

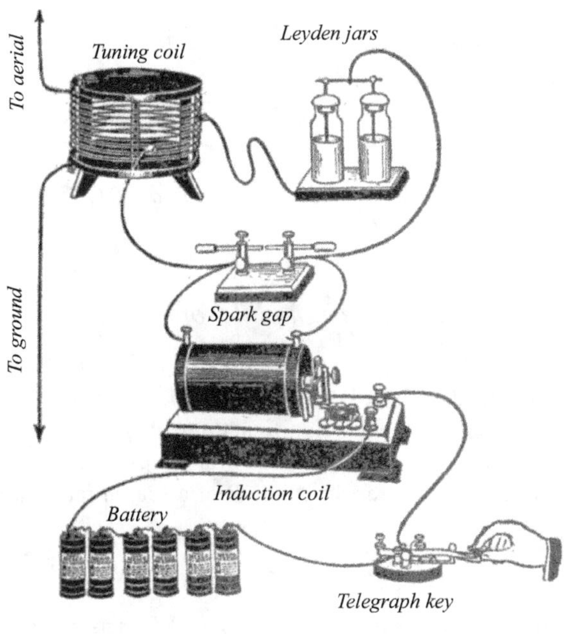

Figure 11.7 *Diagram of a spark-gap used by radio amateurs and hobbyists in the early days of radio [4]*

Figure 11.8 35 kW FM radio transmitter operating at 95.5 MHz [5]

Figure 11.9 15 mW transmitter module operating at 60 GHz (package < 1 in)

11.2.1 Current transmitter design philosophy: radar vs. communication

Historically, radar transmitters have been large and expensive and consume substantial amounts of prime power to generate significant amounts of emitted RF power [6], with the transmitter design evolving from trade-offs in the overall radar system design. The useful theoretical equation for the range R of a search radar to an illuminated target, as given by

$$R^4 \propto PA_e, \tag{11.1}$$

varies as the fourth root of the product of the average RF transmitted power P and the effective area of the antenna aperture A_e, which is related to the antenna gain. This quantity is the well-known power-aperture product, with (11.1) an abbreviated form of the radar range equation. The range varies as the fourth root of power because both the outgoing transmitted power spectral density and the returning scattered energy density from the target are attenuated as the square of the distance traveled. One can contrast this two-way path against the one-way path experienced in communications, where the inverse-square relationship with distance is only encountered once.

Attempting to obtain greater radar range by increasing transmitter power is expensive. For example, doubling the range requires a 16× increase in transmit power. It is also possible to increase the antenna's physical area and thus the effective antenna aperture A_e. However, constraints in doing so are the allowable frequency of operation, the cost, and the intended antenna platform (aircraft, ship, etc.). The power-aperture product is thus fundamental as a measure of radar performance. Although the average transmitter power is only one of the factors in the radar range equation, it can cause the design of the radar to become expensive rather quickly.

Radar transmitters would be much less complex and more cost effective if they could operate in a CW mode like radio broadcast stations. However, the ability to detect targets depends on average power, which is the product of the duty cycle (the pulse width divided by the pulse repetition rate) and the peak power. For many years and even today, many radars, particularly legacy long-range air-search and ATC radars, utilize microwave tubes like magnetrons and klystrons. These tubes have duty cycles of less than 1%. For example, legacy ATC radars use magnetron transmitters at S-band with a duty cycle of 0.1% and a peak power of 1.0 MW, thereby yielding an average power of 1 kW that is sent to the antenna for transmission.

Having to generate very high pulsed RF peak power also leads to higher operating voltages (both DC and RF), energy storage issues, and the necessity for high-power switching devices. RF PAs that are Class C (tube or SS) are self-pulsing and draw current only when the RF drive is applied, but most microwave tubes require some type of pulse modulator so that they economize on power and do not generate interfering noise during the interval between transmit pulses when the radar is listening for extremely faint echoes. While some CW radars are in use, they function by using separate transmit and receive antennas (albeit still co-located), and in some cases employ radar absorbing material between them to try to isolate the sensitive receiver from experiencing direct-path coupling with the transmitter.

11.2.2 Future transmitter architectures

11.2.2.1 Software-defined radio

Emerging transmitter architectures are currently moving in several directions, which may potentially converge in the future into a flexible system capable of doing things that currently require multiple systems. One trend is the increasing development of what is called software-defined radio (SDR). This approach replaces many of the analog portions of receivers and transmitters with a combination of digital circuitry and software. Not only is the software used to replace analog functions with signal processing equivalents, but it is also meant to allow the radio to be reconfigured so that it can operate across multiple bands and modes.

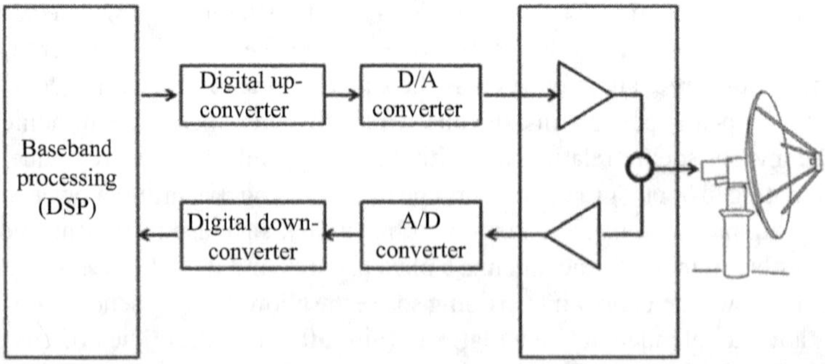

Figure 11.10 Ideal SDR transmitter–receiver—analog only in last stage(s)

Figure 11.10 illustrates an idealized block diagram of an SDR system, where most of the signal flow occurs using digital and signal processing techniques, and it is only in the RF front end where the need for high power requires analog PAs to drive the antenna for cases such as radar and electronic warfare (EW). The use of these techniques provides significant flexibility, thereby allowing the receiver and transmitter not only to be upgraded as requirements change, but also to receive and transmit in different RF bands with different modulation/demodulation schemes to provide multiple capabilities from the same equipment.

Over the past few years, many implementations of this approach have made it to the marketplace, where the flexibility and ability to work across frequency bands or with new spectral allocations has justified the higher price that usually accompanies these types of radios. The biggest technical challenges for greater use of this approach is the development of components that cover very wide frequency ranges, the generation of higher power across a wide frequency range without requiring multiple amplifiers, and the design of antennas that have both broadband performance and high gains, especially if high effective radiated power is required. Because of these issues, it is easier to design SDR receivers than transmitters. However, recent advances in SS amplifiers and broadband design techniques are making software-designed transmitters more feasible.

11.2.2.2 Adaptive/Cognitive/Distributed transmitters

Current research is extending the SDR concept by providing additional components to the transmitter–receiver system to achieve even greater flexibility and capability out of the basic transmitter–receiver system. These approaches have various names, including adaptive radio, cognitive radio, etc. In all cases, the approaches attempt to make transmit–receive systems even more flexible by marrying the traditional RF hardware with databases and software to allow the RF system to interact better with its environment. Thus, the base system in Figure 11.10 can be expanded into a system, such as the one in Figure 11.11, where multiple SDR sub-systems can be interconnected within a system that "simultaneously" monitors its spectral environment, evaluates the environment to determine the best portion of available bandwidth

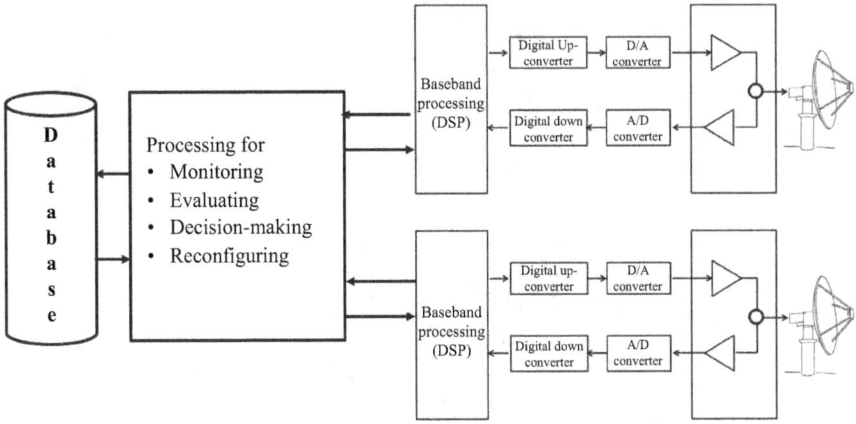

Figure 11.11 Extension of the SDR concept: an adaptive, cognitive reconfigurable system

in which to transmit and receive, and reconfigures itself to implement such a choice. Depending upon the construction and capabilities of the various subsystems, multiple functions could be implemented from the same core hardware, including communication, radar, and EW functions, all working together to make sure that each function works cooperatively with the others while avoiding self-interference and maintaining situational awareness of its EM environment. Also, the functions can be distributed over multiple subsystem implementations to provide both capability and survivability if various pieces are compromised.

To develop this type of system requires hardware that goes beyond that used in a normal RF or radar transmitter, as well as advances in software algorithms and computing power. However, work is progressing on many fronts, and many of the required capabilities are being addressed in current research efforts (see the last section of this chapter).

11.3 Transmitter components

11.3.1 Microwave tubes

The evolution of radar is closely tied to the development of and advances in microwave vacuum tubes. As discussed above, the main reason for this linkage is the very high peak power that tubes can provide, thus facilitating the means to achieve long-range surveillance capabilities. While SS alternatives continue to emerge, tubes will remain in use for the foreseeable future due to the prevalence of legacy systems, the costs and trade-offs involved with upgrades, and ongoing advancements in tube technology. Thus, the following surveys the common types of microwave tubes used in radar applications, and also discusses a promising new form.

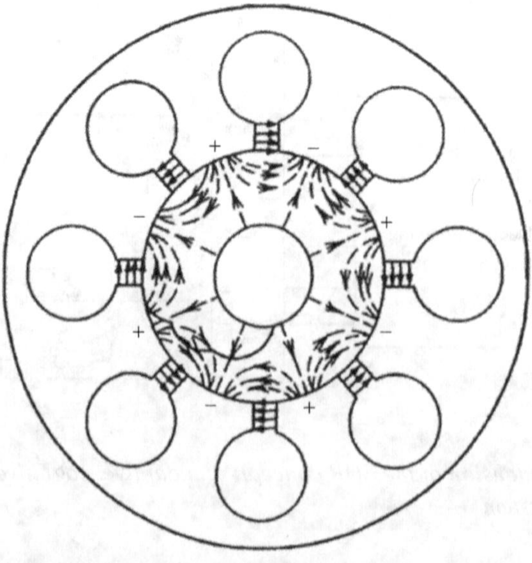

Figure 11.12 Magnetron multicavity interaction space with the magnetic ***B*** *field into the page*

11.3.1.1 Magnetron

The magnetron is the oldest of the microwave tubes, and its application to radar was a result of work in 1939 by Boot and Randall in the United Kingdom [7]. A magnetron is a high-power microwave oscillator in which the potential energy of an electron cloud near the cathode is transformed into RF energy through a series of cavity resonators. Figure 11.12 shows the basic cavity magnetron configuration [8].

From a circuit point of view, a magnetron is basically a diode, possessing a cathode and an anode. However, in this case the anode contains a series of resonant cavities situated symmetrically around the cathode. Magnetrons are actually oscillators and not amplifiers. Figure 11.12 shows a top-down drawing of a complete magnetron (not shown is the permanent magnet, with the **B** field into the page). Magnetrons are crossed-field devices, meaning that the electric and magnetic fields present in the device are orthogonal. Magnetrons are particularly useful as pulsed oscillators whether used in a low-power marine radar or high-power (1.0 MW peak power) ATC radar.

The magnetron functions in the following manner. A large DC voltage is applied by a power supply between the cathode and the anode plate. The cathode and anode plates are negatively and positively charged, respectively. A DC electric field is thus directed from the anode plate to the cathode. A permanent magnet supplies a DC magnetic field that is orthogonal to the plane of the page. Without the DC magnetic field, the electrons emitted by the cathode would traverse a straight-line path through the interaction space to the anode plate. The addition of the DC magnetic field interacting with the orthogonal DC electric field causes the electrons to travel in a circular path

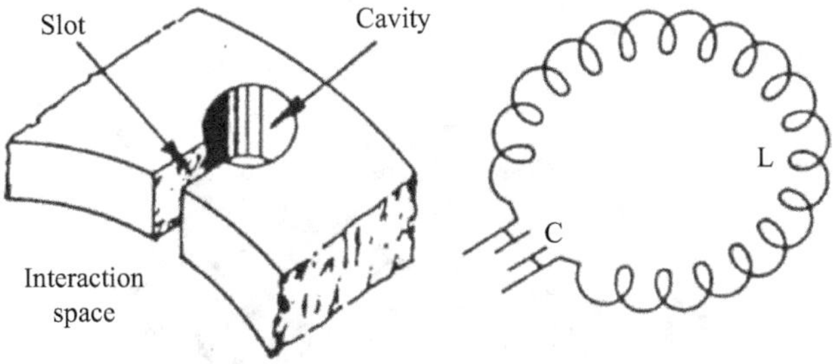

Figure 11.13 Cavity circuit representation

throughout the interaction space toward the anode plate to interact with the cavities. The rotating dense circular cloud of electrons makes its way to the cavities, where it induces an RF EM field within the cavities.

From the circuit standpoint, each cavity may be represented as a parallel resonant circuit as shown in Figure 11.13. The physical sizes of the cavities correlate to a given set of frequencies that are representative of specific electric and magnetic field structures known as modes. The mode shown in Figure 11.12 is the most stable one, known as the π mode, where the RF field structure alternates in polarity with respect to its neighboring cavity, with the cavity approximating a parallel LC circuit whose physical dimensions support the desired RF frequency.

A probe is usually inserted into one of the cavities to conduct away the RF energy. Sometimes a slot from the cavity to a waveguide structure is used as the conduit for RF energy. Most practical radars, such as magnetron transmitters in ATC radars, transmit pulsed waveforms. The magnetron by itself is a CW oscillator; however, if a negative going pulse is supplied to the cathode, the magnetron acts as if it is also a modulator, thus producing a modulated magnetron emission. Figure 11.14 shows a modern X-band magnetron.

11.3.1.2 Cross-field amplifier

The cross-field amplifier (CFA) is a cross-field tube like the magnetron in that it employs a DC magnetic and DC electric field that are perpendicular to one another [8]. CFAs have high efficiency (40–60%), use lower voltages than linear-beam tubes, and are lighter in weight and smaller in size. They have wide fractional bandwidths in the neighborhood of 10–20%. In addition, they exhibit high peak and average power and provide good phase stability, though their gain of 10 dB or so is relatively low. They are sometimes connected in cascade with a high-gain, lower-power traveling wave tube (TWT) preceding the CFA, which serves as a driver amplifier.

The different types of CFAs all use a slow-wave circuit, a cathode, and input and output ports. For radar applications, CFAs usually have a circular configuration as shown in Figure 11.15(a). Many of them are classified as "re-entrant with distributed emission." Re-entrant means that electrons that are not absorbed by the anode slow-wave structure are allowed to return to the input. Distributed emission means that, like the magnetron, the cathode is adjacent to the full length of the RF structure. Electrons

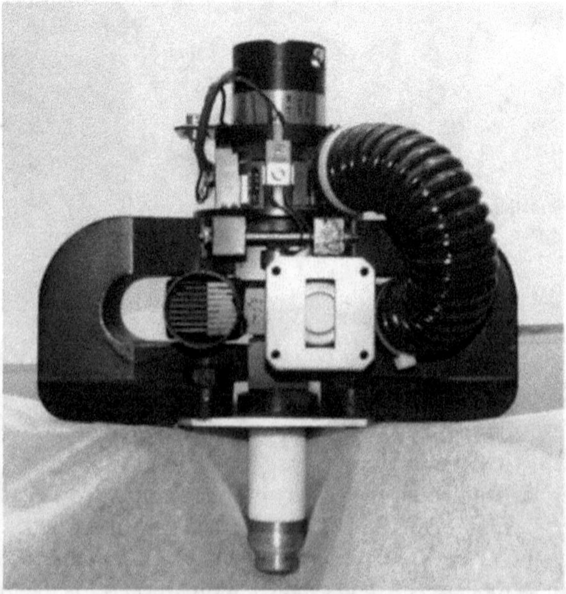

Figure 11.14 Example of X-Band magnetron (Courtesy of CPI)

are emitted from the cylindrical cathode, which is coaxial to the RF slow-wave circuit that serves as the anode. Since the circular CFA is the one most commonly used for radar transmitters, it is described here.

The electrons, under the action of the crossed electric and magnetic fields, form into rotating electron (space-charge) bunches, or spokes, similar to what occurs in the magnetron. These bunches drift along the slow-wave circuit in phase with its RF signal and transfer their energy to the RF wave resulting in amplification. Those electrons that have expended their energy to the RF wave are collected by the slow-wave anode structure, while those electrons that are not collected by the slow–wave anode structure are allowed to re-enter the RF interaction area at the input. Some of the re-entering electrons contain modulation (bunched electrons) that will be amplified in the next pass around the RF circuit. A drift space is situated between the output and input ports. In the drift space, the electric field forces the electron bunches to disperse, thus eliminating any modulation contained within the re-entering electrons.

A CFA has a low insertion loss on the order of 0.5 dB or less. If it is desired to prevent the radar from causing in-band or out-of-band interference to wireless systems, then it is desirable to have low- and high-power modes in the radar transmitter. CFA transmitters can achieve extra gain by operating with a TWT driving a CFA. If it is desired to operate in a low-power mode, then the DC voltage to the final CFA stage can be turned off, resulting in the TWT driving the CFA slow-wave structure as if it is simply a section of low-loss transmission line. The lower-level RF drive can be passed directly through the final stage with little attenuation. Figure 11.15(b) shows an actual S-band CFA.

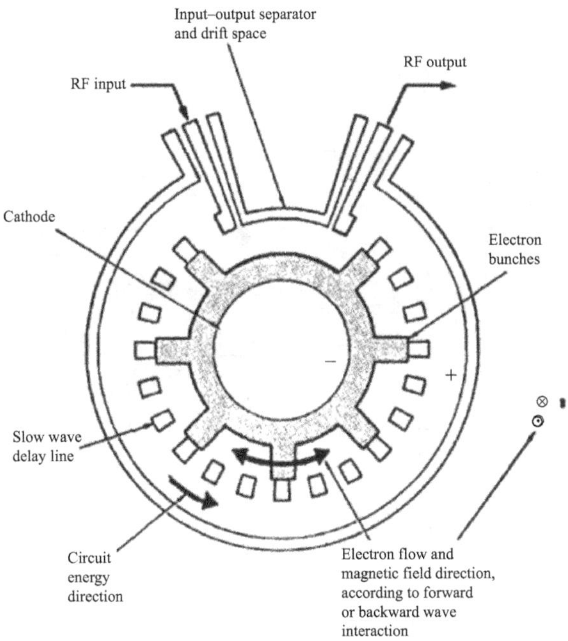

Figure 11.15(a) Circular format CFA

Figure 11.15(b) Actual S-band CFA

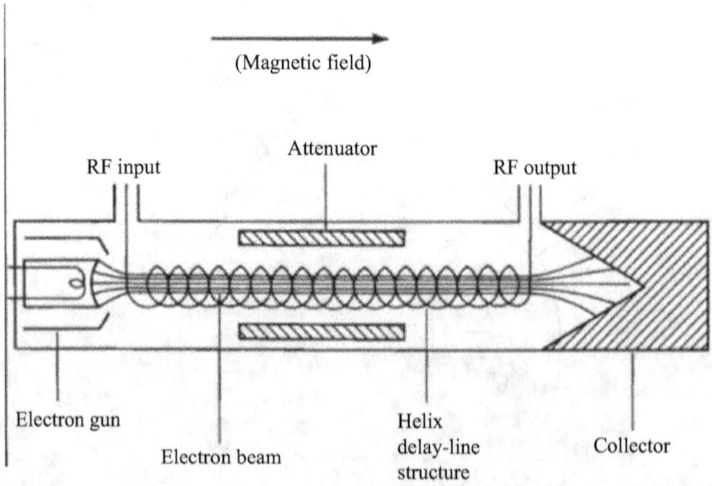

Figure 11.16(a) Helix TWT

Figure 11.16(b) Example of TWT

11.3.1.3 Traveling wave tube

The TWT supports high gain, broad bandwidth, and very low noise when used as an amplifier in radar transmitters. Figure 11.16(a) is a simplified schematic diagram of a helix TWT, and Figure 11.16(b) is an actual TWT with a power supply [8]. The basic TWT is comprised of an electron gun, a filament, a cathode, a collimated electron beam that transitions longitudinally through the helix for supporting a slowed traveling wave, a collector acting as a sink for the electrons returning them to the power supply, a DC magnetic field usually supplied by an electromagnet, an attenuator, and an absorber that reduces the reflected wave to minimize undesirable oscillations. The TWT is structurally similar to the multi-cavity klystron, except for the signal arrangement, which employs a slow-wave structure instead of cavities. A slow-wave structure is a transmission line where the forward signal propagation is slower than normal. A wire in the form of a helix comprises the slow-wave structure in low-power

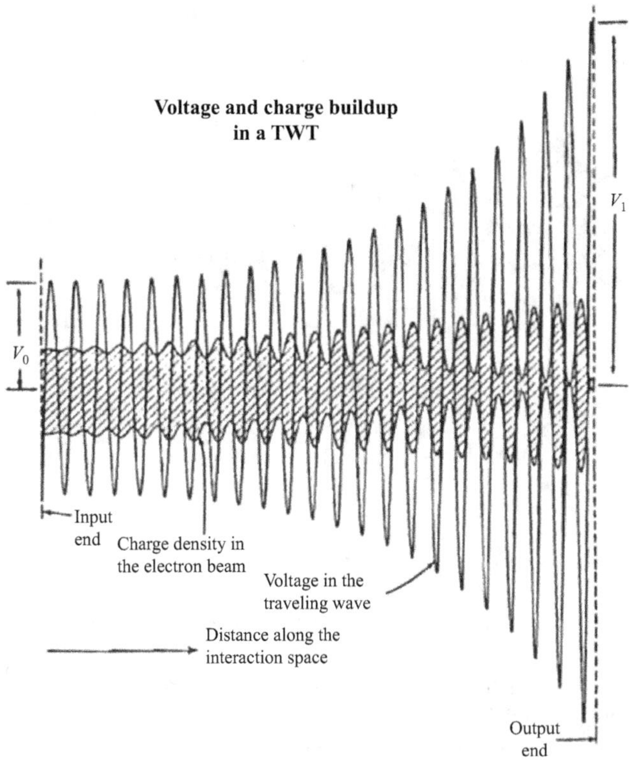

Figure 11.17 Electron beam versus traveling wave buildup on the TWT helix

TWTs. High-power TWTs employ cavities with an aperture. The propagation of the signal to be amplified along the length of the helix is effectively slowed down as the signal must travel farther along the extent of the helical wire. The reduction in propagation velocity of the signal through the helix effectively allows energy from the longitudinal electron beam to be imparted to the signal (amplification) when transitioning through the helix. The methodology supporting amplification by a TWT is different from that of a klystron, in that the interaction between the signal and the electron beam occurs continuously, as opposed to occurring at discrete locations in the resonant cavities. Figure 11.17 shows the interaction between the helical slow-wave structure and the electron beam, resulting in TWT amplification.

11.3.1.4 Klystron

The klystron is a linear-beam tube that is used as an amplifier in the transmitter of radar systems. As shown in Figure 11.18 [8], the klystron is composed of (1) a filament or heater; (2) a cathode coated with a chemical such that when heated by the filament electrons are emitted; (3) a positively charged anode that accelerates the electron beam forward toward the right; (4) the first cavity known as a buncher; (5) a

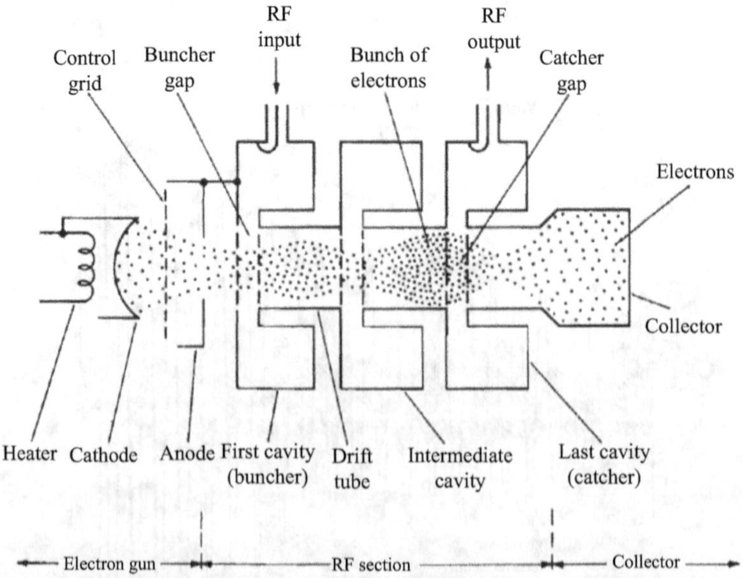

Figure 11.18 Diagram of klystron showing electron bunching

drift tube, also known as the interaction space between the cavities; (6) a final cavity or catcher where an amplified version of the signal is obtained; and (7) a collector that recovers the electrons that surrendered spent kinetic energy in the form of EM energy as the RF output. A permanent magnet or an electromagnet (not shown) surrounds the length of the tube to collimate the electrons into a tight beam.

The operation of the klystron tube starts with heating the thermionic material on the cathode, causing a density of electrons to be emitted. A high-voltage negative-going pulse applied to the cathode propels electrons toward the positively charged anode. These electrons migrate through an aperture to the first cavity (the buncher), which is resonant at the desired radar frequency. The buncher cavity is excited in the form of an EM wave by a microwave input on the inner conductor of a transmission line that is terminated on the inside surface of the cavity. Once within the confines of the cavity, the time-varying EM field at the resonant frequency modulates the electron beam. This field alternates in both the positive and negative polarities. A positive polarity causes electron bunching (i.e. an increase in electron density), while during the peak of the negative excursion, the electrons become less bunched with a decrease in electron density. The sinusoidal wave increases in magnitude, showing an increase in electron density (bunching) within the drift space moving from left to right toward the final cavity known as the catcher. The bunching is caused by the faster electrons in the positive-going EM field catching up with the slower moving electrons due to the negative polarity of the EM field. Both the buncher and the catcher cavities are at the same positive potential as that of the anode. The electron beam with its increased electron density and kinetic energy arrives at the catcher cavity, where it sets up an amplified version of the input signal in the form of an amplified EM field

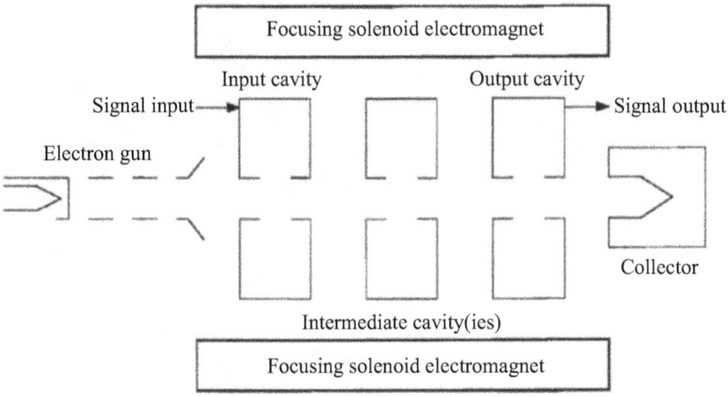

Figure 11.19 Klystron diagram showing the electron focusing electromagnet

within the cavity. A portion of the inner conductor of a transmission line forms a loop, similar to the buncher cavity, where an amplified version of the input signal is induced. The electrons that have lost their kinetic energy in the form of an amplified version of the EM wave then continue out of the catcher cavity through an aperture to a positively charged collector.

Per Figure 11.19, practical klystrons employ multiple or intermittent cavities between the input (buncher) and output (catcher) cavities. These additional cavities provide more gain and if stagger-tuned (i.e. each cavity tuned by having its Q reduced within the passband), they deliver greater instantaneous bandwidth than can be achieved by a single cavity.

11.3.1.5 Multiple beam klystron

Now consider a more recent innovation in tube technology. The single-beam klystron tube is known for its high power and lower in-band and out-of-band noise products when compared to cross-field tubes such as the CFA [8]. Klystrons do require high voltages, particularly when the size is greater to accommodate multicavities. The ability to convert the DC power of the electron beam to RF power in an EM wave is a function of the DC beam current and beam voltage. It is always preferred to operate a tube at the lowest voltage possible because high-voltage power supplies can suffer from breakdown, can be less reliable, and are expensive. A reduction in beam voltage means an increase in beam current to obtain the same amount of power. However, an increase in beam current density causes the electrons to repel one another, resulting in a lack of coherence in the bunching process.

Lower beam-voltage limitations can be overcome by having multiple beams (also known as beamlets) instead of a single beam, where each of the individual beams has a small enough current density to avoid the repulsive effects of a higher current beam. Each of the individual beams also has a separate drift channel, which is a metallic-walled tube. Figure 11.20 shows a multiple beam klystron (MBK) configuration with eight drift channels capable of transporting eight current beams.

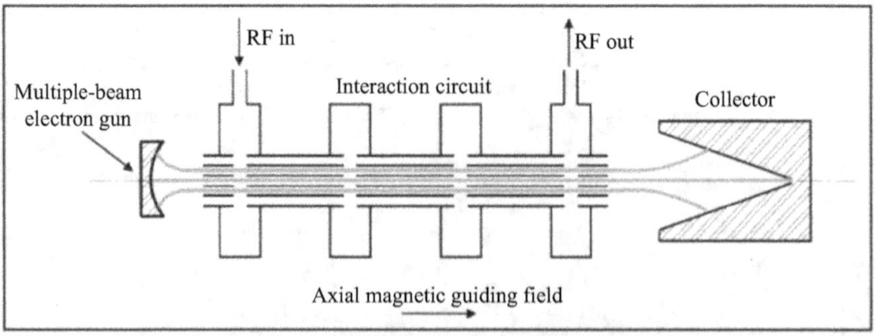

Figure 11.20 MBK diagram

Figure 11.21 600 kW peak power MBK

The MBK is thus capable of peak power outputs greater than 100 kW, lower in-band and out-of-band phase noise products (e.g. 30 dB less than cross-field tubes), a 30% to 60% reduction in length compared to a single beam tube having comparable peak power output, and instantaneous fractional bandwidths greater than 10%. With the explosion of wireless devices in S-band, the ability to minimize adjacent-channel interference from a radar to a wireless base station or handset makes the MBK an attractive option for legacy and developmental radars. Figure 11.21 depicts an early 600 kW peak power MBK tube designed and built by the U.S. Naval Research Laboratory.

11.3.2 Pulse modulators

Modulators functionally turn the transmitter on and off to generate a desired pulsed waveform. Thus, the modulator is also known as a "pulser" [8]. Every RF power source determines the specific characteristics that establish the type of modulator to be used. The magnetron modulator, for example, is required to operate with full

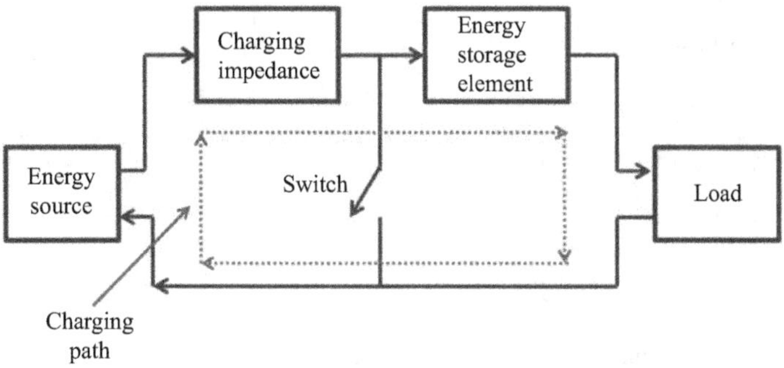

Figure 11.22 Basic modulator configuration

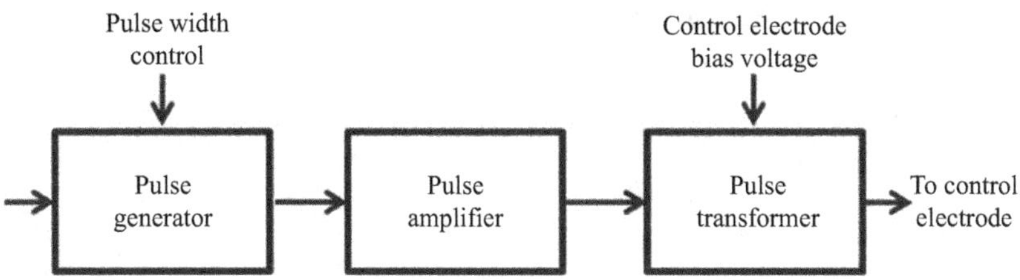

Figure 11.23 Low-level modulator

pulse power. On the other hand, transistor amplifiers require no modulator since the transistor is biased such that it is turned on and off by the occurrence or absence of an input pulse.

As illustrated in Figure 11.22, a modulator consists of an energy source which is usually a high-voltage DC power supply; a charging impedance such as an inductor; a switch, which for many years was a gas-filled thyratron tube but in recent years is a silicon-controlled rectifier; an energy-storage element comprised of a combination of series inductors and parallel capacitors that are electrically similar to a section of transmission line; and the load, which is usually the cathode of the driven microwave tube. Not shown is the pulse transformer that supplies the stepped voltage to the cathode as well as filament voltages to the tube filament.

Modulators can be classified as either low level or high level. Low-level modulators form a pulse of a given width and level at the proper time by applying the required DC bias voltages, which then drives the control electrode of a tube in a transmitter amplifier. Typical control pulse values are a few hundred volts at an ampere. Figure 11.23 contains a simplified diagram of a low-level modulator. The pulse, upon initiation by a transmit trigger, is developed in the pulse generator, where the required pulse repetition rate and width are set. It is then amplified in a pulse amplifier and passed through a pulse transformer where its voltage is stepped up. It is at the pulse transformer that bias voltages for the control electrode, such as a grid, are added. The pulse and associated bias voltages are then applied to the control electrode, such as a grid in a TWT.

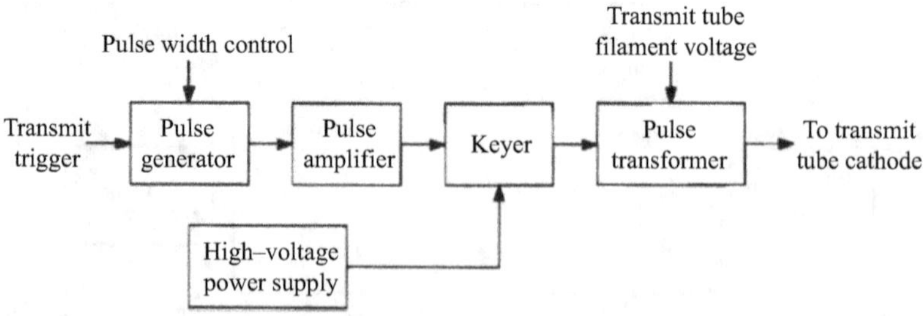

Figure 11.24 Simplified hard-tube modulator

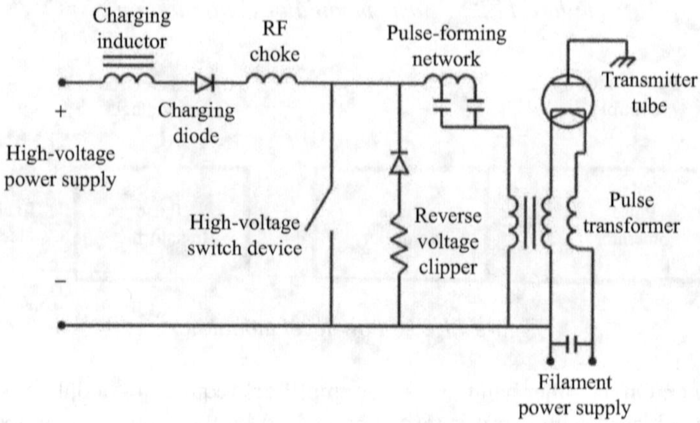

Figure 11.25 Soft-tube modulator

Two types of high-level modulators exist, hard tube and soft tube [9], both of which can be configured with vacuum tubes, SS devices, or a hybrid of both. Hard-tube modulators (Figure 11.24) are basically low-level modulators with added amplification to increase the pulse amplitude high enough to provide the proper final DC drive level in the PA. The pulse length of proper width is formed by a pulse generator. The high level of transmit energy is derived from high-voltage power-supply capacitors that supply a keyer and an amplifier. The high-energy pulse from the output of the keyer is inputted to a pulse transformer, where adjustment of the pulse voltage and impedance matching with respect to the tube takes place. In addition, the bias and filament voltages of the transmit tube are added. The pulse out of the transformer secondary is usually negative going and is applied to the cathode of the transmit tube.

Soft-tube modulators (Figure 11.25) have used thyratron tubes as a high-voltage switch device for many years. A thyratron is a gas-filled tube that can be used as a high-power switch and controlled rectifier. Recently, with advances in high-power semiconductor devices, silicon-controlled rectifiers (SCRs) have taken the place of thyratrons to switch the energy taken from the high-voltage power supply stored in a pulse-forming network (PFN). Unlike the hard-tube modulator, in this configuration the switch does not control the pulse width, which is instead set by the PFN. As can

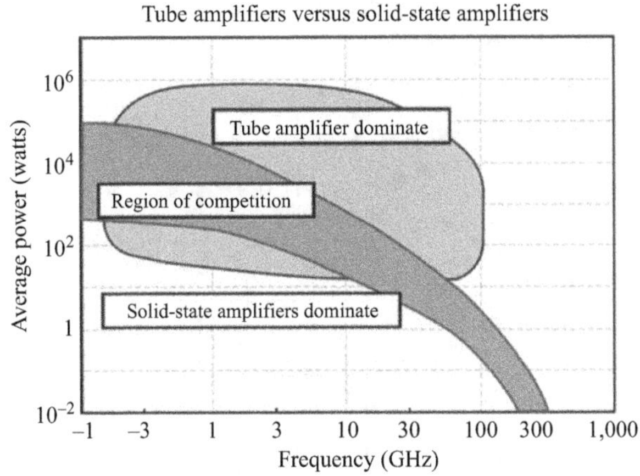

Figure 11.26 Comparison of tube versus SS amplifiers with respect to frequency

be seen, the PFN resembles a section of transmission line with series inductors and parallel capacitors. A characteristic of thyratrons and SCRs is that a control electrode closes the switch, but the current through the device must drop below a certain level before it can be closed. Thus in a soft-tube modulator, the active device only starts the pulse; the PFN shuts it off.

11.3.3 Solid-state transmitter architectures

SS radar transmitter architectures have become very popular, both as a means to replace legacy tube transmitters and for new radar designs. Figure 11.26 shows the competition between tubes and SS devices as a function of average power versus frequency.

One of the first legacy radars to have its vacuum-tube tetrode transmitter redesigned into an SS equivalent was the U.S. Navy's AN/SPS-40 UHF long-range air-search radar, which took place in the 1980s. However, the overall transition from tube to SS radar transmitters has been very gradual. In military radars, for example, tubes such as CFAs, TWTs, klystrons, and even magnetrons have provided the necessary performance objectives. SS device adoption has been far more gradual due to limited power available from each transistor, cost, and the requirement of a higher duty cycle with the associated range-time sidelobes associated with pulse compression. However, the SS devices of today do have advantages, which are described below. Consider a magnetron S-band ATC radar operating from 2700 to 2900 MHz that has a peak power of 1.0 MW and a duty cycle of 0.1%, which implies an average power of 1.0 kW. In contrast, a notional SS radar providing the same requisite average power for detection might have a duty cycle of 10% and a peak power of 10 kW, which can be achieved by a combination of multiple 100 W GaN transistors (discussed later). Consequently, some SS transmitter upgrade efforts have been directed toward ATC radars like the ASR-11.

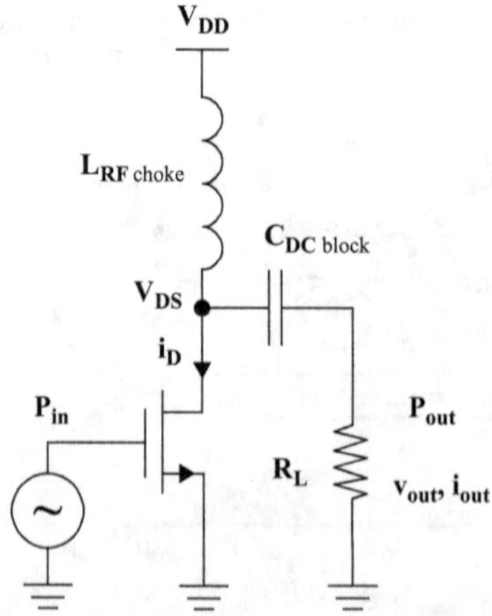

Figure 11.27 SS amplifier in common source configuration for Classes A, B, AB, and C

It is also worth mentioning that SS radars are not direct replacements for tube radars in terms of duty cycle. Tubes, such as CFAs and klystrons, operate at very low duty cycles on the order of 0.1% to avoid destroying the elements inside the tube due to overheating. However, since average power is equal to duty cycle times the peak power, tube radars are able to achieve the required average power by operating with very high peak powers. In contrast, because thermal constraints of SS devices limit the peak-power handling ability of SS radars, they must operate at longer duty cycles of 5% or more.

11.3.3.1 Solid-state devices

SS devices for radar transmitters are constructed from semiconductors, which are neither insulators nor conductors. The ability to transport an electrical charge can be modified by adding small amounts of impurity ions to the semiconducting material. Materials that have been used to fabricate transistors are (1) gallium arsenide; (2) indium phosphide; (3) silicon carbide; (4) gallium nitride; and (5) silicon germanium. Gallium arsenide (GaAs) has had great utility in microwave transistors because of ease in controlling crystal-lattice defects and repeatability in manufactured performance. Most RF transistors currently employed in radars are manufactured with two or more elements. The energy bandgap is often used as a descriptor for semiconductors by assessing the energy difference between the lowest conduction band and the highest valence band, where valence bands are those energy states in which electrons are bound to a particular atom. The conduction band contains electrons that can move freely from atom to atom.

In simple terms, an electric field generated by the DC bias within a transistor imposes the proper condition such that an RF signal experiences amplification [10]. Figure 11.27 shows a simple example of this phenomenon with DC biases applied to the gate, which is the control electrode, and to the drain or output electrode, both with respect to the grounded source. The circuit is supported by two inductors on the gate

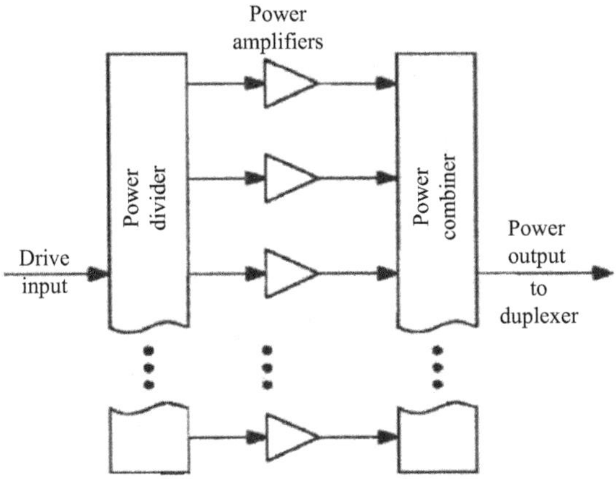

Figure 11.28 Amplifier divider/combiner topology

and drain, which allow a DC bias current to flow to the gate and drain but prevent RF from flowing back to the power supply, and input and output blocking capacitors that prevent DC bias currents from flowing to the RF source and load. The matching networks on the gate and drain help to insure maximum power transfer to the gate and load. Some actual transistor device configurations that are used for radar PAs are (1) bipolar junction transistors; (2) metal-oxide semiconductor field-effect transistors; (3) high electron mobility transistors; (4) heterojunction bipolar transistors; (5) laterally diffused metal oxide semiconductors; and (6) metal semiconductor field effect transistors.

11.3.3.2 Combining solid-state power amplifiers

The radar range equation calls for high peak and average radiated power from the transmitter-antenna combination to extract the best resulting signal-to-noise ratio in the receiver. Individual SS amplifiers do not provide adequate power to satisfy the requirements of a high-power radar. These individual power transistors must be combined to achieve the desired power levels. Combining two or more signals calls for coherency, which means that the signals must be at the exact same frequency and be very similar in phase and amplitude.

A multiple amplifier configuration is shown in Figure 11.28. The drive input signal to be amplified is split using a power divider, with the amplified signals from the individual PAs then added in a power combiner. Figure 11.29 is an example of a hybrid splitter/combiner. Functionally, with all ports matched, the signal power to be amplified entering port 1 is divided evenly between ports 2 and 3, but with a 90° phase shift between the output ports 2 and 3. Port 4 is the isolated port.

11.3.4 Microwave power module

The microwave power module (MPM) was conceived and designed for use in active array antennas, where their small size allows for placement of a large number of modules behind the radiating face of the antenna. Most MPMs, because of their relatively

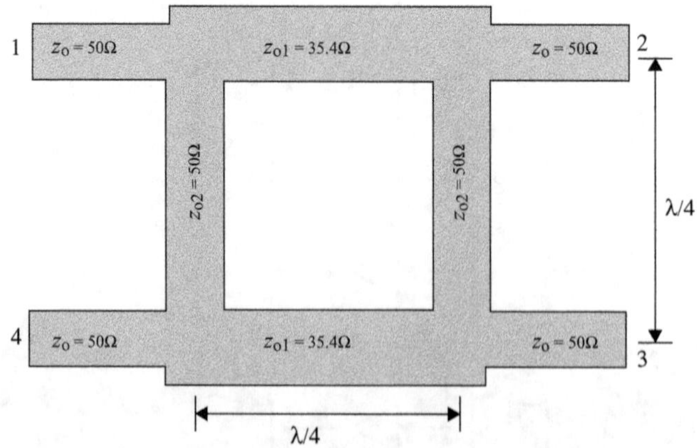

Figure 11.29 Hybrid splitter/combiner

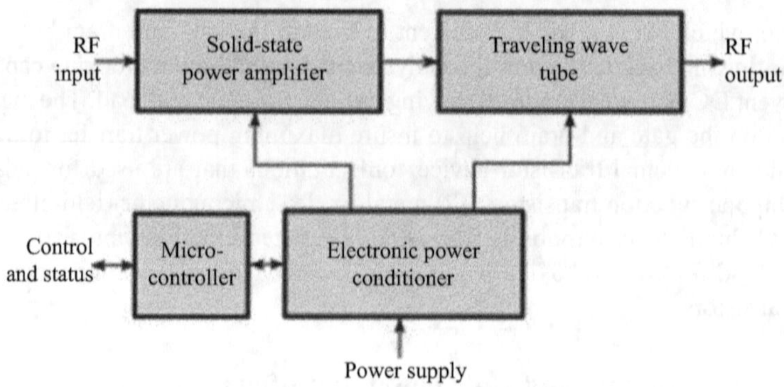

Figure 11.30 Microwave power module

large size, are only useable for linear arrays, partially distributed arrays, and single-module applications. An MPM system-level configuration (Figure 11.30) takes an RF input and applies amplification via an SS PA. The output from the PA is then applied to the helix of a TWT for final HPA. The high-voltage power supply required by the TWT is provided by an electronic power conditioner that insures high-quality DC power. For pulsed-mode MPMs, the power conditioner provides a pulsed high voltage that is controlled by a trigger input. A microcontroller directs the operation of the module, insuring that the timing and sequencing of the appropriate power supply voltages are maintained to prevent damage to the TWT. In addition, the microcontroller provides the module status, which includes the various voltages, currents, and temperatures of the module. Figure 11.31 contains a photograph of an MPM.

11.3.5 Classes of power amplifiers

Amplifier classes are related to the time that the active amplifier device (such as a transistor) is passing current, expressed as a fraction or percentage of the period of a

Figure 11.31 Example of an MPM, courtesy of dB Control

signal waveform applied to the input. This condition is typically based on what occurs over one cycle of operation when a sinusoidal input signal is used. The amplifier classes are also related to the efficiency of the output related to the input, as well as the linearity of the output with respect to the input. The amplifier classes range from highly linear operation, often with a corresponding low efficiency, to highly nonlinear operation and a much higher efficiency, with many variations in between.

Amplifier classes can be divided into two main groups. The first group consists of the "linear" amplifiers that are controlled by varying the conduction angle, which is defined by the length of the conduction state over some portion of the output waveform. Here, the output stage varies in a more or less linear manner between the fully-off state and the fully-on state. The common amplifier classes in this group are classes A, B, AB, and C. Another amplifier related to this group is the Doherty amplifier, which is a combination of a class-AB amplifier and a class-C amplifier.

The other group is comprised of the switching amplifiers, which includes classes D, E, F, G, H, I, J, R, S, and T. The class-D amplifier functions as a nonlinear switching amplifier, where the active element is either fully-on or fully-off and is pulse-width modulated by the input signal. Because of the fully on/off operation, the efficiency of these amplifiers is quite high, approaching 100%. Other switching amplifiers use RLC resonators or multiple power supply voltages to reduce power loss and increase efficiency.

Note that there is no official definition of some of these switching classes, and in fact, some amplifier classes seem to be more like marketing creations than a technical classification. Some of the switching amplifier classes are based on some of the linear classifications, or are simply variations of class D. For instance, the class-J amplifier combines a class-D amplifier with aspects of a class B amplifier, while the class-S amplifier is now more correctly referred to as a class-D amplifier.

Figure 11.32 illustrates the range of efficiencies over which various amplifier classes operate. This figure also shows the amount of time that the output of the active device is conducting during a single cycle of the applied input. Upon consideration of the results of injecting a sinusoidal signal, we see that class-A amplifiers provide

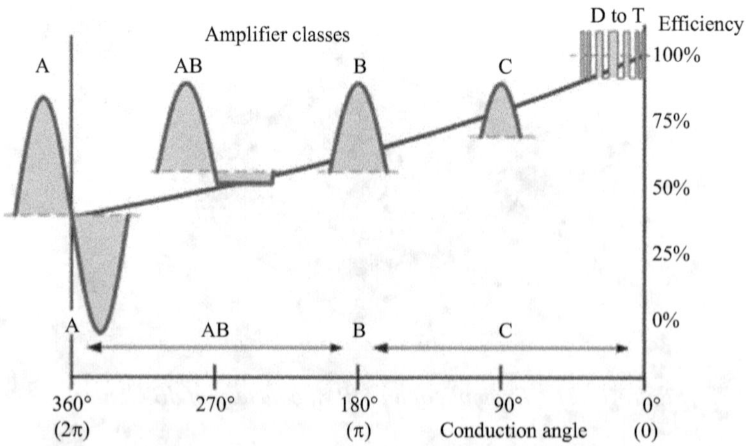

Figure 11.32 Amplifier classes vs. efficiency

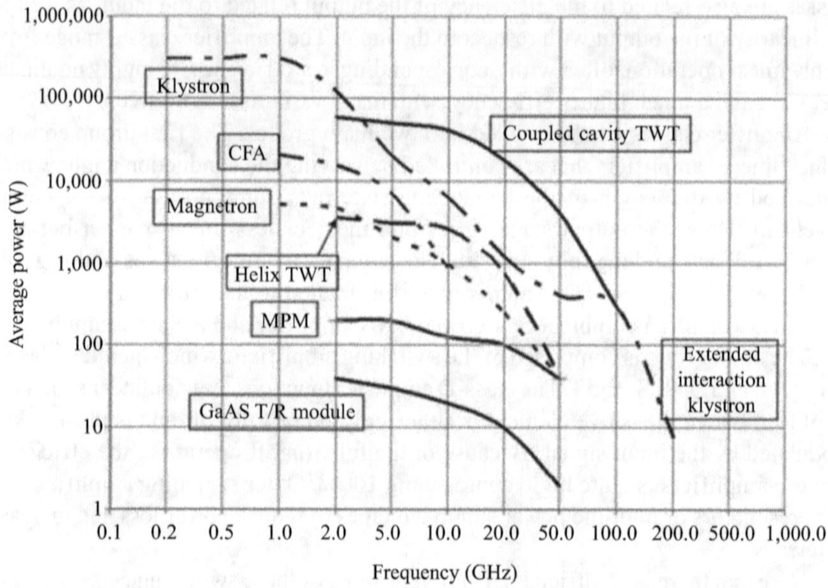

Figure 11.33 Comparison of radar amplifiers and sources

a highly linear output, but a low efficiency (20%–30%), while the switching class amplifiers provide an output characterized by high efficiency (>90%), but a very nonlinear output.

Another consideration is the frequency range over which a system is required to operate. Figure 11.33 provides a frequency versus average power comparison of different microwave tube amplifiers, power oscillators (magnetron), and a gallium arsenide (GaAs) MPM.

There are many excellent papers and texts concerned with the design of PAs (see the "Further reading" section at the end of the chapter) and it is not the intention to get into that level of detail here. Rather, this chapter focuses on the impact that amplifier attributes such as linearity and efficiency have on the overall design of radar systems that are intended to coexist with communication systems. We examine these effects later in the chapter.

11.4 Radar transmitters and the future of radar

11.4.1 Legacy versus new radar developments

The design, development, and deployment of radar transmitters, especially those for high-power long-range search, is now at a crossroads because the primacy of microwave tubes is being challenged by the performance of SS PAs due to the advancement of device technologies like gallium nitride. This situation is compounded by the fact that both legacy and new radars must be considered. The ensuing discussion addresses future transmitter development within this context.

Vacuum and microwave tube radars have been around since the early days of WWII, beginning with VHF and later evolving into microwave radars with the introduction of the magnetron. Many of the current tube-based radar transmitters employ CFAs or klystrons. Attempting to replace the tube-based transmitters with SS ones does not result in a one-to-one replacement. For example, many transistor PAs must be combined in parallel and phase matched to achieve the desirable peak power. Recall that radar detection is based on average power, which is peak power times the duty cycle. Therefore, to achieve an average power commensurate with a tube radar having a 0.1% duty cycle, a 1 microsecond pulse width, and a peak power of 500 kW or more, the corresponding SS radar must have a duty cycle of 5% or greater. These parametric choices imply the use of pulse compression to achieve the range resolution of the tube radar while maintaining similar average power. In addition, there are losses through the SS power combining networks that must be considered. The cost trade-offs can also be difficult to address due to the nature of such a complex system. For instance, when considering whether to transform a tube-based radar to SS, it must be determined whether the phase shifters, one for each antenna element, can safely operate with a significantly increased duty cycle or if they should be replaced.

11.4.2 Transmit/Receive modules

In new radar designs, there has been a great deal of interest in active electronically steered array antennas. In this arrangement, transmit/receive (T/R) modules are used at each antenna element. Figure 11.34 illustrates a typical T/R module circuit topology [8] while Figure 11.35 shows an example of a G-band (150 GHz) T/R module. Starting from the antenna element in Figure 11.34, move backwards to the circulator, which nominally provides about 20 dB of isolation between the transmit and receive portions of the module. At extreme scan angles with respect to antenna boresight, the voltage standing wave ratio (VSWR) can increase, which leads to load-pulling effects on the

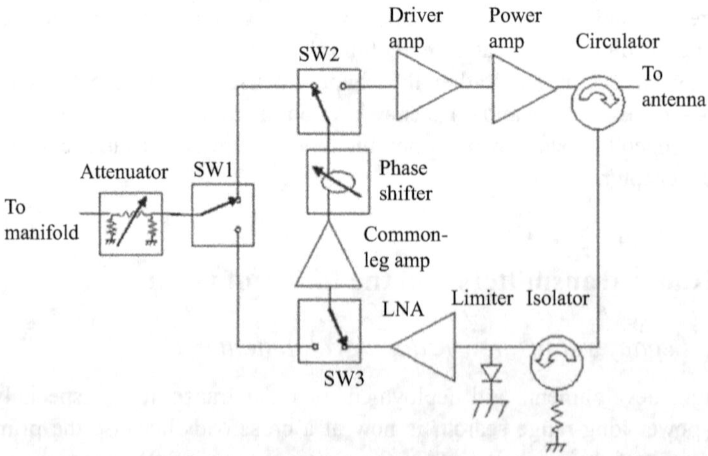

Figure 11.34 Block diagram of a T/R module

PA. An isolator is therefore used to present a matched load to the antenna element and PA regardless of the effect of the low-noise amplifier (LNA) and limiter on VSWR. The phase shifter provides the incremental phase shift for both receive and transmit paths. The limiter provides protection to the LNA from its own transmitter and other external emitters, and also sets the noise figure of the receiver. It is placed as close to the antenna element as possible to minimize ohmic losses. The driver amplifier raises the transmit signal to a level sufficient to operate the PA, which provides a transmit radar signal of sufficient wattage to be radiated by the antenna element.

11.4.3 Power amplifier linearization

With the ever more crowded EM spectrum, there has been a great deal of interest in reducing the emission of unwanted in-band and OOB spectral products. Reducing OOB emissions is particularly problematic for radar since the transmitter is already the most expensive part of the system. Moreover, a radar operating with greater efficiency does not necessarily mean that it is a good neighbor in the sidebands. In fact, radars can and have caused interference in adjacent channels to other users. This adjacent-channel interference problem has become particularly worrisome with the proliferation of wireless in the bands above 3500 MHz.

To minimize adjacent-channel interference, there has been interest in the radar design community to consider linearization and cancellation techniques as part of PA design. One of the earliest of these techniques was used in the design of amplitude modulation (AM) broadcast transmitters and is known as the Doherty amplifier [11,12]. The Doherty amplifier configuration (Figure 11.36) is composed of two amplifiers: a “carrier amplifier” that is operated at Class AB or B, and a “peaking amplifier” operated at Class C. As the carrier amplifier output rises to within

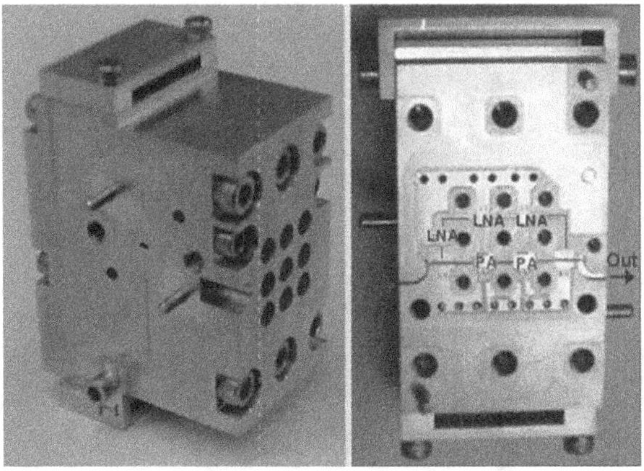

Figure 11.35 150 GHz T/R module

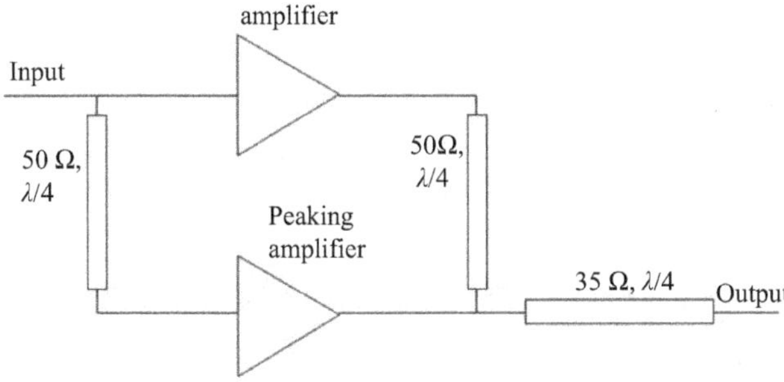

Figure 11.36 Example of a Doherty amplifier configuration

about 6.0 dB of the composite power of both amplifiers, the peaking amplifier goes into conduction, thereby extending the overall linear range of the combination. An important feature of the Doherty configuration is the concept of active load-pull. The quarter wavelengths of transmission line shown at the outputs of the carrier and peaking amplifiers in Figure 11.36, in concert with the load, facilitate active load pulling whereby during conduction of only the carrier amplifier, the overall complex impedance of the 50 and 30 Ω transmission lines effectively decrease the impedance of the load. As the peak amplifier conducts, the overall impedance of the load increases.

The next approach that has drawn attention from radar transmitter designers is the linear amplification with nonlinear components (LINC) configuration depicted in Figure 11.37. By expressing the amplitude excursions of an AM signal $M(t)$ as a phase deviation, $M(t)$ is converted into the phase-modulated signal $\cos^{-1}(M(t))$ that is applied to a divider with upper and lower paths having separate PAs. The two constant-envelope signals are amplified by efficient PAs with gain G, and then the two amplified phase-modulated signals are reintegrated with a combiner. In practical

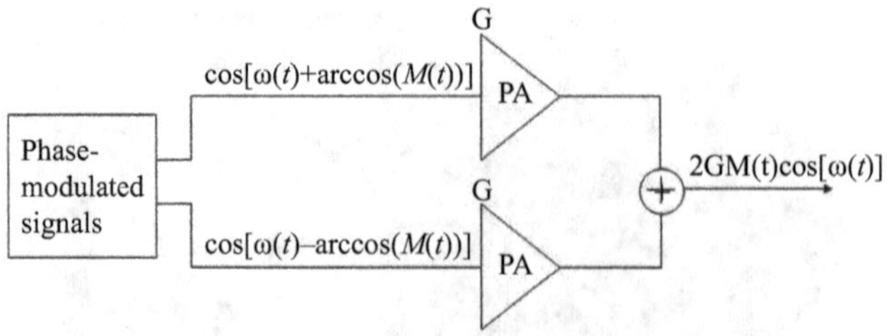

Figure 11.37 LINC configuration

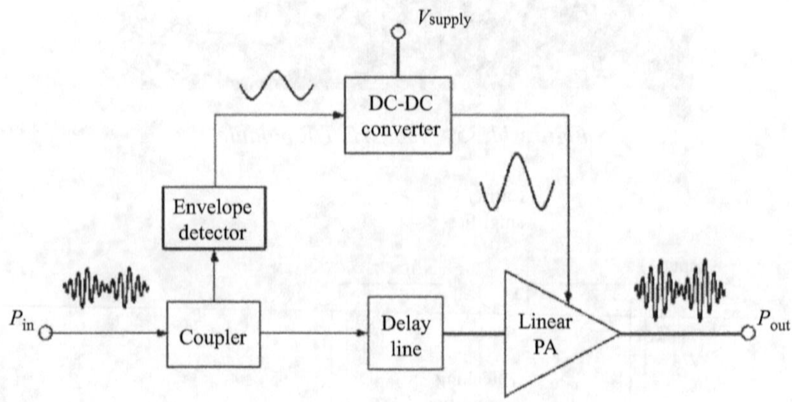

Figure 11.38 Envelope-tracking configuration

implementations, the upper signal has a positive phase shift of 135° with respect to the lower signal prior to each being applied to their respective PAs. The two signal paths are designed to produce maximum phase shifts of 45° and −45° each. When the phase shift of the upper and lower signals is 45° and −45°, respectively, the two inputs to the combiner swing between 90° and 180°. At 90°, the desired in-band signal components are produced; whereas at 180°, the undesired OOB components cancel.

A third technique that has received interest is envelope tracking (Figure 11.38), where the signal to be amplified is input to a coupler that splits it into two paths. One path is directly connected to a PA biased for class B, while the other path goes to a detector in which the signal envelope is extracted. The amplitude excursions of the extracted signals control the duty cycle of a DC-to-DC converter that can be either a buck (lowers the applied voltage) or a boost (raises the applied voltage) converter—the greater the duty cycle, the greater the DC voltage decrease or increase from the output of the converter. These DC voltage fluctuations are thus proportional to the fluctuations of the original signal envelope. The drain of the PA field-effect transistor receives the proper amount of DC bias so that it conducts as a Class B in proportion to input signal levels, thereby achieving a degree of linearity and minimizing OOB spectral products.

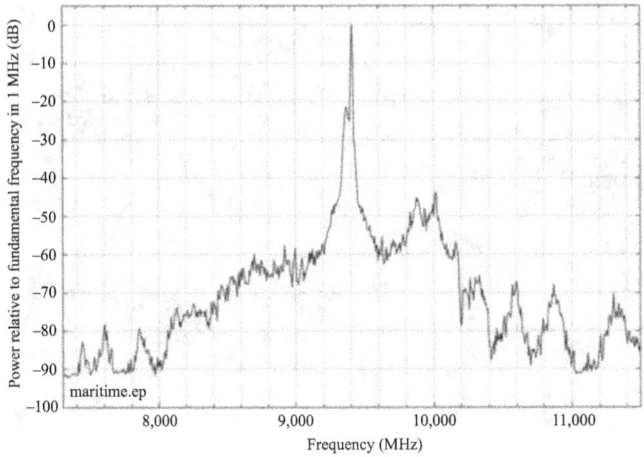

Figure 11.39 X-Band magnetron emission (Courtesy of Frank Sanders, NTIA/ITS)

11.5 Transmitter emission measurements

Modern radar design has created complicated pulses that present significant measurement challenges. Improvements to range, resolution, and immunity to interference have required phase modulated pulses, frequency chirped pulses, and very narrow pulses, with many of these exhibiting high bandwidths. Further, different types of radar systems are used in a variety of applications: avionic, military, automotive, law enforcement, astronomy, mapping, weather, and more. Within this broad range of uses, several radar transmitter technologies have emerged to meet specific needs in terms of performance, cost, size, and capability.

The spectrum analyzer is the instrument that has traditionally been used to assess frequency-dependent power or power spectral density (W/Hz). For example, the spectrum analyzer is used to ascertain whether a radar transmitter has a symmetrical pulse spectrum, thus insuring that the transmitter is supplying the required energy for detection on receive, as well as minimizing the potential for interference in adjacent frequency bands that may be inhabited by other services such as other radars or wireless services. In contrast, an asymmetrical spectrum wastes power and generates undesirable spurious emissions, thereby degrading the performance of the radar system. Figure 11.39 depicts an asymmetrical emission from an X-band magnetron.

For spectrum measurements made with a conventional (swept superheterodyne) spectrum analyzer on pulsed signals, one needs to be aware of various idiosyncrasies related to the settings for the resolution bandwidth (RBW) and sweep time to ensure correct results [13]. For example, Figure 11.40(a) depicts a pulsed 160 MHz carrier signal having $\tau = 2\ \mu\text{s}$ pulse widths and a pulse repetition interval (PRI) of $T = 10\ \mu\text{s}$, which together yield a 20% duty cycle. While legacy tube-type radars have had duty cycles of as low as 0.1%, future SS active arrays could have duty cycles that are this high to provide sufficient average power for target detection.

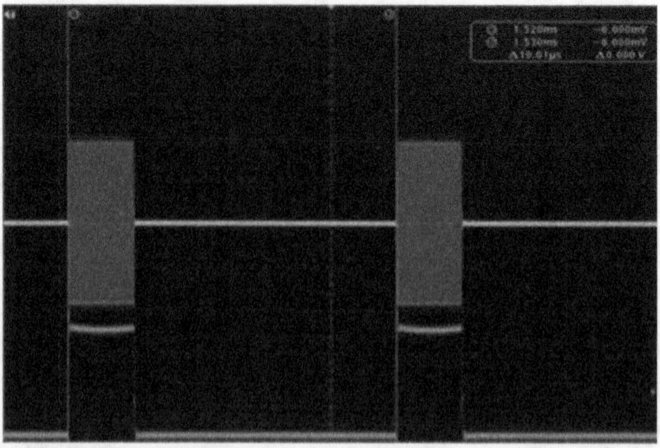

Figure 11.40 (a) Pulsed 160 MHz carrier signal

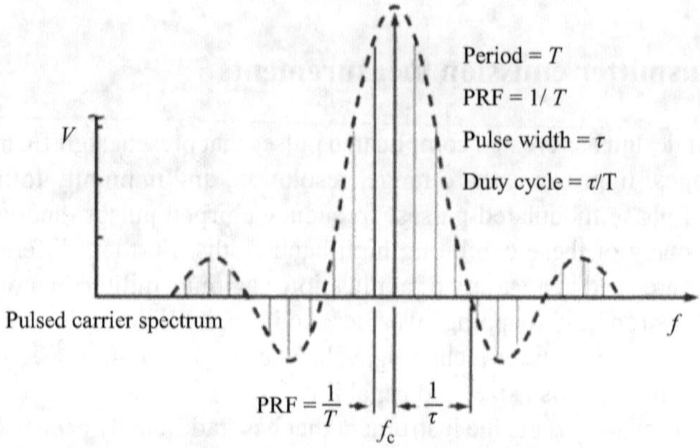

Figure 11.40 (b) Frequency-domain waveform of a pulsed carrier

Upon examining the frequency domain characteristics of the simple pulsed carrier signal from Figure 11.40(a), the discrete spectral lines with frequency spacing of $1/T$ are observed (Figure 11.40(b)). The amplitudes of these spectral lines are determined by the envelope centered on the carrier (fundamental) frequency f_c. In other words, the envelope is a (sin x)/x shape with the nulls at integer multiples of the duty cycle about the carrier ($f_c \pm n/\tau$), where n is an integer. To view the power spectral density in the line spectrum mode so that the spectral lines are visible, the RBW must be adjusted lower (RBW $< 0.3/T$) with respect to the pulse repetition frequency $1/T$. Figure 11.41 illustrates a spectrum analyzer display showing a line spectrum of the pulsed carrier signal in Figure 11.40(a).

When viewing the line spectrum (Figure 11.41), the signal power is dispersed into its spectral components [13]. The desensitization factor of the line spectrum permits

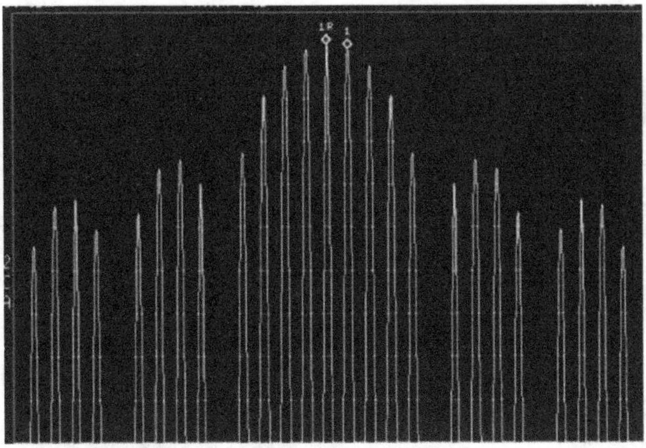

Figure 11.41 Spectrum analyzer line spectrum at 160 MHz with RBW of 100 kHz

determination of the actual peak power from that shown on the display. A marker is placed on the highest component in the line spectrum display, and the pulse width and interval are measured on the screen. The desensitization factor α depends on the duty cycle only and is calculated as $\alpha = 20\log(\tau/T)$. Then α is subtracted from the apparent peak power (highest peak in the line spectrum display). The average power can then be calculated by adding the duty cycle in dB, 10*log(τ/T), to the calculated peak power in dBm. While the line spectrum mode is generally applicable to all spectrum analyzers, whether using conventional superheterodyne or fast Fourier transform (FFT) processing, the user is cautioned to validate through bench testing with signals like or similar to those that will be measured as part of the emission test of a radar transmitter.

If the RBW of the spectrum analyzer is increased until it is larger than the reciprocal of the pulse period, but is still smaller than the reciprocal of the pulse width, the spectrum analyzer will display the pulse or envelope spectrum. The amplitude of the envelope increases linearly with the RBW; for example, doubling the RBW causes a 6 dB increase in the amplitude. In addition, the sweep time setting on the analyzer should be much greater than the reciprocal of the pulse period (at least 1,000×). The desensitization factor α for a pulse spectrum is calculated by $\alpha = 20\log(\tau KB)$, where K is the RBW correction factor (normally 1.5 for a Gaussian filter) and B is the RBW. Taking the peak of the mainlobe of the pulse spectrum and subtracting the desensitization factor provides the peak power.

In addition to supporting measurements in the frequency domain, modern spectrum analyzers are configured for time-domain measurements using the zero-span mode. In zero-span mode, a spectrum analyzer works as a receiver having fixed-frequency tuning and a set RBW. The zero-span mode is useful in transmitter emission measurements because it allows for the determination of the pulse width by applying markers.

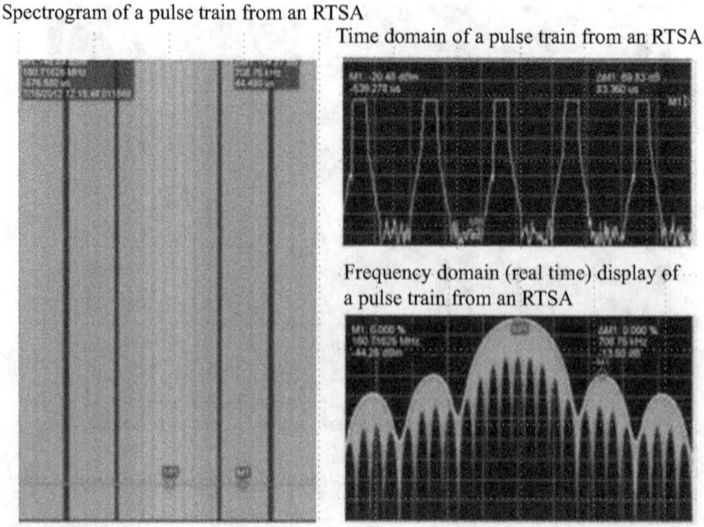

Figure 11.42 Multiframe displays from an RTSA

In recent years, there has been a great deal of interest in spectrum occupancy during given periods of time. The real-time spectrum analyzer (RTSA) is one type of technology that has gained wide acceptance in the transmitter design and spectrum communities [14] because it can perform rapid spectrum measurements within a specified time window. Consequently, with the RTSA, one can capture the transient behaviors of signals that are not easily observed by swept-tuned superheterodyne or FFT-based receivers [8]. Digital signal processing is the heart of the RTSA and permits it to perform millions of FFTs over bandwidths in excess of 100 MHz. This capability is important for radars because large instantaneous bandwidths are a high priority today and will be even more so in the future. Figure 11.42 shows an example of a display from a Tektronix RTSA that is measuring a pulse train in one instant of time. The left frame displays a spectrogram of this pulse train with its nulls and peaks as they evolve in time. The upper right frame shows a time-domain display of the region that contains a group of pulses to be analyzed. The lower right frame shows the signal magnitude versus frequency. This last display is particularly noteworthy because the RTSA capture mode permits transient spectral events that are not observable on conventional swept superheterodyne and FFT analyzers to be easily visible on the RTSA through a multicolor format.

The authors wish to convey some final thoughts on transmitter emission measurements. First, it is preferable, if possible, that radiated emission measurements be made as opposed to measuring from a test port connected to the waveguide through a directional coupler. The reason is that the antenna acts as a bandpass filter, and thus the radiated emissions are what victim receivers will experience in terms of spectral content. It is very important to understand the environment in which the radar is operating in terms of multipath and intervening reflection and diffraction effects of structures and vegetation between the radar antenna and the emission test antenna. In addition, the height of the test antenna relative to the radar antenna is important. In the case of a phased array where the mainbeam position travels from the horizon toward the zenith, it is important that the test antenna be at a sufficient height to allow capture of both the mainbeam and sidelobes to allow for determination of the latter relative to the former. Finally, before emission measurements are taken, knowledge of other sources of potential interference should be known and ambient measurements should be collected beforehand to ensure that the emission measurements of the radar transmitter are not corrupted.

11.6 Future research

Future transmitter research must address a number of disciplines and issues. Complicating any research is the unique spectrum requirements for a given geographic area where radar operates. At both the system and component levels, research needs to address radar transmitters that are able to operate in dynamic and often severe and challenging EM environments. Finally, there needs to be active research into all major pieces of equipment that comprise a transmitter, as well as the software glue that will connect and enable the transmitters of the future. It is the seamlessly integrated combination of hardware and software that will allow various transmitters and receivers to coexist in the crowded spectrum of the future with a high capacity and efficiency.

Given the preceding comments, the authors feel that future research for cooperatively coexisting transmitters and receivers can be grouped into the following areas: (1) RF power generation and amplification; (2) modulation techniques capable of adapting to the spectral environment, as well as other transmitters in the operating area, be they intentional or unintentional; (3) high-speed and high-bandwidth RF sampling techniques; (4) signal processing capability that can detect and identify signals in the operating area; (5) software that is adaptive and can assess its spectral environment to direct the transmitter's own operation with cooperating and coordinating systems in its vicinity; (6) integrated computing power that can handle the aforementioned tasks with sufficient expansion potential to address reasonable advances in any of the preceding areas; and (7) antennas that are optimized to support the hardware and software developments for future transmitter architectures.

An area of active research involving SS PAs is on the subject of adaptive PAs [15], which is also the subject of the next chapter (12). Consider Figure 11.43, where the plot on the left displays a set of conventional load-pull contours on a Smith chart that can be obtained with an adjustable load-pull tuner. These contours represent the real and imaginary components of the complex reflection coefficient Γ and can also be obtained using the load-pull simulation found in RF/microwave design programs such as the Keysight Advanced Design System. The Smith chart presentation on the right shows three queries of a steepest accent search algorithm to determine the shortest path to the particular complex reflection coefficient with the highest power-added efficiency (PAE) that also provides the lowest adjacent channel power ratio

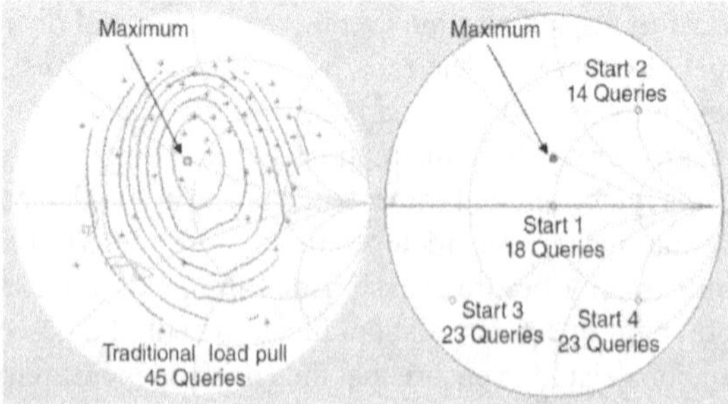

Figure 11.43 Conventional load pull (left) vs. search algorithm (right)

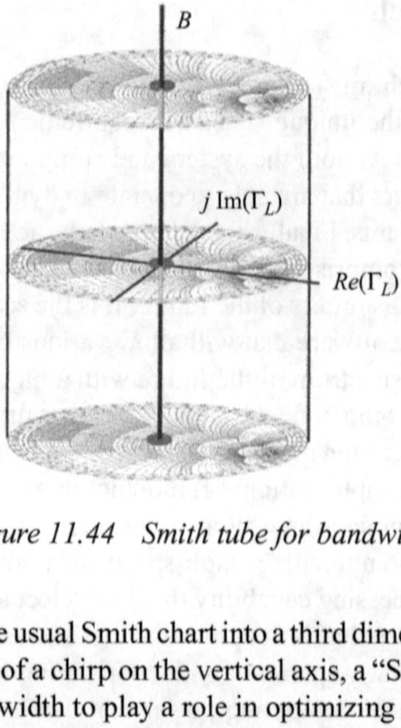

Figure 11.44 Smith tube for bandwidth

(ACPR). By extending the usual Smith chart into a third dimension and by utilizing the adjustment of bandwidth of a chirp on the vertical axis, a "Smith tube" (Figure 11.44) is formed, allowing bandwidth to play a role in optimizing for PAE and ACPR.

To evolve this work into realizable radar transmitters, research will have to provide rapidly converging algorithms that facilitate determination of the optimum complex reflection coefficient that provides for maximum PAE and minimum ACPR at a given transmit signal bandwidth. Load-pull tuners of sufficiently small volume need to be developed that can withstand the required RF power and commensurate heat dissipation in order to support rapid impedance matching between the RF PA and the load (such as an antenna element). In addition, advancements in microelectromechanical systems (MEMS) techniques are critical if high-speed load-pull tuners with small footprints and high power-handling capability are to be realized. Other research efforts to increase PAE, reduce ACPR, and enhance spectrum compatibility involve digital computation and lookup tables based on predistortion and envelope tracking.

Interest has arisen recently in the application of MPMs for transmitters in phased arrays. For example, an new MPM under consideration has a 6-inch, S-band,

1 kW TWT, an SS pre-amplifier driver, and a power supply [16]. The research challenge is to design and fabricate highly reliable, cost-effective MPMs that would offer peak output powers of 1.0 kW or greater to support advanced phased array antennas.

With large numbers of capable and expensive legacy tube-based radar systems still operating, microwave tubes that contribute less in-band and OOB noise are a very attractive alternative to converting to solid state. Research into such tubes is ongoing. The MBK, discussed earlier, is being examined as a possible replacement for more noisy tubes such as CFAs. These activities have received much attention because they offer a possible means to upgrade existing systems as well as of being the basis for future transmitter designs. Of course, in considering a tube like the MBK for legacy and future radars, care must be exercised in evaluating prototype versus long-term development costs in comparison to an SS transmitter.

Important and often overlooked components of radar transmitters are the high-voltage tubes and high-current power supplies necessary for providing reliable, noise-free DC power to tube-based and SS PAs, respectively. While high-voltage power supplies may support duty cycles of 0.01% to 5%, a high-current power supply may be needed to maintain duty cycles of 20% or higher, which could place greater demands on these devices to address heat dissipation. Continued research by the power electronics community needs to be encouraged and supported to ensure that both high-voltage and high-current power supplies provide the necessary reliability and EM compatibility over time. With the increased emphasis on spectrum compatibility, power supplies must achieve EM compatibility instead of causing or being the victim of adjacent-channel interference.

Another area of active research is the development of modulation techniques that are capable of adapting to the spectral environment, as well as to other transmitters (intentional and unintentional) in the operating area. Many of these techniques depend on special waveforms, thereby falling under the umbrella of waveform diversity.

Much effort is underway to increase the speed and bandwidth of analog-to-digital (A/D) and digital-to-analog (D/A) converters and boards suitable for direct sampling of RF signals, both at baseband and at ever higher frequencies, to minimize the losses in power and time required when frequency conversion techniques are needed to achieve high-frequency operation. This field of research and development has received a big boost from the many varied applications beyond communications requiring A/D and D/A conversion.

The development of radar signal processing capabilities that detect and identify interfering signals is an activity that has also received significant attention in areas such as image processing, data fusion, and predictive analytics. From a radar standpoint, there is still much room for improvement in signal identification via comparison against known signals in databases, processing information into a form that allows machine-based decision-making, and the evaluation and updating of those decisions.

Software that is adaptive and can assess its spectral environment and then can direct the transmitter's operation in cooperation with other nearby spectrum users is a key component of systems that will employ advanced transmitter architectures in the future. Not only must this software be maintainable, verifiable, modular, and expandable, it must also address cybersecurity concerns such as authentication and attribution.

Research must also concentrate on integrated processing power that can handle the above tasks with sufficient expansion potential to address reasonable advances in any of the preceding areas. Fortunately, this research area is one that is actively being pursued to facilitate advances in many other fields and should be available for implementation in advanced transmitter architectures.

There is also a continued need for advances in antenna technology to support the many approaches to future transmitter architectures. Because of the many different approaches to transmitter design, depending on what aspect is to be emphasized, no one antenna formulation is optimal. For instance, the antennas of a distributed transmitting array can take many forms, depending on whether it is distributed in time, space, or frequency.

Finally, the ability to build a transmitter that is cognitive and adaptive will depend upon tying all the above hardware and software together into a framework that permits radar operation in conjunction with supporting environmental sensing, decision-making, and adaptation to the surrounding environment. As such, the reader should become familiar with the research activity in the cognitive sensing area. A good place to start is the special issue of the *IEEE Signal Processing Magazine* on *Knowledge-based Systems for Adaptive Radar* [17]. An excellent example of the structure of a cognitive radar is provided in two papers by Wicks [18,19]. Because an adaptive radar is a dynamic system, employing control and feedback concepts, one should become familiar with several papers by Haykin *et al.* [20–22] that provide a review of the framework defining a cognitive and adaptive radar system, of which an adaptive transmitter is a key component.

References

[1] Cohen, L. and Sanders, F., "The current state of spectrum usage," *IEEE International Symposium on Electromagnetic Compatibility*, Raleigh, NC, Aug. 2014.

[2] United States Department of Commerce, National Telecommunications & Information Administration, *Manual of Regulations and Procedures for Federal Radio Frequency Management*, September 2015 Revision of the May 2013 Edition. Located at www.ntia.doc.gov/page/2011/manual-regulations-and-procedures-federal-radio-frequency-management-redbook

[3] Wikipedia article, *Heinrich Hertz*. Located at https://en.wikipedia.org/wiki/Heinrich_Hertz

[4] Wikipedia article, *Spark-gap transmitter*. Located at https://en.wikipedia.org/wiki/Spark-gap_transmitter

[5] Wikipedia article, *Transmitter*. Located at https://en.wikipedia.org/wiki/Transmitter

[6] Skolnik, M., *Radar Handbook, 2nd Edition*. Chapter 4. McGraw-Hill, New York, NY. 1990.

[7] Wilkins, M.H.F., "John Turton Randall (23 March 1905–16 June 1984)," *Biographical Memoirs of Fellows of the Royal Society*, 33, 1987, pp. 493–535.
[8] Baylis, C., Cohen, L., and Jost, R., "Introduction to radar transmitters," *IEEE Radar Conference*, Kansas City, MO, May 2011.
[9] Edde, B., *Radar: Principles, Technology, Applications*. Chapter 8. Prentice-Hall, Upper Saddle River, NJ. 1993.
[10] Kingsley, N. and Guerci, J., *Radar RF Circuit Design*. Chapter 4. Artech House, Boston, MA. 2016.
[11] Lee, T., *Planar Microwave Engineering*. Chapter 20, Cambridge University Press, New York, NY. 2004.
[12] Cripps, S., *RF Power Amplifiers for Wireless Communications, 2nd Edition*. Chapters 1 and 14. Artech House, Boston, MA. 2006.
[13] Keysight, *Radar Measurements*, Application Note, pp. 30–32, Keysight Technologies, 2014.
[14] Cohen, L. and Jost, R., "Basics of performing radar spectrum measurements in a congested electromagnetic environment," *Antenna Measurements and Techniques Association Symposium*, Belleview, Washington, Oct. 2012.
[15] Baylis, C., Fellows, M., Cohen, L., and Marks, R., "Solving the spectrum crisis," *IEEE Microwave Magazine*, July/Aug. 2014.
[16] Song, H., Tekamp, L., Everleigh, C., *et al.*, "A new class of microwave power module for phased array antenna radar applications," *IEEE International Conference on Plasma Science*, Edinburgh, UK, July 2012.
[17] Special Issue on "Knowledge-based systems for adaptive radar," F. Gini, ed., *IEEE Signal Processing Magazine*, vol. 23, no. 1, Jan. 2006.
[18] Wicks, M., "Spectrum crowding and cognitive radar," *2nd International Workshop on Cognitive Information Processing*, pp. 452–457, Elba Island, Tuscany, Italy, June 2010.
[19] Wicks, M., "Cognitive radar: a way forward," *IEEE Radar Conference*, Kansas City, MO, May 2011, pp. 12–17.
[20] Haykin, S., "Cognitive dynamic systems: radar, control, and radio," *Proceedings of the IEEE*, vol. 100, no. 7, pp. 2095–2103, July 2012.
[21] Haykin, S., Fatemi, M., Setoodeh, P., and Xue, Y., "Cognitive control," *Proceedings of the IEEE*, vol. 100, no. 12, pp. 3156–3169, Dec. 2012.
[22] Fatemi, M. and Haykin, S., "Cognitive control: theory and application," *IEEE Access*, vol. 2, pp. 698–710, 2014.

Further reading

C.A. Balanis, *Antenna Theory: Analysis and Design*, 3rd ed., Wiley-Interscience, Hoboken, NJ, 2005.

C.A. Balanis, ed. *Modern Antenna Handbook*, Wiley-Interscience, Hoboken, NJ, 2008.

C. Baylis, M. Moldovan, L. Wang, and J. Martin, "LINC power amplifiers for reducing out-of-band spectral re-growth: a comparative study," *IEEE Wireless and Microwave Technology Conference*, Melbourne, Florida, Apr. 2010.

C. Baylis, L. Wang, M. Moldovan, *et al.*, "Designing for spectral conformity: issues in power amplifier design," *IEEE Waveform Diversity & Design Conference*, Niagara Falls, Ontario, Canada, Aug. 2010.

C. Baylis, L. Dunleavy, S. Lardizabal, R.J. Marks II, and A. Rodriguez, "Efficient optimization using experimental queries: a peak-search algorithm for efficient load-pull measurements," *Journal of Advanced Computational Intelligence and Intelligent Informatics*, Sept. 2010.

M. Borkowski, "Solid-state transmitters," Chapter 11 in *Radar Handbook*, 3d ed., ed. M. Skolnik, McGraw-Hill, New York, 2008.

L. Cohen and R. Jost, "Introduction to modern radar transmitters," *IEEE Radar Conference*, Cincinnati, OH, May 2014.

J. de Graaf, H. Faust, J. Alatishe, and S. Talapatra, "Generation of spectrally confined transmitted radar waveforms," *IEEE Radar Conference*, 2006, pp. 76–83.

G. Ewell, *Radar Transmitters,* McGraw-Hill, New York, 1981.

R. Gilmore and L. Besser, *Practical RF Circuit Design for Modern Wireless Systems Vol. I*, Artech House, Norwood, MA, 2003.

R. Gilmore and L. Besser, *Practical RF Circuit Design for Modern Wireless Systems Vol. II*, Artech House, Norwood, MA, 2003.

A.S. Gilmour, *Klystrons, Traveling Wave Tubes, Magnetrons, Cross-Field Amplifiers, and Gyrotrons*, Artech House, Norwood, MA, 2011.

F. Gini and M. Rangaswamy, *Knowledge Based Radar Detection, Tracking and Classification*, John Wiley & Sons, Hoboken, NJ, 2008.

J.R. Guerci, *Cognitive Radar: The Knowledge-aided Fully Adaptive Approach*, Artech House, Norwood, MA, 2010.

L-3 Communications, *TWT/TWTA Handbook*, 13th ed., L-3 Communications Electron Technologies, Inc., Torrance, CA, 2009.

R. Ludwig, and G. Bogdanov, *RF Circuit Design: Theory & Applications*, 2nd ed., Prentice Hall, Upper Saddle River, NJ, 2008.

K. Mekechuck, W. J. Kim, S. Stapelton, and J. Kim, "Linearizing power amplifiers using digital predistortion, EDA tools, and test hardware," *High Frequency Electronics*, April 2004, pp. 18–27.

T.A. Milligan, *Modern Antenna Design*, 2nd ed., Wiley-IEEE Press, Hoboken, NJ, 2005.

E. Ostroff, M. Borkowski, and H. Thomas, *Solid-State Radar Transmitters*, Artech House, Norwood, MA, 1985.

D.M. Pozar, *Microwave Engineering*, 3rd ed., Wiley, Hoboken, NJ, 2004.

F. Raab, P. Asbeck, S. Cripps, *et al.*, "Power amplifiers and transmitters for RF and microwave," *IEEE Transactions on Microwave Theory and Techniques*, vol. 50, no. 3, pp. 814–826, Mar. 2002.

M.A. Richards, J.A. Scheer, and W.A. Holm, (eds.) *Principles of Modern Radar: Basic Principles*, SciTech Publishing, Raleigh, NC, 2010.

D. Root, J. Horn, L. Betts, C. Gillease, and J. Verspecht, "X-parameters: the new paradigm for measurement, modeling, and design of nonlinear RF and microwave components," *Microwave Engineering Europe*, 2008.

D. Root, "A new paradigm for measurement, modeling, and simulation of nonlinear microwave and RF components," Presentation at Berkeley Wireless Research Center, Apr. 2009.

G. Simpson, "High power load pull with X-parameters – a new paradigm for modeling and design," *IEEE Wireless and Microwave Technology Conference*, Melbourne, Florida, April 2010.

M. Steer, *Microwave and RF Design: A Systems Approach*, SciTech Publishing, Raleigh, NC, 2009.

W.L. Stutzman and G.A Thiele, *Antenna Theory and Design*, 2nd ed., Wiley, New York, NY, 1997.

P. Tasker, "Practical waveform engineering," *IEEE Microwave Magazine*, vol. 10, no. 7, pp. 65–76, Dec. 2009.

T.V. Wallace, R.J. Jost and P.E. Schmid, "Radar transmitters," Chapter 10 in *Principles of Modern Radar*, ed. M.A. Richards, J.A. Scheer, and W.A. Holm, SciTech Publishing, Raleigh, NC, 2010.

M. Wicks, E. Mokole, S. Blunt, R. Schneible, and V. Amuso, *Principles of Waveform Diversity and Design*, SciTech Publishing, Raleigh, NC, 2010.

G. Wimpenny, "Improving multi-carrier PA efficiency using envelope tracking," *RF Engineer Network*, http://www.rfengineer.net, Apr. 2009.

J. Volakis, ed., *Antenna Engineering Handbook*, 4th ed., McGraw-Hill, New York, NY, 2007.

Chapter 12

Adaptively reconfigurable radar: real-time optimization of the transmitter amplifier and waveforms

Charles Baylis[1]

In the ever-changing spectral environment, dynamic spectrum allocation is becoming more prevalent in frequency bands previously assigned wholly to radar systems. The increasing use of dynamic spectrum allocation indicates that future radar transmitters will need to be frequency agile and spectrally sensitive. Based on knowledge of surrounding spectrum users, these transmitters should reconfigure their operating frequency and spectral output to coexist with communications and other spectral occupants. To do so, high-power radar transmitters need to be *adaptive* and *reconfigurable*, which means the radar must be able to optimize its range/Doppler resolution according to the desired operating mode, maximize its power efficiency, and ensure that it meets spectral requirements dictated by the current electromagnetic (EM) environment. While this chapter addresses issues with transmitter circuits, it should be noted that adaptivity and circuit reconfigurability are also useful in the radar receiver to eliminate potentially interfering adjacent-band and in-band signals (see Chapter 10).

The focus here is on circuit and waveform optimization approaches that can be used by the radar to tailor its emissions according to the interference environment while still seeking to maximize detection performance. In particular, physical hardware implementations are emphasized, including experimental assessment. In the chapter, a brief overview of present and future radar spectrum allocations is first presented, followed by algorithms that can be used to optimize the power-amplifier circuitry of future radar transmitters, along with a subsequent discussion of waveform optimization techniques. Combining the circuit and waveform optimization techniques is a present research objective, leading to potential joint circuit and waveform optimization for the radar. The formation of a dynamic spectral mask is then considered as a means to use these optimization approaches in a dynamic EM environment. Finally, a possible path forward for the design of adaptive and reconfigurable radar transmitters is discussed.

[1]Department of Electrical and Computer Engineering, Baylor University, USA

12.1 Why an adaptive and reconfigurable radar transmitter?

Adaptive and reconfigurable radar transmitters are needed to facilitate flexible, real-time sharing with wireless communications. Such sharing is already being required in bands previously dedicated only to radar, and it is expected that shared allocations will continue to become more widespread.

12.1.1 Traditional radar spectrum regulations

The Federal Radar Spectrum Requirements published by the National Telecommunications and Information Administration (NTIA) in the United States show several bands primarily allocated to radar in the U.S. under 10 GHz. These are: 420–450 MHz, 902–928 MHz, 1215–1300 MHz, 1350–1390 MHz, 2310–2390 MHz, 2700–3650 MHz, 4200–5925 MHz, and 8500–10,550 MHz [1]. This arrangement shows that a significant percentage of available frequencies are allocated to radar operation. The Radar Spectrum Engineering Criteria (RSEC), also published by the NTIA, gives examples of how spectral masks are used to limit the bandwidth of radar emissions [2].

12.1.2 Dynamic spectrum access

A robust surge in wireless communication devices over the last decade has caused traditional spectrum management approaches to come into question. The pre-allocation of spectrum is rather inefficient, as considerable allocated spectrum may go unutilized for significant amounts of time. A dynamic, real-time allocation of spectrum is more efficient and can provide access to a greater number of users. However, systems designed under the pre-allocation paradigm can have difficulty operating in this new arrangement. Radars have specifically proven difficult to the deployment of dynamic spectrum access due to their large contiguous bandwidths needed for adequate range resolution, the "listen" time following a radar transmission during which information is actually collected, and the expense of replacing legacy radar transmitters that are not frequency agile.

Despite the issues that exist with regard to dynamic spectrum access in the vicinity of legacy radars, recent regulatory moves have forced these radars into real-time sharing with communications. In the United States, for example, the 3550–3650 MHz S-band radar-only allocation has been changed to now permit sharing between radar and communications; likewise for the Unlicensed National Information Infrastructure (U-NII) bands above 5 GHz. The growth of wireless communication demand is expected to continue. Thus, a new radar transmit/receive paradigm is necessary to facilitate operation in this increasingly more dynamic spectrum access environment.

To enable dynamic spectrum access, radar transmitters must become *adaptive* and *reconfigurable*. Specifically, a radar may adapt its operating frequency, the waveform(s) it emits, its pulse repetition frequency, etc. Reconfigurability, on the other hand, is a specific sub-attribute of adaptability that can be viewed as an enabling capability. Reconfigurability means that the analog microwave circuitry in the radar

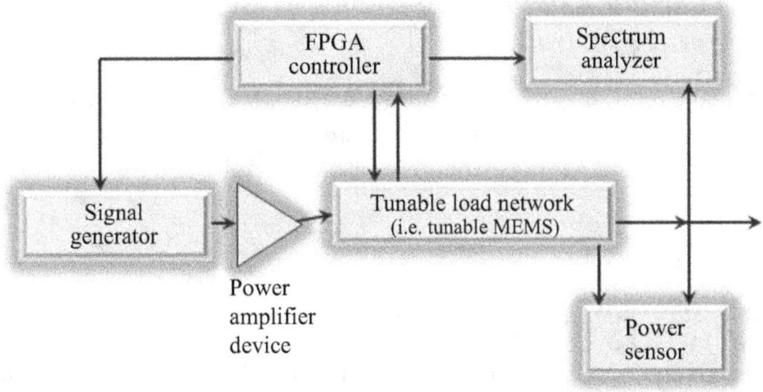

Figure 12.1 Conceptual block diagram of an adaptive and reconfigurable radar transmitter, reprinted from [3]

transmitter and receiver must be able to change to optimize performance for different dynamically assigned frequency bands and spectral masks in real time or "on the fly."

The radar must be able adjust in real time to accomplish three main objectives: (1) perform target detection with the requisite range and Doppler resolution, (2) maximize transmit power efficiency, and (3) meet the spectral requirements dictated by the environment and operating mode. Thus, the waveform and circuitry of the transmitter and receiver must also be optimizable in real-time to meet these demands. A prospective next-generation adaptive and reconfigurable radar transmitter is shown in Figure 12.1 [3].

This chapter focuses on real-time optimization of the transmitter circuit and waveform according to the block diagram of Figure 12.1. Specifically, algorithms for fast circuit optimization address the load impedance (using a tunable load network), input power, and bias voltages. In addition, the radar waveform can be optimized to provide desired ambiguity-function properties. In all cases, the spectral content of the resulting emission is a key consideration. These considerations ultimately lead to a roadmap for joint transmitter/waveform optimization.

12.2 Algorithms for circuit optimization

This section focuses on fast circuit optimization techniques with the goal of achieving high power-added efficiency (PAE) while meeting spectrum requirements. These techniques were developed for the particular application of reconfigurable radar, with laboratory experimental test results on scaled-down, lower power devices used to demonstrate their efficacy. The future application of these algorithms to high-power amplifiers within each tuning element of a phased array will provide even more insights into the design and capabilities of a rapidly reconfigurable radar.

12.2.1 Optimization of load impedance to maximize PAE under spectral constraints

The PAE and adjacent-channel power ratio (ACPR) are useful metrics to describe the operation of a transmitter power amplifier, and both depend on the load reflection coefficient Γ_{L} [4]. The PAE is defined for a power amplifier as

$$\mathrm{PAE} = \frac{P_{\mathrm{out,RF}} - P_{\mathrm{in,RF}}}{P_{\mathrm{DC}}} \times 100\%, \tag{12.1}$$

where $P_{\mathrm{in,RF}}$ and $P_{\mathrm{out,RF}}$ are the input and output power of the radio frequency (RF) signal, respectively, and P_{DC} is the direct current (DC) power added by the power supply. In other words, PAE measures how much of the supplied power is converted into additional RF signal power.

The ACPR is defined as the ratio of the total measured power in a defined adjacent spectrum channel to the total measured power in the defined main channel, or simply

$$\mathrm{ACPR} = \frac{P_{\mathrm{adjacent}}}{P_{\mathrm{main}}}. \tag{12.2}$$

In practice, the ACPR limit is determined by the amount of interference a receiver within the adjacent channel can tolerate while still being able to receive and detect its own desired signal successfully. ACPR is also a useful metric for assessing the degree of linearity of nonlinear devices such as power amplifiers [5] because the nonlinearities of these devices induce odd-order intermodulation products between in-channel frequency components that manifests in the form of spectral spreading into adjacent channels [6–8].

Constrained optimization of these parameters using a variable load impedance tuner therefore provides the means with which to reconfigure a radar transmitter in an adaptive manner. For example, in [9] Sun and Lau demonstrated the use of a genetic algorithm for real-time transmitter impedance tuning with reconfigurable matching. However, in [10], du Plessis and Abrie discuss genetic algorithms as an option for real-time impedance matching, but conclude that they are too slow for many applications.

More recently, a dual search was developed [11] to find the load reflection coefficient Γ_{L}that provides the maximum PAE while meeting ACPR constraints. The real-time optimization goal is to reach this solution on operational time scales. The search space for this endeavor is the Smith Chart, which is the complex representation of Γ_{L} that can also be expressed in terms of the two real-valued parameters $\Gamma_{\mathrm{r}} = \mathrm{Re}(\Gamma_{\mathrm{L}})$ and $\Gamma_{\mathrm{i}} = \mathrm{Im}(\Gamma_{\mathrm{L}})$, where $\mathrm{Re}(\bullet)$ and $\mathrm{Im}(\bullet)$ extract the real and imaginary parts of the argument, respectively. The search is initialized at a user-defined starting point on the Smith Chart and proceeds toward the optimum solution according to the nature of the particular optimization algorithm being employed. Here, we consider a modified gradient approach.

For the first candidate solution, measurements are taken to estimate the gradient, as shown in Figure 12.2 [12]. Each gradient-estimation measurement is made at a

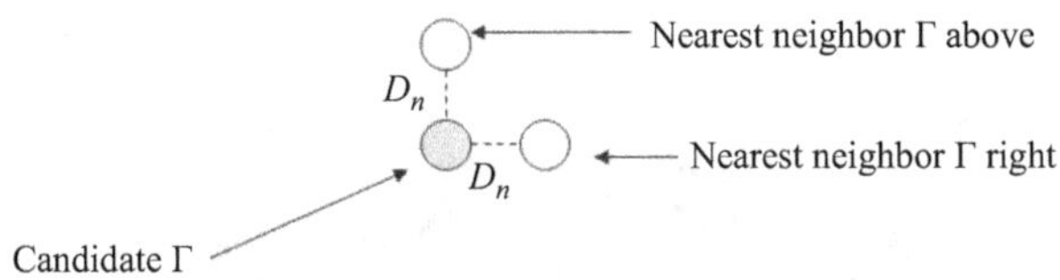

Figure 12.2 Gradient estimation procedure, reprinted from [12]

neighboring point that is distance D_n from the candidate point, varying each dimension. If PAE is represented by the variable p, then the associated gradient vector is defined as

$$\nabla p = \hat{\Gamma}_r \frac{\partial p}{\partial \Gamma_r} + \hat{\Gamma}_i \frac{\partial p}{\partial \Gamma_i} \tag{12.3}$$

in terms of the $\hat{\Gamma}_r$ and $\hat{\Gamma}_i$ coordinates. The derivative portions of (12.3) can be estimated by taking points on the Smith Chart separated from the candidate solution by small increments in the Γ_r and Γ_i directions, such that

$$\frac{\partial p}{\partial \Gamma_r} \approx \frac{(\Delta p)_{\text{right}}}{\Delta \Gamma_r} = \frac{(\Delta p)_{\text{right}}}{D_n} \tag{12.4}$$

$$\frac{\partial p}{\partial \Gamma_i} \approx \frac{(\Delta p)_{\text{up}}}{\Delta \Gamma_i} = \frac{(\Delta p)_{\text{up}}}{D_n}. \tag{12.5}$$

From the estimated gradient, a unit vector in the direction of PAE steepest ascent can then be obtained by dividing the gradient by its magnitude:

$$\hat{p} = \frac{\nabla p}{|\nabla p|}. \tag{12.6}$$

The ACPR gradient ∇a can be expressed in terms of $\hat{\Gamma}_r$ and $\hat{\Gamma}_i$ coordinates similar to (12.3) and can likewise be estimated using ACPR measurements at the same points on the Smith Chart for comparison with the resulting ACPR values at the candidate solution. A unit vector can also be constructed for ACPR as

$$\hat{a} = -\frac{\nabla a}{|\nabla a|}, \tag{12.7}$$

which is the direction of ACPR steepest descent.

A search vector can be constructed based on the gradients as illustrated in Figure 12.3 [12]. The search vector is based on the PAE and ACPR optimum-direction vectors $\hat{p}$ and $\hat{a}$, respectively, where the goal of the search is to determine the location that provides the best PAE that meets specified ACPR constraints. This location resides on the Pareto "trade-off" optimum locus that stretches between $\hat{a}$ and $\hat{p}$, and is also expected to be on the boundary of the ACPR-acceptable region. The Pareto optimum locus is the curve containing the best-compromise solutions between PAE and ACPR, with each point on this locus a constrained optimum for PAE according to a different ACPR constraint. At all points along this locus, it is expected that the

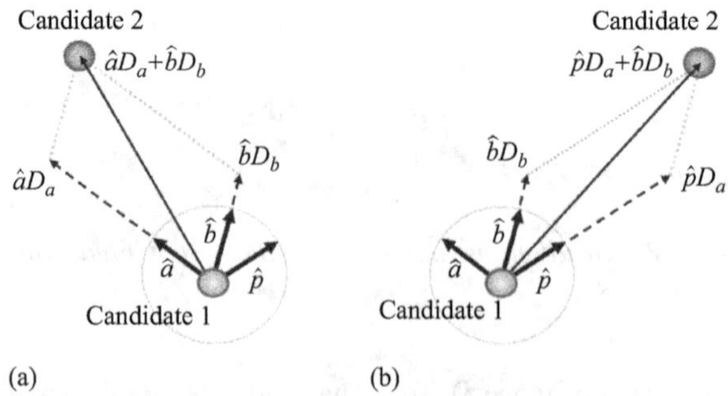

Figure 12.3 Estimation of the search vector in cases where (a) ACPR > $ACPR_{limit}$ (out of compliance) and (b) ACPR ≤ $ACPR_{limit}$ (in compliance). Reprinted from [12]

PAE and ACPR gradients will be generally oppositely directed [11]; for example, see the orientations of $\hat{p}$ and $\hat{a}$ in Figure 12.3.

When the ACPR is greater than an established limit denoted $\mathrm{ACPR}_{\mathrm{limit}}$, the search vector is constructed so as to take the fastest route to the constrained PAE optimum using the following considerations: (1) the constrained PAE optimum is expected to lie on the boundary of the ACPR constraint, and (2) the $\hat{p}$ and $\hat{a}$ vectors have essentially opposite signs relative to this solution. Figure 12.3(a) shows the case where the starting candidate solution (or simply, Candidate 1) is out of compliance, i.e. $\mathrm{ACPR} > \mathrm{ACPR}_{\mathrm{limit}}$. In this case, the next step toward the constrained PAE optimum contains a component in the best ACPR direction to pull the search toward the ACPR-acceptable region, and another component toward the Pareto optimum locus. The best-estimate direction that reaches the nearest point on the Pareto optimum locus is in the direction of a bisector of $\hat{a}$ and $\hat{p}$, which is shown as vector $\hat{b}$ in Figure 12.3. Under this condition the search vector is defined as [12]

$$\overline{v} = \hat{a}D_{\mathrm{a}} + \hat{b}D_{\mathrm{b}} \tag{12.8}$$

using the weights

$$D_{\mathrm{a}} = \frac{D_{\mathrm{s}}}{2}\frac{|\mathrm{ACPR}_{\mathrm{meas}} - \mathrm{ACPR}_{\mathrm{limit}}|}{|\mathrm{ACPR}_{\mathrm{worst}} - \mathrm{ACPR}_{\mathrm{limit}}|} \tag{12.9}$$

and

$$D_{\mathrm{b}} = \frac{D_{\mathrm{s}}}{2}\frac{|\theta_{\mathrm{meas}} - \theta_{\mathrm{target}}|}{\theta_{\mathrm{target}}}, \tag{12.10}$$

where $\mathrm{ACPR}_{\mathrm{meas}}$ is the measured ACPR at Candidate 1, $\mathrm{ACPR}_{\mathrm{worst}}$ is the highest (worst) value of ACPR measured during the search, θ_{meas} is the angle measured between $\hat{a}$ and $\hat{b}$ at Candidate 1, θ_{target} is the desired value of the angle between $\hat{a}$ and $\hat{b}$, which is 90°, and D_{s} is a user-specified search-distance parameter. Relative to the estimated total search travel, (12.9) and (12.10) provide estimates concerning

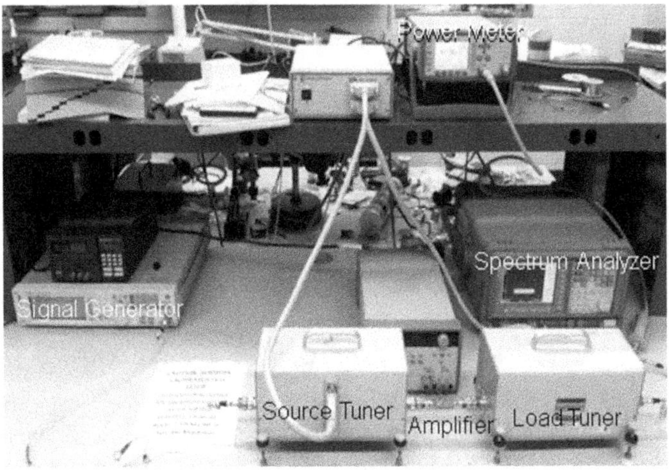

Figure 12.4 Nonlinear test bench setup, reprinted from [12]

the percentage of travel remaining to reach the ACPR limit and the Pareto optimum locus, respectively. As such, each component of $\overline{v}$ is adjusted at each candidate location based on the estimated closeness to the objective it seeks.

Alternatively, if a candidate is inside the ACPR compliance region, i.e. ACPR $\leq$ $\text{ACPR}_{\text{limit}}$, then the search proceeds toward the PAE optimum rather than the ACPR optimum so as to reach the intersection of the Pareto optimum locus with the ACPR constraint boundary. In this case, the search vector is

$$\overline{v} = \hat{p}D_{\text{a}} + \hat{b}D_{\text{b}}, \tag{12.11}$$

where the ACPR unit vector $\hat{a}$ from (12.8) is replaced by the PAE unit vector $\hat{p}$.

It should be noted that this search algorithm uses the characterized reflection coefficient values at the center frequency and does not consider the variation of reflection coefficient over the entire operating main channel band (or of the adjacent channel). While the reflection coefficient could show significant variation with frequency, particularly for wideband operation, experimental measurements of PAE and ACPR should still be accurately assessed and used for optimization. Of course, the modality and/or convexity of the search may be affected, as variation in the reflection coefficient as a function of frequency could cause different effects in the load-pulling of the device. Similarly, the harmonic terminating impedances are also of interest, with the tuning of harmonic frequency reflection coefficients a means to achieve additional performance improvement. This topic is beyond the scope of this chapter, but is certainly one worth further investigation.

This search algorithm was experimentally demonstrated using the nonlinear load-pull test bench setup shown in Figure 12.4. The setup is controlled by MATLAB, thereby allowing for the determination of the values of Γ_{L} to present to the device as the search progresses. A Skyworks-packaged amplifier was used as the device

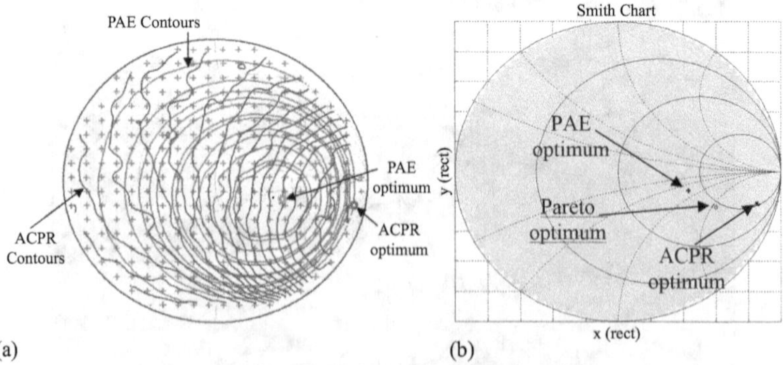

Figure 12.5 (a) Traditionally measured PAE and ACPR contours, with the PAE and ACPR optima indicated, and (b) constrained Pareto optimum for ACPR < −28.2 dBc relative to PAE and ACPR optima. Reprinted from [12]

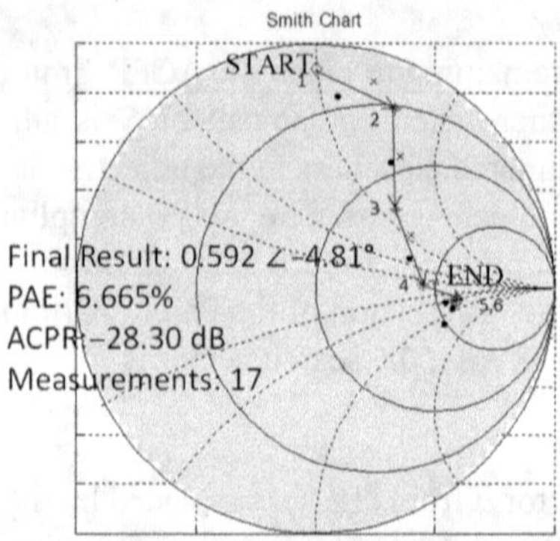

Figure 12.6 Fast measurement search results for Skyworks-packaged amplifier at 3.3 GHz using starting point $\Gamma_L = 0.9\underline{/90^\circ}$, reprinted from [12]

under test, with testing performed using a modified chirp waveform having a center frequency of 3.3 GHz and with an ACPR limit of −28.2 dBc. Figure 12.5 depicts traditionally measured load-pull results for this device while Figure 12.6 shows the results of the fast search algorithm. Here a constrained optimum PAE of 6.67% was found at $\Gamma_L = 0.9\underline{/90^\circ}$, requiring only 17 measurements to locate. Table 12.1 shows the results of the search when performed from multiple starting values of Γ_L. For the five starting points presented in the table, the number of necessary measurements ranges from 11 to 22, which seems reasonable for implementation in a prospective real-time reconfigurable radar transmitter.

A slight adjustment allows this algorithm to make direct use of spectral mask criteria as a constraint instead of the ACPR metric [13], though doing so necessitates

Table 12.1 Algorithm measurement search results for multiple starting points [12]

Start Γ_L	End Γ_L	End PAE (%)	End ACPR (dBc)	# Pts.
$0.9\,\underline{/-90^\circ}$	$0.623\,\underline{/-36.2^\circ}$	6.55	−28.31	13
$0.9\,\underline{/90^\circ}$	$0.592\,\underline{/-4.8^\circ}$	6.67	−28.30	17
$0.9\,\underline{/180^\circ}$	$0.621\,\underline{/-17.2^\circ}$	6.53	−28.28	22
$0.9\,\underline{/0^\circ}$	$0.584\,\underline{/-9.0^\circ}$	6.74	−28.32	11
0	$0.580\,\underline{/-17.7^\circ}$	6.88	−28.28	13

that an associated spectral compliance metric be quantified. Davidson [14] discusses the application of convex optimization based on spectral mask criteria for filter and beamformer design. In [15], Parr describes the generation of time-domain pulses meeting ultra-wideband system spectral mask constraints using the eigenvalues of a transmission matrix equation to quantify the percentage of waveform power inside the spectral mask. Likewise, Sheng [16] shows how waveform parameters can be adjusted based on measurement data to meet spectral mask requirements and Luo [17] discusses pulse shaping for optimization according to spectral mask criteria. The metric defined in [13] quantifies the maximum difference between the spectrum power s_i and the spectral mask m_i via

$$S_{\max} = \max_i (s_i - m_i) \tag{12.12}$$

over all i measured frequencies. Based on (12.12), $S_{\max} > 0$ if the spectrum is out of compliance and $S_{\max} \leq 0$ if the spectrum is compliant.

Optimization using the spectral mask is beneficial for real-time optimization because the metric automatically declares compliance or non-compliance. Search vector construction in this case is similar to that used above and is illustrated in Figure 12.7. Unit vectors $\hat{p}$ and $\hat{m}$ are generated based on the gradients of PAE and $S_{\max}$, respectively. If the amplifier output waveform is out of compliance with the spectral mask for the present candidate solution, the search vector is given by

$$\overline{v} = \hat{m}D_{\max} + \hat{b}D_{\mathrm{b}}, \tag{12.13}$$

where

$$D_{\max} = \frac{D_{\mathrm{S}}}{2}\frac{|S_{\max,\mathrm{meas}}|}{|S_{\max,\mathrm{worst}}|} \tag{12.14}$$

and with D_{b} as defined in (12.10). This search vector is directed toward the point on the Smith Chart boundary of spectral mask compliance that is also on the Pareto optimum locus between PAE and the condition of best spectral mask compliance.

Experimental validation of this search approach was also performed on the Skyworks packaged amplifier using the same load-pull nonlinear test setup as in Figure 12.4. Figure 12.8 shows the result of a search starting from $\Gamma_{\mathrm{L}} = 0.9\,\underline{/-90^\circ}$, where the constrained optimum solution is obtained after only 19 measurements and the final spectral metric value is $S_{\max} = -0.057$ dBc, meaning the closest spectral

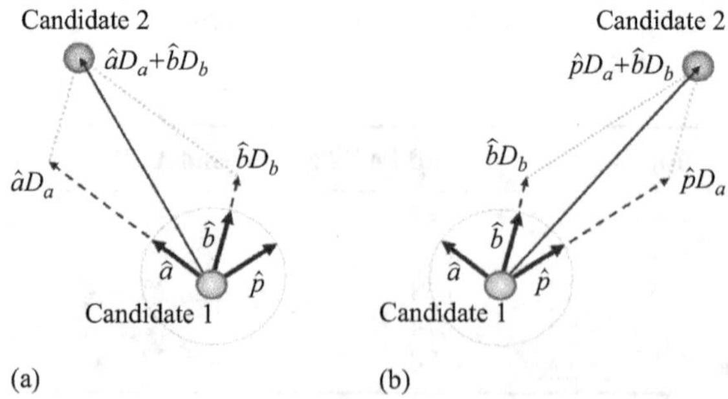

Figure 12.7 Search vector construction for spectral-mask based search when (a) the amplifier output waveform is out of compliance with the spectral mask and (b) the amplifier output waveform is in compliance with the spectral mask Reprinted from [13]

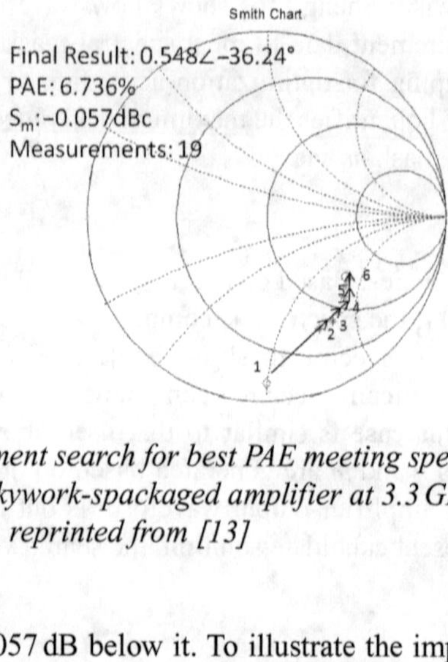

Figure 12.8 Measurement search for best PAE meeting spectral mask compliance using a Skywork-spackaged amplifier at 3.3 GHz and starting at $\Gamma_L = 0.9\underline{/-90^\circ}$, reprinted from [13]

point to the mask is 0.057 dB below it. To illustrate the impact of the optimization process, Figure 12.9 shows that the waveform spectrum is out of compliance with the spectral mask at the start of the search while Figure 12.10 shows that at the search endpoint of $\Gamma_L = 0.548\ \underline{/-36.24^\circ}$ the waveform spectrum is now in compliance, i.e. below the mask.

Table 12.2 shows measurement search results for multiple different starting Γ_L values. The final PAE values obtained are very similar across all the cases, with all search endpoints realizing a small negative dBc value of S_{max} indicating that the resulting waveform spectra closely approach the mask in at least one spectral location but all still achieve compliance with the spectral mask.

The results in this section have shown the usefulness of fast optimization of load impedance (using the load reflection coefficient Γ_L) to maximize efficiency

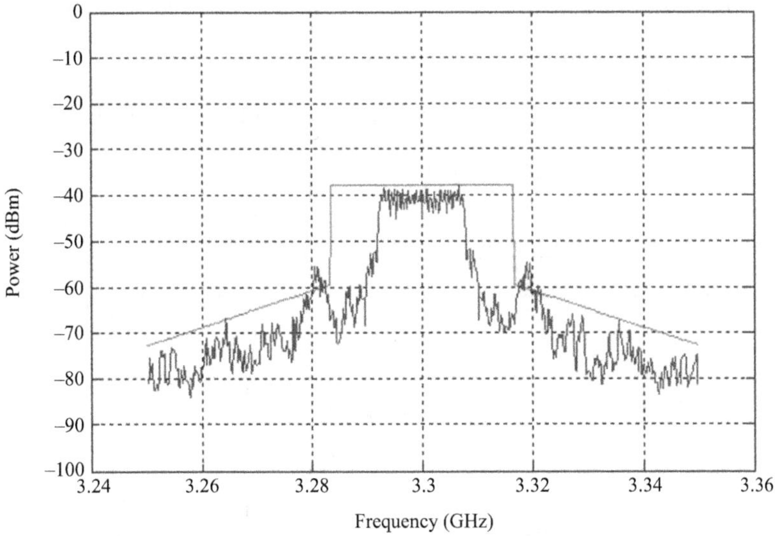

Figure 12.9 Spectral mask and amplifier output waveform spectrum for starting point $\Gamma_L = 0.9\,\underline{/-90^\circ}$, with the spectrum clearly in violation of the mask. Reprinted from [13]

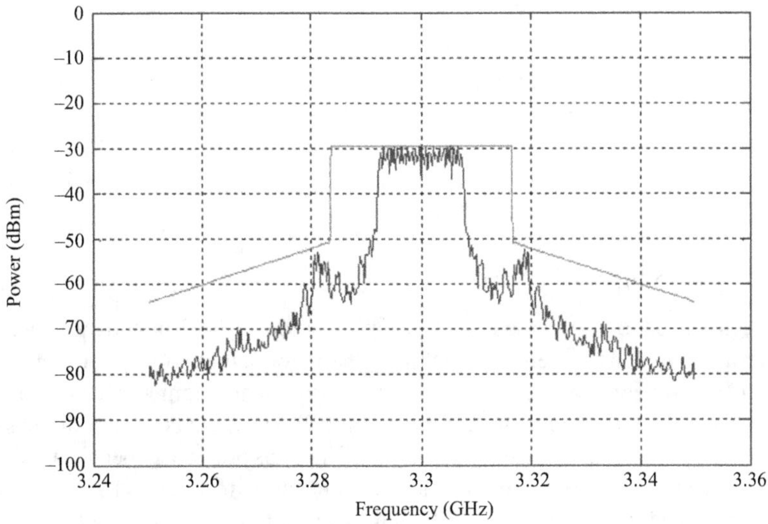

Figure 12.10 Spectral mask and amplifier output waveform spectrum for endpoint value of $\Gamma_L = 0.548\,\underline{/-36.24^\circ}$ after the search in Figure 12.8. The waveform spectrum now complies with the spectral mask. Reprinted from [13]

Table 12.2 Spectral-mask constrained search results for different starting Γ_L [13]

Start Γ_L	End Γ_L	End PAE (%)	End S_{max} (dBc)	# Pts.
$0.9\angle 0°$	$0.570\angle -14.43°$	6.780	−0.322	11
$0.9\angle 90°$	$0.545\angle -17.05°$	6.917	−0.089	26
$0.9\angle 180°$	$0.586\angle -15.54°$	6.659	−0.322	22
$0.9\angle -90°$	$0.548\angle -36.24°$	6.736	−0.057	19
	$0.571\angle -17.44°$	6.772	−0.487	10

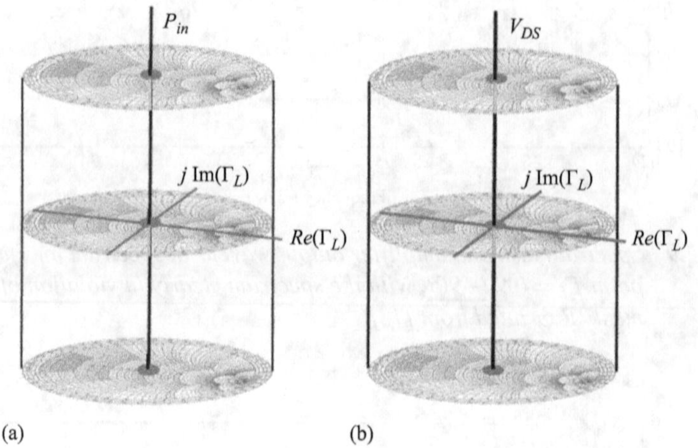

Figure 12.11 Smith Tube representations for (a) power and (b) bias (reprinted from [23] and [24], respectively)

while meeting spectral requirements. The optimization of additional parameters can be accomplished using the Smith Tube as a visualization tool.

12.2.2 The Smith tube

The previous section addressed the fast optimization of load impedance, which is realized as a point on the well-known Smith Chart plot of reflection coefficient, while meeting spectral constraints. The extension to additional parameters can be accomplished using the more recently conceived Smith Tube, which is a three-dimensional (3D) extension of the Smith Chart. The Smith Chart has previously been extended for purposes such as envisioning impedances with negative real parts [18,19], accounting for fractional circuit elements [20], and as a way to represent lossy transmission lines [21]. However, the particular concept of the Smith Tube provides a useful visualization tool for the optimization of additional parameters related to the power amplifier. For example, the vertical axis (the “tube” portion) can be used to represent another parameter such as waveform bandwidth [22], input power [23], or bias voltage [24].

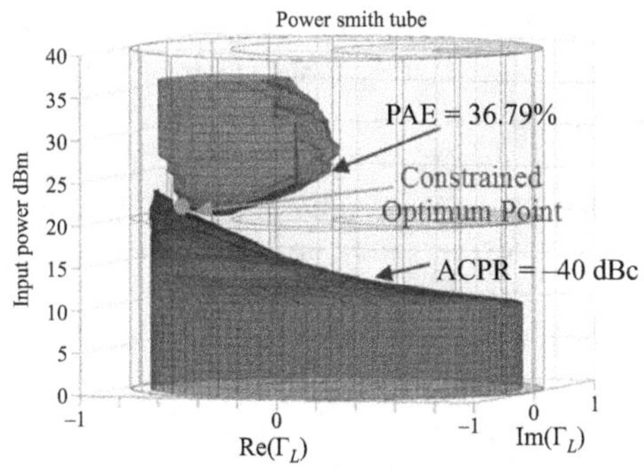

Figure 12.12 Nonlinear FET model for ACPR = −40 dBc and PAE = 36.79% (the constrained optimum value) plotted in the Power Smith Tube, reprinted from [23]

The Power Smith Tube is illustrated in Figure 12.11(a) [23]. Here, the vertical axis represents amplifier input power P_{in}, while the circular cross-section of the tube is the standard Smith Chart. Alternatively, the Bias Smith Tube is shown in Figure 12.11(b) [24], with the vertical axis now representing either a transistor bias voltage (such as V_{GS}, V_{DS}, V_{BE}, or V_{CE}) or an amplifier supply voltage.

The Power Smith tube is a useful design tool because both PAE and ACPR vary according to input power and load reflection coefficient. Snider [25] discussed the variation of the optimal-efficiency load impedance with changing input power level, and Nemati [26] showed that the load impedance that provides optimum PAE is strongly dependent on input power. Likewise, Hajji [27] showed that input power can be changed to meet intermodulation rejection objectives and Nam [28] described load modulation techniques to provide efficiency for Doherty amplifiers.

Figure 12.12 shows the solution to a typical power-amplifier design problem using the Power Smith Tube [23] in which load-pull simulations were performed for a field-effect transistor (FET) model in Keysight Technologies' Advanced Design System at different values of input power P_{in} ranging from 0 to 40 dBm. A sample design goal would be to determine the combination of Γ_L and P_{in} that provides the highest PAE while satisfying the constraint ACPR ≤ -40 dBc. To visualize the design problem, a 3D surface of constant ACPR $= -40$ dBc was fit to the load-pull data, thereby providing the 3D equivalent of a two-dimensional (2D) ACPR contour (such as depicted in Figure 12.5(a)). This 3D surface also illustrates the locus of points over different Γ_L and P_{in} values that meet this ACPR constraint. The highest value of PAE contained within this surface is the constrained optimum solution of 36.79 percent, as shown in Figure 12.12. Figure 12.12 also shows that the constant-PAE surface corresponding to this value and the ACPR $= -40$ dBc surface meet at only one point, which is the constrained optimum point.

This type of visualization is useful to power amplifier designers as well as for real-time reconfigurable optimization. In a typical power amplifier design, the designer would perform a 50 Ω power sweep to gain a general understanding of the output power, efficiency, and ACPR performance as a function of input power variation. The designer would then select a reasonable value of input power (as pertains to these objectives) and perform a load-pull measurement to select a value of Γ_L that optimizes output power, efficiency, or ACPR. Using this Γ_L, another power sweep would likely be performed because the gain compression characteristic is different for different values of Γ_L. The alternating process would likely be repeated until the designer is satisfied with the results. The Power Smith Tube removes the need for iterative power-sweep and load-pull measurements by providing a visualization of the entire design problem in one picture.

12.2.3 Simultaneous optimization of load impedance and input power using the power Smith tube

One disadvantage of performing power-amplifier design using the Power Smith Tube is the time required to measure the data necessary to populate the plot. Certainly, obtaining an exhaustive collection of power-dependent load-pull data is out of the question for real-time transmitter reconfigurability. However, performing a fast search within the Power Smith Tube can quickly determine the optimum solution, thereby speeding up both design and measurement.

This search algorithm is constructed using an extension of the 2D optimization of Γ_L [29]. To construct the Power Smith Tube, input power values on the interval $P_{in,min} \leq P_{in} \leq P_{in,max}$ are selected for the optimization. Normalized input power is then computed via

$$p_{in} = 2\left(\frac{P_{in} - P_{in,min}}{P_{in,max} - P_{in,min}}\right) - 1, \tag{12.15}$$

which maps $P_{in} = P_{in,max}$ to $p_{in} = 1$ and $P_{in} = P_{in,min}$ to $p_{in} = -1$. This normalization is important because it scales the optimization range to fit a cylinder of radius 1 and with the vertical axis defined from -1 to $+1$.

Extending the previous 2D optimization to 3D, the PAE gradient is now determined by expanding (12.3) as

$$\nabla p = \hat{\Gamma}_r \frac{\partial p}{\partial \Gamma_r} + \hat{\Gamma}_i \frac{\partial p}{\partial \Gamma_i} + \hat{p}_{in} \frac{\partial p}{\partial p_{in}}, \tag{12.16}$$

which is in terms of the $\hat{\Gamma}_r$, $\hat{\Gamma}_i$, and $\hat{p}_{in}$ coordinates. The derivatives in (12.16) can be estimated by taking measurements with small separation D_n from the candidate in the Γ_r, Γ_i, and p_{in} directions, as shown in Figure 12.13, such that [29]

$$\frac{\partial p}{\partial \Gamma_r} \approx \frac{(\Delta p)_{\Gamma_r}}{\Delta \Gamma_r} = \frac{(\Delta p)_{\Gamma_r}}{D_n} \tag{12.17}$$

$$\frac{\partial p}{\partial \Gamma_i} \approx \frac{(\Delta p)_{\Gamma_i}}{\Delta \Gamma_i} = \frac{(\Delta p)_{\Gamma_i}}{D_n} \tag{12.18}$$

$$\frac{\partial p}{\partial p_{in}} \approx \frac{(\Delta p)_{p_{in}}}{\Delta p_{in}} = \frac{(\Delta p)_{p_{in}}}{D_n}. \tag{12.19}$$

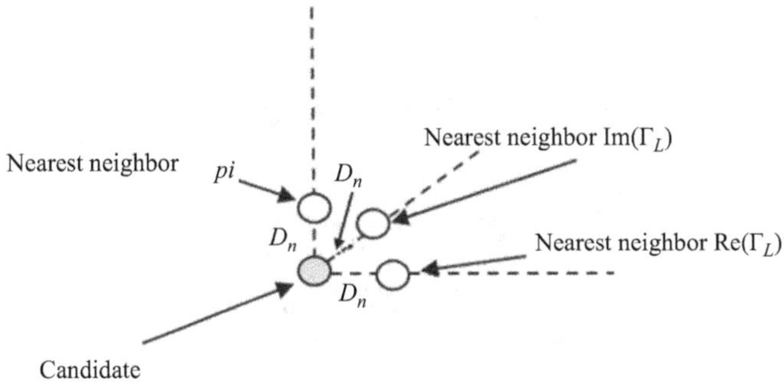

Figure 12.13 Gradient estimation for the Smith Tube, reprinted from [29]

The ACPR gradient ∇a can be estimated in a similar manner. The PAE and ACPR optimum directions are then ascertained similar to the 2D case via

$$\hat{p} = \frac{\nabla p}{|\nabla p|} \tag{12.20}$$

$$\hat{a} = -\frac{\nabla a}{|\nabla a|}. \tag{12.21}$$

The 3D search vector is constructed in the same way as before, albeit in three dimensions. The bisector unit vector $\hat{b}$ is the average of the 3D unit vectors $\hat{a}$ and $\hat{p}$ as [29]

$$\bar{b} = q\frac{1}{2}(\hat{a} + \hat{p}), \tag{12.22}$$

thus resulting in the bisector unit vector

$$\hat{b} = \frac{\bar{b}}{|\bar{b}|}. \tag{12.23}$$

Equations (12.8) through (12.11) are then applied to find the 3D search vector.

Experimental assessment of this algorithm was performed using the load-pull test bench for a Skyworks SKY5017-70LF InGaP packaged amplifier, with the ACPR limit set to −27.5 dBc. Figure 12.14 illustrates the constant ACPR = −27.5 dBc surface along with the measured PAE optimum that satisfies the constraint [29]. Figure 12.15 shows the search progression for starting location $\Gamma_L = 0.9 \underline{/180^\circ}$ and $P_{in} = -4$ dBm.

Table 12.3 demonstrates that good convergence was also obtained for a variety of different starting combinations of Γ_L and P_{in} throughout the Smith Tube. All final values of ACPR are just below the limit of −27.5 dBc and all final PAE values are between 7.0% and 7.5%. In other words, the convergence is consistent regardless of the starting point.

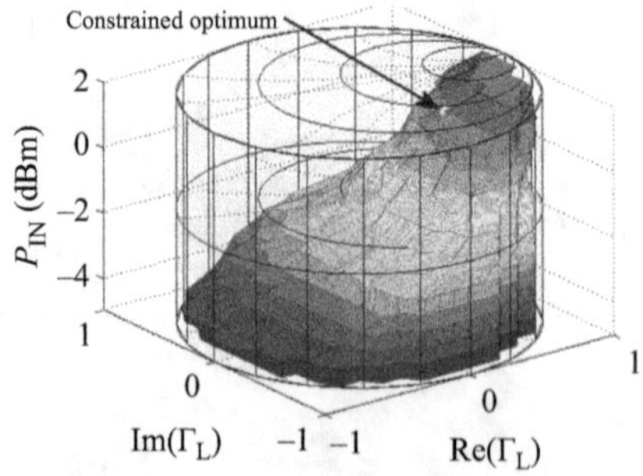

Figure 12.14 Constant-ACPR surface for the limiting ACPR value of −27.5 dBc and the corresponding maximum PAE = 7.39% found at $\Gamma_L = 0.47 \underline{/-41.5^\circ}$ and $P_{in} = 1.5$ dBm. Reprinted from [29]

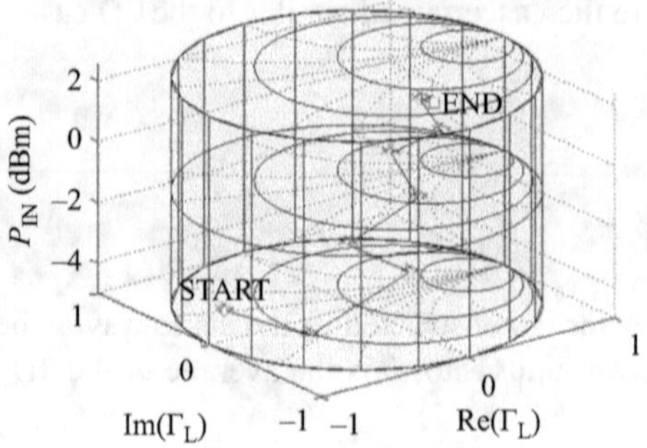

Figure 12.15 Search algorithm path with search parameters $D_s = 1$ and $D_n = 0.05$, input power range $-5\text{ dBm} \le P_{in} \le 2\text{ dBm}$, and starting location $\Gamma_L = 0.9 \underline{/180^\circ}$ and $P_{in} = -4$ dBm. The search converged to $\Gamma_L = 0.42 \underline{/-40.33^\circ}$ and $P_{in} = 1.67$ dBm, providing PAE = 7.23% and ACPR = −27.80 dBc after 39 measured points. Reprinted from [29]

Table 12.3 Power Smith tube search results for different initial Γ_L and P_{in} [29]

Start Γ_L	Start P_{in} (dBm)	End Γ_L	End P_{in} (dBm)	End ACPR (dBm)	End PAE (%)	# Meas
$0.6 \underline{/-135^\circ}$	2.0	$0.42 \underline{/-44^\circ}$	1.60	−27.55	7.32	31
0	0.0	$0.45 \underline{/-47^\circ}$	1.58	−27.86	7.05	23
$0.8 \underline{/0^\circ}$	−3.0	$0.43 \underline{/-24^\circ}$	1.57	−27.51	7.49	35
$0.9 \underline{/180^\circ}$	−4.0	$0.42 \underline{/-40^\circ}$	1.67	−27.80	7.23	39
$0.75 \underline{/45^\circ}$	1.5	$0.39 \underline{/-38^\circ}$	1.21	−27.51	7.29	24

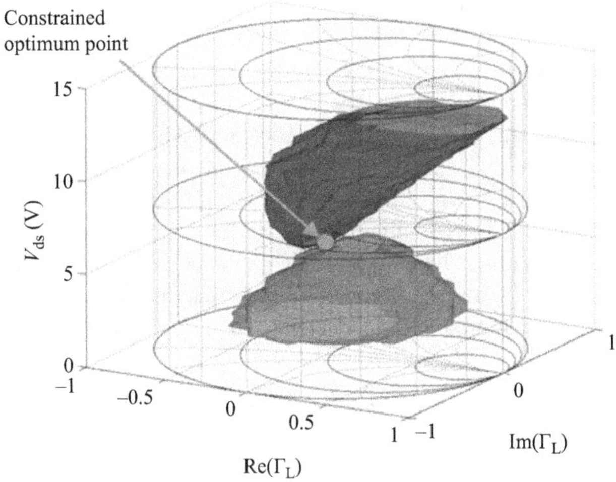

Figure 12.16 Simulated ACPR = −45 dBc surface (top) and maximum constrained PAE surface (bottom) for a Modelithics model of an Excelics EFA060BS5 MESFET. The constrained optimum PAE was found to be 35.79% at $\Gamma_L = 0.1\ \underline{/165^\circ}$ and $V_{ds} = 5.5$ V. Reprinted from [24]

12.2.4 Simultaneous optimization of load impedance and bias voltage using the bias Smith tube

The Bias Smith Tube depicted in Figure 12.11(b) facilitates simultaneous optimization according to load reflection coefficient Γ_L and a given bias voltage [24], which is useful due to the dependence of both PAE and ACPR on bias voltages. Examples of these dependencies include techniques for gate voltage envelope modulation [30], improvement of PAE by tuning drain voltage and current [31], evaluation of nonlinear performance as a function of input and output bias voltage [32], and optimization of bias and RF input power for efficiency [33]. The load resistance has also been shown to have a significant impact on the bias setting and power efficiency [34]. The optimum load impedance in terms of output power and power efficiency has been found to vary on the Smith Chart as bias voltage is adjusted [35], and the optimum PAE has been demonstrated to be a significant function of the bias setting [36].

For instance, Figure 12.16 presents Bias Smith Tube simulation results for a Modelithics nonlinear transistor model of an Excelics EFA060BS5 metal-semiconductor field-effect transistor (MESFET) according to the drain-to-source voltage V_{ds} (shown on the vertical axis). The design goal is to maximize PAE while maintaining ACPR ≤ -45 dBc.

In general, the PAE and ACPR surfaces in the Smith Tube match intuitive expectations. For a constant gate-to-source voltage V_{GS}, the PAE is expected to be highest when V_{DS} is low because increased V_{DS} leads to increased DC power, which subsequently decreases PAE. As such, constant-PAE surfaces are expected to have large

Smith Chart coverage at the bottom of the V_{DS} Smith Tube and then decrease coverage higher on the Smith Tube. However, linearity is expected to improve as V_{DS} is increased, thus allowing more of the Smith Chart to meet an ACPR requirement. Consequently, a constant-ACPR surface tends to increase its coverage higher on the Smith Tube. In addition, the center of the constant-ACPR surface depicted in Figure 12.16 migrates to higher resistance areas as V_{DS} is increased, meaning that the load-line slope that best provides linearity is lowered [24].

Both the Power and Bias Smith Tubes allow for additional design parameters to be optimized jointly with the load impedance. As in the case of the Power Smith Tube, an efficient algorithm is required to achieve real-time reconfigurability using the Bias Smith Tube. One such approach was developed in [37].

12.2.5 Optimization of PAE under multiple constraints based on output power and ACPR

Some reconfigurable transmitter situations may impose multiple constraints on the optimization. For instance, in addition to ACPR, output power is often a significant consideration for radar operation. Specifically, the transmitted power produced by the radar, which is directly related to the output power from the power amplifier, must be large enough to ensure sufficient "energy on target" for subsequent detection in noise and interference. Therefore, the amplifier may also be constrained to achieve a minimum delivered power P_d in addition to a maximum ACPR constraint [38].

Output power is often used as a metric for nonlinear power-amplifier performance. In [39], Sevic uses load-pull measurements to minimize ACPR and maximize PAE for a given output-power requirement. Iwamoto [40] shows that PAE and ACPR vary with output power for a CDMA-excited Doherty amplifier. Zhang [41] and Noh [42] state that PAE, ACPR, and output power are often used as metrics for linearization, while Kusunoki [43] presents the relationship between power back-off and ACPR, as well as the relationship that output power has with both PAE and ACPR. Reveyrand [44] shows how fast load-pull optimization can be used to maximize PAE under constraints on output power and dissipated power through the use of models of Smith Chart dependent behavior.

To perform this type of search on the Smith Chart, the gradients for PAE, ACPR, and delivered power can be estimated using the approach from Figure 12.2. In this multi-criteria optimization, the search vector is constructed according to the location of the initial candidate reflection coefficient Γ_L [38]. If the candidate possesses ACPR and P_d values that are both noncompliant with the constraints, then the search vector is chosen to be a combination of the ACPR direction $\hat{a}$ and the delivered-power steepest-ascent unit direction $\hat{d}$ via

$$\bar{v} = \hat{a}D_a + \hat{d}D_d, \tag{12.24}$$

where D_a is given by (12.8) and

$$D_d = \frac{D_s}{2}\frac{|P_{d,cand} - P_{d,limit}|}{|P_{d,worst} - P_{d,limit}|}, \tag{12.25}$$

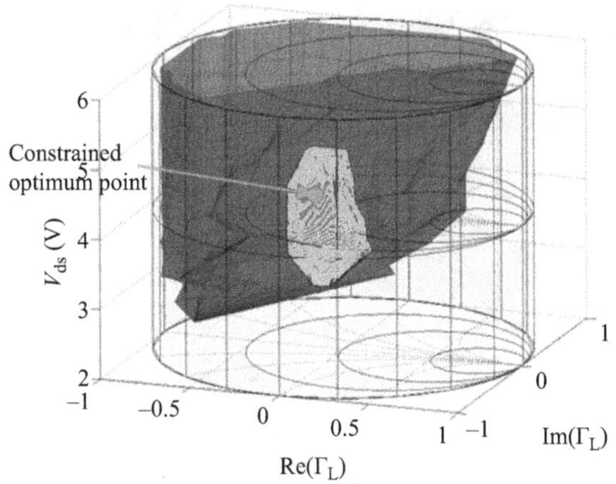

Figure 12.17 Search vector construction when the originating candidate is outside both the ACPR and delivered-power acceptable regions, reprinted from [38]

for $P_{d,cand}$ the delivered power value at the candidate point, $P_{d,limit}$ the P_d constraint value, and $P_{d,worst}$ the lowest value of delivered power over all measurements performed in the search. In (12.24), D_a and D_d are estimates of the fraction of remaining distance in the search for the respective ACPR and P_d constraint values, scaled by the search-distance parameter D_s. The resultant vector $\overline{v}$ directs the search toward the mutually acceptable region for both ACPR and P_d, as illustrated by Figure 12.17 [38].

If a candidate point is in the region of acceptable P_d but is outside the ACPR limit, then the search vector is constructed as [38]

$$\overline{v} = \hat{a}D_a + \hat{b}_{ad}D_{b,ad}, \tag{12.26}$$

where $\hat{b}_{ad}$ is the bisector between $\hat{a}$ and $\hat{d}$, which points toward the Pareto optimum locus between P_d and ACPR. Here $D_{b,ad}$ is an estimate of the relative remaining distance to the Pareto optimum locus for ACPR and P_d, and is defined as

$$D_{b,ad} = \frac{D_s}{2}\frac{|\theta_{ad} - 90°|}{90°}, \tag{12.27}$$

where θ_{ad} is the angle between $\hat{a}$ and $\hat{b}_{ad}$. This case is illustrated in Figure 12.18 [38].

Alternatively, if a candidate meets the ACPR condition but has an unacceptable P_d, the search vector is constructed as

$$\overline{v} = \hat{d}D_d + \hat{b}_{ad}D_{b,ad} \tag{12.28}$$

for $D_{b,ad}$ defined in (12.27). The vector in (12.28) will guide the search toward the acceptable P_d region and also the Pareto optimum locus between the acceptable ACPR and P_d regions, as depicted in Figure 12.19 [38].

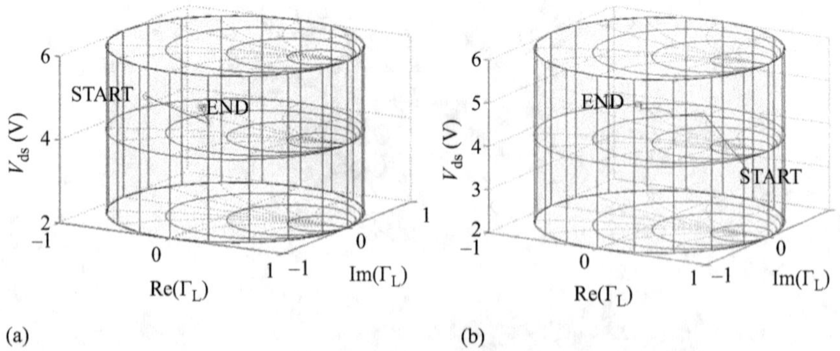

Figure 12.18 Search vector construction when the originating candidate meets the P_d constraint but is outside the ACPR-acceptable region, reprinted from [38]

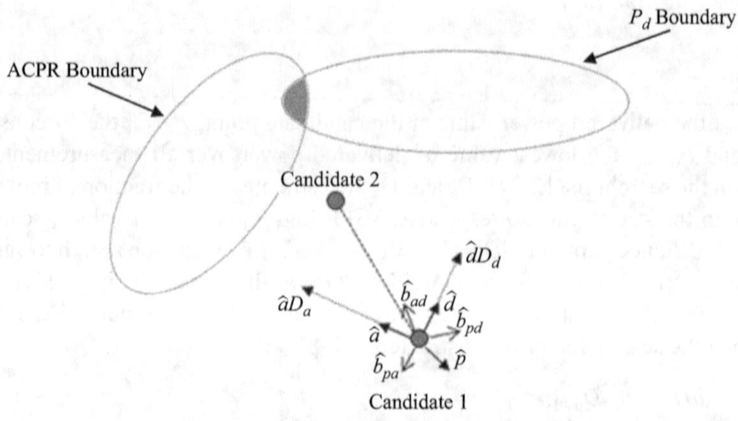

Figure 12.19 Search vector construction when the originating candidate meets the ACPR constraint, but is outside the P_d-acceptable region, reprinted from [38]

Finally, if the candidate point lies in both the ACPR- and P_d-acceptable regions, then the search proceeds toward the constraint boundary that produces the higher PAE. Note that if the smallest angle between $\hat{a}$ and $\hat{p}$ is less than the smallest angle between $\hat{d}$ and $\hat{p}$, then it is expected that the constrained optimum will lie on the boundary of the P_d-acceptable region, as shown in Figure 12.20. The search progresses toward the intersection of the Pareto front joining the P_d and PAE optima with the boundary of the P_d-acceptable region. In this case, the search vector is

$$\overline{v} = \hat{p}D_\mathrm{d} + \hat{b}_\mathrm{pd}D_\mathrm{b,pd}, \tag{12.29}$$

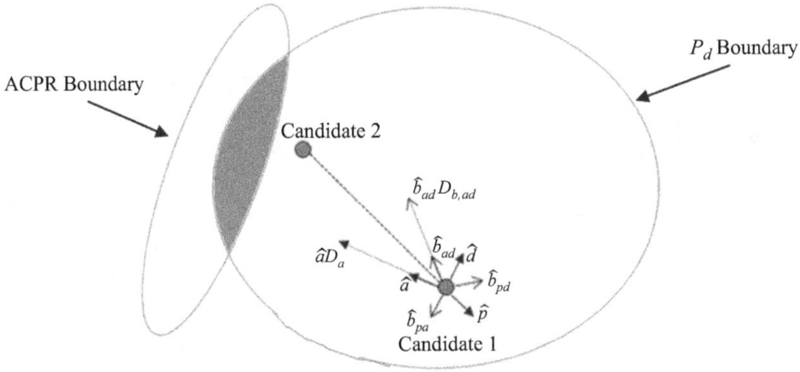

Figure 12.20 Search vector construction when the originating candidate meets both the P_d and ACPR constraints, where the smallest angle between $\hat{a}$ and $\hat{p}$ is less than the smallest angle between $\hat{d}$ and $\hat{p}$. Reprinted from [38]

where

$$D_{\text{b,pd}} = \frac{D_{\text{s}}}{2} \frac{|\theta_{\text{pd}} - 90^\circ|}{90^\circ} \tag{12.30}$$

for θ_{pd}, the angle between $\hat{d}$ and $\hat{b}_{\text{pd}}$ [38].

Conversely, if the candidate point lies in both the ACPR- and PAE-acceptable regions, but the smallest angle between $\hat{a}$ and $\hat{p}$ is greater than the smallest angle between $\hat{d}$ and $\hat{p}$, then it is expected that the constrained optimum will lie on the boundary of the ACPR-acceptable region, as shown in Figure 12.21. In this case the search moves toward the intersection of the Pareto front joining the ACPR and PAE optima with the boundary of the ACPR-acceptable region. The search vector is thus

$$\overline{v} = \hat{p}D_{\text{a}} + \hat{b}_{\text{pa}}D_{\text{b,pa}}, \tag{12.31}$$

with

$$D_{\text{b,pa}} = \frac{D_{\text{s}}}{2} \frac{|\theta_{\text{pa}} - 90^\circ|}{90^\circ}, \tag{12.32}$$

and θ_{pa} the angle between $\hat{a}$ and $\hat{b}_{\text{pa}}$ [38].

Penalty functions can be designed to keep the search inside the intersection of the P_{d}- and ACPR-acceptable regions. For example, if the search is following a compliant trajectory and then becomes non-compliant, one could return to the previous compliant candidate, divide the step-size D_{s} by 2, and then perform the move again. The same strategy can be employed if the value of PAE decreases from one candidate to the next. The search is stopped when $|\overline{v}|$ is less than a pre-specified resolution distance D_{r}.

This multi-criteria search was evaluated on the load-pull measurement bench for the Skyworks packaged amplifier at a center frequency of 3.3 GHz, using constraints

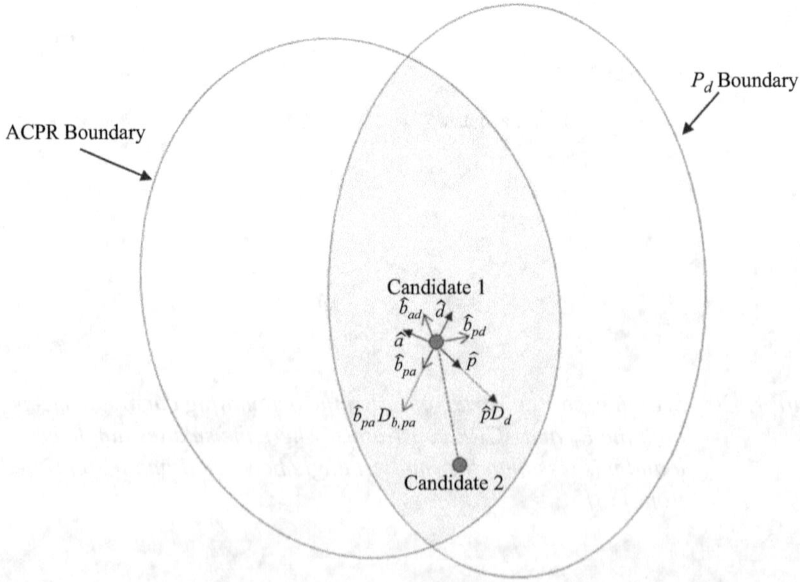

Figure 12.21 Search vector construction when the originating candidate meets both the P_d and ACPR constraints, where the smallest angle between â and p̂ is greater than the smallest angle between $\hat{d}$ and p̂. Reprinted from [38]

ACPR ≤ −27.5 dBc and $P_d \geq 17$ dBm. The search was tested using multiple different starting Γ_L values throughout the Smith Chart, for a modified chirp waveform of 16 MHz bandwidth as the amplifier excitation. The main and adjacent channels were defined as 10 MHz wide, with the adjacent channel separated from the center frequency by 20 MHz [38].

Figure 12.22 shows traditional load-pull results measured for this device, with the intersection of the ACPR and P_d-acceptable regions shaded [38]. In contrast, Figure 12.23 shows two fast load-pull searches from the different starting locations of $\Gamma_L = 0.9\ \underline{/180^\circ}$ (in Figure 12.23(a)) and $\Gamma_L = 0.9\ \underline{/90^\circ}$ (in Figure 12.23(b)) [38]. It is observed that the search makes some adjustments due to entering regions with different search-vector definitions based on the acceptable region(s) in which a candidate resides. Table 12.4 summarizes these results from multiple starting Γ_L values within the Smith Chart [38]. The table shows that all endpoints are near the ACPR limit of −27.5 dBc while the final P_d values exceed the minimum by at least 1.4 dB, thus indicating that ACPR is the limiting constraint in this search. The end values of PAE correspond well for the different starting points. That said, the number of measurements range from 10 to 28, showing an expected dependence on the number of steps required based on the proximity of the starting point to the constrained optimum.

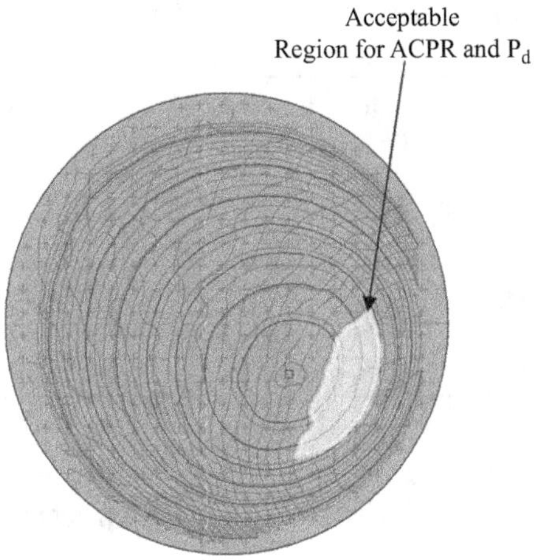

Figure 12.22 PAE, ACPR, and P_d contours measured using a traditional load-pull approach. The intersection of the ACPR and P_d acceptable regions is shaded. Reprinted from [38]

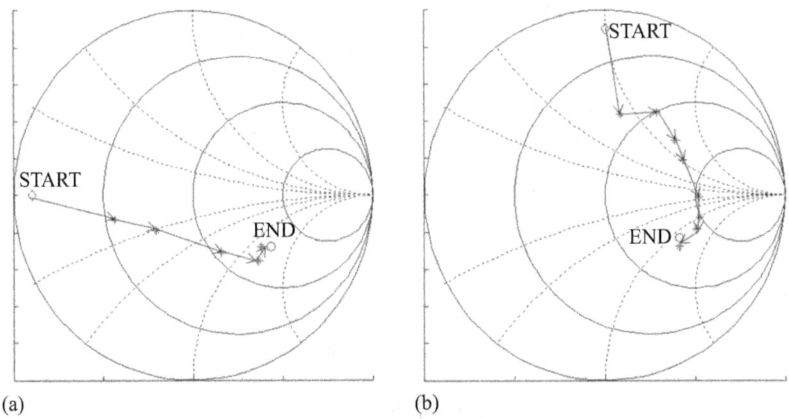

Figure 12.23 (a) Fast measurement search results starting from $\Gamma_L = 0.9\,\underline{/180^\circ}$. The constrained optimum is found at $\Gamma_L = 0.52\,\underline{/-33^\circ}$, providing PAE = 7.59%, ACPR = −27.85 dBc, and P_d = 18.44 dBm after 19 measurements. (b) Fast measurement search results starting from $\Gamma_L = 0.9\,\underline{/90^\circ}$. The constrained optimum is found at $\Gamma_L = 0.47\,\underline{/-29^\circ}$, providing PAE = 7.82%, ACPR = −27.58 dBc, and P_d = 18.57 dBm after 28 measurements. Reprinted from [38]

Table 12.4 Dual-constrained search results for different initial Γ_L values [38]

Start Γ_L	Start ACPR (dBc)	Start P_d (dBm)	Start PAE (%)	End Γ_L	End ACPR (dBc)	End P_d (dBm)	End PAE (%)	# Meas
0	−24.33	17.90	6.70	$0.47\angle -37^\circ$	−27.77	18.54	7.78	17
$0.9\angle 0^\circ$	−28.69	10.31	1.02	$0.46\angle -33^\circ$	−27.51	18.62	7.92	11
$0.9\angle 90^\circ$	−22.66	8.25	0.57	$0.47\angle -29^\circ$	−27.58	18.57	7.82	28
$0.9\angle 180^\circ$	−20.63	9.25	0.75	$0.52\angle -33^\circ$	−27.85	18.44	7.59	19
$0.9\angle -90^\circ$	−23.23	11.79	1.44	$0.51\angle -51^\circ$	−27.56	18.38	7.46	10

12.3 Radar waveform optimization with spectral mask compliance

While circuit optimization techniques provide the means to quickly reconfigure the transmitter to maximize power efficiency and linearity according to spectral requirements, it is really the characteristics of the resulting emitted waveform that dictates radar performance. The waveform plays a significant role in all three objectives of the adaptive radar. First, the waveform determines the radar's range and (fast-time) Doppler resolution, as illustrated by the well-known ambiguity function (AF). Second, the waveform impacts spectral performance, with a narrowband waveform being much more likely to remain within a given spectral mask than a wideband waveform. Finally, the waveform significantly impacts the PAE based on the associated peak-to-average power ratio (PAPR), where waveforms with low PAPR permit much greater efficiency than the converse. In this section, the waveform AF is briefly described, followed by an alternating-projections approach to optimizing the radar waveform for certain range-Doppler ambiguity objectives, all while meeting spectral mask requirements and providing acceptable PAPR.

12.3.1 Woodward's ambiguity function

First derived by Woodward [45], the AF associated with a time-domain signal $x(t)$ describes the resulting matched filter response as a function of relative delay τ and Doppler frequency shift u according to [46,47]

$$\chi(\tau,u) = \int_{-\infty}^{\infty} x(t)x^*(t-\tau)e^{-j2\pi ut}\,dt. \tag{12.33}$$

The AF can also be calculated from the Fourier transform of $x(t)$, denoted as $X(f)$, via

$$\chi(\tau,u) = \int_{-\infty}^{\infty} X(f+u)X^*(f)e^{j2\pi f t}\,df. \tag{12.34}$$

Radar waveform design has employed the AF since the 1960s. In [48], Wilcox outlined an ambiguity-function synthesis technique using least-squares estimation to approximate a desired cross-ambiguity function. Sussman [49] subsequently removed some of the function-type limitations used in Wilcox's approximation. Wolf [50] then showed how a pattern search can be used with basis functions consisting of phase-modulated waveforms to optimize the AF. Costas [51] also demonstrated methods to estimate the desired 2D impulse of the AF.

More recently, Gladkova and Chebanov [52,53] extended Wilcox's method to demonstrate Hermite waveforms, though they can be difficult to use [54,55]. Sebt [54] extended the work of Sussman to include orthogonal frequency division multiplexing (OFDM) with PAPR as a criterion. In [56], Patton discussed joint optimization of the waveform and receive filter according to autocorrelation and cross-correlation constraints, as well as how a waveform can be synthesized based on a desired spectrum [57]. He also described optimization for constraints imposed both on the autocorrelation and modulus of a radar waveform [58].

12.3.2 Alternating projections optimization

Waveform synthesis methods described in the literature have often incorporated the concept of projections. Selesnick and Pillai performed alternating projections for the synthesis of notched chirp waveforms to cancel intermodulation [59,60]. Kassab [61] used alternating projections to adjust the autocorrelation properties of the radar waveform, albeit without taking the entire ambiguity plane into account. Jakabosky and Blunt [62–65] showed how projections can be used for radar waveform design, even with an amplifier "in the loop." Jakabosky, Blunt, and Higgins [66] demonstrated the use of projections to design amplitude-tapered waveforms for use in a linear amplification using a nonlinear components (LINC) amplifier configuration. Nobakht and Civinlar [67] demonstrated the use of projections to place signal energy strategically in both the time and frequency domains. Bin [68] developed a projection-based algorithm that provides a waveform with constant modulus based on a desired Fourier transform magnitude. Much earlier than all of these, yet quite relevant, is the Gerchberg–Saxton method [69] that facilitates waveform synthesis based on auto- and cross-correlation features.

Here we consider the projection of the AF onto an ambiguity minimization template function [70,71] that specifies regions of the range-Doppler ambiguity plane that are to be minimized. A sample of such a minimization template is shown in Figure 12.24 [72]. To meet the requirements of the minimization template, the AF $\chi(\tau, u)$ must satisfy

$$\frac{|\chi(\tau,u)|}{|\chi(0,0)|} \leq M(\tau,u), \quad \begin{array}{c} -B \leq u \leq B \\ -T \leq \tau \leq T \end{array}, \tag{12.35}$$

where $M(\tau, u)$ is the minimization template, $\pm B$ denotes the limits of Doppler frequency under consideration, and $\pm T$ likewise denotes the limits of delay under consideration. The minimization template can be derived from considerations such as required range resolution and Doppler tolerance.

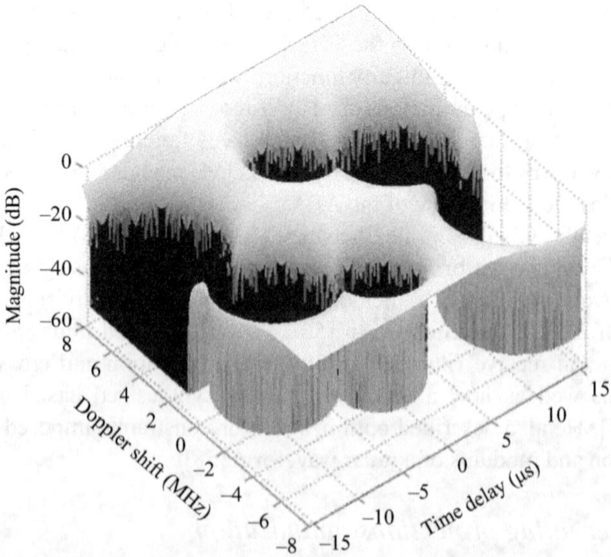

Figure 12.24 Example range-Doppler minimization template function. When used for optimization, the ambiguity is placed in the regions that are not nulled (the circular areas). Reprinted from [72]

In general, the alternating projections approach involves successive interleaved projections onto different sets. Using an initial waveform and AF, the implementation here involves iterative projections onto the following sets:

1. The set of waveforms complying with the minimization template (by removing the minimized parts of the present AF) given by

$$\Phi_i(\tau,u) = P_{\mathcal{M}}(\chi_i) = \begin{cases} \chi_i(\tau,u)\dfrac{M(\tau,u)}{|\chi_i(\tau,u)|}, & (\tau,u)\in\mathscr{B} \\ \chi_i(\tau,u), & (\tau,u)\notin\mathscr{B} \end{cases}, \tag{12.36}$$

   where $\mathscr{B}$ is the set containing combinations of (τ,u) in which the condition in (12.35) is not satisfied;
2. The set of waveforms with valid AFs, obtained by integrating the AF along the τ axis and along the u axis via

$$x(\tau) = \frac{1}{x^*(0)}\int_{-\infty}^{\infty}\chi^*(-\tau,u)du \tag{12.37}$$

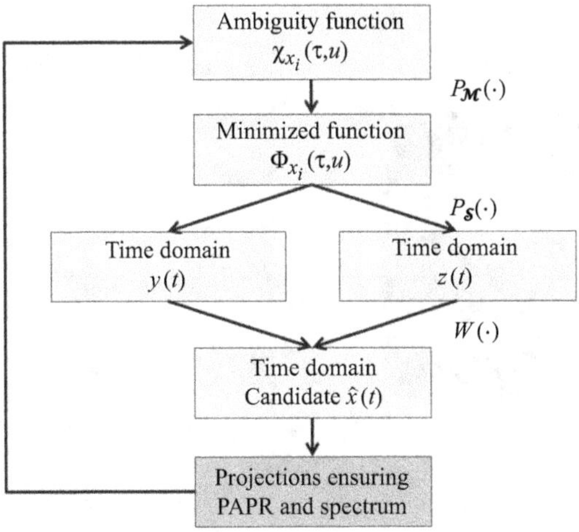

Figure 12.25 Alternating projections waveform optimization involving projections onto the sets of waveforms providing AFs, minimized template functions, acceptable PAPR, and spectral mask compliance. Reprinted from [72]

and

$$X(u) = \frac{1}{X^*(0)} \int_{-\infty}^{\infty} \chi(t,u)dt, \tag{12.38}$$

respectively, where the inverse Fourier transform of $X(u)$ can be performed and then averaged with the time domain version using an appropriate weighting as described in [72];

3. The set of waveforms meeting the energy, PAPR, and spectrum requirements, in which the spectral mask projection is the same as using the spectral mask as a filter for the waveform.

This alternating projections process is outlined in Figure 12.25 and is performed until some stopping criterion is reached. The success of the algorithm in meeting AF requirements can be assessed using either a least-squares or minimax comparison with the minimization template.

Amplifier distortion can cause an optimized waveform to deviate from its optimal structure. A simple way to compensate for this distortion is through the use of predistortion models, of which there are many different kinds. The method considered here uses the hyperbolic tangent relationship [72]

$$y(t) = a_1 \tanh(a_2 x(t)). \tag{12.39}$$

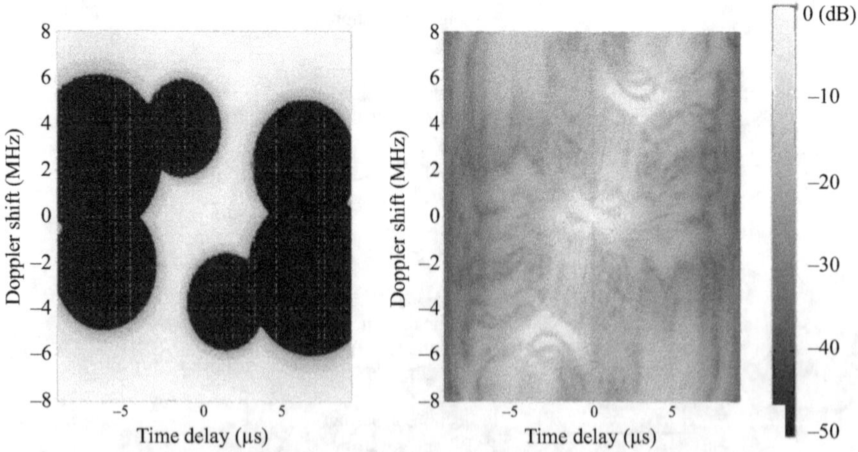

Figure 12.26 Example of waveform optimization using the alternating projections process where (a) is the desired minimization template and (b) is the resulting optimized AF. Reprinted from [72]

After measurements to estimate a_1 and a_2 for a given amplifier, the input waveform $x(t)$ needed to obtain a desired output waveform $y(t)$ can be calculated using the inverse equation [72]

$$x(t) = \frac{1}{a_2}\tanh^{-1}\left(\frac{1}{a_1}y(t)\right). \tag{12.40}$$

Figure 12.26(b) shows the AF that results from applying the optimization process above using the minimization template in Figure 12.26(a). Comparison shows that the resulting ambiguity does a good job of avoiding the minimization regions. The minimization template in (12.35) can also change dynamically with time, with the waveform updated accordingly, which provides the flexibility to account for changing target positions and velocities.

12.4 Adjusting the radar spectral mask based on surrounding wireless devices

Future radar systems will almost certainly exist in an environment that permits dynamic spectrum access for a significant portion of the available spectrum. The incorporation of sharing in the upper part of the S-band radar allocation has already illustrated that transmitter/waveform flexibility will be needed to quickly respond to changing spectral environments containing wireless communication systems. To that end, a dynamic spectral mask can be constructed for the radar emission based on the relative locations and maximum interference power levels of nearby wireless systems [73]. This knowledge would need to be made available to the radar through some type of spectrum sensing capability or via side-channel information conveyed by the wireless network. While outside the scope of this chapter, this manner of infrastructure based on cognitive radio networks [74] shows great promise.

12.4.1 Calculation of a dynamic radar spectral mask

The concept of spectrum shaping using a flexible spectral mask is discussed by Mahmoud [75] for the purpose of enabling or disabling subcarrier sets in OFDM communications. Poston [76] describes how OFDM signals can be inserted between television channel allocations, and Cabric [77] discusses the use of an OFDM transmission mask to protect neighboring channel users from interference. In [78], Srinivasa proposes similar requirements on secondary spectrum users to maintain interference below the acceptable noise floor for the primary users.

The well-known Friis transmission equation establishes the received power as [79]

$$P_{\mathrm{r}} = \frac{G_{\mathrm{t}} G_{\mathrm{r}} \lambda^2}{(4\pi R)^2} P_{\mathrm{t}} \tag{12.41}$$

based on wavelength λ, the distance from the transmitter R, the transmitter antenna gain G_{t}, and the receiver antenna gain G_{r}. Consider free-space propagation where f_n is the frequency of the nth communication receiver, c is the speed of light, and assume an isotropic antenna pattern simplification for both the transmitter and receiver, i.e. $G_{\mathrm{t}} = G_{\mathrm{r}} = 1$. Under this simplifying condition (12.41) can be easily solved in terms of the maximum allowed transmit power in frequency f_n given a maximum allowed interference power p_n at the nth receiver as

$$P_{\mathrm{t}}(f_n) \leq \left(\frac{4\pi R_n f_n}{c} \right)^2 p_n. \tag{12.42}$$

which can be practically applied using a power-per-bandwidth approach that is easily measurable via a spectrum analyzer or with a filter and power sensor.

Given the determination of the maximum transmitted power-per-bandwidth based on (12.42) at the operating frequencies of neighboring wireless devices, a spectral mask could be formed for the radar transmitter like the example in Figure 12.27. As shown, a power margin could also be included to mitigate the prospect of constructive interference due to multipath effects. This arrangement could enable radars to operate as a secondary spectrum user in bands that have been allocated to wireless communications for primary use. Of course, the radar would very clearly have considerably lower performance capabilities in this instance relative to cases in which it is the primary user.

12.4.2 Radar waveform optimization using a dynamic spectral mask

Waveform optimization, such as via the method described above, can be performed based on real-time updates of a dynamic spectral mask to account for other spectrum users. As an example, consider a simulated scenario (see Figure 12.28) involving a geographical distribution of multiple mobile wireless handsets that move relative

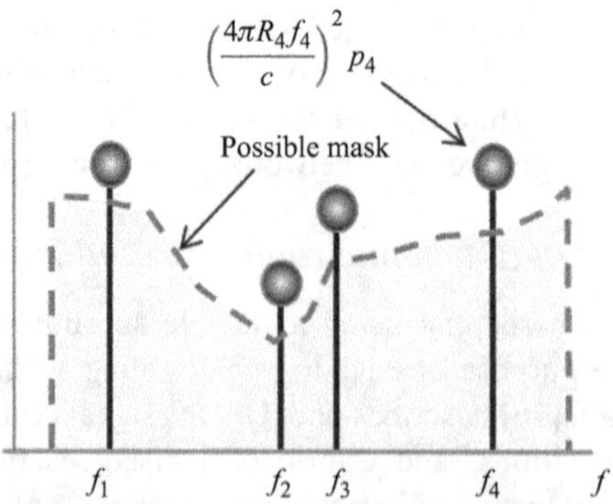

Figure 12.27 Dynamic spectral mask construction based on maximum acceptable transmitted power-per-bandwidth levels. Reprinted from [73]

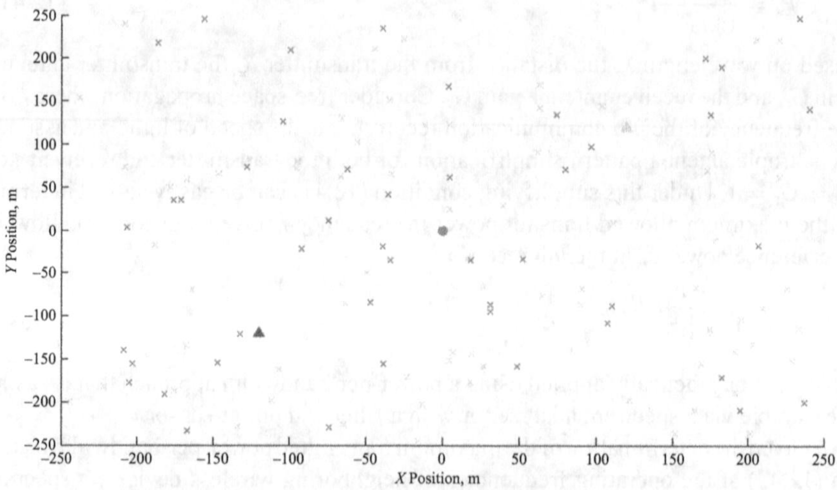

Figure 12.28 Snapshot of a simulated geographical scenario with the radar transmitter the circle in the center, the light and dark triangles two communication base stations, and the light and dark x's the mobile handsets associated with these base stations. Reprinted from [80]

to both the radar transmitter and two associated base stations [80]. Each handset is assumed to be communicating with one of these two base stations. As the distance from the nth handset to its base station increases, which therefore translates into less received power from the base station, the maximum amount of radar interference that the handset could accept without experiencing performance degradation would likewise decrease (i.e. lower p_n).

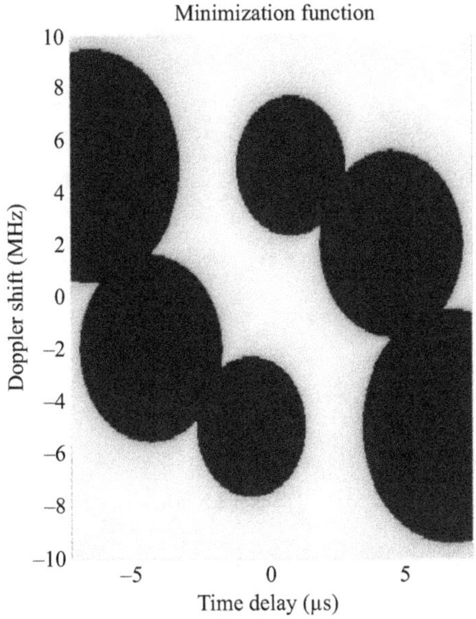

Figure 12.29 Ambiguity function minimization template for dynamic spectral mask simulation, reprinted from [80]

Figure 12.29 shows the ambiguity-function minimization template used for waveform optimization in this scenario. Figure 12.30 shows a time snapshot of the dynamic spectral mask computed based on the above distribution of mobile handsets as well as the resulting optimized waveform spectrum. A comparison between the optimized waveform AF and an overlay of the minimization template is illustrated in Figure 12.31. The narrow optimized waveform spectral content observed in Figure 12.30 clearly translates into a rather broad mainlobe for the delay dimension in Figure 12.31, i.e. rather poor range resolution. The least-squares fit between the AF and minimization template is 5.32×10^{-6}, which shall be useful for comparison in a moment.

A second optimization was performed using the geographical distribution in Figure 12.28 and the minimization template in Figure 12.29, albeit with the additional requirement that PAPR ≤ 2 dB. In this case the waveform spectrum in Figure 12.32 is realized, which unlike the narrow waveform spectrum observed in Figure 12.30, now expands out to the edges of the spectral mask. Subsequently, the delay dimension of the AF mainlobe depicted in Figure 12.33 shows a much narrower mainlobe that achieves better range resolution. The least-squares fit between the AF and minimization template for this case is 508.53×10^{-6}, two orders of magnitude higher than the previous case, which is not unexpected due to the inclusion of the PAPR constraint.

This waveform optimization approach was also examined for a minimization template aimed at achieving high-range resolution (HRR) with low-range sidelobes

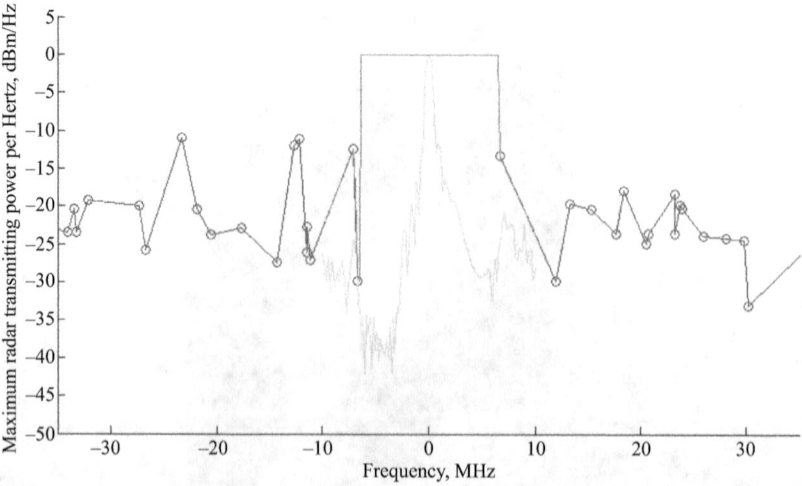

Figure 12.30 Dynamic spectral mask and spectrum of the resulting optimized waveform, reprinted from [80]

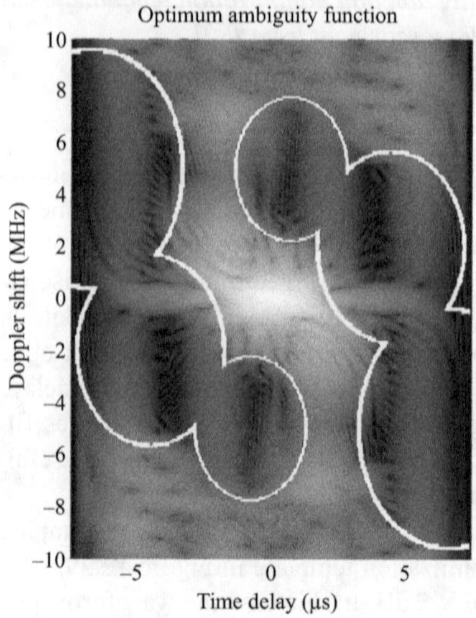

Figure 12.31 Ambiguity function of the optimized waveform, shown with an overlay of the minimization template. Reprinted from [80]

while providing some tolerance to Doppler. A simple way to achieve this goal is via the minimization template in Figure 12.34. First, consider the simulated waveform that results when no spectral mask is applied, i.e. the impact on the surrounding wireless users is ignored. The spectrum of this waveform, depicted in Figure 12.35, possesses

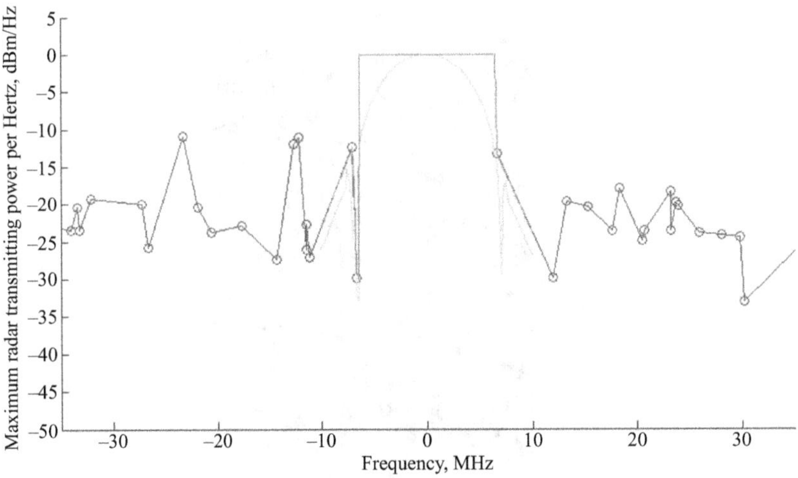

Figure 12.32 Dynamic spectral mask and spectrum of the resulting optimized waveform after inclusion of PAPR ≤ 2 dB constraint, reprinted from [80]

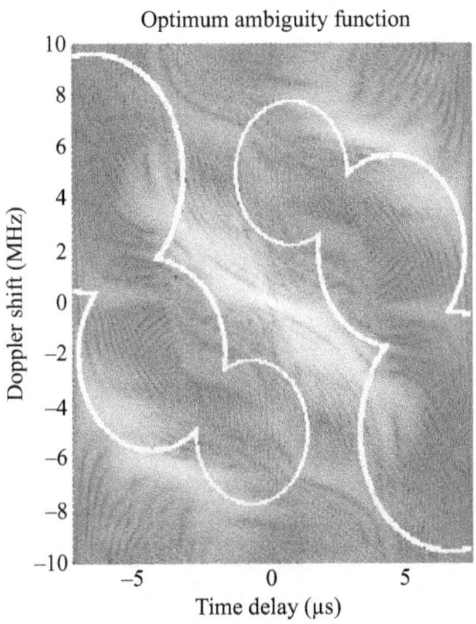

Figure 12.33 Ambiguity function of the optimized waveform for PAPR ≤ 2 dB constraint and the minimization template, reprinted from [80]

a wide bandwidth that is to be expected to achieve the desired HRR. Figure 12.36 shows the AF of the optimized waveform, where all ambiguity is aligned neatly along the Doppler axis. The least-squares distance in this case is 180.35×10^{-6}.

In contrast, the HRR waveform optimization was repeated with the inclusion of the spectral mask generated by the geographical distribution of nearby mobile

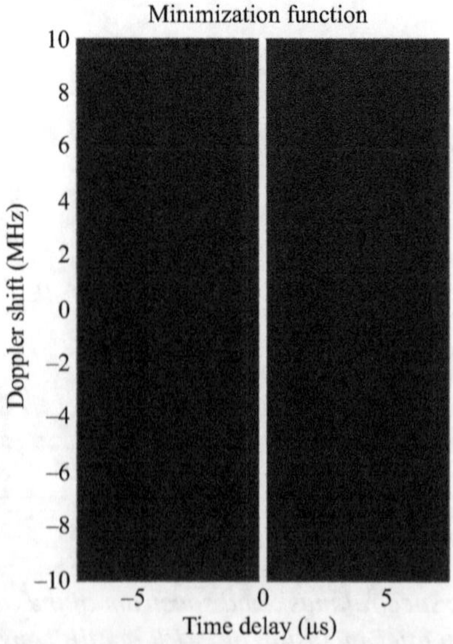

Figure 12.34 HRR minimization template, reprinted from [80]. The ambiguity is minimized everywhere in the range-Doppler plane except along the Doppler axis

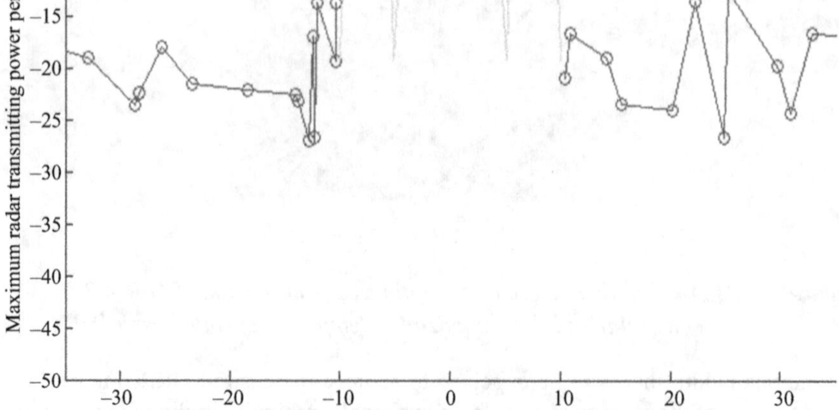

Figure 12.35 Spectrum resulting from waveform optimization without a spectral mask using the HRR minimization template from Fig. 12.34

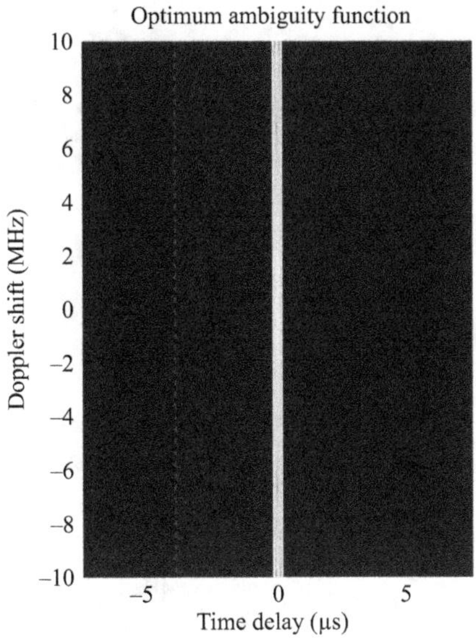

Figure 12.36 Ambiguity function of the optimized waveform using the HRR minimization template from Fig. 12.34

handsets. Figures 12.37 and 12.38 show the resulting waveform spectrum and AF, respectively. As expected, the spectral mask constraint degrades the least-squares error, which increases by roughly a factor of two to 377.13×10^{-6}. The waveform range–Doppler sidelobes in Figure 12.38 are also observed to increase relative to Figure 12.36. While a PAPR constraint was not imposed here, one can easily infer that further deviation from the result in Figure 12.36 would be incurred.

12.4.3 Joint transmitter and waveform optimization

Joint optimization of the transmitter circuit and the waveform can be expected to yield a better overall result relative to separately optimizing these components. This expectation is justified because the transmitter power amplifier operates nonlinearly, and thus the principle of superposition does not apply. Because it establishes the impact on the neighboring EM environment, the construction of a dynamic spectral mask may be critical to achieving joint circuit and waveform optimization for a real-time, reconfigurable radar transmitter. The circuit optimization methods presented earlier based on the spectral mask compliance metric S_{max} can be readily integrated into waveform optimization that is likewise constrained by a spectral mask. This topic is an area of ongoing investigation.

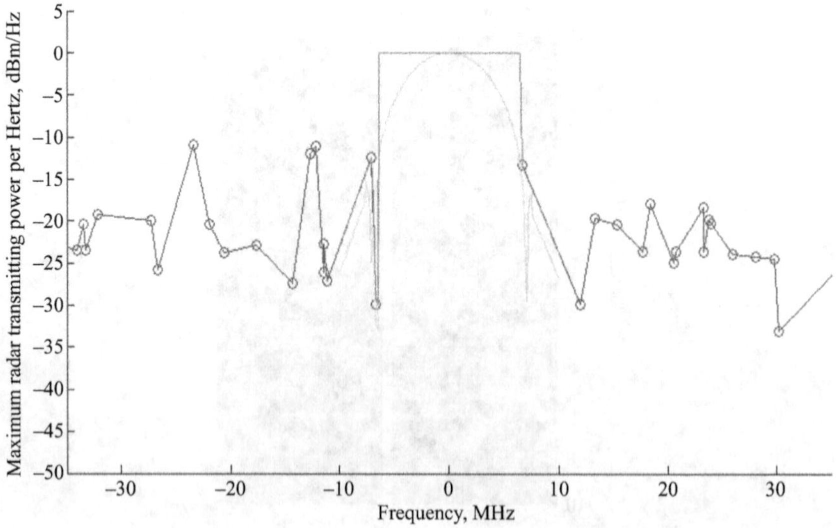

Figure 12.37 Dynamic spectral mask and resulting optimized waveform spectrum using the spectrum-sharing scenario from Figure 12.28 and the HRR minimization template from Figure 12.34, reprinted from [80]

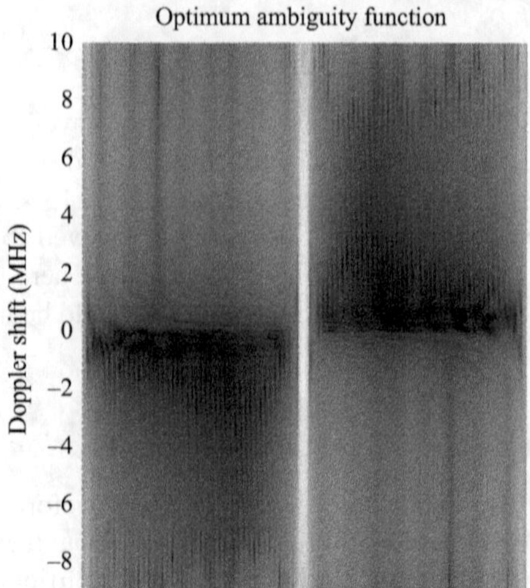

Figure 12.38 Ambiguity function of the optimized waveform based on the spectral mask template from the spectrum-sharing scenario of Figure 12.28 and the minimization template from Figure 12.34, reprinted from [80]

12.5 Looking ahead

What is a realistic path forward to realize reconfigurable radar transmitters? Significant enablers to this technology include: (1) regulatory freedom, (2) inter-system cooperation, and (3) the development of high-power tunable components. While only briefly discussed below, certainly each of these topics is an area of research all by itself.

12.5.1 Creating a regulatory environment favoring adaptive radar transmission

In recent years the previously quite rigid spectrum allocations have begun to loosen, with dynamic spectrum access and cognitive radio being employed in the 3.5 GHz radar band and in television "white space" spectrum. Spectrum sharing of communication applications with radar is also expected to be a significant component of fifth-generation (5G) wireless systems [81]. In fact, the actual motivation for constructing reconfigurable radar transmitters comes from this expectation of greater spectrum sharing in the future. Further, greater overall freedom for both operating modes means that while communication systems are increasingly allowed into what have historically been radar-only bands, the converse of radar systems operating in wireless bands could likewise occur.

As this more relaxed regulatory environment emerges, it is important to remember that the "silence" between radar pulses is when the radar is actually listening for echoes, i.e. the spectrum during this time period is not being underutilized. In fact, with radar duty cycles varying from roughly 25% on the high end to much lower than 1% on the low end, the radar spends far more time listening than transmitting [82]. Thus any sharing paradigm between radar and communications must be carefully provisioned (see the "lessons learned" throughout Chapter 4). The development of greater transmit adaptivity for radar will therefore permit regulatory bodies to realize further spectrum sharing opportunities while ensuring robustness for critical sensing applications related to safety, security, and weather monitoring.

12.5.2 Cooperation between radar and communication systems

Sharing between radar and communication systems will be most effective in situations where these systems can operate in a cooperative manner. Such a scenario would be facilitated by the distribution of data such as position, transmit frequencies, and acceptable levels of interference power, between the different devices. Obviously, in some cases radar systems would not be able to share this information due to safety or security concerns. For example, entities involved with managing air travel such as the U.S. Federal Aviation Administration may not wish to share this information for fear that it could be exploited by nefarious reasons.

The question then becomes: how can such systems still cooperatively share spectrum in an effective way? One possible answer is that the cooperation could be radar-centric [83]. In other words, the radar could receive information from surrounding communication devices via a wireless or wired network and then make decisions regarding its own operating frequency and power level without informing others. Mandating private cellular companies to provide this type of information in real-time to government-operated radar systems in exchange for utilizing radar spectrum could be an acceptable arrangement. Of course, the possible drawback to the radar is that unfettered access to its spectrum could create situations in which there is insufficient remaining spectrum for the radar to meet its mission. Clearly, regulatory involvement is necessary in this context as well.

12.5.3 Development of high-power-tunable components

To facilitate reconfigurable radar transmitters, continued advances in tunable passive components that can handle 10 W up to 1,000 W of power are needed, depending on the particular application. Such a capability would be a significant expansion over the power-handling abilities of presently available tunable components. Having said that, recent developments show great promise in this direction, such as the use of evanescent-mode cavity technology to achieve a 90 W impedance tuner capable of providing matching impedances over most of the Smith Chart [84]. With sustained support from government and industry, continued further advances are expected.

12.6 Conclusions

Future radar systems will need to be adaptive and reconfigurable to allow for adequate sharing of the spectrum with wireless communication systems. This chapter has described several algorithms that facilitate the optimization of transmitter circuitry and radar waveforms. The goals of these approaches include (1) ensuring desired radar range/Doppler resolution characteristics, (2) achieving high power- PAE, and (3) spectral compliance to minimize the interference to other spectrum users. Algorithms for the fast tuning of the power amplifier load impedance, as well as simultaneous tuning of load impedance with input power or bias voltage, have been demonstrated to locate the optimal solution within a feasible number of required measurements.

Optimization of the waveform AF based on an ambiguity minimization template has also been presented. To connect these optimization algorithms with regulatory situations, the ability to derive a dynamic spectral mask for radar transmission has been demonstrated based on the knowledge of the locations, frequencies, and maximum acceptable interference power levels of nearby communication receivers. Collectively, these methods illustrate ways in which future radar systems may themselves become more amenable to spectrum sharing while ensuring that sensing capabilities vital to safety and security are maintained.

Acknowledgments

I would like to thank collaborator Robert J. Marks II of Baylor University, who has contributed greatly to this chapter. He has worked tirelessly with our students and I in developing the alternating-projections waveform optimization algorithm, as well as contributing greatly to the construction of this chapter. I also thank our past and present student research assistants, including Matthew Fellows, Joseph Barkate, Dylan Eustice, Casey Latham, Lucilia Lamers, Zach Hays, Sarvin Rezayat, Matthew Flachsbart, Jennifer Barlow, and Christopher Kappelmann, who have innovated the developments shown in this work and have originally published many of these results. Roberto Gomez-Garcia has also provided helpful comments that have improved the presentation of this chapter. This work has been funded by the National Science Foundation (Grant No. ECCS-1343316) and the Naval Research Laboratory. Gratitude is expressed to Keysight Technologies for cost-free loan of the Advanced Design System software used for many of the simulations, and to Modelithics for their donation of simulation model libraries through the Modelithics University Program. I am grateful for my wife Rebekah who has been an encourager and supporter throughout the writing of this chapter. Finally, I wish to thank the Lord Jesus Christ, who provides an eternal purpose and meaning that transcends the pursuits and concerns of this present life. It is my sincere hope that all who read these words will be drawn to Him and experience the joy, peace, and eternal life that He offers.

References

[1] "Federal Radar Spectrum Requirements." National Telecommunications and Information Administration (NTIA) Special Publication 00-40, U.S. Department of Commerce, May 2000. Available at https://www.ntia.doc.gov/files/ntia/publications/ntia00-40.pdf.

[2] "Radar Spectrum Engineering Criteria." National Telecommunications and Information Administration (NTIA).

[3] Baylis, C., Fellows, M., Cohen, L., and Marks II, R.J. "Solving the spectrum crisis: intelligent, reconfigurable microwave transmitter amplifiers for cognitive radar" *IEEE Microwave Magazine*, vol. 15, Jul. 2014, no. 5, pp. 94–107.

[4] Sevic, J.F., Burger, K.L., and Steer, M.B. "A novel envelope-termination load-pull method for ACPR optimization of RF/microwave power amplifiers," *IEEE International Microwave Symposium*, vol. 2, Jun. 1998, pp. 723–726.

[5] Raab, F., Asbeck, P., Cripps, S., *et al.*, "RF and microwave power amplifier and transmitter topologies, part 1," *High Frequency Electronics*, May 2003.

[6] Baylis, C., Wang, L., Moldovan, M., *et al.*, "Designing transmitters for spectral conformity: power amplifier design issues and strategies." *IET Radar, Sonar & Navigation*, vol. 5, no. 6, Jul 2011, pp. 681–685.

[7] Baylis, C., Wang, L., Moldovan, M., *et al.*, "Designing for spectral conformity: issues in power amplifier design," *International Waveform Diversity & Design Conference*, Niagara Falls, Ontario, Canada, Aug. 2010.

[8] Wu, Q., Xiao, H., and Li, F., "Linear power amplifier design for CDMA signals: a spectrum analysis approach," *Microwave Journal*, 1998.

[9] Sun, Y. and Lau, W.K., "Antenna impedance matching using genetic algorithms," *IEE Conference on Antennas and Propagation*, York, United Kingdom, Aug. 1999, pp. 31–36.

[10] du Plessis, W. and Abrie, P., "Lumped impedance matching using a hybrid genetic algorithm," *Microwave and Optical Technology Letters*, vol. 37, no. 3, May 2003, pp. 201–212.

[11] Martin, J., Baylis, C., Cohen, L., and Marks II, R.J., "A peak-search algorithm for load-pull optimization of power-added efficiency and adjacent-channel power ratio," *IEEE Transactions on Microwave Theory and Techniques*, vol. 62, no. 8, Aug. 2014, pp. 1772–1783.

[12] Fellows, M., Baylis, C., Martin, J., Cohen, L., and Marks II, R.J., "Direct algorithm for the Pareto load-pull optimization of power-added efficiency and adjacent-channel power ratio," *IET Radar, Sonar & Navigation*, vol. 8, no. 9, Dec. 2014, pp. 1280–1287.

[13] Fellows, M., Baylis, C., Cohen, L., and Marks II, R.J. "Real-time load impedance optimization for radar spectral mask compliance and power efficiency," *IEEE Transactions on Aerospace and Electronic Systems*, vol. 51, no. 1, Jan. 2015, pp. 591–599.

[14] Davidson, T.N., Luo, Z.-Q., and Sturm, J.F., "Linear matrix inequality formulation of spectral mask constraints with applications to FIR filter design," *IEEE Transactions on Signal Processing*, vol. 50, no. 11, Nov. 2002, pp. 2702–2715.

[15] Parr, B., Cho, B., Wallace, K., and Ding, Z., "A novel ultra-wideband pulse design algorithm," *IEEE Communications Letters*, vol. 7, no. 5, May 2003, pp. 219–221.

[16] Sheng, H., Orlik, P., Haimovich, A., Cimini Jr., L., and Zhang, J., "On the spectral and power requirements for ultra-wideband transmission" *IEEE International Conference on Communications*, May 2003, pp. 738–742.

[17] Luo, X., Yang, L., and Giannakis, G.B., "Designing optimal pulse-shapers for ultra-wideband radios," *IEEE Conference on Ultra Wideband Systems and Technologies*, Nov. 2003, pp. 349–353.

[18] Zelley, C., "A spherical representation of the Smith chart," *IEEE Microwave Magazine*, vol. 8, no. 3, Jun. 2007, pp. 60–66.

[19] Wu, Y., Zhang, Y., Liu, Y., and Huang, H., "Theory of the spherical generalized Smith chart," *Microwave and Optical Technology Letters*, vol. 51, no. 1, Nov. 2008, pp. 95–97.

[20] Shamin, A., Radwan, A.G., and Salama, K.N., "Fractional Smith chart theory," *IEEE Microwave and Wireless Components Letters*, vol. 21, no. 3, Mar. 2011, pp. 117–119.

[21] Kretzschmar, J., and Schoonaert, D., "Smith charts for lossy transmission lines," *Proceedings of the IEEE*, vol. 57, no. 9, Sep. 1969, pp. 1658–1660.

[22] Fellows, M., Flachsbart, M., Barlow, J., Baylis, C., and Marks II, R.J., "The Smith tube: selection of radar chirp waveform bandwidth and power amplifier load impedance using multiple bandwidth load-pull measurements," *IEEE Wireless and Microwave Technology Conference*, Tampa, FL, USA, June 2014.

[23] Barkate, J., Barlow, J., Fellows, M., Baylis, C., Cohen, L., and Marks II, R.J., "The power Smith tube: joint optimization of power amplifier input power and load impedance for power-added efficiency and adjacent-channel power

ratio," *IEEE Wireless and Microwave Technology Conference*, Cocoa Beach, Florida, Apr. 2015.

[24] Fellows, M., Rezayat, S., Barlow, J., *et al.*, "The bias Smith tube: simultaneous optimization of bias voltage and load impedance in power amplifier design," *IEEE Radio and Wireless Symposium*, Austin, TX, USA, Jan. 2016.

[25] Snider, D.M., "A theoretical analysis and experimental confirmation of the optimally loaded and overdriven RF power amplifier," *IEEE Transactions on Electron Devices*, vol. ED-14, no. 12, Dec. 1967, pp. 851–857.

[26] Nemati, H.M., Fager, C., Gustavson, U., Jos, R., and Zirath, H., "Design of varactor-based tunable matching networks for dynamic load modulation of high power amplifiers," *IEEE Transactions on Microwave Theory and Techniques*, vol. 57, May 2009, no. 5, pp. 1110–1118.

[27] Hajji, R., Beauregard, F., and Ghannouchi, F.M., "Multitone power and intermodulation load-pull characterization of microwave transistors suitable for linear SSPA's design," *IEEE Transactions on Microwave Theory and Techniques*, vol. 45, no. 7, Jul. 1997, pp. 1093–1099.

[28] Nam, J., Shin, J.-H., and Kim, B., "A handset power amplifier with high efficiency at a low level using load-modulation technique," *IEEE Transactions on Microwave Theory and Techniques*, vol. 53, no. 8, Aug. 2005, pp. 2639–2644.

[29] Barkate, J., Flachsbart, M., Hays, Z., *et al.*, "Fast, simultaneous optimization of power amplifier input power and load impedance for power-added efficiency and adjacent-channel power ratio using the power Smith tube." *IEEE Transactions on Aerospace and Electronic Systems*, vol. 52, no. 2, Apr. 2016, pp. 928–937.

[30] Cha, J., Yang, Y., Shin, B., and Kim, B., "An adaptive bias controlled power amplifier with a load-modulated combining scheme for high efficiency and linearity," *IEEE Microwave Theory and Techniques Society International Microwave Symposium*, Philadelphia, PA, USA, June 2003, pp. 81–84.

[31] Hau, G., Nishimura, T.B., and Iwata, N., "A highly efficient linearized wide-band CDMA handset power amplifier based on predistortion under various bias conditions," *IEEE Transactions on Microwave Theory and Techniques*, vol. 49, no. 6 Jun. 2001, pp. 1194–1201.

[32] Forestier, N., Bouysse, P., Quere, R., Mallet, A., Nebus, J.-M., and Lapierre, L., "Joint optimization of the power-added efficiency and the error-vector measurement of 20-GHz pHEMT amplifier through a new dynamic bias-control method," *IEEE Transactions on Microwave Theory and Techniques*, vol. 52, no. 4, Apr. 2004, pp. 1132–1141.

[33] Weiss, M.D., Raab, F., and Popovic, Z., "Linearity of X-band class-F power amplifiers in high-efficiency transmitters," *IEEE Transactions on Microwave Theory and Techniques*, vol. 49, no. 6, Jun. 2001, pp. 1174–1179.

[34] Asbeck, P., Larson, L.E., and Galton, I.G., "Synergistic design of DSP and power amplifiers for wireless communications," *IEEE Transactions on Microwave Theory and Techniques*, vol. 49, no. 11, Nov. 2001, pp. 2163–2169.

[35] Raab, F.H., "High-efficiency linear amplification by dynamic load modulation," *IEEE International Microwave Symposium*, Philadelphia, PA, USA, June 2003, vol. 3, pp. 1717–1720.

[36] Yang, Y., Choi, K., and Weller, K.P., "DC boosting effect of active bias circuits and its optimization for class-AB InGaP-GaAs HBT power amplifiers," *IEEE Transactions on Microwave Theory and Techniques*, vol. 52, no. 5, May 2004, pp. 1455–1463.

[37] Fellows, M., Rezayat, S., Lamers, L., *et al.*, "Smith tube optimization of drain voltage and load reflection coefficient to maximize power-added efficiency under ACPR constraints," To appear in *IEEE Transactions on Aerospace and Electronic Systems*.

[38] Fellows, M., Lamers, L., Baylis, C., Cohen, L., and Marks II, R.J., "A fast load-pull optimization for power-added efficiency under output power and ACPR constraints," *IEEE Transactions on Aerospace and Electronic Systems*, vol. 52, no. 6, Dec. 2016, pp. 2906–2916.

[39] Sevic, J., Burger, K., and Steer, M., "A novel envelope-termination load-pull method for ACPR optimization of RF/microwave power amplifiers," *IEEE International Microwave Symposium*, vol. 2, June 1998, Baltimore, MD, USA, pp. 723–726.

[40] Iwamoto, M., Williams, A., Chan, P-F., Metzger, A.G., Larson, L.E., and Asbeck, P.M., "An extended Doherty amplifier with high efficiency over a wide power range," *IEEE Transactions on Microwave Theory and Techniques*, vol. 49, no. 12, Dec. 2001, pp. 2472–2479.

[41] Zhang, X., Saycocie, C., Munro, S., and Henderson, G., "A SiGe HBT power amplifier with 40% PAE for PCS CDMA applications," *IEEE International Microwave Symposium*, June 2000, Boston, MA, USA, pp. 857–860.

[42] Noh, Y.S. and Park, C.S., "PCS/W-CDMA dual-band MMIC power amplifier with a newly proposed linearizing bias circuit," *IEEE Journal of Solid-State Circuits*, vol. 37, no. 9, Sept. 2002, pp. 1096–1099.

[43] Kusunoki, S., Yamamoto, K., Hatsugai, T., *et al.*, "Power-amplifier module with digital adaptive predistortion for cellular phones," *IEEE Transactions on Microwave Theory and Techniques*, vol. 50, no. 12, Dec. 2002, pp. 2979–2986.

[44] Reveyrand, T., Gasseling, T., Barataud, D., Mons, S., and Nebus, J.-M., "A smart load-pull method to safely reach optimal matching impedances of power transistors," *IEEE International Microwave Symposium*, June 2007, pp. 1489–1492.

[45] Woodward, P.M., *Probability, Information Theory, and Radar*. Oxford, United Kingdom: Pergamon, 1953.

[46] Cheney, M., "A mathematical tutorial on synthetic aperture radar," *SIAM Review*, vol. 43, no. 2, pp. 301–312, 2001.

[47] Eustice, D., Baylis, C., and Marks II, R.J., "Woodward's ambiguity function: from foundations to applications," *Texas Symposium on Wireless and Microwave Circuits and Systems*, Waco, TX, USA, Apr. 2015.

[48] Wilcox, C.H., "The synthesis problem for radar ambiguity functions," Tech. Summary Report No. 157, Mathematics Research Center, University of Wisconsin, Madison, Wisconsin, 1960.

[49] Sussman, S.M., "Least-square synthesis of radar ambiguity functions," *IRE Transactions on Information Theory*, vol. 8, pp. 246–254, 1962.

[50] Wolf, J.D., Lee, G.M., and Suyo, C.E., "Radar waveform synthesis by mean-square optimization techniques," *IEEE Transactions on Aerospace and Electronic Systems*, vol. AES-5, no. 4, 1968, pp. 611–619.

[51] Costas, J.P., "A study of a class of detection waveforms having nearly ideal range–Doppler ambiguity properties," *Proceedings of the IEEE*, vol. 72, no. 8, 1984, pp. 996–1009.

[52] Gladkova, I. and Chebanov, D., "On a new extension of Wilcox's method," *5th WSEAS International Conference on Applied Mathematics*, no. 31, Miami, FL, USA, Apr. 2004.

[53] Gladkova, I., and Chebanov, D., "On the synthesis problem for a waveform having a nearly ideal ambiguity surface," *IEEE Radar Conference*, Philadelphia, PA, USA, Apr. 2004.

[54] Sebt, M.A., Sheikhi, A., and Nayebi, M.M., "Orthogonal frequency-division multiplexing radar signal design with optimised ambiguity function and low peak-to-average power ratio," *IET Radar, Sonar, and Navigation*, vol. 3, no. 2, 2008, pp. 122–132.

[55] Byrnes, J., "Computational Noncommutative Algebra and Applications." *Proceedings of the NATO Advanced Study Institute on Computational Noncommutative Algebra and Applications*. New York: Springer Science & Business Media, 2006, p. 351.

[56] Patton, L.K., and Rigling, B.D., "Autocorrelation constraints in radar waveform optimization for detection," *IEEE Transactions on Aerospace and Electronic Systems*, vol. 48, no. 2, Apr. 2012, pp. 951–968.

[57] Patton, L.K., and Rigling, B.D., "Phase retrieval for radar waveform optimization," *IEEE Transactions on Aerospace and Electronic Systems*, vol. 48, no. 4, Oct. 2012, pp. 3287–3302.

[58] Patton, L.K., and Rigling, B.D., "Autocorrelation and modulus constraints in radar waveform optimization," *International Waveform Diversity and Design Conference*, Kissimmee, FL, USA, Feb. 2009, pp. 150–154.

[59] Selesnick, I.W., and Pillai, S.U., "Chirp-line transmit waveforms with multiple frequency-notches," *IEEE Radar Conference*, Kansas City, MO, USA, May 2011.

[60] Selesnick, I.W., and Pillai, S.U., "An iterative algorithm for the construction of notched chirp signals," *IEEE Radar Conference*, Washington, D.C., USA, May 2010.

[61] Kassab, M., Lesturgie, M., and Fiorina, J., "Alternate projections technique for radar waveform design," *International Radar Conference*, Bordeaux, France, 2009.

[62] Jakabosky, J., Blunt, S.D., Cook, M.R., Stiles, J., and Seguin, S., "Transmitter-in-the-loop optimization of physical radar emissions," *IEEE Radar Conference*, Atlanta, GA, USA, May 2012.

[63] Jakabosky, J., Lane, R., and Blunt, S., "Transmitter-in-the-loop optimization of distorted OFDM radar emissions," *IEEE Radar Conference*, Ottawa, Ontario, Canada, Apr./May 2013.

[64] Blunt, S., Cook, M., Jakabosky, J., de Graaf, J., and Perrins, E., "Polyphase-coded FM (PCFM) radar waveforms, part I: implementation," *IEEE Transactions on Aerospace and Electronic Systems*, vol. 50, no. 3, Jul. 2014, pp. 2218–2229.

[65] Blunt, S., Jakabosky, J., Cook, M., Stiles, J., Seguin, S., Mokole, E., "Polyphase-coded FM (PCFM) radar waveforms, part II: optimization," *IEEE Transactions on Aerospace and Electronic Systems*, vol. 50, no. 3, Jul. 2014, pp. 2230–2241.

[66] Jakabosky, J., Blunt, S., and Higgins, T., "Ultra-low sidelobe waveform design via spectral shaping and LINC transmit architecture," *IEEE Radar Conference*, Arlington, VA, May 2015.

[67] Nobakht, R. and Civanlar, M.R., "Optimal pulse shape design for digital communication systems by projections onto convex sets," *IEEE Transactions on Communications*, vol. 43, no. 12, Dec. 1995, pp. 2874–2877.

[68] Bin, W., Wenfang, Y., and Jinkuan, W., "Adaptive waveform design for multiple radar tasks based on constant modulus constraint," *Journal of Applied Mathematics*, vol. 2013, Jul. 2013, pp. 1–6.

[69] Gerchberg, R. and Saxton, W., "A practical algorithm for the determination of phase from image and diffraction plane pictures," *Optik*, vol. 35, no. 22, 1972, pp. 237–248.

[70] Eustice, D., Baylis, C., Cohen, L., Latham, C., and Marks II, R.J., "Optimizing radar waveforms using generalized alternating projections," *Texas Symposium on Wireless and Microwave Circuits and Systems*, Waco, TX, USA, Apr. 2015.

[71] Eustice, D., Baylis, C., Cohen, L., and Marks II, R.J., "Waveform synthesis via alternating projections with ambiguity function, peak-to-average power ratio, and spectrum requirements," *IEEE Radio and Wireless Symposium*, Austin, TX, USA, Jan. 2016.

[72] Eustice, D., Latham, C., Baylis, C., Cohen, L., and Marks II, R.J., "Amplifier-in-the-loop adaptive radar waveform synthesis in a dynamic environment," *IEEE Transactions on Aerospace and Electronic Systems*, vol. 53, no. 2, Apr. 2017, pp. 826–836.

[73] Latham, C., Baylis, C., Cohen, L., and Marks II, R.J., "Dynamic spectral mask construction for radar transmission based on communication receiver locations," *Texas Symposium on Wireless and Microwave Circuits and Systems*, Waco, TX, USA, Mar. 2016.

[74] Liu, Y., and Dong, L., "Spectrum sharing in MIMO cognitive radio networks based on cooperative game theory," *IEEE Transactions on Wireless Communications*, vol. 13, no. 9, Sep. 2014, pp. 4807–4820.

[75] Mahmoud, H., Yucek, T., and Arslan, H., "OFDM for cognitive radio: merits and challenges," *IEEE Wireless Communications*, vol. 16, no. 6, Apr. 2009, pp. 6–15.

[76] Poston, J.D., and Horne, W.D., "Discontiguous OFDM considerations for dynamic spectrum access in idle TV channels," *First IEEE International Symposium on New Frontiers in Dynamic Spectrum Access*, Baltimore, MD, USA, Nov. 2005, pp. 607–610.

[77] Cabric, D., O'Donnell, I.D., Chen, M.S.-W., and Broderson, R.W., "Spectrum sharing radios," *IEEE Circuits and Systems Magazine*, vol. 6, no. 2, Jul. 2006, pp. 30–45.

[78] Srinivasa, S., and Jafar, S.A., "Cognitive radios for dynamic spectrum access—the throughput potential of cognitive radio: a theoretical perspective," *IEEE Communications Magazine*, vol. 45, no. 5, May 2007, pp. 73–79.

[79] Pozar, D.M., *Microwave Engineering*, Fourth Edition. Hoboken, New Jersey: John Wiley & Sons, Inc., 2012.

[80] Latham, C., Fellows, M., Baylis, C., Cohen, L., and Marks II, R.J., "Radar waveform optimization for ambiguity function properties and dynamic spectral mask requirements based on communication receiver locations," *IEEE Radio and Wireless Symposium*, Phoenix, AZ, USA, Jan. 2017.

[81] Mitola III, J., Guerci, J., Reed, J., *et al.*, "Accelerating 5G QoE via public-private spectrum sharing," *IEEE Communications Magazine*, vol. 52, no. 5, May 2014, pp. 77–85.

[82] Griffiths, H., Cohen, L., Watts, S., *et al.*, "Radar spectrum engineering and management: technical and regulatory issues," *Proceedings of the IEEE*, vol. 103, no. 1, Jan. 2015, pp. 85–102.

[83] Khawar, A., Abdel-Hadi, A., and Clancy, T.C., "Spectrum sharing between S-Band radar and LTE cellular system: a spatial approach," *IEEE International Symposium on Dynamic Spectrum Access Networks*, McLean, VA, USA, Apr. 2014, pp. 7–14.

[84] Semnani, A., Abu Khater, M., Wu, Y.-C., and Peroulis, D., "An electronically-tunable high-power impedance tuner with integrated closed-loop control," *IEEE Microwave and Wireless Components Letters*, vol. 27, no. 8, Aug. 2017, pp. 754–756.

Part III

Addressing interference

Chapter 13

Radar/Wi-Fi spectrum sharing: evaluation of radar protection regions

Farzad Hessar[1]*, Anish Ashok*[1]*, and Sumit Roy*[1]

13.1 Background

In this chapter, an aspect of spectrum sharing between a monostatic radar and 802.11 wireless local area networks (WLANs) (Wi-Fi) is explored. Specifically, it focuses on the determination of *protection regions* around the radar to keep the aggregate interference caused by active Wi-Fi (secondary) sources to within acceptable limits. Of particular interest is the impact of the deployment geometry of the Wi-Fi networks—modeled via suitable random spatial processes—on the statistics of the aggregate interference. The impact of various stochastic geometries on the resulting protection region is explored through both (Monte Carlo) simulation and analytical approaches.

A critical challenge facing regulators is the so-called spectrum crunch (lack of sufficient available spectrum within the usable range) for the roll-out of new services [1]. A primary contributor to this issue is the escalating pressure on mobile cellular access networks due to high-end mobile devices running bandwidth hungry applications. Accordingly, various strategies for increasing the network efficiency of next-generation systems are under active consideration, including new notions of *spectrum sharing*. Unlike sharing in unlicensed bands, the significant novelty of the proposed concepts center around sharing between dissimilar (heterogeneous) systems [2], with a designated *primary* (licensed incumbent) sharing with a *secondary* (unlicensed) network via principles of dynamic spectrum access. The first instance of such primary/secondary sharing was the unlicensed reuse of local TV spectrum pursuant to the *TV White Space* ruling by the US FCC in 2008 [3], for which the IEEE 802.22 working group developed standard technology [4,5].

Here, we study the coexistence of radars and WLAN. The reemergence of interest in this topic is based on the large amount of licensed spectrum allocated to radar operations (over 1.7 GHz of spectrum in the range of 225 MHz–3.7 GHz) that has significant potential for reuse due to the temporally and spatially sporadic nature of how many search radars operate. Pursuant to a Presidential directive [6], the FCC has required that 150 MHz of spectrum between 3550 and 3700 MHz be made available

[1]Department of Electrical Engineering, University of Washington, United States

Table 13.1 NTIA federal spectrum requirement forecast for radar bands

Frequency band	Federal government use
92–100 GHz	Airborne fire-control, beacons, atmospheric research, cloud detection, and synthetic vision radars
31.8–36 GHz	Airborne navigational, mapping, weather, beacon, terrain following and avoidance; aircraft carrier PAR, test range, atmospheric and oceanic research, altimeter, scatterometer, and synthetic vision radars
24.05–24.65 GHz	Doppler radiolocation, vehicle speed detection, scatterometer, and precipitation radars
15.4–17.3 GHz	Airborne and shipborne multimode search, battlefield, aircraft carrier PAR, fire-control, test range, ASDE, scatterometer, precipitation, atmospheric research, and spaceborne radars
13.25–14.2 GHz	Airborne and shipborne search and acquisition Doppler, airborne weather, altimeters, scatterometer, precipitation, environmental research, and spaceborne radars
8.5–10.55 GHz	Airborne and shipborne surveillance and navigation, fire-control, battlefield, maritime, weather, test range, airborne radionavigation, ATC, SAR's, altimeters, ASDE, scatterometer, vehicle speed detection, and spaceborne radars
5250–5925 MHz	NOAA weather radars, FAA TDWR, surveillance and air defense (airborne, shipborne, land-based), fire-control, maritime, test range, SAR's, altimeters, scatterometer, airborne, and spaceborne radars
4200–4400 MHz	Aircraft radar altimeters
3100–3650 MHz	DOD surveillance and air defense (airborne, shipborne, land-based), ATC, SAR's, altimeters, test range, and spaceborne radars
2700–3100 MHz	ATC, maritime, and weather radars; DOD shipborne, airborne, ground air surveillance radars; range control, and spaceborne radars
2310–2385 MHz	Planetary and lunar radar
1215–1390 MHz	ATC, SAR's, and DOD early warning air defense, battlefield, shipborne long-range surveillance, and spaceborne radars
890–942 MHz	Navy shipborne long-range surveillance, test range, NASA research, and wind profiler radars
420–450 MHz	DOD early warning and long-range surveillance radars; and wind profiler radars
216–220 MHz	DOD space surveillance radar
3–30 MHz	DOD OTH and surface wave radar

for wireless broadband applications for the Citizens Band Radio Service [7]. A comprehensive listing of US Federal spectrum allocations for various radars is shown in Table 13.1.

This arrangement is not, however, the first instance of spectrum sharing between a (licensed) radar and an unlicensed network. The globally unlicensed bands according

Table 13.2 Unlicensed bands per Article 5 of ITU regulations (Primary allocation in capital, secondary allocation in small letters) Region 1: Europe, Africa, the former Soviet Union, Mongolia, and the Middle East west of the Persian Gulf. Region 2: Americas including Greenland, and some of the eastern Pacific Islands

Frequency range	**Availability**	**Licensed users**
6.765–6.795 MHz	Subject to local acceptance	FIXED SERVICE and Mobile service
13.553–13.567 MHz	Worldwide	FIXED and Mobile services except Aeronautical mobile (R) service
26.957–27.283 MHz	Worldwide	FIXED and MOBILE SERVICE except Aeronautical mobile service, CB Radio
40.66–40.7 MHz	Worldwide	Fixed, Mobile services and Earth exploration-satellite service
433.05–434.79 MHz	Only in region 1, subject to local acceptance	AMATEUR SERVICE and RADIOLOCATION SERVICE
902–928 MHz	Region 2 only	FIXED, Mobile except aeronautical mobile and Radiolocation service
2.4–2.5 GHz	Worldwide	FIXED, MOBILE, RADIOLOCATION, Amateur and Amateur-satellite service
5.725–5.875 GHz	Worldwide	FIXED-SATELLITE, RADIOLOCATION, MOBILE, Amateur and Amateur-satellite service
24–24.25 GHz	Worldwide	AMATEUR, AMATEUR-SATELLITE, RADIOLOCATION and Earth exploration-satellite service (active)
61–61.5 GHz	Subject to local acceptance	FIXED, INTER-SATELLITE, MOBILE and RADIOLOCATION SERVICE
122–123 GHz	Subject to local acceptance	EARTH EXPLORATION-SATELLITE, FIXED, INTER-SATELLITE, MOBILE, SPACE RESEARCH (passive) and Amateur service
244–246 GHz	Subject to local acceptance	RADIOLOCATION, RADIO ASTRONOMY, Amateur and Amateur-satellite service

to Article 5 of the International Telecommunication Union (ITU) regulations are shown in Table 13.2. Wi-Fi operation in the 5 GHz (U-NII unlicensed bands in North America but not elsewhere) are already subject to dynamic frequency select requirements (necessitating channel sensing and avoidance) to avoid interfering with various radars operating worldwide [8,9]. Currently, only 36% of 5 GHz channels is unencumbered by radar protection requirements.

Figure 13.1 The original protection regions computed by the NTIA (inner traces) were revised recently in [12] (outer traces). Even the revised region encompasses many population centers such as San Francisco and New York

Any coexistence scenario must consider two important factors:

1. The impact of interference from radar systems on the performance of unlicensed wireless networks and
2. Measures needed to protect radars (primary users) from unlicensed wireless (secondary).

In this chapter, we focus exclusively on the second objective, which is typically achieved by enforcing a *protection region* that represents a spatial domain surrounding the protected primary location in which cochannel transmission by secondary users (SUs) is forbidden.[1] We assume all radars are monostatic. The protection regions are determined by regulatory policy-making through a collaboration between NTIA and FCC, representing the need to balance incumbent protection and the needs of secondary civilian networks to set up an effective service [10]. Often, these protection regions (Figure 13.1) can extend for hundreds of kilometers from the radar location, which if located near population centers, makes deployment of secondary systems unattractive. The NTIA in its Fast Track report [11] initially inferred that large geographical separation and frequency offsets were required to minimize the interference higher power macrocell networks would cause to federal radar systems in the 3.5 GHz band. However, as shown in Figure 13.1 this analysis was later revised because the initial assumptions were overly conservative (e.g. 4G and beyond networks use smaller cells with lower power Base Stations). The above example and our own

[1] Outside such regions, secondary operation is allowed since the resulting interference is deemed acceptable to primary receivers.

subsequent analysis emphasizes this important underlying point—that the determination of protection regions is significantly impacted by a multiplicity of factors on both the secondary (notably, the transmit power, deployment geometry, transmit antenna patterns) and primary (receiver sensitivity and operational margins, receive antenna patterns) sides as well as the channel characteristics (distance and frequency dependent path loss, and any statistical effects such as small-scale fading and large-scale shadowing).

13.2 Search radar: noise limited operation

Consider a rotating search radar, with all interfering (Wi-Fi) sources in the same plane as the radar and the desired target at some elevation.[2] Radar performance characteristics for the noise-limited case are quantified in terms of the (desired) probability of detection P_D and acceptable false alarm rate P_{FA}. In turn, this performance determines the minimum signal-to-noise ratio (SNR) requirements at the radar receiver input (corresponding to maximum operating range) and sets the baseline for comparison with any spectrum sharing regime. In order to enable cognitive reuse by secondary transmitters outside the protection region, it is necessary to define rules for "acceptable" interference at the radar. A recent DARPA program [13] suggested that a drop of 5% in P_D for fixed P_{FA} at the edge of the radar operating range is acceptable, which we adopt here.[3] This margin can be allocated to aggregate interference produced by a (random) spatial distribution of multiple Wi-Fi networks that are operational outside of the protection region.

At the monostatic radar receiver, the power of the reflected signal from the target, assuming free space propagation, is given by the *Radar Equation* [14]

$$P_R = \frac{P_T G^2 \lambda^2}{(4\pi)^3 d^4}\sigma. \tag{13.1}$$

Here, P_T is the radar's instantaneous transmit power, G is the radar's antenna gain on both transmit and receive in the direction of the target, λ is the wavelength corresponding to the radar's center frequency, d is the distance from the radar to the target of interest, and σ represents the target's *radar cross section*. Consequently, the effective SNR (denoted SNR_{eff}) for reliable operation can be computed for a given probability of detection P_D and probability of false alarm P_{FA}. The minimum required SNR per pulse (denoted SNR_{min}) is determined by the nature of any pulse integration employed in the detection chain, namely *coherent* or *non-coherent*. Using SNR_{min}, one can determine the maximum interference level that can be tolerated by the radar, which we denote I_{max}. In turn, this level of interference determines the minimum separation between the radar and SUs.

[2]Hence, in the sequel, consider a general radar antenna pattern $G(\theta, \phi)$ where these represent the azimuth and elevation angles, respectively; all interference calculations use the horizontal antenna pattern corresponding to the $\phi = 0$ plane.

[3]The current version suggests a 5% range reduction for fixed P_D and P_{FA} is acceptable.

13.2.1 Noise-limited radar operation: design equations

For a radar transmitting pulse train $x(t) = \sum_n \sqrt{P_T}\, s(t - (n/f_R))$ with pulse repetition frequency (PRF) of f_R, the power of the reflected signal from the target at the radar receiver is given by (13.1). For a single received pulse, the SNR at the receiver input is

$$\mathrm{SNR_p} = \frac{P_T G^2 \lambda^2}{(4\pi)^3 d^4 N_0 B}\sigma, \tag{13.2}$$

with B the pulse bandwidth and

$$N_0 = FKT_E \tag{13.3}$$

the one-sided noise spectral density, where F is the receiver noise figure and T_E is the ambient temperature. Radar detection typically involves the processing of multiple pulsed echoes received from the target. For a *coherent* receiver that integrates M reflected pulses, the detector SNR is increased by a factor of M, i.e.

$$\mathrm{SNR_{eff}} = M\frac{P_T G^2 \lambda^2}{(4\pi)^3 d^4 N_0 B}\sigma, \tag{13.4}$$

where $M = T_I f_R$ is a product of illumination time T_I and the PRF. The target illumination time T_I depends on the radar scan rate as well as the antenna pattern.

Let θ_V and θ_H (in radians) denote the vertical and horizontal antenna beam widths, respectively. The antenna gain can therefore be approximated as

$$G \approx \frac{4\pi}{\theta_H\,\theta_H}\rho_A \tag{13.5}$$

for ρ_A the antenna efficiency. In other words, the radar concentrates the emitted power into an area of $\theta_V\,\theta_H$ with efficiency ρ_A, where the latter is typically 0.5. If the radar scans an area of Ω (steradians), where a full search of azimuth/elevation implies $\Omega = 4\pi$, within a scan time of T_S, the illumination time is determined as

$$T_I \approx T_S\frac{\theta_H\,\theta_V}{\Omega} \approx T_S\frac{4\pi\rho_A}{\Omega\, G}. \tag{13.6}$$

Using (13.4)–(13.6) subsequently yields

$$\mathrm{SNR_{eff}} = \frac{T_S}{\Omega}\frac{P_T G \lambda^2 f_R}{(4\pi)^2 d^4 N_0 B L}\sigma, \tag{13.7}$$

where the antenna efficiency ρ_A has been replaced by the equivalent value L that represents total losses in the system (e.g. transmission line mismatch, loss of perfect coherence in pulse echo integration, etc.).

The relationship between P_D, P_{FA}, and SNR_p for a single pulse is fairly accurately expressed by [15]

$$\mathrm{SNR_p} = \ln\left(\frac{0.62}{P_{FA}}\right) + 0.12\ln\left(\frac{0.62}{P_{FA}}\right)\ln\left(\frac{P_D}{1-P_D}\right) + 1.7\ln\left(\frac{P_D}{1-P_D}\right) \tag{13.8}$$

for ln(•) the natural log. For a coherent receiver integrating M pulses, $\mathrm{SNR_p}$ in (13.8) should be replaced with ($\mathrm{SNR_{eff}}$) from (13.7). Overall, the baseline performance of a noise-limited radar can be evaluated in two ways:

- If the maximum operational range of the radar is known, calculate SNR from (13.2) or (13.7) for maximum distance d. Then, using (13.8), one can trade-off between P_D and P_FA.
- If target P_D and P_FA are specified, estimate the required SNR from (13.8) and then determine maximum range from (13.2) or (13.7).

As discussed subsequently, the interference from a Wi-Fi transmitter can be modeled as a Gaussian process, which is additive to the (thermal) additive white Gaussian noise (AWGN). Thus, radar system performance in the presence of (Wi-Fi) interference is determined by the signal-to-interference-plus-noise ratio (SINR) at the receiver input, which based on (13.7) is given by

$$\mathrm{SINR} = \frac{T_\mathrm{S}}{\Omega}\frac{P_\mathrm{T}G\lambda^2 f_\mathrm{R}}{(4\pi)^2 d^4 L\left(N_0 B + I_\mathrm{aggr}\right)}\sigma \tag{13.9}$$

for I_aggr the aggregate interference power from all SU(s). Therefore, the maximum additional interference level I_max that can be tolerated is determined by using the minimum required SINR ($\mathrm{SINR_{min}}$) as

$$\begin{aligned} \mathrm{SINR_{min}} &\leq \frac{T_\mathrm{S}}{\Omega}\frac{P_\mathrm{T}G\lambda^2 f_\mathrm{R}}{(4\pi)^2 d^4 L\left(N_0 B + I_\mathrm{aggr}\right)}\sigma \\ I_\mathrm{aggr} &\leq \frac{T_\mathrm{S}}{\Omega}\frac{P_\mathrm{T}G\lambda^2 f_\mathrm{R}}{(4\pi)^2 d^4 L\,\mathrm{SINR_{min}}}\sigma - N_0 B = I_\mathrm{max}. \end{aligned} \tag{13.10}$$

For a given deployment of Wi-Fi networks, the resulting aggregate interference I_aggr at the radar receiver input is modeled (statistically) in the following section. The difference between the received interference from SUs (I_aggr) and the maximum tolerable interference at the radar receiver (I_max) establishes the available margin, where a positive margin implies secondary operation is possible.

For a single Wi-Fi access point (AP), the interference power induced at the radar is given by

$$I_{\mathrm{SU}\rightarrow\mathrm{Radar}} = \frac{P_\mathrm{SU}\, G(\theta)}{L_\mathrm{Rd-SU}(d_\mathrm{Rd-SU})\,\mathrm{FDR}(\Delta f)} \tag{13.11}$$

where $L_\mathrm{Rd-SU}(\,\cdot\,)$ is the path-loss between the radar and SU as a function of distance, P_SU is the SU transmit power, and $G(\theta)$ is the radar's antenna gain in the direction of the SU (function of azimuth angle). Further, $\mathrm{FDR}(\Delta f)$ is the *frequency dependent rejection* factor that accounts for the mismatch between the spectral support of the transmitted Wi-Fi signal and the radar receiver input band. Depending on the spectral shape of the transmitted signal $P(f)$ and the receiver input filter $H(f)$, this factor is computed as

$$\mathrm{FDR}(\Delta f) = \frac{\int_0^\infty P(f)\,df}{\int_0^\infty P(f)H(f+\Delta f)\,df}. \tag{13.12}$$

Since the Wi-Fi and radar spectra have been assumed (here) to be flat and overlapping, the FDR is calculated as the ratio of corresponding bandwidths $FDR = (20\ \text{MHz}/653\ \text{kHz}) = 30.6$ [16], based on the characteristics of radars operating in the 2700–2900 MHz band. Note that we have used the intermediate frequency 3-dB bandwidth for the radar's receiver, which is significantly smaller than the radio frequency 3-dB bandwidth of 10 MHz, since radar detection occurs after down-conversion.

13.3 Cumulative interference to radar from randomly deployed Wi-Fi APs

Wi-Fi transmissions use orthogonal frequency division multiplexing (OFDM) signals, where each OFDM symbol is a linear combination of many randomly modulated sub-carriers. Hence, using the central limit theorem, the transmitted OFDM signal in the time-domain is well approximated as a Gaussian random process [17]. Thereafter, the transmitted signal undergoes linear time-invariant filtering—the equivalent of the channel and any receiver front-end processing such as the matched filter—that preserves the Gaussianity of the resulting interference process at the radar detector input [18]. Consequently, the power (variance) of any interference (I_{aggr}) at the detector input is additive to the AWGN noise power, effectively raising the noise floor and with system performance determined by the corresponding SINR.

If multiple SUs coexist with the radar simultaneously, the accumulated interference power at the radar's location can only be characterized statistically. The aggregate interference encountered by the radar receiver is determined by several factors, the most notable of which are the stochastic geometry of the deployment of networks (which impacts distances and hence the component contributed to I_{aggr}) and the multiple access protocol employed (which determines the users that are simultaneously active). Since users within a single network share the common channel in time based on carrier-sense multiple access with collision avoidance, usually a *single* source can be active at any instant within one network. However, different networks can simultaneously operate in the vicinity of a radar. Here, we assume that each network is simultaneously active (i.e. there is an active user), which represents a worst-case scenario and acts as an upper bound on I_{aggr}, and focus exclusively on the impact of the network deployment geometry.

The stochastic geometry [19] of network deployments impacts the aggregate interference at the radar receiver. Let the APs and their associated users be randomly distributed in 2-D space according to a point process. Without loss of generality, assume a radar located at the 2-D origin (Figure 13.2), with APs randomly distributed over Cartesian locations $\mathbf{x} \in R^2$. The aggregate interference from SUs at the radar receiver is therefore described as [20]

$$I_{\text{aggr}} = \sum_{\mathbf{x} \in \Phi} \frac{P_{\text{SU}}\, G(\theta_{\mathbf{x}})}{\text{FDR}(\Delta f)}\, l(||\mathbf{x}||) \tag{13.13}$$

where $||\mathbf{x}||$ is the distance between radar and secondary AP, $l(\cdot)$ describes the impulse response function that models signal attenuation (inverse of path loss) between the

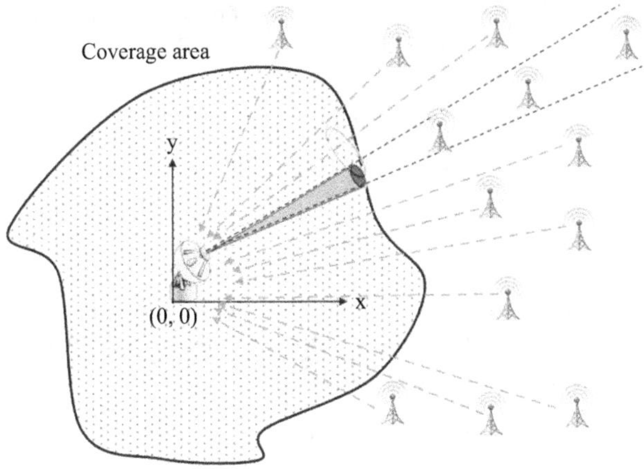

Figure 13.2 Aggregate interference to the radar receiver caused by multiple access points

source and receiver, and $G(\cdot)$ is the radar's antenna gain in the azimuthal direction of interferer $\theta_{\mathbf{x}}$. A flat world model is considered for this exercise. Note that (13.13) is the aggregate interference from multiple sources that are simultaneously active and is the summation of components (13.11) that represent the interference from a single user. The random spatial deployment of sources in turn implies the aggregate interference I_{aggr} at the radar must also be statistically characterized and plays an important role in the performance analysis of radar detection in the presence of interference. When the Wi-Fi deployment can be modeled as a homogeneous Poisson point process (PPP) with intensity λ, assuming propagation with pathloss exponent[4] $\alpha = 4$, a Levy probability density function (pdf) is realized for I_{aggr}, which takes the form [20]

$$f_{I_{\text{aggr}}}(x) = \frac{\pi \lambda}{2x^{3/2}} \exp\left(-\frac{\pi^3 \lambda^2}{4x}\right). \tag{13.14}$$

It is worth noting that the Levy pdf has a heavier tail than the Gaussian distribution and was derived [20] for the scenario involving *no protection region* around the radar, i.e. nearby sources contribute significantly to I_{aggr}.

13.3.1 Average and variance of aggregate interference (μ_I, σ_I^2): Campbell's theorem

Given the fact that the pdf of I_{aggr} is available in closed form for only a few specific cases, we now resort to the use of a basic result that allows characterization of the 1st

[4]There exists no closed form solution other than for $\alpha = 4$; the pdf for these must be obtained via numerical simulations.

and 2nd moments of I_{aggr}. For a broad class of point processes, the mean and variance of the random sum $\sum_{\mathbf{x}\in\Phi} f(\mathbf{x})$ is given by Campbell's theorem [20] as

$$E\left[\sum_{\mathbf{x}\in\Phi} f(\mathbf{x})\right] = \int_{\mathcal{R}^2} f(\mathbf{x})\Lambda(\mathbf{x})\,d\mathbf{x} \tag{13.15}$$

$$var\left[\sum_{\mathbf{x}\in\Phi} f(\mathbf{x})\right] = \int_{\mathcal{R}^2} f^2(\mathbf{x})\Lambda(\mathbf{x})\,d\mathbf{x} \tag{13.16}$$

where $\Lambda(\mathbf{x})$ is the intensity function. Identifying $f(\mathbf{x}) = ((P_{\text{SU}}G(\theta_{\mathbf{x}}))/(\text{FDR}(\Delta f)))l(||\mathbf{x}||)$ from (13.13) allows us to compute the statistics of the aggregate interference at the radar input, as derived below.

13.3.1.1 Homogeneous Poisson point process (PPP)

We first consider the homogeneous PPP, for which $\Lambda(\mathbf{x}) = \lambda$. The average interference incident at the radar is calculated from (13.15) by integrating over the $\mathcal{R}^2$ plane, i.e.

$$\mu_I = E[I_{\text{aggr}}] = \frac{\lambda P_{\text{SU}}}{\text{FDR}(\Delta f)} \int_{\mathcal{R}^2} G(\theta_x) l(||x||)\,dx. \tag{13.17}$$

To incorporate the protection region, it is assumed that there is a minimum separation distance between the radar and active SUs, as a function of azimuth angle θ, denoted as $d(\theta)$. Using polar coordinates and considering $l(r) = K_0 r^{-\alpha}$ for a general path loss model with radial symmetry, for K_0 a constant, we obtain

$$\mu_I = \frac{\lambda P_{\text{SU}} K_0}{\text{FDR}(\Delta f)} \int_{\theta} \int_{d(\theta)}^{\infty} G(\theta) r^{1-\alpha}\,dr\,d\theta = C_{\mu_I} \int_{\theta} G(\theta) d^{2-\alpha}(\theta)\,d\theta$$

$$C_{\mu_I} = \frac{\lambda P_{\text{SU}} K_0}{(\alpha - 2)\,\text{FDR}(\Delta f)}, \tag{13.18}$$

where $\alpha > 2$ is necessary to guarantee convergence of the integral over θ and C_{μ_I} is a constant of the integral.

It is evident that the mean aggregate interference is dependent on the choice of the function $d(\theta)$, which in turn depends on the radar antenna pattern $G(\theta)$. As shown in the next section, the choice of $d(\theta)$ can be optimized subject to some constraints. The variance of aggregate interference—calculated from (13.16)—follows a similar approach to that for μ_I, resulting in

$$\sigma_I^2 = \frac{\lambda P_{\text{SU}}^2 K_0^2}{\text{FDR}^2(\Delta f)} \int_{\theta} \int_{d(\theta)}^{\infty} G^2(\theta) r^{-2\alpha} r\,dr\,d\theta = C_{\sigma_I^2} \int_{\theta} G^2(\theta) d^{2-2\alpha}(\theta)\,d\theta$$

$$C_{\sigma_I^2} = \frac{\lambda P_{\text{SU}}^2 K_0^2}{(2\alpha - 2)\,\text{FDR}^2(\Delta f)}, \tag{13.19}$$

where $\alpha > 1$ ensures convergence of the integration over r and $C_{\sigma_I^2}$ is a constant of the integral.

13.3.1.2 Nonhomogeneous Poisson process

Now consider the Thomas cluster process, which belongs to the class of Neyman–Scott cluster processes [20] and is characterized by

$$\Lambda(\mathbf{x}) = \lambda_{\mathrm{p}}\, \bar{c},$$

where λ_{p} is the intensity of a parent standard Poisson process that identifies a cluster head (AP) and the number of (client) points associated with a cluster is independently Poisson distributed with density $\bar{c}$. Around every parent point, each daughter point is distributed according to a symmetric normal distribution in 2-D with variance ζ^2 as

$$f_{\mathrm{cl}}(\mathbf{x}) = \frac{1}{2\pi\zeta^2}\exp\left(-\frac{||\mathbf{x}||^2}{2\zeta^2}\right) \tag{13.20}$$

where $\lambda = \lambda_{\mathrm{p}}\bar{c}$, and $||\mathbf{x}||$ is the radial distance of the client with respect to a cluster head (AP). The average interference is given by [20]

$$E[I_{\mathrm{aggr}}] = E[I_{\mathrm{aggr-PPP}(\lambda)}] + \frac{\bar{c}\, P_{\mathrm{SU}}}{(4\pi\zeta^2)\,\mathrm{FDR}(\Delta f)} \int_{\mathcal{R}^2} G(\theta_{\mathbf{x}})\, l(||\mathbf{x}||) \exp\left(-\frac{||\mathbf{x}||^2}{4\zeta^2}\right) d\mathbf{x} \tag{13.21}$$

where $E[I_{\mathrm{aggr-PPP}(\lambda)}] = \mu_I$ is the average interference at the radar receiver when the Wi-Fi nodes are distributed according to a homogeneous PPP with density λ. The solution for the above integral is generally not available in closed-form and must be computed numerically.

Note that the average interference at the radar is identical at each azimuth when the Wi-Fi deployment follows a homogeneous PPP or the Thomas cluster process. We will later consider an inhomogeneous process in azimuth for the Wi-Fi deployment and explore its impact on the aggregate interference via simulation.

13.4 Simulation results for omni antenna gain pattern

For a radar receiver located at the origin (per Figure 13.2), nodes are scattered according to a PPP with density $\lambda = 10^{-6}$. We assume an omnidirectional antenna pattern for the radar with a gain of $G(\theta) = 10$ dBi. The SU power (including antenna gain) is set to $P_{\mathrm{SU}} = 1.0$ W, which is in line with commercial devices. The path-loss component is $\alpha = 4$ and the maximum outage probability is $P_{\mathrm{out,max}} = 0.1$, where the outage probability is defined as

$$P_{\mathrm{out}} = \Pr\{I_{\mathrm{aggr}} > I_{\max}\}.$$

The aggregate interference power I_{aggr} at the radar receiver is simulated via the following steps:

1. Initially, for a given omniprotection region ($d(\theta) = R$), the mean and variance of I_{aggr} at the radar is calculated using Campbell's theorem.
2. For each $d(\theta)$, the numerically computed mean (μ_{theory}) and variance ($\sigma^2_{\mathrm{theory}}$) based on Campbell's theorem are validated by Monte-Carlo simulations. For

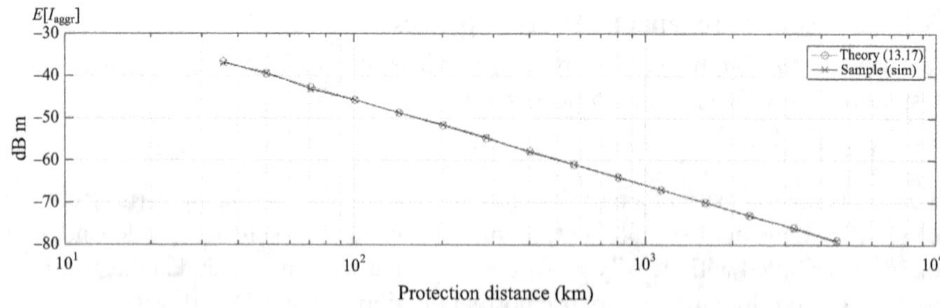

Figure 13.3 Mean of aggregate interference (1,000 iterations)

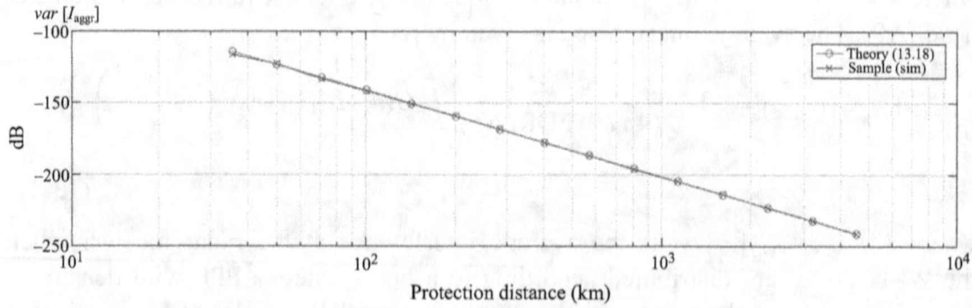

Figure 13.4 Variance of aggregate interference (1,000 iterations)

every simulated case, a total of 1,000 snapshots are generated and the sample mean/variance of the aggregated interference is calculated.

13.4.1 Campbell's theorem verification

The protection region in this case is circular, i.e. $d(\theta) = R$, where R is the radius of the protection region. Figures 13.3 and 13.4 show the calculated mean and variance, respectively, for 1,000 snapshots. It is observed that Campbell's theorem is corroborated for a range of distances from 35 km up to 4,500 km. The results are tabulated in Table 13.3.

13.4.2 Optimizing the protection region for homogenous PPP

Now consider how one might optimize $d(\theta)$ subject to a constraint on I_{aggr} and the outage probability, i.e.

$$\Pr\{I_{\text{aggr}} > I_{\max}\} \leq P_{\text{out,max}}\,, \tag{13.22}$$

where a typical value is $P_{\text{out,max}} = 0.1$. We choose specific values of $I_{\max}$ (e.g. -50 and -60 dBm) for which the protection region is optimized, i.e. $d_{\text{opt}}(\theta)$ is determined such that this constraint is satisfied. In the following, three different methods are used

Table 13.3 Campbell's theorem verification table

$d(\theta) = R$ (km)	μ_{theory} (dB m)	$\mu_{\text{sample}-1,000}$	σ^2_{theory} (dB)	$\sigma^2_{\text{sample}-1,000}$	Time (s)
35.6	−36.7733	−37.0256	−114.3099	−116.2947	22.22
50.3	−39.7836	−39.4451	−123.3408	−122.9496	21.81
71.1	−42.7939	−43.2891	−132.3717	−133.4190	22.24
100.6	−45.8042	−45.8034	−141.4026	−141.7799	21.73
142.3	−48.8145	−48.8520	−150.4335	−150.7537	21.87
201.2	−51.8248	−51.5716	−159.4644	−159.2130	22.07
284.5	−54.8351	−54.6039	−168.4953	−168.2962	22.10
402.4	−57.8454	−58.0629	−177.5262	−177.6571	22.11
569.0	−60.8557	−60.8903	−186.5571	−186.6209	22.00
804.8	−63.8660	−63.9500	−195.5880	−196.1037	21.81
1,138.1	−66.8763	−66.8855	−204.6189	−204.6284	21.79
1,609.5	−69.8866	−69.8694	−213.6498	−213.5399	21.69
2,276.2	−72.8969	−72.9489	−222.6807	−222.8702	21.70
3,219.0	−75.9072	−76.0084	−231.7116	−231.7877	20.88
4,552.4	−78.9175	−79.1378	−240.7425	−240.7864	20.40

to calculate $d(\theta)$: brute-force simulation, iterative Monte-Carlo, and an analytical method based on the Gaussian assumption.

13.4.2.1 Numerical brute-force approach for omniprotection region

For the brute-force method, we use the same parameters as in Section 13.4, along with $I_{max} = -50$ dBm. The following steps are followed to calculate the protection distance:

1. Initially, select $d(\theta) = 100$ km. For a given value of $d(\theta)$, a histogram of I_{aggr} is generated for 1,000 snapshots.
2. Using the I_{max} value, the outage probability $P_{out} = \Pr\{I_{aggr} > I_{max}\}$ is calculated and compared against $P_{out,max}$. If $P_{out} \leq P_{out,max}$, then stop; the current value of $d(\theta)$ is the optimum $d_{opt}(\theta)$. Otherwise, go to next step.
3. If $P_{out} > P_{out,max}$, then increment $d(\theta)$ by $\Delta = 0.5$ km and repeat the first step.

Figure 13.5 shows the variation of P_{out} vs. $d(\theta)$, from which it is observed that P_{out} oscillates around the true value. We then use a cubic spline function with a smoothing parameter of 0.001 and also tabulate the results in Table 13.4. The brute-force method takes substantial time to complete. The next section demonstrates an iterative Monte-Carlo method that is significantly faster.

13.4.2.2 Iterative Monte-Carlo approach to determine $d_{opt}(\theta)$

Now consider an adaptive on-line approach to update the values of $d(\theta)$ based on the Adam algorithm proposed in [21]. Adaptive moment estimation (Adam) computes adaptive learning rates for each parameter by using an exponentially decaying average

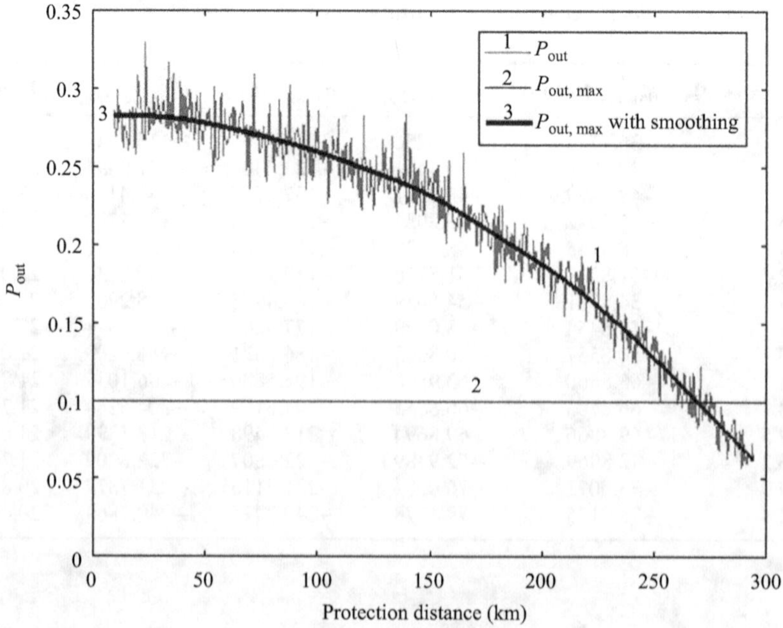

Figure 13.5 P_{out} *vs.* $d(\theta)$*—Numerical brute force approach* ($I_{max} = -50$ *dBm)*

of past gradients (m_t) and past squared gradients (v_t). The steps of this algorithm are outlined below:

1. The gradient is chosen as $g_t = P_{out} - P_{out,max}$.
2. Initialize: Step-size $\alpha = 100, \beta_1 = 0.9, \beta_2 = 0.999$, and $\epsilon = 10^{-8}$.
3. The first moment vector (m), second moment vector (v), and time-step vector (t) are initialized as
   $m \leftarrow 0, v \leftarrow 0, t \leftarrow 0$ and $d(\theta) = 10$ km.
4. **while** $d(\theta)$ has not converged, **do**
   $t \leftarrow t + 1$
   run Monte-Carlo experiment 1,000 times to obtain P_{out-t}
   $g_t \leftarrow P_{out-t} - P_{out,max}$
   $m_t \leftarrow \beta_1 m_{t-1} + (1 - \beta_1)g_t$
   $v_t \leftarrow \beta_2 v_{t-1} + (1 - \beta_2)g_t^2$
   $\alpha_t \leftarrow \alpha(1 - \beta_2^t)^{\frac{1}{2}}/(1 - \beta_1^t)$
   $\hat{m}_t \leftarrow \beta_1 m_{t-1} + (1 - \beta_1)g_t$
   $\hat{v}_t \leftarrow \beta_2 v_{t-1} + (1 - \beta_2)g_t^2$
   $d_t(\theta) \leftarrow d_{t-1}(\theta) - \alpha_t \hat{m}_t/(\hat{v}_t)^{\frac{1}{2}} + \epsilon)$
   end while
5. stop: $d_t(\theta)$ is optimal protection distance

To check convergence at each iteration, a histogram of I_{aggr} is generated using 1,000 snapshot results for a given value of $d(\theta)$. Using I_{max}, the outage probability

Table 13.4 Effect of smoothing in Figure 13.5

	$d_{opt}(\theta)$ (km)	P_{out}	Time (s)
Without smoothing	252.5	0.100	4,082.4
With smoothing	269.9	0.098	4,082.4

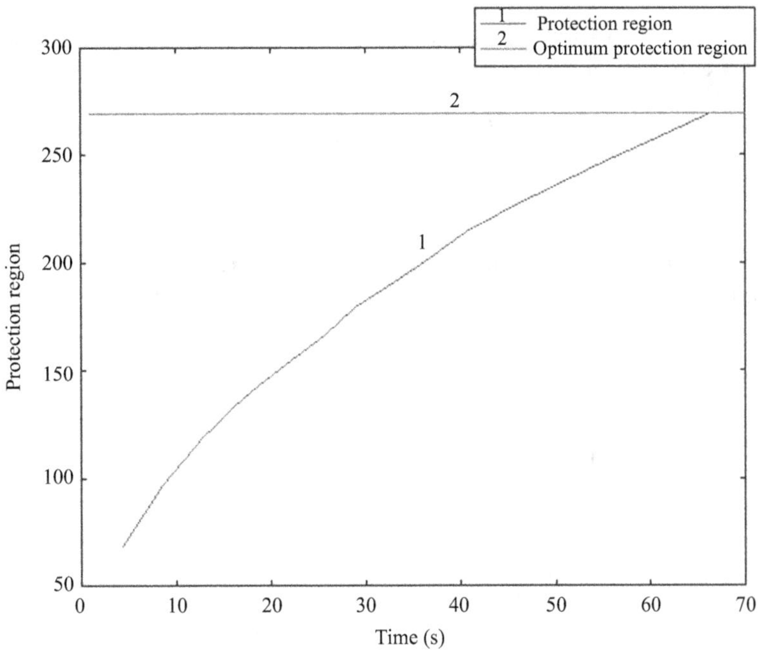

Figure 13.6 Update of $d(\theta)$ vs. time (s)—Iterative Monte-Carlo simulation ($I_{max} = -50$ dBm)

Table 13.5 Comparison of processing time to find d_{opt}

Algorithm	$d_{opt}(\theta)$ (km)	P_{out}	Time (s)
Numerical brute-force	269.9	0.100	4,082.4
Iterative Monte Carlo	268.8	0.099	66.2

$P_{out} = \Pr\{I_{aggr} > I_{max}\})$ is calculated and compared against $P_{out,max}$. If $|P_{out} - P_{out,max}| \leq \delta = 0.005$, then the algorithm stops; keeping the current value of $d(\theta)$ as $d_{opt}(\theta)$. Figure 13.6 illustrates the updating process of $d(\theta)$ in time (s) from initialization to convergence. The benefit, in terms of time to converge, of using the iterative update is captured in Table 13.5, which shows significant gains.

13.4.2.3 Gaussian assumption for distribution of I_{aggr}

In [22], the distribution of I_{aggr} was approximated as a normal distribution $N(\mu_I, \sigma_I^2)$ for mathematical tractability to realize an analytical determination of the optimum protection region. With the Gaussian assumption, the outage probability can be determined as

$$P_{outage} = \Pr\{I_{aggr} > I_{max}\} = Q\left(\frac{I_{max} - \mu_I}{\sigma_I}\right), \tag{13.23}$$

where $Q(\cdot)$ is the tail probability of the standard normal distribution. It is desired to set an upper bound for probability of outage, $P_{out,max}$, which is expressed as

$$\begin{aligned} &P_{outage} \le P_{out,max} \rightarrow Q\left(\frac{I_{max} - \mu_I}{\sigma_I}\right) \le P_{out,max} \\ &I_{max} \ge \mu_I + \sigma_I Q^{-1}\left(P_{out,max}\right). \end{aligned} \tag{13.24}$$

Here, the coexistence criteria is defined by (13.24). This inequality is trivially achieved by letting $d(\theta) \rightarrow \infty$ [apparent from (13.18), for example]. In order to avoid this trivial condition, we *minimize the total protection area* subject to a net interference limit as formulated by the optimization problem

$$d_{opt} = \arg\min_{d(\theta)} \int_0^{2\pi} d^2(\theta)d\theta \tag{13.25}$$

subject to

$$\mu_I + \sigma_I Q^{-1}\left(P_{out,max}\right) \le I_{max}. \tag{13.26}$$

For a general antenna pattern $G(\theta)$, it is proven in Appendix A of this chapter that the optimum protection distance $d_{opt}(\theta)$ is proportional to $G^{1/\alpha}(\theta)$ with a constant γ that is determined by numerically solving

$$\begin{aligned} &d_{opt}(\theta) = \gamma G^{(1/\alpha)}(\theta) \qquad \text{for} \\ &\mathcal{A}\gamma^{2-\alpha} + \mathcal{B}\gamma^{1-\alpha} - I_{max} = 0, \end{aligned} \tag{13.27}$$

in which $\mathcal{A}$ and $\mathcal{B}$ are determined by

$$\begin{aligned} &\mathcal{A} = C_{\mu_I} \int_0^{2\pi} G^{(2/\alpha)}(\theta)d\theta \\ &\mathcal{B} = Q^{-1}\left(P_{out,max}\right)\sqrt{C_{\sigma_I^2} \int_0^{2\pi} G^{(2/\alpha)}(\theta)d\theta}\,. \end{aligned} \tag{13.28}$$

Table 13.6 Comparison of calculated protection distance and resulting outage probability for Gaussian vs. Monte-Carlo

I_{max} (dBm)	−40	−50	−60	−70
Iterative Monte Carlo				
d_{opt} km	41.41	261.49	654.7	1,787.5
(P_{out})	(0.098)	(0.102)	(0.100)	(0.101)
Gaussian assumption				
d_{opt} km	112.08	262.45	659.0	1,809.0
(P_{out})	(0.057)	(0.098)	(0.106)	(0.089)

13.4.2.4 Comparison of the Gaussian assumption with Monte-Carlo simulation

To evaluate the accuracy of the Gaussian assumption, the calculated protection distance is compared with that of the Monte-Carlo simulations. The maximum interference level I_{max} is varied from −70 to −40 dBm to cover both large and small cases of protection distance. Table 13.6 quantifies the resulting distances as well as the simulated outage probabilities for both methods.

The Monte-Carlo method achieves the desired outage probability of $P_{out} \approx 0.1$ for all cases of I_{max}. For smaller values of I_{max}, the Gaussian assumption results in similar P_{out} and d_{opt}, providing an accurate analytical solution. For large values such as $I_{max} = -40$ dBm, however, the Gaussian assumption results in an outage probability of $P_{out} \approx 0.057$ that is significantly smaller than the allowed bound. Therefore, the corresponding protection distance is significantly larger than that which is determined by the Monte-Carlo method (112 vs. 41 km).[5]

In Figures 13.7 and 13.8, the histogram of I_{aggr} from the above simulation scenario (for the cases of $I_{max} = -50, -70$ dBm) is plotted to compare with the Gaussian, Inverse Gaussian, and Levy pdfs. In [20], the Inverse Gaussian distribution is shown to be a better fit than the Levy distribution for cases involving active nodes outside the protection region, with the pdf given by

$$f_I(x) = \left[\frac{\kappa}{2\pi x^3}\right]^{1/2} \exp\left(-\frac{\kappa(x-\nu)^2}{2\nu^2 x}\right) \quad (13.29)$$

where $E[I] = \nu$ and $var(I) = \nu^3/\kappa$.

13.4.3 Poisson cluster process

Now consider Wi-Fi deployments that are modeled as *spatially inhomogeneous* point processes. Here, we focus on Poisson Cluster Processes for the spatial distribution of

[5]The approximate value from the Gaussian method in such cases can be used to initialize the iterative Monte-Carlo simulation, which can then find the accurate value rather quickly.

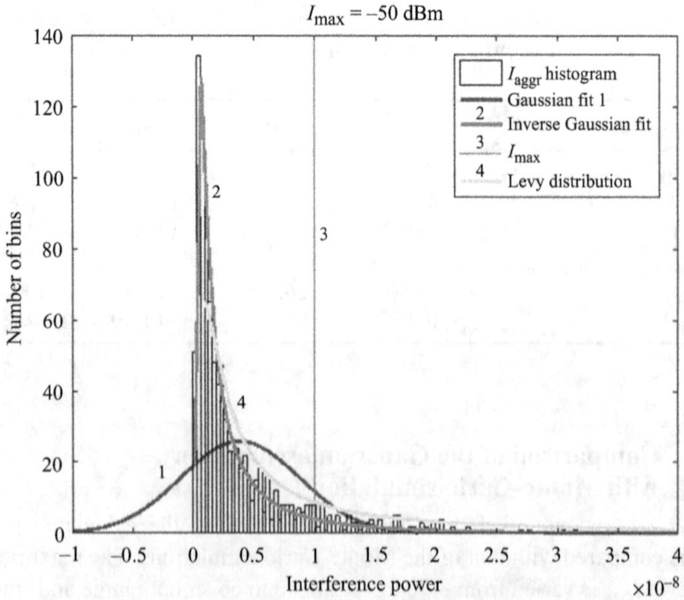

Figure 13.7 Gaussian fit is not a good approximation for I_{max} relatively large (here −50 dBm)

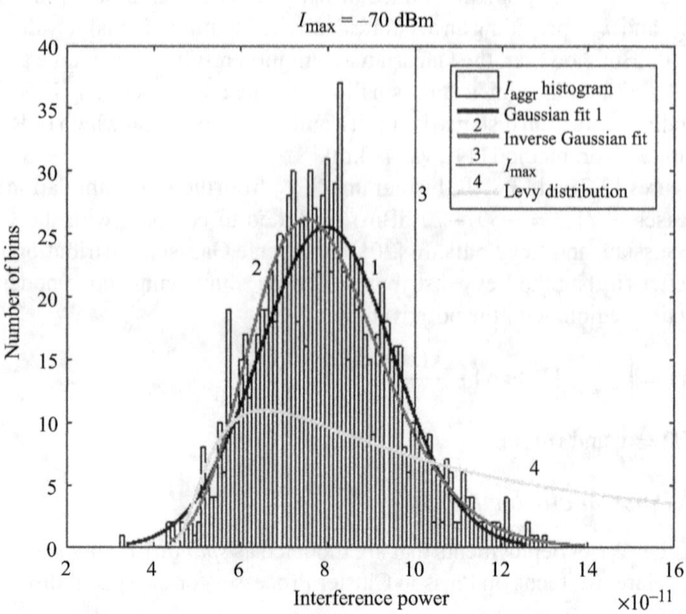

Figure 13.8 Gaussian fit is a reasonable approximation for I_{max} relatively small (here −70 dBm)

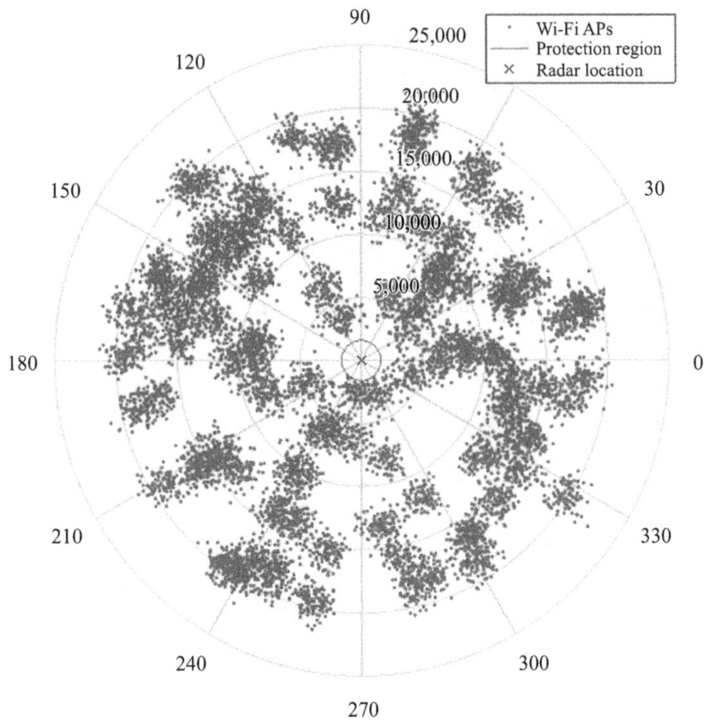

Figure 13.9 One realization of the Thomas cluster process

nodes. A cluster process is created using a parent process with density λ_p that defines the location of the clusters, and a child process that is created around each cluster point. The number of points in a cluster is a Poisson random variable with mean $\bar{c}$. These points are scattered around the cluster center, with a Gaussian distribution of zero mean and variance ζ.

Similar to previous results, the radar antenna pattern is selected to have $G(\theta) =$ 10 dBi, along with $\lambda_p = 10^{-7}$ km^{-2}, $\bar{c} = 100$, and $\zeta = 800$ for the Poisson cluster (Thomas) process. The path-loss exponent is set to $\alpha = 4$, $I_{max} = -60$ dBm, and $P_{out,max} = 0.1$ is used. The simulations are again conducted using 1,000 snapshots for averaging. The protection distance is still a circle with radius $d_{opt}(\theta)$ and results from the fact that the spatial process, while inhomogeneous, exhibits radial symmetry of cluster/node locations (as illustrated in Figure 13.9) and all sources use an omnidirectional antenna pattern.

13.4.3.1 Campbell's theorem verification

Similar to Section 13.4.1, Campbell's theorem is verified numerically using the new spatial distribution and for various protection distances. Figure 13.10 shows

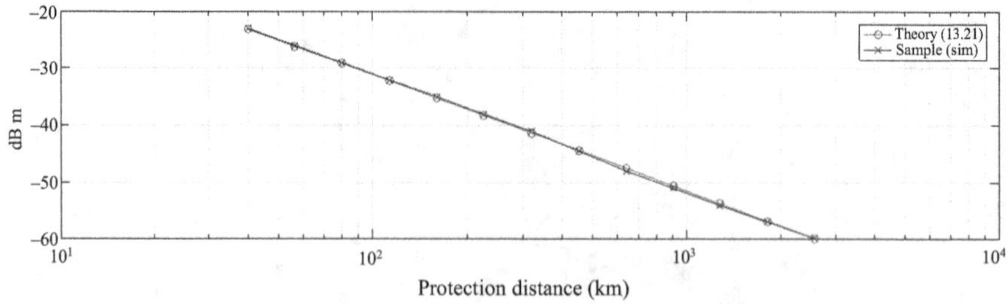

Figure 13.10 Mean of aggregate interference (1,000 iterations)

Table 13.7 Comparison of protection distance ($I_{max} = -60$ dBm, $\lambda_p = 10^{-7}$, $\bar{c} = 10$)

Point process	**Intensity**	d_{opt}
Homogeneous PPP	$\lambda = \lambda_{\text{p}}\bar{c}$	662.8
Thomas cluster process	$\lambda_{\text{p}}, \bar{c}$	663.2

the average interference measured at the radar receiver from nodes with a Poisson cluster distribution. A perfect match is observed between what Campbell predicts for average interference and these simulation results.

It is insightful to compare the average interference from a Poisson cluster process with a corresponding homogeneous PPP. From (13.21), we know that the increase in average interference power, compared to the homogeneous PPP, is given by

$$\frac{\bar{c}\,P_{\text{SU}}}{\text{FDR}(\Delta f)\,(4\pi\zeta^2)} \int_{\mathcal{R}^2} G(\theta_{\mathbf{x}})\,l(||\mathbf{x}||)\exp\left(-\frac{||\mathbf{x}||^2}{4\zeta^2}\right) d\mathbf{x}\,.$$

This value tends to 0 for large values of protection regions, and therefore the increase in interference power due to clustering is negligible. A cluster process using $\lambda_{\text{p}} = 10^{-7}$ and $\bar{c} = 2{,}000$ has been simulated for $\zeta \in (0, 300]$, with no visible difference observed when compared to a corresponding homogeneous PPP.

The optimum protection region can be found using the same methods described for the homogeneous PPP. A fair comparison of the Thomas cluster process with the homogeneous PPP is made by setting $\lambda = \lambda_{\text{p}}\bar{c}$. Table 13.7 quantifies the optimum protection regions using the iterative Monte-Carlo simulation for both cases of spatial distributions. The resulting distances are very close, which is in line with the previous observation about average I_{aggr}.

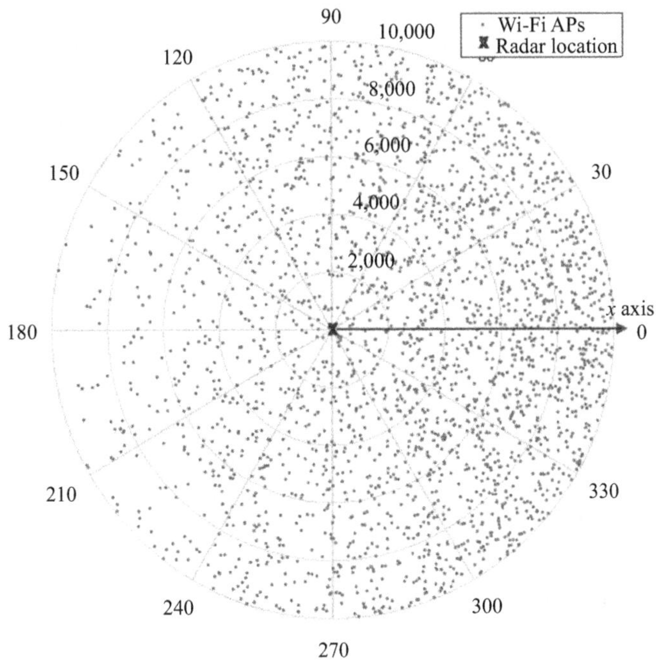

Figure 13.11 One realization of an asymmetric inhomogeneous PPP

13.4.4 Inhomogeneous PPP

In contrast to the findings above, we now examine the impact of an inhomogeneous PPP that is *not* symmetric in azimuth angle. Per [19], a homogeneous PPP with intensity λ_0 is initially simulated. Then points are randomly removed with a *spatially dependent* probability $1 - ((\lambda(\mathbf{x}))/\lambda_0)$ by assigning every point a random number uniformly distributed in [0, 1]. Let random number u_n be assigned to the nth point, with the given point being deleted if $u_n < 1 - ((\lambda(\mathbf{x}))/\lambda_0)$. The resulting point process has an intensity $\lambda(\mathbf{x})$ and an example realization is shown in Figure 13.11. The resulting density function $\lambda(x)$ of the point process is depicted in Figure 13.12.

For such an inhomogeneous point process, the protection region is found for small sectors based on a local assumption of a homogeneous PPP. For a given sector, Campbell's theorem is verified and a protection distance is obtained. This approach is repeated for all (sufficiently small) sectors to obtain the overall protection region. The result is shown in Figure 13.13 and compared with the protection region for a corresponding homogeneous PPP with intensity $\lambda_0/2$. In this case, the protection region is not circular. However, the total area of the protected region for both cases is very similar, as detailed in Table 13.8.

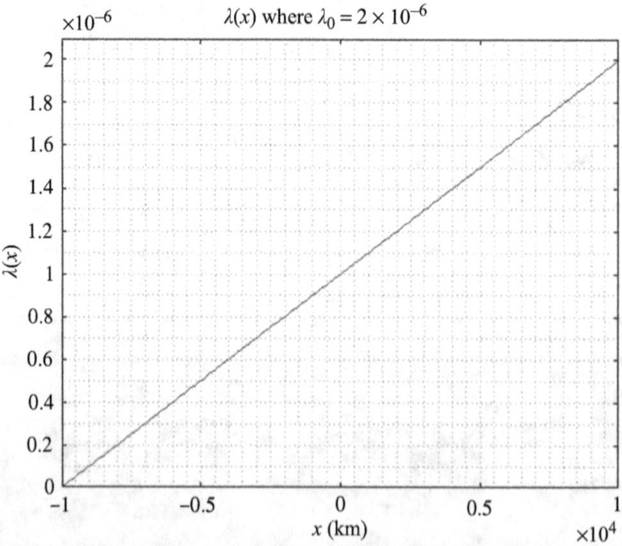

Figure 13.12 Resulting density function $\lambda(x)$ for asymmetric inhomogeneous PPP

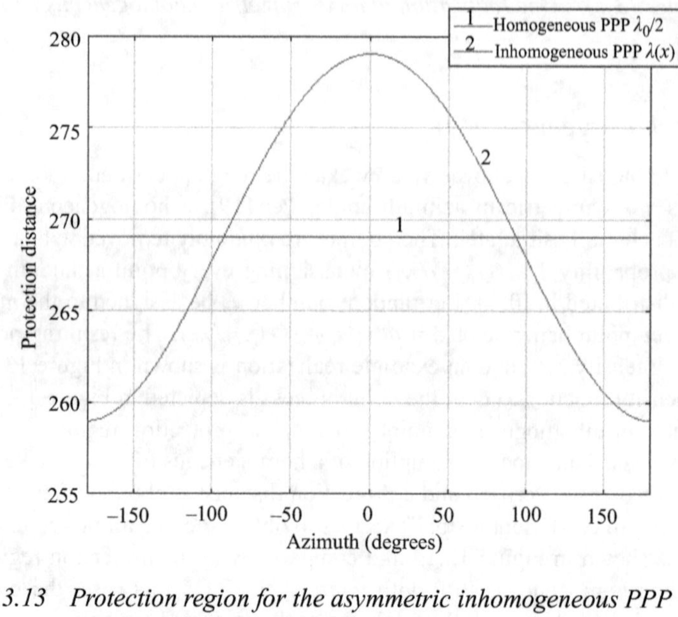

Figure 13.13 Protection region for the asymmetric inhomogeneous PPP Wi-Fi deployment, compared with the homogeneous PPP deployment ($\lambda_0 = 2 \times 10^{-6}, I_{max} = -50$ dBm)

Table 13.8 Fair comparison with a homogeneous PPP having $\lambda_0/2$ intensity for $I_{max} = -50$ dBm

Poisson point process	Intensity	Area of protection region (sq. km)
Inhomogeneous	$\lambda(x)$	2.2733×10^5
Homogeneous	$\lambda_0/2$	2.2770×10^5

13.5 Simulation results for typical radar gain pattern

Here we investigate the case of realistic (non-omni) antenna gain patterns for the radar receiver using the model in [23]. The maximum gain of the antenna (mainlobe) is typically in the range of $22 < G_{\max} < 48$ dBi. The antenna gain $G(\theta)$ vs. azimuth angle can be described by the piece-wise function

$$G(\theta) = \begin{cases} G_{\max} - 0.0004\,\theta^2 \times 10^{G_{\max}/10} & \theta \in [0, \theta_{\mathrm{M}}] \\ 0.75G_{\max} - 7 & \theta \in [\theta_{\mathrm{M}}, \theta_{\mathrm{R}}] \\ 53 - G_{\max}/2 - 25\log(\theta) & \theta \in [\theta_{\mathrm{R}}, \theta_{\mathrm{B}}] \\ 11 - G_{\max}/2 & \theta \in [\theta_{\mathrm{B}}, 180] \end{cases} \tag{13.30}$$

where $\theta_{\mathrm{M}} = 50\sqrt{(0.25G_{\max} + 7)}/(10^{G_{\max}/20})$, $\theta_{\mathrm{R}} = 250/(10^{G_{\max}/20})$, and $\theta_{\mathrm{B}} = 48$. Figure 13.14 shows the antenna gain vs. azimuth with a mainlobe value of $G_{\max} =$ 33.5 dBi.

The results for the homogeneous PPP are again used as a baseline for the non-omni gain pattern. It is important to note that the protection region is now *not* radially symmetric but is instead larger along the mainbeam direction of the radar and significantly smaller in the other directions, such as is depicted in Figure 13.15. By itself, clustering in the deployment geometry is observed to have little effect on the protection region (as observed in Section 13.4.3), though inhomogeneous PPP models that are not radially symmetric do shift the protection region toward the denser portion of the Wi-Fi deployment, as one would intuitively expect.

13.5.1 Homogeneous PPP

The protection distance $d_{\mathrm{opt}}(\theta)$ has been calculated using the antenna pattern in (13.30), with Table 13.9 providing simulation results for different cases of $I_{\max}$. Comparing the theoretical mean from (13.15) with sample averages verifies Campbell's theorem for the case of a directional antenna as well. In addition, Table 13.9 shows the outage probability and the minimum/maximum protection distance d_{opt} (corresponding to minimum/maximum $G(\theta)$) for both Monte-Carlo simulation as well as the Gaussian assumption method.

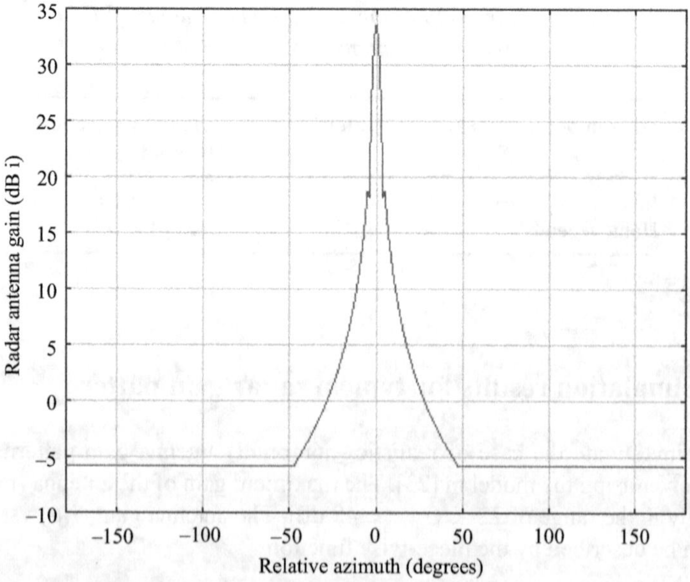

Figure 13.14 Radar antenna gain vs. azimuth

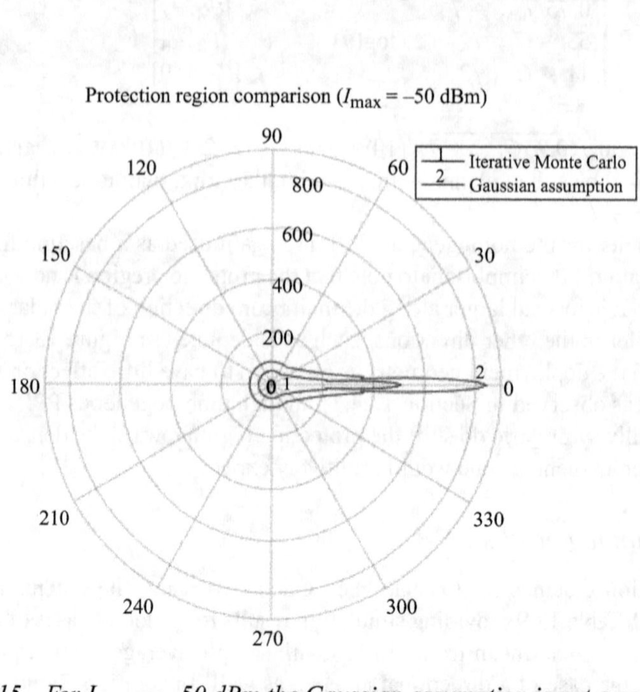

Figure 13.15 For $I_{max} = -50$ dBm the Gaussian assumption is not a good approximation for the distribution of I_{aggr}

Table 13.9 Comparing d_{opt}, P_{out} for Monte-Carlo and Gaussian methods

I_{max}	**−50 dBm**	**−60 dBm**	**−70 dBm**
Iterative Monte Carlo			
d_{opt} km (mainlobe)	509.14	2,055	5,173.1
d_{opt} km (sidelobe)	53.16	214.6	540.1
(P_{out})	(0.100)	(0.103)	(0.104)
Gaussian assumption			
d_{opt} km (mainlobe)	845.75	2,023.9	5,242.8
d_{opt} km (sidelobe)	88.32	211.31	547.41
(P_{out})	(0.057)	(0.125)	(0.098)
Theoretical mean (dBm) per (13.15)	−55.76	−63.33	−71.60
Sample mean (dBm)	−56.38	−63.26	−71.63

Figures 13.15 and 13.16 present $d_{opt}(\theta)$ as a function of azimuth for the two cases of $I_{max} = -50, -70$ dBm. In each figure, Monte-Carlo results are compared with analytical results from the Gaussian assumption. For $I_{max} = -70$ dBm, it is observed that both methods have rather similar outcomes and the Gaussian assumption provides an accurate estimate of $d_{opt}(\theta)$. However, for $I_{max} = -50$ dBm, the Gaussian assumption leads to a more conservative protection region that has larger distances and lower outage probability.

The conservative result from the Gaussian assumption can also be understood from the point of view of the I_{aggr} pdf. Figures 13.17 and 13.18 depict histograms of I_{aggr} (calculated from the simulation results) that are compared with the Gaussian and Inverse-Gaussian pdfs (with same mean and variance). It is observed that the Gaussian pdf mimics the histogram of I_{aggr} with high accuracy when $I_{max} = -70$ dBm. In contrast, the Gaussian pdf is a poor approximation when $I_{max} = -50$ dBm. The inverse-Gaussian pdf, however, provides an accurate model for the pdf of I_{aggr} in both cases.

For smaller protection distances (where Gaussian is a poor model), the heavier tails in the distribution of I_{aggr} arise from the contributions from Wi-Fi users that are at (or close to) the boundary of the protection region, which are possibly orders of magnitude larger than the far-off sources. As the protection region increases, the effect of interference from Wi-Fi sources in the vicinity of the protection region (relative to the rest) reduces and the distribution becomes less heavy tailed (i.e. more symmetric).

13.5.2 Inhomogeneous PPP

Following the approach in Section 13.4.4, the protection region is calculated for an inhomogeneous distribution of nodes and with a directional antenna pattern at the radar. In Figure 13.19, the result is compared with the protection region of a

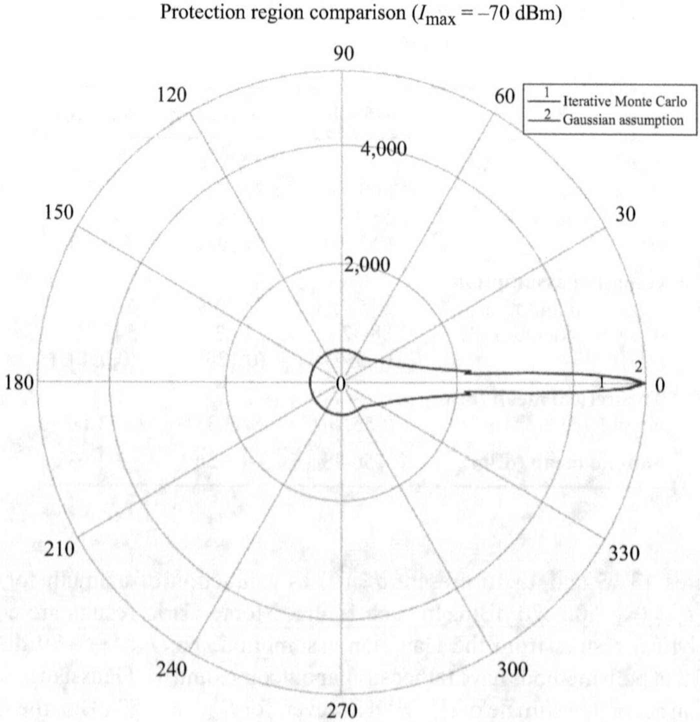

Figure 13.16 For $I_{max} = -70$ dBm the Gaussian assumption is a good approximation for the distribution of I_{aggr}

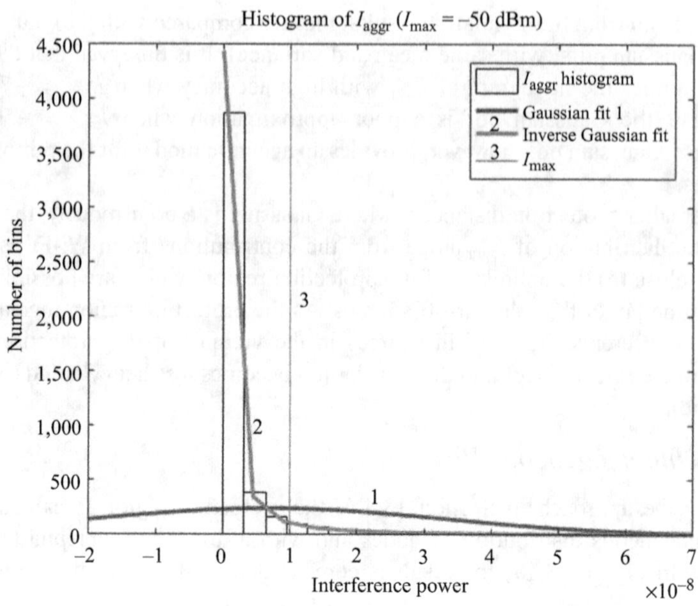

Figure 13.17 Gaussian fit is not a good approximation when $I_{max} = -50$ dBm

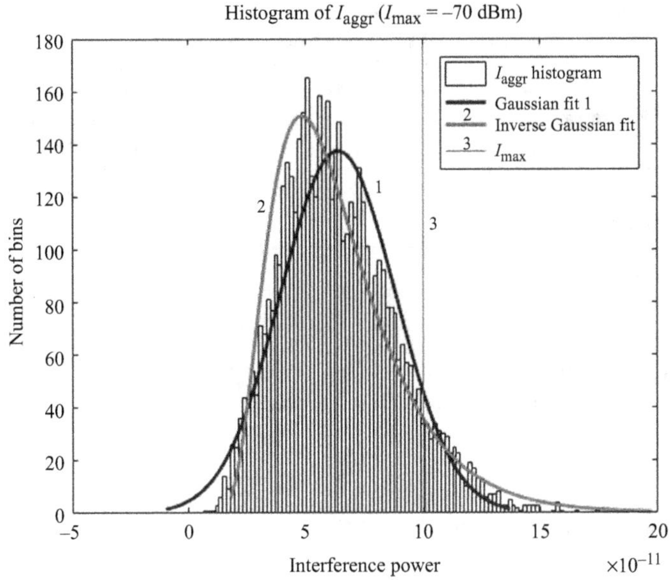

Figure 13.18 Gaussian fit is a reasonable approximation when $I_{max} = -70$ dBm

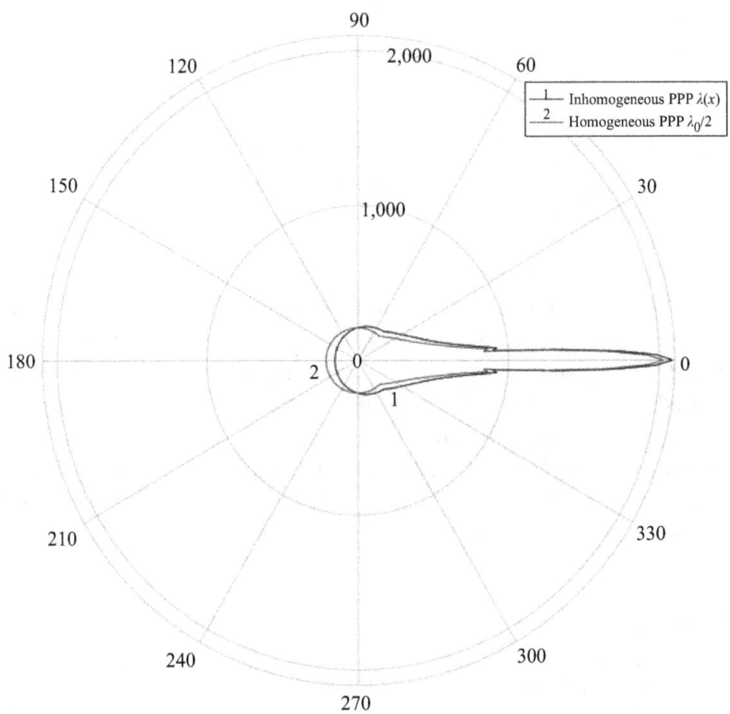

Figure 13.19 Protection region for the inhomogeneous PPP Wi-Fi deployment, compared with the homogeneous PPP Wi-Fi deployment ($\lambda_0 = 2 \times 10^{-6}$, $I_{max} = -60$ dBm)

Table 13.10 Fair comparison with a homogeneous PPP having $\lambda_0/2$ intensity

Poisson point process	Intensity	Area of protection region (sq. km)
Inhomogeneous	$\lambda(x)$	4.1664×10^5
Homogeneous	$\lambda_0/2$	4.1105×10^5

homogeneous PPP having intensity $\lambda_0/2$. The computed protection region area for both cases is quantified in Table 13.10. It is observed that the two cases are in strong agreement.

13.6 Looking ahead: future work

This chapter has established a framework for more precise computation of the protection region for radars operating close to metropolitan areas in which a high density of Wi-Fi networks are deployed. However, there remain several unresolved issues and a few of those are highlighted here.

The primary objective of this analysis has been to explore the impact of various *stochastic point process models* on the aggregate interference I_{aggr} experienced by the radar. Since map data for actual real-world deployments of Wi-Fi APs is available, a useful sanity check would be to explore which of the above models (if any) adequately represent these real deployments. This assessment will potentially result in actual prescriptions for the protection region for a specific metro area, which would likely be of great interest to mobile operators. Field measurements would help determine the validity of this analysis.

Further, as already noted, I_{aggr} is also impacted by the probability of a wireless network transmitting at a given time instant, which is related to the medium access control (MAC) protocol employed. Integrating the probability of transmission into the above models would lead to a more complete analysis. Likewise, we have assumed that various parameters regarding radar operation (e.g. the location of the radar and mainbeam direction) are known exactly. Extending the analysis to situations in which such parameters are unknown and must be estimated—and the consequent effect on the protection region—is another path for future work.

Acknowledgment

This work was supported in part under award AFRL CERFER FA8650-14-D-1722.

Appendix A

Based on (13.25) and (13.26), the optimal protection distance based on limiting the maximum outage probability is obtained as

$$d_{\mathrm{opt}} = \arg\min_{d(\theta)} \int_0^{2\pi} d^2(\theta)\, d\theta$$

$$\mu_I + \sigma_I Q^{-1}\left(P_{\mathrm{out,max}}\right) \leq I_{\mathrm{max}}$$

where μ_I and σ_I are determined in (13.18) and (13.19) to be

$$\mu_I = C_{\mu_I} \int_\theta G(\theta)\, d^{2-\alpha}(\theta)\, d\theta$$

$$\sigma_I = \sqrt{C_{\sigma_I^2} \int_\theta G^2(\theta)\, d^{2-2\alpha}(\theta)\, d\theta}\,.$$

The optimal $d(\theta)$ is attained by converting the inequality constraint in (13.22) to equality, which follows because for any $d(\theta)$ for which the strict inequality constraint holds, we can scale down $d(\theta)$ accordingly to increase μ_I, σ_I and achieve the equality constraint (note that $2 - \alpha < 0$ and $2 - 2\alpha < 0$). This result will clearly produce a smaller objective function.

With equality the constraint, the Lagrange multiplier method is applied with dummy variable ϵ to redefine the objective function as

$$d_{\mathrm{opt}} = \arg\min_{d(\theta)} \int_0^{2\pi} d^2(\theta) d\theta + \epsilon[\mu_I + \sigma_I Q^{-1}\left(P_{\mathrm{out,max}}\right) - I_{\mathrm{max}}].$$

Taking partial derivatives of the new objective function with respect to $d(\theta)$ then results in

$$d(\theta) + \epsilon \left[\frac{\partial \mu_I}{\partial d(\theta)} + Q^{-1}(P_{\mathrm{out,max}}) \frac{\partial \sigma_I}{\partial d(\theta)} \right] = 0,$$

and subsequently replacing μ_I and σ_I yields

$$d(\theta) + \epsilon \left[(2-\alpha) C_{\mu_I} G(\theta) d^{1-\alpha}(\theta) + \frac{(2-2\alpha) Q^{-1}(P_{\mathrm{out,max}}) C_{\sigma_I^2} G^2(\theta) d^{1-2\alpha}(\theta)}{2\sqrt{C_{\sigma_I^2} \int_\theta G^2(\theta) d^{2-2\alpha}(\theta) d\theta}} \right] = 0.$$

Let $X = G(\theta)\,d^{-\alpha}(\theta)$ such that the above equation can be written as $1 + \epsilon\left[\Gamma X + \Lambda X^2\right] = 0$, where Λ and Γ are constants. Solving for X results in

$$G(\theta)d^{-\alpha}(\theta) = \frac{-\epsilon\Gamma \pm \sqrt{\epsilon^2\Gamma^2 - 4\epsilon\Lambda}}{2\epsilon\Lambda}.$$

Thus, $d(\theta) = \gamma G(\theta)^{(1/\alpha)}$ and the proportionality constant γ is found from the constraint equation

$$C_{\mu_I}\int_\theta G(\theta)\gamma^{2-\alpha}G^{((2-\alpha)/\alpha)}(\theta)d\theta + Q^{-1}\left(P_{\text{out,max}}\right) \times \sqrt{C_{\sigma_I^2}\int_\theta G^2(\theta)\gamma^{2-2\alpha}G^{((2-2\alpha)/\alpha)}(\theta)d\theta} = I_{\max},$$

that simplifies to

$$\gamma^{2-\alpha}C_{\mu_I}\int_\theta G^{2/\alpha}(\theta)d\theta + \gamma^{1-\alpha}Q^{-1}\left(P_{\text{out,max}}\right)\sqrt{C_{\sigma_I^2}\int_\theta G^{2/\alpha}(\theta)d\theta} = I_{\max}.$$

References

[1] Federal Communications Commission, "The National Broadband Plan: Connecting America," March 2010. http://www.broadband.gov/plan/.

[2] O. Ileri and N.B. Mandayam, *Dynamic spectrum access models: Toward and engineering perspective in the spectrum debate*, IEEE Communications Magazine, vol. 46, no. 1, pp. 153–160, January 2008.

[3] FCC, ET Docket 08-260, "Second Report and Order and Memorandum Opinion and Order in the Matter of Unlicensed Operation in the TV Broadcast Bands," November 14, 2008.

[4] L. Lu, S. Chang, J. Zhang, *et al.*, "Channel Sensing and Radio Resource Allocation Algorithms for WRAN Systems," IEEE802.22 WG, 2006.

[5] N. Han, S. Shon, J.H. Chung, and J.M. Kim, *Spectral correlation based signal detection method for spectrum sensing in IEEE 802.22 WRAN Systems*. ICACT 2006, Seoul, South Korea, February 1, 2006.

[6] "Presidential Memorandum: Unleashing the Wireless Broadband Revolution," June 2010. http://www.whitehouse.gov/the-press-office/presidential-memorandum-unleashing-wireless-broadband-revolution.

[7] FCC, "Amendment of Commission Rules with Regard to Commercial Operation in 3550–3650 MHz Band," Report&Order 15-74A1, 2015.

[8] U.S. Dept. of Commerce, "Fast Track Evaluation of the 5350–5470 MHz and 5850–5925 MHz Bands Pursuant to Section 6406(b) of the Middle Class Tax Relief and Job Creation Act of 2012," Nat. Telecom. Inf. Admin., Washington,

DC, USA, Tech. Rep., January 2013. https://www.ntia.doc.gov/report/2013/evaluation-5350-5470-mhz-and-5850-5925-mhz-bands.

[9] ETSI, "EN 301 893, document REN/BRAN-0060011," pp. 1–26, 2012. [Online]. Available: http://www.etsi.org/deliver/etsi_en/301800_301899/301893/01.07.02_20/en_301893v010702a.pdf

[10] E.F. Drocella, Jr., J.C. Richards, R.L. Sole, F. Najmy, A. Lundy, and P.M. McKenna, "3.5 GHz Exclusion Zone Analyses and Methodology," NTIA, Washington, DC, USA, Tech. Rep. 15-517, June 2015.

[11] L. Gary, L.E. Strickling, and A. Secretary, "An Assessment of the Near-Term Viability of Accommodating Wireless Broadband Systems in the 1675–1710 MHz, 1755–1780 MHz, 3500–3650 MHz, and 4200–4220 MHz, 4380–4400 MHz Bands," U. S. Dept. of Commerce, Washington, DC, October 1, 2010.

[12] "NTIA Letter to FCC on Commercial Operations in the 3550–3650 MHz Band," Nat. Telecommun. Inf. Admin., Washington, DC, USA, Tech. Rep. 12-354, GN 12-354, April 2015. [Online]. Available: https://www.ntia.doc.gov/fcc-filing/2015/ntia-letter-fcc-commercial-operations-3550-3650-mhz-band.

[13] DARPA STO, "Shared Spectrum Access between Rotating Radar and Communications (SSPARC)," BAA 15-24, 2013. https://www.fbo.gov/spg/ODA/DARPA/CMO/DARPA-BAA-15-42/listing.html.

[14] M.I. Skolnik, Introduction to radar. *Radar Handbook*. 2nd ed., 1962.

[15] N. Levanon, *Radar Principles*. 1st ed., John Wiley and Sons, United States of America, 1988.

[16] ITU, "Characteristics of Radiolocation Radars, and Characteristics and Protection Criteria for Sharing Studies for Aeronautical Radionavigation and Meteorological Radars in the Radiodetermination Service Operating in the Frequency Band 2700–2900 MHz", Tech. Rep., 2000–2003.

[17] R. van Nee and R. Prasad, *OFDM for Wireless Multimedia Communications*. Artech House, Norwood, MA, 2000.

[18] H. Ochiai and H. Imai, *On the distribution of the peak-to-average power ratio in OFDM signals*. IEEE Transactions on Communications, vol. 49, no. 2, pp. 282–289, February 2001.

[19] D. Stoyan, W.S. Kendall, and J. Mecke, *Stochastic Geometry and Its Applications*. 2nd ed., John Wiley & Sons, Hoboken, NJ, 1995.

[20] M. Haenggi and R.K. Ganti, *Interference in large wireless networks*. Foundations and Trends® in Networking, vol. 3, no. 2, pp. 127–248, February 2009. [Online]. Available: http://dx.doi.org/10.1561/1300000015.

[21] D.P. Kingma and J. Ba, *Adam: A method for stochastic optimization*. International Conference on Learning Representations, pp. 1–13, 2014.

[22] F. Hessar and S. Roy, *Spectrum sharing between a surveillance radar and secondary Wi-Fi networks*, IEEE Transactions on Aerospace and Electronic Systems, vol. 52, no. 3, pp. 1434–1448, June 2016.

[23] E.F. Drocella, L. Brunson, and C.T. Glass, "Description of a Model to Compute the Aggregate Interference from Radio Local Area Networks Employing Dynamic Frequency Selection to Radars Operating in the 5 GHz Frequency Range," National Telecommunications and Information Administration (NTIA), Tech. Rep., May 2009.

[illegible]

[illegible] 2005, [illegible]

[illegible] Available [illegible]

[illegible]

[illegible] and [illegible] in [illegible]

[illegible] Seybold, [illegible] and J. [illegible]

[illegible]

[illegible]

[illegible] and [illegible] [illegible]

Chapter 14

Spectrum sharing via interference tolerant transform domain waveform design

Vasu Chakravarthy[1], Zhiqiang Wu[2], and John Ellinger[1]

14.1 Introduction

A recent President's Council of Advisors on Science and Technology (PCAST) report calls for a new Federal spectrum architecture that brings a paradigm shift to spectrum usage based on spectrum sharing. The new architecture should enable multiple users to share the spectrum without interfering with one another [1]. It is worth noting that the PCAST study suggests the sharing of government-held spectrum to meet the growing commercial demand. In other words, commercial systems will share spectrum primarily allocated to military systems, which include radar and communication systems. It is obvious that the present practice is static in the sense that fixed spectrum is allocated to new systems. Furthermore, spectrum coexistence is limited to interference management that involves defining an interference contour for each system. Present approaches are limited to interference avoidance performed at the receiver, but to achieve dynamic spectrum sharing, innovative systems and waveform designs are needed that are interference-tolerant and proactively account for the interference from the transmit side.

Traditionally, radars and communication systems have been designed to operate in allocated frequency bands that are isolated from other systems to avoid interference. The new paradigm calls for sharing the spectrum, which translates to the design of interference tolerant systems. This approach requires new spectrum management policies/regulations as well as enabling technologies to realize the necessary agile/adaptive radio/radar systems. Four synergistic enabling capabilities that will facilitate dynamic spectrum access and sharing are flexibility, agility, radio frequency (RF) sensing, and networking [2].

- *Flexibility* provides the ability to adapt and reconfigure. Software-defined radio platforms are being considered as a key enabler of flexibility.
- *Agility* is the ability to change system or signal processing parameters based on spatial, spectral, and temporal features to minimize inference to other incumbent users.

[1]U.S. Air Force Research Laboratory, United States
[2]Electrical Engineering Department, Wright State University, United States

- *RF sensing*, probably one of the most important functions, is the ability of a system or network to understand other users, channel conditions, and interference sources.
- *Networking* provides the ability to communicate between multiple user groups to coordinate dynamic spectrum allocation and management functions and to exchange RF spectrum knowledge between both homogenous and heterogeneous users.

In general, any type of interference degrades performance and requires additional processing to compensate. The ideal approach is to adopt space, time, and spectral diversities to avoid interference altogether. This chapter focuses on the application of spectral diversity via transform domain (TD) waveform design instead of the traditional time domain formulation. The basic idea behind TD waveform design is to address interference at the transmitter and thereby minimize signal-processing complexities at the receiver [3]. This approach is generalized within the spectrally-modulated-spectrally-encoded (SMSE) framework and includes all multicarrier waveform designs such as orthogonal frequency division multiplexing (OFDM) and multicarrier code-division multiple access (MC-CDMA) [4,5] as special cases. These multicarrier SMSE waveforms are utilized in the design of overlay and underlay cognitive radio waveforms.

The vast majority of existing research on physical layer spectrum coexistence addresses cognitive radio systems, where the focus is mainly on the sharing of spectrum between multiple communication systems [2,6,7]. In this chapter, we introduce the foundational work for a transform domain communication system (TDCS) and SMSE framework, and then discuss the design of multicarrier communication and radar waveforms based on SMSE.

14.2 Foundations and framework

14.2.1 Transform domain communication system

Traditionally, communication waveforms are synthesized in the time domain using assigned frequency allocation(s) to the user(s). If interference is present, it can be mitigated using real-time TD filtering techniques to provide interference suppression. Such techniques can be traced back to [8,9] where the primary responsibility for achieving the requisite signal-to-noise ratio (SNR) improvement resides with the receiver. Subsequent advances in processing power have enabled more computationally intense techniques [10,11], whereby SNR improvement is achieved synergistically through transmit/receive waveform diversity to provide interference avoidance.

The basic idea behind the generation of a TDCS fundamental modulation waveform (FMW) is to avoid existing users (or jammers) by operating dynamically over a given bandwidth to share the spectrum. In 1988, German [10] proposed a system that uses spectral information to modify a direct-sequence (DS) spread-spectrum waveform to avoid jammed frequencies. Subsequently, in 1991, Andren at Harris

Corporation patented a conceptual low probability of intercept (LPI) communication system to hide transmitted signals in noise using TD signal processing [11], though the patent does not provide theoretical analysis or address implementation issues associated with functional processing. The Air Force Research Laboratory and Air Force Institute of Technology (AFIT) subsequently adopted Andren's framework for environmental sampling and waveform generation along with German's transmit signal processing [12], using conventional time-domain matched filtering and maximum likelihood (ML) detection estimation at the receiver.

The present TDCS architecture assumes that both the transmitter and receiver are observing the same electromagnetic environment (EME) and thereby produce similar spectral estimates and notches (identical estimates in the ideal case). This assumption is suitable for "localized" short-range data links where the transmitter and receiver reside in the same jamming/interference environment. There are a number of scenarios where this "local" assumption is valid, e.g. aircraft flying in tight formation with the interference remotely located outside the formation. However, since spectral estimation is performed independently at geographically separated locations, the estimates are generally not identical. This condition can impact transmitted symbols such that (1) they contain energy in spectral regions avoided by the receiver, resulting in a loss of desired signal energy or (2) they have no energy in regions retained by the receiver, which yields an increase in received noise. The overall result is decreased detection SNR and increased symbol error rate [13]. An alternate approach to independent spectral estimation would be to use a dedicated feedback channel between the transmitter and receiver. This channel could be used to convey the receiver spectral environment and performance of the forward link to the transmitter [6].

A functional TDCS implementation involves environmental sampling, spectral estimation, thresholding, notching, phase generation, phase mapping, and inverse transformation to obtain the time-domain FMW. Figure 14.1 shows the functional flow of TDCS signal generation and transmission beginning with environmental sampling and spectral estimation. Once "clean" or interference-free spectral regions are

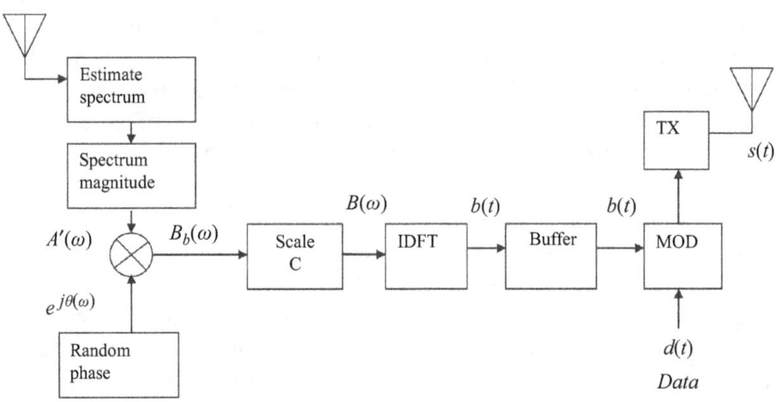

Figure 14.1 TDCS transmitter block diagram

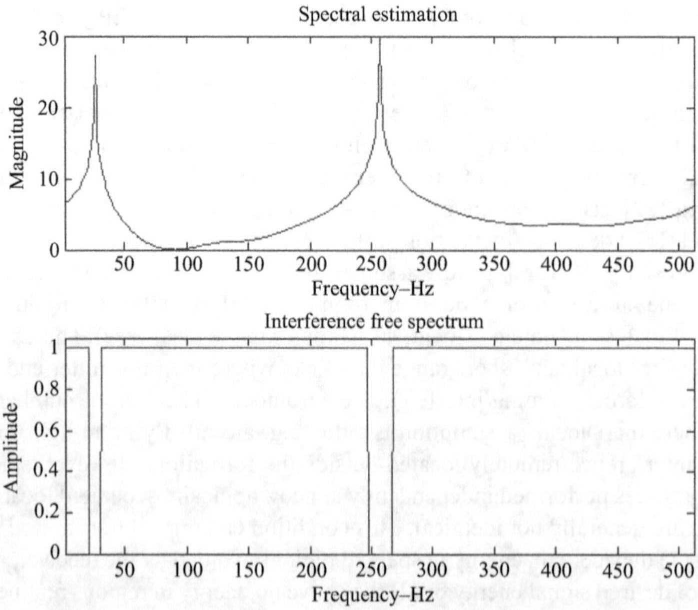

Figure 14.2 Illustration of environmental spectral estimation (top) and corresponding notches (bottom)

established, the FMW $b(t)$ is generated, stored, data modulated, and transmitted. Each functional block of the TDCS transmitter and receiver, and signal processing therein, is explained sequentially in the following.

Estimate spectrum: Spectrum estimation/identification determines the interference-free spectral regions. If interference occurs due to other cooperative systems, prior knowledge can be used to establish their spectral characteristics. In the case of non-cooperative interference, spectral estimation techniques can be used to determine the interference-free bands. A cooperative system is defined here as a legitimate user operating in the assigned spectral region, while a noncooperative system is a jammer. Spectral estimation techniques used previously in TDCS research include the periodogram, autoregressive modeling, and wavelets [13–15].

Spectrum magnitude: The spectrum magnitude is calculated from the spectral estimate. To avoid frequencies containing interference, a hard-limiting threshold is applied (the threshold value is usually based on the mean power in the spectrum). Applying a threshold to the estimated spectrum generates a "clean" or interference-free spectrum $A'(\omega)$ to the extent that the spectral regions containing interference have been avoided. The amplitudes of interfering frequency components exceeding the threshold are set to zero ("nulled"), and the remaining spectral components are assigned a value of one. For example, the top plot in Figure 14.2 shows the spectral

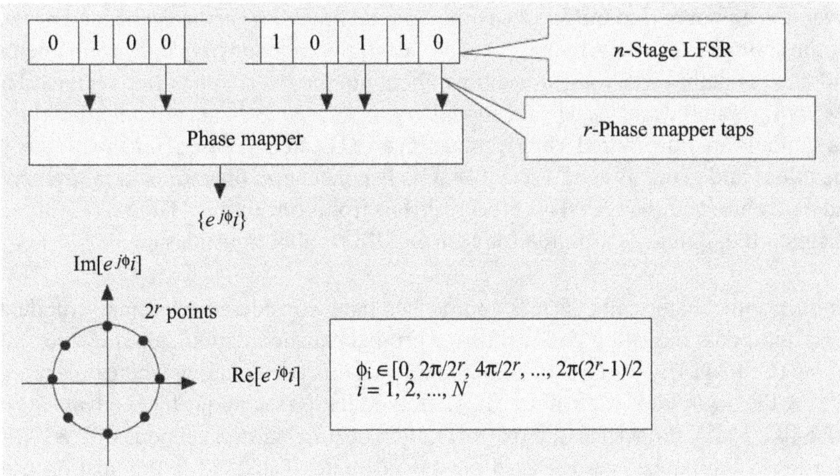

Figure 14.3 TDCS phase mapping process

estimate of a representative environment consisting of two narrow band, single-tone jammers (or other narrowband users) that produce the corresponding-notched response (interference-free spectrum) shown in the bottom plot.

Random phase: A multi-valued, complex, pseudo-random (PR) phase sequence is generated and element-by-element multiplied with $A'(\omega)$ to produce $B_b(\omega)$. This randomization of phase as a function of frequency ensures that the time domain FMW has correlation properties similar to that of sampled noise. As shown in Figure 14.3, each r-bit snapshot of an m-sequence (generated from a n-stage linear feedback shift register (LFSR)) is mapped to one of the 2^r complex phase values. The m-sequence serves two important functions: (1) the PR phasing is critical in the development of noise-like TDCS symbols, which is discussed below, and (2) multiple access is implemented by assigning each user pair a unique primitive polynomial for a different m-sequence [14]. The use of PR coding and the noise-like property of FMW can lead to an erroneous impression that TDCS processing is identical to DS-CDMA. However, PR coding in DS-CDMA serves to spread the spectrum, while TDCS uses the PR code to randomize the phase of the spectral components.

Magnitude scaling: The complex spectrum $B_b(\omega)$ is then scaled to provide the desired energy in the signal spectrum $B(\omega)$. This scaling effectively permits all communication symbols to be transmitted with equal energy, since energy is distributed evenly across the spectral components outside of the notches. Note that for applications in which the peak-to-average-power ratio (PAPR) of the resultant time domain waveform is a concern, several coding techniques have been developed to provide the desired power relationships [16].

Inverse transform and buffer: The time-domain FMW $b(t)$ is then generated by taking the appropriate inverse transform of the spectrally encoded frequency components, with the resultant FMW only containing energy in the interference-free regions. The FMW $b(t)$ is stored and used by the modulator for subsequent data modulation. Multiple symbols are transmitted using a single Inverse Discrete Fourier Transform (IDFT) operation, and generation of a new FMW is dependent on operational requirements and environmental changes. This process differs from conventional OFDM techniques where an IDFT stage is required for each OFDM symbol transmission.

Modulation: Historically, TDCS approaches have considered two binary modulations: antipodal signaling and a form of orthogonal modulation called cyclic shift keying (CSK) [14,17]. Antipodal modulation is a form of signaling where binary signals are the negative of each other. The CSK modulation technique takes advantage of noise-like FMW properties, i.e. the correlation of time-shifted versions of the FMW with itself approaches zero. Based on this attribute, TDCS CSK modulation uses circular shifts of the FMW to represent different symbols. For binary CSK (BCSK), the first symbol $s_1(t)$ is the FMW itself and the second symbol $s_2(t)$ is generated by circularly shifting the FMW over one-half its symbol period T_s. This circular shift in the time domain induces a linear phase shift in the frequency domain without affecting the magnitude. In the case of M-ary CSK (MCSK), the first symbol $s_1(t)$ is again the FMW. The jth symbol $s_j(t)$ for $j = 1, 2, 3, \ldots, M$ is generated by circular shifting the FMW by $(j/M)T_s$ seconds. Thus, the M-ary formulation can be viewed as an extension of the binary case [17].

14.2.1.1 Multiple access in TDCS

TDCS uses phase mapping generated from an LFSR configured to output a maximum-length binary sequence or m-sequence. An LFSR may be configured to provide several different binary sequences depending on the generator polynomial (feedback taps). Figure 14.4 shows two FMWs (top plots) generated by different polynomials that represent different users. The autocorrelation response is indeed an impulse, as expected for noise, and the reduced crosscorrelation response between users is indicative of the quasi-orthogonal behavior required for multiple access [14].

14.2.1.2 TDCS receiver

Figure 14.5 is a block diagram of the TDCS receiver, where the received signal $r(t)$ is comprised of the transmitted signal $s(t)$, channel noise $n(t)$, and possibly interference $i(t)$. As with any communication system, the first step in TDCS receive processing prior to demodulation is acquisition, detection, and synchronization. The received signal $r(t)$ is preprocessed using either direct time correlation or German's technique [10]. The preprocessor output is then passed to the detector to determine signal presence and perform alignment. If a signal is detected and adequately aligned, the receiver continues with estimation followed by frame and symbol synchronization [15]. The TDCS receiver demodulation structure is highlighted by the dotted line in Figure 14.5 that encloses the identical FMW generation process used by the transmitter. Signal $r(t)$ is correlated with locally generated reference signals $c_j(t)$,

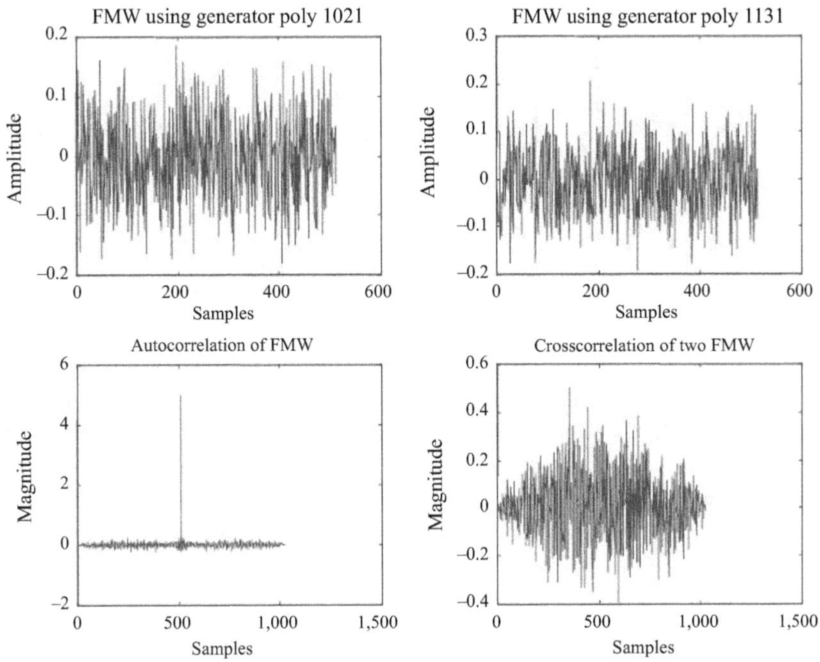

Figure 14.4 TDCS multiple access: two distinct FMWs (top plots) and resultant auto and crosscorrelations (bottom plots)

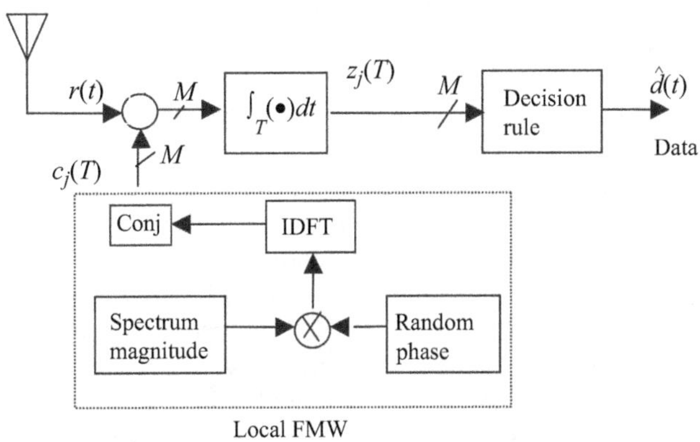

Figure 14.5 TDCS receiver block diagram

where $M = 2$ for binary modulation, there is one locally generated reference for each possible symbol.

Each of the M correlators generates a test statistic $z_j(T)$. An ML decision rule is applied to the test statistic to produce the data estimate $\hat{d}(t)$, with the decision rule dependent on the data modulation. For antipodal modulation, if $z(T)$ is

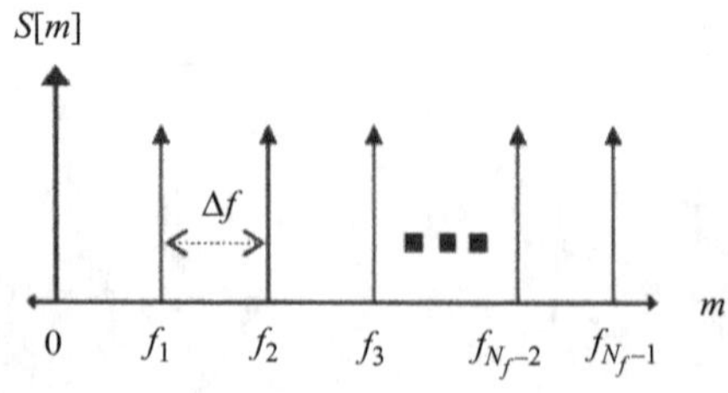

Figure 14.6 Discrete spectral components

positive, then $s_1(t)$ is estimated, otherwise $s_2(t)$ is declared. For BCSK modulation, if $z_1(t) - z_2(t) > 0$, then $s_1(t)$ is estimated, otherwise $s_2(t)$ is declared. Finally, for MCSK, the received signal is correlated with M reference signals, and the test statistic having the largest magnitude is chosen as the estimated symbol. In summary, TDCS provides a communication system that has the capability of sharing spectrum with narrowband transmissions residing in the same frequency band by notching out the frequency components occupied by those signals.

14.2.2 SMSE framework

Developed at AFIT, the spectrally-modulated–spectrally-encoded (SMSE) framework is a general mathematical structure with which to generate a rich family of multicarrier waveforms for RF communication and radar signals [18–20]. The first consideration in developing such a general framework is the number and type of variables required to ensure that the desired level of diversity is achieved. For the purpose of spectrum sharing using SMSE, the most important variable is the number of frequency components available to the TDCS transmitter within a given bandwidth that contains other narrowband emissions. Figure 14.6 illustrates a starting point where the bandwidth is divided into a number of equal subcarriers or spectrum bins.

The identification of usable spectral components involves two variable classes. The N_f assigned frequency components in Figure 14.6 are defined by the vector $\mathbf{a} = [a_1, a_2, \ldots, a_{N_f}]$, for $a_i \in \{0, 1\}$. From this assigned frequency vector, certain frequencies may be unavailable due to interference or by system design. Therefore, the spectral environment can be described by the available frequency vector $\mathbf{u} = [u_1, u_2, \ldots, u_{N_f}]$, also with $u_i \in \{0, 1\}$. In both cases, zeros indicate unused frequencies, and there are $P \leq N_f$ occupied frequencies.

The other variables to consider are the code $\mathbf{c} = [c_1, c_2, \ldots, c_{N_f}]$, for $c_i \in \mathbb{C}$, the data modulation $\mathbf{d} = [d_1, d_2, \ldots, d_{N_f}]$, for $d_i \in \mathbb{C}$, and the windowing function $\mathbf{w} = [w_1, w_2, \ldots, w_{N_f}]$ for spectral shaping, where $w_i \in \mathbb{C}$. These complex variables account for component-wise amplitude and/or phase variations applied to the frequency components. One final "phase only" variable to provide orthogonality among users is also considered and is represented by $\theta_\mathbf{o} = [o_1, o_2, \ldots, o_{N_f}]$, for $\theta_{o_i} \in [0, 2\pi)$.

A continuous time-domain expression for the discrete frequency-domain signal represented in Figure 14.6 is found via an inverse fast Fourier transform (IFFT) of the positive and negative frequency components of $S[m]$, respectively denoted as r_m^{pos} and r_m^{neg}, which yields [19]

$$s_k(t) = \sum_m \text{Re}\left[\mathcal{F}^{-1}\left\{r_m^{\text{pos}}\,\delta(f - f_m) + r_m^{\text{neg}}\,\delta(f + f_m)\right\}\right], \tag{14.1}$$

for time interval $t_i \leq t \leq t_i + T$ with $T = N_f \Delta t$, and Re[·] extracts the real part of the argument. Using (14.1) as a starting point, an SMSE waveform can be developed by applying the data modulation, coding, and windowing factors. The resultant expression for the kth data modulated symbol $S_k[m]$ can be written as

$$S_k[m] = \left\{c_m\, d_{m,k}\, w_m\, e^{-j(2\pi f_m t_n + \theta_{d_{m,k}} + \theta_{c_m} + \theta_{w_m})}\right\}_{m=0}^{N_f-1}, \tag{14.2}$$

where m is the index of frequency component $f_m = m/(N_f \Delta t)$, the discrete-time variable is $t_n = n\Delta t$ for $n = 0, 1, \ldots, N_f - 1$ [18], and $\mathbf{c}$, $\mathbf{d}$, and $\mathbf{w}$ denote the complex code, data modulation, and windowing vectors (magnitudes and phases, as appropriate). By design, the coding and windowing factors only vary with frequency index m, while the data modulation factors are varied according to symbol index k as well. The expression in (14.2) can be further modified to incorporate *orthogonality* and to account for frequency *assignment* and *usage* via

$$S_k[m] = \left\{a_m\, u_m\, c_m\, d_{m,k}\, w_m\, e^{-j(\theta_{d_{m,k}} + \theta_{c_m} + \theta_{w_m} + \theta_{o_{m,k}})}\right\}_{m=0}^{N_f-1}, \tag{14.3}$$

where $a_m u_m \in \{0, 1\}$ and $\theta_{o_{m,k}} = 2\pi k\,(a_m u_m/P)\sum_{i=1}^{m} a_m u_m$ with $P \leq N_f$. The phase term $\theta_{o_{m,k}}$ is used to ensure orthogonality between users.

The SMSE spectral content expression in (14.3) contains the data modulation, encoding magnitude, and phase factors. A time-domain version of the SMSE symbol is generated by applying an IDFT, resulting in

$$s_k[n] = \frac{1}{N_f}\text{Re}\left\{\sum_{m=0}^{N_f-1} a_m\, u_m\, c_m\, d_{m,k}\, w_m\, e^{j(2\pi f_m t_n + \theta_{d_{m,k}} + \theta_{c_m} + \theta_{w_m} + \theta_{o_{m,k}})}\right\} \tag{14.4}$$

for $t_k \leq t_n \leq t_k + T$ and $f_m = f_c + m\Delta f$, with Δf the frequency resolution [20].

14.3 SMSE-based overlay multicarrier communication waveforms

The SMSE framework provides a unified expression for generating and implementing a host of multicarrier type waveforms (e.g. OFDM, MC-CDMA, CI/MC-CDMA[1], TDCS, etc.) and satisfies the spectrum-sharing goal of exploiting *unused* spectral bands, as shown in Table 14.1. We show next how to generate the OFDM and TDCS waveforms from the general SMSE framework.

[1] CI = carrier interferometry.

Table 14.1 Spectrally encoded waveform variables

Operation	Basic OFDM	MC-CDMA	CI/MC-CDMA	TDCS
Data modulation	MPSK M-QAM relies on m,k	MPSK M-QAM relies on k	MPSK M-QAM relies on k	MPSK M-QAM relies on k
Coding	$\mathbf{c}=\mathbf{1}$ $\theta_{\mathbf{c}}=\mathbf{0}$	$\mathbf{c}=\mathbf{1}$ $\theta_{c_i} \in \{0,\pi\}$	$\mathbf{c}=\mathbf{1}$ $\theta_{c_i} \in [0,2\pi)$	$\mathbf{c}=\mathbf{1}$ $\theta_{c_i} \in [0,2\pi)$
Windowing	$\mathbf{w}=\mathbf{1}$ $\theta_{\mathbf{w}}=\mathbf{0}$	$\mathbf{w}=\mathbf{1}$ $\theta_{\mathbf{w}}=\mathbf{0}$	$\mathbf{w}=\mathbf{1}$ $\theta_{\mathbf{w}}=\mathbf{0}$	$\mathbf{w}=\mathbf{1}$ $\theta_{\mathbf{w}}=\mathbf{0}$
Orthogonality	$\theta_{\mathbf{o}}=\mathbf{0}$	$\theta_{\mathbf{o}}=\mathbf{0}$	$\theta_{\mathbf{o}}=\mathbf{0}$	$\theta_{\mathbf{o}}=\mathbf{0}$
Frequency assignment	$\mathbf{a}$	$\mathbf{a}$	$\mathbf{a}$	$\mathbf{a}$
Frequencies used	$\mathbf{u}=\mathbf{a}$	u_i depends on subcarrier spacing	u_i depends on subcarrier spacing	u_i depends on notching process

14.3.1 OFDM via SMSE

Using the variables defined in Table 14.1 for a basic OFDM implementation, the analytic expression can be simplified by setting $u_i = a_i$, $w_i = 1$, and $\theta_{w_i} = 0\ \forall i$. Since there is no windowing or coding in basic OFDM, the expression in (14.3) simplifies to

$$\begin{aligned} S_k[m] &= \left\{ a_m\, d_{m,k}\, e^{-j\left(\theta_{d_{m,k}}\right)} \right\}_{m=0}^{N_f-1} \\ &= \left\{ \alpha_{m,k} + j\beta_{m,k} \right\}_{m=0}^{N_f-1}, \end{aligned} \tag{14.5}$$

where $d_{m,k}\, e^{-j\theta_{d_{m,k}}} = \alpha_{m,k} + j\beta_{m,k}$. For example, $\alpha_{m,k}, \beta_{m,k} \in \{\pm 1\}$ for QPSK and $\alpha_{m,k}, \beta_{m,k} \in \{\pm 1, \pm 3\}$ for 16-QAM. Finally, applying an IFFT to (14.5) results in the discrete-time domain OFDM expression

$$s_k[n] = \frac{1}{N_f} \mathrm{Re} \left\{ \sum_{m=0}^{N_f-1} (\alpha_{m,k} + j\beta_{m,k})\, e^{j(2\pi f_m t_n)} \right\}. \tag{14.6}$$

14.3.2 TDCS via SMSE

The SMSE framework also facilitates the generation of a TDCS waveform that resides in multiple noncontiguous spectral fragments as a means to share spectrum via the avoidance of multiple narrowband users. The analytic expression of a basic TDCS implementation is realized by setting the variables $w_i = 1$ and $\theta_{w_i} = 0\ \forall i$ (no windowing). Coding is applied through random (or PR) phase variations between 0 and 2π, and thus $c_i = 1\ \forall i$. Data variables only depend upon symbol k because of spreading. The orthogonality variable $\theta_{o_i} = 0\ \forall i$ does not play a role in TDCS because the application of PR phase provides orthogonality between users [3]. Substituting these variables into (14.3) results in

$$S_k[m] = \left\{ a_m\, u_m\, d_k\, e^{-j\left(\theta_{cm} + \theta_{d_k}\right)} \right\}_{m=0}^{N_f-1}. \tag{14.7}$$

Spreading is performed in the frequency domain by using the PR phase distribution $\theta_{c_m} \in [0, 2\pi)$. The $a_m\, u_m$ term depends on the spectrum environment estimation as well as the adaptive TDCS notching process that deems specific frequency components as usable/unusable based on the characterization of channel interference. For these "notched" components, $a_m\, u_m = 0$. Applying these conditions results in

$$S_k[m] = \left\{ d_k\, e^{-j\left(\theta_{cm} + \theta_{d_k}\right)} \right\}_{m=0}^{N_f-1}. \tag{14.8}$$

After performing an IFFT on (14.8), the transmitted TDCS signal is

$$s_k[n] = \frac{1}{N_f} \mathrm{Re} \left\{ \sum_{m=0}^{N_f-1} d_k\, e^{j\left(2\pi f_m t_n + \theta_{cm} + \theta_{d_k}\right)} \right\}, \tag{14.9}$$

which is consistent with the TDCS discrete-time analytic expression from the previous section.

14.4 SMSE-based multicarrier radar waveforms

Inspired by the success of multicarrier waveforms such as OFDM in radio communications, multicarrier-based radar waveform designs have gained strong interest recently [21–34]. Reasons for this include their resistance to multipath fading, ability to overcome the limitations of a congested spectrum, ability to exploit frequency diversity gains stemming from the fact that target scattering centers inherently resonate differently at different frequencies, and the potential to perform radar and communication functions simultaneously within the same hardware using the same waveform [22].

In this section, we take an example from the seminal work of Levanon [23], and Levanon and Mozeson [24,25] on multicarrier waveforms for radar and explain that such multicarrier radar waveforms can also be generated by exploiting the SMSE framework. These waveforms can also be easily extended to form a spectrally noncontiguous waveform for spectrum sharing purposes.

Specifically, the work in [23–25] examined the use of multicarrier communication technologies and multi-frequency complementary phase-coded (MCPC) sequences to increase the range resolution of radar systems. MCPC signals are sequences that form a complementary set, which is defined as cyclically shifted versions of a phase-coded sequence having an ideal periodic *autocorrelation function*. Two such codes (that form the basis of MCPC signals) are the polyphase P3 and P4 code sequences [26]. The phase terms of a P-length P3 code are constructed as

$$\phi_p = \frac{\pi}{P}(p-1)^2 - \pi\,(p-1) \tag{14.10}$$

Table 14.2 Phases (in degrees) of the elements in a P3-based 5 × 5 MCPC signal

0	−144	−216	−216	−144
−144	−216	−216	−144	0
−216	−216	−144	0	−144
−216	−144	0	−144	−216
−144	0	−144	−216	−216

for $p = 1, 2, \ldots, P$. Likewise, the phase terms for a P-length P4 code are constructed as

$$\phi_p = \begin{cases} \dfrac{\pi}{P}(p-1)^2 & p \text{ is even} \\ \dfrac{\pi}{P}(p-1)p & p \text{ is odd} \end{cases} \tag{14.11}$$

for $p = 1, 2, \ldots, P$.

The MCPC signal is then a $P \times P$ matrix of cyclically shifted versions of either a P-length P3 code or a P-length P4 code. It is also possible to stack MCPC sequences to realize a $P \times N$ sequence, where N is an integer multiple of P. Such a stacked sequence likewise forms a complementary set. Table 14.2 provides an example of a P3-based 5 × 5 MCPC signal.

Using a multicarrier signaling scheme, the P sequences are transmitted on N subcarriers separated in frequency by

$$\Delta f = \frac{B}{n} = \frac{1}{t_b}, \tag{14.12}$$

where B is the overall signal bandwidth and t_b is the time duration of each bit in the sequence. The total duration of the transmitted pulse is then $P\,t_b$. The transmitted pulse is mathematically defined as

$$X(t) = \begin{cases} \sum_{m=0}^{N-1} \exp(j2\pi\, m\, \Delta f\, t) \sum_{p=1}^{P} u_{m,p}(t - (p-1)\,t_b) & 0 \le t \le Mt_b \\ 0 & \text{otherwise} \end{cases} \tag{14.13}$$

where

$$u_{m,p} = \begin{cases} \exp(j\phi_{m,p}) & 0 \le t \le t_b \\ 0 & \text{otherwise} \end{cases} \tag{14.14}$$

and $\phi_{m,p}$ is the pth phase element of the mth sequence (i.e. the mth column and pth row of Table 14.2). It is observed that this OFDM modulation scheme can be efficiently implemented with the IFFT operation, and that upon reception, the transmitted sequence can be demodulated by applying the FFT.

Note that the existing work on this multicarrier radar waveform only generates a signal that occupies a contiguous band of frequencies. However, the SMSE framework facilitates the extension to noncontiguous operation.

Using the SMSE framework a multicarrier MCPC radar signal can be generated by setting $w_i = 1$ and $\theta_{w_i} = 0\,\forall i$ (no windowing). The orthogonality variable $\theta_{o_i} = 0\,\forall i$ does not play a role here either. Coding is clearly applied via use of the P3 or P4 codes. Since this waveform is a pure radar waveform, no data modulation exists, and thus $d_k = 1$ and $\theta_{d_k} = 0$. It is important to note that a joint multicarrier MCPC radar/communication waveform can also be realized through the manipulation of d_k and θ_{d_k}.

The available frequency vector **a** enables the SMSE framework to generate a spectrally noncontiguous multicarrier MCPC radar signal. Similar to TDCS, the $a_m\, u_m$ term depends not only on the estimation of the spectrum environment but also upon the adaptive spectrum notching process that deems specific frequency components as usable/unusable based on channel interference characterization. For these "notched" components, $a_m\, u_m = 0$. Applying these conditions, along with the variable settings above, result in

$$S_k[m] = \left\{a_m u_m e^{-j(\theta_{cm})}\right\}_{m=0}^{N_f-1}, \tag{14.15}$$

where θ_{c_m} corresponds to the MCPC sequence in the available spectrum bins and 0 in the unavailable spectrum bins. After application of an IFFT to (14.15), the transmitted multicarrier MCPC radar signal is thus given by

$$s_k[n] = \frac{1}{N_f}\text{Re}\left\{\sum_{m=0}^{N_f-1} a_m\, u_m\, e^{j(2\pi f_m t_n + \theta_{cm})}\right\}. \tag{14.16}$$

14.5 Coexistence of radar and communications

By exploiting the SMSE-based spectrally noncontiguous multicarrier waveform, overlay communication and radar waveforms can be easily generated as a means to coexist with other users in the same spectrum. The following sections discuss a few scenarios involving this manner of spectrum sharing.

14.5.1 Scenario 1: Communication coexistence with narrowband emissions

Figure 14.7 illustrates the scenario where an SMSE-based spectrally noncontiguous communication waveform shares the spectrum with two distinct narrowband signals. The total number of subcarriers N_f is 64 and the narrowband signals occupy subcarriers 17–32 and 49–64. By notching out these bands, the TDCS waveform can be transmitted while avoiding mutual interference.

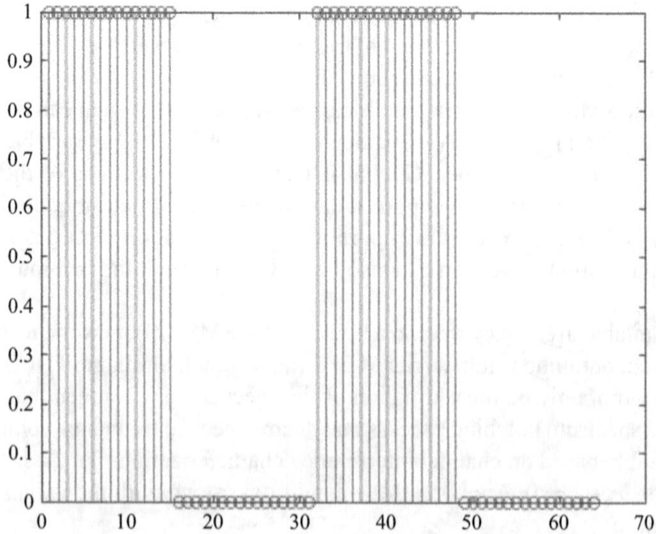

Figure 14.7 Spectrum profile for scenario 1

In general, the total received signal in such a spectrum-sharing scenario is

$$r(t) = \sum_{k=1}^{K} s_{\mathrm{NB},k}(t) + \sum_{\ell=1}^{L} s_{\mathrm{TDCS},\ell}(t) + n(t) \tag{14.17}$$

where $S_{\mathrm{NB},k}(t)$ is the kth of K narrowband emissions, $S_{\mathrm{TDCS},\ell}(t)$ is the ℓth of L TDCS transmitters, and $n(t)$ is additive Gaussian noise.

When the TDCS transmission operates as a secondary overlay waveform only the unoccupied spectrum "holes" are used, thus avoiding mutual interference. Here, it is assumed that one secondary user is operating over all the available spectrum holes. Hence, the bit error rate (BER) performance of a BPSK modulated TDCS system in an additive white Gaussian noise (AWGN) channel is

$$\mathrm{BER} = Q\left(\sqrt{\frac{2E_\mathrm{b}}{N_0}}\right), \tag{14.18}$$

for E_b the energy per bit and N_0 the noise power spectral density.

In the presence of multipath fading the ℓth received TDCS signal is

$$r_{\mathrm{TDCS},\ell}(t) = h(t) * s_{\mathrm{TDCS},\ell}(t), \tag{14.19}$$

where $h(t)$ is the channel impulse response and $*$ denotes convolution. In the TDCS receiver, the signal is recombined across all available subcarriers to exploit frequency diversity via

$$R^{(n)} = \sum_{i=0}^{M-1} w_i\, r_i^{(n)} \tag{14.20}$$

where w_i is the ith element of the weight vector and

$$r_i^{(n)} = \alpha_i b^{(n)} \sqrt{\frac{E_b}{M}} + \alpha_i \sqrt{\frac{E_b}{M}} \sum_{\ell=1,\ell\neq n}^{L} b^{(\ell)} \beta_i^{(\ell)} \beta_i^{*(n)} + n_i \tag{14.21}$$

for α_i the fading gain of the ith subcarrier, $\beta_i^{(\ell)}$ the ith component of the ℓth TDCS user's PR spreading code, and $b^{(n)}$ is the nth TDCS user's FMW. Maximum likelihood combining (MLC) is employed since it offers the best BER performance in multipath fading channels, and consequently

$$w_i = \frac{\alpha_i \sqrt{E_b/M}}{(L-1)\alpha_i^2 (E_b/M) + (N_0/2)}. \tag{14.22}$$

Note that when there is only one secondary user (i.e. $L = 1$), MLC reduces to maximum ratio combining which is known to be the optimal frequency diversity combining scheme in which

$$w_i = \alpha_i. \tag{14.23}$$

After frequency combining, the final decision variable is

$$\begin{aligned} R^{(n)} &= \sum_{i=0}^{M-1} w_i r_i^{(n)} \\ &= \sum_{i=0}^{M-1} w_i \alpha_i b^{(n)} \sqrt{\frac{E_b}{M}} + \sum_{i=0}^{M-1} w_i \alpha_i \sqrt{\frac{E_b}{M}} \sum_{\ell=1,\ell\neq n}^{L} b^{(\ell)} \beta_i^{(\ell)} \beta_i^{*(n)} + \sum_{i=0}^{M-1} w_i n_i, \end{aligned} \tag{14.24}$$

where the first term is the desired signal, the second term is the multiple access interference (MAI), and the third term is the noise contribution. The power of the desired signal is

$$P_{\text{Signal}} = \left(\sum_{i=0}^{M-1} w_i \alpha_i b^{(n)} \sqrt{\frac{E_b}{M}} \right)^2 = \frac{E_b}{M} \left(\sum_{i=0}^{M-1} w_i \alpha_i \right)^2. \tag{14.25}$$

Likewise the power of the MAI is

$$P_{\text{MAI}} = E\left[\sum_{i=0}^{M-1} w_i \alpha_i \sqrt{\frac{E_b}{M}} \sum_{\ell=1,\ell\neq n}^{L} b^{(\ell)} \beta_i^{(\ell)} \beta_i^{*(n)} \right]^2 = (L-1)\frac{E_b}{M} \sum_{i=0}^{M-1} w_i^2 \alpha_i^2 \tag{14.26}$$

and the power of the noise contribution is

$$P_{\text{Noise}} = E\left[\sum_{i=0}^{M-1} w_i n_i \right]^2 = \frac{N_0}{2} \sum_{i=0}^{M-1} w_i^2. \tag{14.27}$$

The instantaneous signal-to-interference-plus-noise ratio (SINR), denoted as γ, is thus

$$\gamma = \frac{P_{\text{Signal}}}{P_{\text{MAI}} + P_{\text{Noise}}}$$
$$= \frac{E_b/M\left(\sum_{i=0}^{M-1} w_i\,\alpha_i\right)^2}{(L-1)E_b/M\sum_{i=0}^{M-1} w_i^2\,\alpha_i^2 + (N_0/2)\sum_{i=0}^{M-1} w_i^2}.$$

The expression for probability of error as a function of instantaneous SINR is then

$$P_{\text{inst}}(e) = Q\left(\sqrt{\gamma}\right) \tag{14.28}$$

and the average error probability is obtained by integrating over the pdf of SINR as

$$P(e) = \int_0^\infty Q\left(\sqrt{\gamma}\right) p(\gamma)\, d\gamma, \tag{14.29}$$

which can be determined via numerical methods.

14.5.2 Scenario 2: Radar coexistence with narrowband emissions

Figure 14.8 illustrates another scenario where an SMSE-based spectrally noncontiguous MCPC radar waveform shares the spectrum with four narrowband signals.

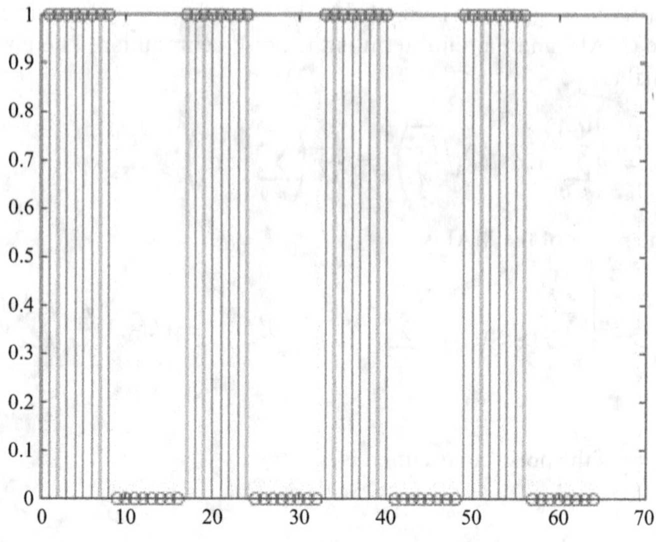

Figure 14.8 Spectrum profile for scenario 2

The total number of subcarriers N_f is 64 and the four narrowband emissions occupy subcarriers 9–16, 25–32, 41–48 and 57–64. By notching out these bands, the MCPC radar waveform is able to operate over the four available spectrum holes by employing MCPC sequences.

A commonly used metric to evaluate the performance of a radar waveform is the delay/Doppler ambiguity function (AF) that measures the response of a matched filter to a finite energy signal as a function of time delay τ and Doppler frequency shift ν. The periodic AF can be expressed as [35,36]

$$|\chi(\tau,\nu)| = \left| \int_0^T u(t)\, u^*(t-\tau)\, e^{2\pi j \nu t}\, dt \right| \tag{14.30}$$

where $u(t)$ is the radar waveform.

Figure 14.9 shows the AF of a contiguous multicarrier MCPC radar waveform, while Figure 14.10 illustrates the AF of a noncontiguous multicarrier MCPC radar waveform corresponding to scenario 2. It is evident that the latter is capable of stitching together multiple noncontiguous spectrum bands as a means to coexist with other in-band signals while only modestly increasing the AF sidelobes.

Further, Figure 14.11 shows the autocorrelation (i.e. zero-Doppler cut of the AF) of the contiguous multicarrier radar waveform, while Figure 14.12 illustrates the autocorrelation of the noncontiguous multicarrier radar waveform. It is evident that the latter maintains the ideal periodic nulls.

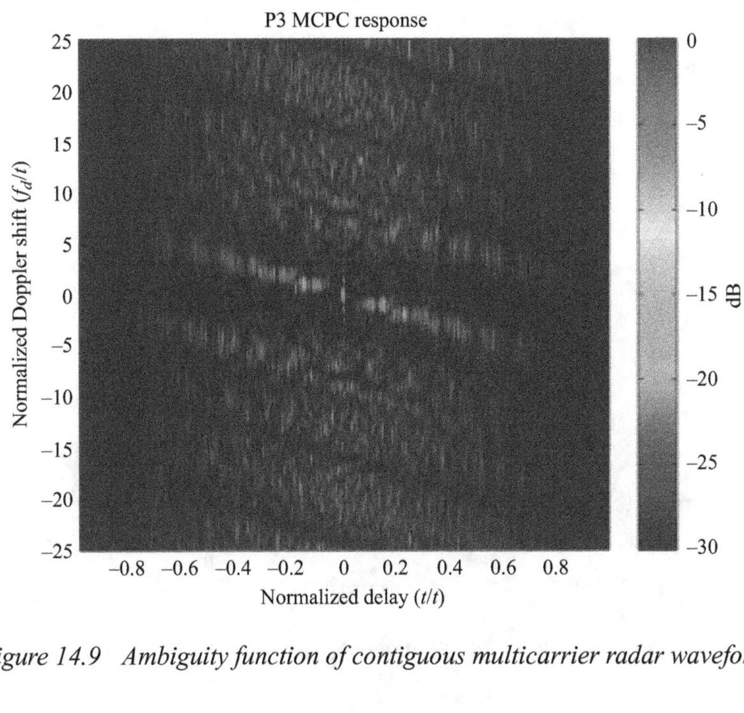

Figure 14.9 Ambiguity function of contiguous multicarrier radar waveform

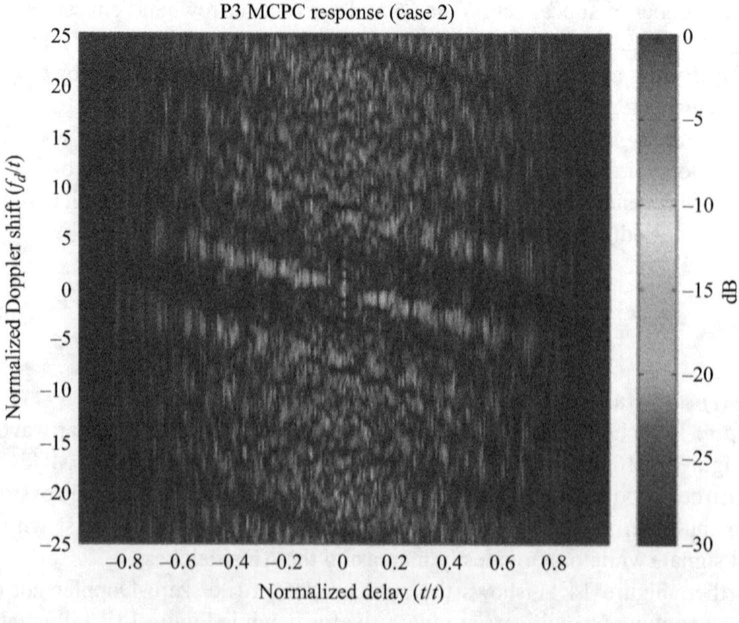

Figure 14.10 Ambiguity function of noncontiguous multicarrier radar waveform for scenario 2

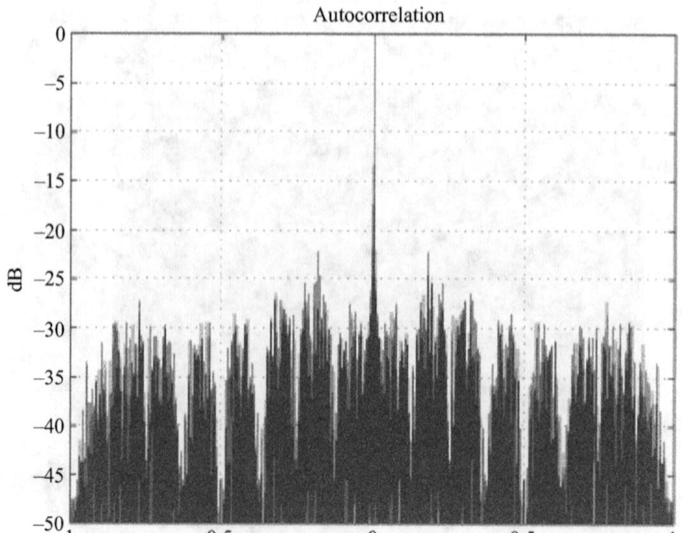

Figure 14.11 Autocorrelation function of contiguous multicarrier radar waveform

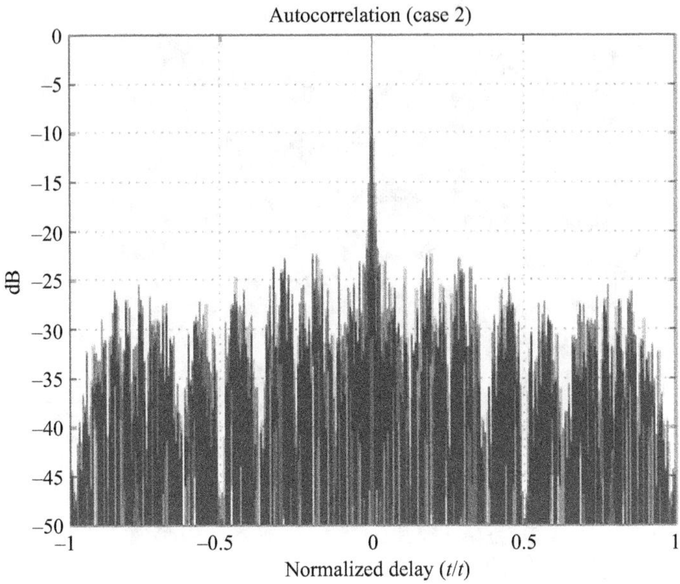

Figure 14.12 Autocorrelation function of noncontiguous multicarrier radar waveform for scenario 2

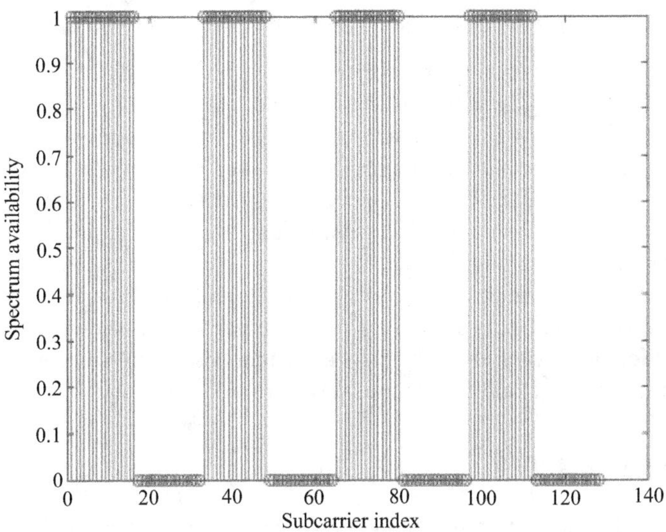

Figure 14.13 Spectrum profile for scenario 3

14.5.3 Scenario 3: Radar/Communication coexistence with narrowband emissions

Figure 14.13 illustrates the third scenario where an SMSE-based spectrally noncontiguous MCPC radar waveform and a TDCS communication waveform coexist with

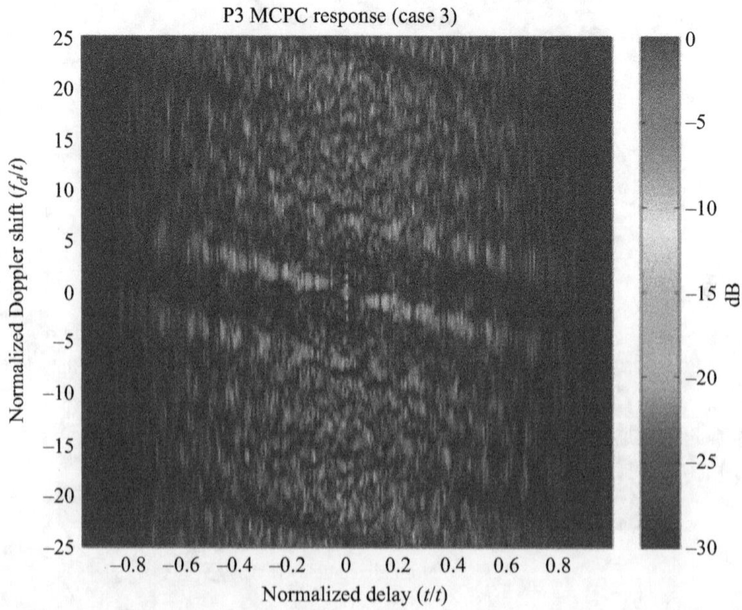

Figure 14.14 Ambiguity function of noncontiguous multicarrier radar waveform for scenario 3

one another as well as four other distinct narrowband signals. The total number of subcarriers N_f is now 128 and the four narrowband signals occupy subcarriers 17–32, 49–64, 81–96 and 113–128. By notching out these bands, there are four spectrum holes available for radar and communication operation. Here, we let the MCPC radar waveform transmit over the first two spectrum holes and the TDCS communication waveform to occupy the last two spectrum holes. As a direct result, both share the spectrum while avoiding interference.

Figure 14.14 illustrates the AF of a noncontiguous multicarrier MCPC radar waveform corresponding to scenario 3, while Figure 14.15 shows the associated autocorrelation. It is evident that this noncontiguous waveform realizes performance similar to that of a spectrally contiguous waveform.

14.6 Looking ahead

Spectrum sharing between radar and communication systems can be accomplished by adopting one of the three methods: coexistence, cooperation, or co-design [37].

- Coexistence: Both radar and communication systems operate and share the same RF spectrum. Each system treats the other as an interference source and tries to mitigate the effect by applying spatial, spectral, or temporal diversity.
- Cooperation: In a cooperative design some knowledge about each system is shared to mitigate mutual interference.

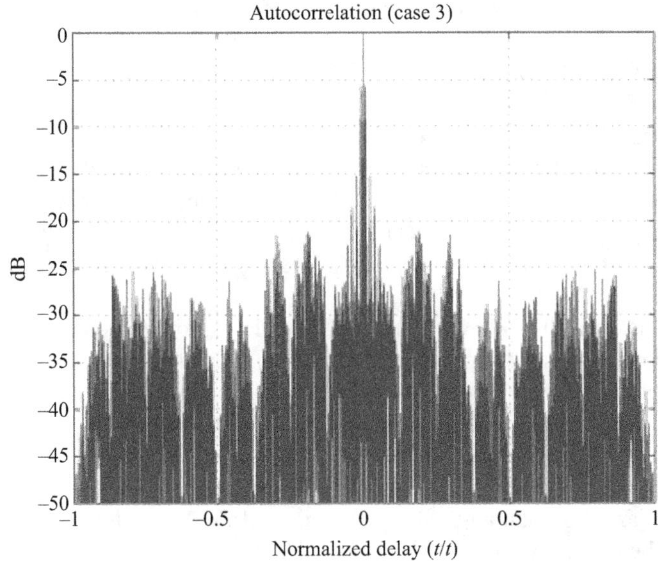

Figure 14.15 Autocorrelation function of noncontiguous multicarrier radar waveform for scenario 3

- Co-design: This arrangement is where both the radar and the communication system are designed to maximize joint performance. Both modes try to utilize features of the other to improve performance. For example, communication systems can use codes for known training sequences that are also beneficial to radar operation. Similarly, radar signals can be used to improve channel estimation.

Even though the SMSE-based overlay TD waveform discussed in this chapter can be applied to all three spectrum sharing methods, its main intent was co-design to maximize joint performance. The analysis above assumed ideal SMSE-based multicarrier waveforms to demonstrate the coexistence of radar and communication systems. The effects of bandlimited nonlinear RF front end component effects such as filters, amplifiers, mixers, analog-to-digital, and digital-to-analog converters have not yet been considered. Based on each system's performance requirements and interference tolerance levels, waveform shaping based on spectral characteristics should also be considered.

References

[1] Executive Office of the President, President's Council of Advisors on Science and Technology, *Report to the President: Realizing the Full Potential of Government-Held Spectrum To Spur Economic Growth*, http://www.whitehouse.gov/administration/eop/ostp/pcast/docsreports., July 2012.

[2] B. Fette, *Cognitive Radio Technology*, 2nd edition, Academic Press, Burlington, MA, 2009.

[3] V. Chakravarthy, A. S. Nunez, J. P. Stephens, A. K. Shaw, and M. A. Temple, "TDCS, OFDM, and MC-CDMA: A brief tutorial," *IEEE Communications Magazine*, vol. 43, no. 9, pp. S11–S16, September 2005.

[4] V. Chakravarthy, Z. Wu, M. Temple, and F. Garber, "Novel overlay/underlay cognitive radio waveforms using SD-SMSE framework to enhance spectrum efficiency—part I: theoretical framework and analysis in AWGN channel," *IEEE Transactions on Communications*, vol. 57, no. 12, pp. 3794– 3804, December 2009.

[5] V. Chakravarthy, Z. Wu, and M. Temple, "Novel overlay/underlay cognitive radio waveforms using SD-SMSE framework to enhance spectrum efficiency—part II: analysis in fading channels," *IEEE Transactions on Communications*, vol. 58, no. 6, pp. 1868–1876, June 2010.

[6] S. Haykin, "Cognitive radio: brain-empowered wireless communications," *IEEE Journal of Selected Areas in Communications*, vol. 23, pp. 1–20, 2005.

[7] S. Hijazi, B. Natarajan, M. Michelini, and Z. Wu, "Flexible spectrum use and better coexistence at the physical layer of future wireless systems via a multicarrier platform," *IEEE Wireless Communications*, vol. 11, no. 2, pp. 64–71, April 2004.

[8] L. B. Milstein, "An analysis of a real-time transform domain filtering digital communication system—part 1: narrow-band interference rejection," *IEEE Transactions on Communications*, vol. 28, no. 6, pp. 816–824, June 1980.

[9] L. B. Milstein, "An analysis of a real-time transform domain filtering digital communication system—part II: wide-band interference rejection," *IEEE Transactions on Communications*, vol. 31, no. 1, pp. 21–27, January 1983.

[10] E. H. German, *Transform Domain Signal Processing Study Final Report, Technical Report*, Reisterstown, MD: Contract: Air Force F30602-86-C-0133, August 1988. DTIC: ADB132635.

[11] C. F. Andren, L. V. Lucas, and J. A. Schachte, "Low probability of intercept communication system," Harris Corporation, U.S. Patent 5029184, 1991.

[12] R. Radcliffe and G. C. Gerace, "Design and simulation of transform domain communication system," *IEEE MILCOM*, Monterey, CA, Oct. 1997.

[13] M. J. Lee, M. A. Temple, R. L. Claypoole, R. A. Raines, and J. P. Stephens, "Wavelet domain communication system: bit error sensitivity characterization for geographically separated transceivers," *IEEE MILCOM*, Anaheim, CA, Oct. 2002.

[14] P. J. Swackhammer, M. A. Temple, and R. A. Raines, "Performance simulation of a transform domain communication system for multiple access application," *IEEE MILCOM*, Atlantic City, NJ, Oct./Nov. 1999.

[15] M. L. Roberts, M. A. Temple, R. A. Raines, and E. P. Magee, "Initial acquisition performance of a transform domain communication system: modeling and simulation results," *IEEE MILCOM*, Los Angeles, CA, Oct. 2000.

[16] L. Hanzo, M. Munster, B. Choi, and T. Keller, *OFDM and MC-CDMA for Broadband Multi-User Communications*, Wiley, West Sussex, 2003.

[17] M. J. Lee, *Wavelet Domain Communication System (WDCS): Packet-Based Wavelet Spectral Estimation and M-ARY Signaling*, Masters Thesis, AFIT/GE/ENG/02M-14, Air Force Institute of Technology, March 2001, DTIC: ADA401433, Approved for Public Release.

[18] M. Roberts, M. A. Temple, R. A. Raines, R. F. Mills, and M. E. Oxley, "Communication waveform design using an adaptive spectrally modulated, spectrally encoded (SMSE) framework," *IEEE Journal of Selected Topics in Signal Processing*, vol. 1, no. 1, pp. 203–213, January 2007.

[19] M. Roberts, M. A. Temple, M. E. Oxley, R. F. Mills, and R. A. Raines, "A general analytic framework for spectrally modulated, spectrally encoded signals," in *Proc. IEEE Waveform Diversity Conf.*, Lihue, HI, USA, January 2006.

[20] M. Roberts, M. A. Temple, M. E. Oxley, R. F. Mills, and R. A. Raines, "A spectrally modulated, spectrally encoded analytic framework for carrier interferometry signals," in *Proc. International Wireless Commun. Mobile Computing Conf. (IWCMC)*, July 2006.

[21] S. Blunt and E. L. Mokole, "Overview of radar waveform diversity," *IEEE Aerospace and Electronics Systems Magazine*, vol. 31, no. 11, pp. 2–42, November 2016.

[22] J. Ellinger, Z. Zhang, M. Wicks, and Z. Wu, "Multi-carrier radar waveforms for communications and detection," *IET Radar, Sonar & Navigation*, pp. 1–9, September 2016, DOI: 10.1049/iet-rsn.2016.0244.

[23] N. Levanon, "Multifrequency complementary phase-coded radar signal," *IEE Proceedings Radar, Sonar and Navigation*, vol. 147, no. 6, pp. 276–284, 2000.

[24] N. Levanon and E. Mozeson, "Multicarrier radar signal-pulse train and CW," *IEEE Transactions on Aerospace and Electronic Systems*, vol. 38, no. 2, pp. 707–720, 2002.

[25] E. Mozeson and N. Levanon, "Multicarrier radar signals with low peak-to-mean envelope power ratio," *IEE Proceedings Radar, Sonar and Navigation*, vol. 150, no. 2, pp. 71–77, 2003.

[26] B.L. Lewis and F.F. Kretschmer, "Linear frequency modulation derived polyphase pulse compression codes," *IEEE Transactions on Aerospace and Electronic Systems*, vol. AES-18, no. 5, pp. 637–641, September 1982.

[27] B. J. Donnet and I. D. Longstaff, "Combining MIMO radar with OFDM communications," in *Proc. 3rd European Radar Conf. 2006*, September 2006, pp. 37–40.

[28] G. Lellouch, P. Tran, R. Pribic, and P. van Genderen, "OFDM waveforms for frequency agility and opportunities for Doppler processing in radar," in *Proc. IEEE Radar Conf.*, May 26–30, 2008, pp. 1–6.

[29] A. Garmatyuk, J. Schuerger, Y. T. Morton, K. Binns, M. Durbin, and J. Kimani, "Feasibility study of a multi-carrier dual-use imaging radar and communication system," in *Proc. 37th European Microw. Conf.*, Munich, Germany, October 2007, pp. 1473–1476.

[30] R. F. Tigrek, W. de Heij, and P. van Genderen, "Multi-carrier radar waveform schemes for range and Doppler processing," in *Proc. IEEE Radar Conf.*, May 4–8, 2009, pp. 1–5.

[31] C. R. Berger, B. Demissie, J. Heckenbach, P. Willett, and S. Zhou, "Signal processing for passive radar using OFDM waveforms," *IEEE Journal of Selected Topics in Signal Processing*, vol. 4, no. 1, pp. 226–238, February 2010.

[32] C. Sturm, E. Pancera, T. Zwick, and W. Wiesbeck, "A novel approach to OFDM radar processing," in *Proc. IEEE Radar Conf.*, Pasadena, CA, May 2009.

[33] J. Jakabosky, L. Ryan, and S. D. Blunt, "Transmitter-in-the-loop optimization of distorted OFDM radar emissions," in *Proc. IEEE Radar Conf.*, Ottawa, Canada, April/May 2013.

[34] S. C. Surender and R. M. Narayanan, "UWB noise-OFDM netted radar: physical layer design and analysis," *IEEE Transactions on Aerospace and Electronic Systems*, vol. 47, no. 2, pp. 1380–1400, April 2011.

[35] M. Nowak, M. Wicks, Z. Zhang, and Z. Wu, "Co-designed radar–communication using linear frequency modulation waveform," *IEEE Aerospace and Electronic Systems Magazine*, vol. 31, pp. 28–35, October 2016.

[36] M. Nowak, Z. Zhang, L. LoMonte, M. Wicks, and Z. Wu, "Mixed-modulated linear frequency modulated radar–communications," *IET Radar, Sonar & Navigation*, pp. 1–20, August 2016, DOI: 10.1049/iet-rsn.2016.0249.

[37] A. R. Chiriyath, B. Paul, and D. Bliss, "Radar–communications convergence: coexistence, cooperation and co-design," *IEEE Transactions on Cognitive Communication and Networking*, vol. 3, no. 1, pp. 1–12, February 2017.

Chapter 15
Radar bandwidth optimization for interference mitigation

Anthony F. Martone[1], Kelly D. Sherbondy[1], and Kyle A. Gallagher[1]

This chapter investigates techniques that modify radar waveforms in order to share the band with communication systems while maximizing radar performance. These techniques are based on a coexistence strategy involving a two-step approach that characterizes the electromagnetic environment via spectrum sensing and then accesses the spectrum dynamically. Spectrum sensing can detect communication systems that are attempting to access the same band as the radar. For the paradigm considered here, the frequency locations of these communication systems are estimated using energy detection. Access to the spectrum is then achieved by identifying a channel, or sub-band, devoid of activity so as to minimize radio frequency interference (RFI). This approach avoids potentially harmful RFI while simultaneously reducing the radar spectral footprint. Proper sub-band selection is key to maximizing radar performance in the presence of RFI.

In this context we consider the potentially conflicting radar performance metrics of signal-to-interference plus noise ratio (SINR) and range resolution. The goal is to identify the most advantageous sub-band that maximizes both SINR and range resolution. However, while the latter would clearly drive the radar bandwidth to be larger, the former is maximized by avoiding interference, consequently driving the bandwidth smaller. A prospective solution to this problem is the spectrum sensing, multi-objective optimization (SS-MO) technique that was developed to permit spectrum sharing in dynamic environments via RFI avoidance. Section 15.1 outlines the need for radar coexistence followed by an introduction of the SS-MO technique. Section 15.2 develops the mathematical formulation of SS-MO and then Section 15.3 discusses its computational cost and the associated latency involved with this decision process and corresponding RF tuning. Section 15.4 presents an information refinement technique to significantly reduce the data input to SS-MO as a means to reduce convergence time, thereby realizing a fast equivalent for low latency operation. Finally, Section 15.5 provides experimental results that compare the

[1]U.S. Army Research Laboratory, USA

performance and computational cost of SS-MO and its fast variant against the standard nonadaptive solution.

15.1 Introduction

Spectrum policy reform throughout the world has generated the need for new and innovative spectrum sharing strategies for radar and communication systems [1,2]. Indeed, the development of software-defined radio (SDR) and cognitive radio technologies in the late 1990s offer an opportunity for radio frequency (RF) systems to adapt to congested electromagnetic environments (EMEs) [3]. The capability for these technologies to mitigate mutual interference has motivated various government organizations to redefine spectrum policy in ways that continue to evolve. For example, in 2003, the U.S. Federal Communications Commission (FCC) identified cognitive radio technologies as having "the potential for more flexible, efficient, and comprehensive use of available spectrum while reducing the risk of harmful interference" [4]. In 2004, the FCC proposed the use of unlicensed radio transmitters in the television (TV) frequency band [5], a regulatory action that led to the development of the IEEE 802.19.1 standard in 2014. This "white space" coexistence standard specifies the requirements for sharing among different RF users in the TV band [6]. Several countries and organizations have adopted similar spectrum policies for coexistence [7], with these policy changes fueling the need for the RF community to develop new and innovative spectrum sharing technologies for radar and communication systems.

Much of spectrum sharing research is divided into either cooperative or coexistence categories, with both including various time, frequency, polarization, and/or space domain approaches [8]. Cooperative strategies favor a co-design between the radar and communication system whereby a policy is implemented to minimize mutual RFI [8]. The cooperative category requires these systems to work together to function in a compatible manner (e.g. [9–16] and Chapters 19–23 in this book). In contrast, the coexistence category does not support explicit cooperation between systems, but rather requires the capability to sense the EME for the presence of other spectrum users. The information sensed is typically in the form of interference. Once sensing is performed, the radar or communication system then decides if and how to access the spectrum.

Methods for coexistence have been extensively investigated for cognitive radio in the research area of dynamic spectrum access (see [3,4]). The cognitive radio, in this context the unlicensed secondary user, attempts to access a frequency band allocated to a licensed primary user. To do so, the cognitive radio must detect the primary user by passively monitoring the EME before dynamically accessing the spectrum. Such spectrum sensing detection techniques include matched filtering, cyclostationary detection, and energy detection [17]. The particular spectrum sensing method employed is dependent upon *a priori* knowledge of the primary user's transmission waveform; for instance, a matched filter can be used if the exact waveform of the primary user is known, whereas energy detection is necessary if little is known. After detection of the primary user, the cognitive radio decides "if" and "how" to access the spectrum. Thus the coexistence approach provides the cognitive radio with the means to exploit underutilized spectrum belonging to inactive primary users [18].

The cognitive radio coexistence paradigm has likewise been explored for cognitive radar, where in this case the radar is operating as the secondary user [19]. In other words, just as commercial communications continues to share more of the spectrum allocated to radar, so too could radar share spectrum allocated for communications. In this arrangement, the cognitive radar monitors a set of prospective sub-bands within a wideband channel that are predefined in terms of bandwidth and center frequency. The cognitive radar implements a spectrum-sensing and channel characterization strategy to mitigate RFI to the primary user while utilizing as much of the available bandwidth as possible. The authors in [20] (and Chapter 16) investigate compressive sensing to limit the required sample rate, thereby improving the response time of the radar for this purpose. A fast response time is necessary to maintain an up-to-date measure of the RFI and to maximize the radar's coherent processing interval (CPI). For example, if the time required to sense the spectrum is on the order of the radar pulse repetition interval (PRI), then the radar would necessarily transmit fewer pulses in the CPI, which translates to a lower SINR due to less "energy on target." This concept is further explored in Section 15.3.

A more structured approach to dynamic spectrum access relies on the use of a management system to regulate access for different RF systems; an example is the TV White Space (TVWS) standard developed in 2014 [6]. This standard was first proposed by the FCC in 2004 as a way to mitigate mutual interference among communication systems [5]. It specifies a system architecture to identify radio systems requesting access to the spectrum, to modify their operating parameters as needed to manage the overall performance, and to serve as an interface for different types of radio systems [21]. Preliminary investigation of the TVWS standard has demonstrated a capability for successful coexistence [21], with this approach serving as a prospective future framework for sharing among radar and communication systems.

Here, we examine a radar coexistence strategy for equal sharing among multiple RF users. The "equal sharing" coexistence strategy grants equal access to a frequency channel for multiple RF systems and facilitates greater spectral efficiency by permitting access to underutilized portions of the spectrum by multiple RF device types [8]. A possible scenario considers sharing among radar, Bluetooth, cellular, and unlicensed radiators whose protocol is unknown and must be learned over time. The radar attempts to access a frequency channel of wide bandwidth B by using spectrum sensing. Ideally, the radar would make use of the entire band B to maximize range resolution [22]. However, doing so degrades the radar's SINR if B is occupied by other RF systems and causes mutual interference. The proposed sharing approach can be achieved with or without the use of the TVWS standard management system discussed above.

An example of an occupied band is illustrated in Figure 15.1. This measured spectrum was collected by the spectrum sensing system depicted in Figure 15.2 and is comprised of an Agilent Technologies N9342CN handheld spectrum analyzer, an AOR DA3200 discone antenna, and a data acquisition controller (laptop) to store the data. This measurement setup was used to collect the data for the experiments discussed

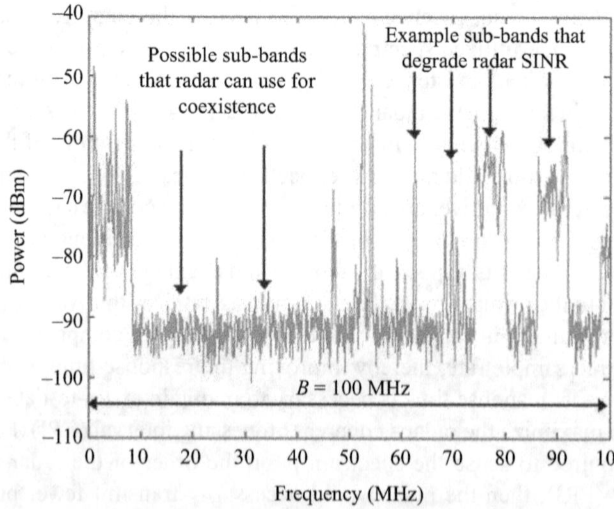

Figure 15.1 Power spectrum collected by the measurement system in Figure 15.2, with B = 100 MHz at a center frequency of 150 MHz, with a frequency resolution of 0.1 MHz. Multiple sub-bands contain interference that would degrade the radar's SINR and would conversely experience interference from the radar

Figure 15.2 Measurement system used to collect RF spectral data, consisting of an AOR DA3200 discone antenna (left), an Agilent Technologies N9342CN handheld spectrum analyzer (middle), and a data acquisition controller (right)

throughout the chapter. The power spectrum in Figure 15.1 was collected using a bandwidth of $B = 100$ MHz at a center frequency of 150 MHz. This spectrum was sampled using $N = 1{,}000$ frequency bins, realizing a frequency resolution of $\Delta_r = 100$ kHz. The average noise level (noise floor) was $P_N = -92$ dBm, corresponding to a noise factor of $N_F = 33$ dB [23].

Figure 15.1 shows that multiple frequency sub-channels within B are occupied, and thus access to the full channel would certainly degrade SINR for the radar, not to mention the interference that would be imposed on the other spectrum users. Instead of operating in the entire band, the radar could occupy an available continuous sub-band within B, which necessitates the joint maximization of SINR and bandwidth. The particular sub-band evaluation/selection approach examined in this chapter is the spectrum sensing, multi-objective optimization (SS-MO) technique [24].

For SS-MO, spectrum sensing is first used to passively sense the channel bandwidth B while the radar remains inactive. An energy detector is then applied to ascertain the presence of other RF systems operating in B. Energy detection is advantageous for the equal sharing paradigm because multiple unknown and unlicensed users could be occupying the channel. A power spectrum, formed via fast Fourier transform, represents an empirical estimate of the EME. Multi-objective optimization then processes the power spectrum to find the "best," or feasible Pareto-optimal, sub-band of B that maximizes SINR and bandwidth objective functions. If a management system exists (e.g. the TVWS standard), then the radar requests use of this sub-band for operation. If no management system exists, the radar would dynamically access the sub-band.

Note that the SS-MO framework supports multiple objective functions in addition to SINR and bandwidth. For example, [25] investigates the use of this approach for joint SINR, bandwidth, and mutual interference optimization. Other useful objective functions in this context include signal-to-clutter ratio and various attributes of the radar waveform's delay–Doppler ambiguity function, with their utility depending on the radar type and operating scenario. Having said that, the remainder of the chapter limits the discussion to the SINR and bandwidth objective functions.

A comprehensive search routine is required to determine the most advantageous sub-band available. This search first merges the frequency bins of the power spectrum according to all possible combinations to generate a set of candidate sub-bands. For N frequency bins in the power spectrum, the number of possible sub-bands is $O(N^2)$, with each sub-band having an associated SINR and bandwidth. These measures are then weighted and combined to form the solution space, the global maximum of which is the feasible Pareto-optimal solution.

The SS-MO search routine does not search some set of predefined, sequential sub-bands; instead every frequency bin, and combination thereof, is evaluated to ensure the actual globally optimal solution is obtained. For example, the case of $N = 1{,}000$ frequency samples over the $B = 100$ MHz overall bandwidth depicted in Figure 15.1 necessitates the evaluation of $O(10^6)$ candidate sub-bands. The only feasibility constraint is that each sub-band is formed from a continuous set of frequency bins (i.e. a nonfragmented spectrum). This approach is capable of adapting to unknown, time-varying RFI simultaneously generated by multiple different RF sources. It has been demonstrated experimentally that in interference-rich environments SS-MO can increase SINR by 25 dB and the peak-to-average sidelobe level by 15 dB for chirp and stepped-frequency waveform types [24,26].

A primary goal of this approach is to support radar coexistence within a given wideband channel by dynamically accessing the spectrum in real time, or near real time, while minimizing mutual interference between the radar and multiple other unknown spectrum users. To achieve real-time (or near real-time) responsiveness, the radar must rapidly obtain a sub-band solution and subsequently tune to the selected sub-band using a reconfigurable RF front-end—Chapters 10 and 12 discuss these capabilities in detail for receivers and transmitters, respectively. Clearly, these steps must be accomplished faster than the RFI changes to be useful in practice.

The overall latency between interference being incident upon the receive antenna and the determination of a radar spectrum-sharing solution depends on (1) the time needed to passively observe the EME and estimate the power spectrum, and (2) the computational time required to perform the multi-objective optimization, with the latter scaled by $O(N^2)$. To that end, a "fast" spectral refinement technique was developed in [27] as a means to significantly decrease the solution space over which SS-MO operates by employing a data reduction step inspired by the human thalamus. The fast variant of this approach, which is discussed in Section 15.5, reduces the computational cost to $O(N)$ with only very minor performance loss.

SS-MO is designed to select a single sub-band that improves SINR (relative to using the entire band B) with as little bandwidth reduction penalty as possible. The reduced bandwidth limits the radar's capability to resolve closely spaced targets. However, future cognitive radars that support multiple modes of operation could exploit this SINR and bandwidth trade-off. For example, a search and track radar would only require the full bandwidth once a target is detected [12]. Otherwise, the radar could operate with a lower bandwidth, thereby sharing the band while also maintaining sufficient SINR for early detection. The functionality of this manner of RFI avoidance also permits the determination of multiple, disjoint sub-bands that realize minimal interference [25], for use with specially designed waveform nulling approaches [13–14,28–35] and RFI mitigation techniques [36–38].

Waveform nulling produces frequency notches on transmit that limit the emissions into occupied frequency bands, thus reducing the radar spectral footprint. The number of notches, the notch size(s), and the notch depth(s) are limited in practice and vary according to the RFI in the band and the hardware used to transmit and receive the waveform. Further details of these nulling approaches and limitations are discussed in Chapters 6, 17, and 18. The RFI mitigation techniques effectively remove the remaining RFI on receive. In many cases, these techniques require *a priori* knowledge of the interference, such as would be facilitated by SS-MO. Further details of RFI mitigation in the receiver analog front-end is considered in Chapter 10.

In the remainder of this chapter, the SS-MO technique is examined in detail. The mathematical framework was originally introduced in [24] and highlights the search strategy used by multi-objective optimization. An examination of the tuning technology, originally introduced in [39], compares the SINR increase provided by SS-MO to the SINR reduction caused by insufficient tuning speed. The subsequent development of the "fast" implementation reduces the computational cost by an order of magnitude, with experimental assessment in the latter part of chapter showing that the trade-off in performance is quite minor.

15.2 Spectrum sensing, multi-objective optimization for RFI avoidance

The block diagram in Figure 15.3 illustrates the SS-MO technique. Since the formulation is applicable to a variety of sensing applications the radar system is considered generic in this development. The process begins by passively sensing the EME at the start of the radar CPI cycle, with the CPI consisting of the echoes produced by a set of pulses. The spectrum sensing receiver digitizes and

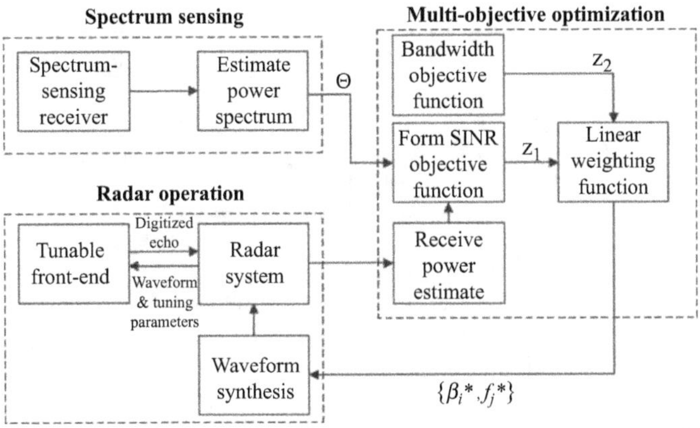

Figure 15.3 SS-MO technique for radar spectrum sharing, which requires empirical measurement of the RF spectrum followed by multi-objective optimization to determine the swath of spectrum in which the radar can operate

processes the observed RF spectrum to generate the power spectrum estimate $\Theta = |\mathcal{F}([x_1, \ldots, x_N])|^2 = \{\theta_1, \ldots, \theta_N\}$ of size N for frequencies $F = \{f_1, \ldots, f_N\}$, where x_n is the nth time domain sample and $\mathcal{F}(\bullet)$ denotes the Fourier transform of the sequence. The frequency resolution Δ_r remains constant so that the noise characteristics remain unchanged between each sensing cycle; otherwise, RFI could be present below the noise floor, resulting in an inaccurate estimate of the EME. For convenience, the frequency resolution used here is the same for the radar and the spectrum sensing receiver. Note that it is possible for the spectrum sensing receiver to use the same RF front-end as the radar or to use a separate front-end. The specific system design depends on the radar application and size, weight, and power (SWaP) limitations.

The SINR objective function is formed using the power spectrum estimate of the interference plus noise and the modeled radar receive signal power. The power spectrum is used to estimate the collective interference plus noise for each of the candidate sub-bands, which is calculated recursively as

$$\Gamma(\beta_i, f_j) = \begin{cases} \theta_j, & i = 1, j = 1, \ldots, N \\ \Gamma(\beta_1, f_j) + \Gamma(\beta_1, f_{1+j}) & i = 2, j = 1, \ldots, N-1 \\ \Gamma(\beta_{i-1}, f_j) + \Gamma(\beta_1, f_{i+j-1}) & i = 3, \ldots N, j = 1, \ldots, N-i+1 \end{cases} \tag{15.1}$$

where $f_j \in F$ is the lower edge frequency (or "start frequency") of the jth sub-band for $j = \{1, \ldots, N\}$, and $\beta_i = i\Delta_r$ is the bandwidth of the ith sub-band for $i = \{1, \ldots, N\}$. For N frequency samples there are

$$N_{\text{cand}} = \sum_{k=1}^{N} (N - k + 1) = \frac{(N^2 + N)}{2} \tag{15.2}$$

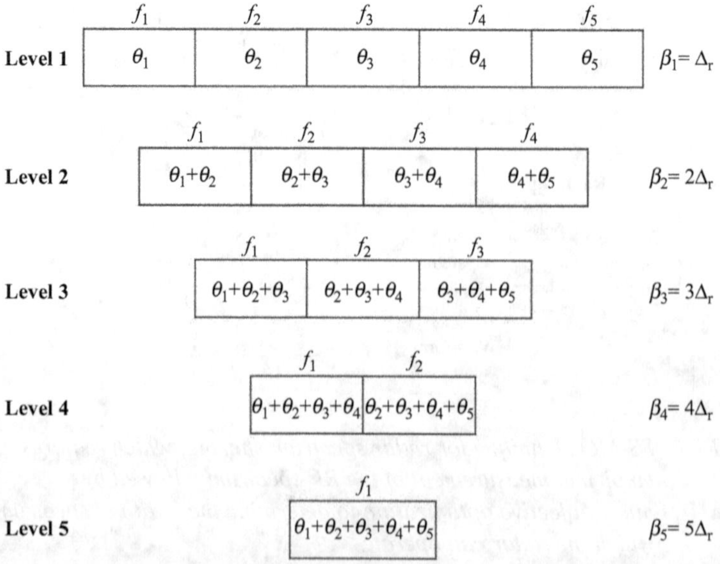

Figure 15.4 Structure of the interference estimate $\Gamma(\beta_i, f_j)$ containing 15 sub-band combinations. The sub-band size increases as the samples are merged together. The lower edge frequency (or "start frequency") of each sub-band is depicted above each block

candidate sub-band combinations of interference and noise (under the nonfragmented spectrum condition).

A simple example of (15.1) is illustrated in Figure 15.4 for the quite modest case of $N = 5$. Here, the power spectrum contains the samples $\{\theta_1, \ldots, \theta_5\}$ corresponding to start frequencies $\{f_1, \ldots, f_5\}$, and there are $N_{\text{cand}} = 15$ possible sub-bands for evaluation. The bandwidth of each sub-band is shown on the right and the start frequency is shown above each block. Each level in Figure 15.4 corresponds to a particular bandwidth and each block in a given level corresponds to a particular start frequency. The Level 1 elements (top row in Figure 15.4) contain the power estimate values having a bandwidth of $\beta_1 = \Delta_r$ (i.e. the resolution of the power spectrum). Level 2 blocks are formed by combining two sequential power estimate values from Level 1, thus realizing individual bandwidths of $\beta_2 = 2\Delta_r$. Continuing in this manner, the single Level 5 block contains the sum of all five power estimate values, with a corresponding bandwidth of $\beta_5 = 5\Delta_r = B$.

The radar receive signal power is modeled using the radar range equation. As shown in Figure 15.3, the radar system provides feedback to the receive power model with updated estimates of the range to target R and the target radar cross section σ. Knowledge of these parameters requires that the radar system be capable of detecting and classifying a given target over time. If a target is not initially detected by the radar, then the radar uses default values such as $R = R_{\max}$, where $R_{\max}$ is the maximum unambiguous range. Based on the well-known radar range equation, the receive signal power is

$$P_{\mathrm{r}} = \frac{P_{\mathrm{t}}\, G^2 \lambda^2 \sigma N_{\mathrm{P}}}{(4\pi)^3 R^4}, \tag{15.3}$$

where P_{t} is the radar peak transmit power, G is the transmit and receive antenna gain, λ is the wavelength of the carrier frequency, and N_{P} is the number of pulses within a CPI. Equations (15.1) and (15.3) are used to form the SINR objective function as

$$Z_1(\beta_i, f_j) = \frac{P_{\mathrm{r}}\, \tau\, \beta_i}{\Gamma(\beta_i, f_j)} \tag{15.4}$$

for τ the pulse width. Equation (15.4) assumes the use of a pulse compression (PC) waveform having a time-bandwidth product of $\tau\beta_i$ for the given sub-band bandwidth. Note that the minimum bandwidth must satisfy a time-bandwidth requirement of $\tau\beta_i \geq 100$, so that the spectrum roll-off is such that 98%–99% of the waveform energy is contained within β_i (under the assumption of a linear frequency modulation (LFM) waveform, which is the most common).

The second objective function, defined simply as

$$Z_2(\beta_i, f_j) = \beta_i, \tag{15.5}$$

is the bandwidth of each sub-band. The goal of the SS-MO approach is to jointly maximize (15.4) and (15.5). Notice that while the numerator of both (15.4) and (15.5) contains β_i, the former also accounts for bandwidth in the denominator via the interference plus noise function $\Gamma(\beta_i, f_j)$, thus establishing a conflict. In other words, while both metrics do benefit from β_i being as large as possible, the denominator term in (15.4) implies an increasing likelihood of encountering SINR-degrading interference as bandwidth increases.

Within this framework, weighted sum multi-objective optimization (WSMO) [40] is used to resolve the conflicting objectives of (15.4) and (15.5) by maximizing the combined objective function

$$Z_{\mathrm{WSMO}}(\beta_i, f_j) = \alpha\, \overline{Z}_1(\beta_i, f_j) + (1-\alpha)\overline{Z}_2(\beta_i, f_j), \tag{15.6}$$

where $0 \leq \alpha \leq 1$ is a user-defined weighting parameter, and $\overline{Z}_1(\beta_i, f_j) = Z_1(\beta_i, f_j)\big/\max\left[Z_1(\beta_i, f_j)\right]$ and $\overline{Z}_2(\beta_i, f_j) = Z_2(\beta_i, f_j)\big/\max\left[Z_2(\beta_i, f_j)\right]$ are the peak-normalized versions of $Z_1(\beta_i, f_j)$ and $Z_2(\beta_i, f_j)$, respectively. The weighting parameter α is selected to emphasize one objective function over the other and is dependent on the radar application. For example, it would be beneficial to set $\alpha \leq 0.5$ for high range resolution (HRR) applications that require large bandwidths. The solution to (15.6) is realized by determining

$$\{\beta_i^*, f_j^*\} = \underset{(\beta_i, f_j)}{\arg\max}\left[Z_{\mathrm{WSMO}}(\beta_i, f_j)\right] \tag{15.7}$$

subject to the constraints

$$Z_1(\beta_i, f_j) \geq Z_{1,\min} \tag{15.8}$$

and

$$Z_2(\beta_i, f_j) \geq Z_{2,\min}, \tag{15.9}$$

where $Z_{1,\min}$ is the minimum SINR required for detection and $Z_{2,\min}$ is the minimum bandwidth needed to meet the range resolution requirement. The constraints of (15.8) and (15.9) require that (15.7) produces a feasible operating solution for the radar. If no feasible solutions exist, then the empty set Ø is returned. The optimal solution selected by WSMO in (15.6) is denoted as

$$Z^*_{\mathrm{WSMO}} = Z_{\mathrm{WSMO}}(\beta^*_i, f^*_j). \tag{15.10}$$

If a feasible solution exists, a PC waveform is then synthesized using the parameters $\{\beta^*_i, f^*_j\}$. For instance, the most straightforward option is the LFM waveform that can be expressed as

$$x(t) = A\cos\left(2\pi f_c\, t + \pi\,[\beta^*_i/\tau]\, t^2\right), \tag{15.11}$$

where $A = \sqrt{P_t}$ is the transmit amplitude, $f_c = f^*_j + \beta^*_i/2$ is the RF center frequency, and the pulse has time support on $-\tau/2 \le t \le +\tau/2$. This waveform is used for the experimental demonstrations later in the chapter. Of course, other waveforms can be used as well, such as stepped frequency [24,26] and digitally tunable waveforms [41]. Once the waveform is synthesized, the radar system likewise tunes the reconfigurable RF front-end to conform to the $\{\beta^*_i, f^*_j\}$ parameters and then subsequently transmits/receives the waveform/echoes (respectively) for the remainder of the CPI cycle.

As an example, consider the application of SS-MO to the power spectrum of Figure 15.1 based on the parameters $Z_{1,\min} = -20$ dB, $Z_{2,\min} = 5$ MHz, and $\alpha = 0.5$, and establishing a baseline SINR of 30 (14.8 dB) when no RFI is present (which would achieve $\beta^*_i = B$). For a given pulsewidth τ, this baseline SINR establishes the received radar signal power P_r via (15.4) under a noise-only condition for the denominator in (15.4) using $\Gamma(\beta_i, f_j) = kT_o N_F B$, where k is the Boltzmann's constant and T_o is the noise temperature.

Based on these parameters, the optimization process selects the sub-band between 8.4 and 46.4 MHz (the dotted trace in Figure 15.5), which does contain lower power narrowband interference at 24.6 MHz that may require interference cancellation in the receiver signal processing. The algorithm selects this sub-band as a trade-off between the degradation in SINR due to the 24.6 MHz interference and the greater range resolution that the larger bandwidth provides. Where a simple hard decision threshold-based spectrum sensing approach—with a threshold at say, 10 dB above the ambient noise floor—would take a binary occupied/non-occupied view of the spectrum, SS-MO considers the amount of interference that is incurred relative to the bandwidth that can be obtained to achieve a better overall solution. Of course, this solution for the radar also means that the RF user at 24.6 MHz may experience interference from the radar (depending on the nature of that user, the spatial separation from the radar, and the spatial orientation relative to the desired look direction of the radar). As discussed above and quantified in [25], additional objective functions can also be used within this framework to account for this mutual interference issue.

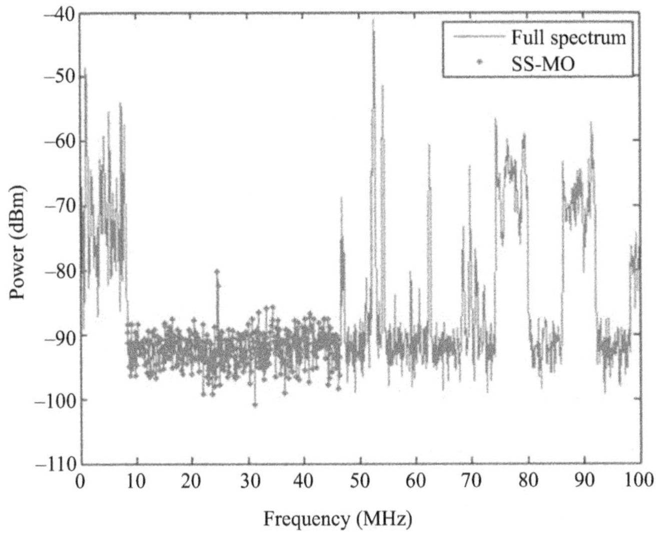

Figure 15.5 Power spectrum from Figure 15.1 with the SS-MO solution highlighted by the dotted trace in the interval from 8.4 to 46.4 MHz. Note that this sub-band does contain narrowband RFI with low power

15.3 Computational cost and RF tuning considerations

The SS-MO technique is applied at the beginning of the CPI so that the radar has an up-to-date measure of the RFI. Ideally, the radar would then transmit N_P pulses for the remainder of the CPI; however, since the radar is devoting resources to determine its appropriate spectral footprint, additional time is needed. If the radar timeline to perform a given sensing task is fixed, this added processing time necessitates reducing the number of pulses in the CPI, which consequently decreases the achievable SINR. Clearly, the computational cost of this spectrum sensing/decision process should be kept to a minimum. The same goes for reconfiguring the RF transmit/receive hardware. For this cognitive spectrum usage arrangement, the radar timeline T to perform a given sensing task can be divided up as

$$T = T_S + T_M + T_F + T_R, \tag{15.12}$$

where T_S is the time needed to measure the current EME and estimate the corresponding power spectrum, T_M is the multi-objective optimization processing time, T_F is the RF tuning time (i.e. the time needed to tune the transmitter and receiver to the optimization-determined bandwidth and center frequency), and T_R is the time remaining to transmit and receive the CPI of pulses. The timing diagram for this arrangement is summarized in Figure 15.6. It is generally expected that $T_S << T_M + T_F$, and thus the processing time and RF tuning time are the bottlenecks that limit the radar CPI, and consequently the radar performance.

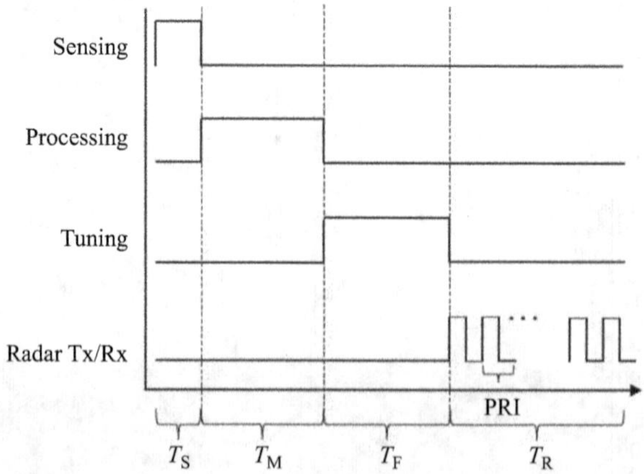

Figure 15.6 Timing diagram of SS-MO spectrum sensing and decision optimization (processing), RF transmit/receive tuning, and radar operation involving a CPI of pulses

15.3.1 Computational cost of WSMO

The WSMO approach discussed in Section 15.2 incurs the following computational costs:

1. Equation (15.1): $N_{\text{cand}} - N = (N^2 - N)/2$ additions to determine each of the candidate interference-plus-noise values.
2. Equation (15.4): Since β_i is size N, the numerator requires N multiplications. Likewise, there are N_{cand} divisions by $\Gamma(\beta_i, f_j)$. The total number of multiplications/divisions is $N_{\text{cand}} + N = (N^2 + 3N)/2$.
3. Equation (15.6): Requires N_{cand} multiplications, N_{cand} divisions, and N_{cand} additions.

Table 15.1 summarizes these contributions, which in total have a computational cost of $O(N^2)$. Section 15.4 discusses a technique to reduce the computational cost to $O(N)$.

15.3.2 Considerations for RF tuning

This form of cognitive radar for RFI avoidance requires reconfigurable RF filter technology that supports rapid tuning speeds (i.e. low latency to keep T_F as small as possible), while maintaining high performance in terms of passband gain, stopband attenuation, linearity, and insertion loss. Further, there is a trade-off between the improved SINR provided by RFI avoidance and diminished SINR resulting from the necessary use of a tunable RF front-end. The following quantifies this trade-off for four RFI cases [39].

Table 15.1 Summary of computational cost for WSMO

	Eq. (15.1)	Eq. (15.4)	Eq. (15.6)	Total
Additions	$(N^2 - N)/2$	0	$(N^2 + N)/2$	N^2
Multiplications and divisions	0	$(N^2 + 3N)/2$	$(N^2 + N)$	$(3N^2 + 5N)/2$

A variety of radar modes operate with a CPI on the order of microseconds (136–680 μs per [42]), including ultra-wideband synthetic aperture radar, airborne-tracking radar, and ground-penetrating radar. These radars require different tuning capabilities depending on the application. Radar applications such as the detection of harmonics generated by electronic components [43,44] also require the RF tuner to have low insertion loss and high linearity. Based on this span of operational CPI values, different categorizations can be made.

Three general groups of tuning technology are considered based on their speeds, categorized as fast, medium, and slow [45]. The first category, C1 (fast), constitutes tuning times from 1 to 300 μs; the next category, C2 (medium), has tuning times between 300 and 600 μs; and the final category, C3 (slow), involves tuning times between 600 and 900 μs. This analysis is broadly applicable to multiple RF tuning technologies including microelectromechanical systems (MEMS), barium-strontium-titanate (BST), Schottky diode, and yttrium-iron-garnet (YIG). The tuner must facilitate a reconfigurable transmitter and receiver to support waveform adaptation (for both narrowband and wideband applications), and be able to reject high-powered interference while not introducing additional intermodulation products.

To evaluate the trade-space between RF tuning latency and RFI rejection, SS-MO is applied to four RF spectra consisting of different interference and noise power levels. These measurements were collected by the system in Figure 15.2 for $B = 100$ MHz and $\Delta_r = 100$ kHz, at center frequencies of 286, 437, 778, and 813 MHz. Optimization was performed using $Z_{1,\min} = -20$ dB, $Z_{2,\min} = 5$ MHz, $\alpha = 0.5$, and a baseline SINR = 30 (14.8 dB). The full radar timeline is 900 μs, which for a PRI of 20 μs corresponds to $N_P = 45$ radar pulses.

The four spectra are illustrated in Figure 15.7 and ordered from least to greatest RFI power, which is computed as $P_a = \sum \theta_i$. The term $P_{a,\text{SSMO}}$ is also used to denote the RFI power within the subsequent optimization-determined bandwidth. The sub-band selected by SS-MO is illustrated by the double-arrow interval. The spectrum in Figure 15.7(a) corresponds to the noise-only condition (no RFI) that is encountered at $f_c = 286$ MHz and has $P_a = -62$ dBm. In this case the full bandwidth $\beta = 100$ MHz is selected, which is an appropriate choice given that no interference is present. Likewise, $P_{a,\text{SSMO}} = P_a = -62$ dBm in this instance.

The spectrum in Figure 15.7(b) was measured at $f_c = 813$ MHz to have an increased P_a of -55 dBm. Here a sub-band bandwidth of $\beta = 76.4$ MHz is selected with $P_{a,\text{SSMO}} = -62$ dBm (equal to noise). In other words, a sub-band is selected that contains no interference. Next, the spectrum in Figure 15.7(c) corresponds to $f_c = 778$ MHz that has $P_a = -46$ dBm. Now a sub-band bandwidth of $\beta = 53.3$ MHz

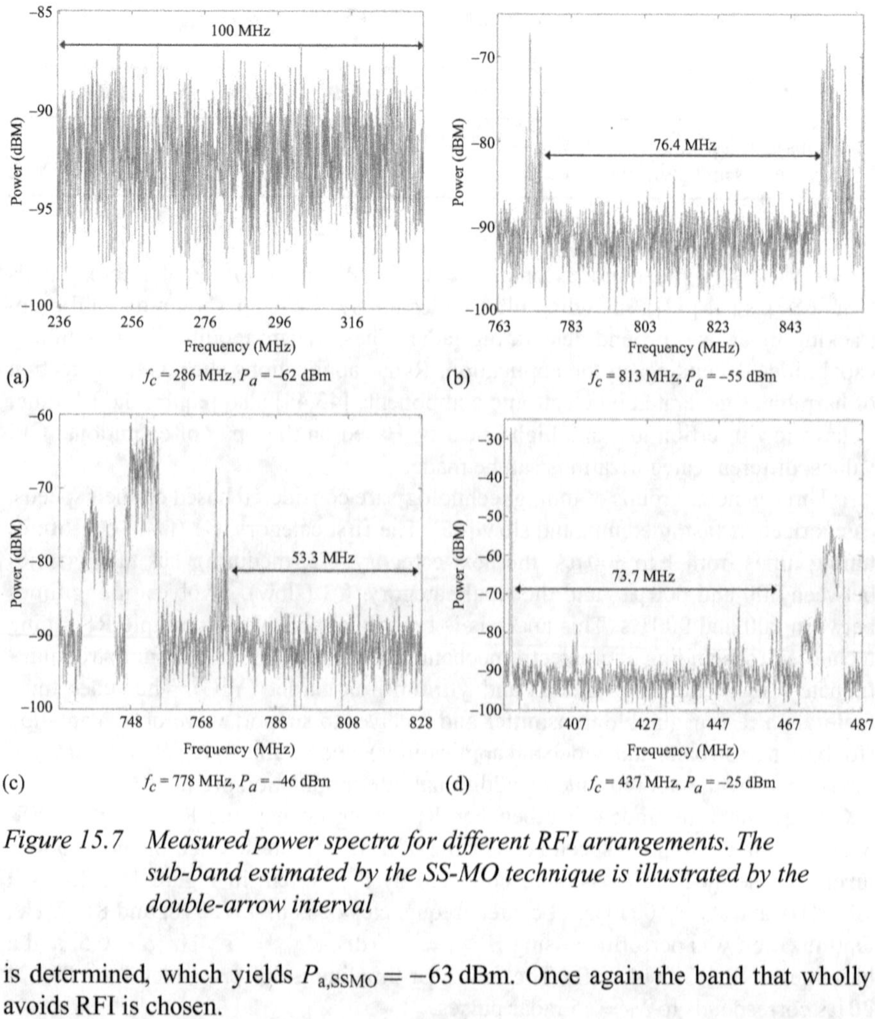

Figure 15.7 Measured power spectra for different RFI arrangements. The sub-band estimated by the SS-MO technique is illustrated by the double-arrow interval

is determined, which yields $P_{a,SSMO} = -63$ dBm. Once again the band that wholly avoids RFI is chosen.

Finally, the spectrum in Figure 15.7(d) is centered at $f_c = 437$ MHz and has the much higher $P_a = -25$ dBm due to the presence of high-power narrowband interference near 390 MHz. In this case a sub-band bandwidth of $\beta = 73.7$ MHz is chosen that has $P_{a,SSMO} = -61$ dBm. While this RFI power still nearly equates to the noise floor measurement, note that some low-power interference is contained in the sub-band. This case once again exemplifies the benefit of multi-objective optimization over simple interference thresholding.

To put the results of Figure 15.7 in the context of RF tuner latency, Figure 15.8 shows the SINR for each tuner category (C1, C2, and C3) along with the baseline case of using the entire bandwidth (no RFI avoidance). As the tuning latency T_F of each filter increases for a fixed overall timeline T, the number of pulses that can be transmitted (with a maximum of $N_P = 45$) decreases, thereby also reducing SINR. The graphs also display a crossover point to indicate the maximum allowable T_F for which

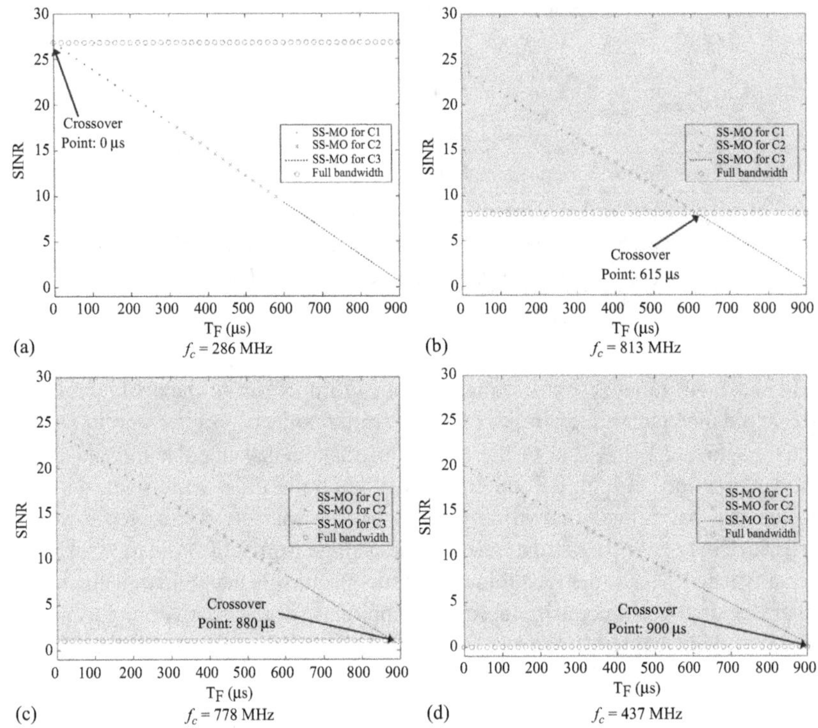

Figure 15.8 SINR trade-space for different RF tuner latencies. The cross-over point (intersection of the horizontal line with the diagonal trace for tuner categories C1, C2, and C2) illustrates where RFI avoidance based on the given RF tuner latency yields the same SINR as using the full band in the presence of RFI

RFI avoidance retains an SINR advantage over simply employing the full bandwidth despite the presence of RFI, with the shaded areas indicating the tuning latencies, and associated tuner categories, where RFI avoidance (based on the use of SS-MO) still provides an advantage.

Figure 15.8(a) shows the SINR for the full-band (FB) usage of Figure 15.7(a), where due to the absence of interference RFI avoidance is unnecessary. In Figure 15.8(b), in which the RFI depicted in Figure 15.7(b) must be addressed, the use of RFI avoidance via SS-MO maintains a higher SINR for tuning latencies up to 615 μs, which includes all of tuner categories C1 and C2.

Figure 15.8(c) illustrates the SINR trade-space for the spectrum of Figure 15.7(c). In this case RFI avoidance provides an advantage over FB operation for tuning times of as much as 880 μs, which includes the entirety of categories C1 and C2 along with nearly all of C3. Finally, Figure 15.8(d) shows that for the spectrum in Figure 15.7(d), all of the RF tuner latencies considered are preferable to operating over the full band.

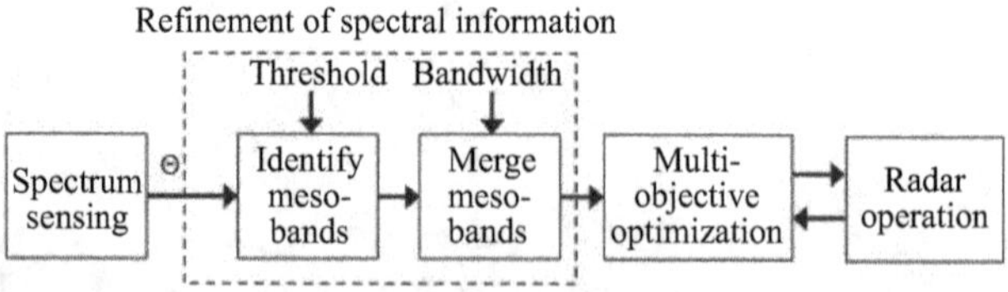

Figure 15.9 fWSMO refinement procedure to reduce spectral information prior to multi-objective optimization. The parameters in the refinement procedure are estimated using the training data

In short, RF tuner response time is critically important to the actual realization of RF avoidance techniques in the cognitive radar context. Higher tuning latencies negatively impact SINR due to the ability to transmit fewer pulses in the remaining radar timeline. In high RFI environments this trade-off is beneficial, while it is expected that operating in environments containing sparsely distributed low-power interference may not necessarily realize an advantage from RFI avoidance. Thus continued advances in RF tuner technology will support further capability enhancements. Further discussion on reconfigurable RF techniques for the receiver and transmitter can be found in Chapters 10 and 12, respectively.

15.4 Fast weighted sum multi-objective optimization for RFI avoidance

The decision portion of RFI avoidance, such as performed above using WSMO, is one of the latency bottlenecks to achieving feasible real-time operation for this manner of cognitive sensing. To reduce the computational cost that drives this latency, a fast version of weighted sum multi-objective optimization denoted as fast weighted sum multi-objective optimization (fWSMO) [27] is discussed that is a bio-inspired processing technique analogous to the operation of the thalamus in the human brain. Specifically, the thalamus regulates sensory input (excluding olfaction) from the external environment as a means to reduce the amount of information sent to the cerebral cortex [46]. An abnormal thalamus disrupts the flow of information to the cerebral cortex, thereby overwhelming a person with external stimuli [47], a condition observed in schizophrenia patients [47,48]. In a manner similar to normal thalamus functioning, fWSMO performs pre-processing to reduce the amount of input to WSMO, thus resulting in a significant computation reduction overall.

A block diagram of fWSMO is shown in Figure 15.9, which is an expansion of SS-MO operation from Figure 15.3. Two additional procedures are added (within the dashed box) to pre-process the information in the power spectrum. The first procedure identifies low-power and high-power interference regions defined as "meso-bands" based on a thresholding operation. The second merges the meso-bands into sub-bands based on how the interference is distributed in frequency.

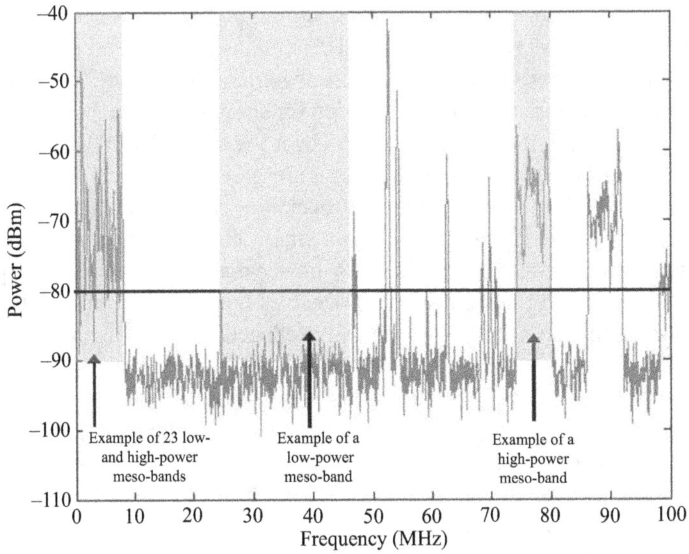

Figure 15.10 Power spectrum from Figure 15.1 where the shaded regions covering 24–46 MHz and 74–80 MHz represent low- and high-power meso-bands, respectively. In contrast, the shaded region from 0–8 MHz contains a combination of several closely spaced meso-bands of both low and high power. The threshold is set to −80 dBm for simplicity

Define the set of $K \leq N$ meso-bands as $\Phi = \{\Phi_1, \ldots, \Phi_K\}$. Meso-band Φ_k has an associated lower edge frequency (or "start" position) denoted as $\overline{S}_k \in \{1, \ldots, N\}$, an upper edge frequency (or "end" position) denoted as $\overline{E}_k \in \{1, \ldots, N\}$, contains $\overline{l}_k \in \{1, \ldots, N\}$ frequency samples, has bandwidth $\overline{B}_k = \Delta_r \overline{l}_k$, and has measured interference power

$$\overline{\Gamma}_k = \sum_{n=\overline{S}_k}^{\overline{E}_k} \theta_n. \tag{15.13}$$

The value Φ_k is deemed to be a "low-power meso-band" when $\theta_n < T_f$ for all $n \in \{\overline{S}_k, \ldots \overline{E}_k\}$, where the interference threshold T_f is determined based on the training procedure discussed in Section 15.4(b). Alternatively, Φ_k is deemed to be a "high-power meso-band" when $\theta_n \geq T_f$ for all $n \in \{\overline{S}_k, \ldots, \overline{E}_k\}$. It should be noted that meso-bands are deemed to be either low or high power, never both; alternating between one and the other when scanning across frequency. Thus the bandwidths of individual meso-bands can vary substantially.

Example meso-bands are illustrated in Figure 15.10 (shaded regions), with the large swath from 24 to 46 MHz identified as a low-power meso-band and the contiguous interference from 74 to 80 MHz a high-power meso-band. Additionally, the shaded spectral interval from 0 to 8 MHz contains a large number of alternating high and low meso-bands that are very close together in frequency. The interference threshold used here is $T_{\rm f} = -80$ dBm (chosen for simplicity), which produces a total of $K = 54$ meso-bands, a dramatic reduction over the original 1,000 samples.

The second refinement procedure in Figure 15.9 merges the large number of alternating high and low meso-bands that occupy a small bandwidth (e.g. the 0-8 MHz shaded region in Figure 15.10) into a single high-power meso-band because it is not useful to consider the narrow low-power regions in between for practical radar operation. This refinement therefore enables the computational cost of subsequent WSMO application to be significantly reduced, even in highly occupied and fragmented spectra.

Specifically, consider Φ_k to be a low power meso-band. Then Φ_k would be merged with Φ_{k-1} and Φ_{k+1} if $\overline{B}_k \leq B_{\rm f}$, where $B_{\rm f}$ is a "merge bandwidth," the valuation of which is determined in Section 15.4.2. To further simplify the meso-band merging operation, the meso-band length in terms of samples can be used by applying $\bar{l}_k \leq B_{\rm f}/\Delta_{\rm r}$ as the merge criterion. Each merged meso-band is defined to be a high-power meso-band since it contains interference.

After the merge procedure, the resulting set of meso-bands are defined as $\hat{\Phi} = \{\hat{\Phi}_1, \ldots, \hat{\Phi}_{\hat{K}}\}$ for $\hat{K} \leq K$. Each meso-band $\hat{\Phi}_{\hat{k}}$ for $\hat{k} \in \{1, \ldots, \hat{K}\}$ has an associated (lower edge frequency) start position $\hat{S}_{\hat{k}} \in \{1, \ldots, K\}$, (upper edge frequency) end position $\hat{E}_{\hat{k}} \in \{1, \ldots, K\}$, number of frequency bins $\hat{l}_{\hat{k}} = \bar{l}_{k-1} + \bar{l}_k + \bar{l}_{k+1}$, bandwidth $\hat{\beta}_{\hat{k}} = \Delta_{\rm r}\hat{l}_{\hat{k}}$, and interference power $\hat{\Gamma}_{\hat{k}} = \overline{\Gamma}_{k-1} + \overline{\Gamma}_k + \overline{\Gamma}_{k+1}$.

After the two refinement procedures are completed, the bandwidth and center frequency pairs $\{\hat{\beta}_{\hat{k}}, f_{\hat{S}_{\hat{k}}}\}$ of the $\hat{K}$ meso-bands are fed into the WSMO process from Section 15.2 to determine the optimal sub-band $\{\hat{\beta}^*_{\hat{k}}, f^*_{\hat{S}_{\hat{k}}}\}$ via (15.7) and the associated

$$\hat{Z}^*_{\rm WSMO} = \hat{Z}_{\rm WSMO}(\hat{\beta}^*_{\hat{k}}, f^*_{\hat{S}_{\hat{k}}}) \tag{15.14}$$

from (15.10). Of course, the refinement process to form and merge meso-bands does result in a reduced set of candidate solutions for WSMO, and thus it is possible that the globally optimal solution is not obtained by fWSMO. However, achieving a "good enough" solution in an operationally feasible amount of time is more practically useful than determining the best solution in an infeasible amount of time.

The computational cost of fWSMO is summarized in Table 15.2, with the individual components incurring:

1. Meso-band formation requires $\sum_{k=1}^{K} \bar{l}_k$ additions to form $\overline{\Gamma}_k$ and K subtractions to compute $\bar{l}_k = \overline{E}_k - \overline{S}_k$.
2. Meso-band merging necessitates computing $\hat{\Gamma}_{\hat{k}}$ and $\hat{l}_{\hat{k}}$, which require $\hat{N} < 2N$ additions. At most, when $K = N$, all bins in the power spectrum are summed. Once $\hat{l}_{\hat{k}}$ is realized, $\hat{K}$ multiplications are needed to determine $\hat{\beta}_{\hat{k}} = \Delta_{\rm r}\hat{l}_{\hat{k}}$.
3. Equation (15.1) for fWSMO: The value $\hat{\Gamma}_{\hat{k}}$ replaces θ_j in (15.1) to form the sub-band interference estimate. The number of additions therefore becomes $N_{\rm cand} = (\hat{K}^2 + \hat{K})/2$ in (15.2), where N is replaced by $\hat{K}$. The calculation of (15.1) requires $(\hat{K}^2 - \hat{K})/2$ additions.

Table 15.2 *Summary of computational cost for fWSMO*

	Form meso-bands	**Merge meso-bands**	**Eq. (15.1)**	**Eq. (15.4)**	**Eq. (15.6)**	**Total**
Additions and subtractions	$\hat{K}+\sum_{k=1}^{K}\bar{l}_k$	$\hat{N}$	$\frac{(\hat{K}^2-\hat{K})}{2}$	0	$\frac{(\hat{K}^2+\hat{K})}{2}$	$\hat{K}^2+\hat{K}+2\sum_{k=1}^{K}\bar{l}_k$
Multiplications and divisions	0	$\hat{K}$	0	$\frac{(\hat{K}^2+3\hat{K})}{2}$	$2(\hat{K}^2+\hat{K})$	$\frac{(5\hat{K}^2+9\hat{K})}{2}$

4. Equation (15.4) for fWSMO: The numerator in (15.4) requires $\hat{K}$ multiplications followed by $(\hat{K}^2+\hat{K})/2$ divisions by $\hat{\Gamma}_{\hat{k}}$. The total number of multiplications and divisions is $(\hat{K}^2+3\hat{K})/2$.
5. Equation (15.6) for fWSMO: Requires $(\hat{K}^2+\hat{K})$ multiplications by α, $(\hat{K}^2+\hat{K})$ divisions to normalize each objective function, and $(\hat{K}^2+\hat{K})/2$ additions to combine each objective function. The total number of multiplications and divisions is $2(\hat{K}^2+\hat{K})$.

15.4.1 Generation of synthetic data for fWSMO training

The input threshold and bandwidth parameters of fWSMO are estimated using training data that can be synthetically produced using randomly generated power spectra representing multiple RFI arrangements. A comprehensive set of training data is needed because the solution determined by fWSMO depends on how the RFI is distributed in terms frequency and power. For instance, the sub-band selection in the presence of high-power, narrowband interference is different than what would be selected if lower power, wideband interference were present. Multiple variations of such interference scenarios must therefore be included in the synthetic data set.

To illustrate the formation of synthetic data, observations of the EME are acquired using the measurement setup in Figure 15.2. The parameters for the data collection are the same as those used to collect the power spectrum in Figure 15.1, namely bandwidth $B=100$ MHz, frequency resolution $\Delta_r=100$ kHz, $N=1{,}000$ frequency bins per spectrum, and average noise power $P_N=-92$ dBm. To capture different arrangements of RFI, the center frequency of the spectrum analyzer was swept from 100 to 900 MHz in 100 kHz increments to collect 8,000 recorded spectra. Each power spectrum in the set contains a variety of RFI types including high-power narrowband interference, wideband interference, no interference (noise only), and combinations of narrow and wideband interference.

Although this measured set is used to produce synthetic data, WSMO and fWSMO can also obviously be applied to each power spectrum as they are in [24]. The 8,000 recorded spectra are defined as $\Psi=\{\Theta_1,\ldots,\Theta_{8{,}000}\}$, where each $\Theta_\psi\in\Psi$ is a

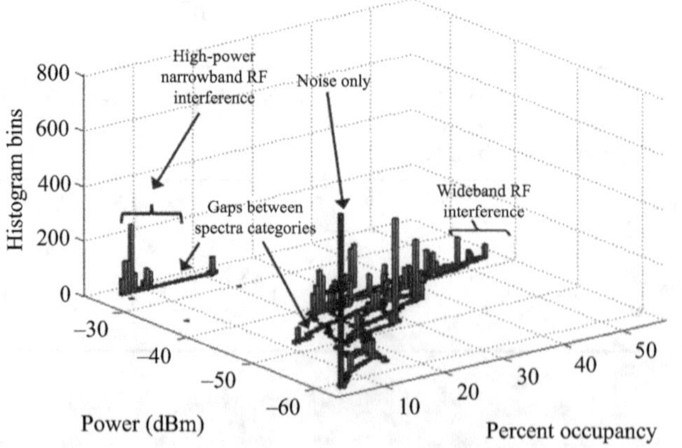

Figure 15.11 Histogram of measured data containing a variety of interference manifestations. Several gaps exist in the histogram; thus, a more comprehensive data set is needed for training

power spectrum that contains the samples $\{\theta_{\psi,1}, \ldots, \theta_{\psi,1,000}\}$, with $\psi \in \{1, \ldots, 8,000\}$ denoting the spectrum index as Θ_ψ. Note that the power spectrum depicted in Figure 15.1 is actually $\Theta_1 = \{\theta_{1,1}, \ldots, \theta_{1,1,000}\}$ in this set, where $\psi = 1$.

The complete set Ψ is quantified based on its percent occupancy (PO) and interference/noise power. The PO is defined as

$$\mu_\psi = \frac{1}{N}\left(\sum_{n=1}^{N} y_{\psi,n}\right) \times 100\% \tag{15.15}$$

for

$$y_{\psi,n} = \begin{cases} 1, & \theta_{\psi,n} \geq T_{\text{po}} \\ 0, & \theta_{\psi,n} < T_{\text{po}} \end{cases}, \tag{15.16}$$

with $T_{\text{po}} = -89$ dBm the threshold (3 dB above the -92 dBm noise floor). The interference-plus-noise power is defined as

$$\overline{\theta}_\psi = \sum_{n=1}^{N} \theta_{\psi,n} \tag{15.17}$$

for the measured spectrum Θ_ψ. Figure 15.11 illustrates a histogram of the measured data in terms of (15.15) and (15.17). It is observed that multiple redundant instances occur for certain types of RFI. For example, 194 spectra contain high-power, narrowband interference with -25.3 dBm power and a 7.2% occupancy, while 439 spectra exhibit the noise-only case with -61.5 dBm power and 2.9% occupancy. The histogram also contains information gaps between different spectral categories, implying that a more comprehensive data set is required to train the parameters of fWSMO.

To address the limited scope of the measured data, synthetic data is generated for the following five occupancy levels: 0% to 10%, 10% to 20%, 20% to 30%, 30% to 40%, and 40% to 50%. Note that occupancy above 50% is not considered since such a small number of the measured spectra was found to fit this category. For each occupancy level, 1,000 randomly generated spectra are produced according to a set of synthetic spectrum templates (SSTs) based on observations of the measured spectra. For instance, narrowband interference is simulated by a sinusoid-based SST defined as

$$x_s(t, Q_s) = \sum_{q=1}^{Q_s} A_s(q) \cos(2\pi f_s(q) t), \tag{15.18}$$

where $A_s(q)$ and $f_s(q)$ are uniform random variables for amplitude and frequency, respectively, Q_s is a uniform random variable that specifies the total number of sinusoids in (15.18), $0 \le t \le 1/\Delta_r$ is the time support of the interference, and $0 \le f_s(q) \le B$ bounds the frequency support. The amplitude of each sinusoid is related to the power by $A_s(q) = \sqrt{P_s(q)}$, which has upper and lower limits according to $P_{s,\text{lower}} \le P_s(q) \le P_{s,\text{upper}}$. Likewise, the number of sinusoids in (15.18) is bounded by $Q_{s,\text{lower}} \le Q_s \le Q_{s,\text{upper}}$.

Wideband interference is synthesized using an LFM chirp SST via

$$x_c(t, Q_c) = \sum_{q=1}^{Q_c} A_c(q) \cos(2\pi f_c(q) t + \pi [B_c(q)/\tau] t^2), \tag{15.19}$$

where $A_c(q)$, $f_c(q)$, and $B_c(q)$ are uniform random variables for amplitude, center frequency, and bandwidth, respectively. Here, Q_c is a uniform random variable that defines the number of wideband chirp signals in (15.19) and is bounded by $Q_{c,\text{lower}} \le Q_c \le Q_{c,\text{upper}}$. The lower and upper limits of the random bandwidth and center frequency variables are $B_{c,\text{lower}} \le B_c(q) \le B_{c,\text{upper}}$ and $B_c(q)/2 \le f_c(q) \le B - B_c(q)/2$, respectively, for B the overall bandwidth. The amplitude of each chirp is related to its power by $A_c(q) = \sqrt{P_c(q)}$, which has the lower and upper limits $P_{c,\text{lower}} \le P_c(q) \le P_{c,\text{upper}}$.

A synthetic spectrum instantiation for each occupancy level is then generated by forming the signal $x(t)$, which is a combination of multiple sinusoid-based SSTs from (15.18), multiple chirp-based SSTs from (15.19), and additive white Gaussian noise, with the latter at the same level as the measured noise floor. The power spectrum of $x(t)$ is then determined by applying the Fourier transform using the same parameters as the measured data, i.e. $B = 100$ MHz, $\Delta_r = 100$ kHz, and $N = 1{,}000$ samples.

For the different occupancy levels, the 1,000 spectra each are divided into various categories representing noise-only environments and a wide variety of narrowband and wideband interference combinations at various amplitudes, center frequencies, and bandwidths. The resulting histogram in Figure 15.12 shows that this synthetic data does indeed provide a much more diverse picture of the possible interference environments relative to Figure 15.11. These 5,000 synthetic spectra are labeled as $\Psi_S = \{\Omega_1, \ldots, \Omega_{5,000}\}$, where each $\Omega_\psi \in \Psi_S$ is a power spectrum containing the power samples $\{\omega_{\psi,1}, \ldots, \omega_{\psi,1,000}\}$ for spectrum index $\psi \in \{1, \ldots, 5,000\}$.

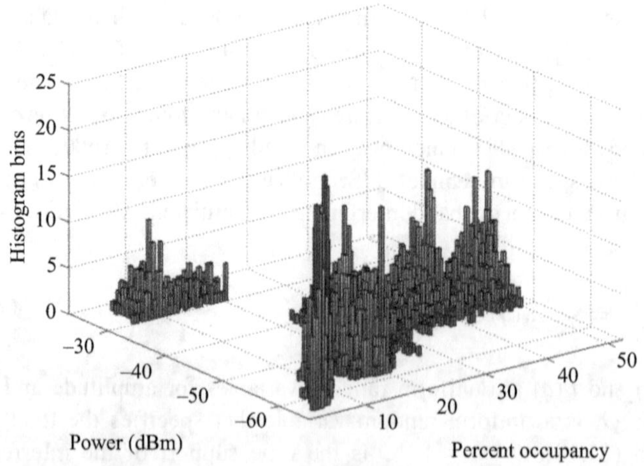

Figure 15.12 Histogram of the synthesized data according based on a variety of interference arrangements modeled on measured data. These data are used to train the fWSMO parameters

15.4.2 Interference threshold and merge bandwidth parameter training for fWSMO

The set Ψ_S is used to train the fWSMO T_f and B_f parameters. For the ψth power spectrum Ω_ψ in Ψ_S fWSMO is applied to determine $\hat{Z}^*_{\mathrm{WSMO}}$ via (15.14), with this solution subsequently denoted as D_ψ. The same parameters of overall bandwidth, frequency resolution, optimization constraints, and multi-objective weighting are used for this determination as are anticipated to be applied to measured data. In addition, the parameters T_f and B_f are swept over a discretized set of reasonable candidate possibilities. From this two-dimensional set of parameter pairs, the best set can be selected.

For example, using the same parameters as in the previous WSMO and fWSMO implementations, and sweeping the fWSMO interference threshold and merge bandwidth through the intervals $T_f \in \{-90, -89, \ldots, -70\}$ dBm and $B_f \in \{0.1, 0.2, \ldots, 4.0\}$ MHz, respectively, 840 different evaluation pairs are obtained for each of the 5,000 synthetic spectra. Define $D_\psi(T_f, B_f)$ as the evaluation of spectrum Ω_ψ for a given $\{T_f,\ B_f\}$ pair, with the average then $D_{\mathrm{ave}}(T_f, B_f) = \sum_{\psi=1}^{5,000} D_\psi(T_f, B_f)/5,000$.

Figure 15.13 plots $D_{\mathrm{ave}}(T_f,\ B_f)$ over these evaluation parameters. A larger T_f produces a smaller $D_{\mathrm{ave}}(T_f,\ B_f)$ value since the majority of the RFI is below the interference threshold. The number of meso-bands is therefore small (i.e. $K << N$), resulting in an inadequate solution space. A high merge bandwidth B_f also decreases the number of meso-bands (since several meso-bands are merged), again resulting in an inadequate solution space. The parameter combination that produces the largest

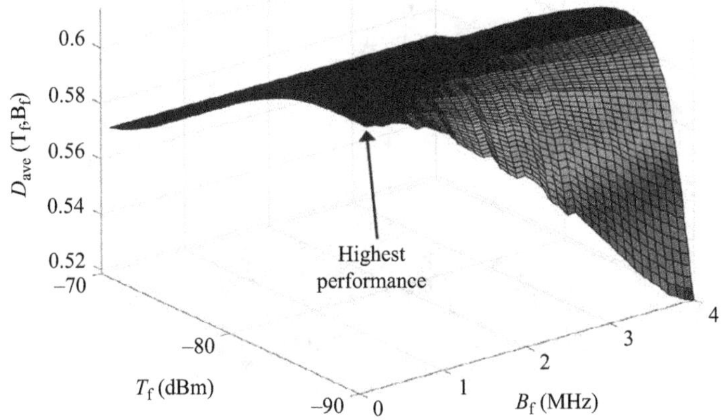

Figure 15.13 $D_{ave}(T_f, B_f)$ *is the objective function evaluation averaged over 5,000 synthetic spectra for prospective* $\{T_f, B_f\}$ *pairs. The highest objective function value occurs for* $T_f = -90$ *dBm and* $B_f = 0.1$ *MHz since a large number of meso-bands exist* ($K \approx N$)

$D_{ave}(T_f, B_f)$ value is $T_f = -90$ dBm and $B_f = 0.1$ MHz, which is expected since a large number of meso-bands exist and $K \approx N$.

An additional criterion used to evaluate the parameterization of fWSMO is computational cost. The evaluation that follows only considers the number of additions since the results for the number of multiplications are rather similar. Like the average objective function metric $D_{ave}(T_f, B_f)$, the number of additions is measured for each combination of T_f and B_f (see Figure 15.14). Clearly, a low interference threshold and a low merge bandwidth realize a high computational cost since a large number of meso-bands exist and $K \approx N$; though, as discussed above, this same arrangement also yields the optimal value of the objective function.

The compromise solution is to jointly evaluate the information in Figures 15.13 and 15.14. The metric in Figure 15.13 is normalized and then converted from a maximization problem to a minimization problem, where the minimum value now corresponds to the highest performance point. This modified metric is then added to the computational cost metric in Figure 15.14 and the minimum found. Figure 15.15 shows that the minimum value of this joint metric is $T_f^* = -86$ dBm and $B_f^* = 1.0$ MHz.

15.5 Evaluation of WSMO and fWSMO

We now consider the application of WSMO and fWSMO to experimental data collected using the measurement system in Figure 15.2. The experimental data

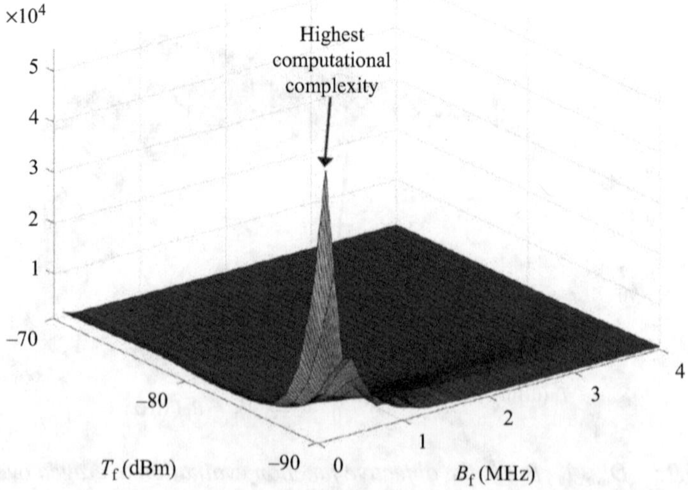

Figure 15.14 The number of fWSMO additions required averaged over 5,000 synthetic spectra for prospective $\{T_f, B_f\}$ pairs. The highest computational cost occurs for $T_f = -90$ dBm and $B_f = 0.1$ MHz, since $K \approx N$

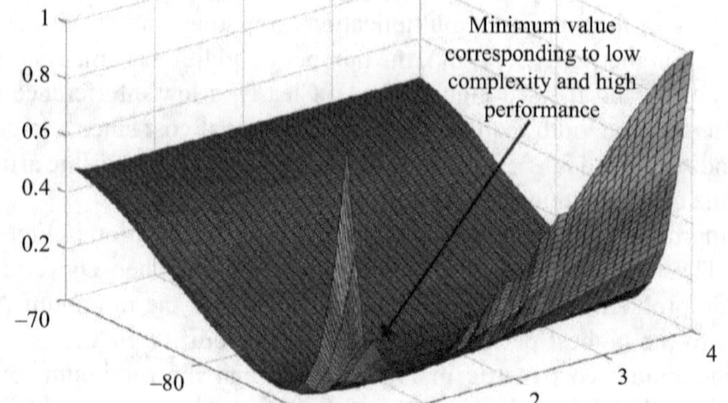

Figure 15.15 Combining the metrics depicted in Figures 15.13 and 15.14 to identify the best trade-off between objective function performance and computational cost. The optimal parameters are $T_f^* = -86$ dBm and $B_f^* = 1.0$ MHz

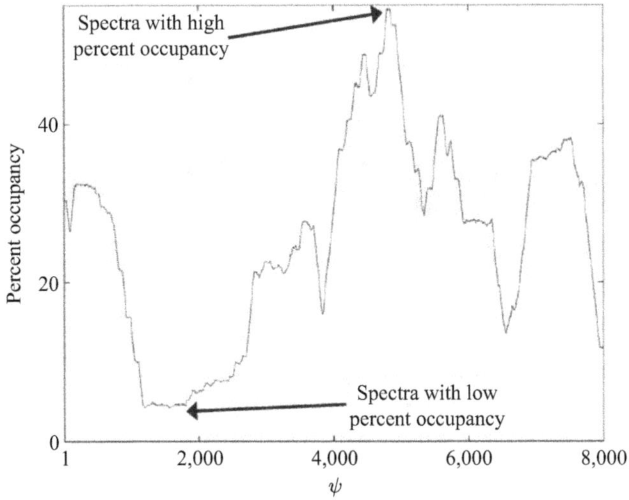

Figure 15.16 Percent occupancy of the recorded spectra in the experiment set, where ψ denotes the spectrum index number. The PO varies over several different interference conditions

are collected in the same manner as the training data discussed in Section 15.4 (similar to Figure 15.11) using the spectrum analyzer parameters of $B = 100$ MHz, $\Delta_r = 100$ kHz, and $N = 1{,}000$ frequency bins per spectrum, with an observed noise floor of $P_N = -92$ dBm. The measurement center frequency is swept from 100 to 900 MHz in 100 kHz increments to collect 8,000 spectra. The sweep for this experiment test set was conducted in the afternoon, whereas the sweep for the previous training data was conducted in the morning of the same day. As before, the algorithm parameters are $Z_{1,\min} = -20$ dB, $Z_{2,\min} = 5$ MHz, baseline SINR = 30 (14.8 dB), and $\alpha = 0.5$, along with the parameters $T_f = -86$ dBm and $B_f = 1.0$ MHz determined from the synthetic training data.

For simplicity, the notation used for the experiment set is the same as the training set. Specifically, the 8,000 recorded spectra are denoted as $\Psi = \{\Theta_1, \ldots, \Theta_{8,000}\}$, where each $\Theta_\psi \in \Psi$ is a power spectrum that contains the power samples $\{\theta_{\psi,1}, \ldots, \theta_{\psi,1,000}\}$ for power spectrum index $\psi \in \{1, \ldots, 8,000\}$. Figure 15.16 shows the PO for these 8,000 power spectra calculated using (15.15). The occupancy clearly varies over many different interference conditions and ranges from a low of 4.2% to a high of 54.5%. The interference plus noise power for this data set using (15.17) is depicted in Figure 15.17, where multiple spectral categories are represented, including the noise-only condition, high-power narrowband interference, wideband interference, and combinations thereof.

The SINR of the processed spectra is shown in Figure 15.18 where three traces are illustrated: (1) FB operation—dashed line; (2) WSMO—starred line; and (3) fWSMO—solid line. The FB, WSMO, and fWSMO approaches realize average

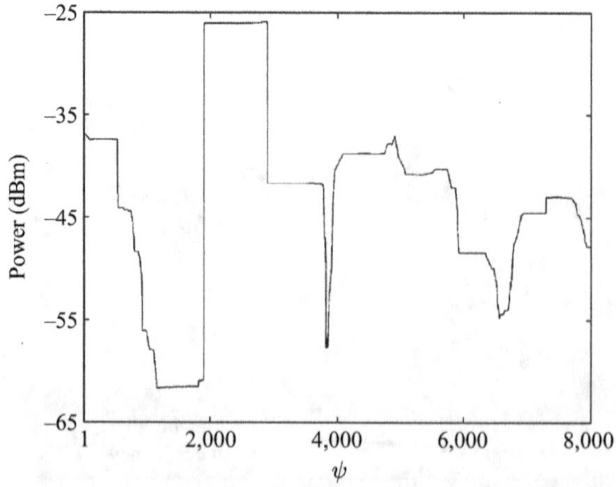

Figure 15.17 Interference and noise power of the recorded spectra in the experiment set with ψ denoting the spectrum index. The spectra contain a variety of interference arrangements, including high-power narrowband interference, wideband interference, and noise-only

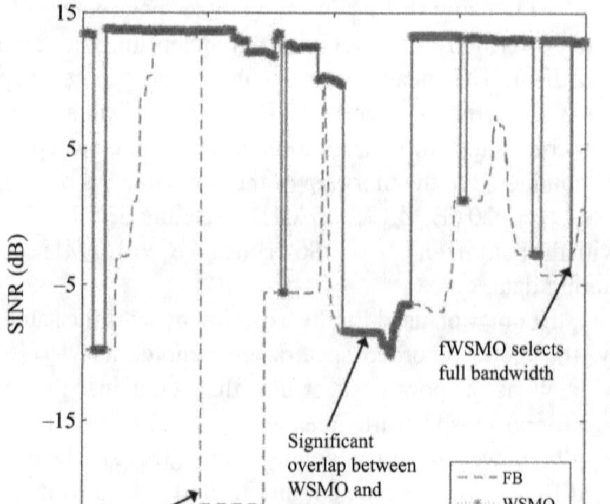

Figure 15.18 SINR results for FB, WSMO, and fWSMO operation on the recorded spectra in the experiment set, for ψ the spectrum index

SINR values of 4.67, 12.10, and 12.09 dB, respectively. The FB method produces some results below the SINR constraint of $Z_{1,\min} = -20$ dB indicating nonfeasible solutions. In contrast, WSMO and fWSMO provide a significant increase in SINR over FB, meeting the SINR constraint for all processed spectra. Both WSMO and fWSMO realize an average SINR that is basically 7.4 dB better than FB. Also observe that the SINR values between fWSMO and WSMO are very similar, with the difference between their average SINRs being a meager 0.01 dB.

We do find that for the small set of spectrum indices $\psi = \{7,766,\ldots,7,781\}$ fWSMO experiences degraded SINR relative to WSMO. Closer examination of these spectra reveals that fWSMO selects the FB solution, whereas WSMO selects a smaller band. For example, for spectrum index $\psi = 7,770$, Figure 15.19(a) shows that WSMO selects the sub-band from 893.9 to 929.4 MHz, while fWSMO selects the FB solution that contains significant RFI. Upon closer examination in Figure 15.19(b) we find that the three meso-bands between 928.8 and 929.9 MHz are merged because the bandwidth of the low-power meso-band is 0.6 MHz $< B_{\mathrm{f}}$. Thus the sub-band from 893.9 to 929.4 MHz is no longer available to fWSMO for a possible solution. By combining meso-bands, the merge process has removed the optimal solution from its candidate sub-bands.

Figure 15.20 shows the optimal value of the objective function for WSMO using (15.10), for fWSMO using (15.14), and for FB via

$$Z^*_{\mathrm{FB}} = \alpha\,\overline{Z}_1(B,f_1) + (1-\alpha)\times 1 \tag{15.20}$$

using $\alpha = 0.5$. The FB objective function in (15.20) employs the entire band (hence the 1 at the end), with $\overline{Z}_1(B,f_1)$ the peak-normalized SINR from (15.6) evaluated under the full band condition. The results in (15.10), (15.14), and (15.20) quantify the combined performance of the SINR and bandwidth objective functions and provide a more accurate comparison between the FB, WSMO, and fWSMO approaches. We observe in Figure 15.20 that WSMO and fWSMO exhibit nearly identical performance, with the average difference being only 0.0002. Both WSMO and fWSMO clearly outperform FB, a result further justifying the SS-MO approach.

The computational cost of WSMO and fWSMO is compared in Figure 15.21, where it is observed that fWSMO is $O(N)$ in terms of additions and multiplications; an order of magnitude reduction relative to WSMO. Specifically, the average number of additions used by fWSMO is 1,638 and is consistent for each spectrum in Ψ. The number of multiplications fluctuates between 85 and 3,043, with an average of 1,360. The multiplications drop significantly for spectra $\psi = \{1,246,\ldots,1,325\}$ (noise only cases), which is due to the small number of merged meso-bands, i.e. only 1 sub-band exists. In contrast, the number of additions and multiplications needed for WSMO are 1×10^6 and 1.5×10^6, respectively. Overall, fWSMO reduces the number of additions by a factor of approximately 610 and the number of multiplications by a factor of 1,100.

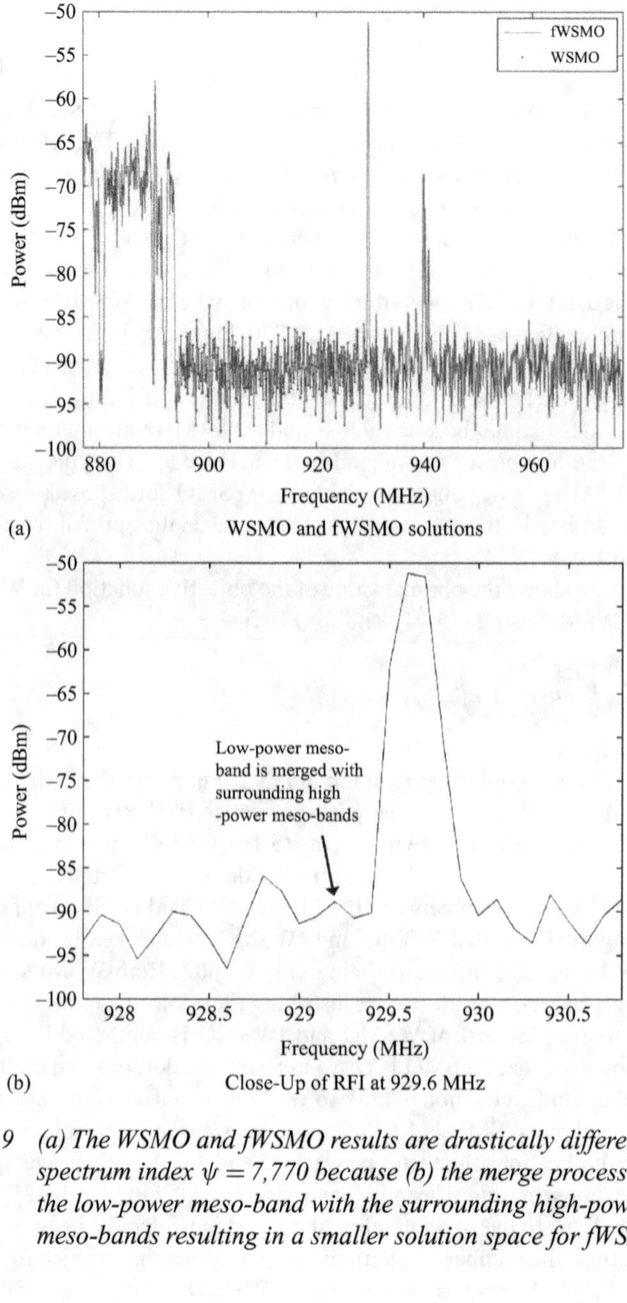

Figure 15.19 (a) The WSMO and fWSMO results are drastically different for spectrum index $\psi = 7{,}770$ because (b) the merge process combines the low-power meso-band with the surrounding high-power meso-bands resulting in a smaller solution space for fWSMO

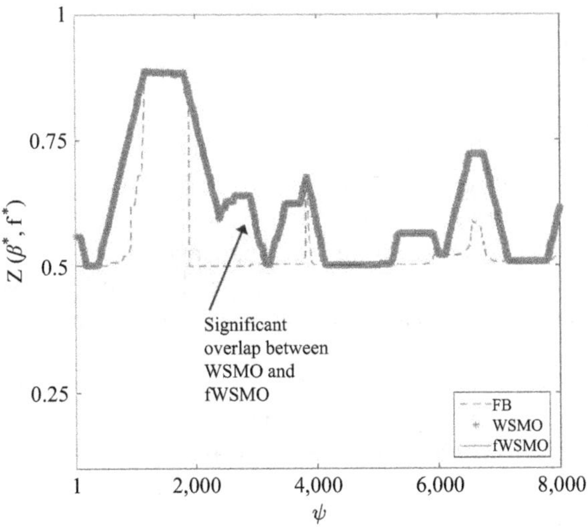

Figure 15.20 Objective function results from (15.10), (15.14), and (15.20) for WSMO, fWSMO, and FB, respectively. The results quantify the combined performance of the individual objective functions of maximizing SINR and bandwidth, and provide a more accurate comparison among the FB, WSMO, and fWSMO approaches

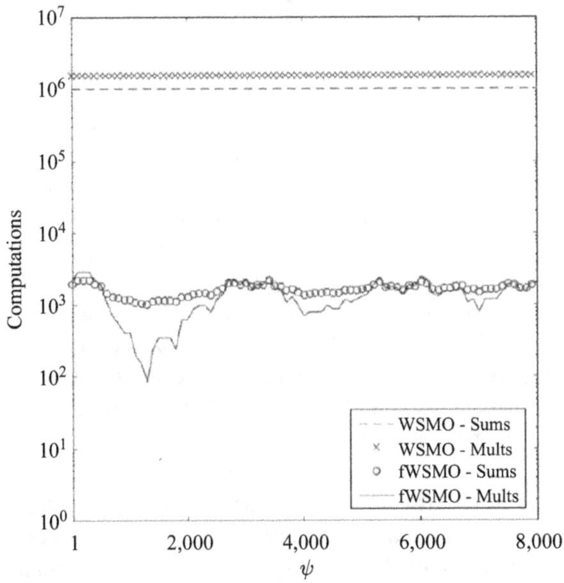

Figure 15.21 Computational cost of WSMO and fWSMO, with the pre-processing data reduction of the latter realizing a significantly lower computational cost

15.6 Conclusions, applications of RFI avoidance, and future work

RFI avoidance techniques such as SS-MO support radar coexistence for congested EMEs in arrangements involving equal sharing of the spectrum without *a priori* knowledge of other RF systems. This process first implements spectrum sensing to estimate the power spectrum of in-band RFI. Multi-objective optimization is subsequently performed to find a sub-band that jointly maximizes SINR and bandwidth. For instance, it was demonstrated using measured data that SS-MO provides an SINR enhancement of more than 7 dB compared to use of the full bandwidth that contains RFI.

This formulation was applied in [49] in the context of a software-defined radar (SDRadar) probing a stationary target in the presence of narrow- and wideband-RFI. The SDRadar implemented SS-MO to sense and assess the RFI, and subsequently an LFM was generated with the appropriate bandwidth and center frequency to avoid the RFI. Compared to using a similar waveform over the full bandwidth, operation based on RFI avoidance was observed to provide over 30 dB of SINR improvement.

This manner of RFI avoidance has also been investigated for a stepped-frequency harmonic radar [26], which is a technology designed to exploit the harmonic properties of a nonlinear target of interest [50]. The harmonic radar probes the target using a specially designed waveform to maximize the harmonic response in a desired frequency band. For example, radio-frequency identification tags contain nonlinear circuit elements and are designed to produce nonlinear signatures [51]. The advantage of harmonic radar is that it separates the target signature from the clutter band; however, this target signature typically has very low received power. Therefore, RFI and corresponding sidelobes greatly impact the detection capability of a harmonic radar. However, the results in [26] indicate that RFI avoidance can achieve a 25 dB improvement in SINR for this manner of operation.

For real-time (or near real-time) implementation of RFI avoidance, the radar must also be capable of tuning to the selected sub-band at a rate faster than the RFI is changing. Here it was shown that a fast variant of SS-MO denoted as fWSMO provides an order-of-magnitude lower computational cost with an essentially negligible degradation in objective function performance. It was also illustrated through experimental examples that the speed of reconfigurable RF tuning technology plays an important role in the SINR/bandwidth trade-space because the latency required to adjust RF filters comes at the cost of fewer pulses that can be emitted and subsequently received as target echoes.

The particular implementation of RFI avoidance also depends on the concept of operation. For example, if a spectrum management system such as proposed for TVWS is employed, then the radar could participate similar to other users. If no such management system exists, or if the radar simply cannot participate, then the radar would independently access the optimal sub-band identified by an RFI sense and avoid technique such as that described here. Independent access to the spectrum allows for radar frequency agility in congested RF environments that cannot be facilitated by a structured management system.

More advanced RFI avoidance could involve investigating the means to address rapidly time-varying spectra through the use of prediction methods and statistical or policy-based objective functions. For instance, RFI behavior could be modeled over time to ascertain advantageous repeated characteristics that could be exploited for radar spectrum usage. Likewise, game-theoretic approaches could facilitate quick look-up table solutions based on the observed EME. Further means to perform data reduction preprocessing like the thalamus would also prove useful. Similar to the discussion in Chapter 12, there is also interest in exploring how RFI avoidance approaches could potentially incorporate the means with which to optimize various waveform characteristics beyond just center frequency and bandwidth.

References

[1] H. Griffiths, L. Cohen, S. Watts, *et al.*, "Radar spectrum engineering and management: technical and regulatory issues," *Proceedings of the IEEE*, vol. 103, no. 1, pp. 85–102, Jan. 2015.

[2] A.F. Martone, "Cognitive radar demystified," *URSI Bulletin,* no. 350, pp. 10–22, Sept. 2014.

[3] J. Mitola and G.Q. Maguire, "Cognitive radio: making software radios more personal," *IEEE Personal Communications*, vol. 6, no. 4, pp. 13–18, Aug. 1999.

[4] Federal Communication Commission, "Facilitating opportunities for flexible, efficient, and reliable spectrum use employing cognitive radio technologies," Washington, DC, FCC 03-322, Dec. 2003.

[5] Federal Communication Commission, "Unlicensed operation in the TV broadcast bands," Wash. D.C., FCC 04-113, May 2004.

[6] IEEE Computer Society, "IEEE standard for information technology, telecommunications and information exchange between systems, local and metropolitan area networks, specific requirements," New York, NY, IEEE 802.19.1, 2014.

[7] C.S. Sum, G.P. Villardi, M.A. Rahman, et al., "Cognitive communication in TV white spaces: An overview of regulations, standards, and technology," *IEEE Communications Magazine*, vol. 51, no. 7, pp. 138–145, July 2013.

[8] J.M. Peha, "Sharing spectrum through spectrum policy reform and cognitive radio," *Proceedings of the IEEE*, vol. 97, no. 4, pp. 708–719, Apr. 2009.

[9] M. Labib, J.H. Reed, A.F. Martone, and A.I. Zaghloul, "Coexistence between radar and LTE-U systems: Survey on the 5 GHz band," in *USNC-URSI Radio Science Meeting*, Boulder, CO, Jan. 2016.

[10] H. Wang, J. Johnson, C. Baker, L. Ye, and C. Zhang, "On spectrum sharing between communications and air traffic control radar systems," *IEEE International Radar Conference*, Arlington, VA, pp. 1545–1550, May 2015.

[11] R. Saruthirathanaworakun, J.M. Peha, and L.M. Correia, "Opportunistic primary–secondary spectrum sharing with a rotating radar," *International*

Conference On Computing, Networking and Communications, Maui, HI, pp. 1025–1030, Jan./Feb. 2012.

[12] S.S. Bhat, R.M. Narayanan, and M. Rangaswamy, "Bandwidth sharing and scan scheduling in multimodal radar with communications and tracking," *IETE Journal of Research*, vol. 59, no. 5, pp. 551–562, Sept. 2013.

[13] J. Jakabosky, S.D. Blunt, and A.F. Martone, "Incorporating hopped spectral gaps into nonrecurrent nonlinear FMCW radar emission," *IEEE International Workshop on Computational Advances in Multi-Sensor Adaptive Processing*, Cancun, Mexico, pp. 281–284, Dec. 2015.

[14] J. Jakabosky, B. Ravenscroft, S.D. Blunt, and A.F. Martone, "Gapped spectrum shaping for tandem-hopped radar/communications & cognitive sensing," *IEEE Radar Conference*, Philadelphia, PA, May 2016.

[15] A. Khawar, A. Abdel-Hadi, and T.C. Clancy, "Spectrum sharing between S-band radar and LTE cellular system: A spatial approach," *IEEE International Symposium on Dynamic Spectrum Access Networks,* Mclean, VA, pp. 1–4, Apr. 2014.

[16] J.G. Metcalf, C. Sahin, S.D. Blunt, and M. Rangaswamy, "Analysis of symbol design strategies for intrapulse radar-embedded communications," *IEEE Transactions on Aerospace & Electronic Systems*, vol. 51, no. 4, pp. 2914–2931, Oct. 2015.

[17] T. Yucek and H. Arslan, "A survey of spectrum sensing algorithms for cognitive radio applications," *IEEE Communications Surveys & Tutorials*, vol. 11, no. 1, pp. 116–130, 2009.

[18] Q. Zhao and B.M. Sadler, "A survey of dynamic spectrum access," *IEEE Signal Processing Magazine*, vol. 24, no. 3, pp. 79–89, May 2007.

[19] P. Stinco, M.S. Greco, and F. Gini, "Spectrum sensing and sharing for cognitive radars," *IET Radar, Sonar & Navigation*, vol. 10, no. 3, pp. 595–602, Feb. 2016.

[20] P. Stinco, M. Greco, F. Gini, and B. Himed, "Cognitive radars in spectrally dense environments," *IEEE Aerospace and Electronic Systems Magazine*, vol. 31, no. 10, pp. 20–27, Oct. 2016.

[21] S. Filin, T. Baykas, H. Harada, F. Kojima, and H. Yano, "IEEE Standard 802.19.1 for TV white space coexistence," *IEEE Communications Magazine*, vol. 54, no. 3, pp. 22–26, Mar. 2016.

[22] M.A. Richards, J.A. Scheer, and W.A. Holm, *Principles of Modern Radar*, 1st ed., Edison, NJ: Scitech Publishing, 2010.

[23] *Agilent N9342C handheld spectrum analyzer (HAS) 7 GHz data sheet*, www.agilent.com/find/N9342C, Agilent Technologies, May 2014.

[24] A.F. Martone, K.D. Sherbondy, K.I. Ranney, and T.V. Dogaru, "Passive sensing for adaptable radar bandwidth," in *Proceedings of the 2015 IEEE International Radar Conference*, Arlington, VA, May 2015, pp. 280–285.

[25] R.K. Pooler, "An analysis of spectrum sharing for radar applications," M.S. thesis, Department of Electrical Engineering, The Pennsylvania State University, State College, PA, Dec. 2016.

[26] A.F. Martone, K.A. Gallagher, K.D. Sherbondy, *et al.*, "Adaptable bandwidth for harmonic step-frequency radar," *International Journal of Antennas and Propagation*, Article ID 808093, July 2015.

[27] A.F. Martone, K.I. Ranney, K.D. Sherbondy, K.A. Gallagher, and S.D. Blunt, "Spectrum allocation for noncooperative radar coexistence," *IEEE Transactions on Aerospace and Electronic Systems*, vol. 54, no. 1, pp. 90–105, Feb. 2018.

[28] M.J. Lindenfeld, "Sparse frequency transmit-and-receive waveform design," *IEEE Transactions on Aerospace & Electronic Systems*, vol. 40, no. 3, pp. 851–861, July 2004.

[29] M.R. Cook, T. Higgins, and A.K. Shackelford, "Thinned spectrum radar waveforms," *International Waveform Diversity & Design Conference*, Niagara Falls, ON, Canada, Aug. 2010.

[30] K. Gerlach, M.R. Frey, M.J. Steiner, and A. Shackelford, "Spectral nulling on transmit via nonlinear FM radar waveforms," *IEEE Transactions on Aerospace & Electronic Systems*, vol. 47, no. 2, pp. 1507–1515, Apr. 2011.

[31] C. Nunn and L.R. Moyer, "Spectrally-compliant waveforms for wideband radar," *Aerospace & Electronic Systems Magazine*, vol. 27, no. 8, pp. 11–15, Aug. 2012.

[32] A. Aubry, A. De Maio, Y. Huang, M. Piezzo, and A. Farina, "A new radar waveform design algorithm with improved feasibility for spectral coexistence," *IEEE Transactions on Aerospace & Electronic Systems*, vol. 51, no. 2, pp. 1029–1038, Apr. 2015.

[33] Z. Geng, H. Deng, and B. Himed, "Adaptive radar beamforming for interference mitigation in radar-wireless spectrum sharing," *IEEE Signal Processing Letters*, vol. 22, no. 4, pp. 484–488, Apr. 2015.

[34] T. Higgins, T. Webster, and A.K. Shackelford, "Mitigating interference via spatial and spectral nulling," *IET Radar, Sonar & Navigation*, vol. 8, no. 2, pp. 84–93, 2014.

[35] K. Gerlach, "Thinned spectrum ultrawideband waveforms using stepped-frequency polyphase codes," *IEEE Transactions on Aerospace and Electronic Systems*, vol. 34, no. 4, pp. 1356–1361, Oct. 1998.

[36] L.H. Nguyen, T. Tran, and T. Do, "Sparse models and sparse recovery for ultra-wideband SAR applications," *IEEE Transactions on Aerospace and Electronic Systems*, vol. 50, no. 2, pp. 940–958, April 2014.

[37] J. Salzman, D. Akamine, and R. Lefevre, "Optimal waveforms and processing for sparse frequency UWB operation," in *Proceedings of the 2001 IEEE Radar Conference*, Atlanta, GA, 2001, pp. 105–110.

[38] A. Reigber and L. Ferro-Famil, "Interference suppression in synthesized SAR images," *IEEE Geoscience and Remote Sensing Letters*, vol. 2, no. 1, pp. 45–49, Jan. 2005.

[39] A.F. Martone, C. Dietlein, M. Govoni, K.D. Sherbondy, and J. Pulskamp, "Tuning technology for adaptable radar bandwidth," in *Proceedings of the 2016 IEEE International Microwave Synposium (IMS) Conference*, San Francisco, CA, May 2016.

[40] A. Konaka, D. Coitb, and A. Smith, "Multi-objective optimization using genetic algorithms: a tutorial," *Reliability, Engineering, and System Safety*, vol. 91, no. 9, pp. 992–1007, Sept. 2006.

[41] M.A. Govoni, H. Li, and J.A. Kosinski, "Low probability of interception of an advanced noise radar waveform with linear-FM," *IEEE Transactions on Aerospace and Electronic Systems*, vol. 49, no. 2, pp. 1351–1356, Apr. 2013.

[42] B.R. Phelan, M.A. Ressler, G.J. Mazzaro, K.D. Sherbondy, and R.M. Narayanan, "Design of spectrally versatille forward-looking ground penetrating radar for detection of concealed targets," in *Proceedings of Radar Sensor Technology XVII*, SPIE, Vol. 8714, 87140B-1, 2013.

[43] G.J. Mazzaro, A.F. Martone, and D.M. McNamara, "Detection of RF electronics by multitone harmonic radar," *IEEE Transactions on Aerospace and Electronic Systems*, vol. 50, no. 1, pp. 477–490, 2014.

[44] A.F. Martone, "Forensic characterization of RF circuits," Ph.D. dissertation, Department of Electrical and Computer Engineering, Purdue University, Dec. 2007.

[45] G.M. Rebeiz, K. Entesari, I. Reines, *et al.*, "Tuning in to RF MEMS," *IEEE Microwave Magazine*, vol. 10, no. 6, pp. 55–72, Oct. 2009.

[46] W.M. Connelly, M. Laing, A.C. Errington, and V. Crunelli, "The thalamus as a low pass filter: Filtering at the cellular level does not equate with filtering at the network level," *Frontiers in Neural Circuits*, vol. 9, no. 89, pp. 1–10, Jan. 2016.

[47] N.C. Andreasen, S. Arndt, V. Swayze, *et al.*, "Thalamic abnormalities in schizophrenia visualized through magnetic resonance image averaging," *Science*, vol. 266, no. 5183, pp. 294–298, Oct. 1994.

[48] A.R. Gilbert, D.R. Rosenberg, K. Harenski, S. Spencer, J.A. Sweeney, and M.S. Keshavan, "Thalamic volumes in patients with first-episode schizophrenia," *American Journal of Psychiatry*, vol. 158, no. 4, pp. 618–624, Apr. 2001.

[49] B. Kirk and J. Owen, "Development of a software defined radar," U.S. Army Research Laboratory, Adelphi, MD, ARL-TR-8152, Oct. 2017.

[50] K.A. Gallagher, R.M. Narayanan, G.J. Mazzaro, A.F. Martone, and K.D. Sherbondy, "Static and moving target imaging using harmonic radar," *Electronics*, vol. 6, no. 2, pp. 1–20, April 2017.

[51] J.L. Osborne, S.J. Clark, R.J. Morris, *et al.*, "A landscape scale study of bumble bee foraging range and constancy, using harmonic radar," *Journal of Applied Ecology*, vol. 36, no. 4, pp. 519–533, 1999.

Chapter 16

Compressed sensing and interference occupancy monitoring for spectrum sharing in spectrally dense environments

Pietro Stinco[1], Maria Greco[1], and Fulvio Gini[1]

16.1 Introduction

Radar technology has been evolving towards higher resolution, high-precision detection instruments with an ever-increasing set of functionalities. Multifunction radar systems have immense potential for combining the benefits of these developments. These systems perform multiple functions simultaneously, including surveillance, tracking, confirmation of false alarm, back-scanning, and clutter and interference cancellation, each of which are traditionally performed by dedicated individual radars. With the introduction of this broader set of requirements, future radar systems will need to operate with wider frequency bands than has traditionally been the case.

From a spectrum perspective, this need is clearly in conflict with the growth of activities in civil communications. The emergence of new communication technologies and services brings with it a strong demand for spectrum allocation, thus putting intense pressure upon the frequency channels currently allocated to radars [1]. For this reason, the availability of spectrum for multifunction radar has been severely compromised, with available frequency bands diminishing or already re-allocated to communication services.

For instance, the International Telecommunication Union (ITU) decided to allocate a portion of C-Band (5150–5350 MHz and 5470–5725 MHz) to wireless access systems [2,3]. Likewise, the National Telecommunications and Information Administration (NTIA) in the United States has devoted efforts to identify bands that could be made available for wireless broadband service provisioning. A total of 115 MHz of additional spectrum in S-Band (1695–1710 MHz and 3550–3650 MHz) has been identified for wireless broadband implementation [4].

Recent work [5] has focused on spectrum sharing between a broadband communication system and an air traffic control (ATC) radar operating in the 2.7–2.9 GHz

[1]Department of Ingegneria dell'Informazione, University of Pisa, Italy

band and a surveillance radar in the 16.7–17.3 GHz band. Another case is analyzed in [6] that involves an L-band radar sharing the same channel as a Joint Tactical Information Distribution System operating in 969–1206 MHz.

To cope with the issue of spectrum crowding and steadily increasing radar requirements, future radar systems will need to coexist with other radio frequency (RF) systems, anticipating their behavior and properly reacting to avoid interference. Consequently, intelligent spectrum utilization coupled with agile waveform diversity has become a necessity [7], and in turn, leads to the need for new methodologies based on cognition as an enabling technology [8–11].

In this chapter, we analyze the general problem of a wideband radar that shares the same channel with a communication system, assuming that the communication band is divided into several frequency channels used for dynamic spectrum access (DSA). We specifically consider the case in which the radar is the secondary user (SU) and the communication system is the primary user (PU).

In cognitive radio terminology, PUs are those who have higher priority or legacy rights to the usage of a specific part of the spectrum. SUs have lower priority and exploit the spectrum in such a way as to not cause interference to PUs. Therefore, SUs must possess some cognitive radio capabilities, such as an ability to sense the spectrum reliably to determine whether or not it is currently occupied by a PU. If the given channel becomes occupied by a PU, the SU must then move to an unused part of the spectrum, often referred to as white spaces or spectrum opportunities.

The decision that white space is available can be made according to various metrics. The traditional approach is to leverage the assumption that the power of a transmitter should not exceed a prescribed noise floor at a certain distance from the transmitter. However, due to the increased mobility and variability of RF emitters, this assumption becomes problematic since unpredictable new sources of interference may appear. To address this issue, the U.S. Federal Communications Commission (FCC) has proposed a new metric on interference assessment denoted as the *interference temperature* to enforce an interference limit as perceived by receivers [12].

The interference temperature is defined as the temperature equivalent of the RF power available at a receiving antenna per-unit-bandwidth. The FCC further established an interference temperature limit that dictates the maximum amount of tolerable interference for a given frequency band. Any secondary transmitter using this band must guarantee that its transmission plus the existing noise and interference will not exceed the interference temperature limit at a PU. Given a particular frequency band in which the interference temperature limit is not exceeded, that band could be made available for radar usage.

To monitor channel occupancy and subsequently access the channel opportunistically, while minimizing interference to the PUs, cognitive radars must adopt a strategy that is conceptually summarized in the cognitive cycle shown in Figure 16.1. Here, a spectrum sharing function controls access to the channel by exploiting feedback from the channel characterization function, while a spectrum sensing function performs the necessary observations of the RF channel to discern the presence of other users and to search for spectrum opportunities where it is possible to transmit without interfering.

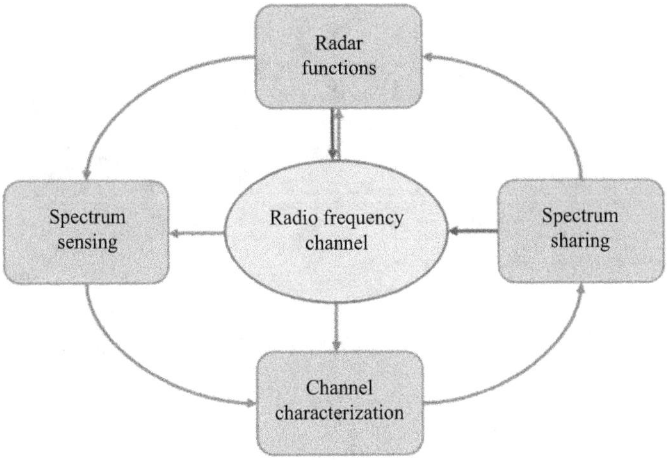

Figure 16.1 Perception-action cycle of a cognitive radar system

In the multifunction radar context, however, where the time required to estimate spectral occupancy is very short and the monitored frequency band can be quite wide, an important limitation is the power consumption and hardware cost involved with high sampling rate analog-to-digital converters (ADCs) that possess large dynamic range and associated high-speed digital signal processing. This chapter therefore investigates the use of compressive sensing (CS) for spectrum sensing, showing how this emerging technology can help to solve important challenges related to hardware complexity that arise when attempting to rapidly sense the presence of other RF users and available spectrum opportunities in a dynamic manner.

16.2 Cognitive spectrum sensing and sharing

Figure 16.2 illustrates a realization of spectrum occupancy as a function of time–frequency blocks. The PUs do not occupy all the bands simultaneously, and thus some bands might be available for an SU. The actual usage of a particular band can be classified into two categories: *black-spaces*, that are time–frequency slots occupied by a PU; and *white-spaces*, or "spectrum opportunities", that are time–frequency slots free of RF interference aside from ambient noise.

To facilitate the discovery of spectrum opportunities in a standardized manner, the FCC [12] proposed the interference temperature metric to enforce an interference limit as perceived by receivers. The interference temperature [13]

$$T_{\mathrm{I}}(f_{\mathrm{C}}, B) = \frac{P_{\mathrm{I}}(f_{\mathrm{C}}, B)}{kB} \tag{16.1}$$

is the temperature equivalent of the RF power available at a receiving antenna per-unit-bandwidth, where $P_{\mathrm{I}}(f_{\mathrm{C}}, B)$ is the average interference power centered at frequency f_{C} that covers bandwidth B, where k denotes Boltzmann's constant.

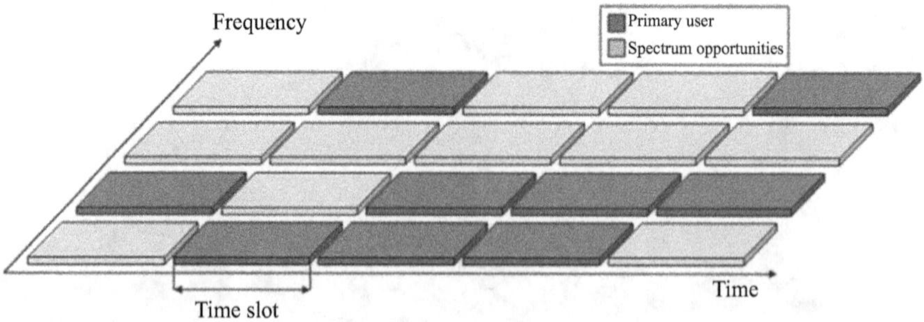

Figure 16.2 Spectrum occupancy in the time–frequency channel

The FCC further established an interference temperature limit that establishes the maximum amount of tolerable interference for a given frequency band. Any secondary transmitter using this band must guarantee that its transmission plus the existing noise and interference will not exceed the interference temperature limit at the receiver of a PU.

If the interference temperature limit is T_L for a particular frequency band with bandwidth B, then a prospective SU must maintain the average interference below kBT_L. Therefore, assuming that an SU is operating with average power P in band $[f_C - B/2, f_C + B/2]$, the interference temperature limit ensures that

$$\left[T_I(f_C, B) + \frac{LP}{kB} \right] \leq T_L(f_C), \tag{16.2}$$

where L is path loss attenuation between the secondary transmitter and the primary receiver.

In the next sections we examine how interference temperature is exploited to model the dynamics of a time–frequency channel to perform spectrum sensing. Various techniques have been proposed in the wireless communication literature to identify the presence of primary signal transmissions such as the energy detector (ED), waveform or matched filter based detectors, feature based detectors, and some emerging eigenvalue-based and wavelet-based detectors [14]. Each of these detectors has advantages and disadvantages with varying detection capabilities, implementation complexity, sensing time, and assumptions/requirements on the PU signal.

Compared to the others, ED is simple to implement and does not make any assumptions on the PU signal structure. On the other hand, it also cannot differentiate the PU signal from other unknown signal sources. Of course, waveform or matched filter detection algorithms make explicit assumptions about a known pilot waveform or the exact transmitted signal to design detectors, and also require some degree of synchronization between the PU and the detector, which can lead to synchronization errors that severely degrade detection performance. Feature-based detectors, which lie between EDs and waveform or matched filter detectors, exploit only structural or statistical properties of a PU signal such as the cyclostationary that is not otherwise present in noise. In general, cyclostationarity-based detectors operate well, but at the cost of high complexity [14].

To further increase the spectrum awareness of a cognitive system (radio or radar), it has also been proposed to use a radio environmental map (REM) [15]. The idea behind the REM is to store and process a variety of data to extract all the available information on transmitter locations, propagation conditions, and spectrum usage in space and time. Exploiting a REM, the radar could become aware of the surrounding electromagnetic environment, and then intelligently use the transmit bandwidth and probing waveforms.

16.2.1 Cognitive cycle

To access the channel opportunistically while minimizing interference to PUs, cognitive radars must adopt a strategy that is conceptually summarized in Figure 16.1. This strategy is known as the *perception-action cycle* and was defined by Haykin in [16].

One aspect of this cycle is *spectrum sensing*. The purpose of spectrum sensing is to acquire the observations about the RF channel that are necessary to determine the presence of other RF users and potential spectrum opportunities where the SU can operating without interfering. These observations can be obtained by exploiting the signal captured at the radar receiver or with a separate receive chain.

Proceeding counter-clockwise from spectrum sensing in Figure 16.1, decisions regarding whether a given channel is available or not then drives *channel characterization*. The role of this task is to characterize statistically the spectrum usage of the PUs, to identify the most appropriate model for the PU dynamics, and to estimate the channel parameters that describe this model.

Channel characterization subsequently drives *spectrum sharing*, the purpose of which is to limit interference from the radar to other services and vice versa. This task exploits the information arising from the channel characterization task to control the various functions that the radar itself must perform.

These functions comprise the typical operation of a radar system, with the exception that in each time-slot, which coincides with the radar pulse repetition interval (PRI), transmit/receive parameters such as transmission power and operating frequency can be dynamically adapted in order to achieve an efficient spectrum utilization and to avoid interference. The two darker arrows in Figure 16.1 indicate active use of the RF channel, while the lighter arrows correspond to observations of the channel and internal data handling of the cognitive system.

16.2.2 System architecture

Figure 16.3 shows the system architecture of a cognitive radar system. In particular, the figure shows how the cognitive functions described in the perception-action cycle of Figure 16.1 are implemented in a cognitive radar architecture.

As shown, the cognition portion is an additional layer that directly controls the transmitter and receiver of a typical radar system. In more complex systems, the spectrum-sensing and channel characterization components directly observe the RF

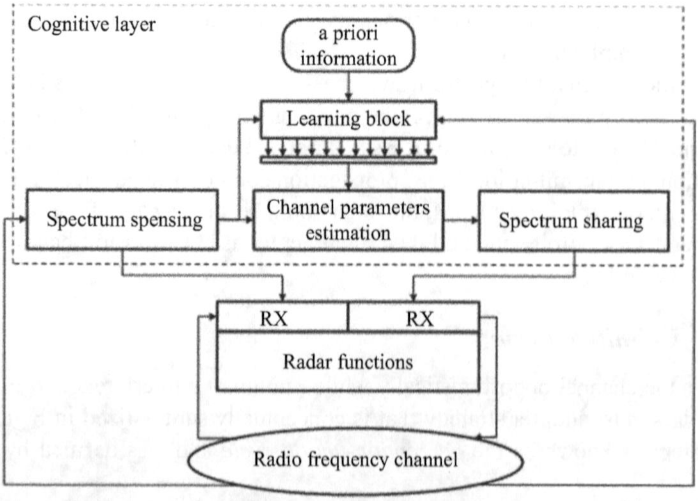

Figure 16.3 System architecture of a cognitive radar

channel with a separate receive chain. On the other hand, in simpler and cheaper systems, the signal captured at the radar receiver can be used instead. In both cases, the radar transmitter is directly controlled by the cognitive layer that makes use of feedback information from the receiver.

16.3 Channel characterization

Not unlike DSA, a cognitive system periodically scans and senses multiple frequency channels to measure the interference temperature. The measured interference temperature for a given channel is compared with the interference limit T_L to determine whether or not the channel is available (i.e. "free" or "busy"). However, due to the ubiquitous presence of noise, a free channel can be incorrectly classified as busy and a busy channel classified as free. These channel dynamics are often modeled with Hidden Markov Models (HMMs) [17–19].

In general, an HMM is comprised of a set S_t of possible states and a set O_t of possible emissions. Here the frequency band used by the PU communication system is subdivided into N frequency channels of bandwidth B used for frequency division multiple access (FDMA) in time-slots of duration Δt. The possible states represent the real activity of the PU in each frequency channel; if the PU is transmitting during time-slot t, the state is $S_t = 1$, otherwise the channel is free and the state is $S_t = 0$. Due to noise, there are also two possible emissions O_t that model the busy/free observations at the output of the Spectrum Sensing detector. Figure 16.4 illustrates the HMM for spectrum occupancy in each frequency channel, where the lower portion describes the PU's dynamics and the upper portion the SU's observations.

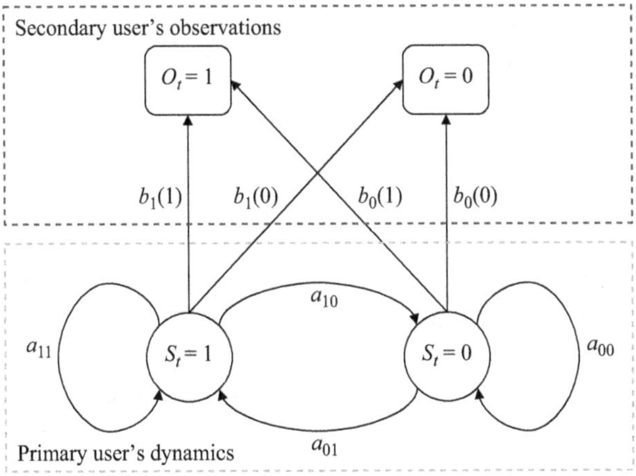

Figure 16.4 HMM representation for spectrum occupancy

The PU's dynamics are characterized by the 2×2 state transition matrix **A** that represents the probabilities associated with changing from one state to another as

$$[\mathbf{A}]_{hk} = a_{hk} = \Pr[S_t = h | S_{t-1} = k] \text{ for } h, k = 0, 1. \tag{16.3}$$

The actual probability values in **A** are specific to each frequency channel.

Independent of the particular modulation of a PU's transmitted signal, if a frequency channel is occupied, the signal observed when performing spectrum sensing will contain a component with that particular frequency whose amplitude depends on the PU distance and the propagation path loss. Therefore, in each frequency channel and time slot, if a PU is transmitting, the received signal samples can be modeled as a discretized sinusoid

$$[\mathbf{f}_i]_n = \alpha \zeta_i \cos\left(\frac{\pi(i-1)(2n-1)}{2N}\right) \text{ for } i, n = 1, \ldots, N \tag{16.4}$$

$$\zeta_i = \begin{cases} 1/\sqrt{N}, & i = 1 \\ \sqrt{2/N}, & 2 \le i \le N \end{cases} \tag{16.5}$$

whose amplitude α is a Gaussian random variable (r.v.) with zero mean and variance $\sigma_{\mathbf{f}}^2$, where i is the frequency channel index and n indicates the nth time sample. The N time samples correspond to the temporal extent of the time slot, which is Δt seconds.

If the channel is free, the received signal $[\mathbf{f}_i]_n$ is zero. In each time slot, the multiband received signal is thus given by the combination

$$\mathbf{f} = \sum_{i=1}^{N} \mathbf{f}_i + \mathbf{w}. \tag{16.6}$$

that contains the signal in each frequency channel and additive white Gaussian noise (AWGN) $\mathbf{w}$ that has zero mean and variance σ_{w}^2. The spectral occupancy $\mathbf{x}$ is then given by the discrete cosine transform (DCT) of $\mathbf{f}$ as

$$\mathbf{x} = \mathbf{\Psi}^T \mathbf{f}, \tag{16.7}$$

where $\mathbf{\Psi}$ is the DCT matrix whose elements are

$$[\mathbf{\Psi}]_{i,j} = \zeta_i \cos\left(\frac{\pi(i-1)(2j-1)}{2N}\right) \text{ for } i,j = 1,\ldots,N. \tag{16.8}$$

Note that it is also possible to consider complex signals by processing the in-phase and quadrature-phase components of the received signal. Having said that, a goal of this work is to reduce hardware complexity of the cognitive portion and so, without loss of generality, we consider only the in-phase component.

Figure 16.5(a) shows the squared absolute value of $\mathbf{x}$ in linear scale, which represents the channel occupancy during an observation time composed of ten time slots. The channel is composed of $N = 256$ frequency bands and the signal to noise ratio, defined as $\mathrm{SNR} = \sigma_{\mathrm{f}}^2/\sigma_{\mathrm{w}}^2$, is 20 dB. In all the simulations σ_{f}^2 was fixed to one while σ_{w}^2 is changed according to the SNR.

According to (16.1), the normalized power of the received signal in each frequency channel and in each time slot is compared with the interference temperature limit T_L to record the observation O_t, which is

$$O_t = \begin{cases} 0, & \text{if } T_{\mathrm{I}}(t) \leq T_{\mathrm{L}} \\ 1, & \text{if } T_{\mathrm{I}}(t) > T_{\mathrm{L}} \end{cases}. \tag{16.9}$$

For a given frequency channel, these observations are periodically recorded over T time slots to form the observation sequence $\boldsymbol{O} = [O_1 \ldots O_T]$. The transitions from states S_t to observations O_t are described by the 2×2 emission matrix $\mathbf{B}$ as

$$[\mathbf{B}]_{hk} = b_h(k) = \Pr[O_t = h|S_t = k], \tag{16.10}$$

which represents the probabilities associated with obtaining a certain output given that the model is currently in state S_t.

Matrix $\mathbf{B}$ is related to the receiver operating characteristic (ROC) of the spectrum-sensing detector. As a matter of fact, $b_0(1)$ is the probability of false-alarm—that is, the probability of incorrectly classifying a free channel as busy—while $b_1(0)$ is the probability of a missed detection—that is, the probability of classifying a busy channel as free. These probabilities depend on the channel noise, the type of signal emitted by the PU, and the spectrum-sensing detector itself. Knowing these characteristics, the elements of $\mathbf{B}$ can either be calculated or evaluated through Monte-Carlo simulations. Also, in the case in which the receiver employs CS, the ROC can be evaluated off-line through Monte-Carlo simulations; hence, without loss of generality, hereafter we assume that $\mathbf{B}$ is known.

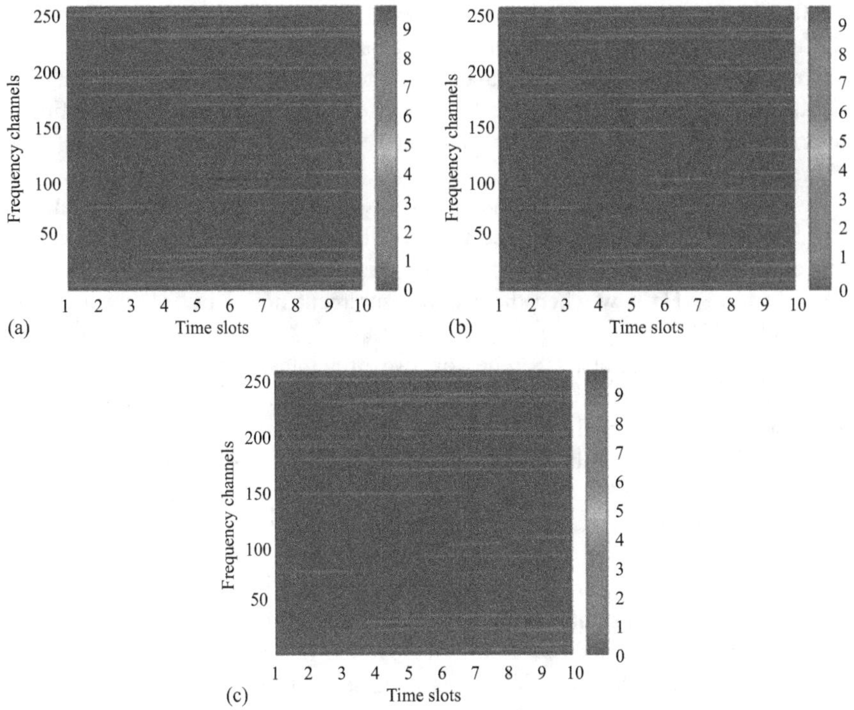

Figure 16.5 Channel evolution in ten time-slots where (a) is the original signal using $N = 256$, (b) is recovered using Gaussian Φ $K = N/2$, (c) recovered using Spiky Φ $K = N/2$

16.4 Compressed spectrum sensing

Spectrum sensing is the first critical step in dynamic spectrum management and it can be performed via two different architectures: single-radio and dual-radio [20,21]. In the single-radio architecture, the radar receiver allocates a specific time slot within which to perform spectrum sensing. Because of this short sensing duration, limited accuracy can be achieved, though the obvious advantage is simplicity and lower hardware cost.

In contrast, the dual-radio architecture involves one receive chain that is dedicated for radar operation and another that is dedicated for spectrum sensing. Note that only one antenna is sufficient for both receive chains. However, for the multifunction radar application, where the required time used to estimate the spectrum occupancy is very short and the monitored frequency band is wide, the main drawback of the dual-radio architecture is increased power consumption and hardware costs. In this context, recent work (e.g. [9,22,23]) proposed the use of CS as a means to perform spectrum sensing with reduced hardware complexity and to facilitate greater responsiveness to dynamically changing channel usage.

To evaluate the time evolution of channel occupancy it is necessary to apply the DCT to the set of received time samples every Δt seconds. When the overall bandwidth to be monitored is wide and/or the time-slot Δt tends to be short, it could be difficult to collect the N time samples at the Nyquist rate. Thus, after a brief introduction to the principles of CS, we consider the notion of compressed spectrum-sensing (CSS) for the problem at hand.

CS is a signal-processing methodology for signal recovery from highly incomplete [24–26] information. The general idea is that a sparse vector[1] $\mathbf{x} \in \mathbb{R}^N$ can be recovered from a small number of linear measurements $\mathbf{y} = \mathbf{H}\mathbf{x}$ for $\mathbf{y} \in \mathbb{R}^K$ and $K \ll N$ (or $\mathbf{y} = \mathbf{H}\mathbf{x} + \mathbf{w}$ when there is measurement noise) by solving a convex program [24,25].

To make this possible, CS relies on two principles: sparsity, which pertains to the signal of interest, and incoherence, which pertains to the sensing modality. Considering the real signal $\mathbf{f} \in \mathbb{R}^N$ defined in (16.6) and letting $\boldsymbol{\Psi} = [\boldsymbol{\psi}_1 \cdots \boldsymbol{\psi}_N]$ be an $N \times N$ orthonormal basis (e.g. the DCT), then the representation of $\mathbf{f}$ on the basis $\boldsymbol{\Psi}$ is given by $\mathbf{f} = \boldsymbol{\Psi}\mathbf{x}$, where $\mathbf{x}$ is a sparse coefficient vector. Given another set of vectors collected into the $K \times N$ "measurement matrix" $\boldsymbol{\Phi} = [\boldsymbol{\varphi}_1 \cdots \boldsymbol{\varphi}_K]^T$, the K measurements are obtained via $\mathbf{y} = \boldsymbol{\Phi}\mathbf{f} = \boldsymbol{\Phi}\boldsymbol{\Psi}\mathbf{x}$ [24–26].

The interest is in under-sampled situations in which the number K of available measurements is much smaller than the dimension N of the signal $\mathbf{f}$. The process of recovering the $N \times 1$ vector $\mathbf{x} = \boldsymbol{\Psi}^T\mathbf{f}$ from the $K \times 1$ measurement vector $\mathbf{y} = \boldsymbol{\Phi}\mathbf{f}$ is, in general, ill-posed when $K < N$. If $\mathbf{x}$ is s-sparse, then the problem can be solved, provided that $K \geq s$. A necessary and sufficient condition for this problem is that, for some small $\delta > 0$, the matrix $\mathbf{H} = \boldsymbol{\Phi}\boldsymbol{\Psi}$ satisfies the restricted isometry property (RIP) [24–26], which can be expressed as

$$(1 - \delta)\|\mathbf{x}\|_2 \leq \|\mathbf{H}\mathbf{x}\|_2 \leq (1 + \delta)\|\mathbf{x}\|_2. \tag{16.11}$$

The RIP implies that matrix $\mathbf{H}$ must preserve the length of s-sparse vectors. A related condition to RIP is referred to as incoherence, where the coherence between the measurement matrix $\boldsymbol{\Phi}$ and the representation matrix $\boldsymbol{\Psi}$ is defined as the largest correlation between any two columns of these matrices, or

$$\mu(\boldsymbol{\Phi}, \boldsymbol{\Psi}) = \sqrt{N} \max_{1 \leq k,j \leq N} \left|\langle \boldsymbol{\phi}_k, \boldsymbol{\psi}_j \rangle\right|. \tag{16.12}$$

It can be shown [24–26] that $\mu(\boldsymbol{\Phi}, \boldsymbol{\Psi}) \in [1, \sqrt{N}]$. The design of $\boldsymbol{\Phi}$ is such that $\mathbf{H} = \boldsymbol{\Phi}\boldsymbol{\Psi}$ having the RIP requires that all possible combinations of s nonzero entries on the length N vector $\mathbf{x}$ must satisfy (16.11). Both the RIP and incoherence can be achieved with high probability by specifying $\boldsymbol{\Phi}$ as a random matrix [24].

It is natural to attempt to recover $\mathbf{x}$ by solving the following optimization problem

$$\hat{\mathbf{x}} = \arg \min_{\mathbf{x} \in \mathbb{R}^N} \|\mathbf{x}\|_0 \text{ such that } \boldsymbol{\Phi}\boldsymbol{\Psi}\mathbf{x} = \mathbf{y}. \tag{16.13}$$

[1]A vector is s-sparse if it has at most s nonzero entries.

In the literature, this minimization is referred as basis pursuit (BP), which can be recast as a linear programming problem. The BP method is guaranteed to provide a reconstruction of a s-sparse signal if there is no measurement noise.

The influence of noise on signal reconstruction can be minimized by applying basis pursuit de-noising (BPDN), which finds a solution to the problem stated as [27]

$$\hat{\mathbf{x}} = \arg\min_{\mathbf{x}\in\mathrm{R}^N} \|\mathbf{x}\|_1 \text{ such that } \|\mathbf{y} - \boldsymbol{\Phi\Psi}\mathbf{x}\|_2 \leq \sigma, \tag{16.14}$$

where the positive parameter σ is an estimate of the noise level in the data. The specific case of $\sigma = 0$ corresponds to the basis pursuit problem. The BPDN method can be solved by linear programming algorithms.

When the spectrum of a PU signal is sparse, it is possible to apply CS to perform spectrum sensing. For the problem at hand, we considered $\boldsymbol{\Psi}$ as the DCT in (16.8) and two measurement matrices $\boldsymbol{\Phi}$. The first is a "Gaussian matrix" that is formed by sampling independent and identically distributed (i.i.d.) entries from a Gaussian distribution having zero mean and variance $1/K$. The second measurement matrix is a "Spiky matrix" obtained by randomly selecting K rows of the $N \times N$ identity matrix.

The latter case is more interesting because, from the definition of $\boldsymbol{\Phi}$, vector $\mathbf{y}$ is obtained by simply selecting K elements of vector $\mathbf{f}$ at random. Compressive sensing thus permits the use an ADC with a rate of $K/\Delta t$ instead of one with the higher rate of $N/\Delta t$. This sampling rate reduction generally translates into a reduction in the cost of the ADC. That said, the physical implementation of a CS receiver is not a simple task, and the reader is referred to [22,28,29] for more details and specifically to the "Xampling" method [23] that can actually lower the cost of the ADC.

Figure 16.5(b) and (c) shows the time evolving channel occupancy of Figure 16.5(a) recovered using the Gaussian and Spiky measurement matrices, respectively. In both cases the sampling rate is reduced by half ($K = N/2$) and SNR = 20 dB.

Figure 16.6 shows the root mean square error (RMSE) after reconstruction of the channel occupancy signal. The RMSE measures the error in reconstructing $\mathbf{x}$ using CSS with respect to the reference signal estimated from all N samples via

$$\mathrm{RMSE} = \sqrt{\frac{1}{MH}\sum\nolimits_{h=1}^{H}\sum\nolimits_{m=1}^{M}\left|\hat{\mathbf{x}}_m^{(h)} - \mathbf{x}_m^{(h)}\right|^2} \tag{16.15}$$

where m is the time slot index and h is the index of the Monte-Carlo run.

The results are shown as a function of K for both measurement matrices and for different values of SNR. The performance using the two matrices is quite similar. It is also apparent that, in the absence of noise, it is possible to reconstruct the signal of interest using a rather low number of samples (30% of N to achieve RMSE of roughly 0.1). With noise present, more samples are needed to minimize its influence on signal reconstruction. When the SNR is relative high (20 dB), the signal can be reconstructed with an RMSE of 0.1 using 60% of N, while an SNR of 15 dB requires a little more than 80% of N to achieve the same performance. The reconstruction RMSE is strictly related to the fact that a high number of samples are needed to reconstruct the entire spectrum with high accuracy.

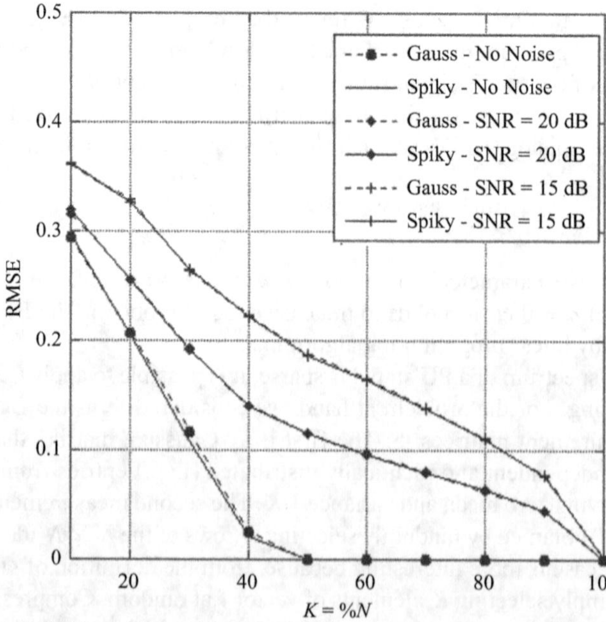

Figure 16.6 RMSE for channel occupancy reconstruction as a function of K (percentage of N) for different SNR values

However, the purpose of the spectrum-sensing task is not to accurately reconstruct the whole spectrum, but rather to determine which channels are busy. Consequently, the classical ED [30] can be applied directly to the imperfectly reconstructed spectrum. This detector compares the squared value of each element in the spectrum occupancy vector (i.e. $r_k = x_k^2$) with a threshold ζ to ascertain whether the channel is busy or free. According to our signal model, the two hypotheses for the elements of the perfectly reconstructed spectrum vector **x** are

$$x_k \sim \begin{cases} N(0, \sigma_{\mathbf{w}}^2), & \text{if } H_0 \\ N(0, \sigma_{\mathbf{f}}^2 + \sigma_{\mathbf{w}}^2), & \text{if } H_1 \end{cases} \tag{16.16}$$

where $\sigma_{\mathbf{f}}^2$ is the variance of the PU's signal and $\sigma_{\mathbf{w}}^2$ is the variance of the noise in (16.6).

Figure 16.7 shows the histogram of x_k under the H_1 hypothesis for SNR = 15 dB. Specifically, histograms are depicted for the perfectly reconstructed spectrum samples and for when the spectrum is reconstructed using the CS technique for $K = 30\%$ of N and $K = 70\%$ of N. For the lower value of K, a disproportionate number of reconstructed samples are zero-valued. In contrast, the higher value of K still produces a Gaussian distribution. Similar results have been observed for the H_0 hypothesis.

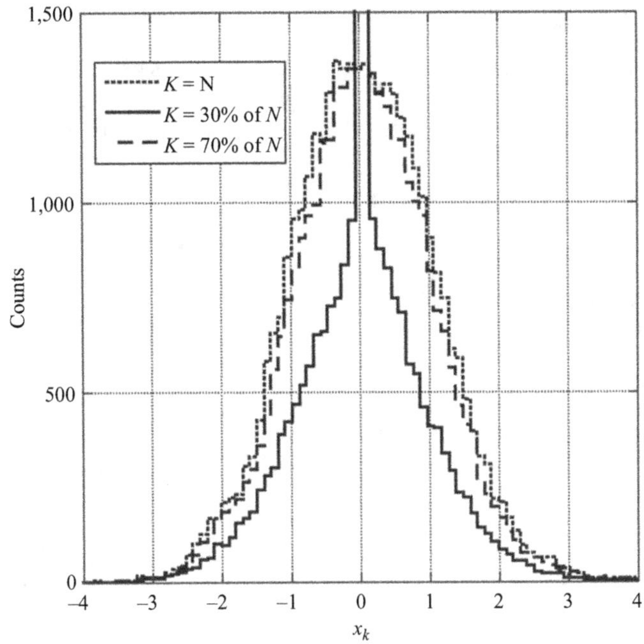

Figure 16.7 Histogram of x_k for different values of K under the H_1 hypothesis, SNR = 15 dB

Squaring a Gaussian r.v. produces a chi-squared r.v. with one degree of freedom, i.e. χ_1^2. Therefore, from (16.16), the binary hypothesis test is

$$r_k \sim \begin{cases} \sigma_{\mathrm{w}}^2 \chi_1^2 & \text{if } H_0 \\ (\sigma_{\mathrm{f}}^2 + \sigma_{\mathrm{w}}^2)\chi_1^2 & \text{if } H_1 \end{cases}. \tag{16.17}$$

Using $P(\bullet)$ to denote the upper incomplete gamma function, the probability of detection and probability of false-alarm are, respectively,

$$P_{\mathrm{D}} = \Pr\{r_k \geq \zeta | H_1\} = \Pr\left\{\chi_1^2 \geq \frac{\zeta}{\sigma_{\mathrm{f}}^2 + \sigma_{\mathrm{w}}^2}\right\} = P\left(\frac{\zeta}{2\left(\sigma_{\mathrm{f}}^2 + \sigma_{\mathrm{w}}^2\right)}, \frac{1}{2}\right) \tag{16.18}$$

and

$$P_{\mathrm{FA}} = \Pr\{r_k \geq \zeta | H_0\} = \Pr\left\{\chi_1^2 \geq \frac{\zeta}{\sigma_{\mathrm{w}}^2}\right\} = P\left(\frac{\zeta}{2\sigma_{\mathrm{W}}^2}, \frac{1}{2}\right). \tag{16.19}$$

Note that without the assumption that the imperfectly reconstructed samples are Gaussian, one must evaluate the distribution for each value of K and SNR and then obtain the detection threshold via Monte-Carlo analysis. Clearly, this assumption introduces error into the decision process. We evaluated the percent error for the channel

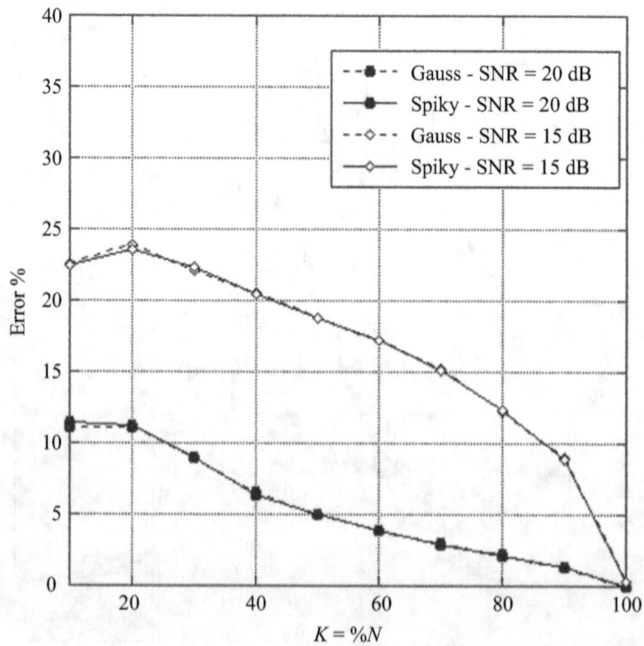

Figure 16.8 Error percentage on the decision of channel occupancy as a function of K (percentage of N) for different SNR values

occupancy decision (i.e. if a free channel is declared as busy and vice versa) by applying the threshold ζ to the imperfectly reconstructed spectrum. These results are shown in Figure 16.8 when ζ is set for $P_D = 0.8$.

Note that radar detectors typically set P_{FA} to a desired value and then maximize P_{D} according to the Neyman–Pearson criterion because false alarms tend to be more problematic than missed detections. In fact, each declared detection invokes other radar processes such as target tracking and identification. Consequently, if there are many false alarms, a significant portion of system resources gets devoted to nonexistent targets.

However, for the application to spectrum sensing where the radar is the SU, missed detections are more problematic because the particular channel is declared as free, when in actuality a PU is transmitting. For this reason, it is more appropriate to set P_{D} to a desired value and minimize P_{FA}.

Also note that, because the threshold is dependent on SNR, P_{FA} likewise depends on SNR. Here P_{D} has been set to 0.8 for each SNR value, resulting in $P_{\mathrm{FA}} = 0.01$ for SNR = 20 dB and $P_{\mathrm{FA}} = 0.15$ for SNR = 15 dB based on (16.19). Thus when the SNR is sufficiently high, the error percentage is reasonably low, such that the busy/free determination can still be made even when the signal is reconstructed from relatively few samples (<30% of N).

16.5 Spectrum sharing

To detect the presence of a PU via the manner discussed above, the spectrum sensing detector would need to process the signal received during the entire time slot. However, if the PRI of the radar and the communication system time slot have roughly the same duration, the radar could not effectively determine if a given frequency channel is occupied by a communication signal until it is already too late for the radar to make use of it. Therefore, to minimize interference to PUs while maximally exploiting spectrum opportunities, the cognitive system should keep track of variations in spectrum availability and subsequently make predictions regarding the upcoming spectral profile.

The statistical parameters that describe each frequency channel are the transition matrix **A**, the emission matrix **B**, and the initial state distribution $\boldsymbol{\pi} = \{\pi_i\}$ that is defined as

$$\pi_i = \Pr[s_1 = S_i] \text{ for } i = 0, 1. \tag{16.20}$$

As discussed previously, **B** is related to the ROC of the spectrum sensing detector and is assumed known. Hence, the problem of channel parameter estimation involves determination of **A** and $\boldsymbol{\pi}$ using a finite observation sequence $\boldsymbol{O} = [O_1 \dots O_T]$ comprised of T elements. This observation sequence is also referred to as a training sequence since it is used to "train" the HMM. There is no way to solve this problem analytically [31]; i.e. given any finite observation sequence there is no optimal way of estimating the model parameters. The most widely adopted iterative procedure is the Baum–Welch method, which is closely related to expectation-maximization and is asymptotically the maximum likelihood estimate of the HMM [32].

The Baum–Welch method selects the parameters **A** and $\boldsymbol{\pi}$ such that $\Pr[O|\mathbf{A}, \boldsymbol{\pi}]$ is locally maximized. To describe this iterative procedure for estimation of the HMM parameters, it is necessary to define some useful variables. First, the forward variable is defined as

$$\alpha_t(i) = \Pr[O_1 O_2 \dots O_t, s_t = S_i | \mathbf{A}, \boldsymbol{\pi}]. \tag{16.21}$$

That is, the probability of the partial observation sequence $O_1 \dots O_t$ and state S_i at time t, given the channel parameters **A** and π. The forward variable can be inductively calculated by initializing as

$$\alpha_1(i) = \pi_i \, b_i(O_1) \text{ for } i = 0, 1, \tag{16.22}$$

and then iterating

$$\alpha_{t+1}(j) = \left[\sum_{i=0}^{1} \alpha_t(i) a_{ij} \right] b_j(O_{t+1}) \text{ for } j = 0, 1 \tag{16.23}$$

and $1 \leq t \leq T - 1$.

In a similar manner, the backward variable is defined as

$$\beta_t(i) = \Pr[O_{t+1} O_{t+2} \dots O_T | s_t = S_i, \mathbf{A}, \boldsymbol{\pi}], \tag{16.24}$$

which is the probability of the partial observation sequence from $t+1$ to T, given state S_i at time t and the channel parameters $\mathbf{A}$ and $\boldsymbol{\pi}$. The backward variable can likewise be solved inductively by initializing as

$$\beta_T(i) = 1 \text{ for } i = 0, 1 \tag{16.25}$$

and iterating

$$\beta_t(i) = \sum_{j=0}^{1} a_{ij}\, b_j(O_{t+1})\beta_{t+1}(j) \text{ for } i = 0, 1 \tag{16.26}$$

and $t = T-1, \ldots, 1$.

Another variable is the probability of being in state S_i at time t, given the observation sequence $\boldsymbol{O}$ and the channel parameters $\mathbf{A}$ and $\boldsymbol{\pi}$, or

$$\gamma_t(i) = \Pr\left[s_t = S_i | O, \mathbf{A}, \boldsymbol{\pi}\right]. \tag{16.27}$$

This probability can be expressed simply in terms of the forward–backward variables according to

$$\gamma_t(i) = \frac{\alpha_t(i)\beta_t(i)}{\sum_{j=0}^{1} \alpha_t(j)\beta_t(j)} \text{ for } i = 0, 1. \tag{16.28}$$

The final variable is the probability of being in state S_i at time t and state S_j at time $t+1$, given the observation sequence $\boldsymbol{O}$ and the channel parameters $\mathbf{A}$ and $\boldsymbol{\pi}$, which can be written as

$$\xi_t(i,j) = \Pr\left[s_t = S_i, s_{t+1} = S_j | O, \mathbf{A}, \boldsymbol{\pi}\right] \text{ for } i,j = 0, 1. \tag{16.29}$$

From the definitions of the forward and backward variables, we can also write (16.29) as

$$\xi_t(i,j) = \frac{\alpha_t(i)\, a_{ij}\, b_j(O_{t+1})\beta_{t+1}(j)}{\sum_{i=0}^{1}\sum_{j=0}^{1} \alpha_t(i)\, a_{ij}\, b_j(O_{t+1})\beta_{t+1}(j)} \text{ for } i,j = 0, 1. \tag{16.30}$$

It is easy to verify using (16.29) that the probability in (16.27) is given by

$$\gamma_t(i) = \sum_{j=0}^{1} \xi_t(i,j) \text{ for } i = 0, 1. \tag{16.31}$$

The summation of $\gamma_t(i)$ over time index t can be interpreted as the expected number of times that state S_i is visited, or equivalently, the expected number of transitions made from state S_i. Similarly, summation of $\xi_t(i,j)$ from $t = 1$ to $t = T-1$ can be interpreted as the expected number of transitions from state S_i to state S_j. Using

(16.28) and (16.30) along with the notion of counting event occurrences, it is possible to estimate the parameters of an HMM iteratively.

The i, j element of state transition matrix $\mathbf{A}$ can be considered as the ratio of the expected number of transitions from state S_i to state S_j to the expected number of transitions made from state S_i. The elements of $\mathbf{A}$ can therefore be estimated using

$$\hat{a}_{ij} = \frac{\sum_{t=1}^{T-1} \xi_t(i,j)}{\sum_{t=1}^{T-1} \gamma_t(i)} \text{ for } i,j = 0, 1. \tag{16.32}$$

Similarly, the initial state distribution π_i can be viewed as the expected number of times to be in state S_i at time $t = 1$, and therefore we can estimate π using

$$\hat{\pi}_i = \gamma_1(i) \text{ for } i = 0, 1. \tag{16.33}$$

Thus the current channel parameters $\mathbf{A}$ and $\boldsymbol{\pi}$ can be used to compute (16.28) and (16.30), which subsequently facilitates re-estimating these channel parameters as $\hat{\mathbf{A}}$ and $\hat{\boldsymbol{\pi}}$ via (16.32) and (16.33). It was proven in [33,34] that the model described by $\hat{\mathbf{A}}$ and $\hat{\boldsymbol{\pi}}$ is more likely to have produced the observation sequence than the model described by $\mathbf{A}$ and $\boldsymbol{\pi}$. In other words, $\Pr[\boldsymbol{O}|\hat{\mathbf{A}}, \hat{\boldsymbol{\pi}}] > \Pr[\boldsymbol{O}|\mathbf{A}, \boldsymbol{\pi}]$.

If this process is repeated using the most recent iterative estimates $\hat{\mathbf{A}}$ and $\hat{\boldsymbol{\pi}}$ in place of $\mathbf{A}$ and $\boldsymbol{\pi}$, the probability of $\boldsymbol{O}$ being observed from the model can be improved until some limiting point is reached. It has been found that 30 iterations allows this process to converge to a stable estimate of $\mathbf{A}$ and $\boldsymbol{\pi}$. Considering that (16.28) and (16.30) measure the expected number of transitions from one state to the other by counting the event occurrences, it is clear that the number of observations T must be sufficiently large to ensure all the possible state transitions; i.e. if T is too small, the estimate of $\mathbf{A}$ will be biased.

As noted above, at the time the radar begins transmitting it is not able to evaluate instantaneously if the channel is currently free or busy. However, using the channel parameter estimates obtained from the most recent T observations, a cognitive radar can calculate the probability that the channel is currently free, i.e. the probability of a spectrum opportunity P_{SO}. If this probability is sufficiently high, the cognitive radar transmits, otherwise it does not.

Figure 16.9 shows how the SU radar processes the continuous sequence of observations at the output of the spectrum-sensing detector. Since the estimation of $\mathbf{A}$ and $\gamma_t(i)$ is time consuming, the receiver obtains these estimates using nonoverlapping blocks of T elements, where in each block the initialization is performed using the channel parameter estimates from the previous block. The estimates obtained for each block are used to evaluate P_{SO} using a sliding window that collects the last T observations captured in the previous time slot.

There are T sliding windows for each block, with the kth sliding window using the estimate of $\mathbf{A}$ and fixing $\pi_i = \gamma_k(i)$ to evaluate the forward and backward variables via

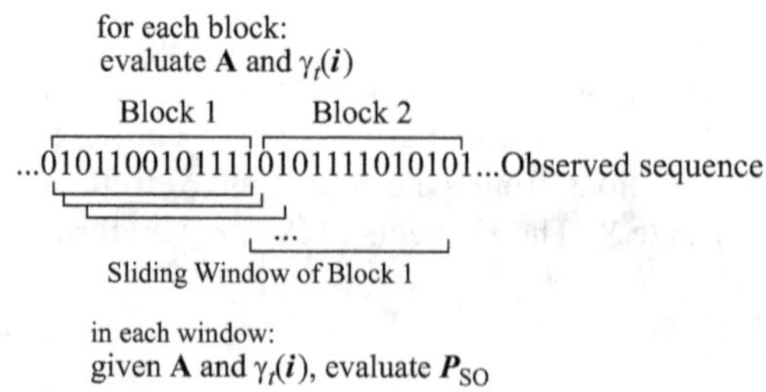

Figure 16.9 How to process the observed sequence

(16.22)–(16.26). Therefore, similar to (16.28), the probability that the last observation in the sliding window corresponds to channel state S_i is

$$\gamma(i) = \frac{\alpha_T(i)\beta_T(i)}{\sum_{j=0}^{1}\alpha_T(j)\beta_T(j)} \quad \text{for } i = 0, 1. \tag{16.34}$$

This probability is used to evaluate P_{SO} as

$$P_{SO} = \gamma(0)a_{00} + \gamma(1)a_{01}, \tag{16.35}$$

i.e. the probability that in the previous time-slot the channel was free and remains free, plus the probability that in the previous time-slot the channel was busy and becomes free. This probability is compared with threshold λ, and the radar transmits only if $P_{SO} > \lambda$.

There are two kinds of errors for this decision process. The event in which the radar does not transmit but the channel is free (i.e. lose a spectrum opportunity) is denoted as e_0. In contrast, the event in which the radar transmits while the channel is already occupied (i.e. a collision) is denoted e_1. Figure 16.10 shows the probability of these two errors as a function of threshold λ, which provides a way to tune the cognitive system to the desired performance. These results were obtained through Monte Carlo analysis by randomly generating a_{00} and a_{11} as independent, uniformly distributed r.v.'s on the range [0,1]. Clearly, when $\lambda = 0$, the radar always transmits and $\Pr[e_1]$ coincides with the probability that the channel is busy, which is 0.5 here. Similarly, when $\lambda = 1$, the radar never transmits and $\Pr[e_0]$ coincides with the probability that the channel is free, which is likewise 0.5.

Figure 16.11 shows $\Pr[e_0]$ and $\Pr[e_1]$ as a function of time for 9,246 time slots (i.e. 9 blocks of 1,024 elements). These results were obtained via Monte Carlo by generating a_{00} and a_{11} as independent, uniformly distributed r.v.'s on [0,1] for each block and by setting $\lambda = 0.65$. These results show that the performance of a cognitive radar that adopts this manner of probabilistic interference avoidance is theoretically constant and far superior to the performance of both (a) a traditional noncognitive system that ignores the presence of a PU when transmitting and (b) a radar that never transmits so as to avoid producing interference.

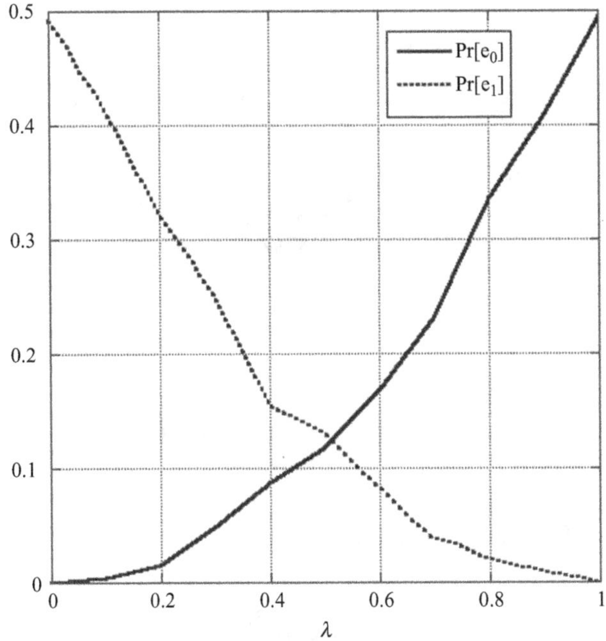

Figure 16.10 Probabilities of e_0 and e_1 as a function of λ

16.6 Conclusions and looking ahead

In recent years, commercial communication systems have proliferated at incredible rates, resulting in an extremely spectrally dense environment and fierce competition for frequency bands that traditionally have been allocated to radar systems as primary legal users. At the same time, with the introduction of new requirements such as advanced imaging capabilities and automatic target recognition, future radar systems need more bandwidth and operational freedom than ever before. To cope with these conflicting demands, one solution is intelligent spectrum utilization coupled with agile waveform diversity, which in turn leads to the formulation of cognitive radar systems that are able to discern and exploit available spectrum opportunities.

Based on the definition of a cognitive system by Haykin [16], where the fundamental structure is the perception-action cycle, we have established a framework involving spectrum sensing, channel characterization, and spectrum sharing that provide direct control over the standard radar functions. Specifically, these tasks measure the spectrum occupancy, estimate the parameters that model the PU operating dynamics as a function of time and frequency channel, and then anticipate the time/frequency arrangement that allows the radar to operate with the least likelihood of inducing interference to the PU.

This manner of learning and anticipating PU behavior, which is based on the use of an HMM and the Baum–Welch method to iteratively estimate channel parameters,

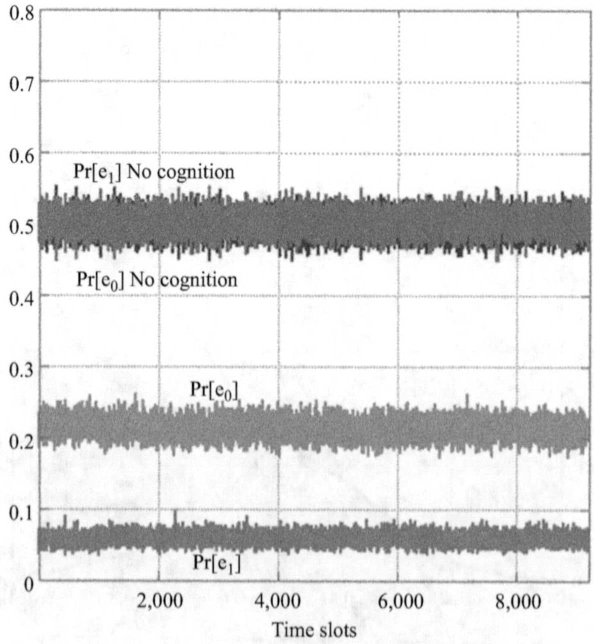

Figure 16.11 Probabilities of e_0 and e_1 as a function of time

facilitates the means with which to reduce mutual interference between the radar and PU in the same channel. Such a capability leverages CS techniques as enabling technology to avoid costly hardware to monitor large bandwidth with high accuracy. Likewise, the notion of interference noise temperature provides a common metric for spectrum sharing systems to ascertain whether they will induce interference to other spectrum users and vice versa.

Interference avoidance approaches such as this can be expected to benefit from the availability of more data regarding typical spectrum usage. For example, use patterns for mobile cellular tend to follow repeatable trends that may be possible to exploit for secondary radar use. Having said that, additional robustness measures are needed to account for RF environments that are or become less predictable.

References

[1] Stinco, P., Greco, M., Gini, F., Himed, B., "Cognitive radars in spectrally dense environments" *IEEE Aerospace and Electronic Systems Magazine*, vol. 31, no. 10, pp. 20–27, Oct. 2016.

[2] ITU-R Resolution 229, "Use of the bands 5150–5250 MHz, 5250–5350 MHz and 5470–5725 MHz by the mobile service for the implementation of wireless access systems including radio local area networks", 2012.

[3] ITU-R RECOMMENDATION M.1652, "Dynamic frequency selection (DFS) in wireless access systems including radio local area networks for the purpose of protecting the radio determination service in the 5 GHz band", 2003.

[4] U.S. Department of Commerce, "An assessment of the near-term viability of accommodating wireless broadband systems in the 1675–1710 MHz, 1755–1780 MHz, 3500–3650 MHz, and 4200–4220 MHz, 4380–4400 MHz bands", 2010.

[5] Obregon, E., Sung, K.W., Zander, J., "Secondary access to the radar spectrum bands: regulatory and business implications", *24th European Regional Conference of the International Telecommunications Society*, Florence, Italy, 2013.

[6] Brick, D., Ellersick, F., "Future Air Force tactical communications" *IEEE Transactions on Communications*, vol. 28, no. 9, pp. 1551–1572, Sep. 1980.

[7] Griffiths, H., Cohen, L., Watts, S., *et al.*, "Radar spectrum engineering and management: technical and regulatory issues," *Proceedings of the IEEE*, vol. 103, no. 1, Jan. 2015.

[8] La Manna, M., Stinco, P., Greco, M., Gini, F., "Design of a cognitive radar for operation in spectrally dense environments", *IEEE Radar Conference*, Ottawa, Ontario, Canada, Apr./May 2013.

[9] Stinco, P., Greco, M., Gini, F., Manna, M.L. "Compressed spectrum sensing in cognitive radar systems" *IEEE International Conference on Acoustics, Speech and Signal Processing*, Florence, Italy, May 2014, pp. 81–85.

[10] Stinco, P., Greco, M., Gini, F., Himed, B., "Channel parameters estimation for cognitive radar systems" *4th International Workshop on Cognitive Information Processing*, Copenhagen, Denmark, May 2014.

[11] Stinco, P., Greco, M., Gini, F., "Spectrum sensing and sharing for cognitive radars" *IET Radar, Sonar and Navigation*, vol. 10, no. 3, pp. 595–602, Feb. 2016.

[12] Federal Communications Commission, "Notice of inquiry and notice of proposed rulemaking: in the matter of establishment of an interference temperature metric to quantify and manage interference and to expand available unlicensed operation in certain fixed, mobile and satellite frequency bands," ET Docket No. 03-237; November 28, 2003.

[13] Clancy, T., "Formalizing the interference temperature model", *Wiley Journal on Wireless Communications and Mobile Computing*, vol. 7, no. 9, pp. 1077–1086, May 2007.

[14] Yücek, T., Arslan, H., "A survey of spectrum sensing algorithms for cognitive radio applications," *IEEE Communications Survey and Tutorials*, vol. 11, no. 1, First Quarter 2009.

[15] Zhao, Y., Morales, L., Gaeddert, J., Bae, K.K., Um, J.-S., Reed, J.H., "Applying radio environment maps to cognitive wireless regional area networks" *2nd IEEE International Symposium on New Frontiers in Dynamic Spectrum Access Networks*, pp. 115–118, Dublin, Ireland, April 2007.

[16] Haykin, S., *Cognitive Dynamic Systems: Perception-action Cycle, Radar and Radio*, Cambridge University Press, Cambridge, UK, 2012.

[17] Bater, J., Tan, H., Brown, K.N., Doyle, L., "Modelling interference temperature constraints for spectrum access in cognitive radio networks," *IEEE International Conference on Communications*, pp. 6493–6498, Glasgow, UK, June 2007.

[18] Sharma, M., Sahoo, A., Nayak, K.D., "Channel modeling based on interference temperature in underlay cognitive wireless networks," *IEEE International Symposium on Wireless Communication Systems*, pp. 224–228, Reykjavik, Iceland, Oct. 2008.

[19] Barnes, S.D., Maharaj, B.T., "Performance of a hidden Markov channel occupancy model for cognitive radio," IEEE Africon, Livingstone, Zambia, Sep. 2011.

[20] Shankar, S., Cordeiro, C., Challapali, K., "Spectrum agile radios: utilization and sensing architectures," *IEEE International Symposium on New Frontiers in Dynamic Spectrum Access Networks*, Baltimore, Maryland, USA, Nov. 2005.

[21] Berbra, K., Barkat, M., Gini, F., Greco, M., Stinco, P., "A fast spectrum sensing for CP-OFDM cognitive radio based on adaptive thresholding", *Signal Processing*, vol. 128, pp. 252–261, Nov. 2016.

[22] Eldar, Y.C., Kutyniok, G., *Compressed Sensing*, Cambridge University Press, Cambridge, UK, June 2012.

[23] Mishali, M., Eldar, Y.C., Dounaevsky, O., Shoshan, E., "Xampling: analog to digital at sub-Nyquist rates," *IET Circuits, Devices & Systems*, vol. 5, no. 1, pp. 8–20, Jan. 2011.

[24] Candes, E.J., Wakin, M.B., "An introduction to compressive sampling," *IEEE Signal Processing Magazine*, vol. 25, no. 2, pp. 21–30, Mar. 2008.

[25] Donoho, D.L., "Compressed sensing," *IEEE Transactions on Information Theory*, vol. 52, no. 4, pp. 1289–1306, Apr. 2006.

[26] Donoho, D.L., Huo, X., "Uncertainty principles and ideal atomic decomposition," *IEEE Transactions on Information Theory*, vol. 47, no. 7, pp. 2845–2862, Nov. 2001.

[27] Chen, S.S., Donoho, D.L., Saunders, M.A., "Atomic decomposition by basis pursuit", *SIAM Journal on Scientific Computing*, vol. 20, no. 1, pp. 33–61, 1998.

[28] Davenport, M.A., Boufounos, P.T., Wakin, M.B., Baraniuk, R.G., "Signal processing with compressive measurements," *IEEE Journal of Selected Topics in Signal Processing*, vol. 4, no. 2, pp. 445–460, Apr. 2010.

[29] Tropp, J.A., Laska, J.N., Duarte, M.F., Romberg, J.K., Baraniuk, R.G., "Beyond Nyquist: efficient sampling of sparse bandlimited signals," *IEEE Transactions on Information Theory*, vol. 56, no. 1, pp. 520–544, Jan. 2010.

[30] Zhao, Q., Sadler, B.M., "A survey of dynamic spectrum access: signal processing, networking, and regulatory policy", *IEEE Signal Processing Magazine*, vol. 4, no. 3, pp. 79–89, May 2007.

[31] Rabiner, L., "A tutorial on hidden Markov models and selected applications in speech recognition," *Proceedings of the IEEE*, vol. 77, no. 2, pp. 257–286, Feb. 1989.

[32] Dempster, A.P., Laird, N.M., Rubin, D.B., "Maximum likelihood from incomplete data via the EM algorithm", *Journal of the Royal Statistical Society*, vol. 39, no. 1, pp. 1–38, Jan. 1977.

[33] Baum, L.E., Sell, G.R., "Growth functions for transformations on manifolds", *Pacific Journal of Mathematics*, vol. 27, no. 2, pp. 211–227, Jan. 1968.

[34] Baker, J., "The dragon system—An overview", *IEEE Transactions on Acoustics, Speech, and Signal Processing*, vol. ASP-23, no. 1, pp. 24–29, Feb. 1975.

Chapter 17
Radar waveform design for spectral coexistence

Aaron M. Jones[1], Francis Watson[2], and Brian D. Rigling[3]

17.1 Introduction

Significant operational challenges exist for modern radar systems, such as scene topography, increasing number of targets, decrease in the size of targets (electrically), and a likely contested/congested radio frequency (RF) spectrum, among others. These concerns lead to the conclusion that simply interrogating the scene is no longer sufficient. The radar (and thus the transmit waveform) must adapt to its environment and understand the impact on performance resulting from the presence of other competing emitters.

In this chapter we discuss design techniques and constraints that facilitate a waveform that is useful for radar. While signal-to-noise ratio (SNR) and signal-to-interference-plus-noise ratio (SINR) have historically been the primary drivers of waveform design and are certainly discussed here, the growing need for spectral coexistence is eliciting new metrics and design approaches that address a broader set of considerations.

This chapter also examines the use of performance prediction models as a means to establish relationships between waveform constraints and SINR performance, which are critical for detection in spectrally crowded environments. An empirical approach is discussed along with models that provide an intuitive way to understand the complex relationships between constraints and SINR performance in dynamic situations.

At its highest level, this chapter has two primary themes: radar-centric and communication-centric waveform designs. After some introductory material in Section 17.1, the optimization of coded radar waveforms for interference-rich environments is discussed in Sections 17.2 and 17.3. This approach is important when the radar is the primary spectrum user and communication systems (or other RF sources) are secondary users. The second theme is discussed in Section 17.4 and views the

[1]Air Force Research Laboratory, Sensors Directorate, United States
[2]Defence Science and Technology Laboratory, United Kingdom
[3]Department of Electrical Engineering, Wright State University, United States

communication system as the primary user (i.e. communication-centric) and radar performance as secondary. Each of these perspectives poses unique challenges and limitations.

Specifically, Section 17.2 presents three interference-plus-noise models that represent spectrally congested environments. These models are necessary to facilitate a complete discussion on how to accomplish waveform optimization in these environments. Such models also provide a relatively easy way to analyse waveform performance, as it is not practical to perform open-air experimental trials for every conceivable scenario.

After discussing the interference and noise models, Section 17.3 considers some of the important criteria for radar waveform design that are valid for both radar- and communication-centric perspectives. While radars are used for many applications (detection, imaging, tracking, and numerous variations thereof), the focus here is on the detection of targets and the practical understanding of how spectral congestion impacts detection performance. For detection, SINR is the highest concern. However, as will be shown, ignoring other practical considerations can provide severely misleading results in expected sensing performance. To investigate this phenomenon, we connect together some of the important requirements for waveform design and the challenging spectral environment into cohesive waveform optimization algorithms. Several techniques are discussed, along with their benefits and limitations. Also in Section 17.3, and a logical next step in the evolution of radar waveform design, is performance prediction of the waveform in complicated spectral environments. Building upon the discussion of interference models in Section 17.2, and considering the modulus and sidelobe attributes of the waveform, we develop multiple performance prediction models for waveform optimization to aid in the selection of what to transmit.

Note that the focus in this chapter is upon the design of waveforms based on a coded structure, which facilitates mathematical tractability for design and analysis. Such codes could be readily deployed in arbitrary operational systems such as through the combination of the waveform implementation scheme discussed in Chapter 6 and arbitrary waveform generation capability.

Finally, while the waveform design problem has evolved into multiple veins of specialties, spectrum sharing being one, note that waveform diversity and design research has witnessed a widespread expansion over the last 10–15 years, which is evidenced by the increase in the number of articles, books, funding opportunities, special sessions in conferences and workshops, and special issues on selected topics within waveform diversity [1–3]. While extensive literature exists on waveform diversity and associated optimization for the purpose of improving many typical functions of radar (e.g. detection, automatic target recognition, imaging, electronic warfare, etc.), it is impractical to review these topics in their entirety. Instead, we focus the discussion on relevant materials to aid in the understanding of key concepts throughout the chapter.

17.1.1 Waveform design for interference mitigation

There has been significant work on radar-centric waveform design, and we build upon these fundamental principles in subsequent sections. The basic problem of designing radar waveforms for SINR maximization has attracted considerable attention, with numerous books being written on different aspects of the subject. Most of these focus on SINR maximization (i.e. detection) and include some of the more popular and fundamental texts in the field, such as [1,4–10]. These references provide a detailed examination of the rich history of radar waveform design, and we employ the definition of waveform diversity as stated in [4].

Definition 17.1. *Adaptivity of the radar waveform to dynamically optimize the radar performance for the particular scenario and tasks.*

There have been several survey papers on various topics in radar waveform design. For example, Stoica *et al.* provide an excellent discussion of the use of transmit code sequences for range resolution and detection [11]. In 2009, Sira [12] wrote about adapting the transmit waveform to match the sensing task and environment. In 2009, Calderbank [13] discussed waveform diversity in signal processing and the control of degrees of freedom (DOF) in the radar illumination pattern. We also suggest the 2009 paper by Benedetto *et al.* [14] that mentions the important role the ambiguity function plays in phase-coded waveforms.

The efforts of many articulate the benefits of constrained waveform design for SINR maximization, including [15–27]. In [28–33], the authors investigate the impact of constraints on SINR performance and formulate performance prediction models to anticipate the degradation of SINR due to spectral coexistence, among other concerns. Other novel waveform design techniques and performance prediction strategies have been investigated in [34–45] and should be examined prior to the development of any new waveform design approaches.

17.1.2 The role of practical constraints

The advent of digital arbitrary waveform generators and modulus tracking technologies provide considerable flexibility for modern radar transmitters. Nevertheless, there are still practical concerns such as finite on-board power supplies, peak-to-average power ratio (PAPR), otherwise known as waveform modulus, and smaller targets being masked by the range sidelobes of nearby large targets. Additional waveform properties can include range resolution (particularly for imaging applications), Doppler tolerance, and the impact of spectral regulations and transmitter distortion effects. Each of these properties imposes particular constraints on the waveform.

As an example of the sidelobe constraint, Figure 17.1 illustrates the impact of ignoring the effect of sidelobes in the waveform design process. The unconstrained waveform realizes high sidelobes that degrade sensitivity to smaller targets. In contrast, the design of the constrained waveform accounts for sidelobes and thus attempts to minimize them.

The suppression of range sidelobes is clearly an important practical aspect of radar waveform design [46]. The process of pulse compression can simultaneously provide good range resolution (bandwidth dependent) while providing the necessary energy on target via the extended pulse length [5,14,47]. It is known that frequency windowing on receive can reduce range sidelobes for chirp waveforms [8,48], though

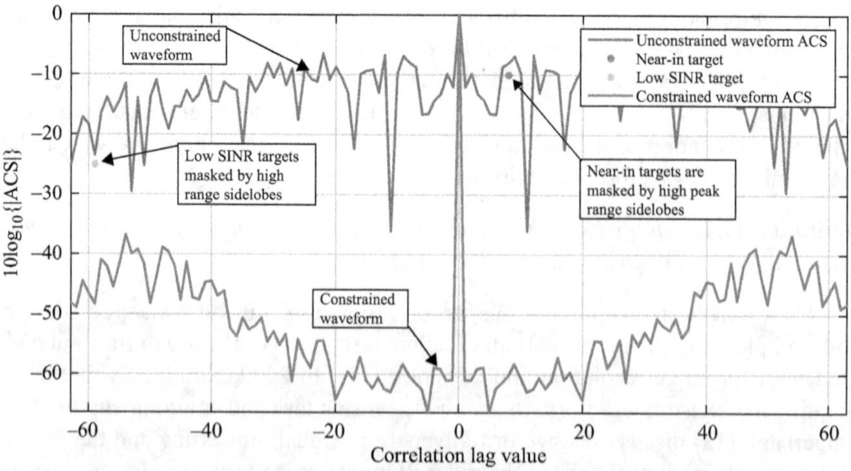

Figure 17.1 Illustration of the consequences of ignoring (and accounting for) range sidelobe effects in the waveform design process. Reproduced, with permission, from Reference [30]

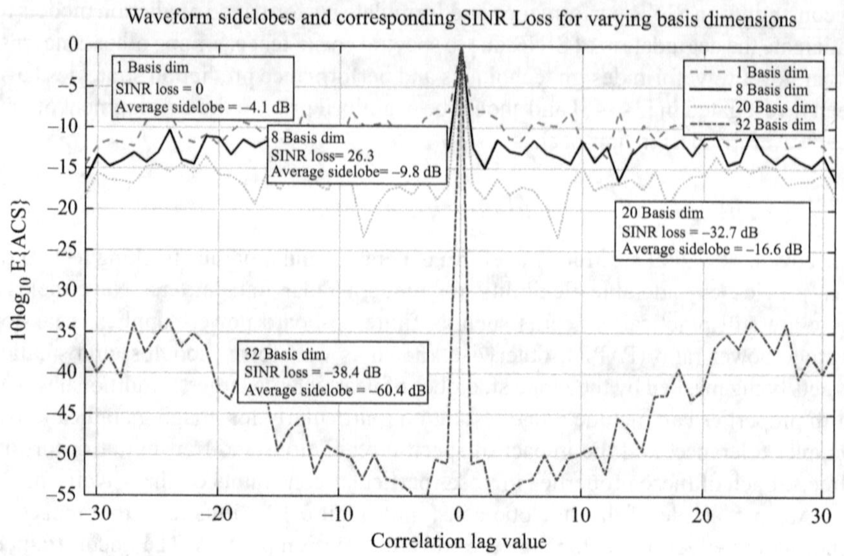

Figure 17.2 Example of SINR loss due to reduced sidelobes in waveform design (length 32 coded waveforms)

here we focus on waveform design to achieve this effect while also minimizing SINR loss (see Figure 17.2 for a conceptual example). Additional work related to the minimization of energy outside the main lobe of the waveform autocorrelation can be found in [25,49–51].

Similarly, the PAPR of the waveform is of critical importance because the transmitter has a maximum supply voltage (i.e. peak power limited). While it is understood that more energy on target increases the likelihood that the echo signal is detected, modulating the amplitude of the waveform limits this energy due to deviation from the peak power. Therefore, efficient use of the available transmit supply voltage is an important waveform design criterion.

In 1973, Mesiya [52] first discussed the need for phase-only waveforms so that the transmitter can operate at peak power to obtain maximum average power. Since then, considerable research has focused on the optimal design of phase-only waveforms given various restrictions [53–56]. For example, in [19,23], Patton discussed the joint role of waveform modulus and the delay–Doppler ambiguity function in phase-only design.

Related literature on optimal waveform design [52,57–59] considers clutter suppression and similarity constraints as the design metric. Waveform design based on PAPR constraints has likewise been investigated (e.g. [54–56]) while also accounting for sidelobes and finite energy. In short, phase-only waveforms have been extensively studied, and the reader is referred to [1,5,10,19,23,60] and the references therein.

By constraining waveforms to be constant modulus, the available design DOF are underutilized. In contrast, while allowing waveform modulus to be perturbed does admit SINR degradation, it also permits the re-purposing of these available DOF to address other constraints, such as those arising in spectrally congested environments. The consequence to SINR resulting from the imposition of these design restrictions is not yet fully understood, though recent work that is reviewed below does alleviate some of this uncertainty.

17.1.3 Performance prediction (empirical vs. analytic)

As the various performance trade-offs involved with operating radars in spectrally crowded environments become better understood, performance prediction models for waveform design will become increasingly important to assist in the decision regarding what to transmit. The performance model use-case considered here is the problem of radar detection in the presence of coloured noise and interference. Most waveform design algorithms provide empirical evidence of their achievable performance (e.g. see [15,19–21]). While such results are valuable as a means to compare different design approaches, it is important to understand the limits of the conclusions, which are usually for highly specific scenarios and potentially not practical. This arrangement motivates a more analytic approach that can generalize across scenarios and algorithms to fully understand the consequences of various constraints and spectral coexistence requirements.

Figure 17.3 depicts a notional example of how performance models can be applied for adaptive radar waveform design to achieve spectral efficiency [61]. The figure illustrates the loss in SINR as the constraint—either modulus or range-sidelobe—is applied for a particular spectral efficiency scenario. Understanding the trade-off between SINR and a given constraint can help avoid over-constraining the waveform design and thereby being penalized with unnecessary SINR loss.

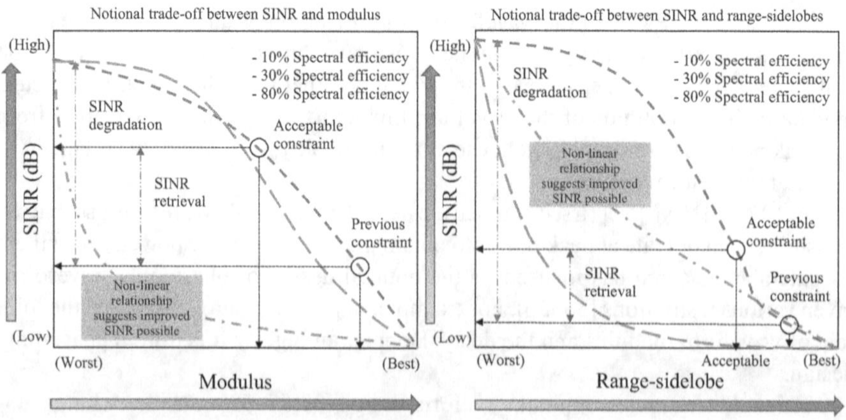

Figure 17.3 Notional example of the trade-off between SINR and modulus/sidelobe constraints for adaptive radar waveform design. Reproduced, with permission, from Reference [30]

Possibly the most convincing argument to develop performance models is the lack of understanding on how constraints influence SINR performance. Intuitively, we understand that enforcing constraints on waveform design limits the DOF available to maximize SINR. It therefore seems misguided to arbitrarily impose constraints without understanding the consequences due to the nonlinearity of SINR performance with respect to the constraints [29,33]. As such, performance prediction models are developed in Section 17.3 and compared against waveform optimization algorithms using simulated and measured data.

17.2 Representing interference and noise for waveform design algorithm development

Now consider the development of contested/congested spectrum models. Evaluation of waveform design algorithms for spectral coexistence requires the generation of representative radio frequency interference (RFI) based on the characteristics of real-world data. One of the challenges in the development of performance prediction models for waveform SINR (as a function of constraints) is the selection of realistic interference and noise. Typically, a toy example of the interference plus noise covariance is used for theoretical development, usually owing to ease of use as opposed to what is actually representative of practical scenarios. In reality, communication power spectra can be rather complicated and temporally non-stationary. Ignoring these considerations in simulated evaluations limits the understanding of how a given radar waveform design approach can perform in practice. A technique that addresses this concern is discussed in Section 17.2.3.

Many waveform design approaches also require the use of Monte Carlo techniques to ascertain their effectiveness across varying conditions. For instance,

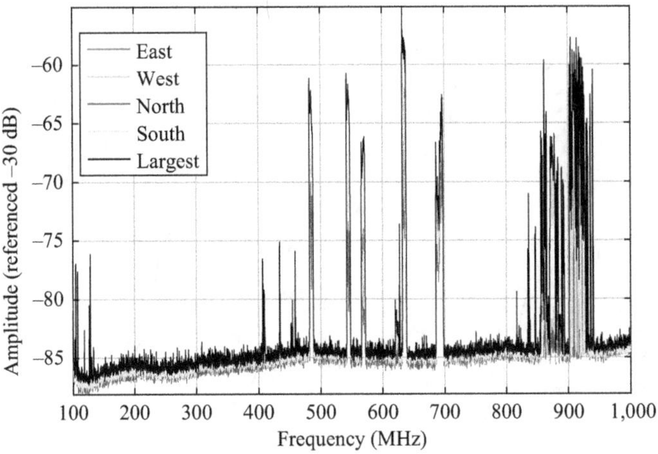

Figure 17.4 Tillman–Pit spectrum survey data. Reproduced, with permission, from Reference [63]

Table 17.1 General interference models for radar waveform design/assessment

Centricity	Model name	Ease of use	Fidelity	Prior knowledge necessary
Radar	AR	Straightforward	Low	No
Radar	MP	Straightforward	Medium	No
Radar	Tillman	Neutral	High	Yes
Comms (discussed in detail in Section 17.4)	PU	Neutral	Medium	Yes

interference-rich environments can degrade radar detection performance and/or limit the effective operating bandwidth [62]. The spectrum survey for a mid-size metropolitan area illustrated in Figure 17.4 exemplifies this type of scenario for the spectral swath from 100 MHz to 1 GHz [63]. It can therefore be inferred that the complexity of the spectral environment must be considered for waveform design and performance prediction to be effective.

For radar waveform design to facilitate spectral coexistence, it is critical that the design process be robust to varying RFI environments. Here we consider three ways in which this design robustness can be achieved. These models and their respective properties are summarized in Table 17.1.

17.2.1 Auto-regressive parametric models

The use of a low model-order auto-regressive (AR) model is one such approach that has been employed extensively in the literature [22–26,64] to represent coloured interference spectra. This model is regarded as having low fidelity relative to practical

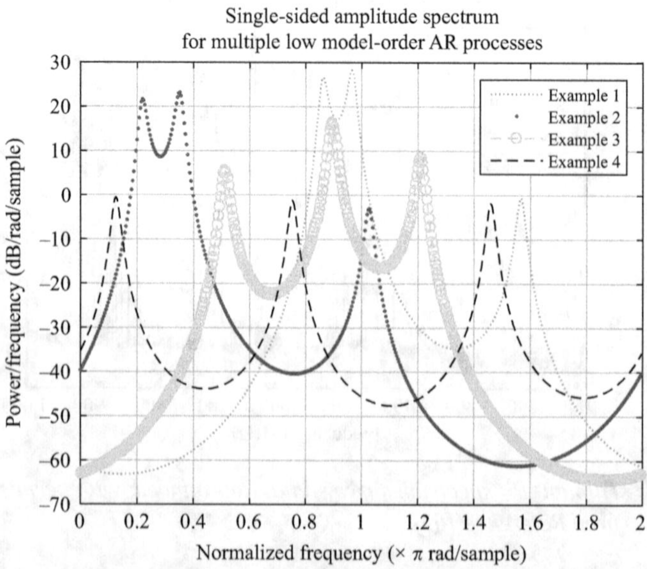

Figure 17.5 Simulated spectrum estimate for multiple AR-3 processes with $M = 256$ samples

power spectra, though it does provide a convenient way to perform Monte Carlo analysis due to the ease with which it can be generated.

Specifically, the received interference and noise is modelled as

$$x_{\mathrm{AR}}(k) = \sum_{p=1}^{P} \phi(p)\, x_{\mathrm{AR}}(k-p) + \varepsilon(k), \tag{17.1}$$

where P is the model order, k the lag position, ϕ_p the model coefficients, and ε_k the driving white noise with distribution $\mathbb{CN}(0, \sigma)$. The time series auto-correlation $R_{xx}(m) = E\{x_{\mathrm{AR}}(k)x_{\mathrm{AR}}(k-m\}$ defines the covariance matrix, which is denoted as $\mathbf{K}_{\mathrm{AR}}$. This all-pole model generates spectrally peaky RF-coloured environments, and the magnitudes of the coefficients ϕ_p describe the correlation.

The low-model order AR power spectrum is a useful tool for testing interference-aware waveform design approaches. Figure 17.5 depicts four power spectra examples for an AR-3 (model-order 3) process. While simple and convenient, it is shown in subsequent sections that this model does not adequately represent the features of measured data and therefore yields misleading SINR performance for waveform design in spectrally contested environments. For a more complete discussion of this model, the reader is referred to [64].

17.2.2 Marčenko–Pastur generalized model

In the absence of prior information about the interference and noise spectrum, the Marčenko–Pastur (MP) model [65] provides a general framework that is predicated on an asymptotic result from random matrix theory. The impetus for this model is the rather common occurrence of having no prior information about a given spectral environment, which arises because many radars are required to operate in unknown environments. This model is important in the analytical development of radar waveform performance prediction models and is used in Section 17.3 to assess multiple waveform design techniques.

Consider the covariance matrix defined as

$$\mathbf{K}_{\mathrm{MP}} = \frac{1}{M}\mathbf{Z}\mathbf{Z}^{H}, \tag{17.2}$$

where the matrix elements $\mathbf{Z}_{ij}$ are $\mathbb{CN}(0, (1/\sqrt{M}))$, and it is assumed that $\mathbf{K}_{\mathrm{MP}}$ has full rank with $\sum_{k=1}^{M} \lambda_k = M$ (equivalent to assuming constant noise power). Therefore, in the limit with regard to the amount of data, the statistical behaviour of eigenvalues of large ($M \gtrsim 32$) random Hermitian matrices conforms to the MP probability density, which is an empirical distribution expressed as [65]

$$p_{\Lambda}(x) = \frac{\sqrt{4x - x^2}}{2x} \quad \text{for } x \in (0, 4]. \tag{17.3}$$

Applying the inverse transform sampling method [66] to (17.3), a closed-form representation of the cumulative distribution function (CDF)

$$\mathscr{M}(x) = \int_{0}^{x} p_{\Lambda}(z)dz, \tag{17.4}$$

$$= \frac{1}{2\pi}\left(\sqrt{4x - x^2} - 2\arcsin\frac{4 - 2x}{4}\right) + \frac{1}{2} \tag{17.5}$$

is obtained that provides a mechanism with which to compute the expected eigenvalues. See Figure 17.6 for a depiction of this density via the probability mass function (PMF) on the left axis and the corresponding CDF on the right axis.

The MP model is unique due to the closed-form expression of the eigenvalue distribution. For the purpose of analytical tractability, this model is used later in the chapter to represent the eigenspectrum of a generic interference and noise covariance matrix, thereby enabling evaluation of SINR performance due to increasing spectrum congestion. Additionally, in the practical case where no information about the noise and interference statistics is available, this model can be used to provide an initial baseline representation of the eigenspectrum for performance prediction.

Using (17.3) to compute the complimentary CDF $\mathscr{S}(x) = 1 - \mathscr{M}(x)$, otherwise known as the survival function, the inverse survival function $\mathscr{Z}(x)$ can be determined to describe the eigenspectrum for $x \in (0, 4]$, which is found to be

$$\mathscr{Z}(x) = \mathscr{S}^{-1}(x) \tag{17.6}$$

$$= \left(1 - \mathscr{M}(x)\right)^{-1} \tag{17.7}$$

$$= \frac{2}{\pi}\left(2\arcsin\left(\frac{4 - 2x}{4}\right) + \pi - \sqrt{4x - x^2}\right). \tag{17.8}$$

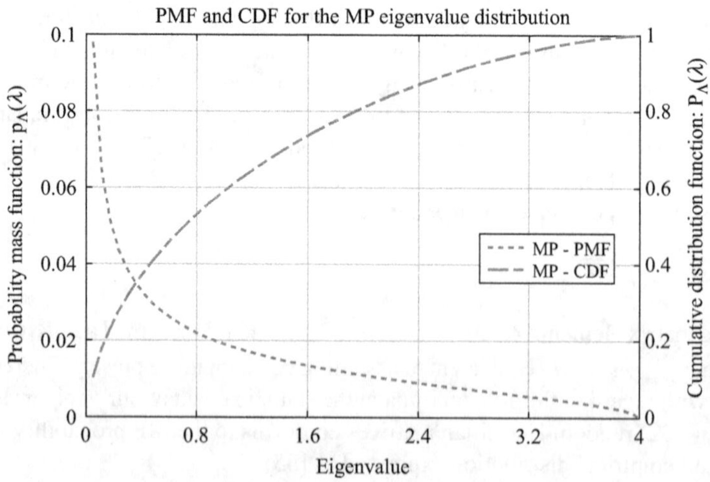

Figure 17.6 (Left axis) Probability mass function (PMF) for the Marčenko–Pastur (MP) asymptotic result using M = 64. (Right axis) The associated cumulative distribution function (CDF) for the MP eigenspectrum. Reproduced, with permission, from Reference [29]

To obtain the eigenspectrum for an M-dimensional covariance, M uniformly sampled values of x are then generated and transformed with the function $\mathscr{L}$ to obtain eigenvalues $\lambda_1 \ldots \lambda_M$ that have an MP-consistent PMF. As an example, Figure 17.7 illustrates 25 realizations of the eigenspectrum of the random matrix along with the theoretical asymptotic distribution for $M = 64$.

Table 17.2 lists the system parameters for measured ambient RF data collected by an air-to-ground radar operating at X-band in an urban area. For $M = 256$, Figure 17.8 shows how the eigenspectrum of this data compares to the theoretical MP model via (17.3) and to a sample eigenspectrum averaged over 25 random instantiations. The MP model is clearly consistent with the eigenspectrum of measured data in general, though it does miss the fine details that are specific to a given data collection.

17.2.3 Empirical models

Finally, we consider an empirical model that permits the generation of unique instantiations of RFI based on real-world data. For dynamic (i.e. temporal non-stationary) environments, where some knowledge may be available, yet the waveform design approach still needs to be evaluated for many possible interference spectra, this model provides a straightforward and highly configurable method to produce representative RFI.

This formulation [32] employs a two-state Markov chain to define the spectral transitions from a 'signal present' condition to a 'noise-only' (or signal absent) condition, and vice versa, in discrete frequency bins. The sequence is then modified with signal-/noise-specific amplitude distributions to arrive at the synthetic power

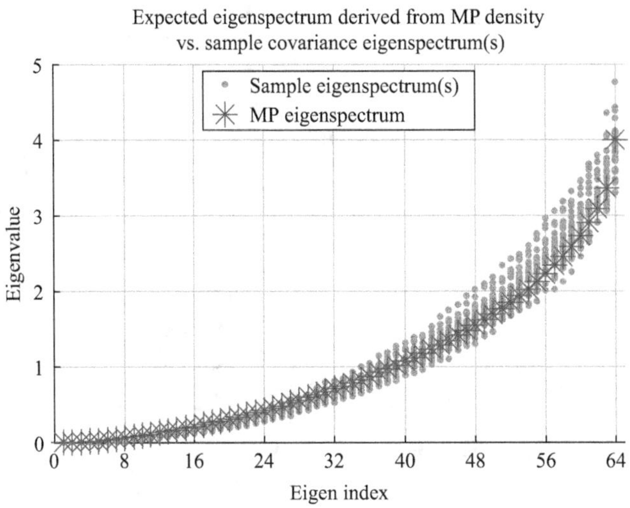

Figure 17.7 Asymptotic Marčenko–Pastur (MP) eigenspectrum derived from the density in (17.3) compared with sample covariance eigenspectra generated from 25 random data matrix instantiations for M = 64. Reproduced, with permission, from Reference [29]

Table 17.2 Interference and noise measurement specifications

Parameter	**Value**
Bandwidth	1 GHz
Minimum frequency	9.2 GHz
Maximum frequency	10.2 MHz
Pulses	256
Fast-time samples	256
Sample rate	3 GHz
Channels	1

spectral density (PSD). This approach is advantageous because it exploits the unique underlying phenomenology of a given band. Further, the technique requires limited computing power and time, thus making it cost effective.

Assuming the signal and noise are complex, independent, stationary, and ergodic Gaussian random processes, a general model can be described as

$$\mathbf{x} = \mathbf{s} + \mathbf{n}, \tag{17.9}$$

where $\mathbf{s} \sim \mathbb{CN}(\mu_s, \sigma_s^2\mathbf{I})$ is the signal component and $\mathbf{n} \sim \mathbb{CN}(0, \sigma_n^2\mathbf{I})$ is the noise component. This model implies that the power spectrum is χ^2 distributed with 1 complex degree of freedom.

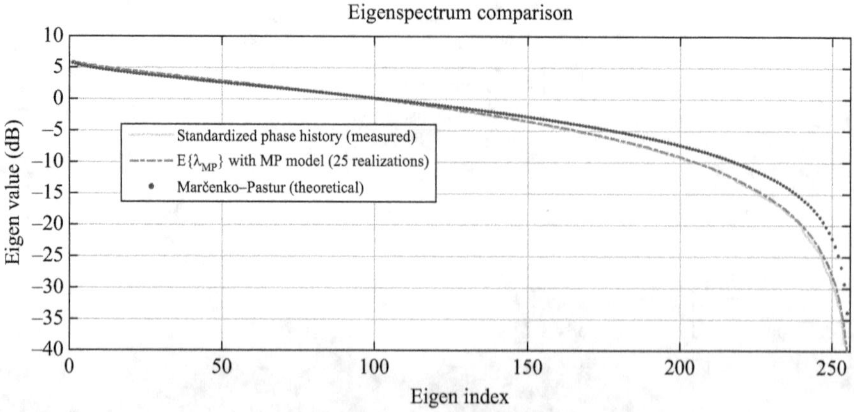

Figure 17.8 Eigenspectrum comparison for $M = 256$ between the theoretical Marčenko–Pastur model of (17.3), the average of 25 random sample covariance instantiations via (17.2), and based on measured data with the parameters in Table 17.2. Reproduced, with permission, from Reference [29]

The objective here is to realize easy to generate models, like in the two previous cases, while improving statistical similarity to real-world interference. This model facilitates effective waveform design for spectral coexistence by accurately representing the impact of additional emitters. However, the signal component is complicated by real-world effects such as channel fading, indirect path contributions, etc. [67]. The noise component may likewise be complicated by the presence of additional noise sources. In a statistical sense, such effects manifest as a deviation from the expected χ^2 distribution. Adequately representing the difference between the modelled and real-world distributions is a major concern.

Given the estimated signal-plus-noise eigenvalues as $\hat{\lambda}_i = l_i$, for $i = 1, \ldots, k$, the minimum description length (MDL) for Gaussian distributed data [68–70] is

$$\mathrm{MDL}(\hat{\lambda}) = -\log\left\{\frac{\rho}{\eta}\right\} + \frac{1}{2}k\log(N_o), \tag{17.10}$$

where $\rho = \prod_{i=k+1}^{g} l_i^{1/(g-k)}$ and $\eta = (1/(g-k))\sum_{i=k+1}^{g} l_i$ are the geometric and arithmetic means, respectively, g is the problem dimensionality, k is the number of free adjusted parameters, and N_o is the number of observations. This strategy can be applied to determine the signal subspace rank of any measurement set where the Gauss–Gauss signal/noise assumption is appropriate (see [32] for details).

Next, a shifted Gamma distribution (for which χ^2 is a special case) is employed to represent the signal and noise amplitudes of the PSD. This distribution has shape

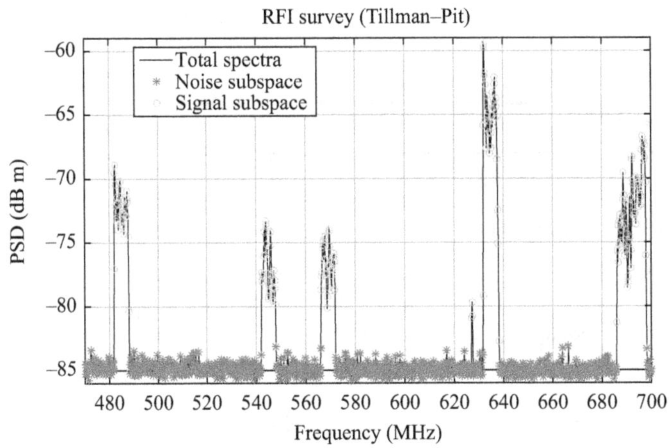

Figure 17.9 PSD survey of Tillman–Pit data (Eastern directivity) in actual power (dB m) from 470 to 700 MHz based on collection parameters in Table 17.3. Reproduced, with permission, from Reference [32]

parameter α, scale parameter β, and shift parameter γ such that it is generically described by the density function

$$G_{\alpha,\beta,\gamma}(x) = \frac{x^{\alpha-1}}{\Gamma(\alpha)\beta^{\alpha}} \exp\left\{\frac{-x}{\beta}\right\} + \gamma, \tag{17.11}$$

where $x \in (0, \infty)$ and $\Gamma(\alpha)$ is the Gamma function evaluated at α. The Gamma distribution is more flexible than a χ^2 distribution because the parameters of the former can be non-integers, thereby permitting higher fidelity to represent possible distributions. Due to this flexibility, the Gamma and shifted-Gamma distributions have been used for probability modelling in many fields [71]. Here the shape, scale, and shift parameters are uniformly distributed.

To illustrate the operation of this approach, the *Tillman–Pit* measured data depicted in Figure 17.9 was used to estimate the model parameters. This data was collected with a spectrum analyser fed by pyramidal horn antennas atop a 50-ft mast, with collection parameters denoted in Table 17.3. This data was collected with a directive antenna in each of the four cardinal directions (North, South, East, and West), providing four data sets to evaluate [63].

Using (17.10), the MDL was computed for the four measured data sets using $N_o = 2{,}000$ observations. Table 17.4 lists the results, where it is observed that the signal subspace rank did not appreciably vary for the different directions [63]. Based on the MDL results, Figure 17.9 likewise illustrates the frequencies corresponding to the noise and signal subspaces.

Table 17.3 Tillman–Pit collection specifications

Parameter	Tillman–Pit
Latitude	39° 46′11.9″
Longitude	−84° 7′25.4″
Bandwidth	230 MHz
Minimum frequency	470 MHz
Maximum frequency	700 GHz
Frequency bin width	181.8 kHz
Samples	1,265

Table 17.4 Minimum description length

Data	North	East	South	West
MDL	201	204	206	219

The CDF and PMF for each of the five distinct signal bands in the East measurement set of Figure 17.9 are depicted in Figures 17.10 and 17.11. In so doing, it becomes evident that, while it can be reasonably assumed that these signals belong to similar distributions, the underlying parameters can vary substantially. As such, now define the new variables $\Theta^s \sim U[\alpha_{\min}, \alpha_{\max}]$ and $\Lambda^s \sim 10^{U[\log_{10}(\gamma_{\min}), \log_{10}(\gamma_{\max})]}$ where the subscripts denote the minimum and maximum of the range for uniform distribution (U). For the shift variable Λ, this arrangement makes intuitive sense because it can be varied from the noise floor to the highest expected signal value uniformly on the log scale. Likewise, Θ controls the power efficiency in each signal band.

To model the possible distributions of these signals, the Gamma scale parameter β is defined to be a function of the two random variables as $\beta = \Lambda^s/\Theta^s$. However, the likelihood function for such a distribution is not mathematically tractable, and therefore we compute the maximum likelihood estimate (MLE) of parameters for the uniform distributions independent of the Gamma distribution. Of course, this approach voids the claim of maximum likelihood on the parameters, but it nonetheless is a reasonable approach that provides desirable features when constructing synthetic PSDs, such as the ability to vary power levels in the signal bands.

To model the signal and noise bands along with the transitions between them in the PSD, a Markov process is used [72]. The simplest Markov process is the two-state model in which the random sequence can assume two discrete values. We define the two possibilities with state vector $\delta = \{s_0, s_1\}$, where $s_0 = 0$ represents the noise state and $s_1 = 1$ the signal state.

The transition matrix **P** is expressed in matrix form (for finite model order) with transition probabilities $Pr[\mathbf{x_P}(m) = s_i \mid \mathbf{x_P}(m-1) = s_j]$, for i,j indicating the possible

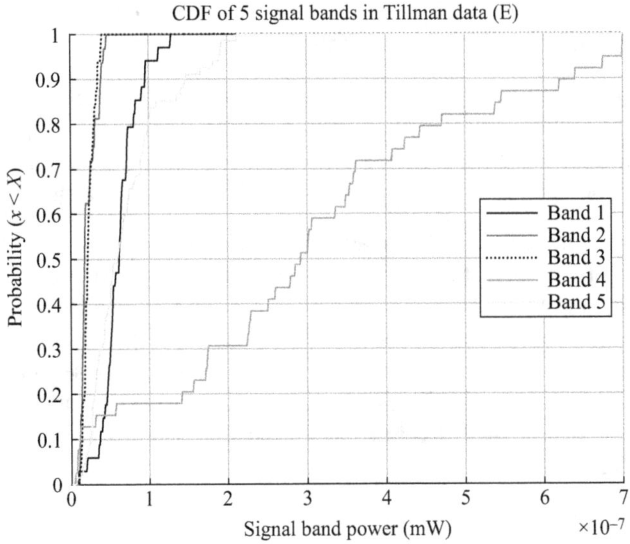

Figure 17.10 CDF for each of the five signal bands in the Tillman–Pitt (East) data collection

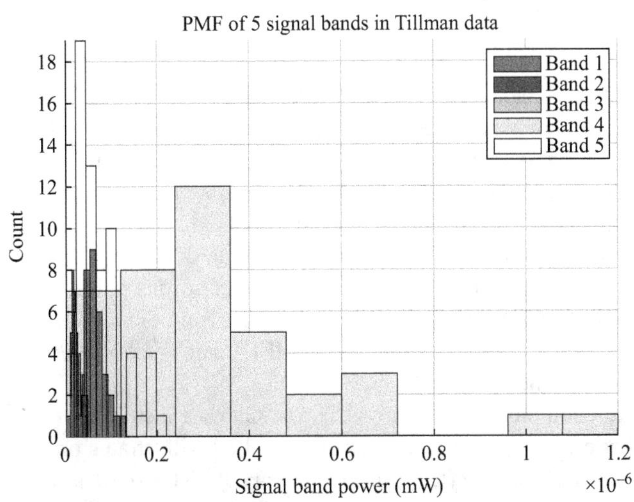

Figure 17.11 PMF for each of the five signal bands in the Tillman–Pitt (East) data collection

states, and $\mathbf{x}_\mathbf{P}(m)$ being the sequence at index m. Here the transition matrix is 2×2 and given by

$$\mathbf{P} = \begin{bmatrix} Pr(s_0|s_0) & Pr(s_1|s_0) \\ Pr(s_0|s_1) & Pr(s_1|s_1) \end{bmatrix}; \tag{17.12}$$

thus we denote a Markov sequence with transition matrix $\mathbf{P}$ and initial state δ_{init} as $\mathbf{x_P} \sim \mathscr{M}(\delta_{\text{init}}, \mathbf{P})$. In this context, each probability corresponds to

- $Pr(s_0|s_0)$, probability of transition from noise bin to adjacent noise bin,
- $Pr(s_0|s_1)$, probability of transition from signal bin to adjacent noise bin,
- $Pr(s_1|s_0)$, probability of transition from noise bin to adjacent signal bin,
- $Pr(s_1|s_1)$, probability of transition from signal bin to adjacent signal bin.

To estimate the state transition matrix from measured data, under an assumption of stationarity (that could be inappropriate in some scenarios), we form a binary sequence using the MDL results such as in Figure 17.9 to assign each frequency bin to either a noise state ($s_0 = 0$) or a signal state ($s_1 = 1$). The estimated transition probabilities are computed by counting the transitions over these sequences.

Once the transition probabilities are estimated, synthetic RFI can be realized via the random placement of RFI centre frequencies and bandwidths by applying the Markov model as

$$\hat{\mathbf{x}} = \begin{cases} (\mathbf{x}_{\hat{\mathbf{P}}} + \mathbf{1}) \odot \hat{\mathbf{f}}^{\text{n}}, & \text{if } \mathbf{x}_{\hat{\mathbf{P}}} = 0 \ \forall m \\ \mathbf{x}_{\hat{\mathbf{P}}} \odot \hat{\mathbf{f}}^{\text{s}}, & \text{for each signal band} \end{cases} \tag{17.13}$$

where $\hat{f}^{\text{n}} \sim G_{\alpha^{\text{n}},\beta^{\text{n}},\gamma^{\text{n}}}(x)$ and $\hat{f}^{\text{s}} \sim G_{\Theta^{\text{s}},\beta^{\text{s}},\Lambda^{\text{s}}}(x)$ with $\Theta^{\text{s}} \sim U[\alpha_{\min}, \alpha_{\max}]$ and $\Lambda^{\text{s}} \sim U[\log_{10}(\gamma_{\min}), \log_{10}(\gamma_{\max})]$ for $\beta = \Lambda^{\text{s}}/\Theta^{\text{s}}$. Note that each signal band is an independent draw of the random variable, which allows for a different mean-amplitude and variance.

Given measured data, this process of (a) computing the MDL, (b) estimating the parameters of the Gamma distribution, and (c) estimating the Markov Chain transition probabilities, allows synthetic RFI data to be generated for use in waveform design. We assume the collected data is prototypical and generalizable to other geographies. However, for the case(s) where this assumption fails, this technique to generate synthetic data is still feasible if the PSD can be estimated. Given a measurement such as Figure 17.9, we can also compute the dimension N of the noise subspace by subtracting the MDL for a particular measurement set from the total dimension. After separating the signal and noise subspaces, MLEs can be computed for the shape, scale, and shift parameters of (17.11).

Using the computed transition matrix and Gamma distribution parameter estimates, synthetic PSD can be generated. Figure 17.12 illustrates one such example derived from the *Tillman–Pit* (East) measurement data. This spectrum shows multiple practical attributes: realistic noise floor, dynamic range, wide noise bands, and multiple randomly placed signal bands. It also contains variation in the mean-amplitude for the signal bands as well as efficient usage of the signal band power.

This practical interference and noise model can be used to evaluate the SINR degradation of waveform design techniques as a function of the complexity of the spectral environment. In so doing, it allows for realistic conclusions to be drawn regarding the effectiveness of a particular waveform design strategy and the impact that practical constraints have upon SINR, thereby facilitating pragmatic waveform selection in dynamic environments.

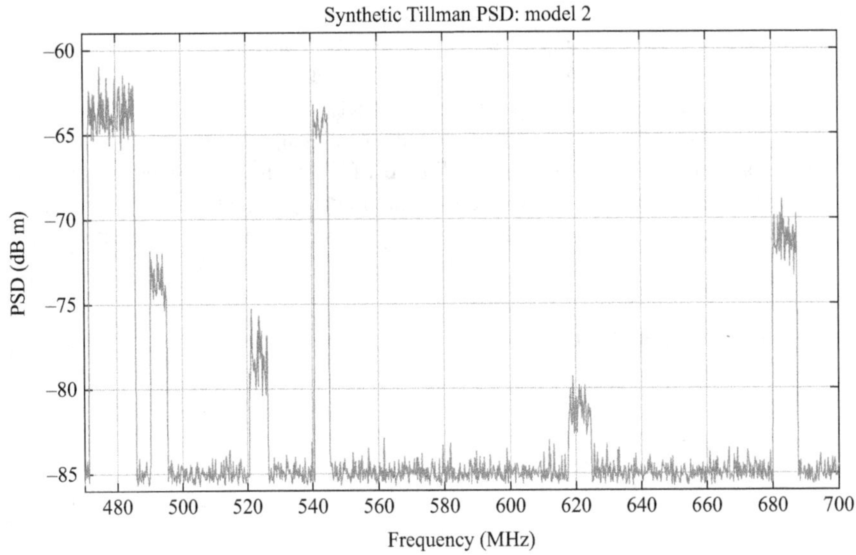

Figure 17.12 Example synthetic PSD derived from the Tillman–Pit (East) data. Reproduced, with permission, from Reference [29]

17.3 Modelling SINR loss for spectral coexistence (radar centric)

In light of the above discussion on modelling the crowed spectral environment in which the radar must operate, now consider the design of coded radar waveforms according to SINR-based optimization with and without constraints. Here the radar is presumed to be the primary user, and all other RF sources are interference, some of which may be secondary users attempting to share spectrum.

Denote the coded transmit waveform as $\mathbf{s} \in \mathbb{C}^{M\times 1}$, such that the received signal resulting from scattering by a unit magnitude point target is

$$\mathbf{y} = \mathbf{s} + \mathbf{n}, \tag{17.14}$$

where $\mathbf{n} \in \mathbb{C}^{M\times 1}$ collectively represents the interference and additive noise. Projecting the received signal onto the transmit signal subspace, otherwise known as matched filtering, then yields

$$\mathbf{s}^H\mathbf{y} = \mathbf{s}^H\mathbf{s} + \mathbf{s}^H\mathbf{n}. \tag{17.15}$$

Based on the average power constraint $\|\mathbf{s}\|_2^2 = 1$, which is equivalently a constraint on pulse energy, the SINR is

$$\text{SINR} = \frac{\mathbf{s}^H\mathbf{s}\mathbf{s}^H\mathbf{s}}{\text{E}\{\mathbf{s}^H\mathbf{n}\mathbf{n}^H\mathbf{s}\}} = \frac{1}{\mathbf{s}^H\mathbf{K}\mathbf{s}}, \tag{17.16}$$

where $\mathbf{K} \in \mathbb{C}^{M\times M}$ is the interference-plus-noise covariance matrix that is positive definite Hermitian. Unless otherwise stated, we assume that this covariance is known. Note that the formulation in (17.16) is not the optimal SINR for joint transmit waveform and receive filter design studied extensively in [73]. This interpretation is intentional to permit investigation of how waveform design constraints affect SINR.

The goal here is to maximize the SINR metric (17.16) in terms of coded waveform $\mathbf{s}$ for fixed transmit energy. Observing that the objective function in (17.16) is fully encapsulated in the denominator, this problem can be equivalently stated as

$$\min_{\mathbf{s}} \left[\mathbf{s}^H\mathbf{K}\mathbf{s}\right] \text{ such that } \|\mathbf{s}\|_2^2 = 1, \tag{17.17}$$

which is a formulation that appears frequently in the literature.

The objective function in (17.17) is easily solved for arbitrary $\mathbf{s}$ using the Lagrange method [74]. Consider the eigendecomposition

$$\mathbf{K} = \mathbf{V}^H\mathbf{L}\mathbf{V}, \tag{17.18}$$

where $\mathbf{V} = [\mathbf{v}_1 \cdots \mathbf{v}_M]$ comprises the eigenvectors and $\mathbf{L} = \text{diag}\{\lambda_1 \cdots \lambda_M\}$ are the eigenvalues with $\lambda_1 \leq \lambda_2 \leq \cdots \leq \lambda_M$, such that maximizing SINR implies use of the smallest number of eigenvectors from the noise-only subspace to construct the waveform. The waveform is thus represented as

$$\mathbf{s} = \sum_{k=1}^{N} a_k\, \mathbf{v}_k = \overline{\mathbf{V}}\mathbf{a}, \tag{17.19}$$

where the scalar a_k is a complex weighting of the kth eigenvector that can be collected into the vector $\mathbf{a} \in \mathbb{C}^{N\times 1}$, and $\overline{\mathbf{V}} \in \mathbb{C}^{M\times N}$ contains the eigenvectors corresponding to the N smallest eigenvalues that comprise the support of $\mathbf{s}$. Note that, per the discussion in the previous section regarding the use of the MDL to estimate the dimensionality of the signal (in this case interference) subspace, N is thus the remaining dimensionality corresponding to the noise-only subspace.

While basis selection is non-unique, it is intuitively understood that the application of constraints onto the waveform design process forces this support to span into higher eigendimensions, which subsequently impacts the resulting SINR. Specifically, the SINR stated in (17.16) that arises from matched filtering the point target response induced by a fixed energy waveform (i.e. $\|\mathbf{s}\|_2^2 = 1$) can likewise be expressed using the eigendecomposition representation from (17.18) and (17.19) as

$$\text{SINR} = \frac{1}{\mathbf{s}^H\mathbf{K}\mathbf{s}}, \tag{17.20}$$

$$= \frac{1}{(\overline{\mathbf{V}}\mathbf{a})^H\mathbf{V}\mathbf{L}\mathbf{V}^H(\overline{\mathbf{V}}\mathbf{a})}, \tag{17.21}$$

$$= \frac{1}{\mathbf{a}^H\overline{\mathbf{L}}\mathbf{a}}, \tag{17.22}$$

$$= \frac{1}{\sum_{k=1}^{N\leq M} |a_k|^2\lambda_k}, \tag{17.23}$$

where $\overline{\mathbf{L}} = \text{diag}\{\lambda_1 \cdots \lambda_N\}$ contains the N smallest eigenvalues.

For known eigenvalues, the determination of SINR simply becomes a matter of selecting the weights, where the 'best' SINR corresponds to $a_1 = 1$ and all others as zero, which is otherwise known as the Rayleigh solution or the minimum eigenvector waveform [75,76]. In contrast, the 'worst' SINR is achieved by using only the Nth eigenvalue via $a_N = 1$ and setting all others to zero. An average can likewise be achieved by setting the N weights to be equal.

Of course, this 'best' solution does not generally produce a desirable radar waveform, since construction from eigenvectors only addresses SINR and offers no guarantees with respect to favourable ambiguity function properties or efficient use of transmit power. In fact, such constructs typically produce terrible radar waveforms in every other sense. Thus, imposing additional constraints is necessary to achieve a useful coded waveform structure.

17.3.1 Cumulative modulus

Finite energy is an obligatory requirement of any waveform. As discussed above, for a coded waveform, this constraint takes the form of the squared L_2 norm as ($\|\mathbf{s}\|_2^2 = 1$), where the unity value is chosen arbitrarily and any value can be used without loss of generality. It is worth noting that for constant modulus, the energy constraint is redundant for a fixed code length. However, when the modulus is allowed to vary, such as to facilitate greater design freedom, it is necessary to impose the energy constraint.

The constant modulus property is rather useful because it maximizes energy-on-target (and thus the numerator of SINR) and, in conjunction with requisite spectral containment, also permits the transmitter power amplifier to operate in saturation (see Chapter 6). However, noting the typical need for a radar to transmit at high power due to propagation and scattering losses, the use of linearizing transmitter architectures such as predistortion and LINC (linear amplification using nonlinear components) that are discussed in Chapters 6, 11, and 12 can facilitate the physical realization of waveforms that possess some degree of amplitude modulation (AM). In so doing, such waveforms can exploit the greater design freedom that results to better achieve sidelobe constraints as well as to avoid interference (i.e. minimize the denominator of SINR).

When allowing a waveform to possess AM characteristics, the cumulative modulus (CM) metric can be used to describe the total point-wise difference between each sample in the waveform and a normalized peak amplitude, described mathematically as

$$C(\mathbf{s}) = \sum_{\ell=1}^{M} \left[1 - |s_\ell|^2\right]^2, \tag{17.24}$$

$$= M - 2\|\mathbf{s}\|_2^2 + \sum_{\ell=1}^{M} |s_\ell|^4, \tag{17.25}$$

where, for $\|\mathbf{s}\|_2^2 = 1$, this quantity can be simplified to

$$C(\mathbf{s}) = M - 2 + \|\mathbf{s}\|_4^4. \tag{17.26}$$

The $M-2$ term in (17.26) is a non-negative constant for code lengths $M \geq 2$, and thus this term can be omitted to realize the simplified function

$$\mathrm{J}_{\mathrm{cm}}(\mathbf{s}) = \|\mathbf{s}\|_4^4. \tag{17.27}$$

Because it is a simplification of (17.24), it can intuitively be understood that $\|\mathbf{s}\|_4^4$ is analogous to minimizing the wasted power provided by a constant power supply when the waveform deviates from constant modulus (i.e. PAPR = 1). It is therefore evident that (17.27) is an alternative formulation to the more well-known PAPR $= \|\mathbf{s}_{\mathrm{peak}}\|^2/\mathbf{s}_{\mathrm{rms}}^2$, where $\mathbf{s}_{\mathrm{peak}}$ and $\mathbf{s}_{\mathrm{rms}}$ are the peak amplitude and root-mean squared (rms) amplitude of **s**, respectively.

Using the technique described in [77], optimization of the CM in (17.27) can be posed as

$$\min_{\mathbf{a}} \|\mathbf{s}\|_4^4 \text{ such that } \|\mathbf{s}\|_2^2 = 1 \text{ and } \mathbf{s} = \sum_{k=1}^{N} a_k \mathbf{v}_k. \tag{17.28}$$

Note that, by imposing the constraint that the waveform reside within a subspace of the eigenvector basis, it is quite possible that a constant modulus solution may not exist.

The optimization problem in (17.28) is non-convex, and thus the solution obtained can be highly dependent on the initialization, without a guarantee of achieving global optimality. However, a convex relaxation of (17.28) may be posed as

$$\min_{\mathbf{A}} \left\{ (\mathrm{vec}(\mathbf{A}))^H \left[\sum_{i=1}^{M} \mathrm{vec}(\hat{\mathbf{V}}_i)\, (\mathrm{vec}(\hat{\mathbf{V}}_i))^H \right] \mathrm{vec}(\mathbf{A}) \right\} \tag{17.29}$$

$$\text{such that } \dim(\mathbf{A})\big[(\mathrm{vec}(\mathbf{I}))^H \mathrm{vec}(\mathbf{A})\big] = 1, \tag{17.30}$$

where $\mathbf{A} = \mathbf{a}\mathbf{a}^H \in \mathbb{C}^{N\times N}$ is positive semi-definite, **I** is the identify matrix, $\hat{\mathbf{V}}_i = \mathbf{v}_i\mathbf{v}_i^H$, and $\dim(\mathbf{A}) = N$. The solution is obtained by determining the vector $\mathbf{a}_{\mathrm{relax}}$ that has the lowest CM. The reader is referred to [77] for the full derivation and implementation strategy.

A useful model to characterize the performance trade-off for CM as a function of eigenbasis dimension N is the Lamẽ curve, otherwise known as a generalized super-ellipse, which takes the form

$$\left|\frac{x - X_{\mathrm{c}}}{a_{\mathrm{L}}}\right|^p + \left|\frac{y - Y_{\mathrm{c}}}{b_{\mathrm{L}}}\right|^q = 1, \tag{17.31}$$

where a_{L} and b_{L} are the minor and major axes, respectively, p and q are parameters to shape the ellipse, and $(X_{\mathrm{c}}, Y_{\mathrm{c}})$ denotes the ellipse centre location. Note that for a curve located at the origin, equal normalized axes, and the powers p and q set to 2, (17.31) reduces to the unit circle.

We seek to fit (17.31) to the expected CM $\mathrm{E}\{\|\mathbf{s}\|_4^4\}$, which is the average solution to (17.28), for a given N. Least-squares estimation is used to determine the parameters p and q based on the solution obtained for (17.28). The parameters x and y are then set to represent the number of eigenvectors used in the design and the CM, respectively.

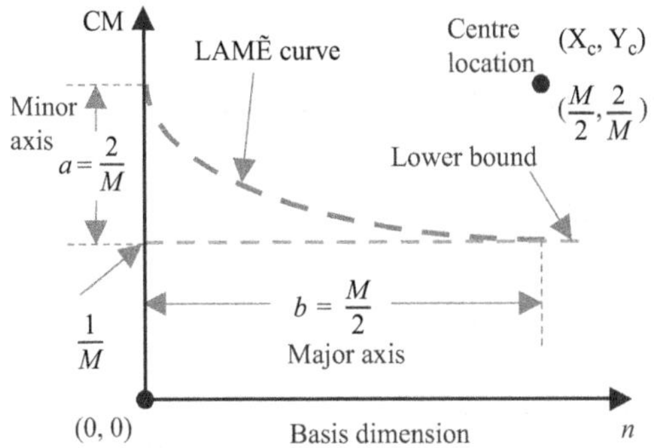

Figure 17.13 Lamẽ model for expected cumulative modulus as a function of the eigenbasis dimension. Reproduced, with permission, from Reference [33]

Solving for y in terms of x subsequently realizes the CM as a function of the eigenbasis dimension, which can be expressed as [78]

$$y(x) = b_{\mathrm{L}} \sqrt[q]{1 - \frac{(x - X_{\mathrm{c}})^p}{a_{\mathrm{L}}^p}} + Y_{\mathrm{c}}. \tag{17.32}$$

The minimum possible solution to (17.28) occurs when the code is constant modulus, thus yielding $\|\mathbf{s}\|_4^4 = 1/M$. In contrast, the condition of $N = 1$ produces the worst performance on average, which is found to be $\mathrm{E}\{\|\mathbf{s}_{N=1}\|_4^4\} = 2/M$, that coincidentally is also the expected value of the fourth-moment of a complex Gaussian random vector with variance $1/M$. Hence, the dynamic range of possible responses for the CM cost function is $(2/M) - (1/M) = (1/M)$.

It is therefore convenient to set $a_{\mathrm{L}} = 1/M$ for the minor axis, such that the major axis b_{L} is the number of eigenvectors required to achieve the optimal solution. Given that a length M code has $2M$ DOF—M for amplitude and M for phase—achieving constant modulus incurs the use of only half of them. Therefore, as depicted in Figure 17.13, only $b_{\mathrm{L}} = M/2$ basis vectors would be required.

Substituting the above parameters into (17.32), the Lamẽ performance model for CM as a function of eigenbasis dimension can be expressed as

$$C \cong \frac{2}{M} - \frac{1}{M} \sqrt[q]{1 - \frac{((M/2) - N)^p}{((M/2) - 1)^p}}, \tag{17.33}$$

with N valid from 1 to $M/2$. Conversely, the eigenbasis dimension required when a modulus constraint is imposed can be written as

$$N(C) \cong \frac{M}{2} - \sqrt[p]{1 - (2 - MC)^q} \left(\frac{M}{2} - 1 \right). \tag{17.34}$$

These models provide insights into the behaviour of the objective function in (17.28).

It is also useful to define a performance model for SINR degradation as a function of N for the MP covariance matrix definition. Based on a power law model, which is attractive due to its simplicity [79], we can therefore write

$$\text{SINR}_{\text{MP}} \cong \frac{1}{\sum_{k=1}^{N\leq M} \frac{1}{k}\mathscr{L}(k)}, \tag{17.35}$$

where $\mathscr{L}(k)$ is the expected eigenvalue, the distribution of which is described in (17.3).

Combining (17.33) and (17.35), using the approximation in (17.34), subsequently permits the development of a performance model for SINR as a function of the CM constraint that takes the form

$$\text{SINR}_{\text{MP}}(C) \cong \frac{1}{\sum_{k=1}^{\lfloor N(C)\rfloor} (1/\sqrt{k})\mathscr{L}(k)} \tag{17.36}$$

where $\lfloor\cdot\rfloor$ is the floor operator. This model implies the rough impact to SINR when applying a CM constraint to waveform design.

17.3.1.1 Numerical results

To validate the CM and SINR models, computer simulations are used to show the effect of the waveform eigenbasis dimension on the modulus and SINR relative to optimal. The covariance matrix is assumed known, and the eigenspectra are approximated with the MP distribution, noting that the MP model is a pessimistic assumption on the covariance statistics. Radar transmit and noise powers are held constant throughout. The MATLAB® Optimization Toolbox [80] and the convex solver CVX [81,82] were used to evaluate 100 Monte Carlo trials. Specifically, we numerically solve for (17.28) and (17.29) and compare the results against the models in (17.33) and (17.35), respectively. Figure 17.14 illustrates this comparison along with the lower bound of $1/M$ and the maximum and minimum computed modulus.

Figure 17.15 also shows the normalized average SINR degradation from optimal. The MP power law model from (17.35) is overlaid and shows good agreement for a highly simplified model when compared against the computationally expensive solution of (17.28).

Using (17.33) and (17.35) in conjunction enables the posing of SINR performance in terms of a modulus constraint on waveform design. Alternatively, given the basis dimension of the noise-only subspace (i.e. avoiding the spectrum associated with interference), one could estimate modulus performance and SINR degradation. More important than the performance models that predict SINR as a function of the eigenbasis, though, is the formulation of a unified model for SINR, CM, and the eigenbasis. Thus, computer simulations can be performed to solve (17.28) and subsequently determine the SINR for a given eigenbasis dimension, which can be compared against the model in (17.36). This comparison is illustrated in Figure 17.16. It is observed that the model predicts the outcome to within a few dB of SINR for a given CM.

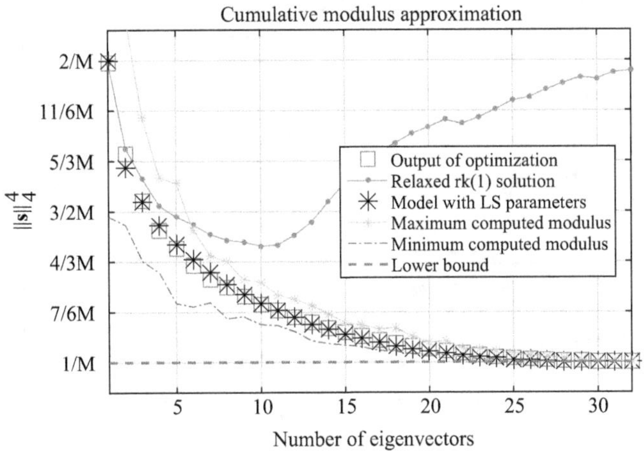

Figure 17.14 Expected cumulative modulus with model comparison and minimum and maximum modulus. Reproduced, with permission, from Reference [33]

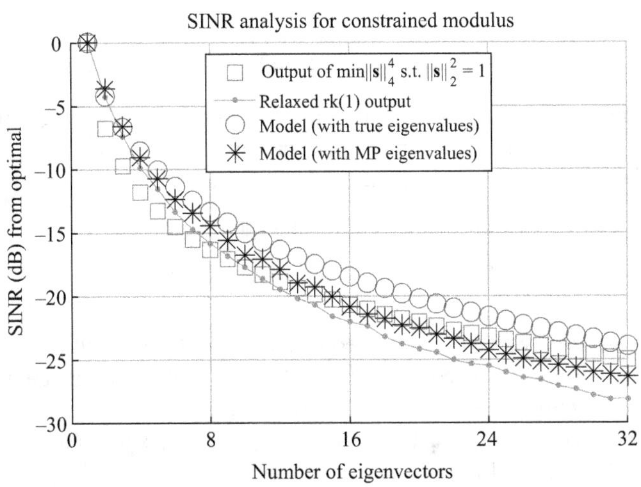

Figure 17.15 Expected SINR with model comparison. Reproduced, with permission, from Reference [33]

17.3.1.2 Application to waveform design

In [53], several techniques are discussed for the design of coded waveforms that maximize SINR while applying PAPR and finite energy constraints for known and unknown normalized Doppler. To illustrate the benefit of the performance prediction

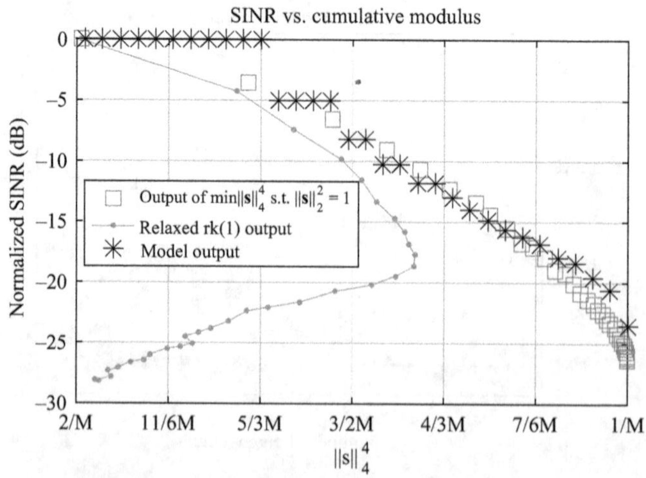

Figure 17.16 SINR vs. CM model compared with the output SINR from (17.28). Reproduced, with permission, from Reference [33]

model in (17.36), a problem similar to the known-Doppler case in [53] is examined. This problem can be stated as

$$\min_{\mathbf{s}} \left(\mathbf{s}^H \mathbf{K}_{\mathrm{MP}}\, \mathbf{s}\right) \quad \text{such that PAPR} \leq \varepsilon \text{ and } \|\mathbf{s}\|_2^2 = 1, \tag{17.37}$$

where ε is the highest acceptable PAPR. It is understood that minimizing PAPR is analogous to minimizing the deviation from constant modulus, and thus a comparable problem employing the CM constraint can be stated as

$$\min_{\mathbf{s}} \left(\mathbf{s}^H \mathbf{K}_{\mathrm{MP}}\, \mathbf{s}\right) \quad \text{such that CM} \leq \varepsilon_{\mathrm{cm}} \text{ and } \|\mathbf{s}\|_2^2 = 1, \tag{17.38}$$

where $\varepsilon_{\mathrm{cm}}$ is allowed to vary from $2/M$ (unconstrained CM) to $1/M$ (constant modulus or fully constrained CM).

Monte Carlo analysis was performed to solve (17.38) for multiple instantiations of the interference-plus-noise covariance matrix $\mathbf{K}_{\mathrm{MP}}$. In this case, no requirement was made that the waveform be comprised of a subset of the eigenbasis, thus permitting a search over the entire vector space. Figure 17.17 compares the Monte Carlo results of (17.38) with the model in (17.36). It is observed that the model provides a good estimation of SINR degradation when the modulus is constrained.

17.3.2 Integrated sidelobe constraints

The next constraint to be considered is based on the waveform integrated sidelobe level (ISL). As a design criterion, ISL has been studied in great depth [21,23,49–51,60] due to its important role in multiple radar functions, including reducing the number of false alarms and improving the resolvability of multiple proximate targets.

The literature mentions many design methodologies for limiting sidelobe levels in coded waveforms, including a periodic autocorrelation function technique possessing

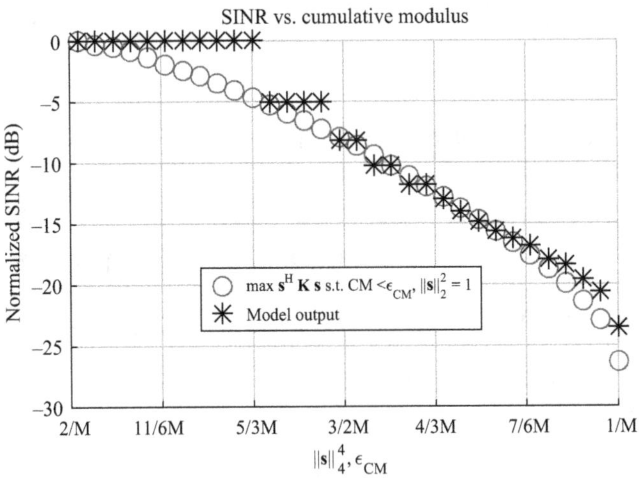

Figure 17.17 Simulation results of 100 Monte Carlo trials for the CM and SINR prediction based on the MP interference model compared to constrained coded waveform optimization via (17.38). Reproduced, with permission, from Reference [33]

theoretically zero sidelobes [83], the cyclic technique PeCAN [50,84], a competing technique denoted as signal waveform's optimal under restriction design (SWORD) [85], and others [86,87]. While such methods propose specific code designs, it is also useful to consider general performance prediction for an arbitrary design methodology that poses SINR as a function of ISL.

For the ISL optimization problem, we begin with the auto-correlation sequence $r_m = \sum_{q=1}^{M} s_m s_{m-q}^*$ for an M length code, where m is the discrete lag index and $(\cdot)^*$ is complex conjugation. The ISL is then defined as

$$\mathrm{J}_{\mathrm{ISL}} = \sum_{m=1}^{M-1} \left| \sum_{q=1}^{M} s_m s_{m-q}^* \right|^2, \tag{17.39}$$

exploiting the complex symmetry of the autocorrelation and assuming the main lobe has a width of 1, which is a reasonable simplification for most codes (note that the continuous definition of ISL instead involves integration over the sidelobes and division by the integral of the main lobe). The notation in (17.39) can be compacted using $\mathbf{E}_m \in \mathbb{R}^{M \times M}$, which has ones on the mth upper super-diagonal and otherwise contains zeros. For example, the $m = 1$ case takes the form

$$\mathbf{E}_1 = \begin{bmatrix} 0 & 1 & 0 & \cdots & 0 \\ 0 & 0 & 1 & \cdots & 0 \\ \vdots & \vdots & \vdots & \ddots & \vdots \\ 0 & 0 & 0 & \cdots & 1 \\ 0 & 0 & 0 & \cdots & 0 \end{bmatrix}. \tag{17.40}$$

Thus, (17.39) can be expressed as $\mathrm{J_{ISL}} = \sum_{m=1}^{M-1} |\mathbf{s}^H \mathbf{E}_m \mathbf{s}|^2$, which allows the ISL minimization problem to be posed as

$$\min_{\mathbf{a}} \sum_{m=1}^{M-1} |\mathbf{s}^H \mathbf{E}_m \, \mathbf{s}|^2 \;\; \text{such that} \;\; \|\mathbf{s}\|_2^2 = 1 \;\; \text{and} \;\; \mathbf{s} = \sum_{n=1}^{N} a_n \mathbf{v}_n, \tag{17.41}$$

which is subject to the finite energy constraint, and the coded waveform is constrained to be a linear combination of the eigenvectors of the interference-plus-noise covariance matrix **K**. While (17.41) does not explicitly maximize SINR, this latter constraint does force the solution to reside in the noise-only subspace, thereby avoiding the dominant interference.

Alternatively, SINR optimization based on the denominator of (17.20) can be posed such that the ISL does exceed $\varepsilon_{\mathrm{ISL}}$ via

$$\min_{\mathbf{s}} \left(\mathbf{s}^H \mathbf{K} \mathbf{s}\right) \;\; \text{such that} \;\; \|\mathbf{s}\|_2^2 = 1 \;\; \text{and} \;\; \mathrm{J_{ISL}} \le \varepsilon_{\mathrm{ISL}}, \tag{17.42}$$

which does not require the restriction that the waveform be composed of the eigenvectors of **K**. Removal of the ISL constraint in (17.42) reduces the problem to that specified in (17.17), the solution of which is simply the eigenvector corresponding to the minimum eigenvalue [75,76]. The purpose of performance prediction in this context is therefore the modelling of the outcome of (17.42) as the ISL constraint is tightened.

17.3.2.1 Expected ISL performance model

Starting with the definition of ISL in (17.41), and recalling that **V** contains the eigenvectors of the interference-plus-noise matrix **K**, the expected ISL can be derived as

$$\mathrm{E}\{\mathrm{J_{ISL}}(M, N, \mathbf{V}, \mathbf{a})\} = \mathrm{E}\left\{\sum_{m=1}^{M-1} |\mathbf{s}^H \mathbf{E}_m \mathbf{s}|^2\right\} \tag{17.43}$$

$$= \mathrm{E}\left\{\sum_{m=1}^{M-1} (\mathbf{s}^H \mathbf{E}_m \mathbf{s})(\mathbf{s}^H \mathbf{E}_m^T \mathbf{s})\right\} \tag{17.44}$$

$$= \mathrm{E}\left\{\sum_{m=1}^{M-1} \left(\sum_{i=1}^{N} a_i^* \mathbf{v}_i^H \mathbf{E}_m \sum_{j=1}^{N} a_j \mathbf{v}_j\right) \left(\sum_{k=1}^{N} a_k^* \mathbf{v}_k^H \mathbf{E}_m^T \sum_{l=1}^{N} a_l \mathbf{v}_l\right)\right\} \tag{17.45}$$

$$= \sum_{i=1}^{N} \sum_{j=1}^{N} \sum_{k=1}^{N} \sum_{l=1}^{N} a_i^* a_j a_k^* a_l \sum_{m=1}^{M-1} \mathrm{E}\{\mathbf{v}_i^H \mathbf{E}_m \mathbf{v}_j \mathbf{v}_k^H \mathbf{E}_m^T \mathbf{v}_l\}. \tag{17.46}$$

To evaluate the expectation, we describe the pth element of each eigenvector **v** as a random variable $v_{p,i} \sim \mathbb{CN}(0, (1/M)) \; \forall \, p, (i, j, k, l)$ whose entries are independent and identically distributed. After manipulation, and under the assumption that M is large ($M \geq 32$ is reasonable), it is found that (17.43)–(17.46) simplifies to [77]

$$\mathrm{E}\{\mathrm{J}_{\mathrm{ISL}}(M,N,\mathbf{a})\} = \frac{1}{2}\sum_{i=1}^{N} |a_i|^4 - \frac{|a_i|}{M}, \tag{17.47}$$

for $\|\mathbf{a}\|_2^2 = 1$. Prediction of ISL performance therefore becomes a matter of weight selection as a function of the eigenbasis dimension N.

We approach modelling the weights from an intuitive perspective by designing the PSD to be as flat as possible for a given eigenbasis dimension, where the eigenvalues are an approximation of the PSD [68]. To flatten the PSD, a reciprocal model of the MP distributed eigenvalues is therefore used, which is modelled via

$$a_{\mathrm{inv},i} = \frac{1}{\hat{\lambda}_i} \quad \text{for } i = 1,\ldots,N \tag{17.48}$$

where $\hat{\lambda} = [\hat{\lambda}_1 \ldots \hat{\lambda}_N]^T$ are the N sorted expected eigenvalues of the interference-plus-noise covariance matrix in (17.2). Thus, the expected ISL becomes

$$\mathrm{E}\{\mathrm{J}_{\mathrm{ISL}}(M,N,\hat{\lambda})\} = \frac{1}{2}\sum_{i=1}^{N} \frac{1}{|\hat{\lambda}_i|^4} - \frac{1}{M|\hat{\lambda}_i|}, \tag{17.49}$$

where $\|\hat{\lambda}\|_2^2 = 1$. Figure 17.18 illustrates the expected ISL from (17.49) for $M = 64$ as a function of the eigenbasis dimension used to the construct the coded waveform.

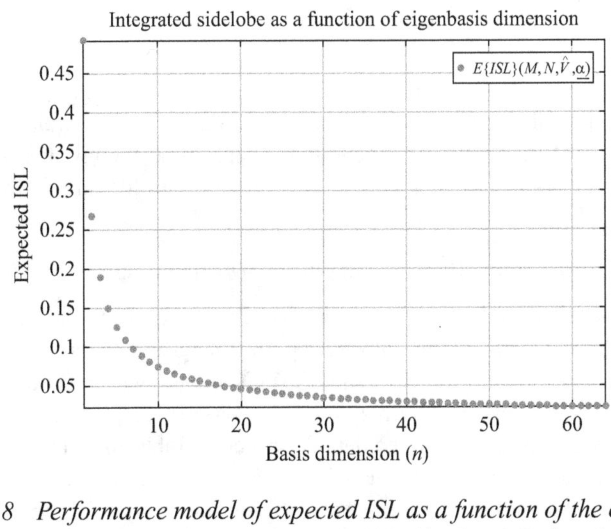

Figure 17.18 Performance model of expected ISL as a function of the eigenbasis dimension of the waveform per (17.49) when MP distributed eigenvalues are employed

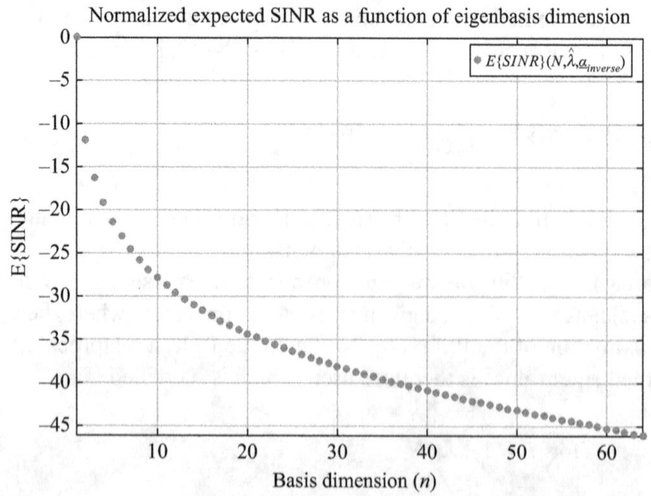

Figure 17.19 Performance model of expected SINR as a function of waveform basis dimension per (17.50) when MP distributed eigenvalues are used

17.3.2.2 Expected SINR performance model

It is useful to relate the expected ISL to expected SINR performance since both are functionally dependent on the weight vector $\mathbf{a}$ via (17.47) and (17.20), respectively. Setting $\mathbf{a} = \mathbf{a}_{\text{inv}}$ from (17.48) and using the MP interference-plus-noise covariance model from (17.2), the expected SINR can be expressed as

$$\mathrm{E}\{\mathrm{SINR}(N, \hat{\lambda}, \mathbf{a}_{\text{inv}})\} = \frac{1}{\sum_{n=1}^{N} |a_{\text{inv},n}|^2 \hat{\lambda}_n} = \frac{1}{\sum_{n=1}^{N} \frac{1}{\hat{\lambda}_n}}. \tag{17.50}$$

Figure 17.19 plots this value for $M = 64$ as a function of the eigenbasis dimension. This result implies that, given the ISL constraint and the inverse weight vector $\mathbf{a}_{\text{inv}}$, the expected SINR is linearly dependent on the basis dimension N. For large M, this relationship is analogous to the amount of bandwidth used for waveform design.

17.3.3 Simulation results and comparison to measured data

For an $M = 64$ length code and the interference-plus-noise covariance model defined in (17.2), 25 Monte Carlo simulations were performed using different realizations of $\mathbf{K}_{\text{MP}}$. In each case, the optimization in (17.41) is solved, incrementing through the entire eigenbasis dimension. Optimization is likewise performed for the covariance matrix estimated from the measured data described in Table 17.2. Figures 17.20 and 17.21 compare these results to the models in (17.49) and (17.50), respectively.

Figure 17.20 shows good agreement between the ISL performance model and the Monte Carlo trials using both simulated and measured data. It is observed that the model deviates somewhat from the solution to (17.41) when N approaches M.

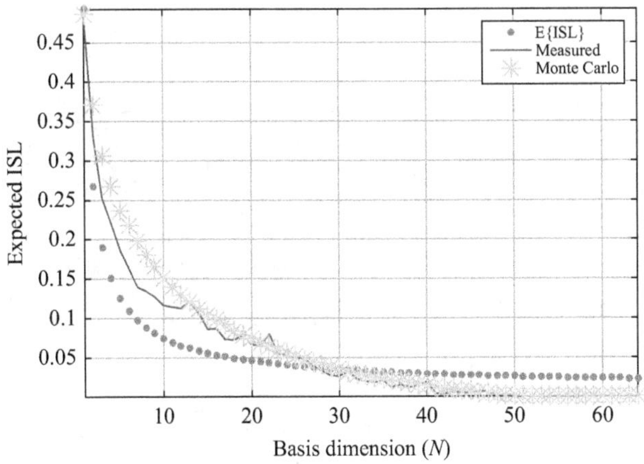

Figure 17.20 ISL model in (17.49) as a function of eigenbasis dimension compared to the outcome of (17.41) for Monte Carlo trials based on covariance $\mathbf{K}_{MP}$ and the covariance from measured data

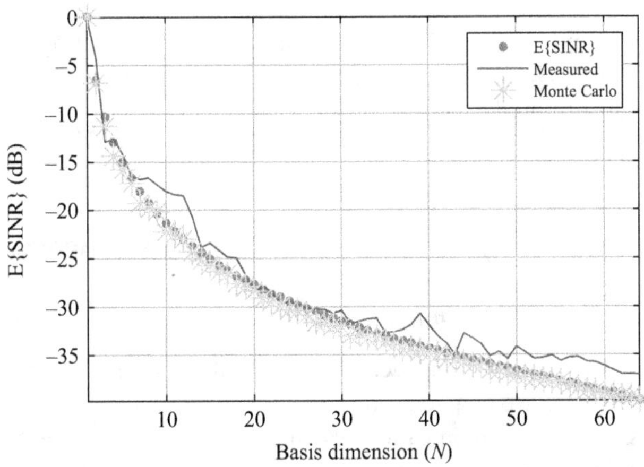

Figure 17.21 SINR model in (17.50) as a function of eigenbasis dimension compared to the outcome of (17.41) for Monte Carlo trials based on covariance $\mathbf{K}_{MP}$) and the covariance from measured data

However, as code length M increases, the model will drive the expected ISL closer to zero, thereby reducing this error.

Figure 17.21 likewise shows good agreement between the SINR performance model and the Monte Carlo trials. Given the direct linkage between N and the frequency support of the waveform, this result confirms the expectation that the primary driver to reduce sidelobes is simply the inclusion of greater spectral content (or in this case, a larger basis dimension).

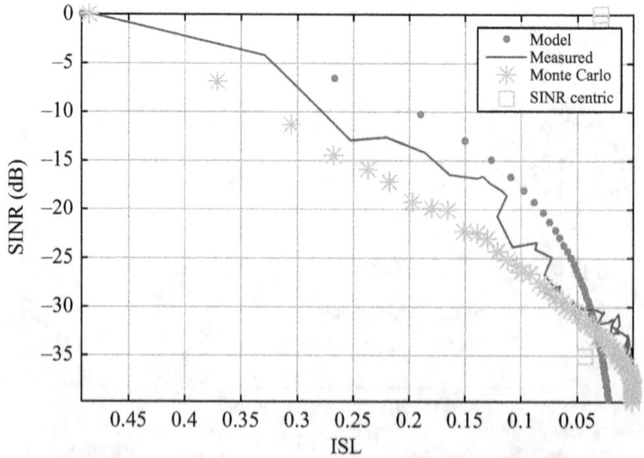

Figure 17.22 Normalized SINR degradation relative to optimal as a function of the ISL constraint on the waveform design

Lastly, we complete the performance model analysis by relating the ISL model in (17.49) to the SINR model in (17.50). Figure 17.22 shows that this modelled relationship is able to predict, within a reasonable error, the SINR degradation as a function of ISL.

17.4 Alternative metrics for spectrally crowded engagements

17.4.1 Maximizing information

Where Sections 17.2 and 17.3 considered waveform design in the context of the radar being the primary user, we now focus on scenarios in which the radar is secondary to other RF users such as communication systems. This arrangement clearly limits the ability of the radar to transmit in certain frequency bands, both to avoid interfering with primary users as well as to avoid bands that have a particularly poor SINR (such occurrences are not necessarily synonymous). Of course, the radar is still required to glean as much information about the illuminated environment as possible, despite the limited available spectrum.

If one aims to maximize the captured information about an illuminated scene, such as to both locate and identify (possibly extended) targets of interest, optimizing a radar waveform to achieve maximal SINR may not always be the most appropriate choice. Indeed, Woodward recognized that [88, p. 62] *there is no general theorem that maximum output SNR insures maximum gain of information*. For example, in the case of extended targets, which present frequency-dependent resonant behaviour, a waveform optimized for SINR may not have the spread of spectral components which would best enable target identification. In fact, for congested and contested environments, it is possible that noise and signal-dependent interference (possibly jamming) could force the waveform into bands that actually inhibit the ability to classify targets in the scene. Thus the motivation for this section is to examine a performance metric that insures maximal gain of information.

The application of information theory to radar was first investigated by Woodward and Davies [88–91], in which they formulated the *a posteriori* probability distribution of measured signal parameters in a radar receiver, such as range, given *a priori* knowledge. It was much later though that Bell [92,93] first applied information theory to waveform design by making use of mutual information (MI). More recently, both SINR and MI were used by Romero and Goodman [94] to design waveforms in the presence of signal-dependent interference, and Bică and Koivunen [95] employed MI to design novel multi-carrier waveforms.

The MI $I(\mathbf{z}, \mathbf{h})$ between observation vector $\mathbf{z}$ (containing interference and noise) and the parameters characterizing the object(s) or scene $\mathbf{h}$ is defined to be the amount of information that $\mathbf{z}$ provides about $\mathbf{h}$. For any decision rule that assigns $\mathbf{h}$ onto a partition based on the observation $\mathbf{z}$, MI has the property that the maximum number of classes that could be identified with an arbitrarily small probability of error p_e is [96]

$$p_\mathrm{e} = \lfloor \exp\{I(\mathbf{z}, \mathbf{h})\} \rfloor. \tag{17.51}$$

Hence, a larger I is necessary, but not sufficient, to identify targets with a larger number of classes.

To specify the MI metric for radar waveforms, we can write the received signal as

$$z(t) = h_\mathrm{t}(t) * x(t) + h_\mathrm{c}(t) * x(t) + r_\mathrm{i}(t) + n(t), \tag{17.52}$$

where h_t and h_c are the impulse responses of the target and clutter, respectively, $x(t)$ is the transmitted waveform, r_i is the interference signal, $n(t)$ is thermal noise, and $*$ is the convolution operator. The terms r_i and h_c could also be considered to be due to active interference. The PSD of the received signal is, in terms of the separate radar and interference/clutter/noise components, therefore

$$S_\mathrm{z}(f) = S_\mathrm{rad}(f) + S_\mathrm{i+c+n}(f) \tag{17.53}$$

$$= |H_\mathrm{t}(f)|^2 |X(f)|^2 + |H_\mathrm{c}(f)|^2 |X(f)|^2 + S_\mathrm{i}(f) + S_\mathrm{n}(f), \tag{17.54}$$

where H_t and H_c are the transfer functions of the target and clutter, respectively. For extended targets (or clutter), these can be described by [97]

$$H_\mathrm{t}(f) = \sum_{m=1}^{M} A_m \left(j \frac{f}{f_\mathrm{c}} \right)^{a_m} \exp\left(j \frac{4\pi f}{c} r_m \right), \tag{17.55}$$

where f_c is the centre frequency, c is the speed of light, M is the number of scatterers in the target (located at different range cells), and A_m, a_m, and r_m are the mth scatterer parameters. For the signal model in (17.52) and (17.54), it can be shown that the MI between $z(t)$ and $h_t(t)$ given that $x(t)$ was transmitted is given by [92, pp. 192–196], [93]

$$I(z(t), h_t(t); x(t)) = T_{obs} \int_B \log(1 + \Psi(f))\, df \tag{17.56}$$

$$= T_{obs} \int_B \log\left(1 + \frac{|H_t(f)|^2 |X(f)|^2}{S_{i+c+n}(f)}\right) df. \tag{17.57}$$

where $\Psi(f) = S_{rad}(f)/S_{i+c+n}(f)$ is the SINR spectral density, T_{obs} is the observation time, and B is the bandwidth, with $H_t(f)$ and $X(f)$ the Fourier transforms of $h_t(t)$ and $x(t)$, respectively.

In order to discretize (17.57) for numerical calculation, we first assume that the observation time is much greater than the pulse width and target impulse response, i.e. $T_{obs} \gg T_p, T_h$. We also assume that the sampling frequency is determined by the passband bandwidth, or $f_s = B$, with $K = T_{obs} f_s$ discrete time samples (note that a finite pulsewidth corresponds to a theoretically infinite bandwidth). The discrete-time expression of MI is thus

$$I(z(t), h_t(t); x(t)) = \frac{1}{K} \sum_{k=0}^{K-1} \log\left(1 + \frac{|H_t(k)|^2 |X(k)|^2}{S_i(k) + \sigma^2}\right). \tag{17.58}$$

For convenience, we shall also use $x_F(k) = X(k)$ to denote the discretized Fourier transform of the radar waveform.

Based on (17.57) and (17.58), it is easy to conceive of a scenario in which SINR is high yet MI is low, because SINR only needs to be large in a narrow band that could convey little information. This arrangement is further exacerbated if the narrow band of high SINR falls outside of the peak response of the targets of interest, which could easily occur in a congested and contested environment. Indeed, an adversary, with knowledge of the resonant characteristics of their platform, might simply deny those parts of the RF spectrum. An adaptive radar that optimizes for SINR could readily be forced to provide minimal information to its operator who may be left unaware of the degraded performance.

17.4.2 Mutual information for cognitive radar

The concept of a cognitive radar, first proposed by Haykin [98], with subsequent work by Guerci [99] amongst others [100–103], can be thought of as [104]

> A radar system that in some sense displays intelligence, adapting its operation and its processing in response to a changing environment and target scene. In comparison to adaptive radar, cognitive radar *learns to adapt* operating parameters as well as processing parameters and may do so over extended time periods.

While a formal definition is yet to be agreed upon, it is clear that this manner of operation involves a *perception–action* cycle [102] and memory (either short or long term), in which the radar learns about its surroundings and adjusts its actions accordingly, making use of any *a priori* knowledge to guide the decision-making process.

Recently, Palamà *et al.* [104,105] have made use of the MI metric in a cognitive radar architecture in which a waveform is designed to fill the gaps in the RF spectrum not employed by primary users (such as communication systems). They consider a scenario in which interference r_i from primary users changes at a regular time period T_PU. For a fixed time slot $kT_\mathrm{PU} \leq t < (k+1)T_\mathrm{PU}$, with N primary users, such an interference signal can be written as

$$r_\mathrm{i} = \sum_{\ell=1}^{N} \gamma_\ell(t) * p_\ell(t) \exp(-j2\pi f_\ell\, t), \tag{17.59}$$

where $p_\ell(t)$ is the signal transmitted by the ℓth primary user, f_ℓ is its centre frequency, and γ_ℓ is the channel impulse response, which can be modelled as a Gaussian stochastic process. Thus, if γ_ℓ has PSD $S_{\gamma_\ell}(f)$ that is assumed constant, the primary user interference signal has PSD

$$S_\mathrm{i}(f) = \sum_{\ell=1}^{N} S_{\gamma_\ell}(f)|P_\ell(f - f_\ell)|^2. \tag{17.60}$$

For simplicity, each waveform $p_\ell(t)$ is assumed to be a sinc function, so that $P_\ell(f)$ is a rectangular window of width B_ℓ.

The waveform optimization problem is constrained such that the total transmitted power is

$$\frac{1}{K}\sum_{k=0}^{K-1} |x_\mathrm{F}(k)|^2 = P_\mathrm{T} \tag{17.61}$$

and that the total power associated with the kth spectral sample must not exceed the maximum value c_k. That is, we wish to solve

$$\tilde{\mathbf{x}}_\mathrm{F} = \arg\max_{\mathbf{x}_\mathrm{F}} \left\{ \frac{1}{K}\sum_{k=1}^{K-1} \log\left(1 + \frac{|H_\mathrm{t}(k)|^2|x_\mathrm{F}(k)|^2}{|H_\mathrm{c}(k)|^2|x_\mathrm{F}(k)|^2 + S_\mathrm{i}(k) + \sigma^2}\right)\right\} \tag{17.62}$$

$$\text{such that } \mathbf{1}^T\mathbf{x}_\mathrm{F} = KP_\mathrm{T} \text{ and } \quad 0 \leqslant x_\mathrm{F}(k) \leqslant c_k \tag{17.63}$$

for $H_\mathrm{c}(k)$ the discretized clutter transfer function and σ^2 the thermal noise power.

Following Palamà *et al.* [104,105], in order to solve the waveform optimization problem (17.63) explicitly it is necessary to satisfy the Karush–Kuhn–Tucker (KKT) conditions. The derivation in the clutter-absent case is given in [104,105]. For the case including clutter, the kth frequency sample of the optimal waveform is given by [104]

$$\tilde{x}_{\mathrm{F},k} = \begin{cases} 0 & A < L_k, \\ -B_k + \sqrt{B_k + \beta_k(A - L_k)} & L_k \leq A \leq U_k, \\ c_k & U_k < A, \end{cases} \tag{17.64}$$

where we have denoted $A = (K\delta)^{-1}$ for δ the Lagrange multiplier, and the bounds and parameters are

$$B_k = \left(\frac{S_{\mathrm{i}}(k) + \sigma^2}{|H_{\mathrm{c}}(k)|^2}\right)\left(\frac{|H_{\mathrm{t}}(k)|^2 + 2|H_{\mathrm{c}}(k)|^2}{|H_{\mathrm{t}}(k)|^2 + |H_{\mathrm{c}}(k)|^2}\right), \tag{17.65a}$$

$$\beta_k = \frac{|H_{\mathrm{t}}(k)|^2(S_{\mathrm{i}}(k) + \sigma^2)}{|H_{\mathrm{c}}(k)|^2(|H_{\mathrm{c}}(k)|^2 + |H_{\mathrm{t}}(k)|^2)}, \tag{17.65b}$$

$$L_k = \frac{S_{\mathrm{i}}(k) + \sigma^2}{|H_{\mathrm{t}}(k)|^2}, \tag{17.65c}$$

$$U_k = \frac{1}{\beta_k}[(c_k + B_k)^2 + B_k]. \tag{17.65d}$$

It remains to solve (17.64) numerically in order to determine the constant level $A = (K\delta)^{-1}$ that satisfies the constraint

$$\frac{1}{K}\sum_{k=0}^{K-1} \tilde{x}_{\mathrm{F},k} = P_{\mathrm{T}}. \tag{17.66}$$

Note that the K-dimensional optimization problem has been reduced to one in a single variable, the Lagrange multiplier δ. The derived result is known as the *water-filling* solution. In the absence of clutter, or $|H_{\mathrm{t}}(k)|^2 = 0$, and without the upper bound c_k, this solution coincides with that found by Bică and Koivunen [95].

17.4.3 Example with primary users

An example target transfer function and interference scenario is shown in Figure 17.23, for which the primary users reside in 30% of the spectrum (baseband representation). The optimal waveform based on (17.62) is depicted in Figure 17.24. In bands where the primary users are operating, the spectral constraints c_k were set to $S_{\mathrm{i}}(k)/5$ to avoid transmitting in those regions. This condition is chosen as an example to illustrate how the method works with a given constraint, rather than simply setting c_k to zero in any bands where primary users are active. The latter could easily result in large parts of the spectrum being unavailable due to the presence of rather distant, low-powered devices. The appropriate value for the constraints c_k may be application specific, or even chosen adaptively (cognitively).

We see that the locally optimal waveform enhances the spectral components for which the target transfer function is high, with lower weights where the interference is high. The effect of the spectral constraints c_k can also be observed at −40 and −20 MHz. While the interference-to-target ratio is low, so too is the waveform weight. We can also see from (17.64) that as the power level P_{T} increases, so too does the constant level A. This relationship results in the output waveform having more spectral components—rather than simply increasing the power of the same components.

The scenario above is repeated with the inclusion of clutter, as shown in Figure 17.25, with the output waveform depicted in Figure 17.26. Here it is observed that wherever the constant level A is above the upper limit U_k or below the lower limit L_k,

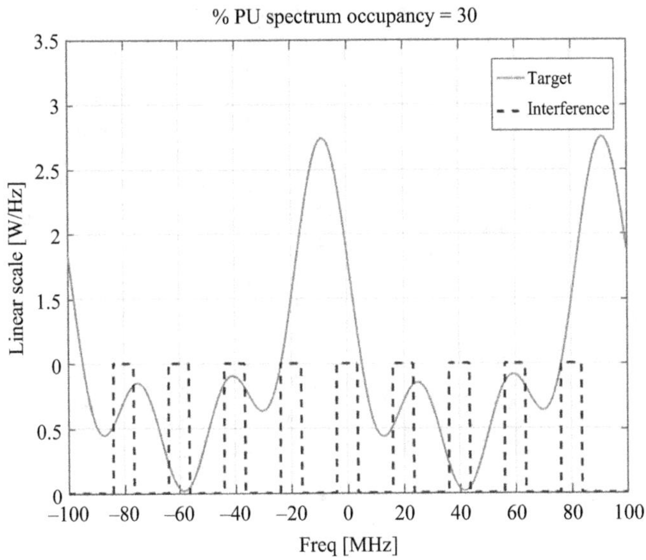

Figure 17.23 Target transfer function (solid) and primary user interference spectrum (dashed). Reproduced, with permission, from Reference [105]

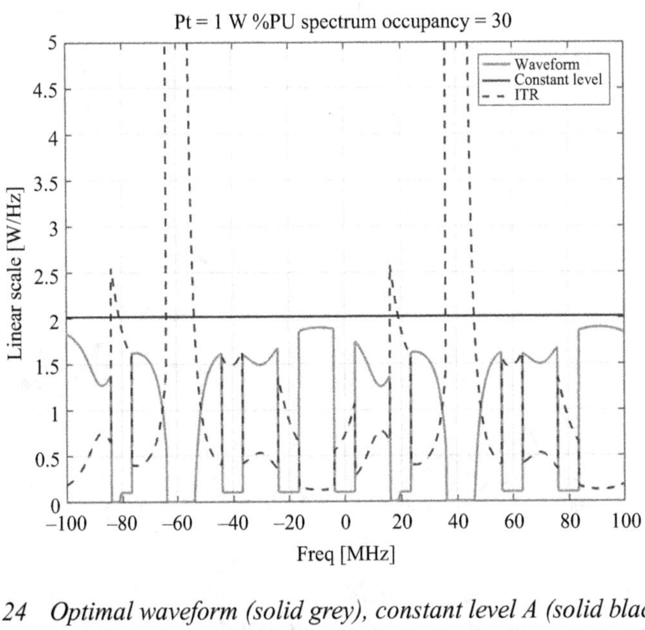

Figure 17.24 Optimal waveform (solid grey), constant level A (solid black), and interference-to-target ratio (dashed), for the scenario in Figure 17.23 with $P_T = 1$ W. Reproduced, with permission, from Reference [104]

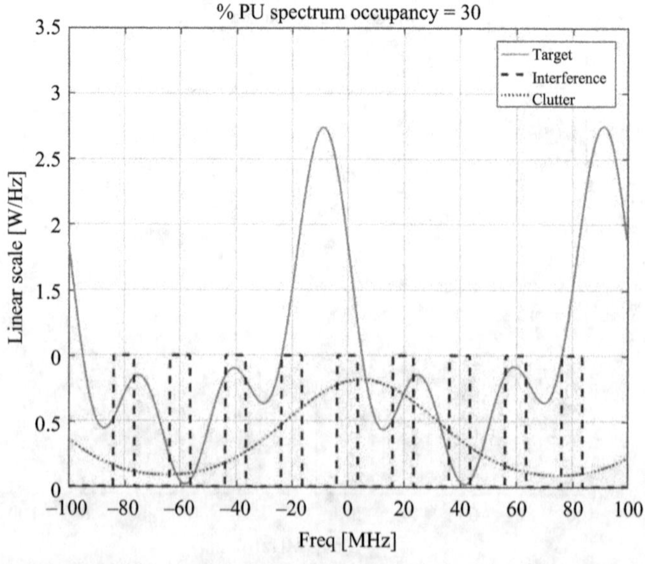

Figure 17.25 Target transfer function (solid), primary user interference spectrum (dashed), and clutter spectrum (dotted). Reproduced, with permission, from Reference [104]

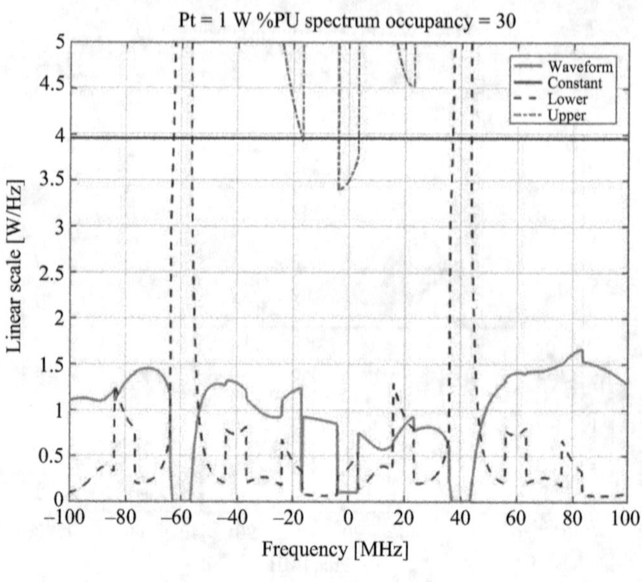

Figure 17.26 Optimal waveform (solid grey), constant level A (solid black), lower limit (dashed), and upper limit (dot-dashed), for the scenario in Figure 17.25 with $P_T = 1$ W. Reproduced, with permission, from Reference [104]

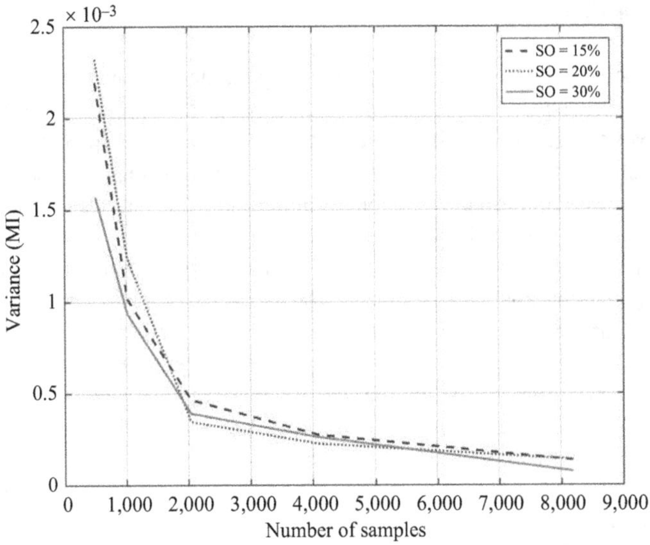

Figure 17.27 Variance of the mutual information values as a function of the number of samples, for a spectrum occupancy of 15% (dashed), 20% (dotted) and 30% (solid). Reproduced, with permission, from Reference [105]

the value given to the waveform weight is zero. We also see how, due to the presence of clutter, the optimal waveform weights are more uniform.

Note that the optimal waveform (17.64) was proposed as part of a perception–action cycle for a 'cognitive' radar architecture. The authors of [104,105] have explored the impact of the observation time in the 'perception' part of the cycle, considering how the number of fast-time samples affects the PSD estimates. In [105], 100 interference samples were generated as a coloured Gaussian random process, with different sample lengths used to estimate the PSD of the interference. For each sample length, the variance of the MI was calculated given the waveform from (17.64) for the estimated spectral densities, with the results shown in Figure 17.27. As expected, if the number of samples is relatively small, the variance of the MI is significantly larger, which would cause greater uncertainty in the classification of multiple targets. This result suggests a minimum requirement on the number of samples needed for a given accuracy level, with the maximum being due to user requirements on the responsiveness of the system.

As a final remark, note that such an architecture could be considered as a *component* of a 'fully cognitive' radar system. Such a fully cognitive architecture might not only adapt its waveform in a perception–action cycle but could also adapt the length of the perception period, other parameters such as the maximum power of spectral components c_k, and even the choice of metric. Thus, the architecture itself could infer when MI is appropriate to its current operation, all aided by memory of past performance.

17.5 Summary

Of the many observations to be made in this chapter, the most significant would have to be the nonlinear behaviour of SINR as constraints are applied. We observed for both the CM and ISL metrics that achieving a high degree of either constraint comes at the cost of significant SINR degradation. This relationship suggests that it may be better to relax very stringent sidelobe or modulus requirements for the benefit of SINR, particular in spectrally congested environments.

Performance prediction for adaptive radar waveform design has been shown to provide new insights that may enable better and faster decision-making for adaptive/cognitive applications that require very low latency. It is clear that waveform adaptation may well become mandatory for future-fielded radar systems, and being able to predict the impact of this manner of operation for a variety of scenarios provides a distinct tactical advantage over static, or non-knowledge-aided, schemes. Such performance predictions could be used in conjunction with spectral occupancy decisions such as those described in Chapter 15 to facilitate the real-time design of RFI-avoiding radar waveforms.

We have also discussed an alternative metric to SINR based on MI that emphasizes the ability to classify extended targets in the illuminated scene. A solution to the MI waveform optimization problem was given in which one must only solve numerically for one parameter. Such a solution may be useful where interference and clutter is varying with frequency throughout the bandwidth, as it often promotes more spectral components than when optimizing for SINR, and can be used as part of a 'cognitive' radar architecture.

17.6 Looking ahead

The area of waveform design for spectral congestion avoidance is assured to be of growing importance in the coming years as RF spectrum usage continues to increase due to wireless access demand and an expanding base of global users. Moreover, changes to spectrum standards and regulations have not in earnest begun to accommodate emerging applications in cognitive RF, including both radar and communications. To bring those technologies to fruition, one will require not only a reliable and effective suite of technologies but also the ability to predict their performance under a variety of operating conditions. Necessarily, one of the earliest barriers to the adoption of cognitive RF techniques will be the ability to guarantee quality of service for primary users. While waveform diversity and design research has been ongoing for in excess of 20 years, performance prediction is still in its infancy, with the most cited examples still being narrowly scoped, empirical studies. Future work in this area will need to give greater emphasis to the prediction of key performance metrics as a function of spectral operating conditions and user needs, independent of the implemented algorithms. Further, specific algorithmic implementations will need effective means for self-monitoring to predict and measure performance in real-time.

References

[1] Wicks, M., Mokole, E., Blunt, S., Schneible, R., Amuso, V., eds. *Principles of Waveform Diversity and Design*, SciTech Publishing, Inc., Raleigh, NC, 2010.

[2] Nehorai, A., Gini, V., Greco, A., Suppappola, A. P., and Rangaswamy, M. Introduction to the issue on adaptive waveform design for agile sensing and communication. *IEEE Journal of Selected Topics in Signal Processing 1*, 1 (June 2007), 2–5.

[3] Suppappola, A. P., Nehorai, A., and Calderbank, R. Waveform-agile sensing and processing. *IEEE Signal Processing Magazine 26*, 1 (January 2009), 10–11.

[4] Gini, V., De Maio, A., and Patton, L., Eds. *Waveform Design and Diversity for Advanced Radar Systems*, The Institution of Engineering and Technology, London, 2012.

[5] Levenon, N., and Mozeson, E. *Radar Signals*. John Wiley and Sons, Inc., Hoboken, NJ, 2004.

[6] Mahafza, B. *Radar Signal Analysis and Processing Using MATLAB*. CRC Press, New Delhi, 2011.

[7] Pillai, S., Li, K. Y., Selesnick, I., and Himed, B. *Waveform Diversity Theory and Applications*, 1st ed. McGraw-Hill, New York, NY, 2011.

[8] Richards, M. *Fundamentals of Radar Signal Processing*. McGraw-Hill, New York, NY, 2005.

[9] Richards, M. *Principles of Modern Radar, Basic Principles*, vol. 1. SciTech Publishing, Inc., Raleigh, NC, 2010.

[10] Skolnik, M. *Introduction to Radar Systems*, 1st ed. McGraw Hill, New York, NY, 1980.

[11] Stoica, P., Li, J., and Xue, M. Transmit codes and receive filters for radar. *IEEE Signal Processing Magazine 25*, 6 (November 2008), 94–109.

[12] Sira, S., Li, Y., Papandreou-Suppappola, A., Morrell, D., Cochran, D., and Rangaswamy, M. Waveform-agile sensing for tracking. *IEEE Signal Processing Magazine 26*, 1 (January 2009), 53–64.

[13] Calderbank, R., Howard, S., and Moran, B. Waveform diversity in radar signal processing. *IEEE Signal Processing Magazine 26*, 1 (January 2009), pp. 32–41.

[14] Benedetto, J., Konstantinidis, I., and Rangaswamy, M. Phase-coded waveforms and their design. *IEEE Signal Processing Magazine 26*, 1 (January 2009), 22–31.

[15] Aubry, A., De Maio, A., Huang, Y., Piezzo, M., and Farina, A. A new radar waveform design algorithm with improved feasibility for spectral coexistence. *IEEE Transactions on Aerospace and Electronic Systems 51*, 2 (April 2015), 1029–1038.

[16] Aubry, A., De Maio, A., Piezzo, M., Naghsh, M. M., Soltanalian, M., and Stoica, P. Cognitive radar waveform design for spectral coexistence in signal-dependent interference. In *Radar Conference, 2014 IEEE* (May 2014).

[17] Davis, M. Frequency allocation challenges for ultra-wideband radars. *IEEE Aerospace and Electronic Systems Magazine 28*, 7 (July 2013), 12–18.
[18] He, H., Stoica, P., and Li, J. Waveform design with stopband and correlation constraints for cognitive radar. In *2nd International Workshop on Cognitive Information Processing* (October 2010).
[19] Patton, L. *On the satisfaction of modulus and ambiguity function constraints in radar waveform optimization for detection*. PhD thesis, Wright State University, 2009.
[20] Patton, L., Bryant, C., and Himed, B. Radar-centric design of waveforms with disjoint spectral support. In *Radar Conference, 2012 IEEE* (2012), IEEE.
[21] Patton, L., Frost, S., and Rigling, B. Efficient design of radar waveforms for optimised detection in coloured noise. *IET Radar, Sonar and Navigation 6*, 1 (2012), 22–29.
[22] Patton, L., and Rigling, B. Nonquadratic regularization for waveform optimization. In *IEEE Radar Conference* (2006), pp. 763–768.
[23] Patton, L., and Rigling, B. Autocorrelation and modulus constraints in radar waveform optimization. In *International Waveform Diversity and Design Conference* (2009), IEEE.
[24] Patton, L., and Rigling, B. Optimal signal/filter design for multi-target detection in colored noise. In *IEEE Radar Conference* (2011), pp. 276–281.
[25] Patton, L., and Rigling, B. Phase retrieval for radar waveform optimization. *IEEE Transactions on Aerospace and Electronic Systems 48*, 4 (October 2012), 3287–3301.
[26] Patton, L., and Rigling, B. D. Autocorrelation constraints in radar waveform optimization for detection. *IEEE Transactions on Aerospace and Electronic Systems 48*, 2 (April 2012), 951–968.
[27] Wicks, M. Spectrum crowding and cognitive radar. In *2nd International Workshop on Cognitive Information Processing (CIP)* (Elba Island, Italy, June 2010), pp. 452–457.
[28] Jones, A. M., Rigling, B., and Rangaswamy, M. Eigen-space analysis of constrained adaptive radar waveform design. In *Proceedings of the 2014 IEEE Radar Conference* (Cincinnati, OH, May 2014), IEEE, pp. 0829–0824.
[29] Jones, A. M., Rigling, B. D., and Rangaswamy, M. Eigen-basis analysis of expected modulus perturbation for constrained signal design. In *Signal, Systems and Computers, 48th Asilomar Conference on* (Pacific Grove, CA, November 2014), IEEE, pp. 2021–2026.
[30] Jones, A. M., Rigling, B. D., and Rangaswamy, M. Motivations to develop performance prediction for adaptive radar. In *IEEE National Aerospace and Electronics Conference and Ohio Innovation Summit (NAECON-OIS)* (Dayton, OH, June 2015), IEEE.
[31] Jones, A. M., Rigling, B. D., and Rangaswamy, M. Performance models for sidelobe constrained signal design with eigen-basis formulation. In *Proceedings of the 2015 IEEE International Radar Conference* (Arlington, VA, May 2015), IEEE, pp. 0828–0833.

[32] Jones, A. M., Rigling, B. D., and Rangaswamy, M. Generation of synthetic UHF RFI in urban North American environments. In *Proceedings of the 2016 IEEE Radar Conference* (Philadelphia, PA, 2016), IEEE.

[33] Jones, A. M., Rigling, B. D., and Rangaswamy, M. Signal-to-interference-plus-noise-ratio analysis for constrained radar waveforms. *Aerospace and Electronic Systems, IEEE Transactions on 52*, 5 (October 2016), 2230–2241.

[34] Aubry, A., De Maio, A., Piezzo, M., and Farina, A. Radar waveform design in a spectrally crowded environment via nonconvex quadratic optimization. *IEEE Aerospace and Electronic Systems Magazine 50* (2014), 1138–1152.

[35] Blunt, S., and Mokole, E. Overview of radar waveform diversity. *IEEE Aerospace and Electronic Systems Magazine 31,* (2016), 2–42.

[36] Frost, S. *Performance analysis of radar waveforms for congested spectrums*. Master's thesis, Wright State University, November 2011.

[37] Frost, S., and Rigling, B. Performance comparison for constrained radar waveform design. In *IEEE RadarCon 2011 Proceedings* (2011), IEEE, pp. 724–728.

[38] Frost, S., and Rigling, B. Sidelobe predictions for spectrally-disjoint radar. *In Radar Conference* (2012), IEEE.

[39] Govoni, M. Enhancing spectrum coexistence using radar waveform diversity. In *Radar Conference, 2016 IEEE* (May 2016).

[40] Rossler, C. W., Patton, L., and Himed, B. Rapid, near-optimal waveform adaptation: precompute and lookup via hash functions. *IET Radar, Sonar Navigation 7*, 8 (2013), 925–932.

[41] Setlur, P., and Devroye, N. Adaptive waveform scheduling in radar: an information theoretic approach. In *SPIE Proceedings* (May 2012), vol. 8361.

[42] Setlur, P., Devroye, N., and Cheng, Z. Waveform scheduling via directed information in cognitive radar. In *Statistical Signal Processing Workshop (SSP)* (Ann Arbor, MI, August 2012), IEEE.

[43] Setlur, P., Devroye, N., and Rangaswamy, M. Radar waveform design with the two step mutual information. In *IEEE Radar Conference* (Cincinnati, OH, May 2014), IEEE.

[44] Setlur, P., Gogineni, S., and Rangaswamy, M. Waveform extraction from reference channel of passive multistatic radar systems. In *49th Asilomar conference on Signals, Systems and Computers* (Pacific Grove, CA, November 2015), IEEE.

[45] Stoica, P., and Li, J. Waveform design with stopband and correlation constraints for cognitive radar. In *2nd International Workshop on Cognitive Information Processing* (June 2010), pp. 344–349.

[46] Skolnik, M. I. *Radar Handbook*, third ed. McGraw-Hill Professional, New York, NY, February 2008.

[47] Soltanalian, M. *Signal design for active sensing and communications*. PhD thesis, Uppsala University, 2014.

[48] Richards, M. A. *Time and Frequency Domain Windowing of LFM Pulses*, radarsp.com, supplementary technical notes for *Fundamentals of Radar Signal Processing*. McGraw-Hill, New York, NY, 2014.

[49] McAulay, R., and Johnson, R. Optimal mismatched filter design for radar ranging, detection, and resolution. *IEEE Transactions on Information Theory 17*, 6 (November 1971), 696–701.

[50] Stoica, P., He, H., and Li, J. On designing sequences with impulse-like periodic correlation. *IEEE Signal Processing Letters 16*, 8 (August 2009), 703–706.

[51] Sussman, S. Least-square synthesis of radar ambiguity functions. *IRE Transactions on Information Theory 8*, 3 (April 1962), 246–254.

[52] Mesiya, M. F., and Mclane, P. J. Design of optimal radar signals subject to a fixed amplitude constraint. *IEEE Transactions on Aerospace and Electronic Systems 9*, 5 (September 1973), 679–687.

[53] De Maio, A., Huang, Y., Piezzo, M., Zhang, S., and Farina, A. Design of optimized radar codes with a peak to average power ratio constraint. *IEEE Transactions of Signal Processing 59*, 6 (June 2011), 2683–2697.

[54] Schweppe, F., and Gray, D. Radar signal design subject to simultaneous peak and average power constraints. *IEEE Transactions on Information Theory 12*, 1 (January 1966), 13–26.

[55] Soltanalian, M., and Stoica, P. Design of perfect phase-quantized sequences with low peak-to-average-power ratio. In *20th European Signal Processing Conference* (August 2012), EUSIPCO.

[56] Tufts, D., and Shnidman, D. Optimum waveforms subject to both energy and peak-value constraints. *Proceedings of the IEEE 52*, 9 (September 1964), 1002–1007.

[57] De Nicola, S., Huang, Y., De Maio, A., Zhang, S., and Farina, A. Code optimization with similarity and accuracy constraints. In *Radar Conference, 2008. IEEE* (2008), pp. 1–6.

[58] DeLong, D., and Hofstetter, E. On the design of optimum radar waveforms for clutter rejection. *IEEE Transactions on Information Theory 13*, 3 (July 1967), 454–463.

[59] Trees, H. Optimum signal design and processing for reverberation-limited environments. *IEEE Transactions on Military Electronics 9*, 3 (July–October 1965), 212–229.

[60] Bergin, J. S., Techau, P. M., Don Carlos, J. E., and Guerci, J. R. Radar waveform optimization for colored noise mitigation. In *Radar Conference* (May 2005), IEEE, pp. 1–6.

[61] Wang, G., and Lu, Y. Designing single/multiple sparse frequency waveforms with sidelobe constraint. *IET Radar, Sonar, Navigation 5*, 1 (2011), 32–38.

[62] Zhou, H., Wen, B., and Wu, S. Dense radio frequency interference suppression in HF radars. *IEEE Signal Processing Letters 12*, 5 (May 2005), 361–364.

[63] Scalzi, G. *RF Range Measurement Capabilities*. U.S. Air Force Research Laboratory – Sensors Directorate, Presentation, October 2014.

[64] Jones, A. M. *Frequency diverse array receiver architectures*. Master's thesis, Wright State University, 2011.

[65] Marcenko, V., and Pastur, L. Distribution of eigenvalues for some sets of random matrices. *Math USSR-Sbornik 1*, 4 (1967), 457–483.

[66] Devroye, L. Sample-based non-uniform random variate generation. In *Proceedings of the 18th Conference on Winter Simulation* (1986), ACM, pp. 260–265.

[67] Proakis, J. *Digital Communications*, 3rd ed. McGraw-Hill, New York, NY, March 1994.

[68] Messerschmitt, D. Autocorrelation matrix eigenvalues and the power spectrum. Tech. Rep. UCB/EECS-2006-90, EECS Department, University of California, Berkeley, June 2006.

[69] Moon, T. K., and Stirling, W. C. *Mathematical Methods and Algorithms for Signal Processing*, vol. 1. Prentice Hall, New York, NY, 2000.

[70] Wax, M., and Kailath, T. Detection of signals by information theoretic criteria. *IEEE Transactions Acoustics, Speech, and Signal Processing 33*, 2 (April 1985), 387–392.

[71] Sunggon, K., Lee, J., and Sung, D. A shifted gamma distribution model for long-range dependent internet traffic. *IEEE Communications Letters 7*, 3 (March 2003), 124–126.

[72] Shanmugan, K. S., and Breipohl, A. M. *Random Signals: Detection, Estimation and Data Analysis*, first ed. John Wiley and Sons, New York, NY, 1988.

[73] Setlur, P., and Rangaswamy, M. Joint filter and waveform design for radar STAP in signal dependent interference. Tech. Rep. AFRL-RY-WP-TP-2015-0170, AFRL Sensors Directorate, October 2015.

[74] Boyd, S., and Vandenberghe, L. *Convex Optimization*. Cambridge University Press, Cambridge, 2004.

[75] Guerci, J. *Space-Time Adaptive Processing for Radar*. Artech House Publishers, Norwood, MA, July 2003.

[76] Kay, S. *Fundamentals of Statistical Signal Processing: Detection Theory*. Prentice Hall, Upper Saddle River, NJ, 1998.

[77] Jones, A. M. *Performance prediction of constrained waveform design for adaptive radar*. PhD thesis, Wright State University, June 2016.

[78] Lame, G. *Examen Des Differentes Methodes Employees Pour Resoudre Les Problemes de Geometrie: Paris, Vve Courcier, 1818*. Paris A. Hermann, 1903.

[79] Newman, M. E. J. Power laws, pareto distributions and zipf's law. *Contemporary Physics 46*, 5 (December 2005), 323–351.

[80] MATLAB. *Version 7.14.0.739 (2013b)*. The MathWorks Inc., Natick, MA, 2012.

[81] Grant, M., and Boyd, S. Graph implementations for nonsmooth convex programs. In *Recent Advances in Learning and Control*, V. Blondel, S. Boyd, and H. Kimura, Eds., Lecture Notes in Control and Information Sciences. Springer-Verlag Limited, London, 2008, pp. 95–110.

[82] Grant, M., and Boyd, S. Cvx: Matlab software for disciplined convex programming, version 2.1. http://cvxr.com/cvx, March 2014.

[83] Chu, D. Polyphase codes with good periodic correlation properties. *IEEE Transactions on Information Theory 18*, 4 (July 1972), 532–532.

[84] Stoica, P., He, H., and Li, J. New algorithms for designing unimodular sequences with good correlation properties. *IEEE Transactions of Signal Processing 57*, (April 2009), 1415–1425.

[85] Li, J., Guerci, J., and Xu, L. Signal waveform's optimal under restriction design for active sensing. In *Sensor Array and Multichannel Processing, 2006. Fourth IEEE Workshop on* (2006), pp. 382–386.

[86] Satyabrata, S. Constant-envelope waveform design for optimal target-detection and autocorrelation performances. In *IEEE International Conference on Acoustics, Speech, and Signal Processing* (Vancouver, Canada, May 2013), pp. 3851–3855.

[87] Wang, G., and Lu, Y. Bounds on generalised integrated sidelobe level in waveforms with stopbands. *IET Radar, Sonar and Navigation, Electronics Letters 46*, 23 (November 2010), 1561–1562.

[88] Woodward, P. *Probability and Information Theory with Applications to Radar*. Pergamon, London, 1953.

[89] Woodward, P., and Davies, I. A theory of radar information. *The London, Edinburgh, and Dublin Philosophical Magazine and Journal of Science 41*, 321 (1950), 1001–1017.

[90] Davies, I. L. On determining the presence of signals in noise. *Proceedings of the IEE – Part III: Radio and Communication Engineering 99*, 58 (March 1952), 45–51.

[91] Woodward, P. M. Information theory and the design of radar receivers. *Proceedings of the IRE 39*, 12 (December 1951), 1521–1524.

[92] Bell, M. R. *Information theory and radar: mutual information and the design and analysis of radar waveforms and systems*. PhD thesis, California Institute of Technology, 1988.

[93] Bell, M. R. Information theory and radar waveform design. In *Proceedings. IEEE International Symposium on Information Theory* (Jan 1993), pp. 428–428.

[94] Romero, R. A., and Goodman, N. A. Waveform design in signal-dependent interference and application to target recognition with multiple transmissions. *IET Radar, Sonar Navigation 3*, 4 (August 2009), 328–340.

[95] Bică, M., and Koivunen, V. Generalized multicarrier radar: models and performance. *IEEE Transactions on Signal Processing 64*, 17 (September 2016), 4389–4402.

[96] Kanaya, F., and Nakagawa, K. On the practical implication of mutual information for statistical decisionmaking. *IEEE Transactions on Information Theory 37*, 4 (1991), 1151–1156.

[97] Potter, L. C., Chiang, D.-M., Carriere, R., and Gerry, M. J. A GTD-based parametric model for radar scattering. *IEEE Transactions on Antennas and Propagation 43*, 10 (1995), 1058–1067.

[98] Haykin, S. Cognitive radar: a way of the future. *IEEE Signal Processing Magazine 23*, 1 (January 2006), 30–40.

[99] Guerci, J. R. Cognitive radar: a knowledge-aided fully adaptive approach. In *2010 IEEE Radar Conference* (2010), IEEE, pp. 1365–1370.

[100] Bell, K. L., Baker, C. J., Smith, G. E., Johnson, J. T., and Rangaswamy, M. Cognitive radar framework for target detection and tracking. *IEEE Journal of Selected Topics in Signal Processing 9*, 8 (2015), 1427–1439.

[101] Charlish, A., and Hoffmann, F. Anticipation in cognitive radar using stochastic control. In *2015 IEEE Radar Conference (RadarCon)* (2015), IEEE, pp. 1692–1697.

[102] Haykin, S. *Cognitive dynamic systems: perception-action cycle, radar and radio*. Cambridge University Press, Cambridge, 2012.

[103] Metcalf, J., Blunt, S. D., and Himed, B. A machine learning approach to cognitive radar detection. In *2015 IEEE Radar Conference (RadarCon)* (2015), IEEE, pp. 1405–1411.

[104] Palamà, R., and Griffiths, H. Beyond the ambiguity function. Tech. Rep., Department of Electronic and Electrical Engineering, University College London, September 2016. Prepared for the Defence Science and Technology Laboratory.

[105] Palamà, R., Griffiths, H., and Watson, F. A radar architecture for joint dynamic spectrum access and target-matched illumination. In *2017 IEEE Radar Conference (RadarConf)* (May 2017), pp. 1472–1477.

Chapter 18
Space–time transmit nulling for RF spectrum interoperability

Tegan Webster[1], Patrick M. McCormick[2], and Thomas Higgins[1]

18.1 Introduction

Transmit nulling can enable interoperability between radar and other RF spectrum users such as communication systems that are nearby in physical location and operational frequency; these nulls reduce the energy transmitted at a particular spatial angle, frequency, or at a combined space–frequency location. The use of nonlinear power amplifiers in radar systems imposes the additional requirement that radar waveforms must be constant amplitude to ensure maximum transmit power, which is achieved when the systems' power amplifiers are operated in the saturated nonlinear region where efficiency is the greatest. In the case of wideband MIMO (multiple-input multiple-output) radar, it is also necessary to optimize the beampattern and spectral content of these emissions jointly because the steering vectors of the array are not constant over the bandwidth.

Numerous methods for creating spatial, frequency, or space–frequency nulls on transmit using constant modulus signals have been published in the literature, e.g. [1–6]. Beampattern optimization for wideband emissions has been investigated in [7,8] and likewise wideband MIMO beampattern optimization has been explored in [9,10].

The reiterative uniform weight optimization (RUWO) algorithm has been proposed as a method to generate phase-only weights that satisfy constant amplitude constraints, as first described in [6]. The RUWO algorithm calculates weights, using either a deterministic interference model or measured interference, via the maximum signal-to-interference-plus-noise ratio (SINR) framework in a reiterative fashion.

The RUWO method has also been incorporated inside of an iterative wideband emission design algorithm to produce physically realizable frequency-modulated (FM) waveforms with large space–frequency nulls [11] and can be implemented via

[1]U.S. Naval Research Laboratory, USA
[2]Electrical Engineering & Computer Science Department, University of Kansas, USA

the polyphase-coded FM (PCFM) framework [12]. This work is an extension to a method for designing wideband MIMO waveforms that shapes the spectrum in certain transmission angles according to a given spectral window that includes a sufficient portion of the spectral roll-off region [13].

This chapter begins with a formulation of the RUWO algorithm in Section 18.2 and its application to producing spatial, frequency, and space–frequency transmit nulls for radio frequency interference avoidance, followed by a description of the Naval Research Laboratory's Space Time Adaptive Nulling (STAN) radar test bed, an eight-channel X-band coherent MIMO radar test bed used to experimentally verify the performance of the RUWO algorithm in Section 18.3. Section 18.4 then provides a description of narrowband loop-back and open-air experimental applications of RUWO that were originally presented in [14,15], respectively. Following the narrowband results is an explanation in Section 18.5 of how the RUWO algorithm is incorporated into a method for producing wideband FM waveforms with large space–frequency nulls as discussed in [11]. The chapter concludes with a summary of presented work and a discussion of future work and open problems on the topic of space–time transmit nulling with the goal of RF spectrum interoperability.

18.2 Overview of reiterative uniform weight optimization (RUWO) algorithm

18.2.1 General formulation

For a given steering vector **b** (e.g. the well-known Vandermonde form observed for a uniform linear array (ULA)) and interference covariance matrix **R**, the maximum SINR solution is given by

$$\mathbf{a} = \mathbf{R}^{-1}\mathbf{b}. \tag{18.1}$$

In general, the solution **a** has arbitrary phase and non-constant amplitude regardless of the properties of **b**, aside from the simple interference-free case of **R** = **I** where **I** is the identity matrix. For the radar applications discussed here, **a** is used to generate constant modulus MIMO waveforms, and thus the RUWO algorithm first removes the amplitude variation from the result of (18.1) to obtain

$$\mathbf{a}_0 = \exp(j\angle\mathbf{a}), \tag{18.2}$$

where $\angle(\cdot)$ determines the phase of the argument. The solution is then modified by iteratively computing the phase-only maximum SINR solution for the output of the previous stage as

$$\mathbf{a}_k = \exp(j\angle(\mathbf{R}^{-1}\mathbf{a}_{k-1})). \tag{18.3}$$

It is shown in [16] that the maximum SINR solution from (18.1) is numerically identical to the projection-based approach proposed in [3,5] for the purpose of forcing precise frequency nulls. The subsequent removal of amplitude variation in steps (18.2) and (18.3) as a means to comply with the constant modulus constraint is equivalent to a projection onto a non-convex set. Consequently, it is not possible to prove convergence of the RUWO algorithm using the method of projection onto convex sets (POCS), although it is possible to show that the algorithm does not diverge [3,5,17]. It has been observed in practice that the iteration in (18.3) generally converges to a stable solution in less than 50 iterations.

While RUWO can be applied to a variety of problems, here it is used to generate spatial, frequency, or space–frequency nulled transmit waveforms that retain the constant modulus property. The covariance matrix in (18.1) can be generated adaptively from received data or deterministically based on known interference sources. In the following derivations, $\mathbf{s}$ denotes the length-N discretized radar waveform and $\mathbf{v}$ is a length-M spatial steering vector based on the desired steering direction relative to boresight. The derivations further assume a linear array with half-wavelength spacing, thus resulting in a spatial steering vector of the form

$$\mathbf{v} = \begin{bmatrix} 1 & e^{-j\pi \sin\theta} & \cdots & e^{-j(M-1)\pi \sin\theta} \end{bmatrix}^T \tag{18.4}$$

where θ is the desired spatial steering angle and $(\cdot)^T$ is the transpose operation. Spatial, frequency, or space–frequency nulls are generated by replacing $\mathbf{b}$ and $\mathbf{R}$ from the RUWO formulation above with different quantities, as described in the following sections.

18.2.2 Spatial nulling

To generate a spatial null, a spatial interference covariance matrix $\mathbf{R}_\text{s}$ is constructed, either deterministically using knowledge of the spatial location of interference sources or adaptively from received data, via the methods described in Section 18.2.5. Substituting $\mathbf{b}_\text{s} = \mathbf{v}$ and $\mathbf{R}_\text{s}$ for the $\mathbf{b}$ and $\mathbf{R}$ terms, respectively, in (18.1) and (18.2), the initial phase-only weight vector becomes

$$\mathbf{a}_{\text{s},0} = \exp(j\angle(\mathbf{R}_\text{s}^{-1}\mathbf{b}_\text{s})), \tag{18.5}$$

and the result after K iterations is

$$\mathbf{a}_{\text{s},K} = \exp(j\angle\mathbf{a}_{\text{s},K-1}). \tag{18.6}$$

The complete set of transmitted waveforms is then expressed as

$$\mathbf{S}_\text{s} = \mathbf{s}\ \mathbf{a}_{\text{s},K}^T, \tag{18.7}$$

in which the mth column of $\mathbf{S}_\text{s}$ corresponds to the waveform transmitted from the mth element of the array. Antenna effects may limit the nulling performance when a deterministic, rather than adaptive, covariance matrix is employed, as shown later in Section 18.4.2.

18.2.3 Frequency nulling

Frequency nulling is achieved by using the deterministic $N \times N$ covariance matrix

$$\mathbf{R}_\text{f} = \mathbf{Q}_\text{f}\mathbf{Q}_\text{f}^H + \delta\mathbf{I}_\text{f}, \tag{18.8}$$

where δ is a diagonal loading term, $\mathbf{I}_{\text{f}}$ is an $N \times N$ identity matrix, and

$$\mathbf{Q}_{\text{f}} = \begin{bmatrix} 1 & 1 & \cdots & 1 \\ e^{j2\pi f_0} & e^{j2\pi f_1} & \cdots & e^{j2\pi f_{P-1}} \\ \vdots & \vdots & \vdots & \vdots \\ e^{j2\pi f_0(N-1)} & e^{j2\pi f_1(N-1)} & \cdots & e^{j2\pi f_{P-1}(N-1)} \end{bmatrix} \tag{18.9}$$

is an $N \times P$ matrix of spectral steering vectors corresponding to P sources of spectral interference, with f_p the frequency of the pth interference source. Here the quantities $\mathbf{b}_{\text{f}} = \mathbf{s}$ and $\mathbf{R}_{\text{f}}$ are substituted into (18.1) and (18.2), and after K iterations of RUWO, the transmitted waveform is expressed as

$$\mathbf{a}_{\text{f},K} = \exp(j\angle\mathbf{a}_{\text{f},K-1}), \tag{18.10}$$

and the complete set of waveforms transmitted from the array is denoted as

$$\mathbf{S}_{\text{f}} = \mathbf{a}_{\text{f},K}\, \mathbf{v}^T. \tag{18.11}$$

In contrast to the spatial nulling case, a deterministic covariance matrix may suffice for producing frequency nulls because radar systems tend to have very accurate timing.

18.2.4 Space–frequency nulling

Combined space–frequency nulling is implemented using the $NM \times NM$ covariance matrix

$$\mathbf{R}_{\text{sf}} = \mathbf{Q}_{\text{sf}}\mathbf{Q}_{\text{sf}}^H + \delta\mathbf{I}_{\text{sf}}, \tag{18.12}$$

where again δ is a diagonal loading term, $\mathbf{I}_{\text{sf}}$ is an $NM \times NM$ identity matrix, and

$$\mathbf{Q}_{\text{SF}} = \begin{bmatrix} \mathbf{q}_{\text{sf},0} & \mathbf{q}_{\text{sf},1} & \cdots & \mathbf{q}_{\text{sf},P-1} \end{bmatrix} \tag{18.13}$$

is an $NM \times P$ matrix of space–frequency steering vectors in which the pth column is

$$\mathbf{q}_{\text{sf},p} = \begin{bmatrix} 1 \\ e^{-j\pi \sin\phi_p} \\ \vdots \\ e^{-j(M-1)\pi \sin\phi_p} \end{bmatrix} \otimes \begin{bmatrix} 1 \\ e^{j2\pi f_p} \\ \vdots \\ e^{j2\pi f_p(N-1)} \end{bmatrix}. \tag{18.14}$$

In (18.14), the terms ϕ_p and f_p are the spatial angle and frequency of the pth interference source, respectively, and $\otimes$ is the Kronecker product. Here the quantities $\mathbf{b}_{\text{sf}} = \text{vec}(\mathbf{s}\,\mathbf{v}^T)$ and $\mathbf{R}_{\text{sf}}$ are substituted into the general RUWO formulation of (18.1) and (18.2), for vec(·) the vectorization operation that involves stacking the columns of a matrix to form a vector.

The resulting space–frequency steering vector after K iterations of RUWO is expressed as

$$\mathbf{a}_{\text{sf},K} = \exp(j\angle\mathbf{a}_{\text{sf},K-1}), \tag{18.15}$$

and discretized versions of the transmit waveforms are given by

$$\mathbf{S}_{\mathrm{SF}} = \begin{bmatrix}\mathbf{I}_N & \cdots & \mathbf{I}_N\end{bmatrix}_{N\times NM} \mathrm{diag}(\mathbf{a}_{\mathrm{sf},K}) \begin{bmatrix} \mathbf{1}_{N\times 1} & \mathbf{0}_{N\times 1} & \cdots & \mathbf{0}_{N\times 1} \\ \mathbf{0}_{N\times 1} & \ddots & \ddots & \vdots \\ \vdots & \ddots & \ddots & \mathbf{0}_{N\times 1} \\ \mathbf{0}_{N\times 1} & \cdots & \mathbf{0}_{N\times 1} & \mathbf{1}_{N\times 1} \end{bmatrix}_{NM\times M} \tag{18.16}$$

where diag($\mathbf{x}$) denotes a matrix with the elements of $\mathbf{x}$ along the diagonal, $\mathbf{1}_{N\times 1}$ is an $N \times 1$ vector of ones, $\mathbf{0}_{N\times 1}$ is an $N \times 1$ vector of zeros, and the mth column of the matrix on the right consists of ones in the $(N(m-1)+1)$th through (Nm)th rows and zeros elsewhere, for $m = 1, \ldots, M$. As with spatial nulling, the space–frequency approach is expected to perform better when the covariance matrices are formed adaptively from received data.

18.2.5 Deterministic and adaptive spatial interference covariance

A deterministic $M \times M$ spatial interference covariance matrix, where M is the number of array elements, can be constructed as

$$\mathbf{R}_{\mathrm{s,det}} = \mathbf{Q}_{\mathrm{s}}\mathbf{Q}_{\mathrm{s}}^{\mathrm{H}} + \delta_{\mathrm{det}}\mathbf{I}_{\mathrm{s}}. \tag{18.17}$$

In (18.17), δ_{det} is a regularization factor, $\mathbf{I}_{\mathrm{s}}$ is an $M \times M$ identity matrix, and $\mathbf{Q}_{\mathrm{s}}$ is an $M \times P$ matrix of spatial steering vectors corresponding to P sources of spatial interference where

$$\mathbf{q}_{\mathrm{s},p} = \begin{bmatrix}1 & e^{-j\pi\sin\phi_p} & \cdots & e^{-j(M-1)\pi\sin\phi_p}\end{bmatrix}^T \tag{18.18}$$

is the pth column of $\mathbf{Q}_{\mathrm{s}}$ and ϕ_p is the spatial angle of the pth interference source.

The interference covariance matrix can also be calculated adaptively from received data as

$$\mathbf{R}_{\mathrm{s,adapt}} = \left(\frac{1}{N}\sum_{n=1}^{N}\mathbf{y}_n^*\,\mathbf{y}_n^T\right) + \delta_{\mathrm{adapt}}\mathbf{I}_{\mathrm{s}}, \tag{18.19}$$

in which $\mathbf{y}_n$ is the nth $M \times 1$ snapshot of data received by the array, N is the number of snapshots used to compute the covariance matrix, δ_{adapt} is a regularization factor, $\mathbf{I}_{\mathrm{s}}$ is an $M \times M$ identity matrix, and $(\cdot)^*$ denotes complex conjugation. Assuming white noise, the sample covariance matrix is naturally loaded by the system noise power. The regularization in (18.19) remains necessary, however, to prevent the phase-only solution from becoming ill-conditioned.

18.3 Experimental MIMO radar test bed

The Naval Research Laboratory's STAN radar test bed is a MIMO configuration consisting of an eight-channel X-band transmitter and receiver as well as a dipole antenna planar array. The antenna array, shown in Figure 18.1(a), consists of 18 cards, and

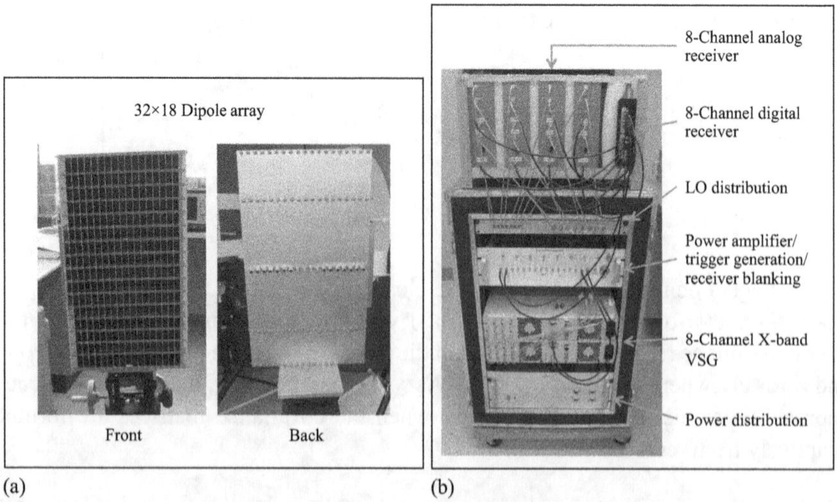

Figure 18.1 STAN radar test bed: (a) front (left) and back (right) of the 32 × 18 dipole antenna planar array and (b) annotated hardware including the eight-channel transmitter and receiver

each card contains a sub-array of 32 dipole antennas stacked vertically with a single corporate feed. Each card can be controlled independently, allowing up to 18 independent elements in azimuth, though in the open-air experimental results presented here only the central eight cards are used. The radiating elements are vertically oriented folded dipoles slightly less than a half-inch in length, with the spacing between dipoles a half-inch in both azimuth and elevation.

The transmitter for the STAN radar test bed is an eight-channel vector signal generator (VSG). Each channel of the transmitter consists of a 16-bit in-phase and quadrature-phase arbitrary waveform generator with a sample rate of 150 MS/s and a two-stage vector up-conversion chain that sets the center frequency to 9.9 GHz.

The analog portion of the receiver has eight channels, each containing a two-stage down-conversion to an intermediate frequency (IF) of 50 MHz. The bandwidth at the IF is 20 MHz. A 16-bit eight-channel digital receiver with a sampling rate of 200 MS/s is used to capture the output of the analog receiver. The system also contains a 2 W power amplifier for each transmit channel, a circulator and blanking switch to protect each channel of the receiver, and an FPGA for generating trigger pulses. Many of the system components are labeled in Figure 18.1(b).

Open-air experiments were conducted in a field using the STAN test bed, the results of which are discussed in Section 18.4.2. The transmitter and receiver were housed in a trailer, and the eight cables used for transmit and receive were run through a hole in the trailer and through a fence to the planar array. Although this configuration had less features than the rooftop setup used for previous experiments [14], the fence located between the trailer and array was unavoidable and may have resulted in multipath. Although the STAN test bed is equipped with a blanking switch, the

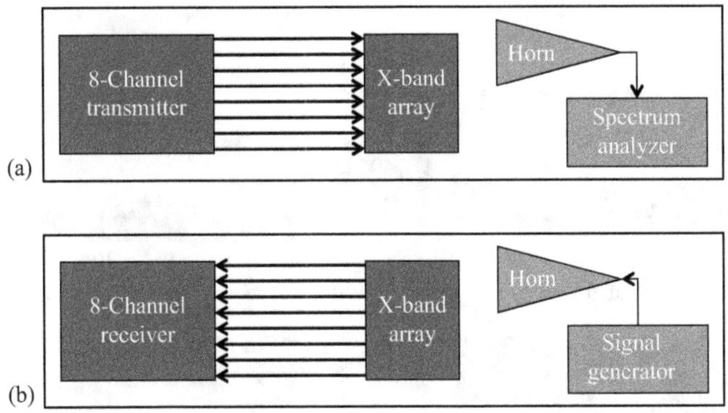

Figure 18.2 Measurement configurations for open-air experiments: (a) transmitting and (b) receiving

test bed was used to perform transmit and receive simultaneously, as the goal of the test was to observe the transmitted antenna patterns with spatial nulls formed deterministically or from recorded interference. Therefore, the test bed was used in two configurations: (1) to transmit from the array toward an X-band horn receive antenna with 27 dBi of gain connected to a spectrum analyzer about 100 ft from the array per Figure 18.2(a) and (2) to receive an interference signal produced by a VSG that was transmitted out of the same horn antenna per Figure 18.2(b).

18.4 Experimental demonstration of transmit nulling

18.4.1 Base-band loop-back demonstration

Prior to conducting open-air experiments with the full STAN radar test bed, preliminary experimental results were collected in a loop-back configuration consisting of the eight-channel VSG and the eight-channel digital receiver components of the test bed, both labeled in Figure 18.1(b). While the VSG does include up-conversion components, there is also an intermediate output of the signal at baseband, which was used in this configuration. The initial transmitted waveform, **s** from Section 18.2, was a 3-μs linear FM (LFM) waveform with a center frequency of 35 MHz and a bandwidth of 15 MHz. The RUWO algorithm was used to generate deterministic phase-only weights that produce spatial, frequency, and space–frequency nulls when applied to calibrated waveforms to generate the signals emitted from each of the eight transmit elements. The resulting transmitted waveforms were recorded at baseband using the eight-channel digital receiver in loop-back, with each transmit channel connected directly to a channel of the receiver.

The measured data were passed through a 20-MHz bandpass filter, digitized, and then collected into a matrix where each column contains the temporal data from the corresponding transmit element. A two-dimensional Fourier transform was applied to the data matrix to examine the spatial and spectral characteristics of the composite

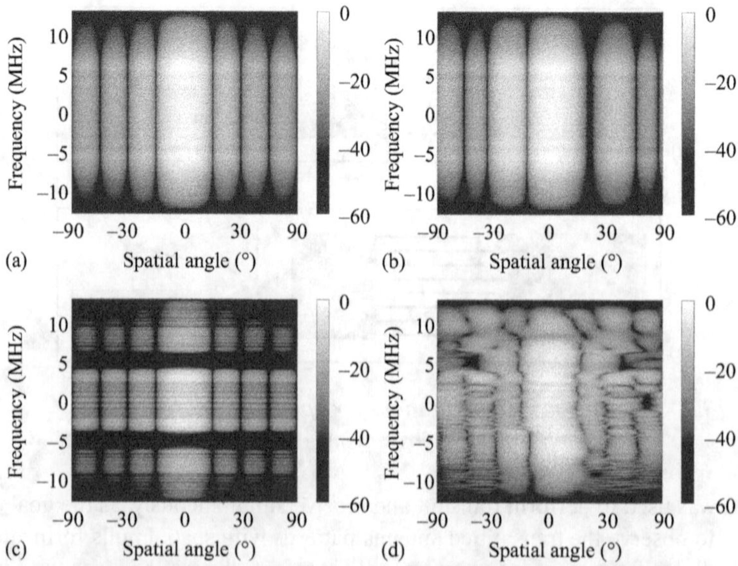

Figure 18.3 Normalized beampattern power (in dB) of (a) the baseline LFM waveform as a function of spatial angle and frequency, (b) the LFM waveform with spatial nulls at 20° and 50°, (c) the LFM waveform with frequency nulls at −5 and 5 MHz, and (d) the LFM waveform with space–frequency nulls at −10° and −5 MHz, 60° and 0 MHz, and both −40° and 40° at 5 MHz

waveform, facilitating visualization of the far-field beampattern of the array as a function of spatial angle and frequency.

Figure 18.3(a)–(d) displays the normalized beampattern power in dB as a function of spatial angle and frequency. The desired steering angle of the emission is 0°, and half-wavelength array element spacing is assumed. Figure 18.3(a) is the space–frequency beampattern of the LFM waveform without nulling, which is used as a baseline for comparison of Figure 18.3(b)–(d). Figure 18.3(b) corresponds to the LFM waveform with spatial nulls placed at 20° and 50°, while Figure 18.3(c) corresponds to the LFM waveform with frequency nulls placed at −5 and 5 MHz. All frequency nulls are 1 MHz in width. It is evident in Figure 18.3(b) that the shape of the mainbeam is perturbed by the spatial nulling. Figure 18.3(d) corresponds to the LFM waveform with four space–frequency nulls placed at −10° and −5 MHz, 60° and 0 MHz, and both −40° and 40° at 5 MHz. The beampattern is distorted by the space–frequency nulls, with nulls closer to the mainbeam generally producing greater distortion.

The classical ambiguity diagrams for the baseline LFM waveform and the three nulling schemes are illustrated in Figure 18.4. The ambiguity diagrams are formed from the sum of the collected data from each channel, i.e. the composite waveform at 0°. There is no observable distortion of the ambiguity diagram for the LFM with spatial nulling in Figure 18.4(b) compared to the baseline LFM case in Figure 18.4(a).

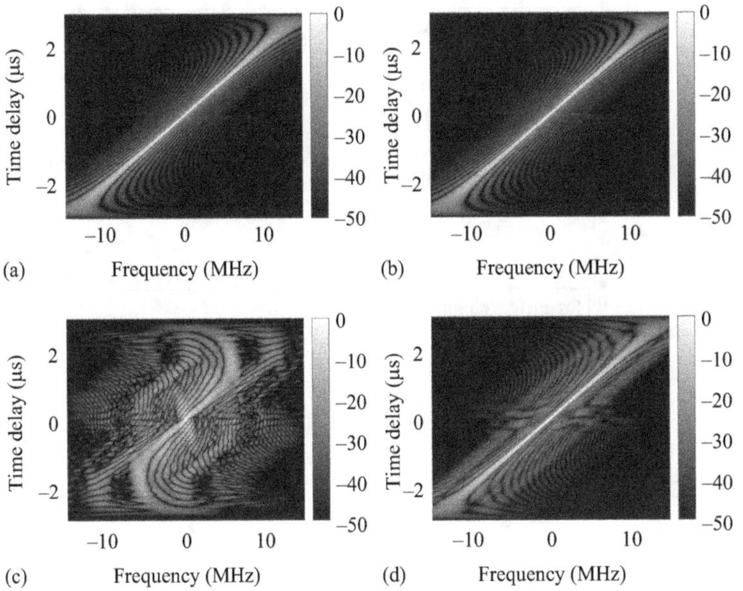

Figure 18.4 *Classical ambiguity diagrams (in dB) of (a) the baseline LFM waveform, (b) the LFM waveform with spatial nulls at 20° and 50°, (c) the LFM waveform with frequency nulls at −5 and 5 MHz, and (d) the LFM waveform with space–frequency nulls at −10° and −5 MHz, 60° and 0 MHz, and both −40° and 40° at 5 MHz*

In contrast, the ambiguity diagram for the LFM with frequency nulling at −5 and 5 MHz shown in Figure 18.4(c) reveals severe distortion, with ambiguity spread away from the time–frequency ridge due to the loss of spectral content in the mainbeam. However, while the ambiguity diagram of the LFM with space–frequency nulling (Figure 18.4(d)) does appear to show more distortion than the spatial nulling case, it is much less than that of the frequency nulling case. In other words, through dimensional coupling, the space–frequency nulling is able to preserve the spectral content in the mainbeam of the transmitted waveform, resulting in an improved ambiguity diagram relative to frequency nulling alone.

Nulling a significant percentage of a radar's bandwidth causes the autocorrelation sidelobes to increase, as seen in Figure 18.4(c), and results in reduced radar performance. Removing a large fraction of the bandwidth also increases the spectral sidelobes, which may violate spectral masks for out-of-band emissions. Frequency nulling, therefore, may not be suitable for some scenarios, such as in the presence of multiple users that span a significant portion of the radar's bandwidth or when strict spectral containment is necessary. In contrast, space–frequency nulling has the ability to null in the spatial directions and frequency bands corresponding to multiple narrowband users. Space–frequency nulling can produce a greater number of narrowband nulls than spatial nulling and does not suffer from the same decrease in radar

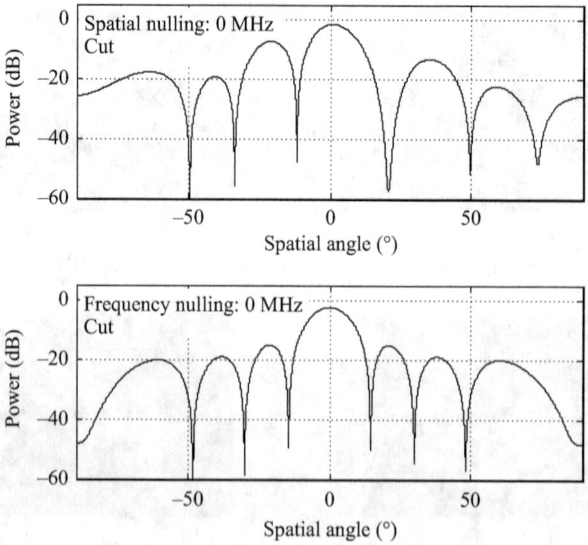

Figure 18.5 Beampattern of the space-nulled waveform from Figure 18.3(b) (top) and the frequency-nulled waveform from Figure 18.3(c) (bottom), both at 0 MHz frequency cuts

performance as does frequency nulling in the presence of multiple narrowband users at different space–frequency locations. An application of space–frequency nulling to a wideband scenario is described in Section 18.5.

Figures 18.5 and 18.6 show beampatterns of the space–frequency emissions from Figure 18.3(b)–18.3(d) at different frequency cuts. The top plot of Figure 18.5 corresponds to a 0 MHz cut of the beampattern of the spatial-nulled LFM from Figure 18.3(b). The nulls at 20° and 50° are clearly visible, but the beampattern shape is distorted. The bottom plot of Figure 18.5 corresponds to a 0 MHz cut of the beampattern of the frequency-nulled LFM from Figure 18.3(c). It is observed that frequency nulling does not impact the beampattern at this frequency because the same waveform is transmitted from each element of the array, although a cut at −5 or 5 MHz would show limited power at any spatial angle.

Figure 18.6 depicts three frequency cuts of the beampattern power for the space–frequency-nulled LFM waveform from Figure 18.3(d). Each plot corresponds to a frequency where the waveform has a null at a particular spatial angle (or angles). The top panel corresponds to the 5 MHz cut, where the nulls at −40° and 40° are visible. The middle panel displays the 0 MHz cut with a spatial null at 60°. The bottom panel corresponds to the −5 MHz cut where a null has been placed in the mainbeam at −10°. The extent of beampattern distortion at each frequency cut is related to the spatial angle of the null, with the null close to the steering direction of 0° causing the most distortion.

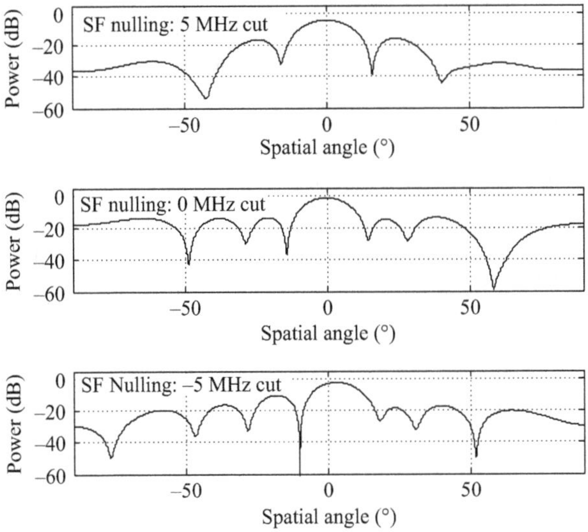

Figure 18.6 Beampattern of the space–frequency nulled waveform from Figure 18.3(d) at 5, 0, and –5 MHz frequency cuts from top to bottom, respectively

18.4.2 Open-air demonstration of space–frequency transmit nulling

The RUWO algorithm has been validated with a variety of open-air experiments, including the most recent efforts that aimed to examine the impact of interference bandwidth on spatial nulling performance. The STAN transmitter and receiver were calibrated according to the method described in [14], and the quiescent pattern of the array was then measured in the configuration shown in Figure 18.2(a) by transmitting an identical waveform from each of the eight transmit elements and recording the power received at the spectrum analyzer as the antenna was rotated. The pattern was measured in azimuth at the highest power elevation angle using an LFM transmitted radar waveform with a 10-μs pulse width, 10 MHz of bandwidth, and a center frequency of 9.9 GHz. Measurements were taken about every 2° in areas of the antenna pattern further from the main beam and away from nulls, and at about every degree or finer elsewhere. Figure 18.7 compares the measured quiescent pattern and the ideal pattern for an eight-element linear array with half-inch element spacing. The patterns are in relatively good agreement, though the measured pattern is not symmetric. All patterns shown here are normalized such that the maximum has a value of 0 dB.

The receive configuration from Figure 18.2(b) was used to record interference from spatial angles of 10° and 25° with a 50 MHz sampling frequency. Three interference signals were considered: a CW tone, a 2 MHz bandwidth LFM, and a 5 MHz

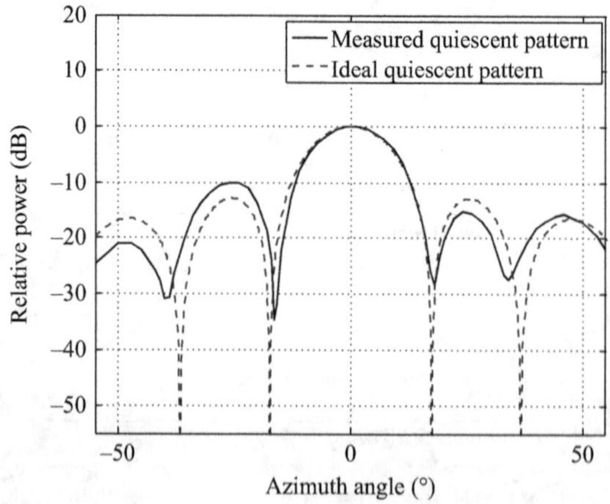

Figure 18.7 Comparison of measured and ideal quiescent antenna patterns

bandwidth LFM. The three interference signals were modulated onto 90 μs pulse trains consisting of 9 identical 10 μs pulses repeated back-to-back. The combination of two spatial angles and three interference signals were recorded using the STAN test bed for a total of six interference scenarios.

The recorded interference was used to adaptively determine waveforms with spatial nulls as described in Section 18.2.2, with 50 μs of training data used to form each data covariance matrix according to the adaptive method described in Section 18.2.5. The interference-to-noise ratio of each of the receiver channels was about 20 dB. Two sets of deterministic weights to produce spatial nulls at 10° and 25° were also formed. The transmitted waveforms were generated by applying the RUWO weights to the LFM test waveform per (18.7), and the patterns corresponding to these sets of waveforms were measured using the same method employed to determine the quiescent pattern.

Figures 18.8 and 18.9 depict data containing pulsed CW interference that was collected by the STAN receiver, before and after adaptive cancellation, using either the phase-only RUWO solution or the phase and amplitude maximum SINR solution. Specifically, Figure 18.8 shows the interference collected at 10° and Figure 18.9 shows the interference collected at 25°. In both figures, the eight channels of data without cancellation have been summed together, i.e. beamformed to 0°. The RUWO solutions in Figures 18.8(a) and 18.9(a) are the result of combining the eight channels of data using the adaptive phase-only weight vector given by (18.6) and the adaptive maximum SINR solutions have been similarly formed using the phase and amplitude maximum SINR weight vector. Figures 18.8(b) and 18.9(b) are provided because the maximum SINR solution is a well-understood standard for comparison. The figures are annotated with the time period that the blanking switch is active, about 13 μs, and the time period after the interference source has finished transmitting when the receiver records only noise with no interference, i.e. the final 10 μs.

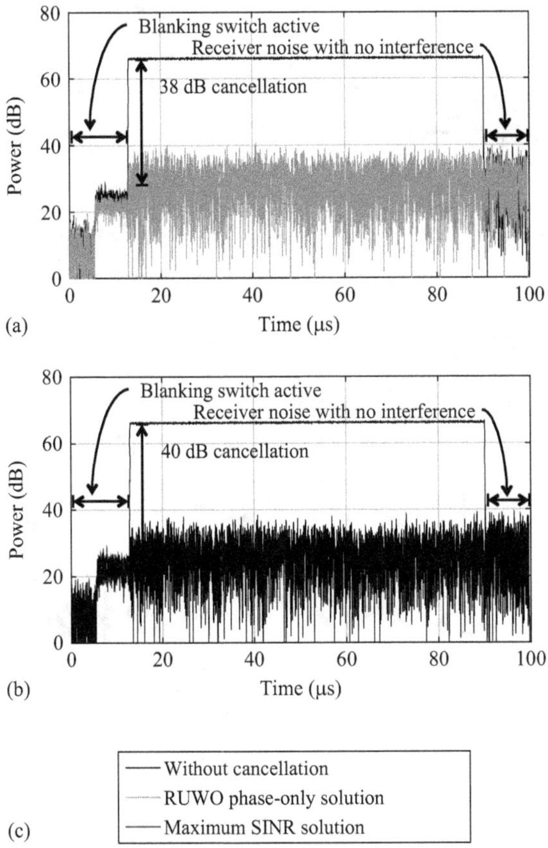

Figure 18.8 Data containing pulsed CW interference at 10° before and after adaptive cancellation using (a) the phase-only RUWO solution and (b) the phase and amplitude maximum SINR solution, with (c) the legend

In Figures 18.8 and 18.9, the adaptive phase-only RUWO weights are able to cancel the interference nearly as well as the adaptive phase and amplitude maximum SINR weights, with RUWO realizing only 2 dB less interference suppression at both 10° and 25°. The maximum SINR solution is expected to perform better because the phase and amplitude weights have more degrees of freedom than the RUWO phase-only weights. However, the ability of the RUWO weights to cancel received interference this well suggests that the RUWO solution can be used to adequately produce a transmit null, assuming the transmitter and receiver are sufficiently calibrated. The RUWO and maximum SINR cancellation performance for data containing LFM interference were quite similar.

Figure 18.10 displays the normalized power of the measured quiescent pattern and the patterns with a spatial null at 10° created by applying RUWO weights obtained

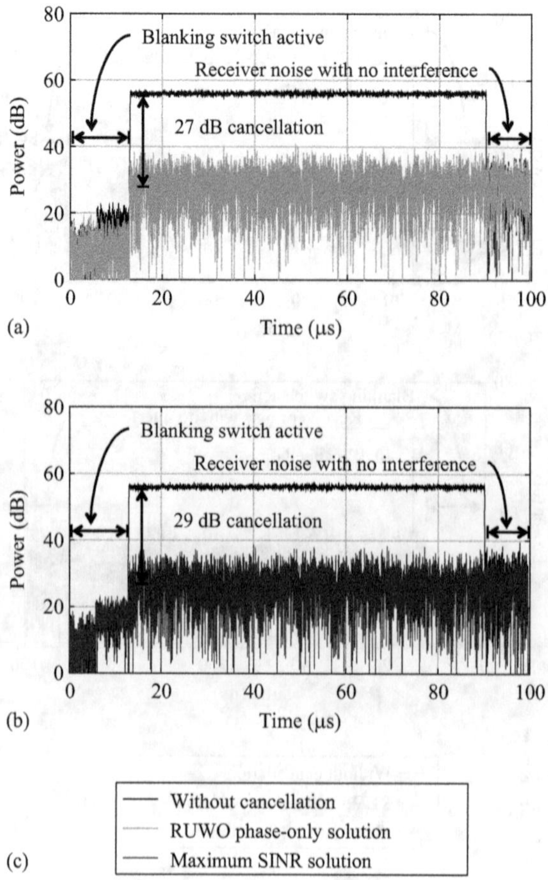

Figure 18.9 Data containing pulsed CW interference at 25° before and after adaptive cancellation using (a) the phase-only RUWO solution and (b) the phase and amplitude maximum SINR solution, with (c) the legend

deterministically or adaptively using each of the three recorded interference signals. Figure 18.10(a) shows the full extent of the measured pattern, while Figure 18.10(b) shows an enlarged view of the null at 10°; both figures use the legend that is given in Figure 18.10(c). The most precise null is obtained at exactly 10° using the tone interference source to produce adaptive weights, but the 2 and 5 MHz bandwidth LFM interference sources produce nulls offset to 9.5° that are 9 and 4 dB deeper, respectively. The deterministic null at 10.5° is both offset in azimuth and is 10 dB less deep than the tone-interference adaptive RUWO null and less deep than both LFM-interference adaptive nulls. The mainbeams of all of the nulled patterns are each offset from 0° to about −6°, but the overall shape of the pattern is unchanged.

Figure 18.11 illustrates the normalized power of the measured quiescent pattern along with the deterministic and adaptive RUWO patterns for a spatial null at 25°.

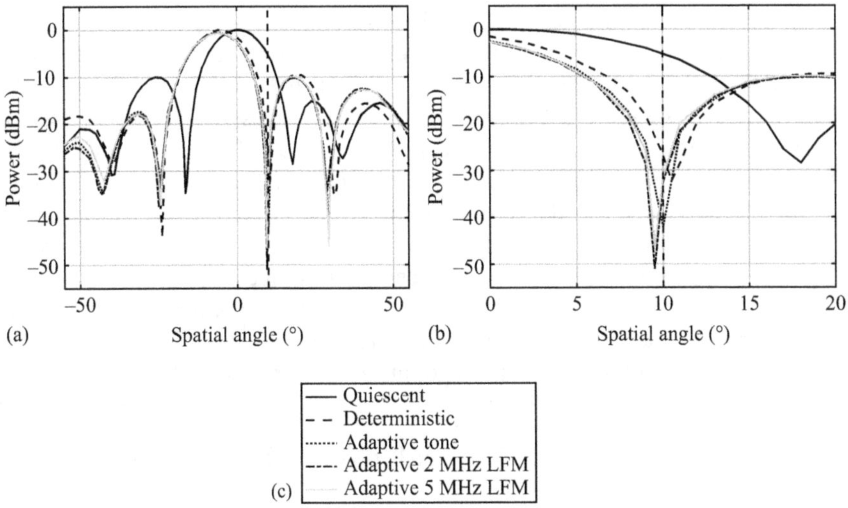

Figure 18.10 *(a) Comparison of the measured quiescent pattern, measured deterministic RUWO pattern, and measured adaptive RUWO patterns using interference sources with three different bandwidths to produce a null at 10°, with (b) an enlarged view of the null, and (c) the legend*

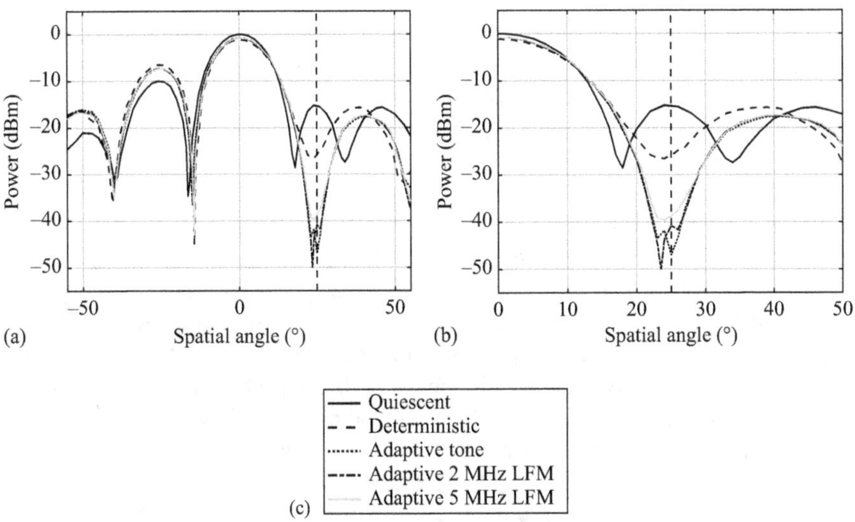

Figure 18.11 *(a) Comparison of the measured quiescent pattern, measured deterministic RUWO pattern, and measured adaptive RUWO patterns using interference sources with three different bandwidths to produce a null at 25°, with (b) an enlarged view of the null, and (c) the legend*

Figure 18.11(a) shows the full extent of the measured pattern, and Figure 18.11(b) shows an enlarged view of the null at 25°. As in the previous example, the most precise null is obtained at exactly 25° using adaptive weights obtained from the pulsed CW interference source. The 2-MHz LFM interference realizes a null that is offset by 1.5° and is 3 dB deeper than that generated by the pulsed CW interference. The 5 MHz LFM interference produces a null that is offset by 1° and is 7.5 dB shallower than the CW case. It is possible that the measurements were not precise enough to determine the true location and depth of the null produced in the 5-MHz interference case, though the results showing the beampattern as a function of frequency would offer more insight into the effects of different bandwidths. The deterministic null is offset by 1° in azimuth and is 20.5 dB shallower than the pulsed CW adaptive null, which is significantly less deep than any of the adaptive pattern nulls.

18.5 Incorporation of RUWO within a wideband emission design problem

Up to this point, RUWO has been applied to a predetermined waveform to form a null or nulled region characterized by some interference covariance matrix **R**. Here, RUWO is combined with an iterative emission design to produce constant amplitude MIMO radar waveforms that match a desired space–frequency allocation of power [11]. The waveforms are emitted in a wideband array configuration, thus the coupling of frequency and spatial transmission angle must be taken into account for both the emission and nulling.

The desired space–frequency allocation of power is defined by shaping the spectrum in certain transmission angles, denoted as "beamlets," according to a predefined spectral window that includes a sufficient portion of the spectral roll-off region. In so doing, the resulting discrete-time signals are spectrally well contained, possess constant amplitude, and can be readily converted into continuous-time waveforms with high fidelity via the PCFM framework [12] discussed in Chapter 6. The beamlets are placed such that the main lobe of the beam (for all operating frequencies) lies entirely within what is known as the "visible domain," which has the effect of minimizing the reactive power that can arise for omni-directional MIMO radar transmissions [18].

The spectra of the beamlets are shaped via an alternating projections algorithm that establishes the amplitude responses of the waveforms in both the time and frequency domains by only retaining the phase in the respective domains, otherwise known as phase retrieval [17,19–23]. This method has been shown to be effective to synthesize polyphase codes via shaping of the power spectrum [23] as well as providing a way to optimize CW and pulsed instantiations of FM noise waveforms [24,25]. However, this method cannot produce nulls of sufficient depth, and thus RUWO is incorporated within the spectral shaping process to facilitate the desired nulls.

18.5.1 Wideband array analysis

For narrowband antenna arrays, the spatially dependent combination of plane waves in the far-field can be approximated as a superposition of phase-shifted versions of the waveforms emitted by the individual array elements. However, in wideband array analysis, the ratio between the antenna spacing and the operating wavelength (sometimes referred to as the electrical spacing) alters significantly over the bandwidth of the emission, and thus the true time delay between the waveforms emitted from different elements must be considered. For a ULA, the frequency dependence of the electrical spacing between elements can cause an array to be spatially oversampled at certain operating frequencies (when the element spacing is less than a half-wavelength). When a ULA is spatially oversampled, transmit power can be placed into what is known as the "invisible" space where no power is actually emitted into the far-field, which leads to reduced power efficiency and possible damage to the RF chain of the transmitter [18,26]. How the invisible space is defined and how it affects a wideband emission is explored below.

18.5.1.1 Wideband beamforming

For an M element, narrowband, ULA with interelement spacing d, the complex-baseband far-field emission as a function of time t and spatial angle θ can be defined as

$$g(t,\theta) = \frac{1}{M}\sum_{m=0}^{M-1} s_m(t)e^{jm\phi(\theta)}, \tag{18.20}$$

where $s_m(t)$ is the complex baseband representation of the continuous, pulsed waveform of duration T emitted by the mth element in the array. The electrical angle $\phi(\theta)$ in (18.20) is defined as

$$\phi(\theta) = \frac{2\pi}{\lambda_{\text{cent}}} d \sin\theta, \tag{18.21}$$

for λ_{cent} the wavelength of center frequency f_{cent}. The spacing d can be defined as half the wavelength λ_{d} corresponding to some frequency f_{d} as

$$d = \frac{\lambda_{\text{d}}}{2} = \frac{c}{2f_{\text{d}}}, \tag{18.22}$$

for c the speed of light. Using this notation, the electrical angle can be represented as

$$\phi(\theta) = \pi\frac{f_{\text{cent}}}{f_{\text{d}}}\sin\theta, \tag{18.23}$$

which is a convenient form for generalizing electrical angle to the frequency dependence that occurs in wideband emissions.

For wideband arrays, the electrical angle becomes dependent on the continuous passband frequency f as

$$\phi(f,\theta) = \pi\frac{f}{f_{\text{d}}}\sin\theta. \tag{18.24}$$

Because of this frequency dependence, the waveforms $s_m(t)$ must be represented in the Fourier domain to facilitate application of the phase shifts defined in (18.24).

Thus, the complex-baseband emission $G(f, \theta)$ as a function of passband frequency f and spatial angle θ becomes

$$G(f,\theta) = \frac{1}{M} \sum_{m=0}^{M-1} \left[\int_0^T s_m(t) e^{-j2\pi(f-f_{\text{cent}})t} dt \right] e^{jm\phi(f,\theta)} \tag{18.25}$$

where the expression inside the brackets $[\cdot]$ represents the Fourier transform of waveform $s_m(t)$. Taking the inverse Fourier transform of (18.25) therefore yields

$$g(t,\theta) = \frac{1}{M} \sum_{m=0}^{M-1} s_m\left(t + \frac{m}{2f_{\text{d}}} \sin\theta\right), \tag{18.26}$$

which is the wideband analog to the far-field, narrowband emission in (18.20). The formulation in (18.26) reveals that for a wideband emission, the true time delay between the waveforms emitted by the elements must be taken into account to represent the complex envelope of the far-field emission. The aggregate (average) beampattern as a function of spatial angle θ displays the power allocation in space by integrating (18.26) over the pulse width as

$$B(\theta) = \frac{1}{T} \int_0^T |g(t,\theta)|^2 dt. \tag{18.27}$$

Now consider an $N \times M$ discretized complex-baseband waveform matrix $\mathbf{S}$ in which the columns correspond to the waveforms emitted by each of the M antenna elements, with N the length of the discretized waveforms in the time domain. The discrete length of the waveform is defined as

$$N = f_{\text{samp}} T, \tag{18.28}$$

where f_{samp} is the sampling frequency of the waveforms contained in $\mathbf{S}$. Because it is theoretically impossible to Nyquist sample a time-limited signal, each waveform is "oversampled" with respect to some bandwidth measure to adequately represent the continuous nature of the waveforms in a discrete form. By doing so, the "oversampling" captures a sufficient portion of the spectral roll-off of the emission such that the power outside the sampled bandwidth can be considered negligible.

The waveform matrix $\mathbf{S}$ can be vectorized into an $MN \times 1$ vector $\mathbf{s} = \text{vec}(\mathbf{S})$ where the waveforms are stacked starting with the leftmost waveform vector. Thus, the complex-baseband emission as a function of passband frequency f and spatial angle θ from (18.25) can be well approximated as

$$G(f,\theta) = \frac{1}{M} [\mathbf{t}(f,\theta)]^H \mathbf{s}, \tag{18.29}$$

where $(\cdot)^H$ is the complex conjugate transpose (or Hermitian) operation. The $MN \times 1$ space–frequency steering vector

$$\mathbf{t}(f,\theta) = \mathbf{v}(f,\theta) \otimes \mathbf{u}(f), \tag{18.30}$$

for $\otimes$ the Kronecker product, is comprised of the $M \times 1$ frequency-dependent steering vector

$$\mathbf{v}(f,\theta) = \begin{bmatrix} 1 & e^{-j\phi(f,\theta)} & \cdots & e^{-j(M-1)\phi(f,\theta)} \end{bmatrix}^T \tag{18.31}$$

and the $N \times 1$ discrete-time Fourier transform (DTFT) vector

$$\mathbf{u}(f) = \left[1 \quad e^{j2\pi\left((f-f_{\text{cent}})/f_{\text{samp}}\right)} \quad \cdots \quad e^{j2\pi(N-1)\left((f-f_{\text{cent}})/f_{\text{samp}}\right)} \right]^T, \tag{18.32}$$

which is a function of continuous passband frequency f and sampling rate f_{samp}.

Naturally, the spectrum must also be discretized as it is infeasible to represent this beamformer in continuous frequency using digital signal processing. Thus, a discrete Fourier transform (DFT) is used instead of the DTFT to transform the emission into the frequency domain. This type of beamforming is known as the DFT beamformer for wideband arrays [27].

Discretizing the frequency spectrum into Q equally spaced points defines the discrete baseband frequencies as

$$f_q = -\frac{f_{\text{samp}}}{2} + \frac{q}{Q} f_{\text{samp}} \tag{18.33}$$

for $q = 0, 1, \ldots, Q-1$. Therefore, the discrete passband frequency becomes

$$f = f_q + f_{\text{cent}}. \tag{18.34}$$

Inserting (18.33) and (18.34) into (18.32) yields the $N \times 1$ DFT vector at discrete baseband frequency f_q as

$$\mathbf{u}(f_q) = \begin{bmatrix} 1 & e^{j2\pi((-1/2)+(q/Q))} & \cdots & e^{j2\pi(N-1)((-1/2)+(q/Q))} \end{bmatrix}^T. \tag{18.35}$$

To prevent aliasing of the complex envelope of the emission due to the true time delay shifts that occur in wideband beamforming, the number of frequency points Q needs to be overspecified relative to the discretized length of the waveform N (i.e. $Q > N$). When the number of frequency points Q is greater than the number of time samples N, the overspecified spectral response represents a zero-padding in the time-domain that accounts for the possible sample delays. Another point of concern is that discrete, time-shifted signals exhibit the well-known Gibbs phenomenon at noninteger sample delays, which produces time sidelobes beyond the extent of the pulsewidth [28] that can also alias in the time-domain. Requiring $Q \gg N$ creates a large enough time window to allow for true-time sample delays and adequately reduces the error due to aliasing of Gibbs sidelobes. It has been observed that $Q = 2N$ frequency points are sufficient to reduce this aliasing-induced error to a negligible amount.

18.5.1.2 Wideband definition and the narrowband assumption

For a signal impinging on or emitted from an array, the methods used to define whether a signal is narrowband depends not only on the spectral content of the signal but also on the geometry of the array. As a result, there are two distinct narrowband definitions: one that we shall call the *spectral* narrowband definition and the other the *array* narrowband definition.

The spectral narrowband definition is based solely on the bandwidth of the signal compared to the center frequency. This definition is based on the fractional (or percent) bandwidth

$$\%\mathrm{BW} = \frac{B}{f_{\mathrm{cent}}}, \tag{18.36}$$

where B is some measure of bandwidth of the emission. Here, we define the bandwidth as the frequency range that contains 98% of the total waveform power centered around f_{cent}, though there are many others (see [29]). The signal is assumed to be narrowband if the fractional bandwidth is below some predefined threshold (10% is commonly used).

In contrast, the array narrowband definition is based on the interplay between the array geometry and the bandwidth of the signal as

$$B \cdot \Delta\tau_{\max} \ll 1 \tag{18.37}$$

where

$$\Delta\tau_{\max} = \frac{M-1}{2f_{\mathrm{d}}} \tag{18.38}$$

is the largest delay term from (18.26), which represents the maximum time shift between waveforms emitted from the array. This relationship allows for the time delay between elements to be treated as a phase shift. Inserting (18.38) into (18.37) yields

$$B\left(\frac{M-1}{2f_{\mathrm{d}}}\right) \ll 1, \tag{18.39}$$

which can be rewritten in terms of fractional bandwidth as

$$\%\mathrm{BW}\left(\frac{M-1}{2f_{\mathrm{d}}/f_{\mathrm{cent}}}\right) \ll 1. \tag{18.40}$$

When the expression in (18.40) approaches 1, the true-time difference between elements must be considered. The number of array elements M, the frequency ratio $f_{\mathrm{d}}/f_{\mathrm{cent}}$, and the fractional bandwidth %BW are the three relative parameters that fully characterize a wideband ULA.

18.5.1.3 Physical considerations and the invisible space

The mutual coupling of power between antenna elements needs to be considered for any emission design involving multiple antenna elements. The driving (or active) impedances of the elements within an array is dependent on the complex weightings (amplitude and phase) of each element in the array [30,31] and directly defines the amount of reflected power due to impedance mismatch. When considering independent control of each waveform in the array, the complex weights across the array become time-varying, thus creating a time-varying driving impedance for each element that can impact the performance of a MIMO radar emission [32,33]. Large reflections can occur if the electrical spacing between elements becomes sufficiently less than a half-wavelength, such that the emission “radiates” into what is known as

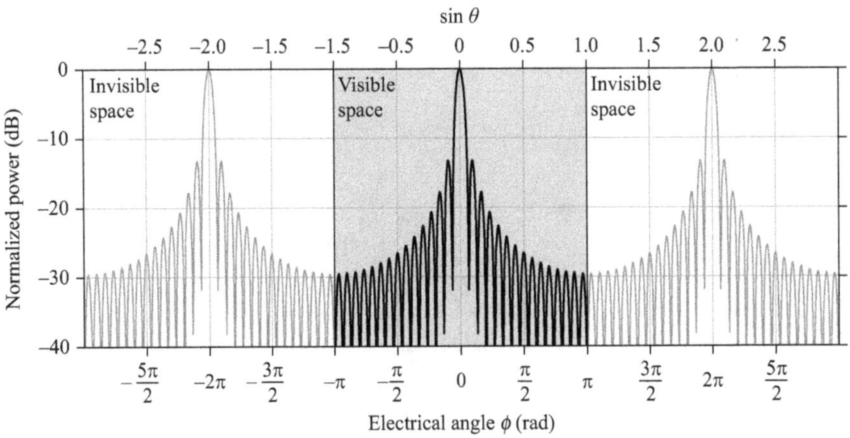

Figure 18.12 Visible and invisible regions for an $M = 30$ element narrowband array with $f/f_d = 1$

the invisible (or imaginary) space [18]. The term "radiate" is really a misnomer in this context as the majority of the power is actually not radiated due to the element drive impedances becoming largely reactive. Instead, the result is storage of additional energy in the near-field of the array, which can lead to large amounts of reflected power that could damage the transmitter [18,26].

The invisible space resides beyond the end-fire direction (in terms of electrical angle) for linear arrays and contains the well-known grating lobe series [30]. This invisible space can be represented as a range of complex-valued spatial angles whereby $|\sin\theta| > 1$. Figure 18.12 defines the visible and invisible regions for an $M = 30$ element narrowband ULA with half wavelength spacing ($f/f_d = 1$). The feasible array excitations reside within electrical angles $\phi \in [-\pi, \pi]$ due to wrapping of the phase beyond these values. For $f/f_d = 1$, these electrical angles perfectly fill the visible region ($|\sin\theta| \leq 1$) that is shaded. Therefore, if the main beam is steered into the invisible space past one end-fire direction, it will be "replaced" by a grating lobe entering the visible space from the opposite end-fire direction.

Now consider the same $M = 30$ narrowband ULA albeit with quarter-wavelength element spacing that equates to $f/f_d = 0.5$. Figure 18.13 illustrates the invisible and visible regions for this array scenario. The visible space is now entirely contained within the electrical angle values $|\phi| \leq \pi/2$. Thus, some percentage of the total available transmit power resides in the region $\pi/2 < |\phi| \leq \pi$ that does not correspond to any physical spatial direction. This region is referred to as the *reactive region* and is associated with reflected power. From an efficiency standpoint, it is not advantageous to "transmit" energy into the reactive region as not all of this energy would actually be emitted. In fact, if large amounts of energy are placed in this region, the amount of reflected power could even damage the RF components of the transmit chain [18].

For wideband arrays, the frequency-dependence of the electrical spacing between antenna elements becomes significant and cannot be assumed a constant over the

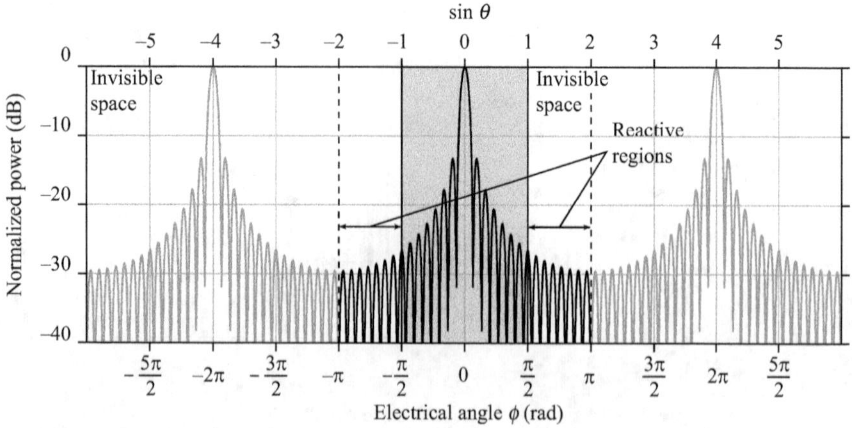

Figure 18.13 Visible, invisible, and reactive regions for an $M = 30$ element narrowband array with $f/f_d = 0.5$

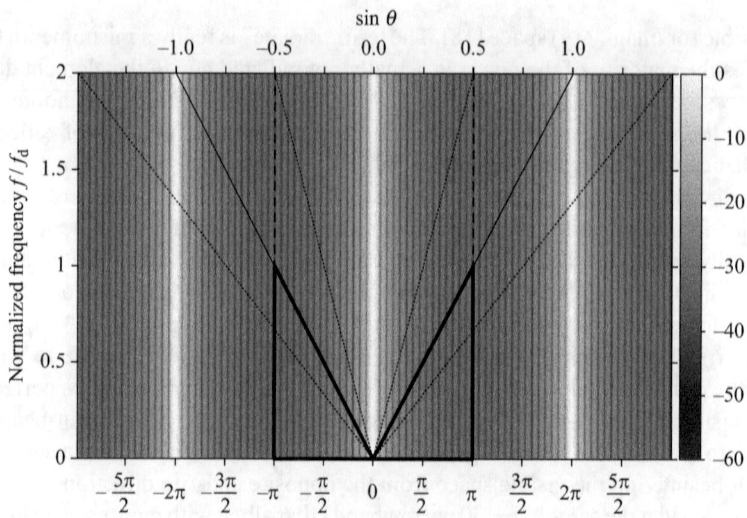

Figure 18.14 Visible ($|\sin\theta| <= 1$), invisible ($|\sin\theta| > 1$), and reactive regions for an $M = 30$ element narrowband array with variable f/f_d. The reactive regions are bounded inside the black triangles

bandwidth. Thus, the ratio of invisible to visible space changes over the bandwidth of the emission. For the example of an $M = 30$ element ULA, Figure 18.14 shows how this ratio changes as the normalized frequency f/f_d is varied. The diagonal traces correspond to particular values of $\sin\theta$, while the vertical dashed lines signify the electrical angles of $\phi = \pm\pi$ just as in Figures 18.12 and 18.13. The reactive region is bounded within the black triangles and only occurs for frequencies $f < f_d$. When designing emissions where any portion of the overall bandwidth occurs at frequencies $f < f_d$, the power distribution of the emission should minimize (to a reasonable extent) the power contained within this region.

An efficiency metric can be defined that takes into account the reduction in visible region power due to "transmission" into the invisible region. Define the fractional reactive power (FRP) as the ratio of average reactive power contained in the invisible domain to the total available power of the emission as

$$\mathrm{FRP} = \frac{\displaystyle\int_0^{f_d}\left[\int_{-\pi}^{-\pi f/f_d}|G(f,\phi)|^2 d\phi + \int_{\pi f/f_d}^{\pi}|G(f,\phi)|^2 d\phi\right] df}{\displaystyle\int_0^{\infty}\int_{-\pi}^{\pi}|G(f,\phi)|^2 d\phi df} \times 100\%. \tag{18.41}$$

The formulation in (18.41) is based on the assumption that the energy contained within the invisible space accounts for all of the reactive power in the emission and that the energy contained within the visible space is fully emitted (no reflected power). This efficiency definition does not account for ohmic losses or reflections due to scan impedance variation within the visible space [30,34].

For uncorrelated MIMO radar emissions (when $\mathbf{S}^H\mathbf{S}$ is approximately proportional to the identity matrix), the energy is spread evenly over all electrical angles, thereby producing a theoretically omni-directional beampattern (in terms of average power) [35]. Thus, if the operating frequency (or portion of the bandwidth for wideband emissions) is less than f_d, a portion of the total available power will be placed inside the reactive region. As an example, consider a discretized waveform matrix $\mathbf{S}$ such that $\mathbf{S}^H\mathbf{S}$ is approximately proportional to the identity matrix and is used to produce a wideband MIMO radar emission having a fractional bandwidth of $\%\mathrm{BW} = 40\%$. The array is comprised of $M = 30$ uniformly spaced antenna elements with interelement spacing d set according to (18.22) with $f_d = 1.2 f_{\mathrm{cent}}$ (the highest frequency within the bandwidth), which is a common practice to prevent grating lobes from appearing [27].

Figure 18.15 shows the frequency content of the described emission versus electrical angle $\phi(f,\theta)$ and normalized frequency f/f_{cent}, where the diagonal traces correspond to different values of $\sin\theta$. Because the MIMO radar waveforms have low crosscorrelations, power is emitted over all $\phi(f,\theta)$ space, which allows a portion of the available power to be contained within the reactive regions. Based on (18.41), the FRP for this emission is 17.4% of the total available power. However, by incorporating knowledge of the number of elements in the array M, the relationship between element spacing and operating frequency f_d/f_{cent}, and fractional bandwidth %BW into the emission design process, the amount of power contained within the reactive regions can be reduced, thereby increasing the overall efficiency of the emission.

18.5.2 RUWO within spectral shaping optimization

Before designing MIMO radar waveforms that meet some set of emission requirements, consider the design of a single waveform to fit to a Gaussian spectral mask with a predefined 98% bandwidth. Further, a large spectral null is to be placed between the normalized frequencies of $f/B = 0.24$ and $f/B = 0.4$. Note that the 98% bandwidth

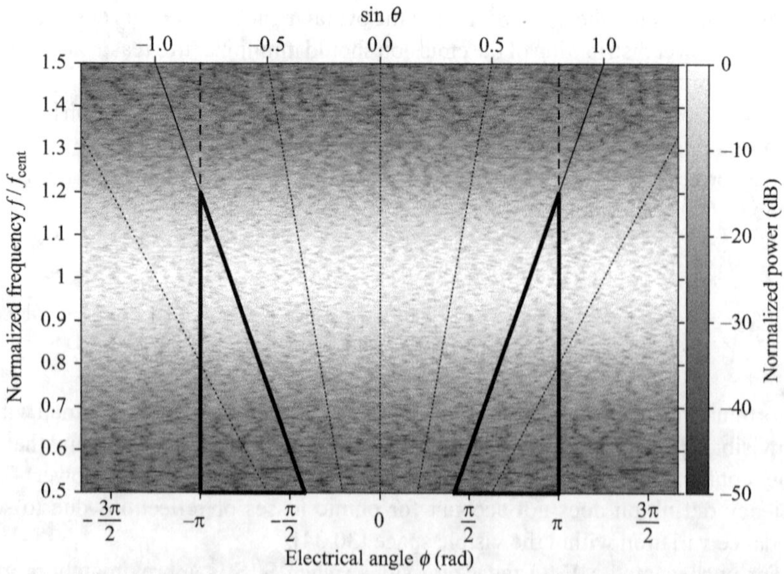

Figure 18.15 Spectrum of wideband (%BW = 40% bandwidth) omni-directional MIMO radar emission versus angle $\phi(f, \theta)$ for an $M = 30$ element ULA with $f_d = 1.2 f_{cent}$. The reactive regions are bounded inside the black triangles and comprise 17.4% of the total available power

is defined according to the base spectral shape and does not alter when the null is introduced. The waveform is designed using an alternating projections approach that leverages phase retrieval by alternating between the time and frequency domains while satisfying some constraint at each stage [17,19–22]. For radar waveform design, the constraints typically involve imposing particular envelopes in the time and frequency domains, commonly known as the Gerchberg–Saxton Algorithm (GSA) [19]. Here, the frequency envelope is set to be the Gaussian window containing a spectral notch in the desired location(s) and the temporal envelope is a rectangular pulse shape. Given some waveform initialization $s_0(t)$ and denoting $X(f)$ as the desired frequency envelope, the process is described below. Note that $\angle(\cdot)$ extracts the angle of the argument, while $\mathscr{F}\{\cdot\}$ and $\mathscr{F}^{-1}\{\cdot\}$ are the Fourier and inverse Fourier transforms, respectively.

GERCHBERG–SAXTON ALGORITHM
Initialize: $s_0(t)$, $i = 0$
repeat
  $S_i(f) = \mathscr{F}\{s_i(t)\}$
  $\tilde{S}_i(f) = X(f)\exp(j\angle S_i(f))$
  $\tilde{s}_i(t) = \mathscr{F}^{-1}\left\{\tilde{S}_i(f)\right\}$
  $s_{i+1}(t) = \exp(j\angle \tilde{s}_i(t))$
  $i = i + 1$
until sufficiently converged
return $s_i(t)$

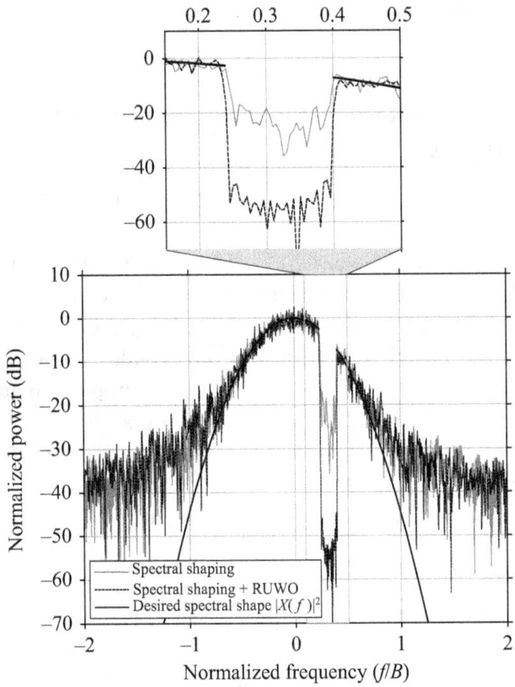

Figure 18.16 Comparison of converged spectra using alternating projections spectral shaping alone or with RUWO

A common problem that occurs with GSA is that higher power areas of the desired frequency window $X(f)$ are prioritized over low power areas (i.e. nulls). As a result, nulled regions tend to be rather shallow. However, the RUWO method described earlier in the chapter can be combined with GSA (or any iterative spectral shaping approach) to ensure that sufficiently deep nulls are enforced. There are two general ways in which RUWO could be incorporated into a spectral shaping algorithm: (1) serially in which RUWO is applied after completion of the spectral shaping optimization or (2) "inside-the-loop" whereby RUWO is applied at each iteration of spectral shaping optimization. When RUWO is applied in each iteration, it is performed until sufficient convergence is achieved. It has been found that the latter scheme tends to provide a better match to a desired notched frequency window $X(f)$, though the former incurs a lower computational cost.

Figure 18.16 shows the desired spectral window $X(f)$ and converged waveform spectra using GSA alone and GSA combined with RUWO (both with same initialization). The RUWO algorithm was implemented inside-the-loop and thus is applied at every iteration of the spectral shaping process. Incorporating RUWO with GSA has driven the null about 25–30 dB deeper relative to the GSA-only approach. In the following section, this incorporation of RUWO with in a spectral shaping algorithm is expanded to design wideband MIMO radar waveforms such that the far-field emission contains space–frequency nulls while adhering to a desired space–frequency allocation of power.

18.5.3 Wideband MIMO radar emission optimization with space–frequency nulling

The design of a wideband emission can be realized by shaping the emitted spectrum in certain predefined angles denoted as "beamlets," where each beamlet adheres to a desired spectral envelope (e.g. Gaussian) with a specified bandwidth. The M elements of a ULA emit M waveforms that are constrained to have constant envelopes in the time domain. For optimization, the waveform is discretized and oversampled (with respect to the specified bandwidth measure) to account for spectral roll-off and to facilitate sufficient fidelity for subsequent continuous implementation. The spectral shaping of the "beamlets" is administered using an iterative algorithm not unlike the GSA method described in Section 18.5.2. The space–frequency nulling is performed using the space–frequency version of RUWO discussed in Section 18.2.4, though the spatial steering vectors are now dependent on frequency f.

18.5.3.1 Parameter initialization

Beamlet steering vectors

Let the ULA have element spacing d as defined in (18.22). The sampling frequency is $f_{\text{samp}} = \alpha B$, where α is a scalar oversampling factor and B is set to be the 98% bandwidth. It has been found that $\alpha = 3$ tends to provide sufficient oversampling to adequately approximate a continuous-time waveform. The discrete length of the waveform defined in (18.28) can thus be written as $N = \alpha BT$, for BT the time–bandwidth product.

Denote $\Theta_P = \{\theta_0, \ldots, \theta_{P-1}\}$ as the set containing the P beamlet directions to be included in the design. From (18.24), (18.33), and (18.34), the pth beamlet spatial angle θ_p corresponds to the discretized frequency-dependent electrical angles

$$\phi(f_q, \theta_p) = \pi \frac{\left(f_{\text{samp}}\left(-(1/2) + (q/Q)\right) + f_{\text{cent}}\right)}{f_{\text{d}}} \sin\theta_p, \tag{18.42}$$

for $q = 0, 1, \ldots, Q-1$ frequency points over the set of $p = 0, 1, \ldots, P-1$ beamlets. Inserting $f_{\text{samp}} = \alpha B$ into (18.42) and dividing the numerator and denominator by f_{cent} yields

$$\phi(f_q, \theta_p) = \pi \frac{(\%\text{BW} \times \alpha(-(1/2) + (q/Q)) + 1)}{f_{\text{d}}/f_{\text{cent}}} \sin\theta_p, \tag{18.43}$$

which can be readily inserted into (18.31) to form the steering vectors $\mathbf{v}(f_q, \theta_p)$ corresponding to the P beamlets, and subsequently the space–frequency steering vectors $\mathbf{t}(f_q, \theta_p)$ using (18.30). The formulation in (18.43) permits direct implementation of the fractional bandwidth %BW and the ratio $f_{\text{d}}/f_{\text{cent}}$ that determines the relative spacing of elements in the ULA. Define

$$\mathbf{T}(\theta_p) = \left[\mathbf{t}(f_0, \theta_p) \quad \cdots \quad \mathbf{t}(f_{Q-1}, \theta_p)\right] \tag{18.44}$$

as the $MN \times Q$ space–frequency transformation matrix that jointly performs a DFT and steers toward spatial angle θ_p.

The beamlet directions in Θ_P should be chosen such that the main lobe of each beamlet is contained entirely within the visible domain (for in-band frequencies) to minimize reactive power as discussed in Section 18.5.1.3. The peak-to-null beamwidth W in $\sin\theta$ space for a ULA with M elements and interelement spacing d is

$$W = \frac{c}{Mfd}.$$

Inserting the representation of d from (18.22) then yields

$$W = \frac{2f_d}{Mf}.$$

The widest beamwidth for in-band frequencies occurs at $f = f_{\text{cent}} - B/2$. The maximum peak-to-null beamwidth for in-band frequencies is therefore

$$W_{\max} = \frac{2f_{\text{d}}/f_{\text{cent}}}{M(1 - (\%\text{BW}/2))}.$$

Thus, the beamlet directions defined in Θ must be chosen such that each main lobe within the set avoids the invisible domain for all in-band frequencies, which is represented by the inequality

$$\sin\Theta_P \leq (1 - W_{\max}) = 1 - \frac{2f_{\text{d}}/f_{\text{cent}}}{M(1 - (\%\text{BW}/2))}.$$

Spectral windows and adaptive beamlet weighting

Each of the P beamlets is shaped according to a desired spectral window $X(f, \theta_p)$. Define the $Q \times 1$ (magnitude) spectral window for the pth beamlet as

$$\mathbf{x}(\theta_p) = \left[X(f_0, \theta_p) \quad \cdots \quad X(f_{Q-1}, \theta_p)\right]^T, \tag{18.45}$$

which is scaled to direct a desired amount of power to the corresponding beamlet. For simplicity, we shall assume that the P spectral windows are identical, though the beamlets can have nonidentical spectral windows.[1] Note that the spectral window extends over the frequency interval $f_{\text{cent}} \pm f_{\text{samp}}/2$, such that the 98% bandwidth is contained in $1/\alpha$ of the total spectral window centered at f_{cent}.

Depending on the number of antenna elements M, the P beamlet steering vectors may be correlated to some extent, with a large number of closely spaced beamlets likely producing ill-conditioning effects. One approach to compensate for this effect is to adapt the relative scaling of the spectral windows $\mathbf{x}(\theta_p)$ to prevent the design process from over-emphasizing (or under-emphasizing) a given beamlet, thus ensuring that the desired spatial beampattern is formed. Define $\beta(\theta_p)$ as the adaptive scaling that dictates the relative contributions of the pth beamlet so as to achieve the desired spatial allocation of power when constructing $\mathbf{x}(\theta_p)$ for $p = 0, \ldots, P-1$. Thus, the spectral window including the adaptive weighting is $\beta(\theta_p)\mathbf{x}(\theta_p)$. Computation of this

[1]The same spectral shape should be used when approximating a wide beam using multiple beamlets.

adaptive scaling is discussed in Section 18.5.3.2. Interestingly, through this beam weighting method, it is shown in Section 18.5.4 that choosing a number of beams larger than the number of antenna elements (i.e. $P > M$) better matches the desired spatial distribution than the case when $P = M$. For further discussion on the benefits of using adaptive scaling, see [11].

Interference covariance matrix

Section 18.2.5 described how deterministic or adaptive interference covariance matrices can be formed for use with the RUWO method. Within the context of this wideband formulation, define L space–frequency pairs (f_ℓ, θ_ℓ) for $\ell = 0, 1, \ldots, L-1$ and denote $\Theta_L = \{\theta_0, \ldots, \theta_{L-1}\}$ and $\mathbf{f}_L = \{f_0, \ldots, f_{L-1}\}$ as the sets containing the spatial angles and corresponding frequencies to be nulled. These points can be condensed into a region to achieve a broad null or can be placed individually. The $MN \times MN$ deterministic wideband, space–frequency interference matrix is then

$$\mathbf{R} = \mathbf{D}\mathbf{D}^H + \delta \mathbf{I}, \tag{18.46}$$

where δ is a loading factor, $\mathbf{I}_{MN}$ is an $MN \times MN$ identity matrix, and

$$\mathbf{D} = \begin{bmatrix} \mathbf{t}(f_0, \theta_0) & \mathbf{t}(f_1, \theta_1) & \cdots & \mathbf{t}(f_{L-1}, \theta_{L-1}) \end{bmatrix} \tag{18.47}$$

is the $MN \times L$ matrix containing the collection of space–frequency steering vectors to be nulled. The loading factor δ supplements the rank of the interference matrix so that it can be inverted. If the loading factor is too large, the null may be too shallow due to the loading factor masking essential eigenvalues.

If the space–frequency null locations defined by the sets Θ_L and $\mathbf{f}_L$ coincide with a value in $X(f_q, \theta_p)$, for $p = 0, 1, \ldots, P-1$ and $q = 0, 1, \ldots, Q-1$, then the spectral window must also be nulled accordingly to prevent the design process from filling in the null when shaping the spectrum. Note that nulling within a spectral window necessitates additional scaling of $\mathbf{x}(\theta_p)$ to achieve the desired power within the beam containing the null.

18.5.3.2 Emission design via alternating projections

Given some initialization of the vectorized waveform matrix $\mathbf{s}_0$ and the initializations discussed in Section 18.5.3.1, the $Q \times 1$ frequency vector of the pth beamlet at the ith iteration can be expressed as

$$\mathbf{g}_i(\theta_p) = \frac{1}{M} \mathbf{T}^H(\theta_p)\, \mathbf{s}_i. \tag{18.48}$$

Using (18.48), the adaptive scaling for the pth beamlet is updated as

$$\beta_{i+1}(\theta_p) = \beta_i(\theta_p) \sqrt{\frac{\left\|\mathbf{x}(\theta_p)\right\|^2 / \gamma_{\mathrm{x}}}{\left\|\mathbf{g}_i(\theta_p)\right\|^2 / \gamma_{\mathrm{g},i}}}, \tag{18.49}$$

where

$$\gamma_{\mathrm{g},i} = \sum_{p=0}^{P-1} \left\|\mathbf{g}_i(\theta_p)\right\|^2 \tag{18.50}$$

and

$$\gamma_{\mathrm{x}} = \sum_{p=0}^{P-1} \left\| \mathbf{x}(\theta_p) \right\|^2 \tag{18.51}$$

are the total powers contained in the beamlets of the current iteration i and the predefined spectral windows, respectively. As the algorithm progresses, the term inside the $\sqrt{\cdot}$ of (18.49) converges to unity, and thus $\beta(\theta_p)$ converges to a beam scaling value that produces the desired spatial beam power specified in $\mathbf{x}(\theta_p)$.

The scaled spectral mask $\beta(\theta_p)\mathbf{x}(\theta_p)$ for the pth beam is applied to $\mathbf{g}_i(\theta_p)$ as

$$\tilde{\mathbf{g}}_i(\theta_p) = \beta_{i+1}(\theta_p)\left[\mathbf{x}(\theta_p) \odot \exp\left(j\angle \mathbf{g}_i(\theta_p)\right)\right] \tag{18.52}$$

for $\odot$ the Hadamard product. The unconstrained (in amplitude) vectorized waveform set at the ith iteration can then be determined as

$$\tilde{\mathbf{s}}_i = \sum_{p=0}^{P-1} \mathbf{T}(\theta_p)\, \tilde{\mathbf{g}}_i(\theta_p), \tag{18.53}$$

which is equivalent to projecting each beamlet back onto the array, taking the inverse Fourier transform, and summing the P responses. This process generates an approximation to the unconstrained waveform matrix that produces the response in (18.52) since it does not account for correlation between the beams when summing the responses. While inversion techniques could be applied here to reduce the correlation between the beamlets, care must be taken because the inversion problem could be ill-conditioned. However, the adaptive scaling method above has been observed to converge to meaningful solutions for a myriad of different simulation scenarios.

The RUWO loop then begins by removing the amplitude variation of the waveforms in (18.53) via $\mathbf{a}_0 = \exp(j\angle\tilde{\mathbf{s}}_i)$. Per (18.3), the update of RUWO at the kth iteration is

$$\mathbf{a}_{k+1} = \exp\left(j\angle(\mathbf{R}^{-1}\mathbf{a}_k)\right) \tag{18.54}$$

based on the interference covariance matrix from (18.46). The process is performed for some predefined number of iterations K (or based on some stopping criterion). The convergence speed of RUWO is dependent on the rank of the interference matrix $\mathbf{R}$, though it typically converges to a stable solution within $K = 50$ iterations [14]. The vectorized $NM \times 1$ waveform set is then updated as

$$\mathbf{s}_{i+1} = \mathbf{a}_K. \tag{18.55}$$

The overall process involving (18.48), (18.49), and (18.52)–(18.55) is repeated a predefined number of times Φ (or likewise based on some stopping criterion). The constant modulus, discrete-time, baseband representation of the optimized waveform set $\mathbf{s}_\Phi$ can be translated into M continuous-time waveforms using the PCFM implementation [12]. Any loss in null sensitivity due to conversion to continuous-time can be mitigated by applying the design process at a higher sampling rate. Table 18.1 provides a summary of the optimization steps. The approach described here involves the design of a wideband emission including space–frequency nulls. The reader is referred to [11] for the implementation that excludes space–frequency nulling.

Table 18.1 Wideband MIMO emission design with space–frequency nulling

1. Establish the number of antenna elements M and discretized waveform length N for "oversampling" factor α and time–bandwidth product BT
2. Select the number of beamlets P and specific directions Θ_P, the number of frequency bins Q, percent bandwidth %BW, and the ratio f_d/f_{cent} that dictates element spacing
3. Determine the L space–frequency points Θ_L and $\mathbf{f}_L$ to be nulled (if any). Form the $NM \times NM$ interference matrix $\mathbf{R}$ using (18.46) and calculate the inverse. Select the number of RUWO iterations K
4. Form the spectral shaping vector $\mathbf{x}(\theta_p)$ for each of the P beamlets. Zero the corresponding values in $X(f_q, \theta_p)$ if they contradict with a nulled region
5. Set the number of optimization iterations to perform Φ. Initialize the vectorized waveform set $\mathbf{s}_0$, the loop index to $i = 0$, and the adaptive scaling coefficients $\beta_0(\theta_p)$ to unity
6. Form $\mathbf{g}_i(\theta_p)$ and update $\beta_{i+1}(\theta_p)$ for $p = 0, \ldots, P-1$ via (18.48) and (18.49)
7. Apply beamscaling $\beta_{i+1}(\theta_p)$ and spectral shaping using $\mathbf{x}(\theta_p)$ to form $\tilde{\mathbf{g}}_i(\theta_p)$ for $p = 0, \ldots, P-1$ via (18.52)
8. Form the unconstrained vectorized waveform set $\tilde{\mathbf{s}}_i$ via (18.53)
9. Initiate RUWO nulling loop. Set loop index to $k = 0$
   a. Form initial vectorized solution $\mathbf{a}_0 = \exp(j\angle\tilde{\mathbf{s}}_i)$
   b. Update vectorized solution $\mathbf{a}_{k+1}$ using (18.54)
   c. Increment $k = k + 1$. Stop if $k = K$. Otherwise, go to step 9b
   d. Set $\mathbf{s}_{i+1} = \mathbf{r}_K$
10. Increment $i = i + 1$. Stop if $i = \Phi$. Otherwise, go to step 6

18.5.4 Emission optimization analysis

Consider the design of a wideband emission having time–bandwidth product $BT = 50$ and fractional bandwidth $\%\text{BW} = 40\%$ for a ULA comprised of $M = 30$ antenna elements. The antenna element spacing is set according to the highest frequency in the 98% bandwidth B to reduce the effect of grating lobes [27]. The normalized frequency f_d/f_{cent} associated with this spacing is

$$\frac{f_d}{f_{\text{cent}}} = \frac{f_{\text{cent}} + B/2}{f_{\text{cent}}} = 1 + \frac{\%\text{BW}}{2} = 1.2. \tag{18.56}$$

The sampling frequency is set to $\alpha = 5$ times the time–bandwidth product of $BT = 50$, and thus the total number of samples in each discrete-time waveform per (18.28) is $N = 250$.

Given the number of elements M, normalized frequency f_d/f_{cent}, and fractional bandwidth %BW, the set of beamlets must be contained within the spatial region defined in (18.5.3.1) as

$$|\sin \Theta_P| \leq 1 - \frac{2f_d/f_{\text{cent}}}{M(1 - (\%\text{BW}/2))} = 0.9 \tag{18.57}$$

to avoid placing main lobe power in the reactive region. The number of frequency points is chosen to be twice the length of the waveform as $Q = 2N = 500$ to minimize aliasing of the complex envelope of the emission that occurs in true-time delay processing.

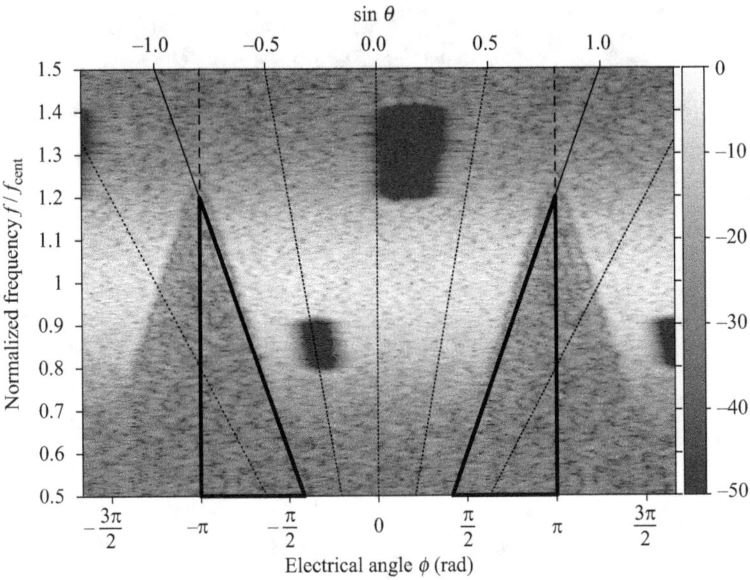

Figure 18.17 Spectrum of wideband (%BW = 40% bandwidth) wide beam MIMO radar emission versus spatial angle $\phi(f, \theta)$ for an $M = 30$ element ULA with $f_d = 1.2f_{cent}$. The reactive regions are bounded inside the black triangles and comprise only 1.4% of the total available power

Using these parameters as a baseline, a wide (nearly omni-directional) beam is generated over the angular interval $\sin\Theta_P \in [-0.9, 0.9]$. A total of $P = 60$ evenly spaced beamlets, all designed to have a nominal Gaussian spectral distribution, are used to generate this wide beam. The desired spatial power distribution is flat across the spatial interval. Two nulled regions are included: the first null is bounded by normalized frequencies $\mathbf{f}_{L,1} \in [0.8, 0.9]$ and spatial interval $\sin\Theta_{L,1} \in [-0.5, -0.42]$, while the second null is placed in frequency interval $\mathbf{f}_{L,2} \in [1.2, 1.4]$ and spatial interval $\sin\Theta_{L,2} \in [0.10, 0.25]$. Therefore, the steering vectors used to construct the interference matrix are defined by normalized frequencies $\mathbf{f}_L = \{\mathbf{f}_{L,1},\ \mathbf{f}_{L,2}\}$ and spatial angles $\sin\Theta_L = \{\sin\Theta_{L,1},\ \sin\Theta_{L,2}\}$.

A total of $\Phi = 100$ design iterations and $K = 20$ RUWO iterations are used to generate the desired far-field emission. These values of Φ and K were observed to provide sufficient convergence of the algorithm, though a more sophisticated stopping criteria could also be used. The waveform matrix $\mathbf{S}_0$ is initialized with white, complex-Gaussian distributed data having unit variance and zero mean so as to start with relatively low crosscorrelation waveforms ($\mathbf{S}^H\mathbf{S}$ is approximately proportional to the identity matrix) which is linked to an omni-directional allocation of power in space [35].

Figure 18.17 shows the converged spectral content of the emission versus electrical angle ϕ after applying the algorithm described in Table 18.1. Notice that only a

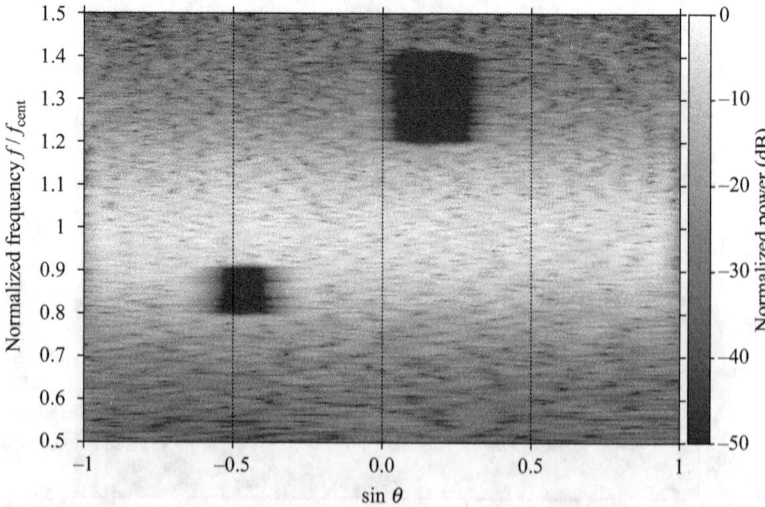

Figure 18.18 Spectrum of wideband (%BW = 40% bandwidth) wide beam MIMO emission versus sin θ for an M = 30 element ULA with $f_d = 1.2f_{cent}$

small amount of power still resides inside the reactive region as compared to the case shown in Figure 18.15 where there was no attempt to minimize the reactive power. Recall that for the emission in Figure 18.15, the FRP was FRP = 17.4% of the total available power. By limiting the emission construction to the visible space, the FRP for the case shown in Figure 18.17 has reduced to FRP = 1.4% of the total available power, more than an order of magnitude reduction.

Figure 18.18 shows the converged emission spectrum versus $\sin\theta$, which is the fully transmitted emission (based on the assumptions of reflected and transmitted power stated in Section 18.5.1.3). Figure 18.19 displays detailed images of both null regions in Figure 18.18. The first null, bounded by normalized frequencies $\mathbf{f}_{L,1} \in [0.8, 0.9]$ and spatial region $\sin\Theta_{L,1} \in [-0.5, -0.42]$, has an average normalized power of −65.44 dB, which is 55 dB below the average power for those frequencies in other spatial directions. The second null, bounded by normalized frequencies $\mathbf{f}_{L,2} \in [1.2, 1.4]$ and spatial region $\sin\Theta_{L,2} \in [0.10, 0.25]$, has an average normalized power of −67.03 dB, which is 45 dB below the average power for that range of frequencies in other spatial directions. Thus, good null depth is being achieved.

Figure 18.20 shows the aggregate beampattern as calculated by (18.27) for the $P = 60$ beamlet emission as well as a $P = 30$ beamlet case optimized using the same initializations. The emitted powers are normalized such that 0 dB is the maximum power that can be produced by the array. The $P = 60$ beamlet case has only 0.74 dB of variation over the prescribed spatial interval $[-0.9, 0.9]$, while the $P = 30$ case has 2 dB of variation. This result exemplifies the fact that the adaptive beam

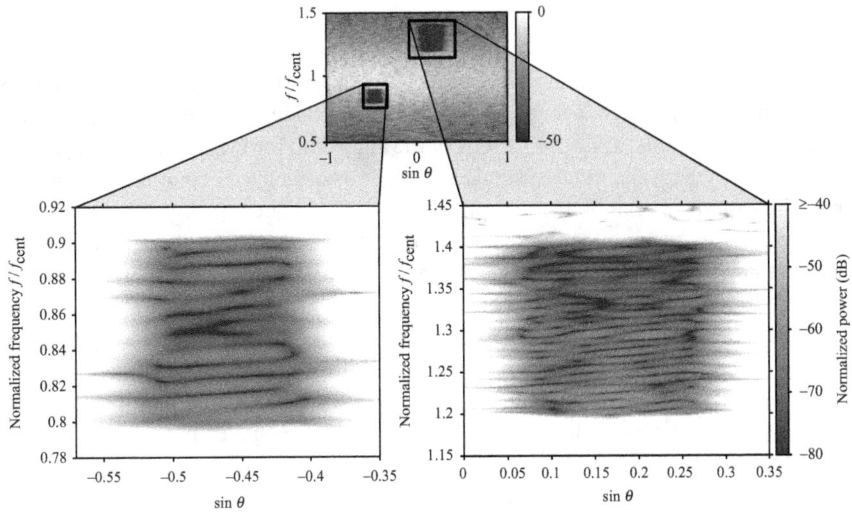

Figure 18.19 Detailed view of the space–frequency nulled regions of Figure 18.18

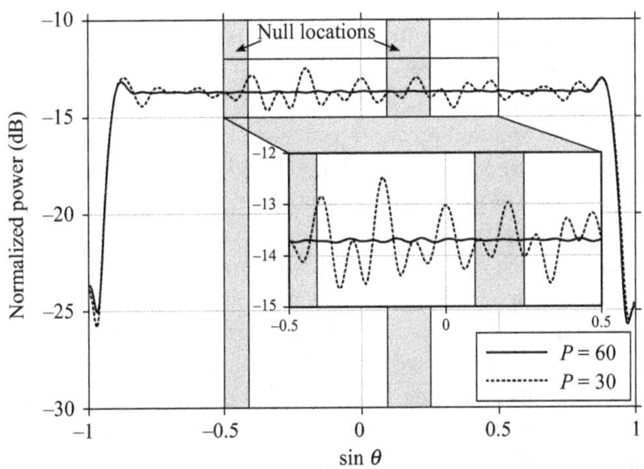

Figure 18.20 Aggregate beampattern calculated via (18.27) for the case of an even power allocation over $\sin\theta = -0.9$ *to* $\sin\theta = 0.9$ *for* $P = 60$ *and* $P = 30$ *beamlets*

scaling allows for precise control of the spatial allocation of power and becomes more accurate with an increased number of beamlets (can essentially be viewed as spatial oversampling). Also note that while large portions of the space–frequency beampattern were removed using the RUWO algorithm, the spatial allocation of power remains smooth over the nulled regions. Further results and discussion on different spatial power distributions within this framework can be found in [11].

18.6 Summary

This chapter provides an overview of the RUWO algorithm, a method based on the maximum SINR framework for generating constant modulus waveform weights that produce transmit nulls, and applications of the algorithm for facilitating RF spectrum interoperability between radar systems and other users, such as communication systems. This approach can be used to produce spatial, frequency, or space–frequency nulls on transmit using either a covariance matrix generated from recorded interference or a deterministic interference covariance matrix produced from knowledge of the space–frequency locations of other RF users.

Experimental demonstrations in both loop-back and open-air configurations using NRL's STAN X-band MIMO radar test bed illustrate the prospect of achieving practical interference avoidance on transmit. While the base-band loop-back experiments using the eight-channel VSG and digital components of the STAN receiver do not include the effects of up-conversion, antenna calibration, or multipath in the scene, the results clearly show the impact of the various nulling schemes on the resulting beampatterns and ambiguity diagrams. As one would expect from removing part of the spatial and/or spectral content of a waveform: spatial nulling distorts the beampattern but has no effect on the ambiguity diagram, frequency nulling does not perturb the beampattern but has undesirable effects on the waveform ambiguity diagram, and space–frequency nulling affects both the beampattern and the ambiguity diagram of the transmitted waveform, albeit to a lesser extent than when either spatial or frequency nulling are used independently as a result of the overall increase in dimensionality due to coupling.

Open-air X-band experiments using the full STAN MIMO radar test bed examined the capability of RUWO to adaptively generate spatial nulls with constant modulus weights for mitigating varying sources of interference with different bandwidths. It was observed that covariance matrices calculated from measured pulsed CW interference produce the most precise nulls, while covariance matrices calculated from wider bandwidth interference produce deeper nulls that are offset in azimuth. Deterministic (model-based) nulls are offset in azimuth and shallower than the adaptive nulls, which is to be expected because modeled interference does not account for hardware calibration effects that a data-adaptive interference covariance matrix naturally contains.

The combination of the RUWO method with a spectral shaping algorithm allows nulls to be enforced when they would otherwise not achieve sufficient depth. The combination of these methods has facilitated an iterative, wideband MIMO optimization scheme for far-field radar emission design that optimizes the spectral content in certain predefined angles denoted as beamlets. The design process avoids placing significant power in the array invisible space by constraining the main lobes of the beamlets to be contained inside the visible domain, thus decreasing reactive power and increasing efficiency. The RUWO method was employed within the spectral shaping formulation to provide wideband space–frequency nulls when the spatial steering vectors are frequency dependent.

18.7 Looking ahead

There are several areas of additional research that would be beneficial to further the performance and utility of transmit nulling approaches such as RUWO and that could be carried out using a small-scale MIMO radar test bed like the STAN setup at NRL. For instance, it would be beneficial to perform field tests with multiple, dynamic interference sources having a variety of different modulation schemes. The evaluation of out-of-band spectral compliance for transmit-nulled waveforms would likewise be instructive. Experimental validation of wideband, wide beam MIMO emissions with space–frequency nulls also warrants additional investigation.

In the broader context of interoperability of radar systems with other RF users, future research should investigate how to best incorporate transmit nulling into future fielded radar systems, including the prospect of MIMO capabilities. The RF environment in the area surrounding a radar system changes over time, but on what time scale should the radar be concerned with ensuring interoperability and with which users? The existence of a trade-off in radar performance to address interoperability with other users is well known, but the exact relationship and the acceptable level of degradation depends on many factors, including the particular radar mission. The amount of distortion in the transmitted waveform's beampattern and ambiguity diagram also depends on whether spatial, frequency, or space–frequency transmit nulling is performed and the radar system's requirements can inform which is more acceptable for a particular situation. The work presented in this chapter also demonstrates that transmit nulling can be accomplished adaptively or, with reduced performance, using deterministic models. However, a mechanism for sensing the interference environment must be included in the radar design to enable adaptive transmit nulling, which necessarily involves the scheduling of an environmental sensing mode. Finally, transmitter and receiver calibration is extremely important to ensure the success of transmit nulling methods and any fielded system would require an integrated calibration mechanism.

Acknowledgments

This work was sponsored by the Office of Naval Research under the Naval Research Laboratory Base Program and the Air Force Research Lab. Some of the work in Sections 18.2, 18.3, and 18.4.1 originally appeared in [14] and was reproduced by permission of the Institution of Engineering & Technology. Some of the work in Section 18.4.2 originally appeared in [15] and was reproduced by permission of the Institute of Electrical and Electronics Engineers. Some of the work in Section 18.5 appeared in [11] and was reproduced by permission of the Institute of Electrical and Electronics Engineers.

References

[1] Gerlach K. Thinned spectrum ultrawideband waveforms using stepped-frequency polyphase codes. IEEE Transactions on Aerospace and Electronic Systems. 1998 Oct;34(4):1356–1361.

[2] Day DA. Fast phase-only pattern nulling for large phased array antennas. In: 2009 IEEE Radar Conference; 2009. p. 1–4.

[3] Selesnick IW, Pillai SU, Zheng R. An iterative algorithm for the construction of notched chirp signals. In: 2010 IEEE Radar Conference; 2010. p. 200–203.

[4] Cook MR, Higgins T, Shackelford AK. Thinned spectrum radar waveforms. In: 2010 International Waveform Diversity and Design Conference; 2010. p. 238–243.

[5] Selesnick IW, Pillai SU. Chirp-like transmit waveforms with multiple frequency-notches. In: 2011 IEEE RadarCon (RADAR); 2011. p. 1106–1110.

[6] Higgins T, Webster T, Shackelford AK. Mitigating interference via spatial and spectral nulls. In: IET International Conference on Radar Systems (Radar 2012); 2012. p. 1–6.

[7] Scholnik DP, Coleman JO. Optimal design of wideband array patterns. In: 2000 IEEE International Radar Conference; 2000. p. 172–177.

[8] Scholnik DP, Coleman JO. Optimal array-pattern synthesis for wideband digital transmit arrays. IEEE Journal of Selected Topics in Signal Processing. 2007 Dec;1(4):660–677.

[9] Antonio GS, Fuhrmann DR. Beampattern synthesis for wideband MIMO radar systems. In: 1st IEEE International Workshop on Computational Advances in Multi-Sensor Adaptive Processing, 2005; 2005. p. 105–108.

[10] He H, Stoica P, Li J. Wideband MIMO systems: signal design for transmit beampattern synthesis. IEEE Transactions on Signal Processing. 2011 Feb;59(2):618–628.

[11] McCormick PM, Blunt SD, Metcalf JG. Wideband MIMO frequency modulated emission design with space-frequency nulling. IEEE Journal of Selected Topics in Signal Processing. 2017 Mar;11(2):363–378.

[12] Blunt SD, Cook M, Jakabosky J, Graaf JD, Perrins E. Polyphase-coded FM (PCFM) radar waveforms, Part I: Implementation. IEEE Transactions on Aerospace and Electronic Systems. 2014 Jul;50(3):2218–2229.

[13] McCormick PM, Blunt SD, Metcalf JG. Joint spectrum/beampattern design of wideband FM MIMO radar emissions. In: 2016 IEEE Radar Conference; 2016. p. 1–6.

[14] Higgins T, Webster T, Shackelford AK. Mitigating interference via spatial and spectral nulling. IET Radar, Sonar Navigation. 2014 Feb;8(2):84–93.

[15] Webster T, Higgins T, Shackelford AK, Jakabosky J, McCormick P. Phase-only adaptive spatial transmit nulling. In: 2015 IEEE Radar Conference (RadarCon); 2015. p. 0931–0936.

[16] Gerlach K, Kretschmer FF. Convergence properties of Gram-Schmidt and SMI adaptive algorithms. IEEE Transactions on Aerospace and Electronic Systems. 1990 Jan;26(1):44–56.
[17] Levi A, Stark H. Image restoration by the method of generalized projections with application to restoration from magnitude. In: Acoustics, Speech, and Signal Processing, IEEE International Conference on ICASSP '84; vol. 9, 1984. p. 88–91.
[18] Frazer GJ, Abramovich YI, Johnson BA. Spatially waveform diverse radar: perspectives for high frequency OTHR. In: 2007 IEEE Radar Conference; 2007. p. 385–390.
[19] Gerchberg RW. A practical algorithm for the determination of phase from image and diffraction plane pictures. Optik. 1972;35:237.
[20] Bauschke HH, Borwein JM. On projection algorithms for solving convex feasibility problems. SIAM Review. 1996;38(3):367–426.
[21] Bucci OM, Franceschetti G, Mazzarella G, Panariello G. Intersection approach to array pattern synthesis. IEE Proceedings H – Microwaves, Antennas and Propagation. 1990 Dec;137(6):349–357.
[22] Fienup JR. Phase retrieval algorithms: a comparison. Applied Optics. 1982;21(15):2758–2769.
[23] Patton LK, Rigling BD. Phase retrieval for radar waveform optimization. IEEE Transactions on Aerospace and Electronic Systems. 2012 Oct;48(4):3287–3302.
[24] Jakabosky J, Blunt SD, Himed B. Waveform design and receive processing for nonrecurrent nonlinear FMCW radar. In: 2015 IEEE Radar Conference (RadarCon); 2015. p. 1376–1381.
[25] Jakabosky J, Blunt SD, Himed B. Spectral-shape optimized FM noise radar for pulse agility. In: 2016 IEEE Radar Conference; 2016. p. 1–6.
[26] Rhodes D. On the stored energy of planar apertures. IEEE Transactions on Antennas and Propagation. 1966 Nov;14(6):676–683.
[27] Van Trees HL. Optimum array processing: Part IV of detection, estimation, and modulation. Wiley, New York, NY; 2002.
[28] Smith S. Digital signal processing: a practical guide for engineers and scientists. Newnes, Boston, MA; 2003.
[29] Blunt SD, Mokole EL. Overview of radar waveform diversity. IEEE Aerospace and Electronics Systems Magazine. 2016 Nov;31(11):2–42.
[30] Allen JL, Diamond B. Mutual coupling in array antennas. DTIC Document; 1966.
[31] Balanis CA. Antenna theory: analysis and design. vol. 1. John Wiley & Sons, Hoboken, NJ; 2005.
[32] Cordill B, Metcalf J, Seguin SA, Chatterjee D, Blunt SD. The impact of mutual coupling on MIMO radar emissions. In: 2011 International Conference on Electromagnetics in Advanced Applications; 2011. p. 644–647.

[33] Babur G, Aubry PJ, Chevalier FL. Antenna coupling effects for space–time radar waveforms: analysis and calibration. IEEE Transactions on Antennas and Propagation. 2014 May;62(5):2572–2586.

[34] Allen J. On array element impedance variation with spacing. IEEE Transactions on Antennas and Propagation. 1964 May;12(3):371–372.

[35] Fuhrmann DR, Antonio GS. Transmit beamforming for MIMO radar systems using signal cross-correlation. IEEE Transactions on Aerospace and Electronic Systems. 2008 Jan;44(1):171–186.

Part IV

Radar/communication co-design

Chapter 19
Communication and radar co-design

Daniel W. Bliss[1]

19.1 Introduction

Wireless communications and radar represent two of the many uses of the radio frequency (RF) spectrum. Historically, these modalities have operated independently in isolated frequency bands, and because hardware was relatively inflexible, this isolationistic design paradigm was reasonable. However, as RF and computational technologies have advanced, a new paradigm is emerging in which more flexible and intelligent use of the RF spectrum is possible [1,2]. In so doing, there are a wide range of important and interesting questions to be answered to enable the RF convergence of communications and radar.

In this chapter, we develop a set of tools to study the co-design of joint communication and radar systems. For the sake of discussion, given a fixed set of resources, one can determine an n-dimensional space of performance metrics such that the best performance of the system can be described by a manifold in this metric space. By changing the parameters the overall system employs, one can therefore move along this surface to different operating points as indicated in Figure 19.1. Thus, an intelligent system can dynamically modify the parameters of a network of joint radar and communication nodes in response to multiple goals that, along with the environment, can likewise be dynamic. Further, in highly dynamic scenarios, distributed (as opposed to centralized) optimization may be required. In fact, dynamic joint radar and communication systems may actually provide a more appropriate example for use of the adjective "cognitive" than other systems that often employ this monicker.

At the simplest level, these RF systems are radiating and receiving energy and subsequently extracting information by analyzing the captured signals. We are motivated by this observation to consider the opportunities for developing joint operations. However, because radars are tasked with detection and estimation, while communication systems are used to transfer information, the nature of their associated performance metrics is different. Here, we rely on the connections between information theory, detection theory, and estimation theory to develop tools for understanding the limits of such a co-designed system.

[1]School of Electrical, Computer and Energy Engineering, Arizona State University, USA

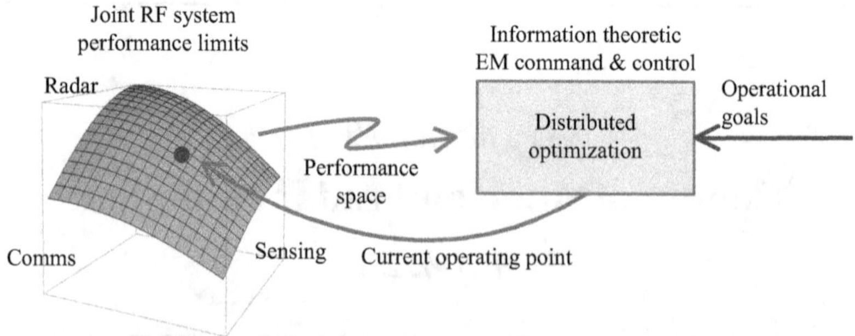

Figure 19.1 Notional diagram of the interaction between system performance bounds, indicated by the multidimensional space of metrics, and operating-point specification that is driven by the optimizer. Here, the notional performance is considered in terms of communications, radar, and spectral occupancy sensing

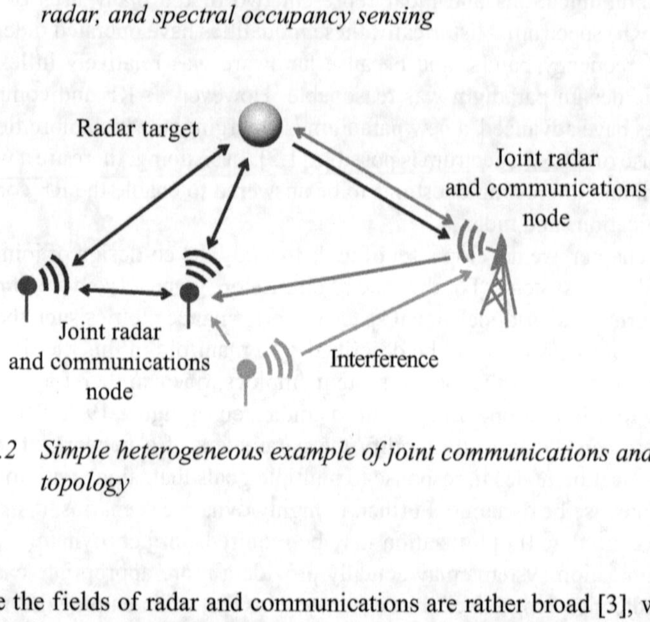

Figure 19.2 Simple heterogeneous example of joint communications and radar topology

While the fields of radar and communications are rather broad [3], we propose that all radar/communication nodes can incorporate activities simultaneously within a heterogeneous network. This integration may include scheduling various functions at different times, though it is important to include the possibility of using the same spectrum simultaneously for multiple functions. Consequently, every emitter can be both a radar illuminator and a communication source. Every receiver can likewise be both a radar and communication receiver. Specifically, we are interested in the reuse of the same spectrum for a variety of such applications, as indicated in Figure 19.2. In developing co-design RF convergence approaches, we draw heavily from two concepts in communication theory [4,5]: (1) successive interference cancellation (SIC) for multiaccess receivers and (2) the "water-filling" approach for evaluating channel capacity.

19.2 Background

There has been significant recent progress in joint radar–communications, although there is admittedly considerable room for continued research, particularly in the area of co-design. While the particular nomenclature with which to classify joint radar and communications has not yet been settled (e.g. see Chapter 8), here we broadly classify systems as belonging to the classes of coexistence or co-design.

It is also worth noting that the observation that RF communication systems can also be used for sensing applications is far from new. For instance, while the early era of wireless research had many important contributors, the earliest direct observation that communication systems simultaneously provide other types of information was made by Alexander Popov around 1897. He observed that the channel of a wireless communication link was affected by nearby ship motion [6]. That said, it is only with modern (and emerging) technology that the real benefits of such observations have begun to be realized.

19.2.1 Coexistence

We define coexistence as the operational paradigm in which independent systems consider the other system(s) as sources of interference. Broadly, such interference is thereby addressed either by avoiding it or by actively mitigating it. Examples of mitigation are given in [7–10]. Under the category of avoidance is the notion of exploiting opportunistic spectrum availability, such as communication systems operating in the nonilluminated intervals of a rotating radar [11,12], or pulsed radars and time-division duplexed communication systems [13,14]. The following briefly highlights the general forms that coexistence can take.

Passive, parasitic, or green radar systems—Passive systems exploit existing RF illumination in the environment to perform radar sensing tasks. The terms passive, parasitic, or green radar systems are often used interchangeably, although some authors imbue slightly different meanings to these terms (see Chapters 7 and 8). Examples of passive bistatic radar concepts include opportunistically using television broadcasts [15] and cellular transmissions [16,17] as illuminators. While appearing to be an obvious and simple solution to the spectral congestion problem, this approach does limit radar performance because important sensing parameters such as the waveform characteristics, beam shape and direction, and transmit power are outside the control of a passive radar. That said, such a system does have access to illumination sources that can be exploited "free of charge" to supplement the scattering information collected from the environment.

Spectral media access control—By using available information or scheduling of users, radar and communication systems may avoid "collisions" in which multiple systems attempt to access the same spectrum and thereby generate mutual interference. Examples of such spectral allocation strategies can be found in [18–23]. In these approaches, the systems do not modify their parameters or properties in a co-design sense but rather actively cooperate to avoid operating in the same frequency band simultaneously. Examples also include communication and radar systems operating as multi-access users [19,24], scheduling and sharing bandwidth [20,25,26], and spatially separating the users [27].

Dynamic "cognitive" media access control—By drawing upon concepts from the cognitive radio literature [28], systems can employ techniques to operate only if adverse effects on another system fall below some interference threshold [29]. With this model, the radar might be a primary user and the communication system is allowed to operate as a secondary user under the condition that it does not degrade radar performance [30]. This arrangement offers a potential improvement to structured spectrum sharing schemes by allowing users to adapt to dynamic spectral availability [31–34].

19.2.2 Co-design of joint radar–communications

In contrast to the coexistence approaches discussed above, this chapter is particularly focused on joint radar–communication co-design whereby a single system is developed to achieve both functions. Such systems potentially enable better performance than that of separate radar and communication systems.

System concepts—Many different co-design system concepts have thus far been explored. Broad surveys of joint radar and communication techniques, including co-design approaches, can be found in [2,3]. Likewise, joint communications and radar specific to vehicular systems is considered in [35]. The following illustrates a cross-section of this work.

In [36], the authors present a joint approach based upon a Hartley capacity measure. The authors in [37] jointly maximized information criteria for radar and communication users to minimize mutual interference by varying the radar waveform and communication orthogonal frequency division multiplexing (OFDM) parameters in response to dynamic bandwidth allocation. In [38,39], Neyman–Pearson-based cooperative metrics were developed that capture both radar detection performance and communication data rate in a joint cost function with parametrizable weighting by considering mutual information and the area under the receiver operating curve, respectively. In [40], the authors suggest an innovative approach to construct radar waveforms by using multiple mobile subscribers in existing cellular networks. Reference [41] provides a detailed system design that employs the resources of what would traditionally be a radar alone to facilitate a system that serves as both a radar and a communication base station.

Our own efforts have largely focused on the theory and design of waveforms and architectures for systems that simultaneously receive radar echoes and communication signals. In [1,42,43], a co-design concept was introduced that involves multiple bounds on performance for a multiaccess radar–communication receiver, and the notion of estimation information rate was proposed. Implicit in the formation of these bounds is control of the communication and radar transmit characteristics. Radar range and Doppler estimation was considered in [44], and these concepts were extended in [45,46] to include the impact of clutter and phase noise. In [47,48], the concept of constant estimation information was employed as a metric to control radar tracking operation, and in [49], preliminary investigations were provided for shaped-spectrum joint communication and radar waveform optimization for multiaccess receivers. Recent observations [50] regarding the connection between mutual information and estimation minimum-mean-squared error (denoted as I-MMSE) have also been employed to evaluate estimation information rates given non-Gaussian distributions [51]. In [39], the multiaccess receiver was considered in the context of detection performance, while [52] developed and demonstrated a multistatic, multiaccess receiver for which the received signals provide both communications and bistatic synthetic aperture radar (SAR) illumination. In simulation, it was shown how the receiver recovers the independent communication streams and forms multiple SAR images.

Waveform design—The co-use of radar illumination as communication signals has been considered by a number of authors. For instance, noncoherent phase encoding approaches were presented in [53], and embedding information by modulating the spatial sidelobe structure of the radar beam pattern was considered in [54]. The authors in [55] observe that OFDM illumination waveforms for multiple-input–multiple-output (MIMO) radar could be used to carry data, and recent vehicle-to-vehicle systems [35,56,57] and software-defined radar [58] implementations also adopt OFDM waveforms. However, this waveform results in numerous problems for radar so research has shifted to designing joint systems to suppress sidelobes [59], maintain a constant envelope [60], or reduce peak-to-average-power ratio [61]. Other authors consider direct relationships between radar estimation sidelobe ambiguity and communications channel coding [33], including specific coding techniques with favorable properties such as finite Heisenberg–Weyl groups [62], Golay waveforms with Doppler resilient properties [63], and complementary sequences [64]. In [52], we consider simultaneous embedding of communication signals and multiaccess receiver examples of multiple bistatic SARs. Co-use waveform design motivated by covert communications is considered in [65–68]. Multiuser joint MIMO communications and radar are considered in [69], while approaches that exploit joint conditioning of the MIMO transmit subspace and sampling approaches are investigated in [70].

19.3 Performance metrics

We construct a framework that permits consideration of a wide variety of joint communication and radar applications. In developing joint performance bounds, a given joint system performance is represented as a vector within a higher dimensional space, where each metric is given its own axis. As indicated in the simple example of Figure 19.1, a surface of bounding performance is established by the set of vectors that jointly satisfy the network performance bounds. Without loss of generality, we assume that the metrics are constructed such that more information is indicated by a larger positive real number. Consequently, moving outward along any ray increases the amount of total information provided by the network. One can imagine dynamically moving along this surface by changing system parameters of the network as indicated by the system's needs at any given moment. The dimensionality and various contributing metrics may then also change as a function of time as the network topology or channel environment changes.

19.3.1 Communication metrics

From a communication link perspective there are a number of useful metrics, the most fundamental of which is given by the Shannon information capacity [4,5]. However, for many scenarios latency and quality of service are of equal or greater importance. For this discussion we shall focus on capacity, though these other metrics could conceivably be incorporated as well. For received signals at complex baseband, the theoretical limit on spectral efficiency c in units of bits per second per Hertz (b/s/Hz)[2] is given by

$$c = \log_2(1 + \mathrm{SINR}), \tag{19.1}$$

where SINR is the signal-to-interference-plus-noise ratio at the receiver after any interference and noise mitigation processing, under the assumption that the noise and residual interference are drawn from a complex, circularly symmetric Gaussian distribution at complex baseband.

The capacity in (19.1) assumes a spectrally flat, known channel. However, if the channel can be decomposed into a set of channels, each with potentially different, yet known SINRs, then a water-filling approach becomes useful [4,5]. If one has access to multiple parallel channels (e.g. multiple frequency subbands) that have different SINRs, then the water-filling solution seeks to distribute the power and rate across the subbands to maximize the total rate, under a constraint on total power. Thus, if the total power P_{com} is partitioned into the power in each subband as P_m, then

$$P_{\mathrm{com}} = \sum_m P_m. \tag{19.2}$$

If we denote the attenuation and noise power in each subband as $\|b_m\|^2$ and $\sigma^2_{n,m}$, respectively, the total capacity in units of b/s is [4,5]

$$C = \sum_m B_m \log_2\left(1 + \|b_m\|^2 \frac{P_m}{\sigma^2_{n,m}}\right), \tag{19.3}$$

where B_m is the spectral width of the mth subband, and the power in subband P_m is obtained via Lagrangian optimization as

$$P_m = \left(\frac{1}{\lambda} - \frac{\sigma^2_{n,m}}{\|b_m\|^2}\right)^+, \tag{19.4}$$

[2]Because the units of "b/s/Hz" simplifies to "b," the form is somewhat superfluous but is useful to remind us that data rate is found by multiplying spectral efficiency by bandwidth.

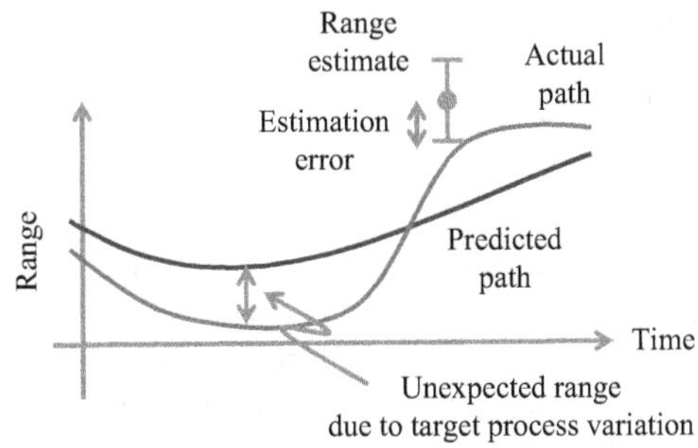

Figure 19.3 For a dynamic target whose range varies as a function of time, the two sources of uncertainty are (1) unexpected changes in target range relative to some predicted path and (2) range estimation error

where $(\cdot)^+$ indicates $\max(\cdot, 0)$. Here, the Lagrangian multiplier λ is varied to satisfy (19.2).

19.3.2 Radar estimation information rate

While the following approach is considered to be more generically applicable to a broad set of sensing applications, for the sake of explanation we develop a metric for estimation of the radar range parameter. To compare performance between the radar and communication functions in terms of information rate, it is convenient to develop a metric that is analogous to data information rate in communications that can be used to measure radar performance. The novel parameterization here is the concept of radar estimation information rate that can be interpreted as the average bit rate required to encode the information produced by target tracking estimation. This information rate is constructed by considering the entropy of the received signal and the entropy of the estimation uncertainty [1].

To estimate the respective entropies, we consider two sources of uncertainty (for target range), which are depicted in Figure 19.3. Assuming that the true target range is not completely predictable—if it were, radar tracking would be unnecessary—the expected squared error between the actual and predicted position is parameterized by the process variance σ^2_{proc}. The second source of uncertainty is due to estimation error. For a given update of the target range estimate, there is a limit to how accurately the range can be determined. This estimation error is parameterized by the variance σ^2_{est}.

We assume that the range estimation achieves the theoretical Cramér–Rao bound (CRB) performance [5,71], which is often a good approximation for waveforms with low range sidelobes. The CRB for estimation of time delay τ is given by [43]

$$\sigma^2_{\text{est}} = \left(\frac{1}{8\pi^2 B^2_{\text{rms}} \, \text{ISNR}} \right), \tag{19.5}$$

where the B_{rms} is the root-mean-squared (RMS) bandwidth defined as

$$B_{\text{rms}} = \frac{\int f^2 \, \|X(f)\|^2 \, df}{\int \|X(f)\|^2 \, df}. \tag{19.6}$$

The term $X(f)$ is the spectral representation of the illumination waveform as a function of frequency f, and ISNR stands for integrated SNR, which is the SNR at the output of any receive processing such as matched filtering.

To evaluate the estimation information rate for a single target, note that the target range evolution from measurement to measurement can be decomposed into a component that is predictable and a component that is random. For the sake of discussion, we assume that the driving component of this random process is drawn from a Gaussian distribution with variance σ^2_{proc}. Denoting T_{pri} as the pulse repetition interval of the illumination, T as the temporal pulse duration, and $\delta = T/T_{\text{pri}}$ as the duty cycle, the mutual information rate (or radar estimation information rate) R_{est} is given by [43]

$$R_{\text{est}} = \frac{h_{\text{rr}} - h_{\text{est}}}{T_{\text{pri}}}. \tag{19.7}$$

Here

$$h_{\text{rr}} = \frac{1}{2} \log_2\left[2\pi\, e\,(\sigma^2_{\text{proc}} + \sigma^2_{\text{est}})\right] \tag{19.8}$$

is the entropy associated with delay estimation, with contributions from both target motion uncertainty and estimation uncertainty, and

$$h_{\text{est}} = \frac{1}{2} \log_2\left[2\pi\, e\,\sigma^2_{\text{est}}\right] \tag{19.9}$$

is the estimation entropy, with contributions from estimation uncertainty only. These entropies are evaluated under the Gaussian assumption [4,5]. Thus, the radar estimation information rate is

$$\begin{aligned} R_{\text{est}} &= \frac{\delta}{2T} \log_2\left(1 + \frac{\sigma^2_{\text{proc}}}{\sigma^2_{\text{est}}}\right) \\ &= \frac{B}{2} \log_2\left[1 + \frac{2\,\sigma^2_{\text{proc}}\,\gamma^2\, B\,(TB)\, a^2\, P_{\text{rad}}}{N_0}\right]^{\delta/(TB)}, \end{aligned} \tag{19.10}$$

where we define $\gamma = B_{\text{rms}}/B$, for B the bandwidth of radar operation (rather than the total bandwidth used later), N_0 is the noise spectral density, a^2 is the total channel attenuation, and P_{rad} is the radar transmit power. With slight modifications the above discussion can be expanded to accommodate multiple pulses.

19.3.3 Detection performance

Another important metric is associated with target detection or model order identification. Multiple approaches have been considered to quantify such performance.

Historically, early target detection efforts employed mutual information to describe performance [72], with a recent joint performance extension [38]. The Kullback–Leibler divergence between the target and null hypotheses may be used as a tool to approximate detection performance such as in [73]. The area under the curve for receiver operating characteristic (ROC) curves [39] and the Akaike information criterion [74] are likewise applicable. However, there is still considerable room for further investigation on this topic.

19.3.4 Practical limitations

It is worth noting that for any metrics based upon theoretical bounds, there are limits to the region of feasibility. As an example, the Shannon capacity limit can be closely approached if one has access to significant computational capabilities and long code words. These conditions are clearly not applicable in all scenarios. For estimation performance, the bounding variance diverges from the CRB at low SNR because of potential nearby ambiguities [5]. Further, real performance diverges from theoretical performance at high SNR due to inaccuracies between the actual and modeled systems. These mismatches can also be a significant source of residual interference for multiaccess receivers, which is discussed in Section 19.4.1.

19.4 Topologies

By considering a heterogeneous network of nodes that incorporate both radar and communication functions, the number of potential topologies is remarkable. Even traditional wireless communications can be a complicated mix of mesh and star networks, where star networks have a central node through which all other nodes communicate and mesh networks are decentralized such that nodes form links with neighbors.

As with networks comprised exclusively of communication nodes, there are an unbounded number of potential topologies for joint radar and communication. A simple example is illustrated in Figure 19.2. Because of the variety of potential roles each node may play, the set of possible radar–communication arrangements is even larger than an equivalent communication network for a given set of nodes. However, as with communication-only networks, there is a small number of representative topologies that captures the essential characteristics of many of the most useful networks. Figure 19.4 depicts three simple example networks that might be used to construct a larger heterogeneous system.

The waveform reuse topology in Figure 19.4(a) assumes that the illumination waveform is used both to communicate with a receiver and for radar target detection and/or parameter estimation. A simple interpretation of this topology is that estimation of the standard communication channel serves the dual purpose of radar sensing. This description, however, underestimates the complexity of waveform optimization, the optimal characteristics of which are often at odds for communications and radar operation (the coding diversity approach in Chapter 23 discusses an instantiation of this approach in detail).

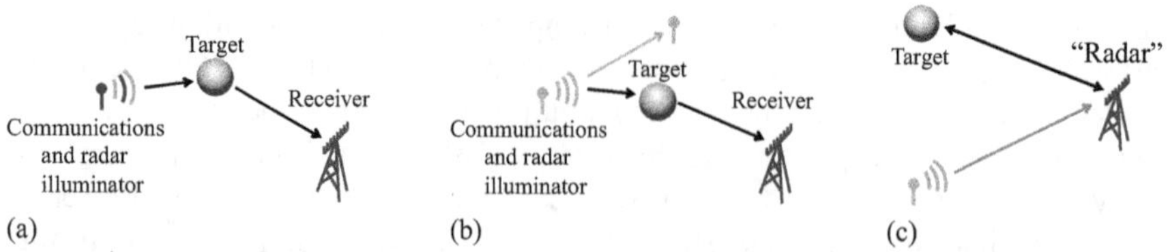

Figure 19.4 Example topologies: (a) waveform reuse, (b) passive radar, and (c) multiaccess for communications and radar

The passive (or parasitic) radar in Figure 19.4(b) relies on a communication broadcast signal for illumination. The radar receiver requires a high-fidelity estimate of the broadcast signal as a reference to perform channel estimation (otherwise known as "pulse compression" in the radar context). This idea is discussed in detail in Chapters 7 and 8.

Finally, the multiaccess communication and radar receiver in Figure 19.4(c) captures both the communication signal and the radar echoes simultaneously [1,43]. By employing a multiuser detection approach, these disparate signals can be separated and evaluated within the receiver.

The number of potential node topologies grows quickly as the number of nodes increases. As an example of how easily the complexity increases, consider how a "radar" node can be readily converted into an extremely capable communication relay through the modification of incorporating data into the radar waveform. This concept is explored in greater detail in Section 19.4.2 and Chapter 23.

19.4.1 Multiaccess sensing and communications

The underlying motivation for the performance bounds in [1,43] is the communication theory behind the multiaccess receiver [4,5]. The channel attenuations for the two communication links are given by a_1 and a_2. Likewise, the associated transmitter powers are P_1 and P_2, along with receiver noise power σ^2_{noise}. The corresponding rates are denoted as R_1 and R_2. The bounding rate region is thus obtained by taking the convex hull of the vertices 1–4 in Figure 19.5.

To determine this region, the concept of SIC is applied. At the receiver, the signals from the two transmitters are superimposed in noise. The rate of transmitter 1 is set such that it is decodable in the presence of interference from the other signal plus noise. After decoding, an exact representation of the signal received from transmitter 1 can be reconstructed and subsequently subtracted from the original received signal. After subtraction, the remaining signal contains only the contribution of transmitter 2 and noise. Consequently, the signal from transmitter 2 can be decoded at a rate commensurate with an environment in which the signal from transmitter 1 is completely absent. This discussion explains vertex 1 of the pentagon in Figure 19.5. Vertex 3 can be achieved in the same manner with the roles of transmitter 1 and 2 reversed. One can also time average between vertices 1 and 3, thus realizing the straight line

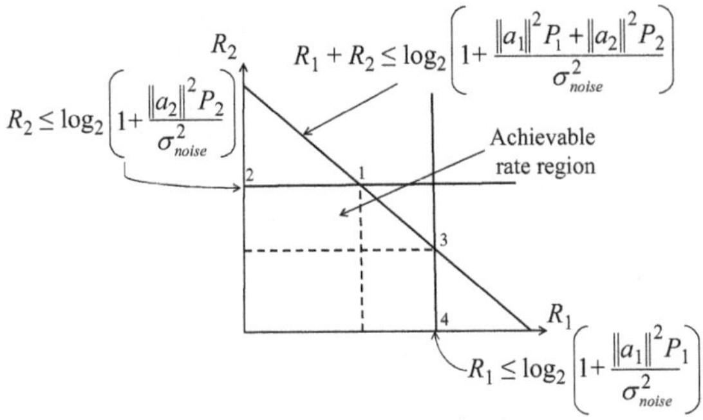

Figure 19.5 The pentagon contains the achievable communication region for two communication transmitters with rates R_1 and R_2, powers P_1 and P_2, and corresponding attenuations $\|a_1\|^2$ and $\|a_2\|^2$, for a single receiver with noise power σ^2_{noise}

between them. Vertices 2 and 4 represent the degenerate cases in which one of the two rates is zero. Finally, subject to the constraints of the capacities of isolated links, it is possible to operate at any point within this pentagon, which represents the inner and outer bounds on capacity.

For the joint radar–communication problem, this pentagon is not expected to be achievable because the set of possible radar echoes does not conform to a countable dictionary of sequences. Thus, the focus here is on achievable inner bounds, with the development of outer bounds left as an open problem.

Consider a multiaccess joint radar–communication performance bound analogous to the communication bound in Figure 19.5. Here, we develop an inner bound on the performance of a joint radar–communication multiaccess system that employs a pulsed radar. It is assumed that the receiver simultaneously observes both the radar echoes and an independent communication signal. This topology corresponds to Figure 19.4(c). As discussed in Section 19.3.2, radar tracking can be measured in terms of the data information rate for the communication function and the estimation information rate for the radar function. A bound similar to the multiaccess communication performance bound is considered to be an unachievable outer bound, though achievable inner bounds are possible. The fundamental system performance limit therefore lies between these inner and outer bounds. To simplify the discussion, only a single target is considered, which is at a range proportional to delay τ and has a channel-attenuation/radar-cross-section product collected into the term a^2. The communication channel has attenuation b^2.

The bounding approach developed here is based upon a simple water-filling formulation where the total bandwidth and total power are split among two subbands [1,43]. Water-filling is known to optimize the power and rate allocation between multiple channels [1,4,5,43]. This mixed-use arrangement operates at the SIC rate vertex (either 1 or 3) from Figure 19.5. The system block diagram of the approach considered in this scenario is shown in Figure 19.6.

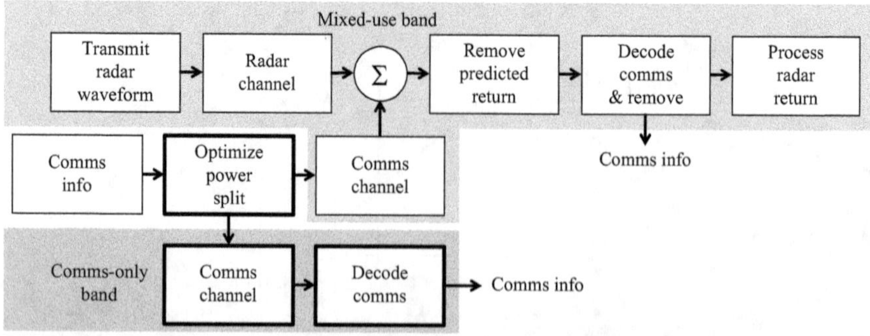

Figure 19.6 System block diagram for communications-only and mixed-use subband operation

For the simple water-filling bound, we assume that the total bandwidth $B = B_{\text{com}} + B_{\text{mix}}$ is split between a contiguous portion with bandwidth B_{com} that is used only for communications and a contiguous portion with bandwidth B_{mix} that is used for both radar and communications simultaneously. The parameter α specifies the fraction of the bandwidth segmentation that is only for communications via

$$B_{\text{com}} = \alpha\, B \qquad \text{and} \qquad B_{\text{mix}} = (1 - \alpha)\, B. \tag{19.11}$$

The communication power utilization parameter β partitions the power P_{com} between the communication-only subband and the mixed-use subband according to $P_{\text{com-only}}$ and $P_{\text{com-mix}}$, respectively. The rates and powers are optimized by using water filling [4,5,43] such that

$$\begin{aligned} P_{\text{com}} &= P_{\text{com-only}} + P_{\text{com-mix}} \\ P_{\text{com-only}} &= \beta\, P_{\text{com}} \\ P_{\text{com-mix}} &= (1 - \beta)\, P_{\text{com}}. \end{aligned} \tag{19.12}$$

The same water-filling argument provides the resulting communication rate bounds. By assuming that estimation error is small, as discussed in [1,43], the contribution of residual interference plus noise $\sigma^2_{\text{int+n}}$ can be evaluated via Parseval's theorem [5], under the assumption that small deviations produce a term that is well approximated by a derivative. Thus the rates for the communication-only subband and the mixed-use subband are [43]

$$R_{\text{com-only}} = \alpha\, B \log_2 \left[1 + \frac{\beta\, P_{\text{com}}\, b^2}{N_0\, \alpha\, B} \right] \tag{19.13}$$

$$R_{\text{com-mix}} = (1 - \alpha)\, B \log_2 \left[1 + \frac{b^2\, (1 - \beta)\, P_{\text{com}}}{\sigma^2_{\text{int+n}}} \right], \tag{19.14}$$

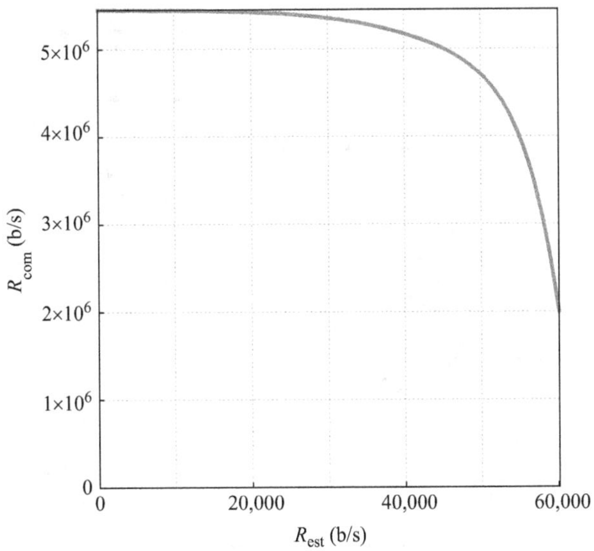

Figure 19.7 Bounding information rates for joint communications and radar based on water-filling, for the system parameters in Table 19.1

where

$$\sigma_{\text{int+n}}^2 = a^2 P_{\text{rad}} (1-\alpha)^2 \gamma^2 B^2 \sigma_{\text{proc}}^2 + N_0 (1-\alpha) B, \tag{19.15}$$

so that

$$R_{\text{com}} = R_{\text{com-only}} + R_{\text{com-mix}}. \tag{19.16}$$

Based on the notion in Section 19.3.2, the corresponding radar estimation rate inner bound is then given by

$$R_{\text{est}} = \frac{\delta}{2T} \log_2 \left(1 + \frac{2\sigma_{\text{proc}}^2 \gamma^2 B_{\text{mix}} (TB_{\text{mix}}) a^2 P_{\text{rad}}}{N_0} \right). \tag{19.17}$$

Figure 19.7 illustrates this joint inner bound for co-design of multiaccess communications and radar based on the water-filling approach [1,43]. This plot was obtained using (19.16) and (19.17) and by varying the bandwidth segmentation parameter α from zero to one. The system operating parameters for this example are listed in Table 19.1. This result shows that when little emphasis is placed on target information extraction (low values of R_{est}), the communication link can operate near full capacity. Conversely, when radar performance is the primary goal (high values of R_{est}), the communications rate is degraded, though it goes no lower than the SIC point.

Table 19.1 System operating parameters for the performance bound example

Parameter	**Value**
Bandwidth	10 MHz
Center frequency	3 GHz
Temperature	1,000 K
Comms range	10 km
Comms power	1 W
Comms transmit antenna gain	0 dBi
Comms receiver sidelobe gain	0 dBi
Radar target range	20 km
Radar antenna gain	30 dBi
Radar power	1 kW
Target cross section	25 m^2
Target process standard deviation	10 m/update
Pulse duration	100 μs
Radar duty factor	1

19.4.2 Multiaccess receiver and relay

The multiaccess concept from the previous section can be extended by including the notion of reuse for the radar illumination waveform as depicted in Figure 19.8. In this arrangement, a node serves as a multiaccess receiver for both radar echoes and an uplink communication signal, while also serving as a dual-function transmitter of radar illumination containing a downlink communication signal. Because the downlink signal is embedded in the radar waveform, this communication relay essentially incurs no cost in terms of spectral occupancy. Furthermore, the capabilities provided by radar platforms are typically more significant than most communication systems, so there may be additional operational advantages beyond the simple fact of extending communication range through higher transmit power. Per Figure 19.8, we assume that the uplink and downlink communication channels reside in spatial sidelobes of the multiaccess node's transmit mainbeam. For the sake of discussion, it is assumed that the co-use of the illumination/downlink signal causes no degradation to radar performance, though this assumption is likely not realistic in general (see Chapter 23 for further discussion of this radar performance trade-off).

The need for such an arrangement may arise because the signal generated by the communication uplink transmitter experiences too much attenuation or has insufficient transmit power for the downlink communication receiver to capture the signal relative to the noise floor. Under this condition, the communication rate from the mixed-use node to the communication receiver is given by

$$R_{\text{relay}} = (1-\alpha)\,B\,\log_2\left(1+\frac{\|b_{\text{down}}\|^2\,P_{\text{rad}}}{N_0\,(1-\alpha)\,B}\right), \tag{19.18}$$

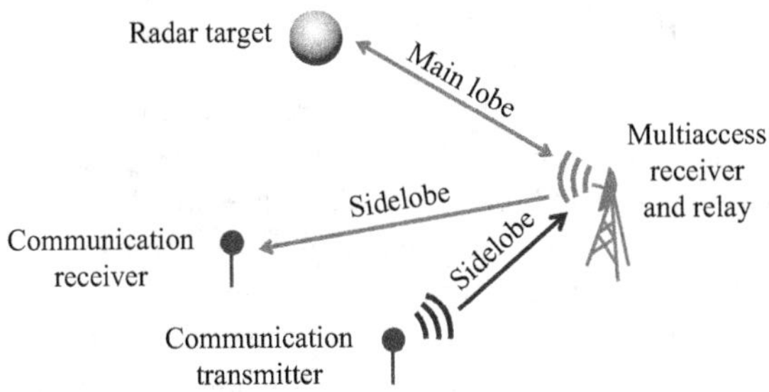

Figure 19.8 Conceptual depiction of a node that serves as a multiaccess receiver for a communication uplink signal and radar echoes. In addition, the radar illumination waveform is used as a relay for a communication downlink

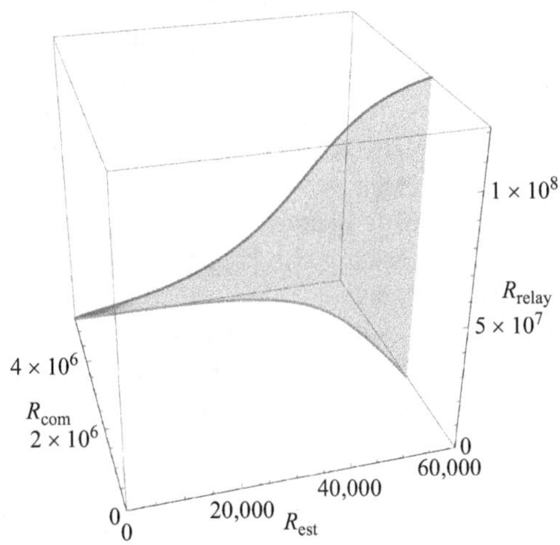

Figure 19.9 Bounding information rates for joint communications (uplink), radar, and relay (downlink), as a function of band splitting factor α for the system parameters in Table 19.1. The manifold is projected onto the $R_{est} - R_{com}$ plane to aid visualization

where B in this context is the bandwidth of the joint waveform that provides both radar illumination and downlink communications.

Figure 19.9 illustrates the manifold of the radar-as-a-relay performance bound, which in this scenario is one dimensional. For the sake of clarity, the manifold is projected onto the $R_{\text{est}} - R_{\text{com}}$ plane, which provides the curve observed in Figure 19.7. It is interesting to note that as the uplink communication rate increases, the downlink communication rate decreases as a natural result of the downlink employing only the mixed-use fraction of the band.

19.5 Operating-point optimization

If we were to imagine that all joint bounds were known, which is not the case for many topologies, there still remains the problem of how to use these bounds for realistic systems as indicated in Figure 19.1. There are two classes of problems in this context. First, if all channel and environmental information are known everywhere and system control parameters can be clairvoyantly distributed instantly across all nodes, then there is a centralized operating-point selection problem. Second, given practical limitations associated with larger networks of nodes, how do we implement distributed solutions that approach the centralized system performance? While recognizing the need for progress on the latter problem, a detailed investigation is beyond the scope of this discussion.

Given access to a complete understanding of joint operational bounds, to all environmental information, and the ability to control all system parameters globally, it is possible to specify a joint optimization metric that is a function of system goals. That said, these relationships are often nonlinear. For example, if some necessary rate is satisfied, additional resources may be of little value (a saturation effect). For static problems, it is expected that joint bounds are convex, and consequently, the optimization is technically easy. The reason for this convex assumption is that any operating points can be functionally averaged, so the joint bounds always conform to the convex hull of any well-performing vertices. However, realistic dynamic systems have memory. As a result, realistic bounds are unlikely to be convex. Once issues associated with distributed control are also included, we have constructed a rather interesting set of technical problems.

For the purpose of preliminary discussion, consider a simple optimization metric within the context of the radar-as-a-relay topology from Section 19.4.2. The metric is constructed from a linear weighting of the three rates, with the resulting weighted rate given by

$$R_{\text{weighted}}(\alpha; w_{\text{est}}, w_{\text{com}}, w_{\text{relay}}) = \frac{w_{\text{est}} R_{\text{est}} + w_{\text{com}} R_{\text{com}} + w_{\text{relay}} R_{\text{relay}}}{w_{\text{est}} + w_{\text{com}} + w_{\text{relay}}}. \tag{19.19}$$

This example is convenient pedagogically because for fixed power and total spectral allocation, the operating point can be specified by the single parameter α, which is the fraction of the band used exclusively by the uplink. Consequently, system operating-point specifications can be determined relative to this single value. The optimal value for α is therefore

$$\hat{\alpha} = \arg\max_{\alpha}\{R_{\text{weighted}}(\alpha; w_{\text{est}}, w_{\text{com}}, w_{\text{relay}})\}. \tag{19.20}$$

To simplify the display for different weighting curves, consider the normalized rate

$$R_{\text{norm,weighted}}(\alpha; w_{\text{est}}, w_{\text{com}}, w_{\text{relay}}) = \frac{R_{\text{weighted}}(\alpha; w_{\text{est}}, w_{\text{com}}, w_{\text{relay}})}{R_{\text{weighted}}(0; w_{\text{est}}, w_{\text{com}}, w_{\text{relay}})}, \tag{19.21}$$

which compares the rate R_{weighted} for a given value of α to that when $\alpha = 0$.

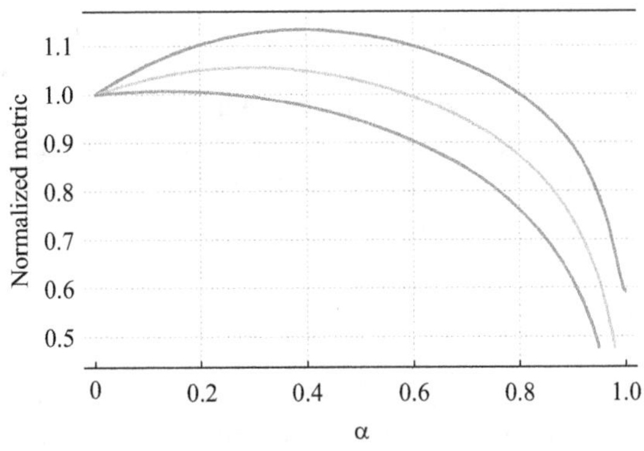

Figure 19.10 *Normalized rate metric from (19.21) for multiaccess radar–communication receiver and relay as a function of band splitting factor α and for the system parameters in Table 19.1. The relative weighting factors are* $w_{com} = w_{relay} = 1$ *and* $w_{est} = \{100\,(top), 200, 400\,(bottom)\}$

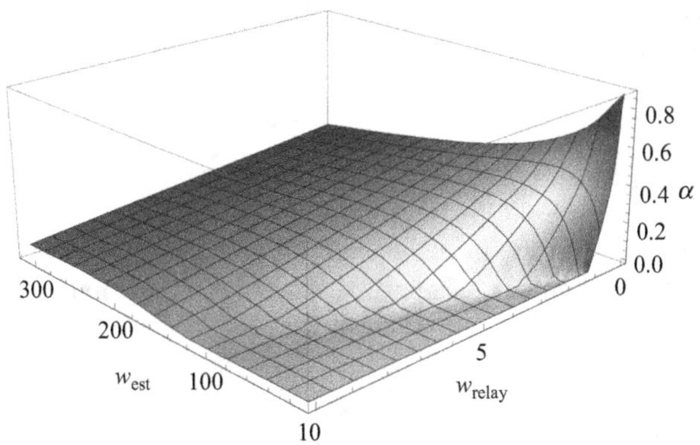

Figure 19.11 *Optimal value of α based on (19.21) as a function of* w_{est} *and* w_{relay} *for multiaccess radar–communication receiver and relay and the system parameters in Table 19.1*

Numerical examples of this radar-as-a-relay topology system optimization are provided in Figures 19.10 and 19.11, where we have again used the system parameters from Table 19.1. In Figure 19.10, the normalized metric is depicted as a function of the uplink-only subband fraction α for equal uplink and downlink weighting and for relative estimation rate weightings of 100, 200, and 400. In each case, the curves are concave, with maxima occurring for different values of α between 0 and 1. As one would expect, as the weight of the radar estimation increases, the optimal α becomes smaller.

In Figure 19.11, the optimal value of α is evaluated as the relative weighting of w_{est} and w_{relay} are varied for fixed $w_{\text{com}} = 1$. As the importance of the downlink communication relay is increased, unsurprisingly, more of the spectrum is given over to mixed-use, which is associated with a smaller value of α. It is interesting, though, that for some values of relay weighting α drops precipitously.

19.6 Conclusions

We have considered the problem of radar–communication co-design from an information theoretic perspective. In so doing, a variety of simple topologies have been proposed that may serve as building blocks for more complex spectrum sharing arrangements. An approach to define performance bounds on a radar–communication multiaccess receiver has been considered that relies on information estimation rate construction combined with more traditional communication information rate metrics. A multiaccess radar-as-a-relay formulation has also been presented, along with a subsequent bounding manifold on performance and an example linear optimization metric that incorporates weights for various functions. For this example, the optimized parameterization as a function of the weighted multidimensional metric was likewise examined.

19.7 Looking ahead

While it is clear that there are many potential advantages to the impending RF convergence of radar and communications, there are still numerous open questions. Performance bounds even for simple topologies are limited, and this understanding becomes even fuzzier as complexity increases. This result is not surprising since even network information theory based only on communications is not solved completely, and the complicated interplay between the myriad different sensing modes means that there is significant work remaining to determine the various performance bounds.

Given an understanding of joint radar–communication co-design bounds, the next step is to develop techniques to use this information to address real system performance goals. This problem necessitates the development of dynamic optimization metrics that capture the key characteristics of evolving goals. An important aspect to this discussion is the integration of relational nonlinearities and system memory. Further, for many of the more complicated topologies, the high-dimensional performance metrics may not be convex, thus leading to optimization challenges.

For larger radar–communication networks, the operational environment and resource status cannot be globally known, which is particularly problematic for highly dynamic scenarios. It is therefore necessary to perform distributed optimization based on partial knowledge. Though as with many engineering problems, exact solutions may not be necessary to produce results that are broadly useful.

Finally, there are a number of challenges with regard to transitioning these concepts to practical systems. An understanding of the implications of model mismatch caused by the nonideal nature of realistic systems needs to be incorporated into operating-point optimization. Likewise, joint signal processing algorithms must be implemented on real-time systems, which includes the development of the appropriate computational capabilities across heterogeneous nodes. Of course, the most significant barrier may ultimately be the traditional stove-piped view of RF system development that will require joint design expertise, system management, and sufficient funding support to overcome.

Acknowledgments

I would like to thank Bryan Paul, Alex Chiriyath, and Andrew Herschfelt for their comments and valuable contributions in investigating references. This work was sponsored in part by the Office of Naval Research. Any opinions, findings, and conclusions or recommendations expressed in this material are those of the author and do not necessarily reflect the views of the Office of Naval Research.

References

[1] D. W. Bliss, "Cooperative radar and communications signaling: the estimation and information theory odd couple," in *IEEE International Radar Conference*. IEEE, May 2014, pp. 50–55.

[2] G. M. Jacyna, B. Fell, and D. McLemore, "A high-level overview of fundamental limits studies for the DARPA SSPARC program," in *IEEE Radar Conference*, May 2016, pp. 1–6.

[3] B. Paul, A. Chiriyath, and D. W. Bliss, "Survey of RF communications and sensing convergence research," *IEEE Access*, vol. 5, pp. 252–270, 2017.

[4] T. M. Cover and J. A. Thomas, *Elements of Information Theory, 2nd Edition*. New York, NY: John Wiley & Sons, 2006.

[5] D. W. Bliss and S. Govindasamy, *Adaptive Wireless Communications: MIMO Channels and Networks*. Cambridge: Cambridge University Press, 2013.

[6] J. D. Taylor, *Ultrawideband Radar: Applications and Design*. Boca Raton, FL: CRC Press, 2012.

[7] A. Elston, "Coexistence of S band radar systems with future mobile services operating in the adjacent 2600 MHz band: A brief," ClearMobitel, Tech. Rep., 2011.

[8] S. Dehnie, V. Chakravarthy, Z. Wu, C. Ghosh, and H. Li, "Spectrum coexistence issues: challenges and research directions," in *IEEE Military Communications Conference*, November 2013, pp. 1681–1689.

[9] S. Sodagari, A. Khawar, T. Clancy, and R. McGwier, "A projection-based approach for radar and telecommunication systems coexistence," in *IEEE Global Communications Conference*, December 2012, pp. 5232–5236.

[10] D. Garmatyuk, Y. Morton, and X. Mao, “On co-existence of in-band UWB-OFDM and GPS signals: tracking performance analysis,” in *IEEE/ION Position, Location and Navigation Symposium*. IEEE, May 2008, pp. 196–202.

[11] R. Saruthirathanaworakun, J. M. Peha, and L. M. Correia, “Gray-space spectrum sharing between multiple rotating radars and cellular network hotspots,” in *IEEE 77th Vehicular Technology Conference (VTC Spring)*, June 2013, pp. 1–5.

[12] R. Saruthirathanaworakun, J. M. Peha, and L. M. Correia, “Opportunistic primary-secondary spectrum sharing with a rotating radar,” in *International Conference on Computing, Networking and Communications*, 2012, pp. 1025–1030.

[13] M. Tercero, K. W. Sung, and J. Zander, “Exploiting temporal secondary access opportunities in radar spectrum,” *Wireless Personal Communications*, vol. 72, no. 3, pp. 1663–1674, 2013.

[14] B. W. Zarikoff and D. Weldon, “Detection of pulsed radar in a time division duplexed system,” in *IEEE 73rd Vehicular Technology Conference (VTC Spring)*, May 2011, pp. 1–5.

[15] H. D. Griffiths and N. R. W. Long, “Television-based bistatic radar,” *IEE Proceedings F – Communications, Radar and Signal Processing*, vol. 133, no. 7, pp. 649–657, December 1986.

[16] S. Searle, J. Palmer, L. Davis, D. W. O’Hagan, and M. Ummenhofer, “Evaluation of the ambiguity function for passive radar with OFDM transmissions,” in *IEEE Radar Conference*, May 2014, pp. 1040–1045.

[17] A. Evers and J. A. Jackson, “Analysis of an LTE waveform for radar applications,” in *IEEE Radar Conference*, May 2014, pp. 200–205.

[18] D. Erricolo, H. Griffiths, L. Teng, M. C. Wicks, and L. L. Monte, “On the spectrum sharing between radar and communication systems,” in *International Conference on Electromagnetics in Advanced Applications (ICEAA)*, August 2014, pp. 890–893.

[19] N. Nartasilpa, D. Tuninetti, N. Devroye, and D. Erricolo, “Let’s share CommRad: effect of radar interference on an uncoded data communication system,” in *2016 IEEE Radar Conference (RadarConf)*, May 2016, pp. 1–5.

[20] H. Shajaiah, A. Abdelhadi, and C. Clancy, “Spectrum sharing approach between radar and communication systems and its impact on radar’s detectable target parameters,” in *IEEE 81st Vehicular Technology Conference (VTC Spring)*, May 2015, pp. 1–6.

[21] H.-A. Safavi-Naeini, S. Roy, and S. Ashrafi, “Spectrum sharing of radar and Wi-Fi networks: the sensing/throughput tradeoff,” *IEEE Transactions on Cognitive Communications and Networking*, vol. 1, no. 4, pp. 372–382, December 2015.

[22] H. Hayvaci and B. Tavli, “Spectrum sharing in radar and wireless communication systems: a review,” in *2014 International Conference on Electromagnetics in Advanced Applications (ICEAA)*, August 2014, pp. 810–813.

[23] M. R. Bell, N. Devroye, D. Erricolo, T. Koduri, S. Rao, and D. Tuninetti, "Results on spectrum sharing between a radar and a communications system," in *2014 International Conference on Electromagnetics in Advanced Applications (ICEAA)*, August 2014, pp. 826–829.

[24] G. Meager, R. A. Romero, and Z. Staples, "Estimation and cancellation of high powered radar interference for communication signal collection," in *2016 IEEE Radar Conference (RadarConf)*, May 2016, pp. 1–4.

[25] S. S. Bhat, R. M. Narayanan, and M. Rangaswamy, "Bandwidth sharing and scheduling for multimodal radar with communications and tracking," in *IEEE Sensor Array and Multichannel Signal Processing Workshop*, June 2012, pp. 233–236.

[26] H. Shajaiah, A. Khawar, A. Abdel-Hadi, and T. C. Clancy, "Resource allocation with carrier aggregation in LTE advanced cellular system sharing spectrum with S-band radar," in *IEEE International Symposium on Dynamic Spectrum Access Networks (DYSPAN)*, April 2014, pp. 34–37.

[27] A. Khawar, A. Abdel-Hadi, and T. C. Clancy, "Spectrum sharing between S-band radar and LTE cellular system: a spatial approach," in *IEEE International Symposium on Dynamic Spectrum Access Networks (DYSPAN)*, April 2014, pp. 7–14.

[28] T. Yucek and H. Arslan, "A survey of spectrum sensing algorithms for cognitive radio applications," *IEEE Communications Surveys & Tutorials*, vol. 11, no. 1, pp. 116–130, 2009.

[29] M. Wicks, "Spectrum crowding and cognitive radar," in *2nd International Workshop on Cognitive Information Processing*, June 2010, pp. 452–457.

[30] L. S. Wang, J. P. McGeehan, C. Williams, and A. Doufexi, "Application of cooperative sensing in radar–communications coexistence," *IET Communications*, vol. 2, no. 6, pp. 856–868, July 2008.

[31] A. Aubry, A. D. Maio, M. Piezzo, M. M. Naghsh, M. Soltanalian, and P. Stoica, "Cognitive radar waveform design for spectral coexistence in signal-dependent interference," in *IEEE Radar Conference*, May 2014, pp. 474–478.

[32] P. Stinco, M. Greco, F. Gini, and B. Himed, "Channel parameters estimation for cognitive radar systems," in *2014 4th International Workshop on Cognitive Information Processing (CIP)*, May 2014, pp. 1–6.

[33] S. D. Howard, W. Moran, A. R. Calderbank, H. A. Schmitt, and C. Savage, "Channel parameters estimation for cognitive radar systems," in *Proceedings IEEE International Conference on Acoustics, Speech, and Signal Processing*, vol. 5, March 2005, pp. 897–900.

[34] Y. Nijsure, Y. Chen, S. Boussakta, C. Yuen, Y. H. Chew, and Z. Ding, "Novel system architecture and waveform design for cognitive radar radio networks," *IEEE Transactions on Vehicular Technology*, vol. 61, no. 8, pp. 3630–3642, October 2012.

[35] C. Sturm and W. Wiesbeck, "Waveform design and signal processing aspects for fusion of wireless communications and radar sensing," *Proceedings of the IEEE*, vol. 99, no. 7, pp. 1236–1259, July 2011.

[36] J. R. Guerci, R. M. Guerci, A. Lackpour, and D. Moskowitz, "Joint design and operation of shared spectrum access for radar and communications," in *IEEE Radar Conference*, May 2015, pp. 761–766.

[37] A. Turlapaty and Y. Jin, "A joint design of transmit waveforms for radar and communications systems in coexistence," in *IEEE Radar Conference*, May 2014, pp. 315–319.

[38] C. D. Richmond, P. Basu, R. Learned, J. Vian, A. P. Worthen, and M. Lockard, "Performance bounds on cooperative radar and communication systems operation," in *2016 IEEE Radar Conference (RadarConf)*, May 2016, pp. 1–6.

[39] A. R. Chiriyath, B. Paul, and D. W. Bliss, "Simultaneous radar detection and communications performance with clutter mitigation," in *IEEE Radar Conference*, May 2017.

[40] J. R. Guerci and R. M. Guerci, "RAST: radar as a subscriber technology for wireless spectrum cohabitation," in *IEEE Radar Conference*, May 2014, pp. 1130–1134.

[41] M. Scharrenbroich and M. Zatman, "Joint radar–communications resource management," in *IEEE Radar Conference*, May 2016, pp. 1–6.

[42] A. R. Chiriyath and D. W. Bliss, "Joint radar–communications performance bounds: data versus estimation information rates," in *2015 IEEE Military Communications Conference, MILCOM*, October 2015, pp. 1491–1496.

[43] A. Chiriyath, B. Paul, G. M. Jacyna, and D. Bliss, "Inner bounds on performance of radar and communications co-existence," *IEEE Transactions on Signal Processing*, vol. 64, no. 2, pp. 464–474, January 2016.

[44] B. Paul and D. W. Bliss, "Extending joint radar–communications bounds for FMCW radar with Doppler estimation," in *IEEE Radar Conference*, May 2015, pp. 89–94.

[45] A. R. Chiriyath and D. W. Bliss, "Effect of clutter on joint radar–communications system performance inner bounds," in *IEEE Asilomar Conference on Signals, Systems and Computers*, November 2015, pp. 1379–1383.

[46] A. R. Chiriyath, B. Paul, and D. W. Bliss, "Joint radar–communications information bounds with clutter: the phase noise menace," in *IEEE Radar Conference*, May 2016, pp. 690–695.

[47] B. Paul and D. W. Bliss, "Constant information radar for dynamic shared spectrum access," in *IEEE Asilomar Conference on Signals, Systems and Computers*, November 2015, pp. 1374–1378.

[48] B. Paul and D. W. Bliss, "The constant information radar," *Entropy*, vol. 18, no. 9, p. 338, September 2016. [Online]. Available: http://www.mdpi.com/1099-4300/18/9/338

[49] B. Paul, A. R. Chiriyath, and D. W. Bliss, "Joint communications and radar performance bounds under continuous waveform optimization: the waveform awakens," in *IEEE Radar Conference*, May 2016, pp. 865–870.

[50] D. Guo, S. Shamai, and S. Verdu, "Mutual information and minimum mean-square error in Gaussian channels," *IEEE Transactions on Information Theory*, vol. 51, no. 4, pp. 1261–1282, 2005.

[51] B. Paul and D. W. Bliss, "Estimation information bounds using the I-MMSE formula and Gaussian mixture models," in *Conference on Information Sciences and Systems (CISS)*, March 2016, pp. 292–297.

[52] A. Herschfelt and D. W. Bliss, "Multi-static space-time-frequency multiple access channel simulation and results," in *IEEE Radar Conference*, May 2017.

[53] A. Hassanien, M. G. Amin, Y. D. Zhang, F. Ahmad, and B. Himed, "Non-coherent PSK-based dual-function radar–communication systems," in *2016 IEEE Radar Conference (RadarConf)*, May 2016, pp. 1–6.

[54] A. Hassanien, M. G. Amin, Y. D. Zhang, and F. Ahmad, "Dual-function radar–communications: information embedding using sidelobe control and waveform diversity," *IEEE Transactions on Signal Processing*, vol. 64, no. 8, pp. 2168–2181, April 2016.

[55] B. J. Donnet and I. D. Longstaff, "Combining MIMO radar with OFDM communications," in *Proceedings of the 3rd European Radar Conference*, pp. 37–40, September 2006.

[56] C. Sturm, T. Zwick, and W. Wiesbeck, "An OFDM system concept for joint radar and communications operations," in *IEEE Vehicular Technology Conference*. IEEE, April 2009, pp. 1–5.

[57] M. Braun, C. Sturm, A. Niethammer, and F. K. Jondral, "Parametrization of joint OFDM-based radar and communication systems for vehicular applications," in *IEEE 20th International Symposium on Personal, Indoor and Mobile Radio Communications*, September 2009, pp. 3020–3024.

[58] C. W. Rossler, E. Ertin, and R. L. Moses, "A software defined radar system for joint communication and sensing," in *IEEE Radar Conference (RADAR)*, May 2011, pp. 1050–1055.

[59] T. Guo and R. Qiu, "OFDM waveform design compromising spectral nulling, side-lobe suppression and range resolution," in *IEEE Radar Conference*, May 2014, pp. 1424–1429.

[60] S. C. Thompson and J. P. Stralka, "Constant envelope OFDM for power-efficient radar and data communications," in *International Waveform Diversity and Design Conference*, February 2009, pp. 291–295.

[61] R. F. Tigrek, W. J. A. de Heij, and P. van Genderen, "Relation between the peak to average power ratio and Doppler sidelobes of the multi-carrier radar signal," in *International Radar Conference – Surveillance for a Safer World*, October 2009, pp. 1–6.

[62] S. D. Howard, A. R. Calderbank, and W. Moran, "The finite Heisenberg-Weyl groups in radar and communications," *EURASIP Journal on Applied Signal Processing*, vol. 2006, pp. 1–12, April 2006.

[63] A. Pezeshki, A. R. Calderbank, W. Moran, and S. D. Howard, "Doppler resilient Golay complementary waveforms," *IEEE Transactions on Information Theory*, vol. 54, no. 9, pp. 4254–4266, September 2008.

[64] L. Xiaobai, Y. Ruijuan, Z. Zunquan, and C. Wei, "Research of constructing method of complete complementary sequence in integrated radar and communication," in *IEEE 11th International Conference on Signal Processing (ICSP)*, vol. 3, October 2012, pp. 1729–1732.

[65] S. Blunt, P. Yatham, and J. Stiles, "Intrapulse radar-embedded communications," *IEEE Transactions on Aerospace and Electronic Systems*, vol. 46, no. 3, pp. 1185–1200, July 2010.

[66] S. Blunt, J. Metcalf, C. Biggs, and E. Perrins, "Performance characteristics and metrics for intra-pulse radar-embedded communications," *IEEE Journal on Selected Areas in Communications*, vol. 29, no. 10, pp. 2057–2066, December 2011.

[67] J. Metcalf, C. Sahin, S. Blunt, and M. Rangaswamy, "Analysis of symbol design strategies for intrapulse radar-embedded communications," *IEEE Transactions on Aerospace and Electronic Systems*, vol. 51, no. 4, pp. 2914–2931, October 2015.

[68] D. Ciuonzo, A. D. Maio, G. Foglia, and M. Piezzo, "Intrapulse radar-embedded communications via multiobjective optimization," *IEEE Transactions on Aerospace and Electronic Systems*, vol. 51, no. 4, pp. 2960–2974, October 2015.

[69] Y. L. Sit, L. Reichardt, C. Sturm, and T. Zwick, "Extension of the OFDM joint radar–communication system for a multipath, multiuser scenario," in *2011 IEEE RadarCon (RADAR)*, May 2011, pp. 718–723.

[70] B. Li, A. P. Petropulu, and W. Trappe. "Optimum co-design for spectrum sharing between matrix completion based MIMO radars and a MIMO communication system," *IEEE Transactions on Signal Processing*, vol. 64, no. 17, pp. 4562–4575, Sept. 2016.

[71] S. M. Kay, *Fundamentals of Statistical Signal Processing: Estimation Theory*. Upper Saddle River, NJ: Prentice Hall, 1993.

[72] P. Woodward and I. Davies, "A theory of radar information," *Philosophical Magazine Series 7*, vol. 41, no. 321, pp. 1001–1017, 1993.

[73] K. W. Forsythe and D. W. Bliss, "MIMO radar: concepts, performance enhancements, and applications," in *Signal Processing for MIMO Radar*, J. Li and P. Stoica, Eds. Hoboken, NJ: John Wiley & Sons, 2008.

[74] H. Akaike, "A new look at the statistical model identification," *IEEE Transactions on Automatic Control*, vol. 19, no. 6, pp. 716–723, 1974.

Chapter 20

Real-time radar/communication spectrum sharing based on information exchange*

Alex Lackpour[1], Joseph R. Guerci[2], and Michal Zatman[3]

20.1 Introduction, background, and the "Three A's" of radar–comms spectrum sharing

The increasing global demand for useful radio frequency (RF) spectrum by wireless communications has led to a conflict with traditional primary users of legacy spectrum such as radar. This arrangement has prompted research into new approaches to sharing spectrum between radar and communications. To that end, the U.S. Defense Advanced Research Projects Agency (DARPA) created a major new project titled Shared Spectrum Access for Radar and Communications (SSPARC) [1–4]. The overall scope of the SSPARC program includes techniques that may be incorporated by existing systems with minimal changes at one end to facilitate spectrum-sharing approaches that entail a "back to the drawing board" approach at the other.

Fundamentally, there are basically three distinct conceptual approaches for combined radar and communications, herein referred to as the "three A's". These are Avoid & Mitigate (A&M), Accept, and Amalgamate, as shown in Figure 20.1.

20.1.1 Avoid & mitigate interference

The A&M approach that forms the basis of most radar–communication (RC) spectrum-sharing techniques is also the basis for the cognitive radio framework that automatically "...changes its transmission or reception parameters to allow more concurrent wireless comms in a given spectrum band at one location" [5]. The broad class of techniques that come under A&M are depicted in Figure 20.2.

The range of techniques that are encompassed by the A&M class may occur in the transmitter or receiver; may avoid or mitigate in time, frequency, and space; and may or may not be coordinated between the transmitter and receiver. Much of the continued research in the A&M class has been directed towards coordination or coupling between communications and radar to eliminate interference [1].

*Distribution Statement "A" (Approved for Public Release, Distribution Unlimited)

[1]Lockheed Martin Advanced Technology Laboratories, USA
[2]Information Systems Laboratories, Inc., USA
[3]SAZE Technologies, USA

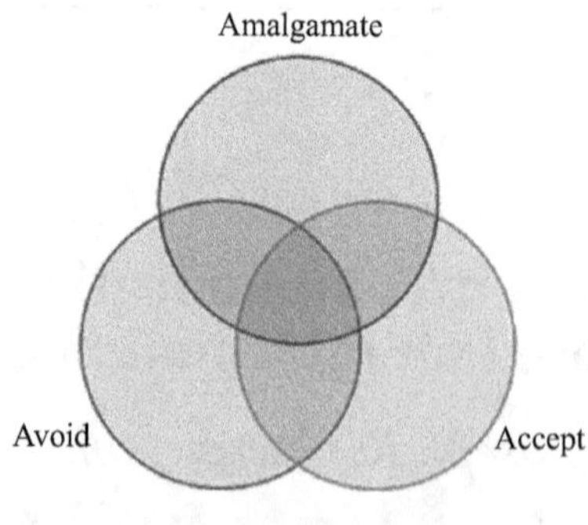

Figure 20.1 The "three A's" of radar–communication spectrum sharing

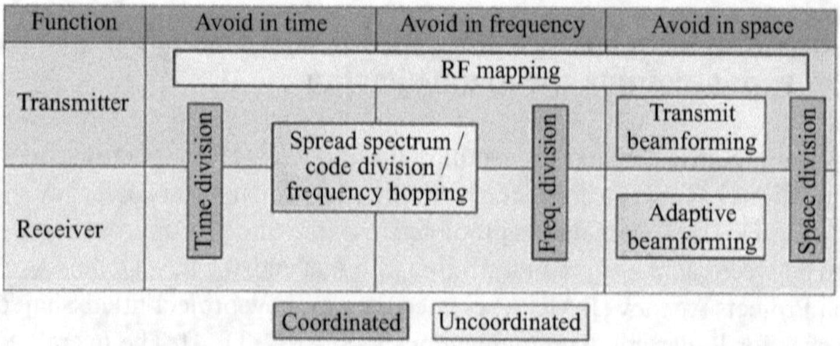

Figure 20.2 Examples of Avoid and Mitigate (A&M) approaches

20.1.2 Accept interference

Many communication systems *accept* co-channel interference as the price that is paid for spatial reuse of the spectrum. In this context the potential channel capacity is governed by Shannon's Theorem [6], which can be expressed as

$$C = B\log_2(1 + \mathrm{SNR}), \tag{20.1}$$

where B denotes the bandwidth and SNR is the signal-to-noise ratio. Consequently, reducing SNR in return for either bandwidth or time is a worthwhile trade for communication systems when the SNR is high. This condition is demonstrated in Figure 20.3, which shows that going from a 30 dB SNR to a 20 dB SNR, while increasing the bandwidth 10-fold, improves the data rate by a factor of 6.7. Accepting co-channel interference in exchange for bandwidth also allows for better spatial reuse—i.e. a higher area spectral efficiency (measured in b/s/Hz/m^2) [7]. Such an arrangement is exemplified by the practical operation of cellular communication systems and Wi-Fi.

In contrast, radar's primary system metric is probability of detection P_d, which (to the first order[1]) is a function of SNR and is independent of bandwidth [8], as shown in Figure 20.4. For the same 10× SINR/bandwidth trades indicated in Figure 20.3

[1]To the extent that the probability of detection depends upon suppressing the ground clutter, a minimal amount of bandwidth is needed for clutter suppression.

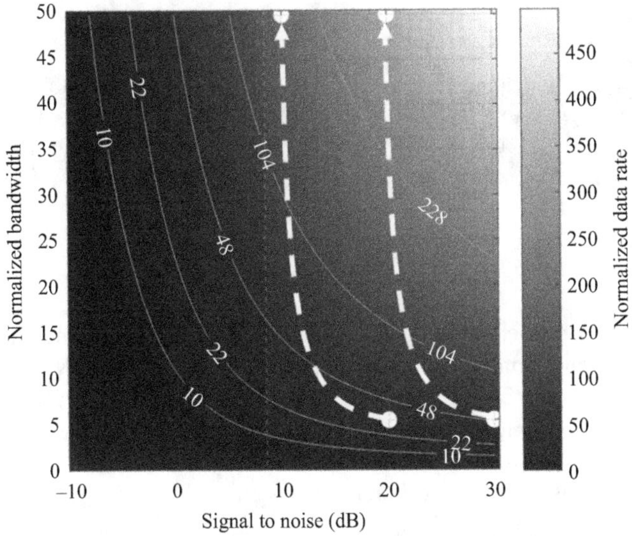

Figure 20.3 Data rate as a function of SNR and bandwidth, demonstrating that bandwidth is more valuable than SNR to a communication system

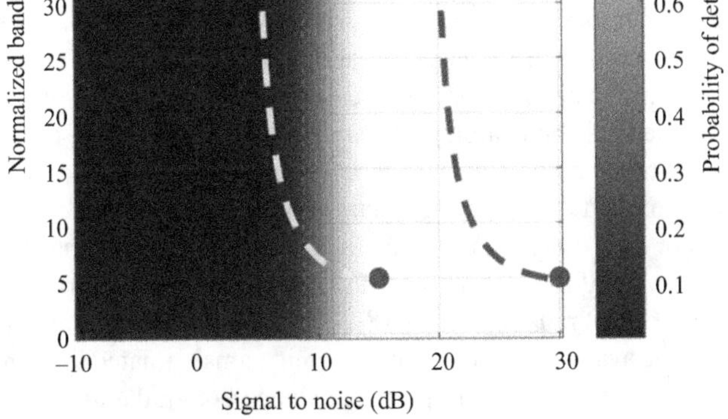

Figure 20.4 Radar probability of detection is independent of the bandwidth (to the first order). The detection threshold was chosen to provide a probability of false alarm of 10^{-6}. The target's radar cross section follows the Swerling 0 model [9]

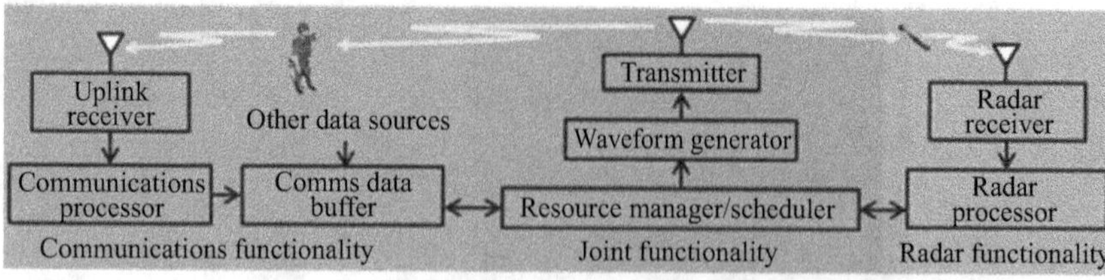

Figure 20.5 Anatomy of an amalgamated RC spectrum-sharing system

for communications there is always degradation in radar performance. Because the SNR/P_d relationship is nonlinear, degradation may be small if the SNR is sufficiently high (dotted line trace on the right), or a significant reduction in P_d may occur if the original SNR operating point is near the transition region (here, 11 dB SNR corresponds to a P_d of 0.9).

This result illustrates a key difference between radar and communications: radar P_d is essentially energy limited, whereas the communication data rate is limited by the bandwidth and the logarithm of energy. Thus the acceptance of interference in exchange for bandwidth that enables better spatial reuse, and hence a higher area spectral efficiency for communication-to-communication spectrum sharing, does not carry over into RC spectrum sharing. The other key aspect of the *accept* solution space is coordination/feedback between radar and communication systems to ensure that the interference level at the radar remains "acceptable."

20.1.3 Amalgamate waveforms

Amalgamating the radar and communication functionalities into a single system results in a single transmitter that blends the radar and communication waveforms. There have been a number of examples of such systems that have been described (see [10–17]), often as a result of the desire to add communications to an existing radar platform rather than the need to share spectrum.

Figure 20.5 depicts the anatomy of an example amalgamated RC system. The transmit chain (with the associated scheduling) is the only joint aspect of the system. The communication and radar receiver chains both interact with the transmitter, but are independent of each other. Amalgamated RC spectrum sharing may be interpreted as a deliberate version of the use of communication signals of opportunity by passive coherent location (PCL) radar systems [18]. While PCL systems must make do with the properties of the available signals of opportunity, amalgamated systems can adjust the waveforms' properties to better optimize/balance both radar and communication performance. Table 20.1 describes how some aspects of the amalgamated waveform affect each modality.

A key issue embedded in the results of Table 20.1 is that radars prefer constant envelope waveforms, thereby maximizing the amount of energy used to illuminate targets. This situation contrasts with many communication systems that accept waveforms with significant peak-to-average power ratios as an aid to achieving higher

Table 20.1 How different waveform properties affect performance

Waveform aspect	Radar	Communications
Amplitude modulation	Reduced SNR	Higher data rate
High duty factor	Reduced coverage	Higher data rate
Multi carrier	Reduced SNR	Lower complexity
MIMO	Better resolution	Higher data rate

data rates (e.g. 64-QAM modulation) and/or simpler receivers (e.g. multi-carrier modulation such as OFDM) [19]. Another area of contention between radar and communications is that of duty factor. A high-power radar is typically unable to receive while transmitting, so a transmit duty factor of around 10% is often preferred [20]. However, a frequency-division duplex (FDD) communication system is most efficient while transmitting 100% of the time and a time-division duplex (TDD) system around 50% of the time (depending upon the proportion of data being uploaded vs. downloaded).

With that being said, there are aspects of the waveform that can be mutually beneficial. One example is multiple-input-multiple-output (MIMO) techniques that both improve the resolution of radar systems [21] and the data rate of communication systems [22].

20.2 Business cases for spectrum stakeholders

To understand the motivations for developing any of the three "A"s of RC spectrum sharing, a taxonomy of the associated business models is required. In their simplest form, such business models consist of radar and communication "service providers" that are independent of one another, providers who are coordinated with one another, or a joint provider.

Independent radar and communication service providers share the same spectrum and adhere to a common set of specifications that determine how their systems operate. However, they are uncoupled, in that there is no real-time coordination between the radar and communication service providers. The use of the Unlicensed National Information Infrastructure (U-NII) band by dynamic frequency selection wireless networks is an example of independent service providers adhering to a common specification, namely, that when wireless networks detect a radar's presence they move to another frequency. One aspect of the independent business model is that the financial success of the communication provider is independent of the radar as long as the former does not interfere with the latter.

Coordinated radar and communication service providers exchange information about each system in "real-time" to ensure the success of both services. DARPA's SSPARC program has investigated both "loosely coupled" and "tightly coupled" communication and radar systems [1,23]. In this case the success of the communication provider is dependent upon the real-time coordination between the

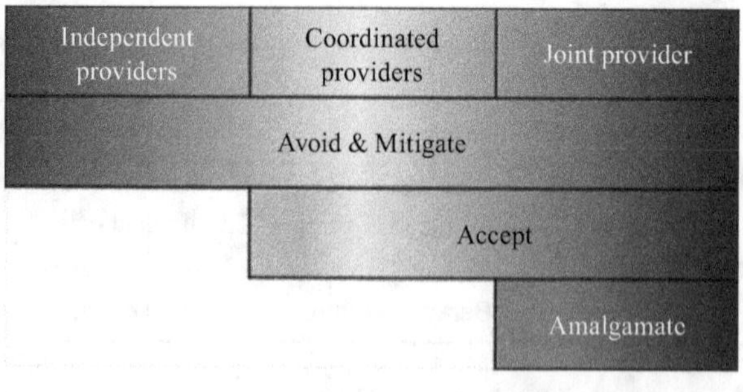

Figure 20.6 The three "As" span successively smaller parts of the service provider spectrum

different modalities. A higher degree of coordination has the potential for pushing larger amounts of data through the communication system while the radar suffers less interference. Finally, a joint radar and communication service provider is a single entity facilitating both services. This arrangement incentivizes the joint service provider to maximize the synergies between the radar and communication systems.

While the taxonomy of business models has been delineated into three discrete types, in reality there is a continuum of service provider types that exist between these Independent, Coordinated, and Joint Provider classes. Figure 20.6 depicts how the three "A"s map into the provider taxonomy. While A&M techniques can be applied across the provider spectrum, the Accept class is only applicable to coordinated and joint service providers because some type of real-time coordination is required to maintain the interference at or below acceptable levels. Similarly, the Amalgamate techniques rely on common transmitters and waveforms for radar and communications, and thus they are only applicable to a joint service provider.

In the next section, we delve deeper into the concept of spectrum sharing via fundamental analysis of *co-design* and *cooperation*, which is an example of a joint service provider developing a tightly coupled A&M spectrum-sharing system. We first develop a basic joint design and operation (JDO) SSPARC theory and develop some performance bounds. A key enabler for JDO SSPARC is a real-time information sharing backbone. To that end, a novel radar/comms spectrum-sharing system (RCS3) is introduced that includes a hierarchical control plane and information sharing subsystem. The timely exchange of state information enables radar and communication networks to apply a variety of coordinated coexistence mechanisms that shrink the minimum required standoff range between systems while sustaining the performance of each.

20.3 Maximizing the joint capacity of a radar–communication network

Radar can be viewed as a special case of communication system in which the target echo bears the information. Consequently, it stands to reason that a measure of capacity directly analogous to that developed by Hartley and Shannon should not be too much of a stretch [6]. Though Woodward's landmark book [24] examined radar from an information theoretic framework (and introduced the delay/Doppler ambiguity function along the way), it did not develop a measure of capacity directly analogous to that of communications. Rather, a general measure of the information gain from a measurement was developed. In this section an intuitively appealing and relatively simple measure of radar capacity is introduced.

For a moving target indication (MTI) radar, one can consider each independent resolution cell (e.g. range-angle-Doppler) as a binary information storage unit, i.e. "0" = target absent and "1" = target present. Viewed in this manner, the maximum capacity of an MTI radar performing a periodic search is given by the Hartley capacity measure [6]

$$C_R = \left(1/T_s\right) \log N \tag{20.2}$$

where T_s denotes the scan or update time so that the capacity is in the conventional "bit rate" form. Here N is the total number of independent MTI resolution cells that possess a minimum prescribed SNR for an assumed target radar cross-section (RCS), denoted as $\mathrm{SNR}_{\min}$, and is given by

$$N \propto (BR_{\max})\left(\frac{2\pi}{\Delta\theta}\right)\left(\frac{\mathrm{PRF}}{\Delta f_D}\right), \tag{20.3}$$

where $\Delta\theta$ and Δf_D denote the angle (e.g. azimuth) and Doppler resolution, respectively, B is the operating bandwidth, and $R_{\max}$ is the maximum range for which $\mathrm{SNR} \geq \mathrm{SNR}_{\min}$ and can be derived from the radar range equation [25] (consistent units assumed). Note that in (20.3) we have assumed the radar scans 360-deg azimuthal (2π radians). This arrangement can obviously be modified to suit the specific application. The maximum range $R_{\max}$ can in turn be expressed in terms of $\mathrm{SNR}_{\min}$ and the total noise σ_n^2 via

$$R_{\max} = \frac{\kappa}{\sqrt{\sigma_n}\ \sqrt[4]{\mathrm{SNR}_{\min}}}, \tag{20.4}$$

where κ is a fixed constant of proportionality that subsumes all of the radar range equation parameters (antenna gain, losses, etc.). Substituting (20.4) into (20.3) and then into (20.2) yields the final basic MTI radar capacity expression

$$C_R = \frac{1}{T_s} \log\left(\frac{\kappa B}{\sqrt{\sigma_n}\ \sqrt[4]{\mathrm{SNR}_{\min}}}\left(\frac{2\pi}{\Delta\theta}\right)\left(\frac{\mathrm{PRF}}{\Delta f_D}\right)\right). \tag{20.5}$$

Note that for a given application, $\mathrm{SNR}_{\min}$ is specified ahead of time and does not usually vary during operation. Typical values of $\mathrm{SNR}_{\min}$ for an MTI radar (after all aperture and signal-processing gains) would be between 10 and 15 dB depending on acceptable false alarm levels, for a given assumed target RCS.

An analogous expression for radar capacity can be readily formulated for synthetic aperture radar (SAR) by recognizing that the equivalent of a "bit" is a SAR "pixel"—again the minimum independent resolution cell with a prescribed minimum SNR (sometimes referred to as terrain-to-noise ratio). Additionally, the most general radar "bit" for either MTI or SAR is inherently an "M-ary" versus binary piece of information since the amplitude (and possibly polarization, monopulse angle, etc.) contains additional information that can be very useful for target identification in MTI, enhanced contrast in SAR, etc. Obviously for the M-ary case, there is an increase in capacity relative to the binary case considered in (20.5).

Finally, for the case in which both radar and communications are operating simultaneously, the combined expression for the total capacity is given by

$$C_{\text{tot}} = \frac{1}{T_{\text{s}}} \log\left(\frac{\kappa B_{\text{r}}}{\sqrt{\sigma_{\text{n}}}\ \sqrt[4]{\text{SNR}_{\text{min}}}}\left(\frac{2\pi}{\Delta\theta}\right)\left(\frac{\text{PRF}}{\Delta f_{\text{D}}}\right)\right) + B_{\text{c}} \log_2\left(1 + \text{SNR}_{\text{c}}\right) \quad (20.6)$$

where B_{r} and B_{c} denote the radar and communication bandwidths, respectively, and SNR_{c} denotes the SNR of the additive Gaussian white noise (AWGN) communication channel. For the assumed co-channel interference model, the coupling between radar and communications occurs in the σ_{n} term for radar (since SNR_{min} is fixed) and in the SNR_{c} term for communications. Thus, in a co-design environment where the cross-coupling terms would depend on operating powers, bandwidths, and relative locations and terrain, one would seek to jointly maximize capacity while maintaining a minimally acceptable level of radar and communication performance. If a given system is capable of performing *both* radar and communication functions, the joint capacity expression in (20.6) realizes an even greater degree of coupling than just the co-channel noise.

The next section introduces a theoretical framework for optimizing the transmit-receive functions of an N-node RF SSPARC network. The resulting design equations provide a tight bound on theoretical performance (as opposed to generally approximate methods such as Cramer–Rao [26]).

20.4 Theory of JDO SSPARC

Consider a notional geographically distributed RC scenario in which there is significant potential for co-channel interference (see Figure 20.7). If the available and limited RF spectrum is to be shared by all parties in a dynamic fashion, knowledge and adaptive functionality is required to avoid highly sub-optimal performance. For any RC node (radar or wireless communication device), at any given moment there is a set of desired channels (target channel for radar, forward link for communication), other channels that are not of interest that might include radar clutter, and other co-channel links.

We begin with the following set of definitions:

$$\begin{aligned} H_{ij} &= \text{Transfer function from the } j\text{th to the } i\text{th nodes;} \\ \mathbf{s}_j &= \text{Signal transmitted by } j\text{th node;} \\ \mathbf{y}_i &= \text{Received signal at } i\text{th node;} \\ \mathbf{n}_i &= \text{Additive receive noise at } i\text{th node.} \end{aligned} \quad (20.7)$$

The dimensionality of the above variables depends on the choice of degrees-of-freedom (DoFs) to be included in the optimization. For example, if each node had say

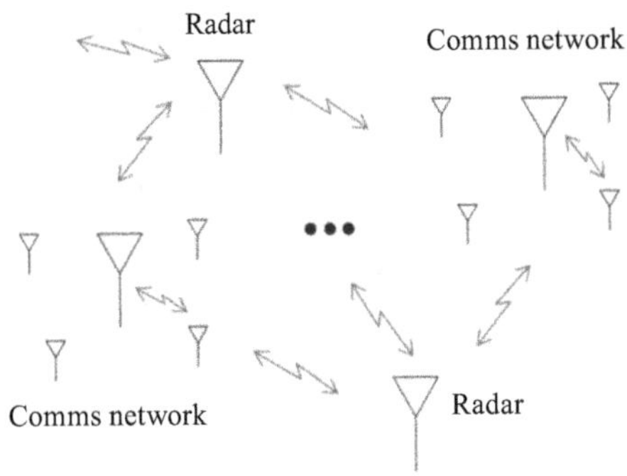

Figure 20.7 Notional combined multi-radar/communication (RC) network

L spatial DoFs on both transmit and receive, the vectors in (20.7) would be L dimensional and complex-valued, and the constituent transfer matrices would be $L \times L$ (and complex valued). We first consider the above model over a short period of time for which the channels can assumed to be nonvarying, though they can still be stochastic. In this way a tight theoretical performance bound can be achieved.

Interpretation of the constituent block matrices $\{H_{ij}\}$ is based on whether the nodes are purposed for radar or communications. For example, it is convenient to define submatrices with equal indices as

$$\begin{aligned} H_{ii} &= \emptyset \text{ for communication nodes} \\ H_{ii} &\neq \emptyset \text{ for monostatic radar nodes,} \end{aligned} \tag{20.8}$$

where $\emptyset$ denotes the null (zero) matrix or vector. In other words, a communication node does not radiate to itself. However, for a monostatic MTI radar $\{H_{ii}\}$ does represent the total signal-dependent radar channel that in general consists of targets and clutter. More specifically,

$$H_{ii} = \sum_{n=1}^{N_{\text{tot}}} H_{\text{T}_{n,i}} + H_{\text{C}_i} \tag{20.9}$$

where $\{H_{\text{T}_{n,i}}\}$ and $\{H_{\text{C}_i}\}$ denote the target and clutter channels for the ith radar.

With the above definitions, the entire composite RC channel matrix is given by

$$H = \begin{bmatrix} H_{11} & H_{12} & \cdots & H_{1N} \\ H_{21} & H_{22} & & \\ \vdots & & \ddots & \vdots \\ H_{N1} & \cdots & & H_{NN} \end{bmatrix}, \tag{20.10}$$

where $H_{ij} = H_{ji}$ (symmetric but not Hermitian) holds if channel reciprocity is assumed. Moreover, some number of diagonal and off-diagonal submatrices will be null matrices, including the diagonal submatrices for communication nodes and the "reverse"

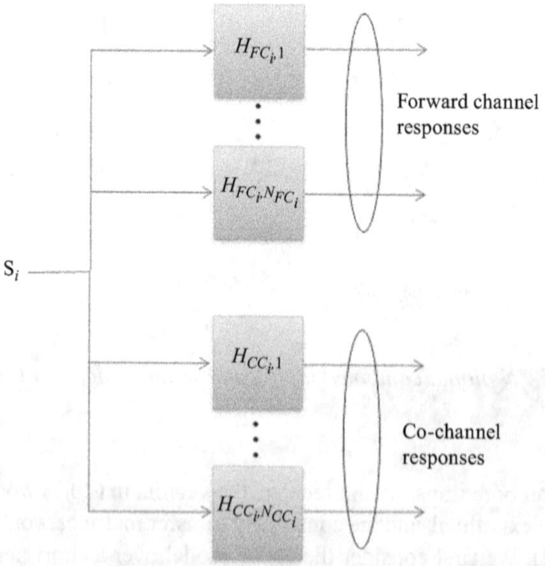

Figure 20.8 Illustration of forward and co-channel signal paths

channel for monostatic radar. As with co-channel communication nodes, radar nodes can interfere with each other and other communication nodes.

The total composite transmit signal defined in (20.7) has the structure

$$\mathbf{s} = \left[\mathbf{s}_1^T \ \mathbf{s}_2^T \ \cdots \ \mathbf{s}_N^T\right]^T \tag{20.11}$$

where $\mathbf{s}_i$ denotes the spatio-temporal transmit signal from the ith RF node and $(\bullet)^T$ is the transpose operation. Note that some nodes may be receive-only, or simply not transmitting at a given moment, in which case the corresponding entries are the null vector.

For both radar and communications there are generally both "forward," or desired, channels and "co-channels" that are undesired. A fundamental goal of JDO SSPARC is to maximize transmission through forward channels, while simultaneously minimizing co-channel interference. Thus, for the ith transmit node there can be as many as N forward channels if the ith node is simultaneously acting as a monostatic radar and/or communication transmitter and/or multistatic radar transmitter (for which the other nodes are passive receivers). In general, the number of forward channels for the ith transmit node is less than N and we shall denote it by N_{FC_i}. Additionally, for the ith transmit channel there will be some number of co-channels (at most $N-1$) for which it is desired to minimize interference. This situation is depicted in Figure 20.8.

Assuming for the moment that all channels have equal "importance" (relative weighting will be addressed later), a common design goal would be to maximize the signal power through the forward channels while simultaneously minimizing the response in the co-channels. For the unconstrained optimization case (other than finite norm), this formulation leads to maximization of the Rayleigh quotient

$$\max_{\mathbf{s}_i} \frac{E\left(\left\|H_{\mathrm{FC}_i}\mathbf{s}_i\right\|^2\right)}{E\left(\left\|H_{\mathrm{CC}_i}\mathbf{s}_i\right\|^2\right)} = \max_{\mathbf{s}_i} \frac{\mathbf{s}_i^H E\left(H_{\mathrm{FC}_i}^H H_{\mathrm{FC}_i}\right)\mathbf{s}_i}{\mathbf{s}_i^H E\left(H_{\mathrm{CC}_i}^H H_{\mathrm{CC}_i}\right)\mathbf{s}_i} \tag{20.12}$$

whose solution is a generalized eigenvector problem. Here H_{FC_i} and H_{CC_i} denote the composite forward and co-channel transfer matrices, respectively, $\|\bullet\|$ is the L_2-norm, $E(\bullet)$ denotes expectation, and $(\bullet)^H$ is the Hermitian operator. For the N-node case, the composite forward and co-channel transfer functions are given by

$$\begin{aligned} H_{\mathrm{FC}_i} &= \sum_{n=1}^{N} w_{\mathrm{FC}_i,n} H_{\mathrm{FC}_i,n} \\ H_{\mathrm{CC}_i} &= \sum_{n=1}^{N} w_{\mathrm{CC}_i,n} H_{\mathrm{CC}_i,n} \end{aligned} \tag{20.13}$$

where $H_{\mathrm{FC}_i,n}$ and $H_{\mathrm{CC}_i,n}$ denote the forward and co-channel transfer matrices from the ith transmit node to the nth receive node, with $w_{\mathrm{FC}_i,n} \in [0, 1]$ and $w_{\mathrm{CC}_i,n} \in [0, 1]$ the relative weighting coefficients.

The solution to (20.12) is based on solving the generalized eigenvalue problem

$$E\left(H_{\mathrm{FC}_i}^H H_{\mathrm{FC}_i}\right)\mathbf{s} = \lambda\, E\left(H_{\mathrm{CC}_i}^H H_{\mathrm{CC}_i}\right)\mathbf{s}, \tag{20.14}$$

where the associated maximum eigenvalue yields the optimum space–time transmit signal. Note that if either (or both) of the matrices in (20.14) are nonsingular, the solution is that of an ordinary eigenvalue problem. It should also be noted that the kernel matrices in (20.14) are positive semi-definite (or positive definite), and thus all eigenvalues are non-negative. In general, additional constraints (beyond finite norm and length) on $\mathbf{s}$ may be imposed such as constant modulus that result in a generally nonlinear programming problem of the form

$$\begin{aligned} &\max_{\mathbf{s}_i} \frac{E\left(\|H_{\mathrm{FC}_i}\mathbf{s}_i\|^2_{w\mathrm{FC}_i}\right)}{E\left(\|H_{\mathrm{CC}_i}\mathbf{s}_i\|^2_{w\mathrm{CC}_i}\right)} \\ &\text{subject to:} \\ &f(\mathbf{s}) \geq \emptyset. \end{aligned} \tag{20.15}$$

An important special case of (20.15) occurs when the constraints are in the form of angular exclusion zones. This situation can arise if, for example, the locations of co-channel RF nodes are known or where there is a desire to achieve a low probability of intercept (LPI) capability. If we let $\mathbf{v}_1, \ldots, \mathbf{v}_J$ for $J < N$ denote the (narrowband) spatial steering vectors associated with these exclusion directions, (20.15) takes the form

$$\begin{aligned} &\max_{\mathbf{s}_i} \frac{E\left(\|H_{\mathrm{FC}_i}\mathbf{s}_i\|^2_{w\mathrm{FC}_i}\right)}{E\left(\|H_{\mathrm{CC}_i}\mathbf{s}_i\|^2_{w\mathrm{CC}_i}\right)} \\ &\text{subject to:} \\ &V^{\mathrm{H}}\mathbf{s} = \emptyset \end{aligned} \tag{20.16}$$

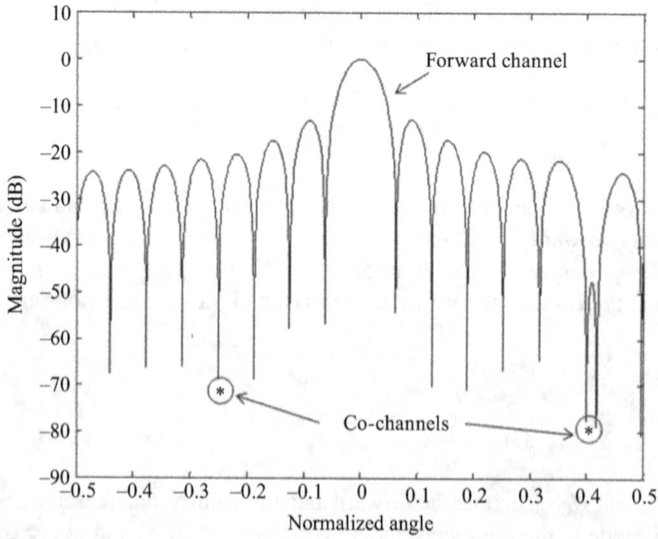

Figure 20.9 Illustration of constrained optimum beamforming with co-channel exclusion directions

where the columns of matrix V are the exclusion zone steering vectors. Figure 20.9 shows an example for the case of one forward channel and two exclusion directions. The antenna consists of a 16-element uniform linear array with half-wavelength element spacing, operating from ±90° (±0.5 normalized angle) [27]. In practice, the beampatterns would be far more complex due to non-line-of-sight (NLOS) multipath propagation.

In general, the constraints in (20.16) can be expected to be nonlinear and not amenable to analytical solution. One such example is a constant modulus constraint on the fast-time waveforms, which allows for Class C (saturation) operation of the transmit amplifiers and thus the most effective radiated power [28]. Besides strictly numerical methods, one approximate solution that is asymptotically optimal is based on the method of stationary phase and nonlinear FM (NLFM) [29].

Once all RC *transmit* space–time waveforms have been designed, the optimal receiver functions can be designed. The reason the transmit waveforms are designed first is because they establish the colored noise component that must be accounted for in the generally colored noise matched filters [26,27].

Although the basic theory for optimal reception is well established, JDO SSPARC leverages recent advances in knowledge-aided (KA), physics-based/model-based, and cognitive signal processing [27,30,31] to provide enhanced knowledge of the colored noise channel, represented by the space–time covariance matrix R_i for the ith receive node. The associated optimal receiver weight vector $\mathbf{w}_i$ that maximizes SINR (signal-to-interference-plus-noise ratio) is given by

$$\mathbf{w}_i = R_i^{-1}\mathbf{p}, \tag{20.17}$$

for $\mathbf{p}$ the desired steering vector of interest (see [26,27] for further details).

20.5 JDO SSPARC channel estimation

The previous section makes it clear that "as goes channel knowledge, so goes performance." Unlike the succinct governing design equations derived in the previous section, high-fidelity channel estimation has many facets and potential solutions, most of which are not mutually exclusive and can thus be combined into an overall effective approach. Therefore, we first provide an overview summary of the different approaches.

The following is a fairly comprehensive listing and brief description of adaptive channel estimation techniques that can be combined into the JDO SSPARC framework:

- *Training sequences and decision feedback:* This classical approach is used in wireless RF communication systems and involves a known signal that is transmitted from one node to another. At the receive node an adaptive estimation problem is solved to derive the channel transfer function. The performance of this approach depends on a number of factors including the complexity of the training sequence, available computational resources at the receiver, and the degree of channel stationarity. The price paid is a reduction in overall capacity relative to the case in which the channel is known. In JDO SSPARC, this concept can easily be extended to allow for channel estimation between both communications, radar, and their combination. This approach can also be combined with MIMO techniques and channel reciprocity (see below). Decision feedback equalization can additionally be used to estimate the channel, though it requires confidence that a previously received and demodulated signal has been accurately obtained [32].
- *MIMO:* Spatial MIMO techniques can enhance the amount of information available by a channel estimator for a given training period. For example, instead of a single transmit training waveform, several simultaneous training sequences can be emitted from different antennas. A receiver would use a bank of matched filters to reconstruct each transmit DoF. An adaptive channel estimate that includes both spatial and temporal information can then be determined, which in turn can provide significant performance improvement over the aforementioned SISO (single-input single-output) approach.
- *Reciprocity:* Except for pathological cases (e.g. ionospheric "ducting"), Maxwell's equations dictate that reciprocity holds for electromagnetic propagation. That is, the forward channel is equal to the time-reversed reverse channel for the same transmit waveform. In theory, this condition means that one forward channel training sequence can be used to estimate the reverse channel. In practice, the main impediment is the generally nonreciprocal nature of the active electronics in the transmit-receiver chains. However, with the advent of "smart RF" electronics (e.g. [33]), it is now possible to have well-calibrated electronics that can support the exploitation of reciprocity.
- *Knowledge-aided (KA) and cognitive RF (CRF)*: These channel estimation methods are far and away the most sophisticated and difficult to implement, yet can in turn yield the biggest benefits [27,28,30,31,34–36]. Though there are a number of

KA/CRF variants, for JDO SSPARC we can assume the most general instantiation that includes:

- *High-fidelity environmental dynamic database (EDDB)*: Environmental databases are increasingly becoming an operational norm. The information they contain can include digital terrain maps, land-cover land-use (LCLU), meteorological, background traffic tracks, and RF "terrain" databases such as the DARPA RadioMap [37]. It is further assumed that mechanisms exist for the continual updating of the database. This is a special case of the assumed supervised learning function of the KA/CRF system.
- *KA/CRF real-time HPEC architecture*: We assume the existence of a real-time high-performance embedded computing (HPEC) architecture that is capable of exploiting the aforementioned database in real-time and perform all necessary online calculations. The DARPA KASSPER project successfully developed and demonstrated several such HPEC architectures. A key enabler is the existence of a "look ahead" scheduler and operating system that uses basic kinematic assumptions and RF scheduler information to overcome the inevitable memory access latencies inherent in KA methods. KA/CRF HPEC architectures are well documented in the literature [27,28,30,31,34–36].
- *Fully adaptive RF*: Lastly, it assumed that adaptivity exists both in the receiver and the transmitter. The full power of KA/CRF in general, and JDO SSPARC in particular, is only realized when both the transmit and receive functions are adaptive [27,31].

Although all of the above could be incorporated in a JDO SSPARC system, the specific instantiations will vary significantly based on scenario specifics and available resources.

20.6 High-fidelity JDO SSPARC design example

In this section, ray tracing was utilized to better gauge the theoretical performance bounds developed in Section 20.3 when applied to adaptive fast-time waveforms. While the previous adaptive transmit spatial nulling example illustrates that there are significant benefits possible with transmit-receive adaptive beamforming, higher-fidelity modeling is required to more accurately gauge fast-time multipath channels.

A five-node JDO SSPARC scenario was modeled using Wireless InSite™ [38], a high-fidelity RF ray tracing software package. The nodes were distributed in a 3 km × 2.5 km set of terrain and building structures from an existing model of the city of Philadelphia, PA, USA. The laydown is shown in Figure 20.10, along with a sample of the propagation paths from node 1 to 4.

Table 20.2 provides a description of the settings used to predict the interference channel impulse responses (CIRs) in the modeled urban scenario. These are the standard model parameters for the full 3D solver that have previously been verified by validating the received power predictions against calibrated measurements for a suburban propagation environment in New Jersey, USA.

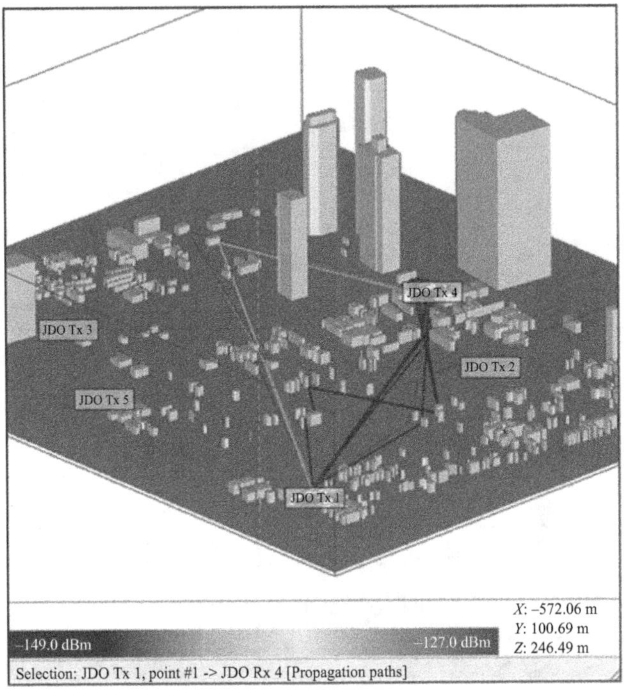

Figure 20.10 View of the simulated urban scenario showing the laydown of 5 RF nodes and propagation paths when node 1 transmits and node 4 receives

Table 20.2 Specific simulation parameters

Wireless InSite™ solver settings	Transmit/receive node configuration
• Solver = 3D shooting and bouncing rays • Ray angle spacing = 0.18° • Six reflections/path (max) • One diffraction/path (max) • 0 transmissions (buildings absorb rays)	• Isotropic antenna (max gain = 0 dBi) • Antenna height = 2 m above the ground level • Waveform = continuous wave (CW) • Carrier frequency = 3 GHz

Figure 20.11 contains the full set of unity power normalized pair-wise CIRs between the five modeled nodes. This result shows that within this relatively small-scale urban scenario, most of the multipath components occur within 2 μs of the arrival of the first ray. It should be noted that the first arriving component is not necessarily a direct line-of-sight path because we have disabled the "transmission" through buildings to account for the relatively high absorption by building materials at 3 GHz.

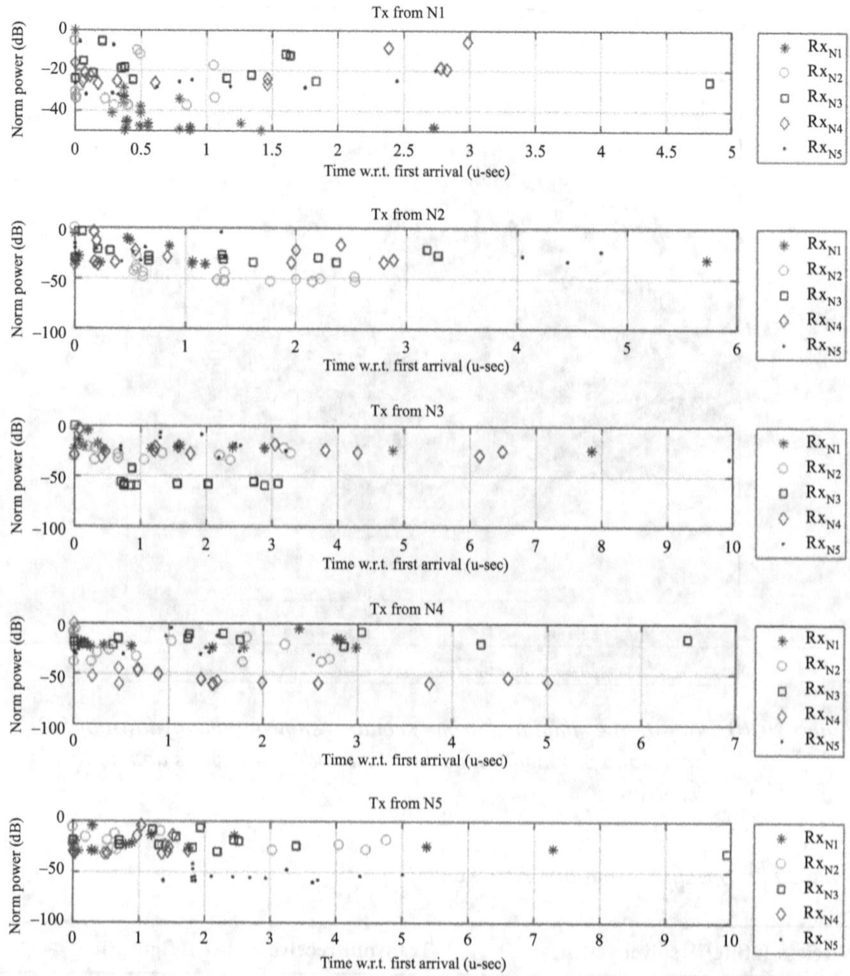

Figure 20.11 Unity power normalized pair-wise channel impulse responses between the five nodes

To tightly bound the performance improvements possible using adaptive JDO SSPARC space–time waveforms, equation (20.14) was solved for a variety of transmit-receive scenarios. In each case significant eigenvalue spread was observed, often greater than 10 to 15 dB. This result in turn implies that there is significant opportunity for performance gains if space–time transmit-receive adaptation is performed via the JDO SSPARC design equations (or constrained variants thereof).

For example, Figure 20.12 shows a typical performance result for the case when node 1 is "broadcasting" to nodes 2 and 3, while simultaneously attempting to minimize its co-channel interference to nodes 4 and 5. The 17 dB eigenvalue spread indicates there is significant opportunity for space-time optimization.

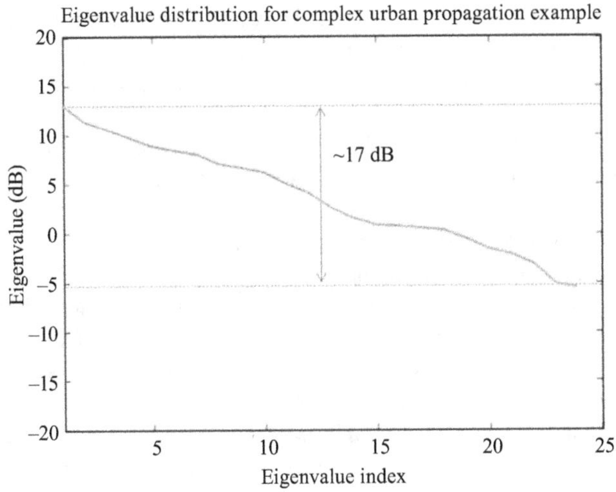

Figure 20.12 Eigenvalue distribution for a site-specific multipath channel realization. The significant spread indicates there is great potential for performance enhancement using adaptive waveforms

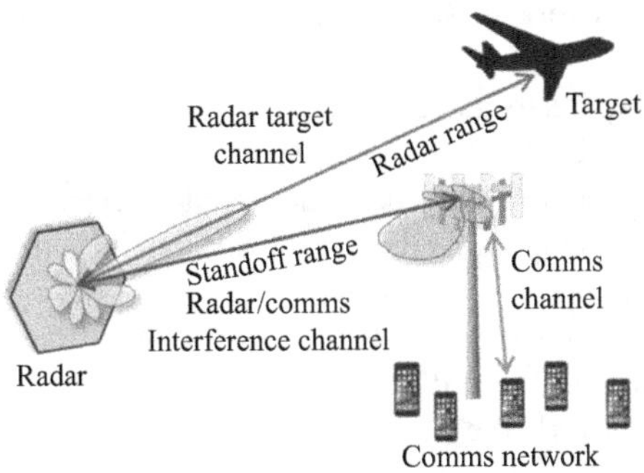

Figure 20.13 Illustration of a radar and communication spectrum-sharing scenario

20.7 A new approach to radar/communication spectrum sharing

Figure 20.13 illustrates the scenario whereby a radar and a civilian cellular communication network share the same RF spectrum at some standoff range. The objective for maximizing shared use of the RF spectrum is to minimize the standoff range between the radar and communication network while simultaneously maintaining a mutually acceptable, minimum performance level for each system. This formulation is a more systems-oriented performance metric than maximizing SINR or capacity, and it was the metric employed by the DARPA SSPARC project [1]. In terms of the three A's discussed at the beginning of the chapter, this arrangement is an example of Coordinated Providers implementing an A&M sharing system.

The first-order performance metric for radar is the reduction in detection range for a fixed P_{d} and probability of false alarm P_{fa} with respect to the standalone case (communications not present). Similarly, the first-order metric for communications is data throughput reduction for a fixed communication range with respect to the baseline case (radar not present).

Table 20.3 contains a partial list of coordinated spectrum coexistence mechanisms and their associated technical challenges and practical solutions. Different types of coexistence mechanisms may be applied in combination to obtain additional RF isolation and further reduce the minimum system standoff range. Not all radars or radios can apply all of these mechanisms due to their internal limitations. Therefore, the sharing system needs to be aware of the capabilities and limitations of the present RF devices and properties of the radar/communication interference channel so that it can intelligently select the set of coexistence mechanisms that maximize sharing performance while meeting other operating goals.

In general, the following three system properties permit coordinated military radar and communication networks to adaptively increase the degree of RF isolation:

1. Military phased array radars create highly directional antenna patterns that rapidly steer across a wide spatial search area—spatial interference rejection allows a communication network to simultaneously operate in the radar's sidelobes at close range.
2. Frequency hopping radars and communication networks possess a relatively narrow instantaneous RF bandwidth that spans a small fraction of their total RF tuning range—frequency dependent interference rejection allows a communication network to temporarily tune to a different RF frequency channel and simultaneously operate at close range.
3. Radars schedule their future dwells in terms of time, frequency, and space. If a representation of that future dwell schedule is shared with a spectrum-sharing communication network, the radios can mitigate severe interference events before they occur.

20.8 Architecture of the spectrum sharing system

A high-level architecture of the proposed Radar–Communication Spectrum Sharing System (RCS3) is shown in Figure 20.14. An authorized and trained spectrum manager uses the operator interface to centrally configure and control the sharing system. The benefit of implementing spectrum-sharing control functions across a multilevel control plane is that it enables the creation of control loops with different time scales so that spectrum sharing can robustly scale to a large number of RF devices while experiencing graceful performance degradation in a dynamic operating environment.

20.8.1 Slow-time control loop via sharing policy composition

At the highest level of the RCS3 control plane, the global coordinator (GC) block is responsible for continuously aggregating information about the performance of

Table 20.3 Partial list of new types of coordinated coexistence mechanisms enabled by the information exchange

Dimension	Description of the coexistence mechanism	Technical challenges and practical solutions
Space	Systems co-schedule their transmissions after exchanging information about the radar's position, scheduled mainbeam pointing angles, and communication node laydown. Measurement and/or prediction of the interference channel is used to identify which radar dwell angles and communication node locations need to be deconflicted. Transmit and receive adaptive nulling/beamforming are also included in this class of coexistence mechanism	Radar and communications need to be time-synchronized to interleave their access to the shared spectrum. In a cluttered environment, RF multipath can lead to unexpected RF interference. Periodic channel sounding is used to measure the RF interference channel for better radar dwell avoidance scheduling. Transmit array calibration is require to achieve adequate transmit nulling.
Frequency	Systems exchange future scheduled frequency channel hopping patterns and use this information to simultaneously access nonoverlapping bands when their transmit/receive filtering provides sufficient RF isolation	Sharing spectrum at relatively short standoff ranges requires improvements to stop-band rejection for existing transmit/receive analog RF filters. Minimize added cost by combining this technique with other sharing mechanisms.
Time	Systems exchange information about scheduled transmit/receive events so that their spectrum access can be temporally interleaved and deconflicted.	Stability and limited duration of predicted future radar schedule creates tight latency constraints on the systems' information exchange communication links (see Figure 20.14).
Coding	Systems exchange information about their code division multiple access (CDMA) waveforms to coordinate selection of code words that offer interference rejection.	CDMA requires transmit power control between users to minimize the near–far interference problem. CDMA communications has been largely replaced by OFDMA in modern civilian systems. Military communications are more likely to use CDMA.
Multipath	Precoding the transmitted MIMO waveforms for radar communications enables a MIMO null space projection interference mitigation technique [39].	Requires MIMO capability. Imperfect channel estimates lower RF isolation between systems and feedback of MIMO channel coefficients to the transmitter requires a low latency channel state information (CSI) link for mobile systems. Enable this mechanism in combination with others when requirements are met.
Polarization	Systems exchange information about their current/future antenna polarizations and then select transmit/receive antenna modes to maximize RF isolation via cross-polarization discrimination.	RF reflections depolarize the transmitted waveform and reduce RF isolation in high clutter environments. Enable this technique in combination with others when the interference channel supports it.

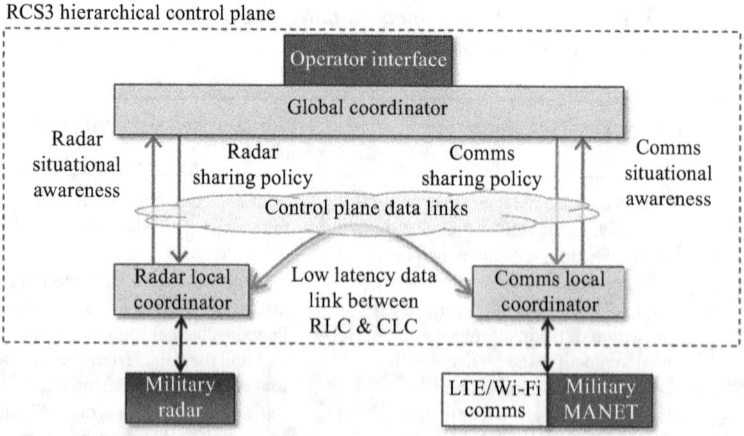

Figure 20.14 High-level architecture of the RCS3

the spectrum-sharing process, information about the location, type, and status of the radar and radios, and the operator's performance goals for sharing the spectrum. The system's sharing performance is periodically presented to the operator for their approval or to optionally select a different course of action. During online operation, the GC iteratively applies an optimization algorithm that generates a sharing policy designed to meet the operator's minimum performance constraints for each system while minimizing the required standoff range.

The spectrum-sharing policy describes the combination of coordinated coexistence mechanisms that the radar and radios need to apply to attain the operator's performance goals. The policy also describes the type and update rate of the intersystem information that is needed to execute the coordinated coexistence mechanisms. Finally, the sharing policy contains a description of what spectrum-sharing circumstance should trigger the various combinations of coexistence mechanisms to meet the operator's performance goals. It is intuitive that combining additional coexistence techniques increases the RF isolation between the systems, thereby reducing the minimum required standoff range.

However, analysis and simulation results have shown that an RF device that applies a spectrum-sharing coexistence mechanism for the sake of reducing RF interference transmitted into the other system will itself experience a performance degradation. This result leads to two important conclusions that drive the design of the RCS3:

1. Coexistence mechanisms should be applied only when absolutely necessary to avoid unnecessary system performance degradation.
2. The spectrum-sharing system needs to perform a continuous optimization process to identify which coexistence mechanism(s) should be executed over a time-varying scenario.

20.8.2 Fast-time control loop via sharing policy enforcement

Once generated, the sharing policy is transferred to the lower level of the control plane where the radar local coordinator (RLC) and comms local coordinator (CLC) rapidly enforce the policy directly with the radar and radios, respectively. The control plane data links between the RLC and CLC are designed to quickly transport delay-sensitive information such as a future radar dwell schedule or other rapidly exchanged control messages. If the deployed system does not include low-latency data links between the local coordinators, then the GC must generate a feasible sharing policy that exclusively uses coexistence mechanisms that are delay tolerant to inter-system information.

Figure 20.15 contains a detailed architecture diagram of the functions of the RCS3 and how the previously mentioned slow-time and fast-time control loops flow through the system. At system initialization, the GC collects information about the location and type of RF systems from existing authoritative spectrum management databases, as well as dynamically collected information about the spectrum-sharing RF devices that register with the RLC and CLC. This situational awareness is then used to construct a detailed RF Environment Map that contains the known/planned locations of radar and communication networks and their spectrum operating requirements. The contents of the RF Map are then transformed in a statistical RF system model that describes the spatial RF activity of the radar and communication networks. If the RF propagation pathloss between the radar and communication networks is unknown, a low complexity RF propagation prediction model is automatically invoked to evaluate the site-specific RF pathloss due to natural effects (e.g. terrain elevation masking and atmospheric ducting) and RF shadows created by human-made objects like urban structures. The final stage of the predictive modeling process takes the statistics of the intersystem RF interference and passes it through a statistical performance evaluation model that quantifies key performance parameters of the radar and communication networks while applying combinations of feasible coexistence mechanisms.

The number of combinations of coexistence mechanisms that need to be evaluated are reduced by first applying heuristic rules that constrain the solution space. The following are some example rules, presented as questions that are evaluated to identify the subset of feasible coexistence mechanisms:

- Does the radar support MIMO?
- Does the radio have a tunable bandpass filter?
- Is the communication network a trusted military radio network?

Based on the results of these queries, an initial spectrum-sharing policy is generated by rank ordering the feasible combinations of coexistence mechanisms by their effectiveness and sharing performance. The highest rank sharing policies are then sent to the RLC and CLC for real-time policy enforcement within the radar and communication networks, respectively. System performance data is periodically collected while the policy is enforced and then forwarded by the RLC and CLC to the GC so that it can trigger a new sharing policy and provide an update on the sharing performance through the operator's interface.

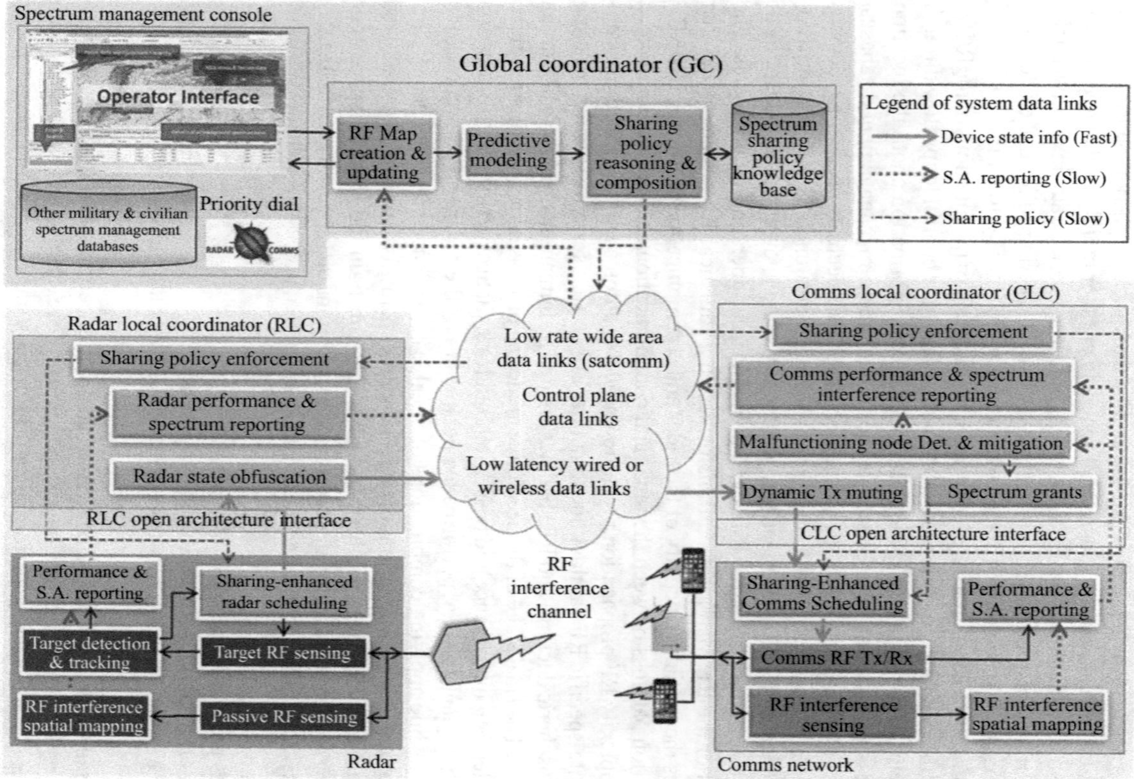

Figure 20.15 Detailed architecture of the RCS3

20.9 Performance of the RCS3

The radar-avoiding communications (RAC) coexistence mechanism is designed to provide RF interference reduction between the radar and radios when sharing RF spectrum with untrusted civilian communication networks. In the RAC concept, a spectrum-sharing radar receives information about the civilian network's frequency hopping pattern and uses that information to determine how to tune the frequency of its radar dwells. The frequency hopping of the civilian communication network ensures that the radar has full access to its entire RF tuning range over relatively short time periods, which allows the radar to maximize frequency diversity and thereby improve target detection while maintaining less predictable operation. While LTE and Wi-Fi protocols and radios were not originally designed to support slow frequency hopping, there is no inherent reason why a synchronized network of radios could not be made to slowly frequency hop together through small modifications to their software and their RF carrier tuning circuits. The radios also need additional RF channelization bandpass filters to prevent receiver saturation when a strong out-of-band radar pulse is received. As shown in Figure 20.16, a practical solution for minimizing the additional hardware cost per radio is to constrain the radio to tune only to two small sub-bands within the radar's tuning range. This arrangement greatly reduces the complexity and additional device cost by adding two low-cost RF channelization filters while still gaining the benefits of the RAC coexistence mechanism.

In contrast, the communication-avoiding radar (CAR) mechanism is designed to reduce RF interference experienced by the radar when sharing spectrum with a communication network. In this case the radar shares its future schedule of dwells with a network of time-synchronized communication nodes so that they can proactively avoid creating RF interference between the systems. Once the communication network receives the radar's future schedule, each communication device can evaluate its standoff distance, spatial angle offset, and frequency offset per dwell to decide if they need to mute their RF emission on a dwell-by-dwell basis. In order to apply CAR with an untrusted communication network, the radios receive an optimized spectrum-sharing policy that specifies the minimum spatial offset angle and/or frequency offset

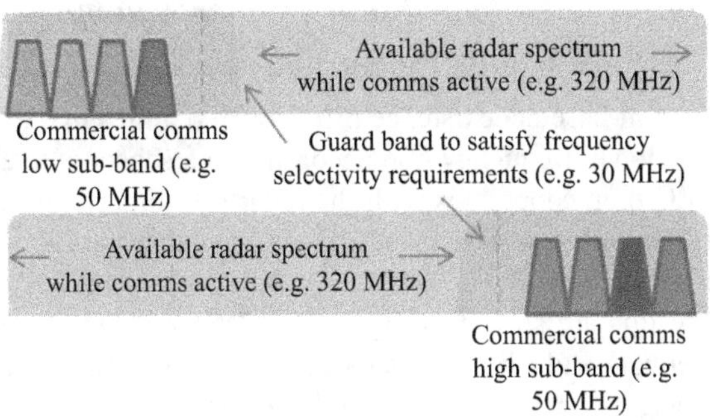

Figure 20.16 The RAC mechanism influences how the radar shares spectrum with LTE and Wi-Fi radio networks

Table 20.4 Simulated spectrum-sharing performance

	Radar and small cell LTE	Radar and MANET
Baseline standoff range	40 km (no coexistence)	52 km (no coexistence)
RCS3 enabled standoff range	5 km (using RAC)	7 km (using CAR)
Standoff range reduction ratio	8×	7.4×

that should cause a radio node to mute its transmission for the duration of the dwell. Alternatively, when applying CAR with a trusted communication network, a radio could receive information about the spatial and frequency-filtering characteristics of the radar so that it can evaluate its spectrum-sharing policy in terms of the minimum RF isolation it experiences per dwell and whether it is sufficiently high for simultaneous transmission with the radar.

20.9.1 Simulation results for radar/communication spectrum sharing

Table 20.4 summarizes simulation results of high-fidelity models of a high-power surface radar-sharing spectrum with either a dense, small-cell LTE network or a military Mobile Ad Hoc Networking (MANET) radio. Recall that the first-order performance metric for the radar is target detection range reduction compared to the standalone radar case for a fixed P_d and P_{fa}. The performance threshold for radar is reached when the target range reduction exceeds 5% for any mainbeam angle within 120 s. The first-order performance metric for the communication network is throughput reduction compared to the standalone case for a fixed link range. The performance threshold for communications is reached when the data throughput reduction exceeds 5%. When a coexistence mechanism is applied, the assumption is that communication performance is further reduced and the standoff range threshold is reached when communication throughput reduction exceeds 50%. In either case, the minimum system standoff range for both systems is set by either the radar or communication network first crossing their respective minimum performance threshold.

20.9.2 Temporally coordinated slow-frequency hopping between radar and communications via RAC

The RAC information-based coexistence mechanism requires that a communication system be able to slowly frequency hop according to a schedule that is shared with the RLC. The RLC then cooperates with the radar (see Figure 20.15) to generate a radar schedule that avoids using frequency bands when they are scheduled to be used by the communication network.

Simulation results are shown in Figure 20.17 for a slowly frequency-hopping small-cell LTE network and a high-power search radar that achieves a standoff range reduction of 8×. RAC offers a unique set of benefits:

1. Sensitive radar internal state information does not need to be shared with an untrusted civilian communication network.
2. The untrusted civilian networks can share spectrum without muting their transmissions so that data throughput is not reduced.

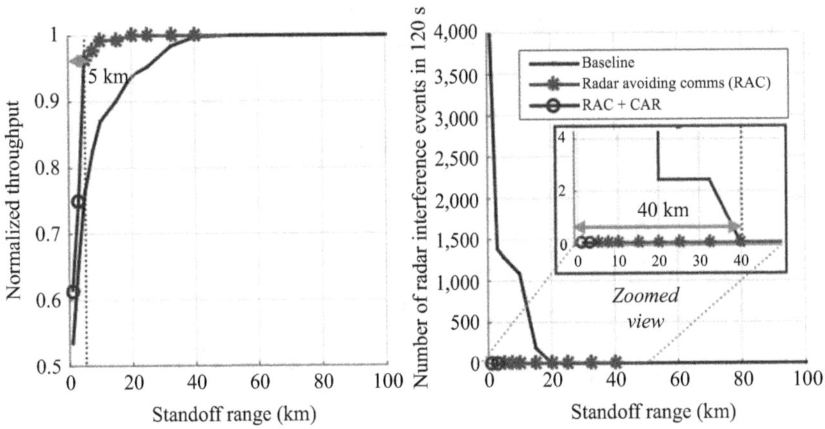

Figure 20.17 Small-cell LTE throughput as a function of system standoff range (left) and number of radar interference events in 120 s (right)

Another valuable aspect of the RAC coexistence mechanism is that it preserves the ability of the radar to access its entire RF tuning range in the shared band over a relatively short time period.

20.9.3 Temporally coordinated transmit muting for military MANETs via CAR

Spectrum sharing between high-power radar and a military MANET creates significant technical challenges at short standoff ranges. This condition arises because both systems transmit with high RF power and frequency hop in unpredictable patterns over the shared RF band as a self-defense mechanism.

Another technical challenge is how to rapidly disseminate the radar's future short-term schedule to the interfering MANET radios before the radar executes the schedule. One approach is to transmit the radar schedule to a single gateway MANET radio and rely on the network to forward the schedule to other interfering radios. However, some MANET medium access control protocols and ad hoc routing protocols introduce significant message latency over network routes with multiple hops. One solution is to co-locate a low latency military radio with the radar so that the encrypted schedule can be broadcast directly to all of the interfering MANET radios. Each MANET radio would need to contain two receivers, one for participating in the MANET and a second for receiving the broadcast radar schedule. This approach does not necessarily increase the cost of the radio equipment because of a new trend to integrate more than one radio receiver in MANET radios.

To share spectrum efficiently, radar and communications need low out-of-band RF emissions or they will cause interference to the other system when their tuned offset frequencies are relatively small. Legacy radars use high-power tube amplifiers on transmit. The problem is that these amplifiers also generate wideband noise during the pulse. Similarly, military MANET radios generate high adjacent band interference when they transmit. Spectrum-sharing efficiency will improve with enhanced RF frequency filtering on transmit because the two systems could be tuned

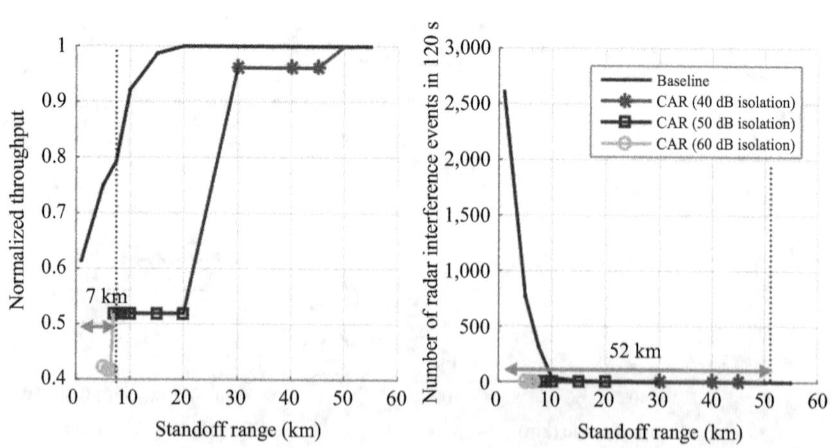

Figure 20.18 Military MANET throughput as a function of system standoff range (left) and number of radar interference events in 120 s (right)

to smaller frequency offsets, consequently allowing for more simultaneous users in the shared band.

Figure 20.18 illustrates the simulated end-to-end performance of a high-power radar and MANET sharing the same spectrum using the CAR coexistence mechanism. The plots show that the minimum standoff range between the MANET and radar is set by the aggregate interference generated by the MANET radios and experienced by the radar. The plot on the left depicts the baseline throughput reduction from the standalone case without CAR enabled and then with CAR enabled for three levels of RF isolation between the systems. As the required RF isolation level provided by CAR is increased, the normalized throughput within the MANET decreases because granting additional RF isolation requires the radios to mute their transmissions at ever-wider spatial angles and frequency offsets in each radar dwell. From these simulation results, it has been determined that the standoff reduction between the MANET and a high-powered radar is 7.4×.

20.10 Summary

Analysis has shown that information exchange between radar and communications greatly enhances sharing performance. Once the current/future system state information is exchanged, radar and communication systems may use coordinated interference avoidance, mitigation, and amalgamation techniques that are otherwise impossible when the systems do not share information and are forced to reactively sense-and-avoid in the spectrum. In addition, information exchange enables these systems to engage in joint co-designed and cooperating operating modes that open the door to amalgamated RC waveforms where the RF equipment can synergistically support each other's operation.

This chapter has covered several key aspects of information-exchange based RC spectrum-sharing systems, including:

- The three A's (Avoid, Accept, and Amalgamate) that describe the mode of operation for RC spectrum-sharing systems.
- A taxonomy of RC service providers that support varying degrees of information-exchange requirements.
- A set of equations for deriving the optimum radar and communication waveforms to maximize joint capacity when sharing spectrum.
- The results of eigen-decomposition of a site-specific simulated multipath channel between RF systems in a typical urban environment. This analysis demonstrates the potential for RC waveform orthogonalization based on exchanged channel state information.
- A description of the RCS3 that uses the practical RAC and CAR coexistence mechanisms to minimize standoff distance while preserving system performance between high-power military radar and MANET or LTE radios that want to share spectrum.
- Simulation results using RAC showing reduced standoff range by 8× over the baseline standoff range for military radar and small-cell LTE radio networks.
- Simulation results using CAR realizing a 7.4× reduction in standoff range over the baseline standoff range for military radar and MANET radios.

20.11 Looking ahead

There is a strong interest by radar and communication service providers to solve the challenges of optimizing RC spectrum sharing using information-exchange-based approaches. The following is a description of a system that is likely to be fielded in the next few years and we conclude with a description of a future research direction based on the three "A"s of spectrum sharing.

A near-term approach for radar/communication spectrum sharing is the Citizens Broadband Radio Service (CBRS). As of the publication of this book, the Federal Communications Commission (FCC) recently approved the CBRS operating rules in the United States as a shared band from 3.55 to 3.7 GHz [40]. The rules call for the use of a spectrum access system (SAS) that contains a combination of spectrum sensing, radio position reporting, and RF physics modeling. When these functions are combined, the SAS can dynamically manage RF interference and dynamic frequency assignments of civilian mobile broadband communication networks that operate in the same band as military radars, such as the SPN-43 air traffic control radar. While this approach is a good pathfinder example for an information-based spectrum-sharing system, the rules for the CBRS minimize the amount of radar information represented in the SAS. Therefore, the SAS is limited to dynamically assigning or revoking short-duration spectrum leases as the ship-borne radar moves away from, or towards, a coastal geographic protection zone [41].

Finally, Figure 20.19 describes the current status of RC spectrum-sharing techniques. Most of the current research is in the A&M [2,4] and Amalgamate areas [10–17]. There is little ongoing work in the Accept area.

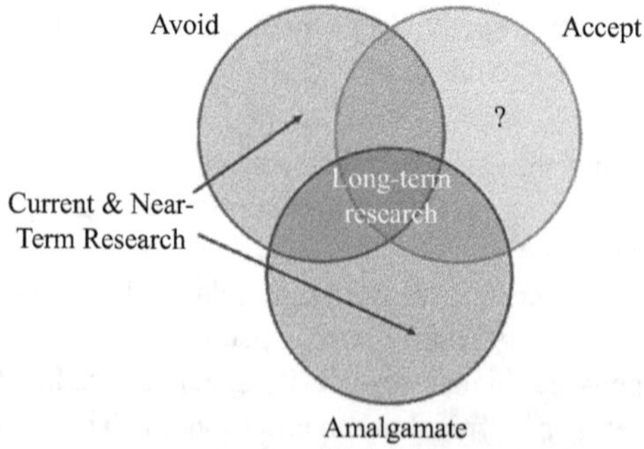

Figure 20.19 Current RC spectrum-sharing research and future directions

Once each of the three "A"s is understood, it is expected that research will then move to their intersections. In particular, there are good reasons why practical systems with realistic business cases may live in these intersections. While "pure" A&M systems assume negligible residual interference between radar and communication systems, it appears likely that the acceptance of some level of interference can provide an additional DoF with which to produce practical and affordable systems. Future Amalgamated waveform systems may also need to share spectrum with other systems, thereby necessitating their operation in the intersections with A&M and Accept. The other communication system could even be a communication uplink to the Amalgamated system. Such an uplink may be implemented more efficiently using A&M techniques other than time- or frequency-division duplexing along with some level of interference acceptance.

Acknowledgments and disclaimer

The authors would like to thank Dr. Joseph Evans and Mr. Barry Fell of DARPA for their encouragement and review of this book chapter. This research was developed with funding from DARPA. The views, opinions, and/or findings expressed are those of the authors and should not be interpreted as representing the official views or policies of the Department of Defense or the U.S. government.

References

[1] Chapin, J., "Shared spectrum access for radar and communications (SSPARC)," *DARPA BAA-13-24* www.darpa.mil,2012.

[2] Guerci, J.R., Guerci, R.M., Lackpour, A., Moskowitz, D., "Joint design and operation of shared spectrum access for radar and communications," *IEEE Radar Conference*, Arlington, VA, USA, May 2015.

[3] Zatman, M., "The three "A's" of Communications—Radar Spectrum Sharing," *WInnComm Dynamic Spectrum Sharing Workshop*, San Diego, CA, USA, Mar. 2015.

[4] Lackpour, A., Rosenwinkel, A., Guerci, J.R., Mody, A., Ryan, D., "Design and analysis of an information exchange-based radar/communications spectrum sharing system (RCS3)," *IEEE Radar Conference*, Philadelphia, PA, USA, May 2016.

[5] "Cognitive Radio", https://en.wikipedia.org/wiki/Cognitive_radio, retrieved March 2017.

[6] Shannon, C.E., Weaver, W., Blahut, R.E., Hajek, B., *The Mathematical Theory of Communication.* vol. 117. University of Illinois Press, Urbana, IL, USA, 1949.

[7] Alouini, M.S., Goldsmith, A.J., "Area spectral efficiency of cellular mobile radio systems," *IEEE Transactions on Vehicular Technology,* vol. 48, pp. 1047–1066, July 1999.

[8] Skolnik, M., *Radar Handbook.* 2nd edition, McGraw-Hill, Boston, MA, USA, 2000.

[9] Meyer, D.P., Mayer, H.A., *Radar Target Detection: Handbook of Theory and Practice*. Academic Press, New York, NY, USA, 1973.

[10] Garmatyuk, D., Schuerger, J., Morton, Y.T., *et al.*, "Feasibility study of a multi-carrier dual-use imaging radar and communication system," *European Microwave Conference*, Munich, Germany, Oct. 2007.

[11] Saddik, G.N., Singh, R.S., Brown, E.R., "Ultra-wideband multifunctional communications/radar system," *IEEE Transactions on Microwave Theory*, vol. 55, no. 7, pp. 1431–1437, Jul. 2007.

[12] Roberton, M., Brown, E.R., "Integrated radar and communications based on chirped spread-spectrum techniques," *IEEE International Microwave Symposium*, Philadelphia, PA, USA, Jun. 2003.

[13] Surender, S.C., Narayanan, R.M., Das, C.R., "Cross-layered resource allocation in UWB noise-OFDM-based ad hoc surveillance networks," *EURASIP Journal on Wireless Communications and Networking*, vol. 2013, no. 4, pp. 1–24, Dec. 2013.

[14] Van Genderen, P., "A communication waveform for radar," *International Conference on Communications*, Bucharest, Romania, Jun. 2010.

[15] Donnet, B., Longstaff, I., "Combining MIMO radar with OFDM communications," *European Radar Conference*, Manchester, UK, Sep. 2006.

[16] Zatman, M., "COMMDAR", special session on Shared Spectrum Access for Radar and Communications at IEEE DYSPAN Workshop, Apr. 2014.

[17] Reed, J., Baxley, R., ""RADCOM", special session on Shared Spectrum Access for Radar and Communications at IEEE DYSPAN Workshop, Apr. 2014.

[18] Griffiths, H.D., Baker, C.J., *An Introduction to Passive Radar*. Boston, MA USA: Artech House, 2017.

[19] Van Nee, R., *OFDM for Wireless Multimedia Communications*. Artech House, Boston, MA, USA, 1999.

[20] Richards, M.A., Scheer, J.A., Holm, W.A., *Principles of Modern Radar, Vol I: Basic Principles*. Raleigh, NC: SciTech, 2010.

[21] Li, J., Stoica, P., *MIMO Radar Signal Processing*. Wiley, Hoboken, NJ, USA, 2008.

[22] Foschini, G.J., "Layered space–time architecture for wireless communication in a fading environment when using multiple antennas," *Bell Labs Technical Journal*, vol. 1, no. 2, pp. 41–59, Autumn 1996.

[23] Chapin, J., "SSPARC", special session on Shared Spectrum Access for Radar and Communications at *IEEE DYSPAN Workshop*, Apr. 2014.

[24] Woodward, P.M., *Probability and Information Theory, with Applications to Radar*. London: Pergamon Press, 1953.

[25] Barton, D.K., *Modern Radar System Analysis*. Norwood, MA: Artech House, 1988.

[26] Van Trees, H.L., *Detection, Estimation and Modulation Theory. Part I*. New York: Wiley, 1968.

[27] Guerci, J.R., *Space–Time Adaptive Processing for Radar, 2nd Edition*. Norwood, MA: Artech House, 2014.

[28] Guerci, J.R., *Cognitive Radar: The Knowledge-Aided Fully Adaptive Approach*. Norwood, MA: Artech House, 2010.

[29] Cook, C.E., Bernfeld, M., *Radar Signals*. New York: Academic Press, 1967.

[30] Legters, G.R., Guerci, J.R., "Physics-based airborne GMTI radar signal processing," *IEEE Radar Conference,* Philadelphia, PA, USA, Apr. 2004.

[31] Guerci, J.R., Guerci, R.M., Rangaswamy, M., *et al.*, "CoFAR: cognitive fully adaptive radar," *IEEE Radar Conference*, Cincinnati, OH, USA, May 2014.

[32] Belfiore, C.A., Park, J.H., "Decision feedback equalization," *Proceedings of the IEEE,* vol. 67, pp. 1143–1156, Aug. 1979.

[33] Kingsley, N., Guerci, J.R., "Adaptive amplifier module technique to support cognitive RF architectures," *IEEE Radar Conference*, Cincinnati, OH, USA, May 2014.

[34] Guerci, J.R., "Cognitive radar: the next radar wave?," *Microwave Journal,* vol. 54, pp. 22–36, Jan. 2011.

[35] Guerci, J.R., Baranoski, E.J., "Knowledge-aided adaptive radar at DARPA: An overview," *IEEE Signal Processing Magazine,* vol. 23, pp. 41–50, Jan. 2006.

[36] Mitola, J., Guerci, J., Reed, J., *et al.*, "Accelerating 5G QoE via public–private spectrum sharing," *IEEE Communications Magazine*, vol. 52, no. 5, pp. 77–85, May 2014.

[37] Chapin, J.M., Evans, J., *DARPA Radiomap Program*. http://www.darpa.mil/program/advance-rf-mapping, retrieved March 2017.

[38] REMCON. *Wireless InSite*. Available: http://www.remcom.com/wireless-insite

[39] Khawar, A., Abdel-Hadi, A., Clancy, T.C., "Spectrum sharing between S-band radar and LTE cellular system: A spatial approach," *IEEE International Symposium on Dynamic Spectrum Access Networks*, McLean, VA, USA, 2014.

[40] FCC 15-47, Report and Order and Second Further Notice of Proposed Rule-making, Amendment of the Commission's Rules with Regard to Commercial Operations in the 3550–3650 MHz Band. Apr. 21, 2015.

[41] Sohul, M.M., Yao, M., Yang, T., Reed, J.H., "Spectrum access system for the citizen broadband radio service," *IEEE Communications Magazine*, vol. 53, no. 7, pp. 18–25, July 2015.

Chapter 21

Embedding communication symbols in radar clutter on an intrapulse basis

Justin G. Metcalf[1], *Cenk Sahin*[1], *and Shannon Blunt*[2]

21.1 Introduction

The radio spectrum is a unique and valuable resource that has enabled some of the most pivotal technological innovations of the past 100 years, particularly with respect to communications and radar remote sensing. However, the demand for high-bandwidth applications, especially digital video, has imposed a severe strain on the existing spectrum allocated for such applications. The exponential growth of the commercial communications industry has created a strong economic incentive to reallocate the spectral resources traditionally assigned to bandwidth intensive applications such as radar, which do not have a comparative economic argument to protect their spectrum. In addition to commercial pressures, the advent of network-centric warfare [1] has led to increasingly stringent requirements for tactical communications capabilities. Unfortunately, the radio spectrum is a fixed resource. Therefore, both defense and commercial industries require innovative techniques to share spectrum between multiple applications, either through the co-design of new multifunction systems or by improving the spectrum sharing capabilities of existing systems. Both cases involve the use of advanced signal processing techniques to mitigate mutual interference [2,3].

Waveform diversity is a key family of techniques enabling the design of shared spectrum systems [4]. There are various waveform diversity approaches that may facilitate spectrum sharing, each with unique requirements and benefits. Research into spectrum sharing tends to take one of two perspectives. One type involves performing two or more different tasks simultaneously, rather than the traditional approach to resource allocation that relies on time multiplexing [5,6]. For example, one could utilize a dual-function waveform to perform concurrent radar and communication operations with a single emission [7–10] or emit multiple waveforms simultaneously in a multiple-input–multiple-output fashion [11–15]. The other type of spectrum sharing considers physically separate entities that contend with the interference to/from

[1]U.S. Air Force Research Laboratory, USA
[2]Electrical Engineering & Computer Science Department, University of Kansas, USA

one another in a responsive manner. Typical approaches include various forms of dynamic spectrum access such as overlay and underlay communications [16–18]. However, these perspectives are not strict. There is growing interest in waveform diversity techniques that can be used by a single system or in tandem with multiple separate systems to share spectrum [3,19–21]. Here, we consider the second type of spectrum sharing in which a communication mode leverages the existence of incumbent radar emissions.

Specifically, consider a communication system capable of remodulating incident radar illumination for the purpose of embedding information into the reflected echoes. The idea of embedding information through remodulation of incident signals is an old one, being first proposed by Stockman in 1948 [22]. Similar systems are referred to as a tag, transponder, retro-directive array, retro-reflector, or radio-frequency identification. Hereafter, we refer to such a system simply as a communication "tag." This manner of communication has been used to assist in calibrating radar systems [23–26] or for identification and localization purposes [27–33]. Recent work on localization/identification also includes modulation of signals of opportunity for passive radar systems [34,35].

These traditional approaches to radar clutter-embedded communication (RCEC) employ *inter*-pulse modulation whereby a sequence of phase shifts is imparted across a large set of pulses incident upon the tag. As such, the data rate is inherently limited and the interpulse modulated signal is embedded into the passband of the radar. Consequently, the tag will readily interfere with a standard radar receiver that does not compensate for the interference from the tag. Recognizing the drawbacks inherent to these interpulse techniques, an *intra*-pulse RCEC waveform design technique was introduced in [36,37] in order to reduce interference with the illuminating radar and provide a higher data rate as compared to the legacy intrapulse methods. This technique has since been expanded, refined, and analyzed in considerable detail [38–46].

In the course of developing and analyzing RCEC symbol designs [38–43,45], the potential spectrum sharing advantages offered by these approaches have become apparent. In contrast to the previous interpulse methods, intrapulse RCEC relies on frequency bands directly adjacent to, and perhaps slightly overlapping with the passband of the radar. The symbol design methods and receive processing strategies that have been developed allow RCEC symbols to effectively transmit information within what would otherwise be considered guard bands of the radar, while also minimizing mutual interference between the radar and communication systems. This low mutual interference, combined with robust communication performance, makes RCEC communications an attractive spectrum sharing technology. This chapter summarizes intrapulse RCEC waveform design based on the notional operation shown in Figure 21.1. Further, multiple performance metrics are discussed that demonstrate the spectrum sharing capability of these intrapulse RCEC waveforms.

21.1.1 Practical considerations

There are a number of practical constraints imposed on the design of RCEC waveforms (henceforth understood to be intrapulse) that are not found in

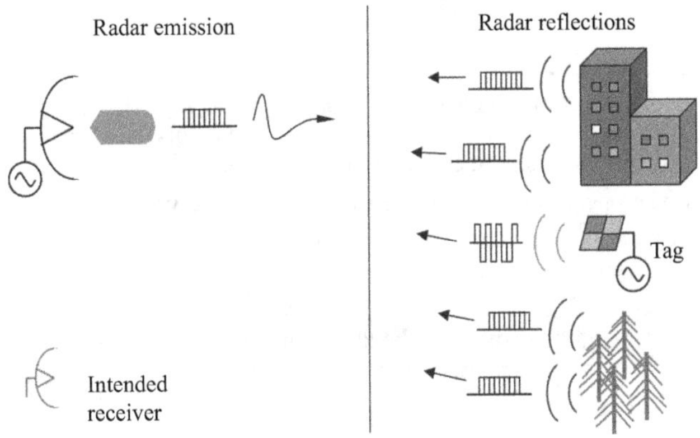

Figure 21.1 Notional intrapulse radar clutter-embedded communications (RCECs)

Table 21.1 Practical constraints for intrapulse RCEC waveform design

1. Radar, tag, and desired receiver are not synchronized
2. Symbols must be determined independently at the tag and the desired receiver based on the observed radar illumination (with possible timing uncertainty and multipath differences)
3. The surrounding clutter structure is not known exactly, though it may be approximately modeled
4. Symbols may be designed to be partially correlated with the clutter but still separable from one another on receive
5. Symbol design must occur in a reasonable time frame (while radar illumination persists)
6. Symbol design must be detectable by the communication receiver while minimizing interference at the radar
7. Minimal (if any) additional spectral resources are devoted to the communication symbols (reuse the radar spectral roll-off region)

traditional communication symbol designs. These constraints are summarized in Table 21.1.

Further, relative to traditional forms of communication, RCEC possesses a set of rather unique operating conditions (see Table 21.2). As noted in Item 4, moving target indicator (MTI) radars mitigate interference from clutter echoes, and thus communication symbols that resemble clutter can be readily suppressed by traditional radar signal processing. This factor certainly underscores the prospective practical utility of this particular form of spectrum sharing since well-designed communication symbols would not incur a performance penalty at the radar in the form of increased residual interference.

Table 21.2 Intrapulse RCEC operating environment

1. Radar waveform can have a high time–bandwidth product (~100 up to several 1,000), while communication symbols have temporal and spectral footprint commensurate with the radar waveform
2. Radar operation is often pulsed at a constant interval
3. The surrounding clutter structure can be modeled as a random signal convolved with the radar waveform
4. A moving target indicator (MTI) radar will attempt to cancel clutter regardless of any embedded communication symbol

Additional constraints may exist according to the specific mode of operation. For instance, the intent may be for the tag to communicate with other tags in the local area or to transmit information back to the illuminating radar. It should be noted that the RCEC symbol design methods described here are agnostic to the radar function. Further work is needed to examine how imaging modes, such as synthetic aperture radar (SAR) and inverse SAR, or tracking modes could be impacted by this form of spectrum sharing.

Subspace techniques form the basis of the RCEC symbol design methods described here, though they are certainly not the only approaches. For example, multiobjective optimization has also recently been proposed [44,46], and there are certainly a variety of other approaches that could be explored.

21.2 RCEC symbol design

Consider a tag and surrounding environment that are being illuminated by a pulsed radar. As is typically the case, the radar is repeatedly transmitting the same waveform over a set of pulses that comprise the coherent processing interval (CPI), though the waveform may vary on a CPI-to-CPI basis. The radar transmits with high power (may be kilowatts to megawatts) and then collects the backscattered echoes. The scattering of the waveform by the environment is called the radar clutter.

The clutter can be modeled as the convolution of a relatively bandlimited signal (the radar waveform) and a white random process. However, the actual spectral content of the radar emission is affected by a variety of factors including the rise/fall time of the pulse characteristics of the radar waveform and nonideal transmitter effects [3,4,47], the collection of which induces an overall spreading effect. As such, the clutter response to a realistic radar emission likewise exhibits this spectral spreading, which can be exploited when designing embedded communication symbols.

For a communication system located within the field of view of an illuminating radar, the clutter may be considered to be a source of interference. One traditional communication scheme that is known to be robust to interference is spread-spectrum [48,49]. A spread-spectrum communication system spreads the information-bearing symbol over a much greater bandwidth than the rate at which the information is being transmitted. Since this spectral spreading reduces the impact of interference sources, it stands to reason that spread-spectrum signals could likewise reduce the interference induced upon an operating radar. There are two general families of spread-spectrum operation: frequency-hopped and direct-sequence [50].

Due to its inherent robustness to interference, direct-sequence spread-spectrum (DSSS) offers an attractive starting point from which to design RCEC symbols. However, the practical constraints imposed in Table 21.1 do require careful consideration.

21.2.1 Signal model

When designing RCEC symbols, both the forward link and reverse link must be considered, borrowing from cellular communication parlance. Here, the forward link is defined as being from the radar to communication tags in the illuminated scene. The reverse link then exists from a tag to the desired receiver, which may or may not be the illuminating radar.

The forward scatter between the radar and the tag can be modeled as a multipath channel, where multiple spatially diverse scatterers cause a superposition of time-delayed, independently scaled copies of the radar emission to impinge on the tag. The multipath-distorted radar emission observed by the tag is therefore

$$\tilde{s}(t) = \alpha_{\mathrm{R}\bullet}\,[s(t) * h_{\mathrm{R}\bullet}(t)] + u(t), \tag{21.1}$$

where $s(t)$ is the transmitted radar waveform with pulsewidth T, $\alpha_{\mathrm{R}\bullet}$ is the attenuation between the radar (R) and some arbitrary receiver (denoted by the "dot" placeholder), $u(t)$ is an additive white Gaussian thermal noise process, and $*$ is the convolution operation. For the forward link from the radar to the tag (subscript T), the attenuation is denoted as $\alpha_{\mathrm{RT}}(t)$ and the particular multipath channel is denoted as $h_{\mathrm{RT}}(t)$. If the tag is communicating with another receiver (i.e. not the illuminating radar), then the desired (subscript D) receiver is subject to an independent forward link attenuation $\alpha_{\mathrm{RD}}(t)$ and multipath process $h_{\mathrm{RD}}(t)$.

If the radar waveform is not known a priori at the tag or the desired receiver (if it is not the radar), then each must independently estimate the illuminating waveform in the possible presence of multipath. Clearly, these independent estimates can affect the performance of the communication system. If the tag and/or the desired receiver have spatial degrees-of-freedom available (i.e. a multi-channel receive antenna array per Figure 21.2), blind deconvolution of the radar signal and the multipath channel may be attempted using adaptive beamforming techniques [39,43,51,52]. The multipath estimate can either be discarded (when the desired receiver is not the radar) or used as a form of predistortion via time-reversal [39,41–43]. Note that the use of time-reversal is specific to the case where the transmitting radar is also the desired receiver. For the remainder of the chapter, we consider an arbitrary desired receiver.

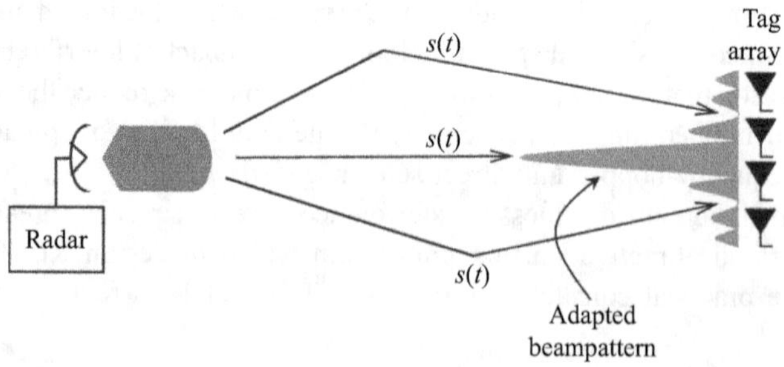

Figure 21.2 Multipath estimation using a spatial array

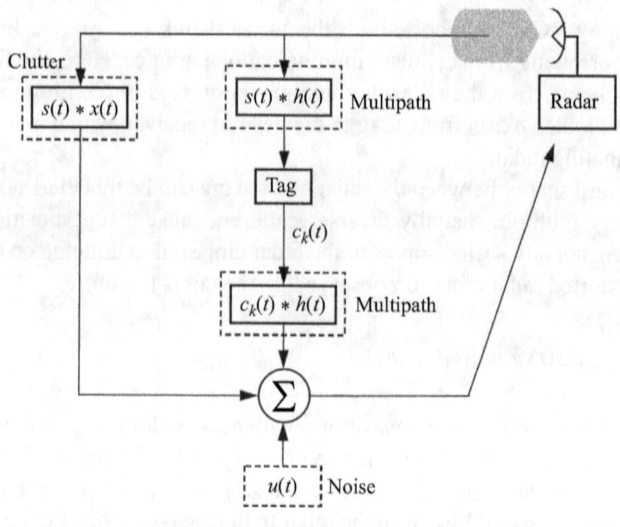

Figure 21.3 Signal model for RCEC when the radar is the desired receiver

The reverse link signal observed at an arbitrary receiver can be expressed as

$$y(t) = s(t) * x(t) + \alpha_{T\bullet} [c_k(t) * h_{T\bullet}(t)] + u(t), \tag{21.2}$$

where $x(t)$ is a random unknown clutter process, $c_k(t)$ is the kth communication symbol that likewise has temporal extent T, $\alpha_{T\bullet}$ is the attenuation between the tag and the given receiver, and $h_{T\bullet}(t)$ is the multipath component between the tag and the particular receiver. Note that in the special case that the tag is communicating back to the illuminating radar one obtains $h_{TR}(t) \equiv h_{RT}(t) \equiv h(t)$ by reciprocity. The signal model for this special case is shown in Figure 21.3.

Now consider discrete-time sampled versions of the signal models in (21.1) and (21.2). For 3 dB bandwidth B and radar pulsewidth T (and symbol extent), let the time–bandwidth product be $BT = N$, which represents the basic dimensionality of the radar waveform in a "fast-time" sense (in contrast, the "slow-time" dimensionality is the number of pulses in a CPI). To allow for additional degrees-of-freedom in the design of RCEC symbols, the radar waveform is oversampled by a factor of M, which captures the roll-off region of the radar spectrum. Note that time-width expansions can also be considered [38]. The baseband discrete-time version of the radar waveform is represented by the length MN vector $\mathbf{s}$. The incident signal at an arbitrary receiver is thus the radar waveform convolved with some random clutter process $\mathbf{x}$, here modeled as white Gaussian $\mathbf{x} \sim \mathscr{CN}(\mathbf{0}, \sigma_x^2\mathbf{I})$, as well as noise that is likewise modeled as additive white Gaussian. Using the $NM \times (2NM - 1)$ Toeplitz discrete-time convolution matrix

$$\mathbf{S} = \begin{bmatrix} s_{NM-1} & s_{NM-2} & \cdots & s_0 & 0 & \cdots & 0 \\ 0 & s_{NM-1} & \cdots & s_1 & s_0 & \cdots & 0 \\ \vdots & \vdots & \ddots & \vdots & \vdots & \ddots & \vdots \\ 0 & 0 & \cdots & s_{NM-1} & s_{NM-2} & \cdots & s_0 \end{bmatrix} \tag{21.3}$$

that is comprised of delay-shifted versions of the discretized waveform $\mathbf{s} = [s_0 \; s_1 \; \cdots \; s_{NM-2} \; s_{NM-1}]^T$, the radar component of the discretized received signal at an arbitrary receiver can be expressed as

$$\mathbf{y} = \mathbf{S}\mathbf{x} + \mathbf{u}, \tag{21.4}$$

where $\mathbf{y}$ has length MN, the vector $\mathbf{u}$ is a length MN noise sequence distributed as $\mathbf{u} \sim \mathscr{CN}(\mathbf{0}, \sigma_u^2\mathbf{I})$, and the product $\mathbf{Sx}$ is the result of scattering (referred to as multipath in the forward link and clutter in the reverse link). While the Gaussian distribution of $\mathbf{x}$ is clearly a simplification of practical scattering, the associated property of being uncorrelated in range is relatively realistic. Therefore, the normalized eigen-decomposition of the fast-time scattering covariance matrix takes the form

$$\begin{aligned} \frac{1}{\sigma_x^2} E[(\mathbf{Sx})(\mathbf{Sx})^H] &= \frac{1}{\sigma_x^2} \mathbf{S} E[\mathbf{x}\mathbf{x}^H]\mathbf{S}^H \\ &= \mathbf{S}\mathbf{S}^H \\ &= \mathbf{V}\boldsymbol{\Lambda}\mathbf{V}^H, \end{aligned} \tag{21.5}$$

where $E[\,\cdot\,]$ is the expectation operator, σ_x^2 is the expected scattering power, and $\boldsymbol{\Lambda}$ is a diagonal matrix of eigenvalues in descending order associated with the respective eigenvectors in $\mathbf{V}$. This eigen-decomposition provides a tool with which to construct a set of communication symbols possessing a controlled degree of separability from the waveform-convolved clutter response and therefore is a means with which to minimize interference with the radar.

Finally, due to the use of nonlinear high-power amplifiers and the desire for maximum power efficiency, radar waveforms are typically constrained to be constant amplitude. Without loss of generality, normalize the radar waveform as $\mathbf{s}^H\mathbf{s} = 1$ by setting $|s(i)|^2 = 1/NM$ for $i = 1, \ldots, NM$. Using this normalization along with (21.5) leads to

$$\begin{aligned} \text{tr}\{\mathbf{S}\mathbf{S}^H\} &= \sum_{i=1}^{NM} \mathbf{s}^H \mathbf{s} = NM \\ &= \text{tr}\{\mathbf{V}\mathbf{\Lambda}\mathbf{V}^H\} = \text{tr}\{\mathbf{\Lambda}\} \end{aligned} \tag{21.6}$$

since $\mathbf{V}$ is unitary and by the cyclic permutation property of the trace operation. This relationship will be invoked for various symbol design strategies.

21.2.2 Direct-sequence spread-spectrum

The first and most obvious set of symbols to consider are those inspired by DSSS [50]. Since the information content of a DSSS symbol is spread across a bandwidth greater than the information bandwidth, DSSS symbols are naturally resistant to narrowband interference. Further, since the purpose of these RCEC symbols is to cohabitate the radar spectrum, DSSS symbol design is an attractive initial approach. However, traditional DSSS symbols require synchronization, which is unlikely to be available (see Table 21.1), and the incumbent radar signal cannot be treated like a narrowband interferer, which rules out traditional DSSS approaches such as code-division multiple access. Therefore, rather than using spreading codes to distinguish between users, as is typically the case in a multiuser system [49,50], the set of spreading codes can themselves be used as a communication alphabet. Given that the eigenvector matrix $\mathbf{V}$ spans the space occupied by the oversampled spectrum, a normalized DSSS symbol can be expressed as

$$\begin{aligned} \mathbf{c}_{\text{DSSS},k} &= \frac{1}{NM}\mathbf{V}^H \mathbf{b}_k \\ &= \frac{1}{NM}\mathbf{q}_k \end{aligned} \tag{21.7}$$

for $k = 1, 2, \ldots, K$, where $\mathbf{b}_k$ is an $NM \times 1$ pseudorandom spreading vector that must be known to both the tag and any intended receivers. Note that the DSSS approach does not account for the interference to/from the radar and thus spreads evenly across the available spectrum.

21.2.3 Eigenvectors as waveforms

In contrast to the DSSS approach above, interference-avoidance based approaches to RCEC symbol design were examined in [36,38,41]. One such example is denoted as eigenvectors-as-waveforms (EaW) in which symbols are defined as

$$\mathbf{c}_{\text{EaW},k} = \mathbf{v}_{NM-k+1}. \tag{21.8}$$

In other words, the kth communication symbol is defined as the eigenvector associated with the kth smallest eigenvalue. In practice, the EaW approach can be approximated by using OFDM subcarriers in the roll-off regions of the radar spectrum. A similar approach to spreading code design for a traditional DSSS communication system

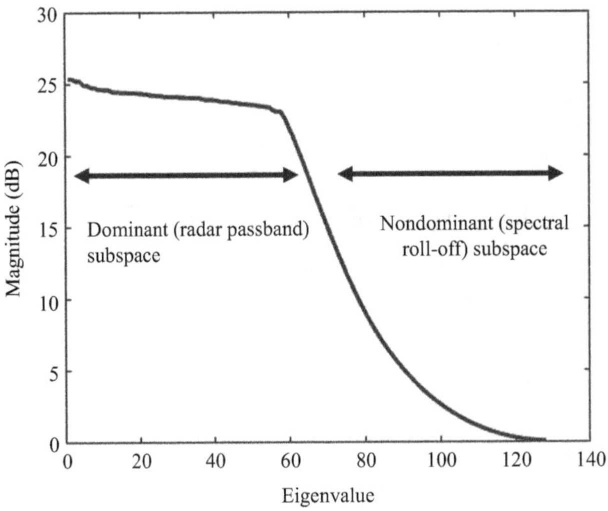

Figure 21.4 Dominant and nondominant subspaces

was explored in [53]. However, it was shown in [53] that for the problem under consideration, the EaW symbol design approach suffers from a large performance degradation in the presence of multipath [38,41].

21.2.4 Dominant projection

While the EaW symbol design uses individual eigenvectors as communication symbols, an alternative approach is to take advantage of the distinct subspaces occupied by the radar passband and spectral roll-off of a received signal, each of which is spanned by a set of mutually exclusive eigenvectors. Specifically, consider the eigenvalues in $\boldsymbol{\Lambda}$ (see Figure 21.4) which are realized for a $N = 64$ linear frequency-modulated (LFM) upchirp waveform oversampled by a factor of $M = 2$. Note that the eigenvalues decrease quite rapidly after approximately the Nth eigenvalue, where $N = BT$ is also the time–bandwidth product of the radar waveform. Therefore, it is natural to designate the passband portion as the dominant subspace, and the roll-off portion as the nondominant subspace [37,38,41] such that

$$\mathbf{V}\boldsymbol{\Lambda}\mathbf{V}^H = \left[\,\mathbf{V}_{\mathrm{D}} \,\middle|\, \mathbf{V}_{\mathrm{ND}}\,\right] \left[\begin{array}{c|c} \boldsymbol{\Lambda}_{\mathrm{D}} & \mathbf{0} \\ \hline \mathbf{0} & \boldsymbol{\Lambda}_{\mathrm{ND}} \end{array}\right] \left[\begin{array}{c} \mathbf{V}_{\mathrm{D}}^H \\ \hline \mathbf{V}_{\mathrm{ND}}^H \end{array}\right]. \tag{21.9}$$

Given that $\mathbf{V}$ contains NM eigenvectors, we can generally ascribe the first m to $\mathbf{V}_{\mathrm{D}}$ and the remaining $NM - m$ to $\mathbf{V}_{\mathrm{ND}}$, where the particular value of m has been shown [42] to be a key parameter when optimizing the design of communication symbols (see Section 21.4.2).

Projecting away from the dominant subspace (or equivalently, onto the nondominant subspace) is accomplished by using the projection matrix

$$\begin{aligned}\mathbf{A} &= \mathbf{I} - \mathbf{V}_{\mathrm{D},m}\mathbf{V}_{\mathrm{D},m}^{H} \\ &= \mathbf{V}_{\mathrm{ND},m}\mathbf{V}_{\mathrm{ND},m}^{H}.\end{aligned} \tag{21.10}$$

The kth dominant projection (DP) communication symbol is then formed as

$$\begin{aligned}\mathbf{c}_{\mathrm{DP},k} &= \beta_{\mathrm{DP},m}^{1/2}\mathbf{A}\mathbf{b}_k \\ &= \beta_{\mathrm{DP},m}^{1/2}\mathbf{V}_{\mathrm{ND},m}\mathbf{V}_{\mathrm{ND},m}^{H}\ \mathbf{b}_k \\ &= \beta_{\mathrm{DP},m}^{1/2}\mathbf{V}_{\mathrm{ND},m}\ \mathbf{q}_{\mathrm{ND},k},\end{aligned} \tag{21.11}$$

where $\beta_{\mathrm{DP},m}^{1/2}$ is a normalization factor (used in Section 21.3 to provide a fair comparison) and $\mathbf{q}_{\mathrm{ND},k} = \mathbf{V}_{\mathrm{ND},m}^{H}\ \mathbf{b}_k$ is the $(NM - m) \times 1$ transformed spreading vector associated with the nondominant subspace. Since $\mathbf{V}$ is a unitary transform and by stipulating $||\mathbf{b}_k||^2 = 1$, the expected power of the individual elements in $\mathbf{b}_k$ and $\mathbf{q}_k$ are

$$\begin{aligned}|b_{k,i}|_{\mathrm{avg}}^2 &= |q_{k,i}|_{\mathrm{avg}}^2 \\ &\approx \frac{1}{NM}.\end{aligned} \tag{21.12}$$

From (21.11) and (21.12), the expected transmit energy for the kth DP symbol is therefore

$$\begin{aligned}S &= ||\mathbf{c}_{\mathrm{DP},k}||^2 = \mathbf{c}_{\mathrm{DP},k}^{H}\mathbf{c}_{\mathrm{DP},k} \\ &= \left(\beta_{\mathrm{DP},m}^{1/2}\mathbf{q}_{\mathrm{ND},k}^{H}\mathbf{V}_{\mathrm{ND},m}^{H}\right)\left(\beta_{\mathrm{DP},m}^{1/2}\mathbf{V}_{\mathrm{ND},m}\mathbf{q}_{\mathrm{ND},k}\right) \\ &= \beta_{\mathrm{DP},m}\mathbf{q}_{\mathrm{ND},k}^{H}\mathbf{q}_{\mathrm{ND},k} \\ &\approx \beta_{\mathrm{DP},m}(NM - m)|q_{k,i}|_{\mathrm{avg}}^2 \\ &= \frac{\beta_{\mathrm{DP},m}(NM - m)}{NM}.\end{aligned} \tag{21.13}$$

Using (21.13) and enforcing a normalized transmit energy of 1 yields a symbol scaling factor dependent on the size of the dominant subspace given as

$$\beta_{\mathrm{DP},m} = \frac{NM}{NM - m}. \tag{21.14}$$

A key property of the DP symbol design is that it projects away from the dominant subspace *as a whole*. Therefore, if the radar waveform is unknown to the tag and intended receiver, estimation errors (due to timing offsets or multipath) tend to be contained within the nondominant subspace. However, for the EaW approach, these estimation errors result in completely different nondominant eigenvectors. This property is why DP symbol design is robust to multipath and timing offset, while the EaW symbols are highly sensitive to these effects [41,42].

21.2.5 Shaped dominant projection

When the DP symbol design projects away from the dominant subspace, it spreads the symbol energy evenly across the nondominant subspace regardless of the radar spectral content. Put another way, the DP symbol design does not utilize the eigenvalue information in $\mathbf{\Lambda}$ (i.e. Figure 21.4). To take advantage of this information, the shaped DP (SDP) symbol design was introduced in [45], for which the kth symbol is

$$\begin{aligned}\mathbf{c}_{\mathrm{SDP},k} &= \beta_{\mathrm{SDP}}^{1/2}\mathbf{V}_{\mathrm{ND},m}\mathbf{\Lambda}_{\mathrm{ND},m}^{1/2}\mathbf{V}_{\mathrm{ND},m}^{H}\mathbf{b}_k \\ &= \beta_{\mathrm{SDP}}^{1/2}\mathbf{V}_{\mathrm{ND},m}\mathbf{\Lambda}_{\mathrm{ND},m}^{1/2}\mathbf{q}_{\mathrm{ND},k},\end{aligned} \tag{21.15}$$

with energy normalization factor

$$\beta_{\mathrm{SDP}} = \frac{NM}{\mathrm{tr}\left\{\mathbf{\Lambda}_{\mathrm{ND},m}\right\}}. \tag{21.16}$$

Comparing (21.15) to the DP approach, here the square-root of the eigenvalues has been inserted into the second line of (21.11), which realizes a weighting of the projection defined in (21.10).

21.2.6 Shaped waterfilling

A final RCEC symbol design was introduced in [45]. The shaped waterfilling (SWF) approach uses the entire eigenspace spanned by $\mathbf{V}$ but employs different shaping in the dominant and nondominant regions. Waterfilling is accomplished by using the partitioned shaped eigenvalue matrix

$$\mathbf{\Lambda}_{\mathrm{P},m} = \left[\begin{array}{c|c}\mathbf{\Lambda}_{\mathrm{D},m}^{-1} & \mathbf{0} \\ \hline \mathbf{0} & \mathbf{\Lambda}_{\mathrm{ND},m},\end{array}\right] \tag{21.17}$$

where the inverse of the large eigenvalues in the dominant subspace serve to minimize the symbol power in the radar passband and the nondominant shaping is the same as for SDP. The kth SWF communication symbol is thus a generalization of (21.15) and takes the form

$$\begin{aligned}\mathbf{c}_{\mathrm{SWF},k} &= \beta_{\mathrm{SWF}}^{1/2}\mathbf{V}\mathbf{\Lambda}_{\mathrm{P},m}^{1/2}\mathbf{V}^{H}\mathbf{b}_k \\ &= \beta_{\mathrm{SWF}}^{1/2}\mathbf{V}\mathbf{\Lambda}_{\mathrm{P},m}^{1/2}\mathbf{q}_k,\end{aligned} \tag{21.18}$$

for energy normalization factor

$$\beta_{\mathrm{SWF}} = \frac{NM}{\mathrm{tr}\left\{\mathbf{\Lambda}_{\mathrm{P},m}\right\}}. \tag{21.19}$$

By examining (21.17) and (21.18), it is observed that the dominant subspace is suppressed to some extent for SWF symbols, though it is not actually nulled as is the case for DP and SDP symbols.

21.3 Receiver design

Receiver detection of RCEC symbols involves multiple interrelated challenges, several of which are noted in Table 21.1. First, the tag and receiver are not synchronized. Therefore, estimation of the particular symbol that is received also requires the determination of whether a symbol is actually present in the radar clutter. However, to determine if a symbol is present, the receiver must be able to separate the symbol from the clutter. Thus, the proper design of receive filters is crucial to enable the detection of an embedded symbol among the clutter interference.

The separability of the communication symbol from clutter and noise can be quantified by the signal-to-interference-plus-noise ratio (SINR) after receive filtering. The processing gain achieved by receive filtering is then the ratio of the SINR after filtering (output SINR) to the SINR before filtering (input SINR), given as [42,43,45]

$$\Delta = \frac{\text{SINR}_\text{o}}{\text{SINR}_\text{i}}. \tag{21.20}$$

The input SINR is independent of symbol design due to the normalization terms β specified in Section 21.2. This SINR is computed from the expected value of the magnitude squared received signal from (21.4), with the addition of the communication symbol $\mathbf{c}_k$, as

$$\begin{aligned} E\left[||\mathbf{y}||^2\right] &= E\left[(\mathbf{Sx} + \alpha\mathbf{c}_k + \mathbf{u})^H(\mathbf{Sx} + \alpha\mathbf{c}_k + \mathbf{u})\right] \\ &= E\left[(\mathbf{Sx})^H(\mathbf{Sx})\right] + E\left[(\alpha\mathbf{c}_k)^H(\alpha\mathbf{c}_k)\right] + E\left[\mathbf{u}^H\mathbf{u}\right] \\ &= R_\text{i} + S_\text{i} + N_\text{i}, \end{aligned} \tag{21.21}$$

where α is the attenuation between the tag and the desired receiver and with the radar clutter, communication symbol, and noise assumed to be independent. With the communication symbols normalized to have $|\mathbf{c}_k|^2 = 1$, the average power of an incident communication symbol is given as

$$S_\text{i} = |\alpha|^2. \tag{21.22}$$

By using (21.5) and (21.6), the average incident interference power is found to be

$$\begin{aligned} R_\text{i} &= \sigma_x^2 \text{tr}\{\mathbf{\Lambda}\} \\ &= \sigma_x^2 NM. \end{aligned} \tag{21.23}$$

Finally, the average input noise power in (21.21) is given as

$$N_{\mathrm{i}} = \sigma_u^2 NM. \tag{21.24}$$

Therefore, combining (21.22)–(21.24) yields the input SINR

$$\mathrm{SINR}_{\mathrm{i}} = \frac{|\alpha|^2}{NM(\sigma_x^2 + \sigma_u^2)} \tag{21.25}$$

which is agnostic to the particular symbol design.

The output SINR is computed using the filtered components of the received signal

$$\begin{aligned} E\left[|\mathbf{w}_k^H \mathbf{y}|^2\right] &= E\left[(\mathbf{Sx} + \alpha\mathbf{c}_k + \mathbf{u})^H \mathbf{w}_k \mathbf{w}_k^H (\mathbf{Sx} + \alpha\mathbf{c}_k + \mathbf{u})\right] \\ &= E\left[(\mathbf{Sx})^H \mathbf{w}_k \mathbf{w}_k^H (\mathbf{Sx})\right] + E\left[(\alpha\mathbf{c}_k)^H \mathbf{w}_k \mathbf{w}_k^H (\alpha\mathbf{c}_k)\right] \\ &\quad + E\left[\mathbf{u}^H \mathbf{w}_k \mathbf{w}_k^H \mathbf{u}\right] \\ &= R_{\mathrm{o}} + S_{\mathrm{o}} + N_{\mathrm{o}}, \end{aligned} \tag{21.26}$$

which are likewise arranged as

$$\mathrm{SINR}_{\mathrm{o}} = \frac{S_{\mathrm{o}}}{R_{\mathrm{o}} + N_{\mathrm{o}}}. \tag{21.27}$$

The terms R_{o} and N_{o} can be further simplified as

$$\begin{aligned} R_{\mathrm{o}} &= E\left[(\mathbf{Sx})^H \mathbf{w}_k \mathbf{w}_k^H (\mathbf{Sx})\right] \\ &= E\left[\mathbf{w}_k^H \mathbf{Sxx}^H \mathbf{S}^H \mathbf{w}_k\right] \\ &= \mathbf{w}_k^H \mathbf{S}\, E\left[\mathbf{xx}^H\right] \mathbf{S}^H \mathbf{w}_k \\ &= \sigma_x^2 \mathbf{w}_k^H \mathbf{SS}^H \mathbf{w}_k \\ &= \sigma_x^2 \mathbf{w}_k^H \mathbf{V\Lambda V}^H \mathbf{w}_k \end{aligned} \tag{21.28}$$

by invoking (21.5) and

$$\begin{aligned} N_{\mathrm{o}} &= E\left[\mathbf{u}^H \mathbf{w}_k \mathbf{w}_k^H \mathbf{u}\right] \\ &= E\left[\mathbf{w}_k^H \mathbf{uu}^H \mathbf{w}_k\right] \\ &= \mathbf{w}_k^H E\left[\mathbf{uu}^H\right] \mathbf{w}_k \\ &= \sigma_u^2 \mathbf{w}_k^H \mathbf{w}_k. \end{aligned} \tag{21.29}$$

Determination of the symbol output power depends on both the particular symbol and receiver filter designs. Receiver designs for RCEC symbols were developed and analyzed in [45] and are summarized in the following sections.

21.3.1 Matched filter

It is well known that the matched filter maximizes SNR in the presence of white Gaussian noise [54]. The matched filter for an arbitrary kth RCEC symbol is simply

$$\mathbf{w}_{\mathrm{MF},k} = \mathbf{c}_k, \tag{21.30}$$

which produces $S_o = |\alpha|^2$ for all symbol designs. Thus, the processing gain from (21.20) is dictated by how the symbol matched filter interacts with the clutter and noise. For the DP symbol design, the processing gain is found to be [45]

$$\Delta_{\mathrm{DP-MF}}(m) = \frac{NM(\sigma_x^2 + \sigma_u^2)}{\sigma_x^2\left((\mathrm{tr}\left\{\mathbf{\Lambda}_{\mathrm{ND},m}\right\})/(NM - m)\right) + \sigma_u^2}, \tag{21.31}$$

recalling that m denotes the dimension of the dominant subspace. Similarly, the processing gain for the SDP and SWF symbol designs are [45]

$$\Delta_{\mathrm{SDP-MF}}(m) = \frac{NM(\sigma_x^2 + \sigma_u^2)}{\sigma_x^2\left(\mathrm{tr}\left\{\mathbf{\Lambda}_{\mathrm{ND},m}^2\right\}/\mathrm{tr}\left\{\mathbf{\Lambda}_{\mathrm{ND},m}\right\}\right) + \sigma_u^2} \tag{21.32}$$

and

$$\Delta_{\mathrm{SWF-MF}}(m) = \frac{NM(\sigma_x^2 + \sigma_u^2)}{\sigma_x^2\left((m + \mathrm{tr}\left\{\mathbf{\Lambda}_{\mathrm{ND},m}^2\right\})/\mathrm{tr}\left\{\mathbf{\Lambda}_{\mathrm{P},m}\right\}\right) + \sigma_u^2}, \tag{21.33}$$

respectively.

21.3.2 Maximum likelihood (ML) decorrelating filter (DF)

While the matched filter provides optimum performance in AWGN, an RCEC system is designed to operate in/near radar clutter. Therefore, the filter design should take into account the correlated nature of the interference. A natural filter structure is the maximum likelihood (ML), or decorrelating (DF), receiver [41,49]

$$\begin{aligned}\mathbf{w}_{\mathrm{DF},k} &= \left(\mathbf{S}\mathbf{S}^H\right)^{-1}\mathbf{c}_k \\ &= \mathbf{V}\mathbf{\Lambda}^{-1}\mathbf{V}^H\mathbf{c}_k.\end{aligned} \tag{21.34}$$

By inspection of (21.34), it can be seen that the DF is a whitening transform followed by a matched filter.

The processing gain when applying the DF to the DP symbols were shown in [45] to be

$$\Delta_{\mathrm{DP-DF}}(m) = \frac{NM(\sigma_x^2 + \sigma_u^2)\left(\mathrm{tr}\left\{\mathbf{\Lambda}_{\mathrm{ND},m}^{-1}\right\}\right)^2}{(NM - m)\left(\sigma_x^2\mathrm{tr}\left\{\mathbf{\Lambda}_{\mathrm{ND},m}^{-1}\right\} + \sigma_u^2\mathrm{tr}\left\{\mathbf{\Lambda}_{\mathrm{ND},m}^{-2}\right\}\right)}. \tag{21.35}$$

Likewise, the DF processing gain for the SDP and SWF symbols are [45]

$$\Delta_{\text{SDP-DF}}(m) = \frac{NM(\sigma_x^2+\sigma_u^2)(NM-m)^2}{\text{tr}\left\{\mathbf{\Lambda}_{\text{ND},m}\right\}\left(\sigma_x^2(NM-m)+\sigma_u^2\text{tr}\left\{\mathbf{\Lambda}_{\text{ND},m}^{-1}\right\}\right)} \tag{21.36}$$

and

$$\Delta_{\text{SWF-DF}}(m) = \frac{NM(\sigma_x^2+\sigma_u^2)\left(\text{tr}\left\{\mathbf{\Lambda}_{\text{D}}^{-2}\right\}+NM-m\right)^2}{\text{tr}\left\{\mathbf{\Lambda}_{\text{P},m}\right\}\left(\sigma_x^2\left(\text{tr}\left\{\mathbf{\Lambda}_{\text{D},m}^{-2}\right\}+NM-m\right)+\sigma_u^2\text{tr}\left\{\mathbf{\Lambda}_{\text{P},m}\mathbf{\Lambda}^{-2}\right\}\right)}, \tag{21.37}$$

respectively.

21.3.3 Diagonally loaded decorrelating filter

Finally, it is useful to consider a diagonally loaded DF (DLDF) [45], which for the kth symbol is defined as

$$\mathbf{w}_{\text{DLDF},k} = \mathbf{V}\tilde{\mathbf{\Lambda}}^{-1}\mathbf{V}^H\mathbf{c}_k \tag{21.38}$$

for the modified eigenvalue matrix

$$\tilde{\mathbf{\Lambda}} = \sigma_x^2\mathbf{\Lambda}+\sigma_u^2\mathbf{I}. \tag{21.39}$$

Note that implementation of the DLDF filters requires knowledge of the relative interference (clutter) and noise powers, which can be readily estimated.

When the DLDF is used, the processing gain for DP symbols is [45]

$$\Delta_{\text{DP-DLDF}}(m) = \frac{NM(\sigma_x^2+\sigma_u^2)\left(\text{tr}\left\{\tilde{\mathbf{\Lambda}}_{\text{ND},m}^{-1}\right\}\right)^2}{(NM-m)\left(\sigma_x^2\text{tr}\left\{\tilde{\mathbf{\Lambda}}_{\text{ND},m}^{-2}\mathbf{\Lambda}_{\text{ND},m}\right\}+\sigma_u^2\text{tr}\left\{\tilde{\mathbf{\Lambda}}_{\text{ND},m}^{-2}\right\}\right)}. \tag{21.40}$$

In like manner, the DLDF processing gain for SDP and SWP symbols are [45]

$$\Delta_{\text{SDP-DLDF}}(m) = \frac{NM(\sigma_x^2+\sigma_u^2)\left(\text{tr}\left\{\mathbf{\Lambda}_{\text{ND},m}\tilde{\mathbf{\Lambda}}_{\text{ND},m}^{-1}\right\}\right)^2}{\text{tr}\left\{\mathbf{\Lambda}_{\text{ND},m}\right\}\left(\sigma_x^2\text{tr}\left\{\mathbf{\Lambda}_{\text{ND},m}^{2}\tilde{\mathbf{\Lambda}}_{\text{ND},m}^{-2}\right\}+\sigma_u^2\text{tr}\left\{\mathbf{\Lambda}_{\text{ND},m}\tilde{\mathbf{\Lambda}}_{\text{ND},m}^{-2}\right\}\right)} \tag{21.41}$$

and

$$\Delta_{\text{SWF-DLDF}}(m) = \frac{NM(\sigma_x^2+\sigma_u^2)\left(\text{tr}\left\{\mathbf{\Lambda}_{\text{P},m}\tilde{\mathbf{\Lambda}}^{-1}\right\}\right)^2}{\text{tr}\left\{\mathbf{\Lambda}_{\text{P},m}\right\}\left(\sigma_x^2\text{tr}\left\{\mathbf{\Lambda}_{\text{P},m}\tilde{\mathbf{\Lambda}}^{-2}\mathbf{\Lambda}\right\}+\sigma_u^2\text{tr}\left\{\mathbf{\Lambda}_{\text{P},m}\tilde{\mathbf{\Lambda}}^{-2}\right\}\right)}, \tag{21.42}$$

respectively. The impact of these different combinations of symbol and filter designs are examined in Section 21.4.2.

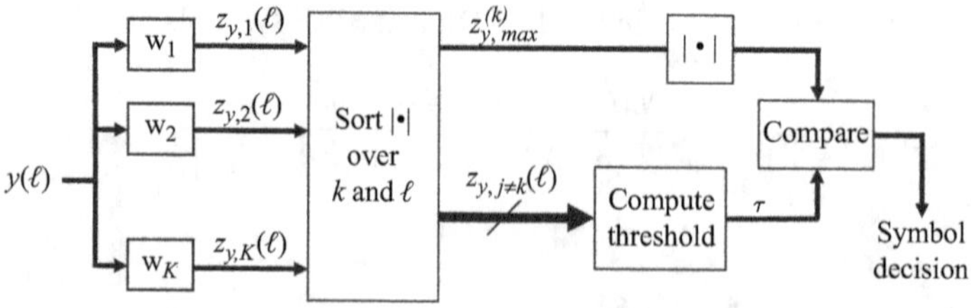

Figure 21.5 Two-stage Neyman–Pearson detector [42]

21.3.4 Two-stage Neyman–Pearson detector

Regardless of the particular receive filter employed, the RCEC receiver must still determine whether or not a symbol is present as part of the symbol estimation process. To that end, an adaptive two-stage Neyman–Pearson detector was developed in [42,43] and further improved in [45]. The receiver is assumed to have knowledge of the oversampling factor M, the particular pseudorandom codes $\mathbf{b}_k$, and the symbol design method (DP, SDP, or SWF) used by the tag.

The determination of the presence of a symbol is incorporated into the symbol estimation process via multiple hypothesis testing with $K+1$ hypotheses. As above, K is the size of the symbol alphabet and

$$\begin{aligned}&\mathscr{H}_0 : \mathbf{y} = \mathbf{S}\mathbf{x} + \mathbf{u}\\ &\mathscr{H}_k : \mathbf{y} = \mathbf{S}\mathbf{x} + \alpha\mathbf{c}_k + \mathbf{u} \quad \text{for} \quad k = 1, 2, \ldots, K,\end{aligned} \tag{21.43}$$

where $\mathscr{H}_0$ is the null hypothesis corresponding to the absence of a symbol, while $\mathscr{H}_k$ is the hypothesis corresponding to the presence of symbol k. This hypothesis test is split into two stages. The first stage determines the most likely symbol present and the second stage subsequently conducts a Neyman–Pearson test between the candidate symbol and the null hypothesis, which is obtained from the other $K-1$ hypotheses. A block diagram of the adaptive two-stage Neyman–Pearson detector is shown in Figure 21.5.

Because the receiver is not synchronized with the RCEC transmitter, symbol estimation begins by observing an interval of $L+NM-1$ range samples, which can be arranged into the L snapshot vectors

$$\mathbf{y}(\ell) = [y(\ell)\; y(\ell-1)\; y(\ell-NM+1)]^T, \; \ell = 0, 1, \ldots, L-1. \tag{21.44}$$

The received snapshots are first filtered with $\mathbf{w}_k$ for $k=1,\ldots,K$ to suppress the radar clutter and to coherently match to a prospective communication symbol that may be present. The output of this filter bank consists of the set of L filtered observations for each of the K filters

$$z_k(\ell) = \mathbf{w}_k^H \mathbf{y}(\ell) \quad \text{for} \quad k = 1, 2, \ldots, K. \tag{21.45}$$

For each symbol, the delay ℓ where the symbol is most likely to be present is then ascertained. By then determining the largest value across this set of K maxima, the most likely symbol over this range interval is estimated. These two operations can be respectively expressed as

$$|z_{\max}^{(k)}| = \max_{k,\ell} |z_k(\ell)| \tag{21.46}$$

and

$$\hat{k} = \arg\max_{k,\ell} |z_k(\ell)|, \tag{21.47}$$

where $\hat{k}$ denotes the mostly likely symbol. Note that this first stage involves testing the K hypotheses

$$\mathcal{H}_k : \mathbf{y} = \mathbf{S}\mathbf{x} + \alpha\mathbf{c}_k + \mathbf{u} \quad \text{for} \quad k = 1, 2, \ldots, K, \tag{21.48}$$

with the accuracy strongly impacted by the crosscorrelation of the symbols, the signal-to-clutter ratio, and the signal-to-noise ratio.

Now that the most likely symbol has been determined, the receiver must decide whether or not that symbol is actually present by comparing the hypothesis $\mathcal{H}_{\hat{k}}$ to the null hypothesis $\mathcal{H}_0$ from (21.43). Using the values estimated in (21.46) and (21.47), and for threshold $\mathcal{T}$, if $|z_{\max}^{(\hat{k})}| > \mathcal{T}$ the receiver decides that the $\hat{k}$th symbol is present in the observation interval. Otherwise, the receiver decides that no communication symbol is present. To obtain this threshold, it is necessary to determine the probability distribution of the null hypothesis in order to properly set the acceptable probability of false alarm [54].

To estimate the distribution under the null hypothesis, L filter responses for each of the $K-1$ rejected symbols are collected. It is assumed that the clutter process samples are zero-mean and independent and identically distributed (i.i.d.). As such, application of the Central Limit Theorem asserts that the linear transformation given by convolution of the radar waveform and the clutter process (i.e. $\mathbf{S}\mathbf{x}$) followed by the filtering $\mathbf{w}_k^H\mathbf{y}(\ell)$ yields a complex Gaussian distributed output [42,43,45]. Therefore, taking the magnitude of the filter output $|z_{y,k}(\ell)|$ realizes i.i.d. Rayleigh samples. If no symbol is present, the probability of false alarm is given by the probability that the maximum value of $(K-1)L$ i.i.d. Rayleigh samples exceeds threshold $\mathcal{T}$, which is given as [54]

$$P_{\text{fa}}(\sigma, \mathcal{T}, K, L) = 1 - \left[1 - \exp\left\{-\frac{\mathcal{T}^2}{2\sigma^2}\right\}\right]^{(K-1)L} \tag{21.49}$$

for σ the Rayleigh scale parameter. Rearranging (21.49), the Neyman–Pearson threshold $\mathcal{T}$ for the second stage of the detector is found to be

$$\mathcal{T}(\sigma, P_{\text{fa}}, K, L) = \sqrt{-2\sigma^2 \ln\left[1 - (1 - P_{\text{fa}})^{1/((K-1)L)}\right]}. \tag{21.50}$$

The ML estimator of σ is given as [55]

$$\hat{\sigma} = \sqrt{\frac{1}{2(K-1)L} \sum_{j=1, j\neq\hat{k}}^{K} \sum_{\ell=1}^{L} |z_{y,j}(\ell)|^2}, \tag{21.51}$$

which has a bias [55]

$$E[\hat{\sigma}] \approx \sigma \left(1 - \frac{1}{8(K-1)L}\right). \tag{21.52}$$

Fortunately, (21.51) is asymptotically unbiased and the number of i.i.d. samples $(K-1)L$ is generally large. The final detector output is thus

$$\text{symbol decision} = \begin{cases} k\text{th symbol}, & \text{if } |z_{\text{max}}^{(k)}| > \mathscr{T} \\ \text{no symbol}, & \text{otherwise.} \end{cases} \tag{21.53}$$

21.4 Performance evaluation

These RCEC symbol/filter designs and symbol detection methods are evaluated by simulation. For these results, an 8 ms, 8 MHz upchirp LFM waveform is used as the radar emission. The communication symbols are allowed to occupy a bandwidth of 16 MHz centered on the radar spectrum and likewise have a duration of 8 ms (i.e. $M = 2$). However, the communication symbols are further upsampled by a factor of 2 in order to illustrate their spectral containment. The pseudorandom spreading vectors $\mathbf{b}_k$ are implemented as random polyphase codes for $K = 8$ symbols.

21.4.1 Spectral content

One method of comparing different symbol design methods is to examine their spectra. As such, Figures 21.6–21.9 illustrate the spectra for the symbol design methods as well as the spectrum of the clutter-plus-noise for particular values of the dominant subspace dimensionality m. Here, the CNR is set to 15 dB, while the signal-to-clutter ratio (SCR) is set to −25 dB. A total of 1,000 instantiations of the pseudorandom spreading vectors $\mathbf{b}_k$ are averaged for each symbol design.

Figure 21.6 shows the three symbol design spectra for $m = 32$. Since only the eigenvectors corresponding to the largest eigenvalues (the dominant subspace) are projected away by the SDP symbol design and suppressed by the SWF symbol design, there is negligible difference between their associated spectra. Further, the SDP and SWF use of the nondominant subspace can be clearly observed in their spectral roll-offs. In contrast, the DP and DSSS designs realize a broader spectral content since no shaping is imposed. It is also interesting to note that the nulls of the DP, SDP, and SWF symbols are aligned with the peaks of the ripple in the clutter spectrum, which arises because these peaks essentially correspond to the largest eigenvalues and

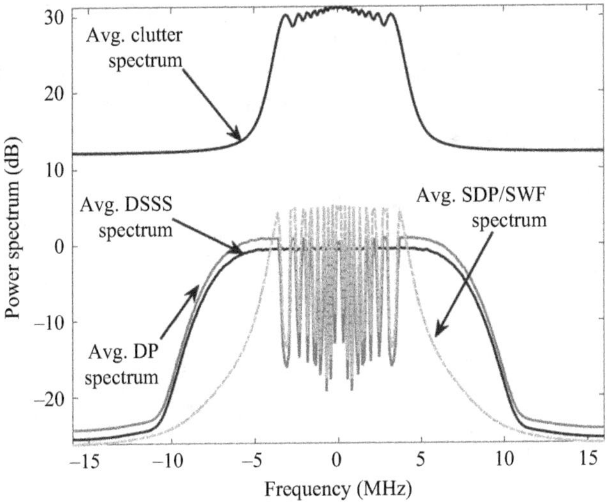

Figure 21.6 Power spectra of symbol design methods for m = 32

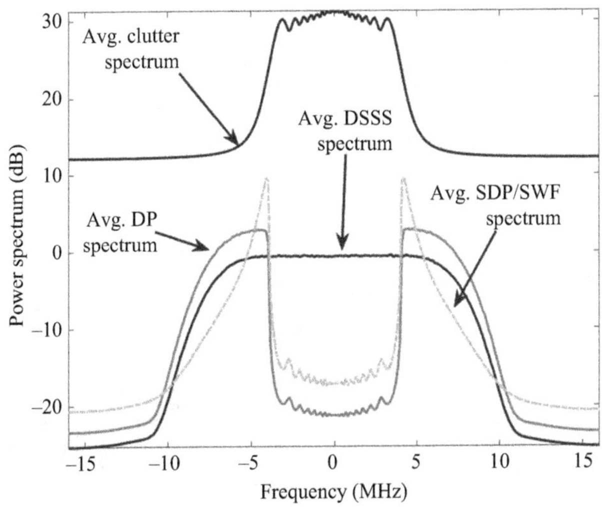

Figure 21.7 Power spectra of symbol design methods with m = 64

setting $m = 32$ for the dominant subspace is insufficient to capture all of the radar clutter passband.

Next, Figure 21.7 shows the spectra for $m = 64$, such that the dimensionality of the dominant subspace closely approximates the time–bandwidth product of the radar waveform (and thus the resulting clutter as well). It is observed that the DP design still follows a similar spectral roll-off as the DSSS design. However, due to the projection

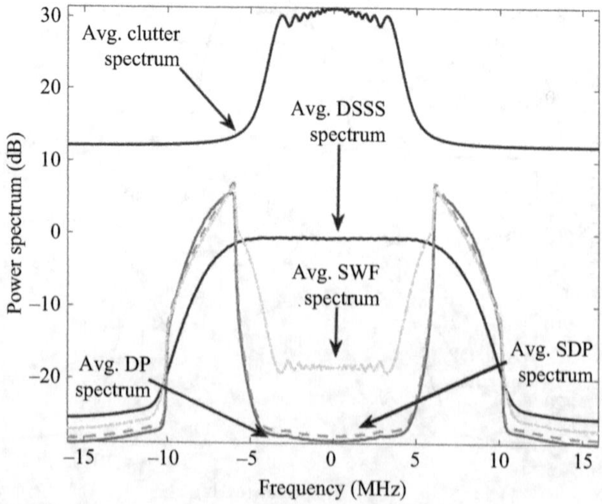

Figure 21.8 Power spectra of symbol design methods with m = 96

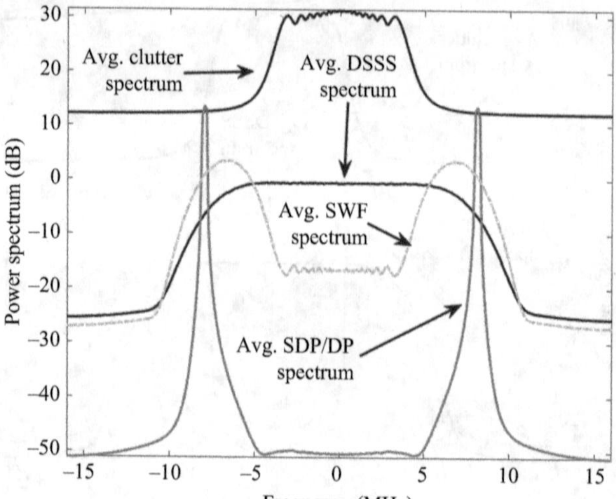

Figure 21.9 Power spectra of symbol design methods with m = 126

away from the radar passband, the DP symbols are observed to allocate power where there is less interference to/from the radar. The SWF and SDP symbols likewise avoid the radar interference, albeit within the spectral roll-off mirroring that of the clutter.

Figure 21.8 considers the dominant subspace dimension of $m = 96$, which exceeds the radar time–bandwidth product by a significant margin. Thus, the nondominant subspace is primarily associated with the noise outside the radar spectral roll-off. As a result, the inherent spectral containment of the symbols now becomes the main factor involved with their spectral roll-off. Additionally, the difference between the projection-based DP and SDP symbol designs and the suppression-based SWF design (i.e. through inversion of the dominant eigenvalues in $\mathbf{\Lambda}_{\mathrm{p},m}$) is evident. Specifically, the SWF symbol design allocates an additional $\sim$ 9 dB of average power across the radar passband relative to the DP and SDP designs. However, the SWF symbol design still realizes $\sim$ 18 dB reduction in the average power allocated in the radar passband compared to DSSS.

Finally, Figure 21.9 considers the extreme case where the dominant subspace dimension is set to $m = 126$ such that the nondominant subspace is spanned by only 2 eigenvectors. The DP and SDP symbols are now virtually identical and they allocate nearly all their power to the frequency bands farthest from the radar passband (as allowed by the RCEC spectral containment). On the other hand, the SWF symbol design spreads the majority of its power across the spectrum outside the radar passband, while reducing (but not nulling) the allocated power within the radar passband.

Examination of the spectrum associated with each symbol design illustrates the effect of the choice of the dominant subspace size. In addition, comparing the spectrum of each design yields a first approximation to the impact on target detection in the radar receiver that can be expected by sharing spectrum with these embedded symbols.

21.4.2 Processing gain

The processing gain equations in Sections 21.3.1–21.3.3 provide a means to examine the interrelated impact of subspace selection and filtering selection. Here, two scenarios are considered: a high CNR of 30 dB and a low CNR of 0 dB, with the RCEC symbols designed according to the high CNR scenario. In other words, the symbols are designed to operate in a spectrum sharing arrangement. However, it is also instructive to examine the processing gain of these symbol and filter designs under the condition of a low CNR to determine the impact of using RCEC communication symbols in a low-interference environment.

Figures 21.10 and 21.11 show the processing gain for the DP symbol design as a function of the dominant subspace size m. For comparison, the coherent gain provided solely by the time–bandwidth product of the symbol is also shown, which is likewise the gain of the DSSS symbols. In the high CNR case for the DP symbol design, the matched filter provides modest gain when the size of the dominant subspace is small because the symbols have to compete with the radar interference. However, as m grows there is less interference in the nondominant subspace and the matched filter becomes as effective as the DF and DLDF. While there is little difference between the performance of the DF and DLDF at high CNR, at low CNR the DLDF is the superior design. It is interesting to note that the processing gain in Figures 21.10 and 21.11 converge to values of $\approx$ 41.5 dB for the high CNR case and $\approx$ 24 dB for the

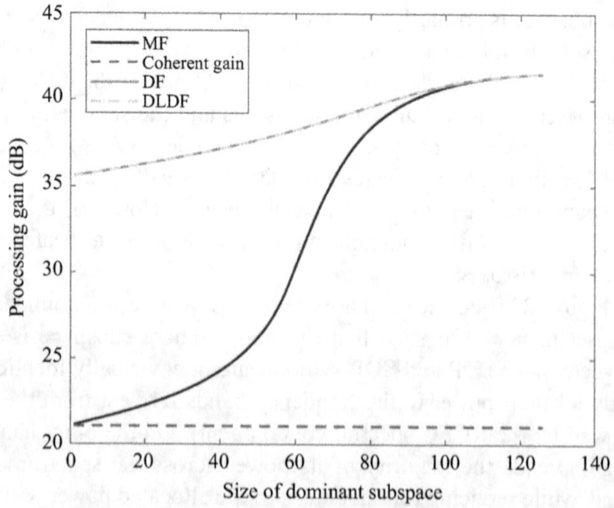

Figure 21.10 Processing gain with DP symbols for the high CNR case

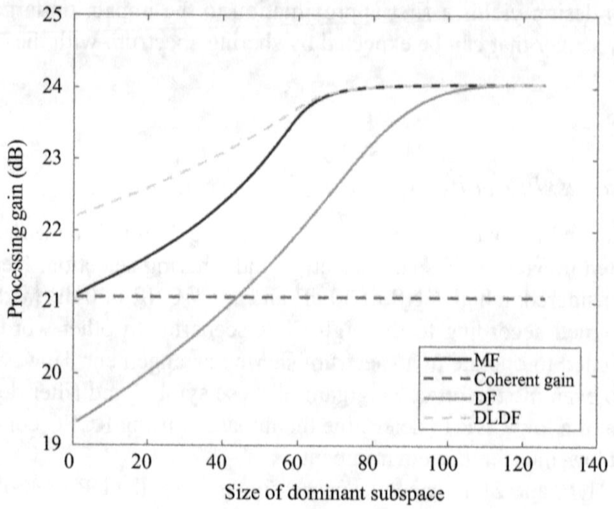

Figure 21.11 Processing gain with DP symbols for the low CNR case

low CNR case. For ease of examination, consider the processing gain offered by a matched filter for the case of dominant subspace size $m = NM - 2 = 126$.

In the high CNR case where $\sigma_x^2 \gg \sigma_u^2$, (21.57) reduces to

$$\Delta_{\text{DP-MF}}(m) \approx \frac{2NM}{\text{tr}\left\{\mathbf{\Lambda}_{\text{ND},m}\right\}}. \tag{21.54}$$

From (21.31) and approximating $\mathrm{tr}\left\{\boldsymbol{\Lambda}_{\mathrm{ND},m}\right\} \approx (NM - m)/NM$ for $m \approx NM$, we obtain

$$\Delta_{\mathrm{DP-MF}}(m) \approx \frac{2\,(NM)^2}{NM - m}. \tag{21.55}$$

It follows that for the simulation parameters used here, the processing gain offered by the matched filter when applied to DP symbols is

$$10\log_{10}\left[\Delta_{\mathrm{DP-MF}}(126)\right] \approx 42.14 \text{ (dB)}, \tag{21.56}$$

which is the approximate convergence value observed in Figure 21.10.

For the low CNR case where $\sigma_x^2 \approx \sigma_u^2$, (21.31) becomes

$$\begin{aligned}\Delta_{\mathrm{DP-MF}}(m) &\approx \frac{2NM}{1 + (\mathrm{tr}\left\{\boldsymbol{\Lambda}_{\mathrm{ND},m}\right\}/2)} \\ &\approx \frac{2NM}{1 + ((NM - m)/2NM)}.\end{aligned} \tag{21.57}$$

For the simulation parameters considered, the denominator is approximately 1 and the processing gain is

$$10\log_{10}\left[\Delta_{\mathrm{DP-MF}}(126)\right] \approx 24 \text{ (dB)}, \tag{21.58}$$

which is the value observed in Figure 21.11. Similar analyses of the convergence of the other filtering strategies are omitted for brevity.

Figures 21.12 and 21.13 show the processing gain for the SDP symbol design. In contrast to the DP symbol design, the dependence on the spectral roll-off provides a disadvantage when m is small. However, once the dominant subspace size exceeds the time–bandwidth product of the radar waveform, the SDP symbol design experiences a rapid processing gain improvement. Like the processing gain result with the DP symbol design, the DF and DLDF have similar performance at high CNR, while the DF suffers greatly at low CNR. Once more, the processing gains converge to ≈ 42 dB for the high CNR case and ≈ 24 dB for the low CNR case.

Figures 21.14 and 21.15 show the processing gain for the SWF symbol design. Comparison of Figures 21.12–21.15 shows that processing gain for the SWF and SDP designs exhibit similar dependencies on the dominant subspace size, filter design, and CNR. However, the matched filter for the high CNR case does not converge to the expected ≈ 42 dB but rather peaks at ≈ 35.6 dB. To examine this effect, once more assume that the dominant subspace size is $m = NM - 2 = 126$ and $\sigma_x^2 \gg \sigma_u^2$.

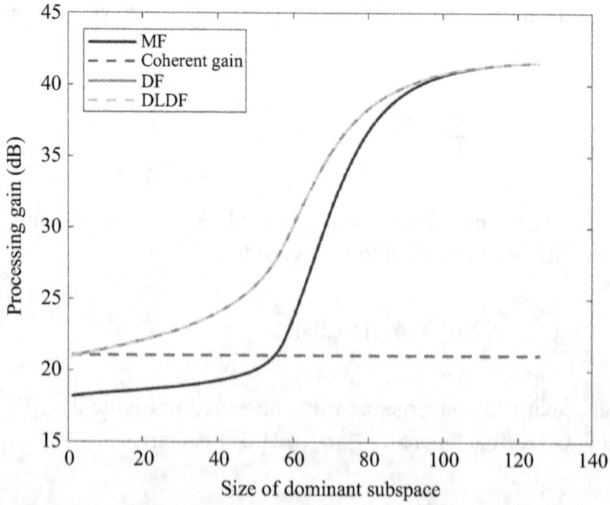

Figure 21.12 Processing gain with SDP symbols for the high CNR case

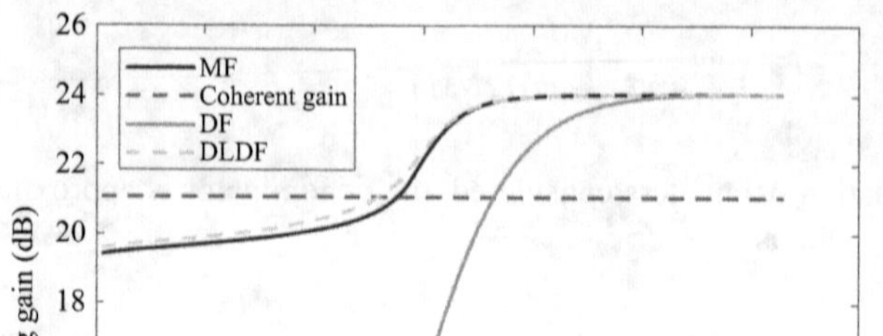

Figure 21.13 Processing gain with SDP symbols for the low CNR case

Applying these assumptions to (21.33) and using the eigenvalues for the scenario considered yields

$$\Delta_{\text{SWF-MF}}(m) \approx \frac{NM\ \text{tr}\left\{\mathbf{\Lambda}_{\text{P},m}\right\}}{m + \text{tr}\left\{\mathbf{\Lambda}^2_{\text{ND},m}\right\}} \approx \frac{NM\ \text{tr}\left\{\mathbf{\Lambda}_{\text{P},m}\right\}}{m}, \tag{21.59}$$

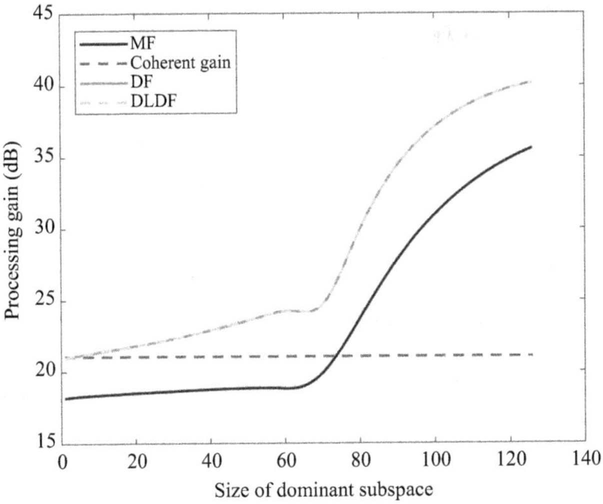

Figure 21.14 Processing gain with SWF symbols for the high CNR case

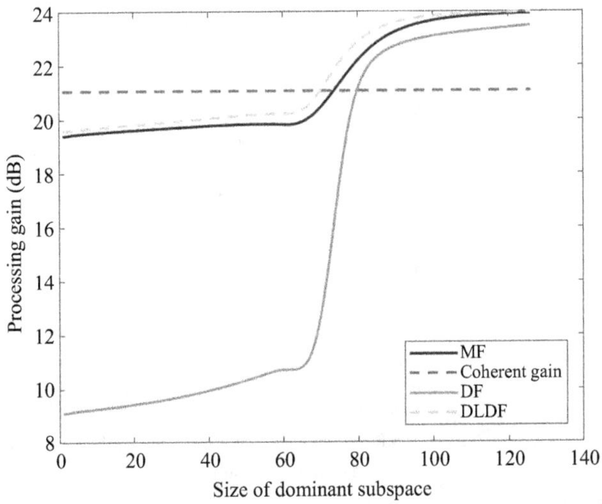

Figure 21.15 Processing gain with SWF symbols for the low CNR case

where for (21.59) we used $\mathbf{\Lambda}^2_{\mathrm{ND},m} \approx 0$ for $m \approx NM$. For $NM = 128$ and $m = 126$, the processing gain is determined to be

$$10 \log_{10} [\Delta_{\mathrm{SWF-MF}}(126)] \approx 35 \text{ (dB)}. \tag{21.60}$$

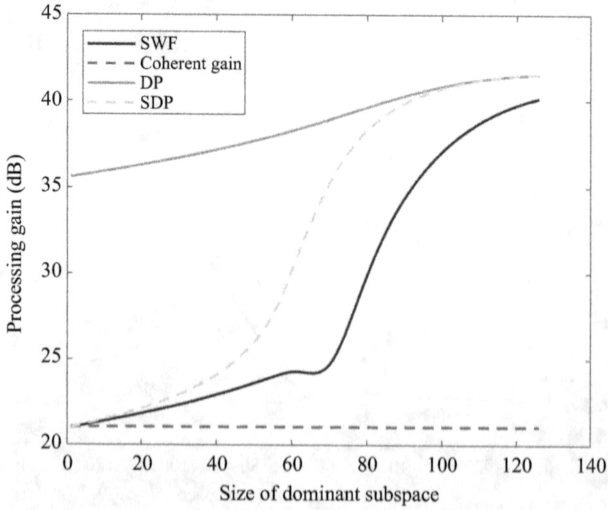

Figure 21.16 Processing gain comparison for the high CNR case

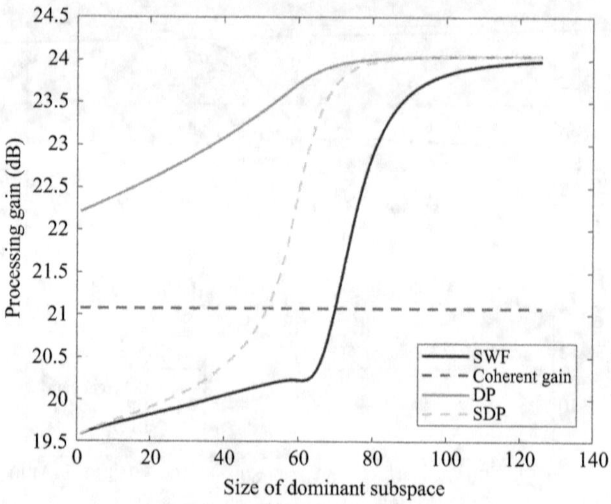

Figure 21.17 Processing gain comparison for the low CNR case

The impact of the partitioned shaped eigenvalue matrix is similarly shown by a loss of ≈ 2 dB in the peak processing gain value for DF and DLDF in the high CNR case when compared to the DP and SDP symbol designs. However, this impact is dependent on the radar waveform used and therefore will not be further explored here.

A comparison of the three symbol designs using the DLDF receive filter is shown in Figures 21.16 and 21.17, where it is observed that the processing gain for DP

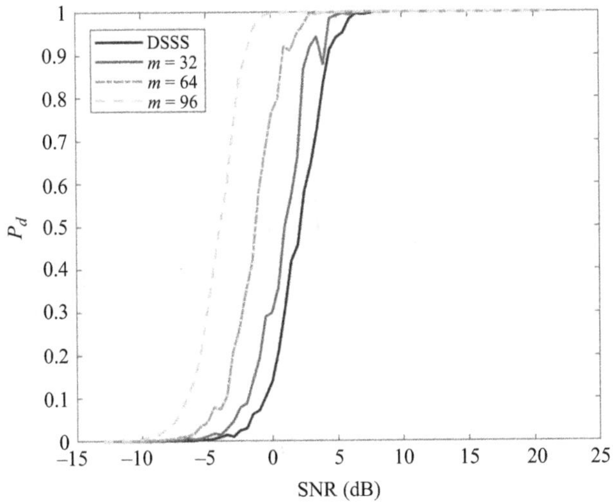

Figure 21.18 Probability of detection for DP and DSSS symbol designs

symbols is superior to that of SDP and SWF symbols. In addition, the processing gain for the DP design is relatively insensitive to the choice of the dominant subspace size for both the low and high CNR conditions. The processing gain analysis presented here provides another metric with which to compare symbol designs. Recall from Figures 21.6–21.9 that the spectral content of the SDP symbols are similar to the SWF symbols for low values of m and to the DP symbols for high values of m. However, Figures 21.16 and 21.17 show a marked difference in the processing gain offered by the three designs that belies the similarities observed in the spectral analysis.

21.4.3 Probability of detection and symbol error rate

Figures 21.18–21.20 show the probability of detection for different symbol designs as a function of input SNR. The symbols are oversampled by a factor of 8 to simulate the continuous nature of the clutter, with the CNR set to 30 dB. The CNR and SNR values are established at the receiver input. Due to the superior processing gain demonstrated in Section 21.4.2, the DLDF receive filter is used here. The probability of false alarm for the two-stage Neyman–Pearson detector is set to 10^{-5}.

Figure 21.18 shows the probability of detection for the DP and DSSS symbols, with the former dependent on the value of m. It is observed that the DP symbols are more likely to be detected for all values of m. In addition, the larger the size of the dominant subspace, the easier the symbols are detected. However, the dependence on the dominant subspace size is not strong since there is only a 3-dB SNR difference between the probability of detection curves of $m = 96$ and $m = 64$. The observed weak dependence on the size of the dominant subspace correlates well with the robust processing gain results from Figures 21.16 and 21.17. Note that an increased subspace size allocated to the communication symbols can be used to enhance the data rate.

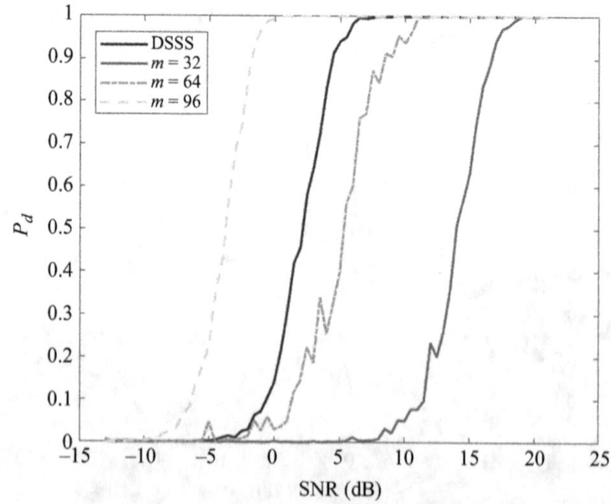

Figure 21.19 Probability of detection for SDP and DSSS symbol designs

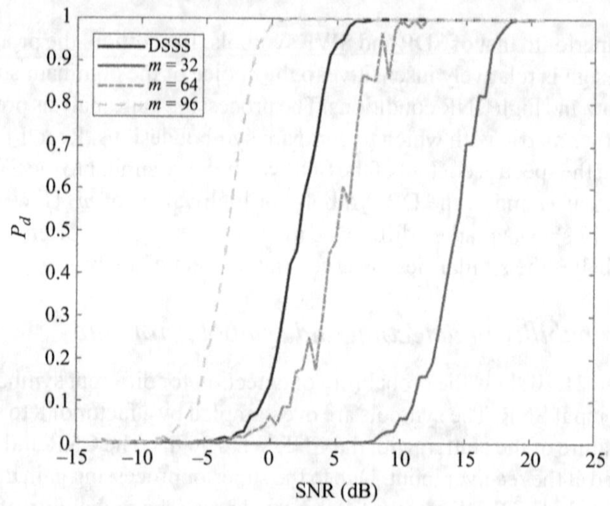

Figure 21.20 Probability of detection for SWF and DSSS symbols

Figure 21.19 shows the probability of detection of the SDP symbols, where it is observed that the probability of detection is degraded when the dominant subspace size is less than or equal to the time–bandwidth product of the radar waveform. Figure 21.20 shows the probability of detection for the SWF symbols, where it is observed that only the $m = 96$ case improves the probability of detection relative to the DSSS symbol design. In addition, the SWF symbols are increasingly difficult to detect as the dominant subspace size decreases.

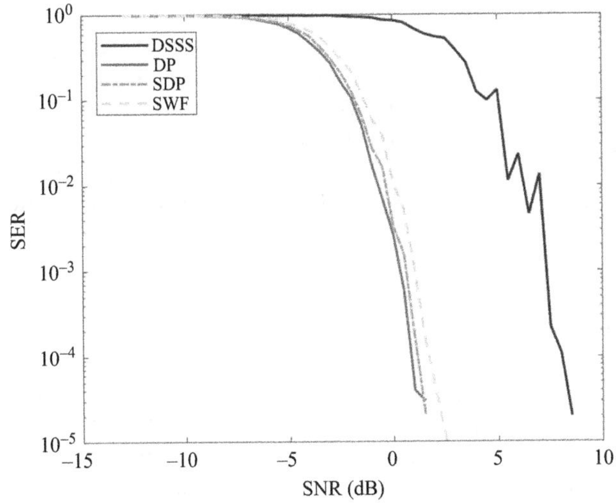

Figure 21.21 Symbol error rates of symbol designs

Figure 21.21 shows the symbol error rate (SER) for different symbol designs, where a symbol error occurs if the detected symbol is not equal to the embedded symbol or fails to pass the second threshold of the two-stage Neyman–Pearson detector. To simplify the comparison, the dominant subspace size is set to $m = 96$ for all designs. It is observed that all symbol designs provide similar SERs, as expected from the processing gain results of Section 21.4.2. However, the SNR improvement over the DSSS symbol design is $\sim$ 7 dB for an SER of 10^{-3}.

21.5 Conclusions

The concept of RCEC has been proposed, with the prospect of various different communication symbol and receive filter design strategies. Several of these symbol/filter designs were examined from the perspective of spectral containment, processing gain, probability of detection, and SER. Of the three proposed symbol designs, it was shown that the DP method of embedding communication symbols into clutter is the most effective from the perspectives of detectability, SER, and interference at the radar receiver. This symbol design offers significant improvements over a traditional DSSS design strategy. Finally, simulation results show that the DLDF provides the most robust performance for all symbol designs, independent of the incident clutter power.

21.6 Looking ahead

A number of technical questions remain for this manner of radar/communication spectrum sharing. First and foremost, the impact of larger alphabets with respect to the size of the nondominant subspace is yet to be examined, which would help to facilitate the selection of the dominant subspace size in an optimal manner. In addition, the work to date has only considered a single user. How the RCEC framework can be generalized to multiuser access remains to be seen. Finally, future work should examine the impact of embedded communication symbols on radar detection performance, which may also require modified symbol designs to compensate. In particular, target masking [56] may occur in cases of low signal-to-interference ratio (SIR) at the radar receiver.

References

[1] Cebrowski AK, Garstka JJ. Network-centric warfare: its origin and future. In: US Naval Institute Proceedings. vol. 124; 1998. p. 28–35.
[2] Strategic Technology Office and Defense Advanced Research Projects Agency. Shared Spectrum Access for Radar and Communications (SSPARC). DARPA-BAA-13-24, Feb. 2013.
[3] Griffiths H, Cohen L, Watts S, *et al.* Radar spectrum engineering and management: technical and regulatory issues. Proceedings of the IEEE. 2015 January;103(1):85–102.
[4] Blunt SD, Mokole EL. Overview of radar waveform diversity. IEEE Aerospace and Electronic Systems Magazine. 2016 November;31(11):2–42.
[5] Cantrell BH, Coleman JO, Trunk GV. Radar Communications. Naval Research Laboratory; 1991. Tech. Rep. 8515.
[6] Hughes PK, Choe JY. Overview of advanced multifunction RF system. In: IEEE International Conference on Phased Array Systems and Technology. Dana Point, CA; 2000. p. 21–24.
[7] Blunt SD, Cook MR, Stiles J. Embedding information into radar emissions via waveform implementation. In: International Waveform Diversity and Design Conference. Niagara Falls, Canada; 2010. p. 195–199.
[8] Sturm C, Wiesbeck W. Waveform design and signal processing aspects for fusion of wireless communications and radar sensing. Proceedings of the IEEE. 2011 July;99(7):1236–1259.
[9] Sahin C, Jakabosky J, McCormick PM, Metcalf JG, Blunt SD. A novel approach for embedding communication symbols into physical radar waveforms. In: IEEE Radar Conference. Seattle, WA; 2017. p. 1498–1503.
[10] Sahin C, Metcalf JG, Blunt SD. Filter design to address range sidelobe modulation in transmit-encoded radar-embedded communications. In: IEEE Radar Conference. Seattle, WA; 2017. p. 1509–1514.
[11] Hassanien A, Amin MG, Zhang YD, Ahmad F. Signaling strategies for dual-function radar communications: an overview. IEEE Aerospace and Electronic Systems Magazine. 2016 October;31(10):36–45.
[12] Hassanien A, Amin MG, Zhang YD, Ahmad F. Dual-function radar-communications: information embedding using sidelobe control and waveform

diversity. IEEE Transactions on Signal Processing. 2016 April;64(8):2168–2181.

[13] Hassanien A, Amin MG, Zhang YD, Ahmad F. A dual function radar-communications system using sidelobe control and waveform diversity. In: IEEE Radar Conference. Arlington, VA; 2015. p. 1260–1263.

[14] McCormick PM, Blunt SD, Metcalf JG. Simultaneous radar and communications emissions from a common aperture, Part I: theory. In: 2017 IEEE Radar Conference. Seattle, WA; 2017. p. 1685–1690.

[15] McCormick PM, Ravenscroft B, Blunt SD, Duly AJ, Metcalf JG. Simultaneous radar and communication emissions from a common aperture, Part II: experimentation. In: IEEE Radar Conference. Seattle, WA; 2017. p. 1697–1702.

[16] Chakravarthy V, Li X, Wu Z, *et al.* Novel overlay/underlay cognitive radio waveforms using SD-SMSE framework to enhance spectrum efficiency – Part I: theoretical framework and analysis in AWGN channel. IEEE Transactions on Communications. 2009 December;57(12):3794–3804.

[17] Chakravarthy V, Li X, Zhou R, Wu Z, Temple M. Novel overlay/underlay cognitive radio waveforms using SD-SMSE framework to enhance spectrum efficiency – Part II: analysis in fading channels. IEEE Transactions on Communications. 2010 June;58(6):1868–1876.

[18] Hessar F, Roy S. Spectrum sharing between a surveillance radar and secondary Wi-Fi networks. IEEE Transactions on Aerospace and Electronic Systems. 2016 June;52(3):1434–1448.

[19] Griffiths H, Darwazeh I, Inggs M. Waveform design for commensal radar. In: IEEE Radar Conference. Arlington, VA; 2015. p. 1456–1460.

[20] Jakabosky J, Ravenscroft B, Blunt SD, Martone A. Gapped spectrum shaping for tandem-hopped radar/communications cognitive sensing. In: IEEE Radar Conference. Philadelphia, PA; 2016.

[21] Ravenscroft B, McCormick PM, Blunt SD, Jakabosky J, Metcalf JG. Tandem-hopped OFDM communications in spectral gaps of FM noise radar. In: IEEE Radar Conference. Seattle, WA; 2017. p. 1262–1267.

[22] Stockman H. Communication by means of reflected power. Proceedings of the IRE. 1948 October;36(10):1196–1204.

[23] Brunfeldt DR, Ulaby FT. Active reflector for radar calibration. IEEE Transactions on Geoscience and Remote Sensing. 1984 March;GE-22(2):165–169.

[24] Freeman A, Shen Y, Werner CL. Polarimetric SAR calibration experiment using active radar calibrators. IEEE Transactions on Geoscience and Remote Sensing. 1990 March;28(2):224–240.

[25] Fujita M. Development of a retrodirective PARC for ALOS/PALSAR calibration. IEEE Transactions on Geoscience and Remote Sensing. 2003 October;41(10):2177–2186.

[26] Fujita M, Nakamura S. A polarization-rotating Van Atta array reflector and its application to polarimetric radar calibration. In: IEEE International Geoscience and Remote Sensing Symposium. vol. 3. Anchorage, AL; 2004. p. 1746–1749.

[27] Axline RM, Sloan GR, Spalding RE. Radar transponder apparatus and signal processing technique; U.S. Patent 5,486,830, January 23, 1996.

[28] Hounam D, Wagel KH. A technique for the identification and localization of SAR targets using encoding transponders. IEEE Transactions on Geoscience and Remote Sensing. 2001 January;39(1):3–7.

[29] Richardson DL, Stratmoen SA, Bendor GA, Lee HE, Decker MJ. Tag communication protocol and system; U.S. Patent 6,329,944, December 11, 2001.

[30] Axline RM, Sloan GR, Spalding RE. Transponder data processing methods and systems; U.S. Patent 6577266, June 10, 2003.

[31] Ormesher RC, Axline RM. Methods and system suppressing clutter in a gain-block, radar-responsive tag system; U.S. Patent 7,030,805, April 18, 2006.

[32] Kaliski B. Microwave tag identification systems. RCA Review. 1983 December;44(4):589–610.

[33] Koelle AR, Depp SW, Freyman RW. Short-range radio-telemetry for electronic identification, using modulated RF backscatter. Proceedings of the IEEE. 1975 August;63(8):1260–1261.

[34] Barott WC. Coherent backscatter communications using ambient transmitters and passive radar processing. In: National Wireless Research Collaboration Symposium. Idaho Falls, ID; 2014. p. 15–20.

[35] Zhang Y, Scott K, Barott WC. Passive backscatter tags as transponders for unmanned aircraft systems. In: IEEE SoutheastCon. Ft. Lauderdale, FL; 2015.

[36] Blunt SD, Yantham P. Waveform design for radar-embedded communications. In: International Waveform Diversity and Design Conference. Pisa, Italy; 2007. p. 214–218.

[37] Blunt SD, Stiles J, Allen C, Deavours D, Perrins E. Diversity aspects of radar-embedded communications. In: International Conference on Electromagnetics in Advanced Applications. Torino, Italy; 2007. p. 439–442.

[38] Blunt SD, Biggs CR. Practical considerations for intra-pulse radar-embedded communications. In: International Waveform Diversity and Design Conference. Kissimmee, FL; 2009. p. 244–248.

[39] Blunt SD, Metcalf JG. Estimating temporal multipath via spatial selectivity: building environmental knowledge into waveform design for radar-embedded communications. In: International Conference on Electromagnetics in Advanced Applications. Torino, Italy; 2009. p. 513–516.

[40] Blunt SD, Metcalf JG. Using time reversal of multipath for intra-pulse radar-embedded communications. In: International Waveform Diversity and Design Conference. Niagara Falls, Canada; 2010. p. 155–158.

[41] Blunt SD, Yatham P, Stiles J. Intrapulse radar-embedded communications. IEEE Transactions on Aerospace and Electronic Systems. 2010 July;46(3):1185–1200.

[42] Blunt SD, Metcalf JG, Biggs CR, Perrins E. Performance characteristics and metrics for intra-pulse radar-embedded communications. Journal of Selected Areas of Communications. 2011 December;29(10):2057–2066.

[43] Metcalf JG. Detection Strategies and Intercept Metrics for Intra-Pulse Radar-embedded Communications. University of Kansas. Lawrence; 2011. [Online]. Available: http://hdl.handle.net/1808/9813.
[44] Ciuonzo D, Maio AD, Foglia G, Piezzo M. Pareto-theory for enabling covert intrapulse radar-embedded communications. In: IEEE Radar Conference. Arlington, VA; 2015. p. 292–297.
[45] Metcalf JG, Sahin C, Blunt SD, Rangaswamy M. Analysis of symbol-design strategies for intrapulse radar-embedded communications. IEEE Transactions on Aerospace and Electronic Systems. 2015 October;51(4):2914–2931.
[46] Ciuonzo D, Maio AD, Foglia G, Piezzo M. Intrapulse radar-embedded communications via multiobjective optimization. IEEE Transactions on Aerospace and Electronic Systems. 2015 October;51(4):2960–2974.
[47] De Graaf J, Faust H, Alatishe J, Talapatra S. Generation of spectrally confined transmitted radar waveforms: experimental results. In: IEEE Radar Conference. Verona, NY; 2006.
[48] Scholtz R. The origins of spread-spectrum communications. IEEE Transactions on Communications. 1982 May;30(5):822–854.
[49] Verdu S. Multiuser Detection. New York, NY: Cambridge University Press; 2003.
[50] Proakis JG, Salehi M. Digital Communications. New York, NY: McGraw-Hill; 2007.
[51] Schmidt R. Multiple emitter location and signal parameter estimation. IEEE Transactions on Antennas and Propagation. 1986 March;34(3):276–280.
[52] Blunt SD, Chan T, Gerlach K. Robust DOA estimation: the reiterative super-resolution (RISR) algorithm. IEEE Transactions on Aerospace and Electronic Systems. 2011 January;47(1):332–346.
[53] Settle TF, Picciolo ML, Goldstein JS. Environmentally-shaped colored spreading sequences for robust wireless communications. In: International Waveform Diversity and Design Conference. Niagara Falls, Canada; 2010. p. 40–44.
[54] Kay SM. Fundamentals of Statistical Signal Processing: Detection Theory. Upper Saddle River, NJ: Prentice Hall PTR; 1998.
[55] Sijbers J, den Dekker AJ, Raman E, Van Dyck D. Parameter estimation from magnitude MR images. International Journal of Imaging Systems and Technology. 1999;10:109–114.
[56] Sanders FH, Sole RL, Bedford BL, Franc D, Pawlowitz T. Effects of RF interference on radar receivers. NTIA; 2006. TR-06-444.

Chapter 22

Dual-function radar–communications using sidelobe control

Aboulnasr Hassanien[1], Braham Himed[2], and Moeness G. Amin[3]

22.1 Introduction

Dual system functionality, in which radar is the primary function, requires strategies for radar-embedded communication signals. Depending upon the quality of service, error tolerance, and accuracy in signal estimation and target localization, dual functionality may require flexibility in beampattern synthesis and amenability to incorporate multiple transmit waveforms. This chapter provides an overview of recent advances in radar-embedded communication signals. Several information-embedding techniques have recently been successful in establishing dual-function systems that simultaneously perform both radar and communication functions. Similar to single-function communication platforms, dual function systems should provide secure communications to protect user signals from being intercepted by other users or unintended receivers. Information embedding into the emission of multiple-input (MIMO) multiple-output radar provides a means for broadcast communications to be secondary to the primary radar function of the dual system.

Recently, the problem of radio frequency (RF) spectrum congestion caused by the explosive growth and accelerating demand of the commercial wireless communications industry has been the focus of intensive research [1–3]. During the last decade, several new concepts with the potential to contribute toward significant improvements in the efficiency of RF spectrum utilization and protection of passive sensing services have been introduced [4–12]. The concept of coexistence between radar and communications offers a promising solution to one of the most pressing problems in the area of spectral congestion and dynamic frequency allocation by providing uncontested shared bandwidth between radar and communications. This situation led to the development of several coexistence solutions based on the simultaneous operations

[1]Department of Electrical Engineering, Wright State University, USA
[2]RF Technology Branch, Air Force Research Lab (AFRL/RYMD), USA
[3]Electrical and Computer Engineering Department, Center for Advanced Communications, Villanova University, USA

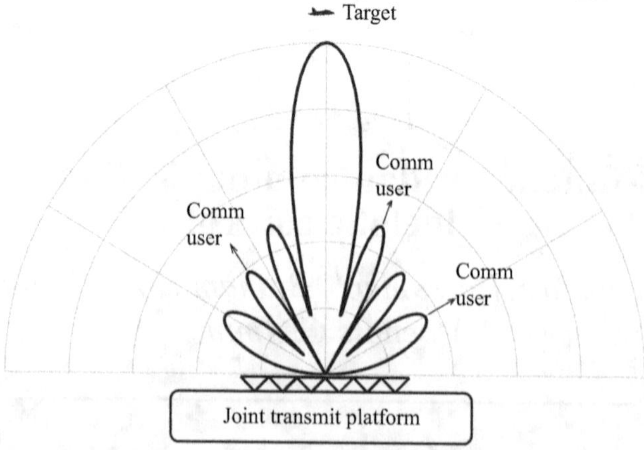

Figure 22.1 Joint transmit platform for a dual-function radar–communication system

of radar target illumination and wireless services within the same band [13–16]. The term coexistence means that radar and communications are two independent systems that are allowed to share the same band and use it either simultaneously or not. This cohabitation arrangement requires some form of bandwidth allocation/scheduling.

A recent approach to coexistence is based on the premise of radar and communication functions that can be deployed from the same platform using the same system resources (see [17] and references therein). This arrangement requires new signaling strategies that permit the same signal and antenna array to be used for both radar and communication operations on transmit, receive, or for the overall system. For instance, joint operation could be achieved with communications being given the higher priority leading to a dual-function communication-radar system whereby the communication operation dictates the joint usage of system resources and available bandwidth, with the radar operation treated as secondary. The commensal waveform discussed at the end of Chapter 23 is one such example.

If priorities are reversed and the radar operation is treated as the primary function, we refer to the joint system as dual-function–radar communications (DFRCs), which is the focus of this chapter. Figure 22.1 provides an illustration of a DFRC system in which the radar operation takes place within the main beam of the transmit radiation pattern, while the communication operation makes use of the otherwise wasted energy in the sidelobe region. Depending upon the quality of service, error tolerance, and accuracy in signal estimation and target localization needed, these strategies may require additional resources and utilize advances in waveform design and multisensor transmit/receive configurations.

During the last few years, several techniques have been developed for embedding communication symbols into radar emissions, clearly demonstrating that information transmission can be achieved as a secondary attribute to the primary radar operation of a DFRC system [18–29]. According to [18], a waveform diversity-based method can be devised that treats waveforms as communication symbols and selects which waveform to be transmitted during a certain radar pulse based on an input stream of communication data. Significant expansions to this approach and others are discussed in greater detail in Chapter 23. Other approaches to DFRC include a sidelobe amplitude modulation (AM) method [19], an amplitude shift keying (ASK)-based method [20,22–24], and a phase-shift keying (PSK)-based approach [24,25,28,29]. The AM signaling scheme of [19] employs time-modulated arrays to modulate the sidelobe levels (SLLs) of the instantaneous pattern toward the communication direction. The ASK signaling scheme embeds information using bilevel sidelobe control and enables the embedding of binary information into each waveform. The number of bits that can be embedded during each radar pulse is proportional to the number of orthogonal waveforms being used. The essence of the PSK-based signaling scheme is to map the binary data into a phase symbol that belongs to an appropriately sized dictionary. During each radar pulse a number of PSK symbols can be embedded into the radar emission based on the number of orthogonal waveforms being simultaneously transmitted.

This chapter considers the problem of embedding information into the radar emission using a common transmit platform. It provides an overview of different approaches to DFRC and the corresponding signaling strategies. The radar waveforms and transmit beamforming weight vector that satisfy the specifications and requirements mandated by the radar are discussed. The chapter also highlights the disturbance caused by the secondary communication function to the primary radar function. Specifically, the effect that information embedding has upon transmit power efficiency is examined. It is shown that, in order to mitigate possible disturbances to the radar functionality, the information embedding strategy can be designed such that the fluctuation in peak-to-average power ratio (PAPR) of the information-bearing signal is minimized. The performance of the discussed techniques is validated using simulation examples.

22.2 System configuration and signal model

This section presents the dual-function system configuration and develops the associated signal models. We assume that the radar operation is the primary function of the dual-function system, and therefore all available resources, including total transmit power and the entire bandwidth, are dedicated to the radar function. A secondary communication capability is embedded into the radar emission as will be explained later in the chapter.

Consider a dual-function system equipped with a joint transmit array comprising M collocated antennas. The system also contains a radar receive array consisting of N collocated antennas and one (or more) single element communication receiver(s). The joint transmit and radar receive arrays are arranged in a linear configuration and, without loss of generality, are assumed to be in the vicinity of each other such that a target located in the far-field would appear at the same spatial angle with respect to both arrays (i.e. pseudo-monostatic). The communication receivers are assumed to be widely separated from each other and from the joint transmit array. For t the fast-time variable, let $\phi(t)$ be a waveform that is optimally designed to satisfy the requirements of radar operation, utilizing all available resources of the dual-function system. It is therefore assumed that $\phi(t)$ occupies the entire bandwidth available to the system and satisfies the range and Doppler resolution requirements mandated for the given radar mode (e.g. surveillance, imaging, tracking, etc.). It is worth noting that the problem of practical waveform design with desirable correlation properties has been thoroughly investigated in the literature (see [30–35], and references therein).

Transmit beamforming is employed by the joint transmit array to focus the total available transmit power toward a specific spatial direction $\theta_{\rm t}$, and thereby discern the presence of targets of interest. We assume that the transmit beamforming weight vector $\mathbf{w}_0$ is optimally designed such that a desired transmit beampattern is achieved for radar operation. Therefore, the $M \times 1$ vector of baseband signals at the input to the joint transmit array is given by

$$\mathbf{s}(t) = \mathbf{w}_0^* \phi(t), \tag{22.1}$$

where $(\cdot)^*$ denotes complex conjugation. The radiated signal observed at a hypothetical target located in spatial direction $\theta_{\rm t}$ is proportional to the inner product $\mathbf{a}^T(\theta_{\rm t})\,\mathbf{s}(t)$, where $\mathbf{a}(\theta_{\rm t})$ is the $M \times 1$ steering vector of the joint transmit array for direction $\theta_{\rm t}$ and $(\cdot)^T$ is the transpose operation. The $N \times 1$ vector of baseband signals at the output of the radar receiver is given by

$$\mathbf{x}(t;\tau) = \gamma_{\rm t}(\tau)\left(\mathbf{w}_0^H \mathbf{a}(\theta_{\rm t})\right)\mathbf{b}(\theta_{\rm t})\,\phi(t) + \mathbf{c}(t;\tau) + \mathbf{z}(t;\tau), \tag{22.2}$$

where τ is the slow-time index (i.e. pulse number), $\gamma_{\rm t}(\tau)$ is the reflection coefficient of the target that is assumed here to obey a Swerling II target model—i.e. the reflectivity remains constant over frequency and during the entire radar pulse, but changes from pulse to pulse—and $\mathbf{b}(\theta_{\rm t})$ is the $N \times 1$ steering vector of the radar receive array in direction $\theta_{\rm t}$. In (22.2), $\mathbf{c}(t;\tau)$ is the $N \times 1$ vector that comprises all interfering signals impinging on the radar receive array from the sidelobe region as well as from targets located at ranges different than the desired target range (i.e. sidelobe ranges), and $\mathbf{z}(t;\tau)$ is the $N \times 1$ vector of additive white Gaussian noise (AWGN) with zero mean and covariance matrix $\sigma_z^2 \mathbf{I}_N$, where $\mathbf{I}_N$ is the $N \times N$ identity matrix and $(\cdot)^H$ is the Hermitian operation.

Assume that the transmit coherent processing gain mandated by the radar requirement is guaranteed, i.e. the transmit beamforming weight vector $\mathbf{w}_0$ is designed such that the condition $\mathbf{w}_0^H \mathbf{a}(\theta_{\rm t}) = 1$ is satisfied. Consequently, matched filtering of the signal vector $\mathbf{x}(t;\tau)$ with the transmitted radar waveform $\phi(t)$ yields

$$\mathbf{y}(\tau) = \int_{T_{\rm p}} \mathbf{x}(t;\tau)\,\phi^*(t)\,dt = M\gamma_{\rm t}(\tau)\mathbf{b}(\theta_{\rm t}) + \mathbf{x}_{\rm i+z}(\tau), \tag{22.3}$$

where $T_{\rm p}$ is the radar pulse width and $\mathbf{x}_{i+z}(\tau)$ is the additive $N \times 1$ vector of interference-plus-noise at the output of the matched filter. Note that in (22.3) it is assumed that the ambiguity function of the radar waveform is sampled at its peak, thus relying on the ambiguity function property $\int_{T_p} \phi(t)\,\phi^*(t)\,dt = \chi(0,0) = 1$, where [36]

$$\chi(\xi,\nu) \triangleq \int_{T_p} \phi(t)\,\phi^*(t-\xi)\,e^{j2\pi\nu t}\,dt \tag{22.4}$$

for ξ and ν are the time delay and frequency shift, respectively.

The model in (22.3) represents the signal at the radar receiver when no communication component is embedded. Therefore, postprocessing of (22.3) yields the best radar performance for a given $\mathbf{w}_0$ and $\phi(t)$. The secondary communication function of the dual-function system can be achieved by slightly modifying $\mathbf{w}_0$ and/or $\phi(t)$ without disturbing the radar operation.

22.3 Transmit beampattern synthesis for dual-function systems

We now consider different techniques for beampattern synthesis in dual-function systems, including design formulations for transmit beamforming to achieve a quiescent main lobe response while varying the SLLs in certain directions. Communication is established by varying the SLL in the direction(s) of intended receivers, and radar performance is maintained by preserving the main lobe signal during the coherent processing interval (CPI). Here it is assumed that the desired SLL takes one of K discrete amplitude values (i.e. symbols) that are denoted as $\mathbb{D}_{\text{AM}} = \{\Delta_1, \ldots, \Delta_K\}$.

22.3.1 Time-modulated array-based synthesis

One possible approach for beampattern synthesis with sidelobe control involves the use of time-modulated arrays. In this context, the beampattern can be synthesized using either sparse time-modulated arrays (STMA) or phase-only synthesis time-modulated arrays (POSTMA). The time-modulated array-based approach optimizes the average array factor (AF) within the radar integration time, which is divided into P time intervals. The average AF for a uniform linear array (ULA) is defined as [19]

$$\text{average AF} = \frac{1}{P}\sum_{p=1}^{P}\sum_{m=0}^{M-1} w_{m,p}\exp\left\{-j\frac{2\pi d}{\lambda}m\sin\theta_0\right\}, \tag{22.5}$$

where $w_{m,p}$ is the complex weight associated with the mth transmit antenna during the pth time interval, λ is the wavelength, d is the interelement spacing, and θ_0 is the spatial direction of the radar main beam. The complex weight is defined as $w_{m,p} = \zeta_{m,p}\,e^{j\varpi_{m,p}}$, where $\zeta_{m,p}$ is the magnitude and $\varpi_{m,p}$ is the phase. The problem of designing these weight coefficients can be solved using one of the following two methods [19].

1. STMA:
   The essence of the STMA method is based on switching the antennas on and off, which corresponds to assigning $\varpi_{m,p} = 0$ and $\zeta_{m,p} \in \{0, 1\}$ to the phase and

magnitude of $w_{m,p}$, respectively. However, due to a reduction in the number of degrees-of-freedom, STMA is capable of achieving only a relatively small number of distinct SLLs that, in turn, reduces the number of communication symbols that can be embedded.

2. POSTMA:
The essence of the POSTMA method is based on fixing the magnitude to $\zeta_{m,p} = 1$ and optimizing the average AF over the phase parameters $\varpi_{m,p} \in [0,\ 2\pi]$. The AF optimization is performed over a total number of *MP* variables using a genetic algorithm.

The main advantage of the STMA and POSTMA techniques is the simple implementation via on/off switching and delay–sum operation, respectively. However, the transmit beampattern synthesis using STMA and POSTMA involves performing highly nonlinear optimization that can be difficult to achieve in a computationally efficient manner. Moreover, while the average AF yields a quiescent main lobe response when averaged over the CPI, the instantaneous AF exhibits large changes from one time interval to another that adversely impacts radar performance.

22.3.2 Convex optimization based design

A computationally efficient method using convex optimization for transmit beamforming design in dual-function systems was developed in [20,22]. Here K distinct SLLs denoted as $\mathbb{D}_{\text{AM}} = \{\Delta_1, \ldots, \Delta_K\}$ can be achieved through the realization of K different transmit beamforming weight vectors. If the K weight vectors are designed such that their individual beampatterns are similar, then the radar function can be achieved using any of them.

Denoting $\boldsymbol{\Theta}$ as the spatial sector of interest for radar operation, one way to design the transmit weight vectors is to minimize the difference between the desired and actual transmit power radiation patterns while enforcing constraints that the sidelobes be bounded by certain predefined levels. This weight vector design criterion can be formulated as the optimization problem [22]

$$\min_{\mathbf{u}_k} \max_{\theta} = \left| |G_{\text{d}}(\theta)| - \left|\mathbf{u}_k^H \mathbf{a}(\theta)\right| \right|, \quad \theta \in \boldsymbol{\Theta} \tag{22.6}$$

$$\text{subject to} = \left|\mathbf{u}_k^H \mathbf{a}(\theta)\right| \leq \varepsilon, \quad \theta \in \bar{\boldsymbol{\Theta}}, \tag{22.7}$$

$$= \mathbf{u}_k^H \mathbf{a}(\theta_{\text{c}}) = \Delta_k, \tag{22.8}$$

where $\mathbf{u}_k$ is the kth $M \times 1$ transmit weight vector, $G_{\text{d}}(\theta)$ is the desired transmit beampattern, $\bar{\boldsymbol{\Theta}}$ is a continuum of spatial angles in the sidelobe region (thus mutually exclusive from the radar spatial sector $\boldsymbol{\Theta}$), the positive number ε is used to upper-bound the SLLs, and θ_{c} is the communication direction.

The objective function in (22.6) is nonconvex, and therefore the optimization problem (22.6)–(22.8) is difficult to solve in a computationally efficient manner.

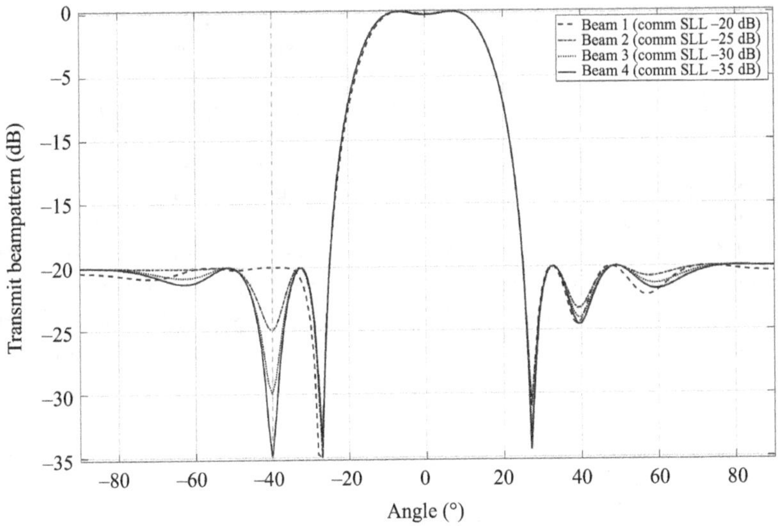

Figure 22.2 Transmit beampattern versus spatial angle for convex optimized dual-function radar–communication system with $M = 10$ antenna elements and $K = 4$ distinct SLLs in the communication direction $\theta_c = -40°$

However, if we set the desired radiation pattern as $G_{\rm d}(\theta) = e^{j\varphi(\theta)}$, then the problem can be reformulated as the convex optimization problem [22]

$$\min_{\mathbf{u}_k} \max_{\theta_i} = \left| e^{j\varphi(\theta_i)} - \mathbf{u}_k^H \mathbf{a}(\theta_i) \right|, \; \theta_i \in \boldsymbol{\Theta}, \; i = 1, \ldots, I \tag{22.9}$$

$$\text{subject to} = \left| \mathbf{u}_k^H \mathbf{a}(\theta_p) \right| \leq \varepsilon, \quad \theta_p \in \bar{\boldsymbol{\Theta}}, \; p = 1, \ldots, P, \tag{22.10}$$

$$= \mathbf{u}_k^H \mathbf{a}(\theta_c) = \Delta_k, \tag{22.11}$$

where θ_i for $i = 1, \ldots, I$ and θ_p for $p = 1, \ldots, P$ are discrete grids of angles used to approximate $\boldsymbol{\Theta}$ and $\bar{\boldsymbol{\Theta}}$, respectively, and $\varphi(\theta)$ is a phase profile selected by the operator. The optimization problem in (22.9)–(22.11) is convex and can be solved in a computationally efficient manner using the interior point method [37].

As an illustrative example, consider a transmit linear array of $M = 10$ elements with half-wavelength spacing. The primary radar operation takes place within the spatial sector $\boldsymbol{\Theta} = [-10° + 10°]$. We assume a single communication receiver is located in spatial direction $\theta_c = -40°$. Four transmit weight vectors are designed by solving (22.9)–(22.11). The highest SLL in the entire sidelobe region is upper bounded by setting $\varepsilon = 0.1$, which establishes the sidelobe attenuation to at least 20 dB. Four distinct SLLs in direction θ_c are achieved by setting $\Delta_1^2 = -20$, $\Delta_2^2 = -25$, $\Delta_3^2 = -30$, and $\Delta_4^2 = -35$ dB. Figure 22.2 shows the synthesized transmit beampatterns versus spatial angle for these four transmit weight vectors. It is observed that the radar main beam is virtually identical for all four beampatterns, while the SLLs in the communication direction θ_c are distinct, which subsequently enables information to be embedded by treating the distinct SLLs as communication symbols.

22.3.3 Beampattern synthesis with minimum disturbance to the radar

The beampattern synthesis technique described in Section 22.3.2 considered the design of K transmit weight vectors independently without relating them to a principal weight vector $\mathbf{w}_0$ predesigned for optimum radar operation. The problem formulation in (22.6)–(22.8) achieves the desired SLL modulation in the communication direction while minimizing variation within the main lobe region of the beampattern. However, there is no guarantee that the SLLs in directions other than those specified for radar and communication operation will remain fixed, and such variation can lead to a modulation of the clutter and/or interference signals that can degrade radar performance.

One way to address the problem of unintended SLL variation and thereby mitigate the effect of clutter modulation is to design the $\mathbf{u}_k$ weight vectors such that they possess the same pattern as $\mathbf{w}_0$ while enforcing constraints to enable SLL variation in the communication direction alone. As such, the desired beampattern in (22.6) can be replaced by the beampattern associated with $\mathbf{w}_0$, resulting in the optimization problem [22]

$$\min_{\mathbf{u}_k} \max_{\theta} = \left|\mathbf{w}_0^H \mathbf{a}(\theta) - \mathbf{u}_k^H \mathbf{a}(\theta)\right|, \quad \forall\theta \in [-\pi, \pi] \tag{22.12}$$

$$\text{subject to} = \mathbf{u}_k^H \mathbf{a}(\theta_c) = \Delta_k \, . \tag{22.13}$$

Using the fact that $\left|\mathbf{w}_0^H \mathbf{a}(\theta) - \mathbf{u}_k^H \mathbf{a}(\theta)\right| = \left|(\mathbf{w}_0 - \mathbf{u}_k)^H \mathbf{a}(\theta)\right| \le M\left|\mathbf{w}_0 - \mathbf{u}_k\right|$, this problem can be reformulated as

$$\min_{\mathbf{u}_k} \left\|\mathbf{w}_0 - \mathbf{u}_k\right\| \quad \text{subject to } \mathbf{u}_k^H \mathbf{a}(\theta_c) = \Delta_k, \quad k = 1, \ldots, K, \tag{22.14}$$

which is convex and thus can be efficiently solved using the interior point method [37].

Consider, for example, the case of a phased array radar with $M = 10$ elements having half-wavelength spacing. The predesigned transmit weight vector mandated for radar operation is $\mathbf{w}_0 = \mathbf{a}(\theta = 0)$. To embed a secondary communication function, we design four weight vectors by solving (22.14) for $\Delta_1^2 = -20$, $\Delta_2^2 = -25$, $\Delta_3^2 = -30$, and $\Delta_4^2 = -35$ dB. The intended communication direction is $\theta_c = -30°$.

Figure 22.3 shows the synthesized transmit beampatterns versus spatial angle for the principal weight vector $\mathbf{w}_0$ and the four designed weight vectors $\mathbf{u}_k$, $k = 1, \ldots, 4$. It is noted that the main lobe for all beampatterns remains essentially identical, while the SLLs in direction θ_c are distinct. It is also observed that the sidelobes in other directions remain rather similar. As a result, the SLL modulation of clutter and/or interference should have little effect on radar performance.

22.3.4 Beampattern synthesis with phase control

Methods that embed information into a spatial sidelobe of the radar emission via the modulation of phase have been reported in a number of papers [21,25, 28,29]. By controlling phase while keeping magnitude invariant, the problem of

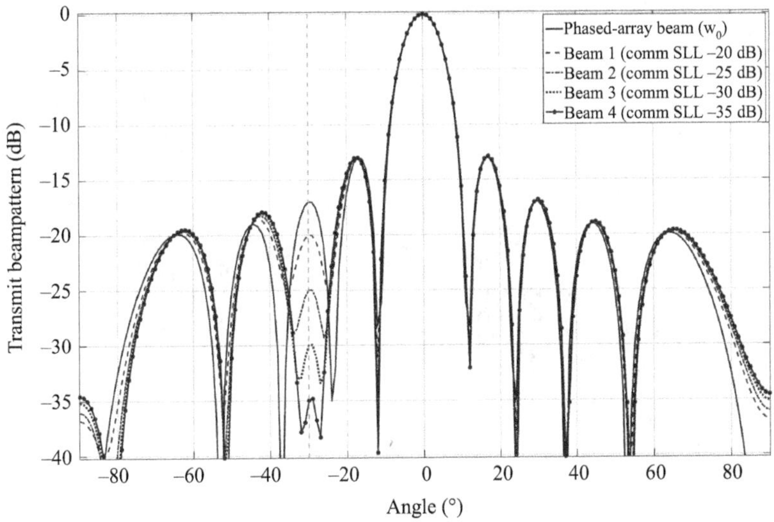

Figure 22.3 Transmit beampattern versus spatial angle for a dual-function radar–communication system with an $M = 10$ element ULA and $K = 4$ distinct SLLs in the communication direction $\theta_c = -30°$

clutter/interference modulation can be avoided in the radar receiver. The notion of beampattern synthesis with phase control again involves the design of K transmit weight vectors denoted as $\mathbf{u}_k$, $k = 1, \ldots, K$, albeit with the communication symbols now defined according to the phase dictionary $\mathbb{D}_{\mathrm{PM}} = \{\Omega_1, \ldots, \Omega_K\}$, the kth value of which is given by

$$\Omega_k = \angle(\mathbf{u}_k^H \mathbf{a}(\theta_c)), \tag{22.15}$$

for $\angle(\cdot)$ the angle of a complex scalar. This formulation permits the construction of PSK constellations that can be emitted in desired directions that may be in sidelobe or main lobe regions without changing the magnitude of the beampattern. Therefore, PSK symbols can be embedded in a manner that is transparent to the primary radar function.

One method to design the $\mathbf{u}_k$ weight vectors such that they satisfy (22.15) while preserving the same pattern as that of the principal weight vector $\mathbf{w}_0$ is to minimize the norm of their difference subject to the particular phase constraint. This condition translates into the convex optimization problem

$$\min_{\mathbf{u}_k} \left\| \mathbf{w}_0 - \mathbf{u}_k \right\| \quad \text{subject to} \;\; \mathbf{u}_k^H \mathbf{a}(\theta_c) = G_0 e^{-j\Omega_k}, \quad k = 1, \ldots, K, \tag{22.16}$$

where $G_0 = |\mathbf{w}_0^H \mathbf{a}(\theta_c)|$ is the magnitude of the transmit gain in direction θ_c. Another advantage of this formulation is that the desired phase rotations (22.15) are satisfied with equality.

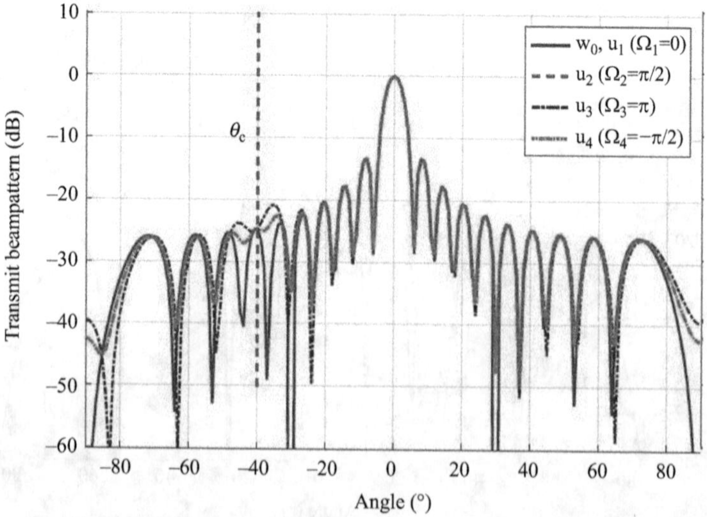

Figure 22.4 Transmit beampattern versus spatial angle for a dual-function radar–communication system with an $M = 20$ element ULA and phase-controlled sidelobes in the communication direction $\theta_c = -40°$

However, it should be noted that the solution yields weight vectors with beampatterns that are very similar, yet not identical, to that of $\mathbf{w}_0$, as illustrated in Figure 22.4 for a phased array with $M = 20$ elements and $K = 4$ weight vectors obtained by solving (22.16). In particular, it is observed that the phase changes in the communication direction of $\theta_c = -30°$ introduce some modest perturbations to the surrounding sidelobes.

Another approach to transmit beampattern synthesis with phase control is based on the beampattern invariance property. This property states that for a ULA and an arbitrary $M \times 1$ weight vector $\mathbf{w} = [w_1, \ldots, w_M]^T$, there exist 2^{M-1} weight vectors that yield beampatterns of the same magnitude as that of $\mathbf{w}$ [38]. The 2^{M-1} weight vectors can be obtained by using the entries of vector $\mathbf{w}$ as the coefficients of a polynomial of order $M - 1$, which is defined as $f(z) = w_1 + w_2 z + \ldots + w_M z^{M-1}$ and has roots r_i, $i = 1, \ldots, M - 1$. It is worth noting that the square of the beampattern magnitude can be obtained from $f(z)$ as

$$|G(\theta)|^2 = f(z)f(1/z^*)\big|_{z=e^{-j(2\pi d/\lambda)\sin\theta}}. \tag{22.17}$$

From inspection of (22.17), it can easily be concluded that reflecting each root about the unit circle does not change the magnitude of the beampattern. Therefore, by taking all possible combinations of roots r_i (or $1/r_i^*$) for $i = 1, \ldots, M - 1$ into account, at most 2^{M-1} different polynomials of the same order can be constructed (for a detailed proof, the reader is referred to [38,39]).

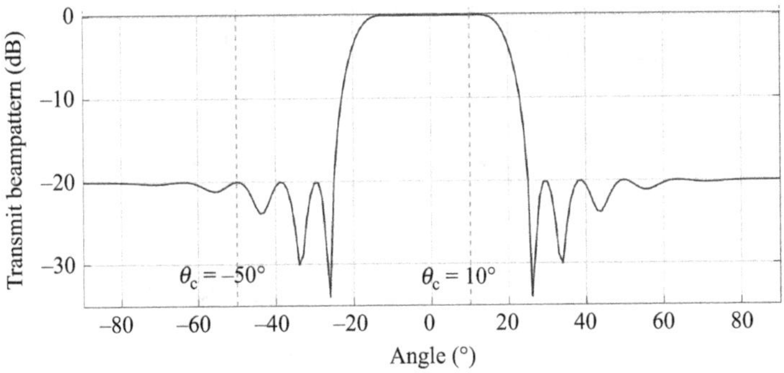

Figure 22.5 Transmit beampattern versus spatial angle for a dual-function radar–communication system based on transmit pattern invariance with $M = 16$ (thus 2^{15} weight vectors yield the same beampattern)

By using this beampattern invariance property, the principal weight vector $\mathbf{w}_0$ can be used to generate a population of 2^{M-1} weight vectors denoted as $\mathbf{W} = \{\mathbf{w}_1, \ldots, \mathbf{w}_{2^{M-1}}\}$ that have the same dimensionality and the same beampattern as that of $\mathbf{w}$. However, the phase of the transmit gain associated with each of these alternative vectors is not necessarily the same as that of the principal weight vector. Therefore, a subset of K vectors $\mathbf{u}_k$, $k = 1, \ldots, K$ (with $K \ll 2^{M-1}$) can be selected from this set of 2^{M-1} vectors by solving

$$\min_{\mathbf{u}_k \in \mathbf{W}} \left|\angle(\mathbf{u}_k^H \mathbf{a}(\theta_c) - \Omega_k\right|, \qquad k = 1, \ldots, K. \tag{22.18}$$

To illustrate the efficacy of this approach, consider an $M = 16$ element ULA with half-wavelength spacing. The principal weight vector $\mathbf{w}_0$ is obtained by solving the optimization problem

$$\min_{\mathbf{w}_0} \max_{\theta} = \left|e^{-j\varphi(\theta)} - \mathbf{w}_0^H \mathbf{a}(\theta)\right|, \; \theta \in [-10°, 10°] \tag{22.19}$$

$$\text{subject to} = \left|\mathbf{w}_0^H \mathbf{a}(\theta)\right| \leq 0.1, \quad \theta \in [-90°, -25°] \cup [25°, 90°], \tag{22.20}$$

where $\varphi(\theta) = 5\pi \sin(\theta)$ is arbitrarily chosen. The corresponding transmit beampattern is shown in Figure 22.5, where it is observed that the transmit gain within the main beam is unity, while the sidelobes are all below −20 dB. Using the beampattern invariance property, the 16×1 vector $\mathbf{w}_0$ enables the generation of 2^{15} other weight vectors that possess the same beampattern. Specifically, Figure 22.6(a) shows the magnitude and phase of the 2^{15} weight vectors evaluated in the communication direction of $\theta_c = -50°$. It is observed that the magnitude has a constant value of 0.1, as dictated by the beampattern in this direction, while the resulting phase in direction $\theta_c = -50°$ inscribes a full circle and is uniformly distributed on $[-\pi, \pi]$. From the 2^{15} possibilities, a set of $K = 4$ weight vectors are selected by solving (22.18) to obtain the phase dictionary $\mathbb{D}_{PM} = \left\{\Omega_1 = 0, \Omega_2 = (\pi/2), \Omega_3 = \pi, \Omega_4 = (-\pi/2)\right\}$.

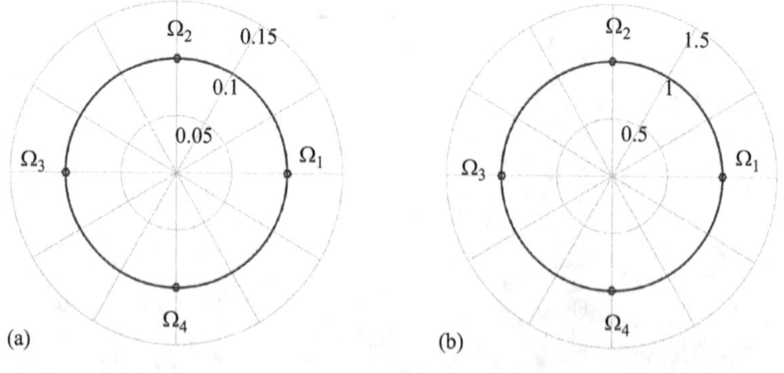

Figure 22.6 (a) Magnitude versus phase for 2^{15} *weight vectors (inscribed circle) and* $K = 4$ *chosen vectors (small dots) for communication direction* $\theta_c = -50°$*; (b) magnitude versus phase for* 2^{15} *weight vectors (inscribed circle) and* $K = 4$ *chosen vectors (small dots) for communication direction* $\theta_c = +10°$

In contrast, Figure 22.6(b) shows the magnitude and phase of the 2^{15} weight vectors evaluated instead for a communication direction of $\theta_c = +10°$. The key difference here is that the magnitude, which is determined by the beampattern in that direction, is now unity. As above, the particular weight vectors that provide the desired PSK symbols in this direction are determined by solving (22.18). Thus, the secondary communication function can be achieved in this manner anywhere in the beampattern, regardless of whether the desired direction is in the main beam or sidelobes.

It is worth noting, though, that the set of $K = 4$ weight vectors obtained in these two circumstances are not the same. Consequently, if the intended communication direction changes with time, then the set of K weight vectors obtained using (22.18) must likewise be updated as needed.

22.4 Transmit signaling strategies for embedding information

Where the previous section discussed various mechanisms by which a secondary amplitude or phase modulation effect could be enabled within the context of a primary radar emission, we now consider how these formulations can be used to realize a communication link. In the following, it is assumed that Q bits are to be embedded into each radar pulse (or a time interval of prespecified duration).

22.4.1 Sidelobe AM-based communications

The idea behind the sidelobe AM DFRC arrangement is to embed communication symbols into the radar emission by modulating the SLL in the spatial angle at which the communication receiver is located [19]. To ensure the primary radar operation is preserved, the main lobe of the beampattern should be kept unchanged during the entire CPI. The secondary communication function requires the construction of an AM constellation of size $K = 2^Q$ denoted as $\mathbb{D}_{\text{AM}} = \{\Delta_1, \ldots, \Delta_K\}$. We assume that the SSLs are preselected such that $\Delta_1 > \Delta_2 > \cdots > \Delta_K$.

During the τth radar pulse, Q bits are first mapped into the corresponding symbol $\Delta(\tau)$ in the AM constellation $\mathbb{D}_{\text{AM}}$, which subsequently determines the associated weight vector $\mathbf{u}(\tau) \in \{\mathbf{u}_1, \ldots, \mathbf{u}_K\}$. In the same manner as (22.1), the resulting transmit signal can be expressed as

$$\mathbf{s}(t;\tau) = \mathbf{u}^{*}(\tau)\,\phi(t). \tag{22.21}$$

The baseband signal at the output of a single-antenna communication receiver can then be modeled as

$$\begin{aligned} y_{\text{com}}(t;\tau) &= \alpha_{\text{ch}}\, \mathbf{u}^{H}(\tau)\, \mathbf{a}(\theta_{\text{c}})\, \phi(t) + n(t;\tau), \\ &= \alpha_{\text{ch}}\, \Delta(\tau)\, \phi(t) + n(t;\tau), \end{aligned} \tag{22.22}$$

where α_{ch} is the channel coefficient of the received signal that includes the propagation effects between the transmit array and communication receiver, and $n(t;\tau)$ is AWGN with zero mean and variance σ_{c}^2. It is assumed that the channel coefficient is known or can be accurately estimated. Matched filtering the received communication signal of (22.22) according to the presumed known transmitted radar waveform yields

$$y_{\text{AM}}(\tau) = \int_{T_{\text{p}}} y_{\text{com}}(t;\tau)\, \phi^{*}(t)\, dt = \alpha_{\text{ch}}\, \Delta(\tau) + n(\tau) \tag{22.23}$$

where $n(\tau) \triangleq \int_{T_{\text{p}}} n(t;\tau)\, \phi^{*}(t)\, dt$ is the filtered additive noise term that has the same statistics as that of $n(t;\tau)$. The strength of the signal at the output of the matched filter is measured as

$$\eta_{\text{AM}}(\tau) = |y_{\text{com}}(\tau)| = |\alpha_{\text{ch}}\, \Delta(\tau) + n(\tau)|. \tag{22.24}$$

Thus, an estimate of the embedded communication symbol can be obtained by comparing $\eta_{\text{AM}}(\tau)$ to a set of thresholds T_k for $k = 1, \ldots, K-1$ via

$$\hat{\Delta}(\tau) = \begin{cases} \Delta_1, & T_1 \le \eta_{\text{AM}}(\tau), \\ \Delta_2, & T_2 \le \eta_{\text{AM}}(\tau) < T_1, \\ \vdots & \\ \Delta_K, & \eta_{\text{AM}}(\tau) < T_{K-1}. \end{cases} \tag{22.25}$$

The threshold values in (22.25) can be preselected such that they divide the K SLLs in an appropriate manner, e.g. to minimize the probability of symbol error. The estimated AM symbol $\hat{\Delta}(\tau)$ can subsequently be converted into the corresponding binary sequence.

Remark: It is worth noting that the number of SLLs required to embed Q bits of information clearly grows exponentially with Q. Unfortunately, symbol estimation performance deteriorates rapidly as the number of SLLs increases due to the higher constellation density. Therefore, the number of bits that can reasonably be embedded

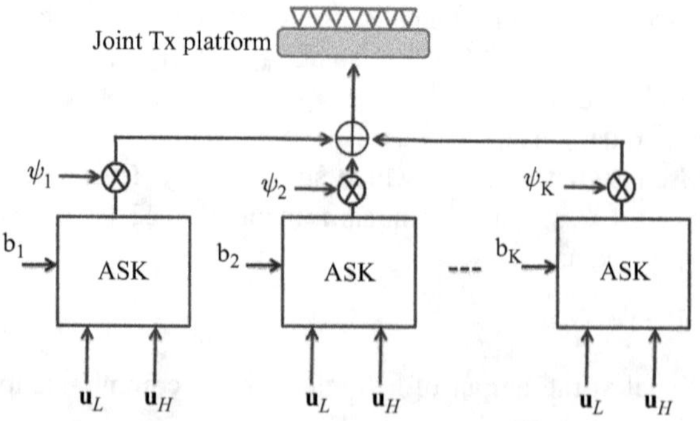

Figure 22.7 ASK-based signaling for embedding information into a radar emission [22]

using the sidelobe AM method is small, and thus the expected achievable data rate is low.

22.4.2 Multi-waveform sidelobe ASK-based communications

The sidelobe AM signaling scheme above has a rather limited achievable data rate. One method that can be used to increase the data rate is to employ multiple orthogonal waveforms, where each can be used to embed Q bits during each radar pulse. In so doing, the number of bits that can be simultaneously embedded during each pulse is proportional to the number of orthogonal waveforms.

The radar waveform $\phi(t)$ can be decomposed into K_{w} orthogonal components such that

$$\phi(t) = \psi_1(t) + \cdots + \psi_{K_{\mathrm{w}}}(t). \tag{22.26}$$

We assume that $\psi_1(t), \ldots, \psi_Q(t)$ satisfy the orthogonality condition at zero time-delay, that is

$$\int_{T_{\mathrm{p}}} \psi_k(t)\, \psi_{k'}^*(t)\, dt = 0, \quad k, k' = 1, \ldots, K_{\mathrm{w}}, \quad k \neq k'. \tag{22.27}$$

It is worth noting that there are an infinite number of solutions to the decomposition in (22.26), and consequently the orthogonality condition in (22.27) is not difficult to satisfy. One way to perform the decomposition is to construct K_{w} orthogonal subspaces (i.e. K_{w} orthogonal bases) and then $\psi_k(t)$ can be obtained by projecting the radar waveform $\phi(t)$ onto the kth subspace. Another way to achieve this waveform decomposition is to apply a subband decomposition in the frequency domain.

The multi-waveform ASK method of [22] employs K_{w} waveforms and two transmit beamforming weight vectors denoted as $\mathbf{u}_{\mathrm{H}}$ and $\mathbf{u}_{\mathrm{L}}$. In this case, the number of bits per waveform is $Q = 1$. Therefore, the number of bits that can be embedded into each radar pulse is K_{w}. The corresponding signaling strategy is illustrated in Figure 22.7.

The weight vectors $\mathbf{u}_{\mathrm{H}}$ and $\mathbf{u}_{\mathrm{L}}$ can be designed by solving the optimization problem in (22.14). Let Δ_{H} and Δ_{L} be the SLLs associated with $\mathbf{u}_{\mathrm{H}}$ and $\mathbf{u}_{\mathrm{L}}$ in the spatial direction of the communication receiver. We assume that the condition $\Delta_{\mathrm{H}} > \Delta_{\mathrm{L}}$ is satisfied while designing these weight vectors. During the τth radar pulse, each of the waveforms, $\psi_k(t)$, $k = 1, \ldots, K_{\mathrm{w}}$, is radiated via either $\mathbf{u}_{\mathrm{H}}$ or $\mathbf{u}_{\mathrm{L}}$ depending on whether the associated bit $b_k(\tau)$ is a 0 or a 1, respectively. Therefore, the $M \times 1$ vector of transmit signals from (22.1) can be expressed as

$$\mathbf{s}_{\mathrm{ASK}}(t;\tau) = \frac{1}{\sqrt{K_{\mathrm{w}}}} \sum_{k=1}^{K_{\mathrm{w}}} \left(b_k(\tau)\, \mathbf{u}_{\mathrm{L}}^{*} + \bar{b}_k(\tau)\, \mathbf{u}_{\mathrm{H}}^{*} \right) \psi_k(t), \tag{22.28}$$

where $\bar{b}_k(\tau) \triangleq 1 - b_k(\tau)$ is the complement of $b_k(\tau)$.

Based on this emission structure, the resulting baseband signal at the output of a single-antenna communication receiver located in the direction θ_{c} is given as

$$\begin{aligned} y_{\mathrm{ASK}}(t;\tau) &= \frac{\alpha_{\mathrm{ch}}}{\sqrt{K_{\mathrm{w}}}} \sum_{k=1}^{K_{\mathrm{w}}} \left(b_k(\tau)\, \mathbf{u}_{\mathrm{L}}^{H} \mathbf{a}(\theta_{\mathrm{c}}) + \bar{b}_k(\tau)\, \mathbf{u}_{\mathrm{H}}^{H} \mathbf{a}(\theta_{\mathrm{c}}) \right) \psi_k(t) + n(t;\tau) \\ &= \frac{\alpha_{\mathrm{ch}}}{\sqrt{K_{\mathrm{w}}}} \sum_{k=1}^{K_{\mathrm{w}}} \left(b_k(\tau)\, \Delta_{\mathrm{L}} + \bar{b}_k(\tau)\, \Delta_{\mathrm{H}} \right) \psi_k(t) + n(t;\tau). \end{aligned} \tag{22.29}$$

The received signal component associated with each transmitted orthogonal waveform can be extracted using matched filtering, with the baseband signal at the output of the kth matched filter expressed as

$$\begin{aligned} y_k(\tau) &= \int_{T_{\mathrm{p}}} y_{\mathrm{ASK}}(t;\tau)\, \psi_k^{*}(t)\, dt \\ &= \begin{cases} \frac{\alpha_{\mathrm{ch}}}{\sqrt{K_{\mathrm{w}}}}\, \Delta_{\mathrm{H}} + n_k(\tau), & b_k(\tau) = 0, \\ \frac{\alpha_{\mathrm{ch}}}{\sqrt{K_{\mathrm{w}}}}\, \Delta_{\mathrm{L}} + n_k(\tau), & b_k(\tau) = 1, \end{cases} \end{aligned} \tag{22.30}$$

where $n_k \triangleq \int_{T_{\mathrm{p}}} n(t;\tau)\, \psi_k^{*}(t)\, dt$ is an additive noise term that has the same statistics as $n(t;\tau)$. The receiver can then estimate the embedded information bits by performing

$$\hat{b}_k(\tau) = \begin{cases} 0, & \text{if } |y_k(\tau)| \geq T_0, \\ 1, & \text{if } |y_k(\tau| < T_0, \end{cases} \tag{22.31}$$

where T_0 is an appropriately selected threshold. The process of embedding and subsequently estimating each bit is performed independently from other bits due to the orthogonality of the component waveforms.

It is worth noting that similar to the sidelobe AM-based scheme in Section 22.4.1, this multi-waveform ASK-based approach enables the embedding of information into the radar emission toward communication receivers located in the sidelobe region. In contrast, the following PSK-based scheme facilitates communication in any direction.

22.4.3 Multi-waveform PSK-based communications

The essence of the PSK-based scheme is to modulate the phase of the signal emitted in a particular direction while keeping the beampattern magnitude fixed. As above, the input binary data are first mapped into PSK symbols that belong to the PSK constellation of an appropriate size, with each PSK symbol representing Q bits. Like the previous multi-waveform formulation, the number of PSK symbols that can be embedded during each radar pulse equals the number orthogonal waveforms being transmitted. In contrast to the AM and multi-waveform ASK methods, the PSK-based method permits embedded information to be emitted in any direction, even within the main beam. It was also shown in a number of papers that the PSK-based signaling scheme realizes a superior bit error rate (BER) compared to the other methods [17,21,29].

Assume that a PSK constellation $\mathbb{D}_{\mathrm{PSK}}$ of size $K = 2^Q$ is constructed using (22.15). Without loss of generality, it is assumed that the PSK symbols are uniformly distributed on the unit circle as

$$\mathbb{D}_{\mathrm{PSK}} = \left\{0, \frac{2\pi}{K}, \ldots, \frac{(K-1)2\pi}{K}\right\}. \tag{22.32}$$

The corresponding set of beamforming weight vectors denoted as $\mathbf{U} = \{\mathbf{u}_1, \ldots, \mathbf{u}_K\}$ can be obtained by solving either (22.16) or (22.18). During the τth radar pulse QK_{w} bits of information are mapped into the PSK symbols $\Omega_k(\tau) \in \mathbb{D}_{\mathrm{PSK}},\ k = 1, \ldots, K_{\mathrm{w}}$. To embed the PSK symbols into the radar emission, the baseband transmit signal vector is expressed as

$$\mathbf{s}_{\mathrm{PSK}}(t;\tau) = \frac{1}{\sqrt{K_{\mathrm{w}}}} \sum_{k=1}^{K_{\mathrm{w}}} \mathbf{u}_k^*(\tau)\, \psi_k(t), \tag{22.33}$$

where $\mathbf{u}_k(\tau) \in \mathbf{U}$ is the weight vector associated with $\Omega_k(\tau) = \angle \mathbf{u}_k^H(\tau)\, \mathbf{a}(\theta_{\mathrm{c}})$ for $k = 1, \ldots, K_{\mathrm{w}}$.

It is assumed that the waveform dictionary used at the transmitter is known to the communication receiver. It is also assumed that phase synchronization is established between the dual-function transmitter and the communication receiver. Hence, the baseband representation of the signal at the output of a single-antenna communication receiver can be expressed as

$$\begin{aligned} y_{\mathrm{PSK}}(t;\tau) &= \alpha_{\mathrm{ch}}\, \mathbf{a}^T(\theta_{\mathrm{c}})\, \mathbf{s}_{\mathrm{PSK}}(t;\tau) + n(t;\tau), \\ &= \frac{\alpha_{\mathrm{ch}}}{\sqrt{K_{\mathrm{w}}}} \sum_{k=1}^{K_{\mathrm{w}}} \mathbf{u}_k^H(\tau)\, \mathbf{a}(\theta_{\mathrm{c}})\, \psi_k(t) + n(t;\tau). \end{aligned} \tag{22.34}$$

Matched filtering the received communication signal (22.34) based on the component waveforms $\psi_k(t),\ k = 1, \ldots, K_w$ therefore yields

$$\begin{aligned} y_k(\tau) &= \int_{T_p} y_{\text{PSK}}(t; \tau)\, \psi_k^*(t)\, dt \\ &= \frac{\alpha_{\text{ch}}}{\sqrt{K_w}} \mathbf{u}_k^H(\tau)\, \mathbf{a}(\theta_c) + n(\tau), \\ &= \frac{\alpha_{\text{ch}}}{\sqrt{K_w}} \left| \mathbf{u}_k^H(\tau)\, \mathbf{a}(\theta_c) \right| e^{j\Omega_k(\tau)} + n_k(\tau). \end{aligned} \tag{22.35}$$

The receiver then obtains estimates of the embedded PSK symbols as

$$\hat{\Omega}_k(\tau) = \angle y_k(\tau), \qquad k = 1, \ldots, K_w, \tag{22.36}$$

which are subsequently compared to the constellation $\mathbb{D}_{\text{PSK}}$ to determine the nearest symbol and corresponding binary data. Since this manner of symbol estimation requires phase synchronization between the transmitter and the communication receiver, we refer to the communication process as being coherent.

It is worth noting that all waveforms are radiated simultaneously and propagate through the same environment. Therefore, any phase synchronization error that occurs due to, for example, propagation distortion yields the same phase error in all waveform components. However, the phase rotation of one waveform with respect to another remains the same. Since measuring the phase associated with one waveform with respect to the phase associated with the another waveform cancels out any common-phase error, the phase synchronization mismatch has no effect on the estimation of the phase rotation at the receiver. This arrangement enables noncoherent operation by making one of the waveforms a known common reference to all others, thus yielding $K_w - 1$ PSK communication symbols each pulse. In this way, each communication symbol is embedded in the phase difference relative to the phase of the reference waveform. The reader is referred to [25,29] for further details.

22.4.4 Embedding information into MIMO radar emissions

Now consider the operation of a DFRC system in which the radar utilizes collocated antennas in a MIMO configuration. Let $\phi_m(t),\ m = 1, \ldots, M$, be orthonormal waveforms such that

$$\int_{T_p} \phi_m(t)\, \phi_{m'}^*(t - \xi) dt = \begin{cases} 1, & m = m', \xi = 0, \\ 0, & \text{otherwise.} \end{cases} \tag{22.37}$$

Note that the orthogonality condition in (22.37) is more restrictive than that of (22.27) because MIMO radar requires that the waveforms be orthogonal for $\xi \neq 0$. On the other hand, the secondary communication function only requires waveform orthogonality at zero time-delay. It is likewise known that (22.37) cannot actually be realized in practice, though waveforms with relatively low cross-correlations can be obtained (see [30–35] and references therein).

It is assumed that L targets are located in the far-field. Thus, the $N \times 1$ complex vector of received observations can be expressed as [40–42]

$$\mathbf{x}(t, \tau) = \sum_{\ell=1}^{L} \gamma_\ell(\tau) \left[\mathbf{a}^T(\theta_\ell)\, \mathbf{\Phi}(t)\right] \mathbf{b}(\theta_\ell) + \mathbf{n}(t, \tau), \tag{22.38}$$

where τ is the pulse number, $\gamma_\ell(\tau)$ is the reflection coefficient of the ℓth target, θ_ℓ is the spatial angle associated with the ℓth target, $\mathbf{a}(\theta)$ and $\mathbf{b}(\theta)$ are the steering vectors of the transmit and receive arrays, respectively, $\mathbf{\Phi}(t) \triangleq [\phi_1(t), \ldots, \phi_M(t)]^T$ contains the orthogonal waveforms, and $\mathbf{n}(t, \tau)$ is the $N \times 1$ vector of zero-mean white Gaussian noise. In (22.38), the reflection coefficients $\gamma_\ell(\tau)$ for $\ell = 1, \ldots, L$ are assumed to obey a Swerling II target model [36], i.e. they remain constant during the pulse duration but vary independently from pulse to pulse.

A bank of M radar matched filters can be used to extract the reflected signal components associated with the individual transmitted waveforms, which results in the $MN \times 1$ extended virtual data vector

$$\begin{aligned} \mathbf{y}(\tau) &= \mathrm{vec}\left(\int_{T_\mathrm{p}} \mathbf{x}(t, \tau)\, \mathbf{\Phi}^H(t)\, dt \right) \\ &= \sum_{\ell=1}^{L} \gamma_\ell(\tau) \left[\mathbf{a}(\theta_\ell) \otimes \mathbf{b}(\theta_\ell)\right] + \tilde{\mathbf{n}}(\tau). \end{aligned} \tag{22.39}$$

Here vec$(\cdot)$ is the vectorization operator that stacks the columns of a matrix into a single column vector, $\otimes$ denotes the Kronecker product, and

$$\tilde{\mathbf{n}}(\tau) = \mathrm{vec}\left(\int_{T_\mathrm{p}} \mathbf{n}(t, \tau)\, \mathbf{\Phi}^H(t)\, dt \right) \tag{22.40}$$

is the $MN \times 1$ additive zero-mean noise term with covariance matrix $\sigma_\mathrm{n}^2 \mathbf{I}_{MN}$.

Let $\mathbf{\Omega} = [e^{j\Omega_1}, \ldots, e^{j\Omega_M}]^T$ be the $M \times 1$ vector of phase rotations, where $\Omega_m \in [0, 2\pi]$ for $m = 1, \ldots, M$ arbitrary phases. The $M \times 1$ vector of phase-rotated waveforms $\mathbf{\Psi}(t)$ can thus be defined as

$$\mathbf{\Psi}(t) = \mathbf{\Pi}\mathbf{\Phi}(t), \tag{22.41}$$

where $\mathbf{\Pi} = \mathrm{diag}\{\mathbf{\Omega}\}$ is an $M \times M$ diagonal matrix and the diag$\{\cdot\}$ operation constructs a diagonal matrix using the elements of the vector argument. Note that if the condition $\mathbf{\Phi}(t)\mathbf{\Phi}^H(t) = \mathbf{I}_M$ is satisfied, then it can easily be verified that phase-rotated waveforms likewise satisfy $\mathbf{\Psi}(t)\mathbf{\Psi}^H(t) = \mathbf{I}_M$.

During the τth radar pulse, transmission of the phase-rotated waveforms $\mathbf{\Psi}(t)$ causes the $N \times 1$ complex received observation vector from (22.38) to be replaced with

$$\tilde{\mathbf{x}}(t, \tau) = \sum_{\ell=1}^{L} \gamma_\ell(\tau) \left[\mathbf{a}^T(\theta_\ell)\mathbf{\Psi}(t)\right] \mathbf{b}(\theta_\ell) + \mathbf{n}(t, \tau). \tag{22.42}$$

Matched filtering this received signal using the phase-rotated orthogonal waveforms in (22.41) yields

$$\breve{\mathbf{y}}(\tau) = \text{vec}\left(\int_{T_\text{p}} \tilde{\mathbf{x}}(t,\tau)\,\boldsymbol{\Phi}^H(t)\,\boldsymbol{\Pi}^H dt\right)$$
$$= \sum_{\ell=1}^{L} \gamma_\ell(\tau)\,[\mathbf{a}(\theta_\ell)\otimes\mathbf{b}(\theta_\ell)] + \breve{\mathbf{n}}(\tau), \tag{22.43}$$

where

$$\breve{\mathbf{n}}(\tau) = \text{vec}\left(\left[\int_{T_\text{p}} \mathbf{n}(t,\tau)\,\boldsymbol{\Phi}^H(t)\,dt\right]\boldsymbol{\Pi}^H\right) \tag{22.44}$$

is the $MN \times 1$ additive noise term with zero mean and covariance matrix $\sigma_\text{n}^2\mathbf{I}_{MN}$.

It can be observed from the comparison between the extended virtual signal models in (22.39) and (22.43) that applying phase rotation to the transmit waveform yields the same data model at the radar receiver except for a phase-rotated additive noise term. In this respect, we can express the phase-rotated noise of (22.44) in terms of the noise term from (22.40) using the relationship

$$\breve{\mathbf{n}}(\tau) = [\boldsymbol{\Pi}^* \otimes \mathbf{I}_M]\,\tilde{\mathbf{n}}(\tau). \tag{22.45}$$

Both sides in (22.45) have the same statistics, and therefore it can be concluded that applying phase rotation to the orthogonal transmit waveforms via (22.41) does not alter the radar receiver operation.

To realize a secondary communication function within this primary MIMO radar context, the phases Ω_m can be used as communication symbols, with one symbol embedded in each of the M waveforms. Without loss of generality, we shall use the PSK dictionary $\mathbb{D}_\text{PSK}$ defined in (22.32) that comprises $K = 2^Q$ unique symbols and each symbol represents Q bits.

During the τth pulse, the PSK symbols are referred to as $\Omega_m(\tau) \in \mathbb{D}_\text{PSK}$ for $m = 1,\ldots,M$. Thus the set of phase-rotated orthogonal waveforms can be expressed as

$$\boldsymbol{\Psi}(t,\tau) = \boldsymbol{\Pi}(\tau)\,\boldsymbol{\Phi}(t), \tag{22.46}$$

where

$$\boldsymbol{\Pi}(\tau) = \text{diag}\left\{\left[e^{j\Omega_1(\tau)},\ldots,e^{j\Omega_M(\tau)}\right]^T\right\}, \quad \Omega_m(\tau) \in \mathbb{D}_\text{PSK}. \tag{22.47}$$

For a single-antenna communication receiver located at an arbitrary direction θ_c, which is assumed known or at least estimable by the receiver, the signal at the output of the communication receiver can be expressed as

$$y_\text{MIMO}(t,\tau) = \frac{\alpha_\text{ch}}{\sqrt{M}}\,\mathbf{a}^T(\theta_\text{c})\,\boldsymbol{\Psi}(t,\tau) + w(t,\tau)$$
$$= \frac{\alpha_\text{ch}}{\sqrt{M}}\,\mathbf{a}^T(\theta_\text{c})\,\boldsymbol{\Pi}(\tau)\,\boldsymbol{\Phi}(t) + w(t,\tau), \tag{22.48}$$

where $w(t,\tau)$ is an additive interference-plus-noise term that is assumed to be Gaussian-distributed with zero mean and variance σ_{w}^2. In (22.48), we assume that α_{ch} remains constant during the entire CPI and that an accurate estimate of it is available to the communication receiver.

Under the condition of phase synchronization between the transmit array and the communication receiver, matched filtering of (22.48) using the orthogonal waveforms $\phi_m(t)$, $m = 1, \ldots, M$, yields

$$\begin{aligned} y_m(\tau) &= \int_{T_{\mathrm{p}}} r(t)\,\phi_m^*(t)\,dt \\ &= \frac{\alpha_{\mathrm{ch}}}{\sqrt{M}}\,\mathbf{a}_{[m]}\,e^{\Omega_m(\tau)} + w_m(\tau), \quad m = 1, \ldots, M, \end{aligned} \tag{22.49}$$

where $\mathbf{a}_{[m]} \triangleq e^{-j2\pi d_m \sin\theta_{\mathrm{c}}}$ is the mth entry of $\mathbf{a}(\theta_{\mathrm{c}})$, for d_m the displacement between the first and mth elements of the transmit array measured in wavelengths, and $w_m(\tau) \triangleq \int_{T_{\mathrm{p}}} w(t,\tau)\,\phi^*(t)\,dt$ is the additive noise term at the output of the mth matched filter. The noise term $w_m(\tau)$ has the same statistics as that of $w(t,\tau)$. Equation (22.49) shows that the received communication signal at the output of the mth matched filter is a phase-shifted and noisy version of the mth entry of the steering vector $\mathbf{a}(\theta_{\mathrm{c}})$, which means that the phase shift $\Omega_m(\tau)$ can be recovered from the received signal $y_m(\tau)$.

Let us assume that an accurate estimate of the channel is available to the communication receiver and also that the communication receiver knows its spatial direction with respect to the dual-function transmitter. In practice, training sequences can be periodically transmitted to update the channel estimate and adjust phase synchronization between the transmit array and communication receiver. The embedded PSK symbols can therefore be recovered from (22.49) using the formula

$$\hat{\Omega}_m(\tau) = \angle(y_m(\tau)) - \varphi_{\mathrm{ch}} + 2\pi d_m \sin\theta_{\mathrm{c}}, \tag{22.50}$$

where φ_{ch} is the phase of the channel coefficient. Once the PSK symbols are estimated the communication receiver proceeds to convert them into the corresponding binary data.

It is worth noting that the number of phase symbols that can be embedded during a single pulse repetition interval (PRI) equals the number of transmit antennas. Consequently, the achievable data rate in bits per second (bps) can be expressed as

$$R = (QM) \times \mathrm{PRF}. \tag{22.51}$$

In radar applications with a high pulse repetition frequency (PRF), such as in X-band radar, a data rate on the order of dozens of Mbps could be achieved.

22.5 Performance analysis

We now evaluate the theoretical performance of the signaling strategies discussed above. Specifically, communication performance is assessed in terms of BER using simulation examples.

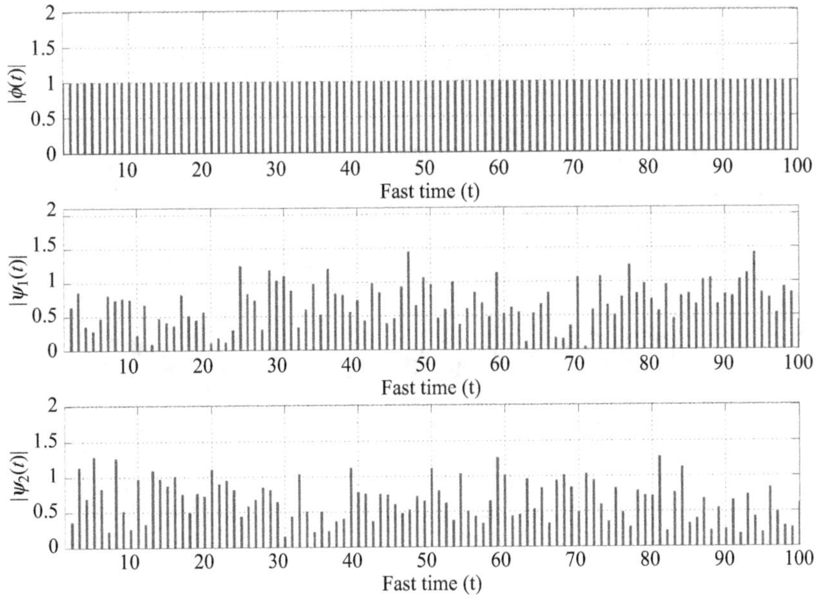

Figure 22.8 Unimodular sequence $\phi(t)$ decomposed into orthogonal components $\psi_1(t)$ and $\psi_2(t)$

22.5.1 Impact of dual-functionality on radar performance

Consider a uniform linear transmit array consisting of $M = 20$ elements at half-wavelength spacing. For maximum transmit power efficiency, the radar waveform $\phi(t)$ is represented by a unimodular sequence. The phase profile of $\phi(t)$ is generated randomly with a uniform distribution on $[0,\ 2\pi]$. Two orthogonal subspaces are formed and used to decompose $\phi(t)$ into the orthogonal components $\psi_1(t)$ and $\psi_2(t)$. Figure 22.8 shows the magnitude of $\phi(t)$, $\psi_1(t)$, and $\psi_2(t)$, respectively. It is observed that, though the radar waveform is unimodular, the two orthogonal components are not.

It is assumed that the radar main beam is pointing in the direction $\theta_t = 0°$ and the transmit weight vector is chosen as $\mathbf{w}_0 = \mathbf{a}(\theta_t)$, i.e. the radar operates in a phased-array mode. Using this principal weight vector, the weight vectors $\mathbf{u}_k,\ k = 1, \ldots, 4$, are designed by solving the optimization problem in (22.16) to realize a QPSK constellation $\mathbb{D}_{\text{QPSK}} = [0,\ \pi/2,\ \pi, -\pi/2]$ for the communication direction $\theta_c = -30°$. The transmit beampatterns for $\mathbf{w}_0$ and $\mathbf{u}_k,\ k = 1, \ldots, 4$, are shown in Figure 22.4. Note that the first symbol in $\mathbb{D}_{\text{QPSK}}$ equals 0° and therefore $\mathbf{w}_0 = \mathbf{u}_1$.

To show the effect of embedding QPSK symbols on the transmit power efficiency of the system, we assume that the signal $\psi_1(t)$ is radiated at all times via $\mathbf{w}_0$. We also assume that each of the weight vectors $\mathbf{u}_k,\ k = 1, \ldots, 4$, is used with equal probability to radiate the signal $\psi_2(t)$. Figure 22.9 shows the magnitude of signals transmitted from the first antenna for four possible cases given as $\mathbf{s}(t) = \mathbf{w}_0\psi_1(t) + \mathbf{u}_i\psi_2(t)$,

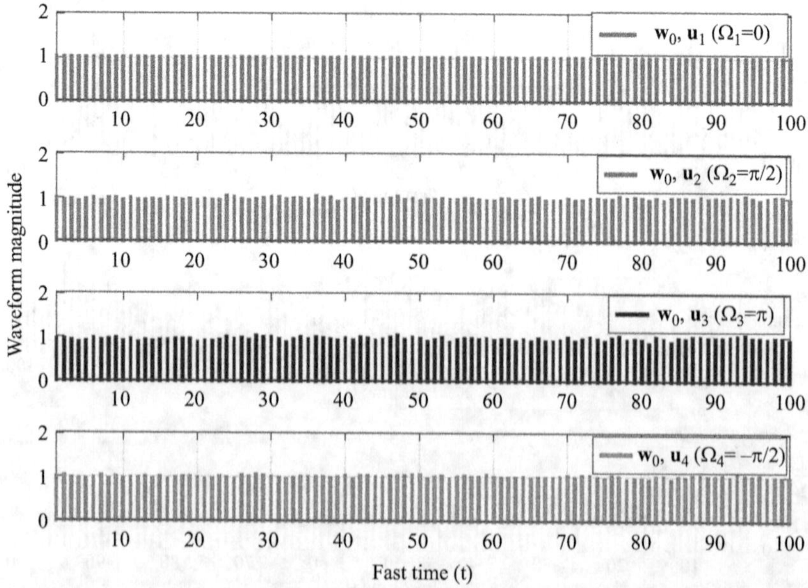

Figure 22.9 Effect of embedding information into the radar emission on the amplitude of a constant-modulus waveform

$i = 1, \ldots, 4$. For case $i = 1$, the transmitted signal is shown at the top of the figure, where it can be seen that the signal remains constant modulus due to the fact that $\mathbf{w}_0 = \mathbf{u}_1$ and $\phi(t) = \psi_1(t) + \psi_2(t)$. For the other three cases, the signal is no longer constant modulus, which in general could negatively impact transmit power efficiency. However, it is observed that the degradation in transmit power efficiency is small and therefore may be tolerable by the radar. Similar behavior has been observed when examining the waveforms of other transmit elements.

22.5.2 Bit error rate performance analysis

Here the communication performance for the above methods is assessed in terms of BER. For the sidelobe AM-based method discussed in Section 22.4.1, the signal at the output of the communication receiver, when $\phi(t)$ is transmitted via $\mathbf{u}_k$, simplifies to

$$y_k(\tau) = \alpha_{\mathrm{ch}}\, \Delta_k + n(\tau) = \tilde{\alpha}_{\mathrm{ch}} + n_k(\tau), \tag{22.52}$$

where $\tilde{\alpha}_{\mathrm{ch}} \triangleq \alpha_{\mathrm{ch}}\, \Delta_k$ is a complex variable that is assumed to remain constant during the CPI, and $n(\tau)$ is white Gaussian noise with zero-mean and variance σ_{n}^2. Let $\eta_k \triangleq |y_k(\tau)|$ be the magnitude of the communication receiver output given in (22.52). Since $\tilde{\alpha}_{\mathrm{ch}}$ is constant and $n_k(\tau)$ is zero-mean Gaussian noise, η_k is a random variable with Rician distribution that follows the probability density function

$$\mathrm{R}(\eta_k|A_k, \sigma_{\mathrm{n}}) = \frac{\eta_k}{\sigma_{\mathrm{n}}^2} e^{-(\eta_k^2 + A_k^2)/2\sigma_{\mathrm{n}}^2} I_0\left(\frac{\eta_k A_k}{\sigma_{\mathrm{n}}^2}\right), \tag{22.53}$$

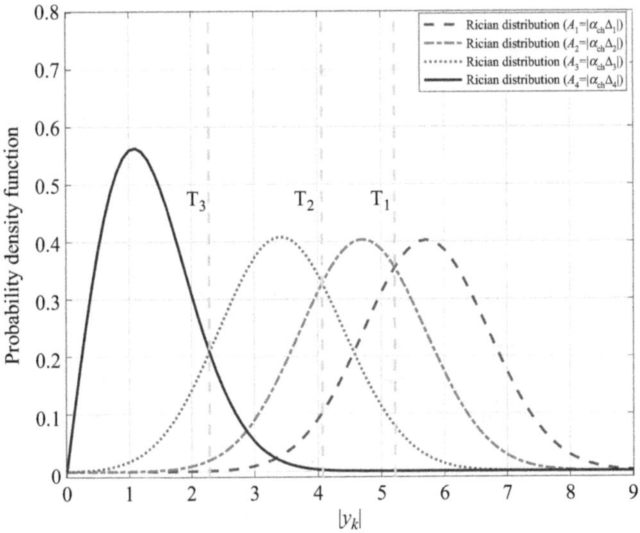

Figure 22.10 Rician probability density function versus magnitude of received signal strength for AM-based signaling strategy

where $A_k \triangleq |\alpha_{\mathrm{ch}}\,\Delta_k|$ and $I_0(\cdot)$ is a modified Bessel function of the first kind with order zero.

Consider the case in which $Q = 2$ bits are embedded in each radar pulse. The sidelobe AM-based method discussed in Section 22.4.1 maps the two bits of information into one of four symbols $\mathbb{D}_{\mathrm{AM}} = \{\Delta_1, \ldots, \Delta_4\}$. Figure 22.10 depicts the Rician probability distribution associated with the four different symbols. The variance of the noise is set to $\sigma_{\mathrm{n}}^2 = 1$. The four SLLs are chosen such that $A_1 = 5.6234$, $A_2 = 3.9364$, $A_3 = 2.2494$, and $A_4 = 0.5623$. The detection thresholds are indicated as vertical dashed lines in the figure, which shows that the tails of the four Rician distributions on either side of the thresholds represent regions where incorrect decisions occur that lead to erroneous decoding of the binary data.

For the sidelobe ASK-based method discussed in Section 22.4.2, the binary data are transmitted via independent waveforms of equal energy. Therefore, the BER analysis for all simultaneously transmitted bits is the same, so we shall focus on $b_1(\tau)$. From (22.35), the signal at the output of the communication receiver when $b_1(\tau) = 0$ is transmitted simplifies to

$$y_k(\tau) = \frac{\alpha_{\mathrm{ch}}}{\sqrt{K_{\mathrm{w}}}}\,\Delta_{\mathrm{H}} + n_k. \tag{22.54}$$

Thus, the probability distribution of $\eta_{\mathrm{H}} = |y_k(\tau)|$ follows the Rician distribution $\mathrm{R}(\eta_{\mathrm{H}}|A_{\mathrm{H}}, \sigma_{\mathrm{n}})$, where $A_{\mathrm{H}} \triangleq \left|(\alpha_{\mathrm{ch}}/\sqrt{K_{\mathrm{w}}})\Delta_{\mathrm{H}}\right|$. Similarly, when $b_1(\tau) = 1$ is transmitted, the probability distribution of $\eta_{\mathrm{L}} = |y_k(\tau)|$ follows the Rician distribution $\mathrm{R}(\eta_{\mathrm{L}}|A_{\mathrm{L}}, \sigma_{\mathrm{n}})$, where $A_{\mathrm{L}} \triangleq \left|(\alpha_{\mathrm{ch}}/\sqrt{K_{\mathrm{w}}})/\Delta_{\mathrm{L}}\right|$.

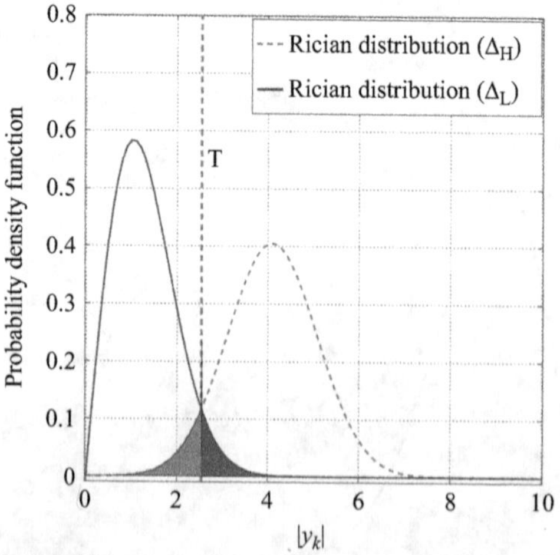

Figure 22.11 Rician probability density function versus magnitude of received signal strength for ASK-based signaling strategy

Figure 22.11 depicts the Rician probability density functions versus the magnitude of the received signal at the communication receiver for the ASK-based signaling strategy. In this case, the high and low SLLs are chosen such that $A_H = A_1/\sqrt{2}$ and $A_L = A_4/\sqrt{2}$. The reason for this choice is that the AM-based method assigns the entire transmit power to a single waveform $\phi(t)$, whereas the ASK-based method divides the total transmit power equally among two orthogonal waveforms $\psi_1(t)$ and $\psi_2(t)$. The threshold used to detect the embedded binary data is indicated as a vertical line in the figure. The tails of the Rician distributions are also observed, which correspond to the regions where incorrect symbol decisions occur.

22.5.3 Simulation results

In the first example, we evaluate the BER performance of the communication receiver versus angle for a ULA of $M = 10$ elements. The main radar operation takes place within the spatial sector $\mathbf{\Theta} = [-10^\circ + 10^\circ]$, while the intended communication receiver is located at $\theta_c = -50^\circ$. We assume perfect knowledge of both the radar waveforms employed and the information-embedding strategy at the receiver. The number of bits embedded in each radar pulse is $Q = 3$.

For the AM-based method, a dictionary of eight symbols is generated, with associated weight vectors and distinct SSLs oriented toward the communication direction. The eight weight vectors are designed by solving the optimization problem (22.9)–(22.11). The highest and lowest SLLs are selected by choosing $\varepsilon = 0.1$ and $\varepsilon = 0.01$, respectively. The other six SLLs are uniformly distributed between these extremes.

For the ASK-based method, $K_w = 3$ orthogonal waveforms are used, with one bit embedded in each waveform. The two transmit beamforming weight vectors used to

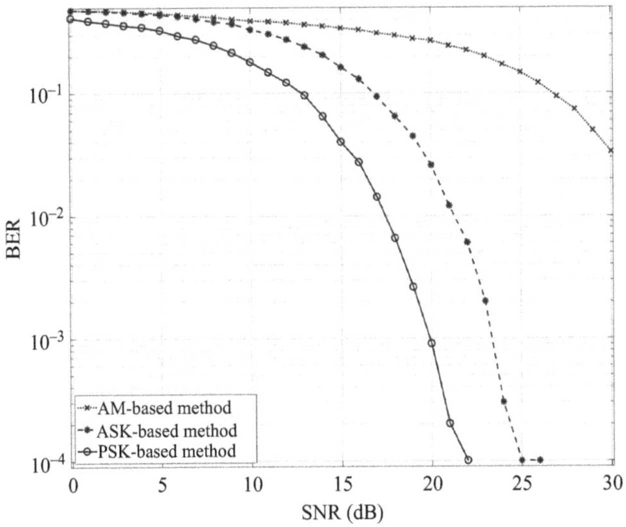

Figure 22.12 Bit error rate versus SNR

implement this approach are chosen as the first and eighth weight vectors employed in the AM-based method.

Finally, the PSK-based method employs an 8-PSK dictionary, where the symbols are obtained by substituting $K = 8$ into (22.32). The corresponding weight vectors $\mathbf{u}_k$, $k = 1, \ldots, 8$, are designed by solving the optimization problem (22.16) in which the principal weight vector $\mathbf{w}_0$ is the first weight vector associated with the AM-based method. The propagation coefficient α_{ch} is modeled as a random variable with a constant unity magnitude and uniformly distributed random phase.

Figure 22.12 depicts the BER curve obtained for each of the three methods versus SNR, where it is observed that the PSK-based methods achieves superior performance relative to the ASK-based and AM-based methods, with the latter realizing the worst performance. Figure 22.13 then shows the BER versus spatial angle, with SNR fixed to 20 dB for all methods. As expected, the AM-based method has poor BER performance for all directions including the true communication direction. The ASK-based method exhibits better BER performance, while the coherent PSK-based method is the best.

In the second example, we consider a MIMO radar operating in X-band with a carrier frequency of $f = 8$ GHz and bandwidth of 500 MHz. The sampling frequency is taken as the Nyquist rate of $f_s = 5 \times 10^8$ complex sample/s. The PRI is $T_P = 10\ \mu$s, which corresponds to a PRF is 100 kHz. The transmit array is an $M = 12$ ULA with half-wavelength spacing and M random waveforms are generated.

We investigate the symbol error rate (SER) performance of the method presented in Section 22.4.4 when embedding information using BPSK, QPSK, 16-PSK, and 256-PSK constellations, which correspond to data rates of $R = 1.2,\ 2.4,\ 4.8$, and 9.6 Mbps, respectively. To assess SER performance, a total of 12×10^7 random symbols are used. Figure 22.14 shows the SER performance versus spatial angle for the various

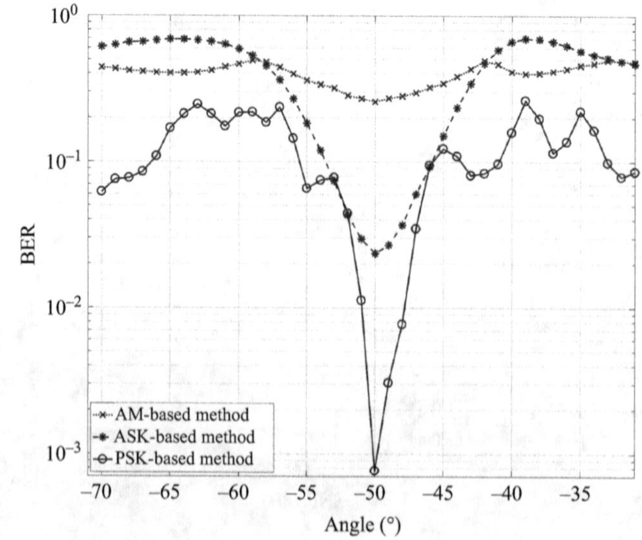

Figure 22.13 Bit error rate versus spatial angle

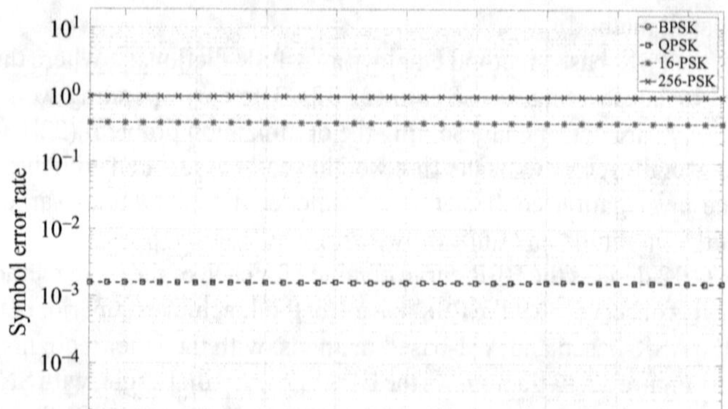

Figure 22.14 SER versus spatial angle of the communication receiver

constellations when SNR is fixed at 10 dB. It is observed that the SER is essentially flat for all constellations, thereby implying that symbol estimation is not sensitive to the direction of the communication receiver. This result can be attributed to the omnidirectional transmission of MIMO radar such that the communication receiver experiences the same power regardless of direction. The figure also shows that, as expected, SER performance degrades as constellation size increases.

22.6 Looking ahead: discussion and future directions

We have considered the problem of dual-function radar–communication system design, which is considered to be a special case of radar–communication coexistence. An overview of different strategies for radar-embedded communication signals was presented. Such strategies are key to realizing dual-function systems that permit simultaneous operation of both radar and communication functions using the same platform.

As a future research direction, new schemes for efficient and high data rate communications need to be developed that leverage waveform diversity while satisfying an overall power constraint. Low probability of intercept and enhanced communication system security against eavesdroppers, especially when the eavesdropper overlaps with the radar main beam, need to be ensured; perhaps by adopting angle-selective signaling schemes for single and multiple users. A full assessment of dual-function system performance is needed, which would involve the interplay between different system parameters, particularly with regard to the implications for different radar modes (e.g. surveillance, imaging, tracking). The prospect of using DFRC systems within a larger radar network is also worth investigation.

References

[1] Griffths, H., Blunt, S., Cohen, L., Savy, L., "Challenge problems in spectrum engineering and waveform diversity," *IEEE Radar Conference*, Ottawa, Canada, Apr./May 2013, pp. 1–5.

[2] Baylis, C., Fellows, M., Cohen, L., Marks, R.J., "Solving the spectrum crisis: Intelligent, reconfigurable microwave transmitter amplifiers for cognitive radar," *IEEE Microwave Magazine*, vol. 15, no. 5, pp. 94–107, July/Aug. 2014.

[3] Griffths, H., Cohen, L., Watts, S., *et al.*, "Radar spectrum engineering and management: Technical and regulatory issues," *Proceedings of the IEEE*, vol. 103, no. 1, pp. 85–102, Jan. 2015.

[4] Goldsmith, A.J., Greenstein, L.J., *Principles of Cognitive Radio*, Cambridge University Press, Cambridge, UK, 2012.

[5] Haykin, S., *Cognitive Dynamic Systems: Perception–Action Cycle, Radar and Radio*, Cambridge University Press, Cambridge, UK, 2012.

[6] Paisana, F., Marchetti, N., DaSilva, L.A., "Radar, TV and cellular bands: which spectrum access techniques for which bands?" *IEEE Communication Surveys & Tutorials*, vol. 16, no. 3, pp. 1193–1220, Third Quarter 2014.

[7] Deng, H., Himed, B., "Interference mitigation processing for spectrum-sharing between radar and wireless communications systems," *IEEE Transactions on Aerospace and Electronic Systems*, vol. 49, no. 3, pp. 1911–1919, July 2013.

[8] Bliss, D.W., "Cooperative radar and communications signaling: the estimation and information theory odd couple," *IEEE Radar Conference*, Cincinnati, OH, May 2014, pp. 50–55.

[9] Geng, Z., Deng, H., Himed, B., "Adaptive radar beamforming for interference mitigation in radar–wireless spectrum sharing," *IEEE Signal Processing Letters*, vol. 22, no. 4, pp. 484–488, Apr. 2015.

[10] Hayvaci, H.T., Tavli, B., "Spectrum sharing in radar and wireless communication systems: a review," *International Conference Electromagnetics in Advanced Applications*, Palm Beach, Aruba, Aug. 2014, pp. 810–813.

[11] Guerci, J.R., Guerci, R.M., Lackpour, A., Moskowitz, D., "Joint design and operation of shared spectrum access for radar and communications," *IEEE International Radar Conference*, Arlington, VA, May 2015.

[12] Khawar, A., Abdelhadi, A., Clancy, C., "Target detection performance of spectrum sharing MIMO radars," *IEEE Sensors Journal*, vol. 15, no. 9, pp. 4928–4940, Sep. 2015.

[13] Wang, L., McGeehan, J., Williams, C., Doufexi, A., "Application of cooperative sensing in radar–communications coexistence," *IET Communications*, vol. 2, no. 6, pp. 856–868, July 2008.

[14] Huang, K.-W., Bica, M., Mitra, U., Koivunen, V., "Radar waveform design in spectrum sharing environment: coexistence and cognition," *IEEE Radar Conference*, Arlington, VA, May 2015, pp. 1698–1703.

[15] Surender, S.C., Narayanan, R.M., Das, C.R., "Performance analysis of communications & radar coexistence in a covert UWB OSA system," *IEEE Global Communications Conference*, Miami, FL, Dec. 2010, pp. 1–5.

[16] Sit, Y.L., Sturm, C., Reichardt, L., Zwick, T., Wiesbeck, W., "The OFDM joint radar–communication system: an overview," *International Conference Advances in Satellite and Space Communications*, Budapest, Hungary, Apr. 2011, pp. 69–74.

[17] Hassanien, A., Amin, M.G., Zhang, Y.D., Ahmad, F., "Signaling strategies for dual-function radar-communications: an overview," *IEEE Aerospace and Electronic Systems Magazine*, vol. 31, no. 10, pp. 36–45, Oct. 2016.

[18] Blunt, S.D., Cook, M.R., Stiles, J., "Embedding information into radar emissions via waveform implementation," *International Waveform Diversity & Design Conference*, Niagara Falls, Canada, Aug. 2010, pp. 8–13.

[19] Euziere, J., Guinvarc'h, R., Lesturgie, M., Uguen, B., Gillard, R., "Dual function radar communication time-modulated array," *International Radar Conference*, Lille, France, Oct. 2014.

[20] Hassanien, A., Amin, M.G., Zhang, Y.D., Ahmad, F., "A dual function radar–communications system using sidelobe control and waveform diversity," *IEEE International Radar Conference*, Arlington, VA, May 2015, pp. 1260–1263.

[21] Hassanien, A., Amin, M.G., Zhang, Y.D., Ahmad, F., "Dual-function radar–communications using phase-rotational invariance," *European Signal Processing Conference*, Nice, France, Aug./Sept. 2015.

[22] Hassanien, A., Amin, M.G., Zhang, Y.D., Ahmad, F., "Dual-function radar–communications: information embedding using sidelobe control and waveform

diversity," *IEEE Transactions on Signal Processing*, vol. 64, no. 8, pp. 2168–2181, Apr. 2016.

[23] Hassanien, A., Amin, M.G., Zhang, Y.D., Ahmad, F., "Efficient sidelobe ASK based dual-function radar–communications," *SPIE Defense + Security, Radar Sensor Technology Conference*, Baltimore, MD, USA, Apr. 2016.

[24] Hassanien, A., Amin, M.G., Zhang, Y.D., "Computationally efficient beampattern synthesis for dual-function radar–communications," *SPIE Defense + Security, Radar Sensor Technology Conference*, Baltimore, MD, USA, Apr. 2016.

[25] Hassanien, A., Amin, M.G., Zhang, Y.D., Ahmad, F., Himed, B., "Non-coherent PSK-based dual-function radar–communication systems," *IEEE Radar Conference*, Philadelphia, PA, USA, May 2016.

[26] BouDaher, E., Hassanien, A., Aboutanios, E., Amin, M.G., "Towards a dual-function MIMO radar–communication system," *IEEE Radar Conference*, Philadelphia, PA, USA, May 2016.

[27] Hassanien, A., Amin, M.G., "Radar system design for dual-functionality platforms," *IEEE Sensor Array and Multichannel Signal Processing Workshop*, Rio de Janeiro, Brazil, July 2016.

[28] Hassanien, A., Amin, M.G., Zhang, Y., Himed, B., "A dual-function MIMO radar–communications system using PSK modulation," *European Signal Processing Conference*, Budapest, Hungary, Aug. 2016.

[29] Hassanien, A., Amin, M.G., Zhang, Y.D., Ahmad, F., "Phase-modulation based dual-function radar–communications," *IET Radar, Sonar & Navigation*, vol. 10, no. 8, pp. 1411–1421, Oct. 2016.

[30] Song, X., Zhou, S., Willett, P., "Reducing the waveform cross correlation of MIMO radar with spacetime coding," *IEEE Transactions on Signal Processing*, vol. 58, no. 8, pp. 4213–4224, Aug. 2010.

[31] Patton, L., Bryant, C., Himed, B., "Radar-centric design of waveforms with disjoint spectral support" *IEEE Radar Conference*, Atlanta, GA, May 2012, pp. 1106–1110.

[32] Aubry, A., De Maio, A., Piezzo, M., Farina, A., "Radar waveform design in a spectrally crowded environment via nonconvex quadratic optimization," *IEEE Transactions on Aerospace and Electronic Systems*, vol. 50, no. 2, pp. 1138–1152, Apr. 2014.

[33] Aubry, A., De Maio, A., Huang, Y., Piezzo, M. Farina, A., "A new radar waveform design algorithm with improved feasibility for spectral coexistence," *IEEE Transactions on Aerospace and Electronic Systems*, vol. 51, no. 2, pp. 1029–1038, Apr. 2015.

[34] Lo Monte, L., Himed, B., Corigliano, T., Baker, C.J., "Performance analysis of time division and code division waveforms in co-located MIMO," *IEEE International Radar Conference*, Arlington, VA, May 2015, pp. 794–798.

[35] Jakabosky, J., Blunt, S.D., Himed, B., "Waveform design and receive processing for nonrecurrent nonlinear FMCW radar," *IEEE International Radar Conference*, Arlington, VA, May 2015, pp. 1376–1381.

[36] Skolnik, M.I., *Introduction to radar systems*, 3rd ed., McGraw-Hill, New York, NY, 2001.

[37] Boyd, S., Vandenberghe, L., *Convex Optimization*, Cambridge University Press, Cambridge, UK, 2009.

[38] Hassanien, A. Vorobyov, S.A., Khabbazibasmenj, A., "Transmit radiation pattern invariance in MIMO radar with application to DOA estimation," *IEEE Signal Processing Letters*, vol. 22, no. 10, pp. 1609–1613, Oct. 2015.

[39] Khabbazibasmenj, A., Hassanien, A., Vorobyov, S.A., "How many beamforming vectors generate the same beampattern?" arXiv preprint, arXiv:1402.1682, 2014.

[40] Li, J., Stoica, P., *MIMO Radar Signal Processing*. Wiley, Hoboken, NJ, 2009.

[41] Ravan, M., Adve, R.S., Riddolls, R.J., "MIMO fast fully adaptive processing in over-the-horizon radar," *IEEE International Radar Conference*, Kansas City, MO, USA, May 2011, pp. 538–542.

[42] Hassanien, A., Vorobyov, S.A., "Transmit energy focusing for DOA estimation in MIMO radar with co-located antennas," *IEEE Transactions on Signal Processing*, vol. 59, no. 6, pp. 2669–2682, June 2011.

Chapter 23
Embedding communications into radar emissions by transmit waveform diversity

Cenk Sahin[1], Patrick M. McCormick[2], Brandon Ravenscroft[2], Justin G. Metcalf[1], and Shannon Blunt[2]

The radio spectrum is a fixed resource with an exponentially increasing demand from commercial communication applications [1,2]. The resulting erosion of radar spectrum to meet this communication demand is creating additional strain on defense applications that must already operate in congested and contested environments. As such, ongoing research is focused on improving spectral efficiency (e.g. dynamic spectrum access [3]) or developing methods to share spectrum between multiple functions (e.g. radar and communications [4]). Generally speaking, spectrum sharing can take two forms: cohabitation or co-design. Where the former tends primarily to address the interference that separately operated systems could cause to one another, the latter can involve cooperative control within the same system. The subject of this chapter is the co-design problem, in particular the realization of a single dual-function system with simultaneous radar and communication capabilities.

The notion of radar/communication shared-spectrum co-design necessitates the use of some manner of waveform diversity [5–8]. As a general principle, waveform diversity can involve the exploitation of the available time, frequency, coding, spatial, and polarization degrees-of-freedom [9–20]. This chapter focuses on distinct methods for coding, spectral, and spatial forms of waveform diversity for this application. Specific design strategies for each are considered, along with the discussion of practical implementation issues.

Section 23.1 explores the coding diversity approach with special attention devoted to spectrally contained constant-envelope radar-embedded communication (REC) waveforms [10] and the inherent range sidelobe modulation (RSM) effect [9] that subsequently arises. The RSM effect, when not addressed, degrades radar detection performance. A REC formulation capable of mitigating RSM by trading-off communication performance [10] is described. Receive filter design to address RSM within this REC system [21] is discussed. Finally, RSM is mathematically characterized [22]

[1]U.S. Air Force Research Laboratory, USA
[2]Electrical Engineering & Computer Science Department, University of Kansas, USA

to quantify its dependence on the REC parameters and to enhance the understanding of REC system design trade-offs.

Section 23.2 describes a spatial diversity technique where radar and communication signals having the same temporal and spectral support are simultaneously transmitted in different spatial directions from an antenna array via a multiple-input–multiple-output (MIMO) arrangement [23–27]. To increase the power efficiency of the emission, an iterative process denoted as far-field radiated emission design (FFRED) [20] is introduced that facilitates the generation of constant amplitude frequency-modulated (FM) waveforms that are naturally amenable to a radar transmitter. The power efficiency of the emission under different scenarios is characterized by specific measures of the spatial distribution of power and the spatially dependent bit error rate (BER) for the communication mode [20,28].

Finally, Section 23.3 introduces a spectral diversity scheme in which a notch is placed in the spectrum of an FM noise radar waveform [29–31] and a communication signal is designed to reside within the spectral notch [19]. Through proper shaping of the communication signal spectrum to blend in with the surrounding radar spectrum, a composite far-field emission with favorable autocorrelation properties is formed [19]. Simulation results are presented showing the efficacy of the proposed approach (experimental measurements are also available in [19]). As a special case of this approach, a type of spectrally shaped communication waveform for secondary usage in a passive "commensal" radar mode [32] is also explored.

23.1 Coding waveform diversity

For the coding perspective on dual-function radar/communications, the radar emission is selected on a pulse-to-pulse basis from a set of radar waveforms that collectively form a communication alphabet [9–12,33]. Each radar waveform in the alphabet is associated with a distinct binary information sequence. Let $\mathbf{b}_{n_\mathrm{p}} = [b_{n_\mathrm{p},0}, \ldots, b_{n_\mathrm{p},N_\mathrm{b}-1}]$, with $b_{n_\mathrm{p},n_\mathrm{b}} \in \{0, 1\}$ and $n_\mathrm{b} = 0, \ldots, N_\mathrm{b} - 1$, denote the information embedded into the radar emission at the n_pth pulse, where $n_\mathrm{p} = 0, \ldots, N_\mathrm{p} - 1$, for N_p is the number of pulses in the coherent processing interval (CPI). This REC system thereby selects a waveform denoted by $s_{n_\mathrm{p}}(t; \mathbf{b}_{n_\mathrm{p}})$ according to $\mathbf{b}_{n_\mathrm{p}}$. While each waveform $s_{n_\mathrm{p}}(t; \mathbf{b}_{n_\mathrm{p}})$ is unique, all waveforms in the communication alphabet have the same time duration (pulse width T) and spectral support.

In practice, the waveform selection may be implemented as a form of waveform generation when the number of distinct radar waveforms 2^{N_b} is large. The N_b bits of information embedded in each pulse may be viewed as an underlying code, hence *coding diversity*. The particular physical waveform $s_{n_\mathrm{p}}(t; \mathbf{b}_{n_\mathrm{p}})$ is generated within the transmitter, where it is amplified, modulated onto a carrier, and emitted in a desired spatial direction. The communication receiver, which is equipped either with knowledge of the set of waveforms or the waveform generation rule, can recover the information sequence by determining which waveform was transmitted.

The primary challenge with this coding diversity approach is the clutter RSM [9] that arises because the matched filtering (pulse compression) of different REC waveforms leads to different range sidelobe structures. When Doppler processing is performed across the CPI of pulsed echoes, the presence of RSM induces a partial loss of sidelobe coherency, the consequence of which is increased residual clutter after cancelation and thus degraded target detection performance. To maintain acceptable radar performance, RSM should be reduced, which can be achieved by designing the set of 2^{N_b} waveforms and the associated set of 2^{N_b} radar receive filters such that all waveform-filter pair responses are driven to be approximately identical [9].

23.1.1 PCFM-based radar-embedded communication (REC)

With this scheme [10], information sequences are implemented using continuous phase modulation (CPM) [34] and phase-attached to a fixed polyphase-coded FM (PCFM) [35,36] radar waveform, the implementation of which is also derived from CPM. The CPM implementation comprises a family of constant-envelope signaling schemes that are used in Bluetooth standards [37], and with two variants standardized for military applications in aeronautical telemetry [38,39]. The resulting REC waveforms retain the CPM structure and, therefore have a constant envelope and a continuous phase. The continuous phase aspect leads to high spectral efficiency and constrains the spectral roll-off, thereby ensuring good spectral containment. The constant envelope attribute allows the radar transmitter power amplifier to be operated in saturation, which maximizes "energy on target." Further, the adjustable parameterization of this PCFM + CPM scheme enables direct control of the degree of RSM by trading off bit error rate (BER) and/or data throughput.

23.1.1.1 System model

With the PCFM-based scheme, the REC waveform during the n_pth pulse is given as

$$\begin{aligned} s_{n_p}(t;\overline{\boldsymbol{\beta}},\boldsymbol{\alpha}_{n_p}) &= \exp\left\{j\left(\psi(t;\overline{\boldsymbol{\beta}})+\phi(t;\boldsymbol{\alpha}_{n_p})\right)\right\} \\ &= \underbrace{\exp\{j\psi(t;\overline{\boldsymbol{\beta}})\}}_{s_r(t;\overline{\boldsymbol{\beta}})}\underbrace{\exp\{j\phi(t;\boldsymbol{\alpha}_{n_p})\}}_{s_{c,n_p}(t;\boldsymbol{\alpha}_{n_p})} \end{aligned} \tag{23.1}$$

where $s_r(t;\overline{\boldsymbol{\beta}})$ is the PCFM waveform implementation of the *fixed* radar code $\overline{\boldsymbol{\beta}} = [\overline{\beta}_0,\ldots,\overline{\beta}_{N_c-1}]$, with $|\overline{\beta}_{n_c}| \le \pi$ for all $n_c = 0,\ldots,N_c-1$ and N_c the time-bandwidth product. Likewise, $s_{c,n_p}(t;\boldsymbol{\alpha}_{n_p})$ is the CPM waveform resulting from modulating the 2^m-ary communication sequence $\boldsymbol{\alpha}_{n_p} = [\alpha_{n_p,0},\ldots,\alpha_{n_p,N_s-1}]$ where $\alpha_{n_p,n_s} \in \{\pm 1, \pm 3,\ldots,\pm(2^m-1)\}$ and $n_s = 0,\ldots,N_s-1$. The communication symbol sequence $\boldsymbol{\alpha}_{n_p}$ is obtained from the binary information sequence $\mathbf{b}_{n_p}$ where $N_b = mN_s$ and m, the number of bits per symbol, is the *modulation order.*

The PCFM [35] radar code $\overline{\boldsymbol{\beta}}$ represents the phase changes between successive chips of the fixed polyphase radar code $\overline{\boldsymbol{\eta}} = [\overline{\eta}_0,\ldots,\overline{\eta}_{N_c}]$. Because the radar code $\overline{\boldsymbol{\beta}}$ is fixed, the radar waveform component $s_r(t;\overline{\boldsymbol{\beta}})$ does not change from pulse to pulse. In contrast, the CPM communication waveform component $s_{c,n_p}(t;\boldsymbol{\alpha}_{n_p})$ does change on a pulse-to-pulse basis as new information is transmitted in each pulse.

The CPM-based PCFM implementation of $s_r(t;\overline{\boldsymbol{\beta}})$ [35] is shown in Figure 23.1. The length-N_c train of impulses, with the n_c-th impulse scaled by $\overline{\beta}_{n_c}$, is convolved

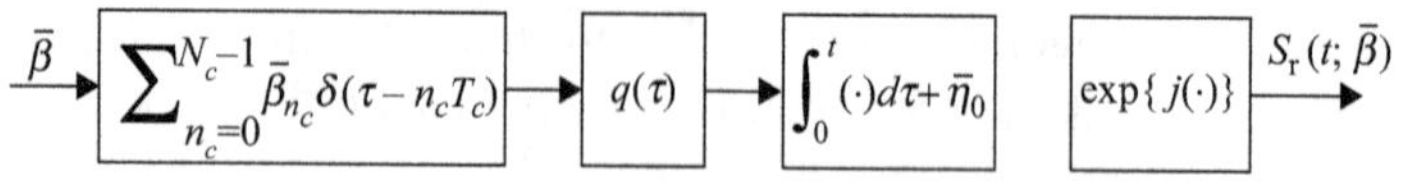

Figure 23.1 Block diagram of CPM-based PCFM implementation for radar [35]

with the *shaping filter* $q(t)$ and subsequently integrated to produce the continuous phase of the radar waveform, denoted by

$$\psi(t;\overline{\boldsymbol{\beta}}) = \int_0^t q(\tau) * \left[\sum_{n_c=0}^{N_c-1} \overline{\beta}_{n_c}\delta(\tau - n_c T_c)\right] d\tau + \overline{\eta}_0, \tag{23.2}$$

where T_c is the time duration of the chip interval, $*$ denotes convolution, and $\overline{\eta}_0$ is the initial phase value. The shaping filter $q(t)$ has time support on $[0, T_c]$ and the area under $q(t)$ is unity. The resulting PCFM radar waveform $s_r(t;\overline{\boldsymbol{\beta}}) = \exp\{j\psi(t;\overline{\boldsymbol{\beta}})\}$ of time duration $T = N_c T_c$ is referred to as the *base radar waveform*.

The communication component $s_{c,n_p}(t;\boldsymbol{\alpha}_{n_p}) = \exp\{j\phi(t;\boldsymbol{\alpha}_{n_p})\}$ is likewise obtained by modulating $\boldsymbol{\alpha}_{n_p}$ as a CPM waveform with symbol interval T_s, modulation index h (a rational number that dictates the degree of possible phase deviation), and shaping filter $q_c(t)$ that has time support on $[0, T_s]$ and unit area. The resulting communication waveform phase is [34]

$$\phi(t;\boldsymbol{\alpha}_{n_p,n_s}) = h\pi \int_0^t q_c(\tau) * \left[\sum_{n_s=0}^{N_s-1} \alpha_{n_s}\delta(\tau - n_s T_s)\right] d\tau. \tag{23.3}$$

The CPM implementation of $\boldsymbol{\alpha}_{n_p}$ is equivalent to the PCFM implementation of $\overline{\boldsymbol{\beta}}$ shown in Figure 23.1 with the mapping $\overline{\beta}_{n_c} \to h\pi\alpha_{n_p,n_s}$, $T_c \to T_s$, $q(\tau) \to q_c(\tau)$, $\overline{\eta}_0 \to 0$. The modulation index h is a key parameter as it controls the total phase change due to a communication symbol transmission, which occurs over T_s. The phase change due to α_{n_p,n_s} is therefore $h\pi\alpha_{n_p,n_s}$, such that the magnitude of the maximum phase change is $h\pi(2^m - 1)$. The communication waveform $s_{c,n_p}(t;\boldsymbol{\alpha}_{n_p})$ has the same duration as the base radar waveform (pulsewidth T), leading to the relationship

$$N_s = \frac{T}{T_s}. \tag{23.4}$$

It follows that for a given base radar waveform, the communication symbol interval can be increased/decreased to realize a conversely decreased/increased symbol rate.

As shown in (23.1), the communication sequence $\boldsymbol{\alpha}_{n_p}$ is embedded into the n_pth REC emission by multiplying the base radar waveform by the communication waveform, which is equivalent to adding the respective phase components. The resulting REC waveforms possess high-power efficiency and spectral efficiency. Moreover, the base radar waveform maintains a degree of commonality across the set of changing REC waveforms in the CPI that is uniquely specified by h, m, and T_s, with greater commonality translating to reduced RSM. In particular, the REC waveforms become more alike as h or m is decreased, and/or as T_s is increased [10], which reduces the severity of RSM. However, these strategies are undesirable from a communication performance perspective since decreasing m or increasing T_s reduces the data throughput of mN_s bits/pulse, while reducing h increases the BER for a given SNR [34]. If forward error correction coding is employed, the coding rate must be lowered as h is decreased to satisfy a given probability of decoding error constraint [40,41].

Related work in the literature involves the phase modulation of a linear FM (LFM) radar waveform [11,12]. In [12], information sequences modulated with minimum-shift keying (MSK) are multiplied by an LFM pulse. In [11], information sequences modulated with phase-shift keying (PSK) [42] having an adjustable phase parameter are multiplied by a higher rate pseudorandom binary sequence (i.e. spread spectrum) and then a discretized LFM pulse. The PCFM-based scheme presented here (from [10]) represents a generalization of such approaches, the spreading of [11] notwithstanding, that is applicable to arbitrary FM radar waveforms. Specifically, MSK can be modeled as a special case of binary CPM. Likewise, PSK with an adjustable phase parameter can be implemented as a CPM waveform with an adjustable modulation index, an impulse function $\delta(\tau)$ as the shaping filter, and a precoder to convert binary PSK (BPSK) symbols to ternary CPM symbols akin to the CPM implementation of shaped-offset quadrature PSK [43].

Assuming knowledge of the base radar waveform, communication system parameters, and ideal synchronization at the communication receiver, demodulation is performed by first multiplying by $\exp\{-j\psi(t;\overline{\boldsymbol{\beta}})\}$, which realizes

$$\begin{aligned} r(t;\boldsymbol{\alpha}_{n_\mathrm{p}}) &= \exp\{-j\psi(t;\overline{\boldsymbol{\beta}})\}(\sqrt{P_\mathrm{r}}\, s_{n_\mathrm{p}}(t;\overline{\boldsymbol{\beta}},\boldsymbol{\alpha}_{n_\mathrm{p}}) + n(t)) \\ &= \sqrt{P_\mathrm{r}}\exp\{j\phi(t;\boldsymbol{\alpha}_{n_\mathrm{p}})\} + \widehat{n}(t), \end{aligned} \tag{23.5}$$

where P_r is the received power, $n(t)$ is the complex white Gaussian receive noise process, and $\widehat{n}(t) = \exp\{-j\psi(t;\overline{\boldsymbol{\beta}})\}n(t)$ is the resulting noise process, which is statistically equivalent to $n(t)$. The optimal detection of $\boldsymbol{\alpha}_{n_\mathrm{p}}$ is then achieved by applying the Viterbi algorithm [44] which operates on the $2v$-state trellis, for v the denominator of h used in the CPM modulator. Since $r(t;\boldsymbol{\alpha}_{n_\mathrm{p}})$ only depends on the communication sequence $\boldsymbol{\alpha}_{n_\mathrm{p}}$, the BER does not depend on the radar system parameters such as the base radar waveform or the number of pulses in the CPI. As an example, consider the BER resulting from the use of binary CPM with a rectangular (RECT) shaping filter and $N_\mathrm{s} = 64$ symbols/pulse as a function of communication receiver SNR for modulation indices $h = 1/2, 1/4, 1/8$, and $1/16$. The resulting BER curves are shown in Figure 23.2, where it is observed that the required SNR to achieve a given BER increases with decreasing modulation index h.

From the radar perspective, Figure 23.3 illustrates the radar range-Doppler response (in dB) to a point target for modulation indices $h = 0, 1/2, 1/4$, and $1/16$. The case with $h = 0$, where all REC waveforms are identical to the base radar waveform and hence no RSM is present, is included as a baseline. The base radar waveform is a PCFM-implemented LFM chirp [36] using a RECT shaping filter, a time-bandwidth product of $N_\mathrm{c} = 64$, and $N_\mathrm{p} = 32$ pulses in the CPI. The CPM-based communication component is also implemented with a RECT shaping filter, a binary symbol alphabet, and $N_\mathrm{s} = 64$ symbols/pulse. A random binary information

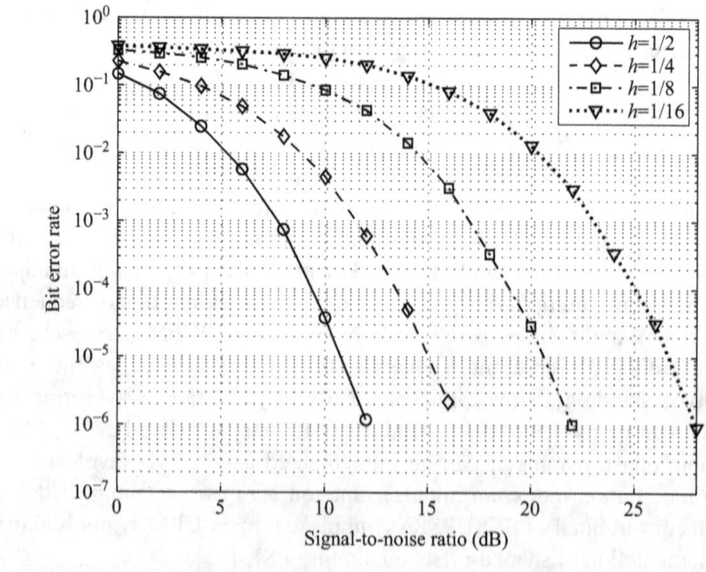

Figure 23.2 The bit error rate (BER) of binary CPM with rectangular shaping filter for various values of modulation index h

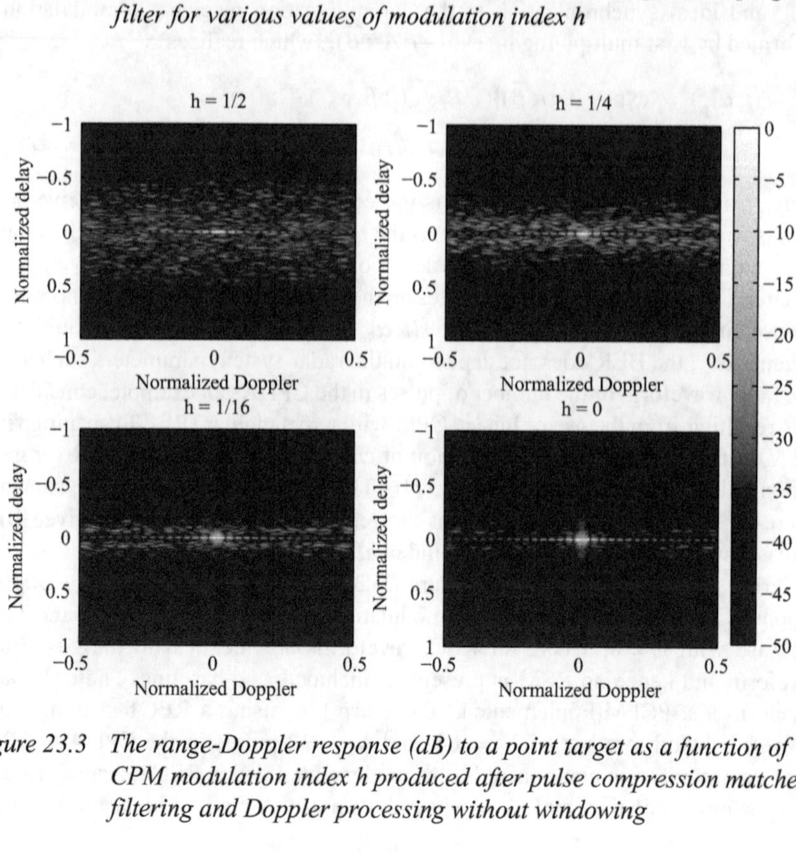

Figure 23.3 The range-Doppler response (dB) to a point target as a function of CPM modulation index h produced after pulse compression matched filtering and Doppler processing without windowing

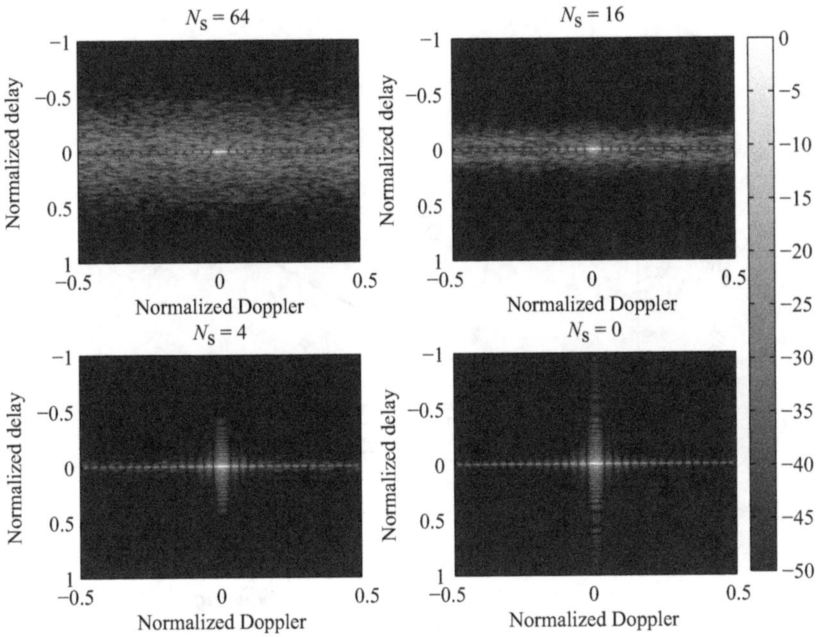

Figure 23.4 The range-Doppler response (dB) to a point target as a function of the number of symbols per pulse N_s produced after pulse compression matched filtering and Doppler processing without windowing

sequence is generated for each pulse. The radar receive processing consists of pulse compression matched filtering (MF) and discrete Fourier transform (DFT) Doppler processing without windowing. Reducing the modulation index clearly reduces the RSM, which is observed by noting that the range-Doppler response for $h = 1/16$ is very similar to the $h = 0$ response, while the RSM for $h = 1/2$ is more severe.

Alternatively, Figure 23.4 shows the range-Doppler response (in dB) to a point target for symbol interval values $T_s = T_c, 4T_c$, and $16T_c$, which correspond to $N_s = 64, 16$, and 4 symbols per pulse, respectively. The case with no communication symbols ($N_s = 0$), where no RSM is present, is again included as a baseline. The modulation index is $h = 1/4$, and all other parameters are the same as those used for Figure 23.3. It is observed in Figure 23.4 that decreasing N_s (i.e. increasing T_s) also reduces the severity of RSM.

23.1.1.2 Spectral content

The spectral broadening relative to that of the base radar waveform is an important aspect of PCFM-based REC waveforms. The broadening effect can be observed from (23.1) by noting that the Fourier transform of $s_{n_p}(t; \overline{\boldsymbol{\beta}}, \boldsymbol{\alpha}_{n_p})$ is the convolution of the Fourier transforms of $s_r(t; \overline{\boldsymbol{\beta}})$ and $s_{c,n_p}(t; \boldsymbol{\alpha}_{n_p})$. The spectrum of $s_{c,n_p}(t; \boldsymbol{\alpha}_{n_p})$, and thus $s_{n_p}(t; \overline{\boldsymbol{\beta}}, \boldsymbol{\alpha}_{n_p})$, broadens with increasing h and m, and with decreasing T_s [42]. However,

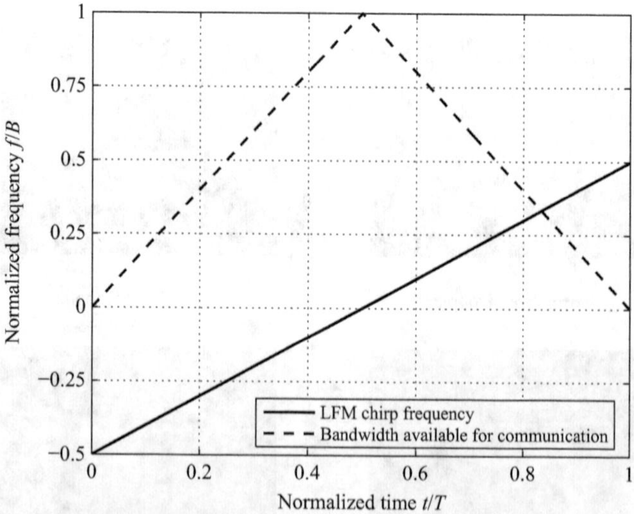

Figure 23.5 LFM frequency and bandwidth available for communication normalized by available bandwidth B as a function of time normalized by pulse duration T

for chirp-like radar waveforms (e.g. LFM and most non-LFM) that traverse the band during the pulse, the spectral broadening can be avoided by the use of *guard symbols* at the beginning and end of the pulse. These guard symbols translate into guard bands at the edges of the base radar waveform spectrum.

The quantity $f_c(t) = d\psi(t; \overline{\boldsymbol{\theta}})/dt$ can be viewed as a time-varying (baseband) *carrier frequency* of the communication waveform $s_{c,n_p}(t; \boldsymbol{\alpha}_{n_p})$. For example, in the case of an up-chirp LFM base radar waveform with bandwidth B, this time-varying carrier frequency is

$$f_c(t) = -\frac{B}{2} + \frac{tB}{T}, \text{ where } 0 \leq t \leq T. \tag{23.6}$$

If the REC waveforms are constrained to occupy the same band as the base radar waveform, then the bandwidth $W(t)$ available for communication as a function of time $0 \leq t \leq T$ is

$$W(t) = 2\min\left\{\left|f_c(t) + \frac{B}{2}\right|, \left|f_c(t) - \frac{B}{2}\right|\right\}. \tag{23.7}$$

The normalized LFM frequency and normalized bandwidth available for communication are shown in Figure 23.5 as a function of normalized time t/T during the pulse. Assume that the communication waveforms are contained within the band $[-B_c/2, +B_c/2]$ (i.e. the out-of-band power is negligible), where B_c is uniquely specified by h, m, and $q_c(t)$ (hence T_s). By transmitting communication symbols only during the time interval $[t_1, t_2]$, where $B_c \leq W(t)$ for $t_1 \leq t \leq t_2$, the spectrum

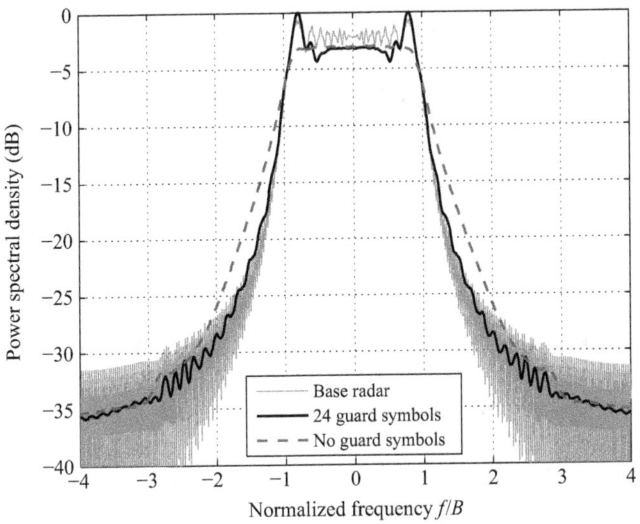

Figure 23.6 PSD of the base radar waveform (dB) and REC PSDs (averaged over different communication sequences) for cases without guard symbols, and with $2N_g = 24$ guard symbols

occupied by the REC waveforms can be limited to $[-B/2, +B/2]$. This arrangement is equivalent to transmitting *null* communication symbols (i.e. guard symbols) at the beginning and end of the pulse, i.e. $\alpha_{n_p,n_s} = 0$ for $0 \leq n_s \leq N_g - 1$ and $N_s - N_g \leq n_s \leq N_s - 1$, for some N_g determined by $h, m, q_c(t)$ and B.

Figure 23.6 compares the power spectral density (PSD) of the base radar waveform with the PSDs realized by REC waveforms (averaged over multiple independent information sequences) with and without guard symbols. The base radar waveform is a PCFM-implemented up-chirp LFM radar waveform [36] with a RECT shaping filter and $N_c = 64$. The CPM parameters are $h = 1/4$, $m = 1$ bit/symbol, and a RECT shaping filter of time support $T_s = T_c$ ($N_s = N_c = 64$ symbols/pulse). There are 64 and 40 communication symbols transmitted, without and with guard symbols, respectively. It is observed that the use of guard symbols eliminates the spectral broadening that would otherwise occur without guard symbols.

Given that a null symbol is equivalent to setting $h = 0$ or $m = 0$, the use of guard symbols can alternatively be formulated as a special case of a time-varying modulation index or modulation order, i.e. $h(t) = h$ or $m(t) = m$ for $t_1 \leq t \leq t_2$, and $h(t) = 0$ or $m(t) = 0$ otherwise. When either quantity is defined as a function of time, the communication bandwidth B_c likewise becomes a function of time denoted by $B_c(t)$. A more spectrally efficient implementation from a communication performance perspective would aim to achieve $B_c(t) \approx W(t)$ by varying $h(t)$ or $m(t)$ throughout the pulse. Such an implementation would allow $h(t)$ or $m(t)$ to increase/decrease before/after the middle of the pulse (an up-down staircase pattern), which would result in reduced BER or increased data throughput, respectively.

23.1.2 Receive filter design to address RSM for coding diversity REC

For coding diversity REC, the objective of receive filter design, in addition to reducing the range sidelobes, is to produce pulse compression responses that are similar despite the pulse-to-pulse change in REC waveforms so that RSM can be mitigated [9,21,33]. The design methods in [9,21,33] focus on the realization of a desired *common filter response* $y(t)$, rather than optimization of the receive filters individually. The joint least squares (JLS) approach is an iterative method introduced in [9] that is applicable to general pulse-agile, or waveform-agile, operation (i.e. changing the waveform on a pulse-to-pulse basis) with a small number of waveforms (around 2–4). The closed-form solution for the JLS method derived in [33] for moving target indication (MTI) is likewise only applicable to a small number of waveforms.

One of the attractive features of the PCFM-based REC is the high-data throughput, which is on the order of the time-bandwidth product per pulse. As a result, the number of distinct REC waveforms 2^{N_b} is very large, which prohibits joint optimization of the desired filter response $\mathbf{y}$ (discrete-time representation of $y(t)$) based on the set of waveforms as in [9,33]. In [45–48], joint range-Doppler processing was considered for arbitrary pulse-agile emissions, though the computational complexity of these approaches is likewise prohibitive when applied to the PCFM-based REC.[1] On the other hand, if the individual pulse-agile waveforms are themselves optimized to provide low sidelobes (see [49]), then simpler mismatched filtering can be employed. However, the PCFM-based REC waveforms are not expected to individually possess low sidelobes.

The methods introduced in [21] address RSM within the PCFM-based REC framework [10] by establishing a *fixed* (i.e. not jointly optimized) desired common filter response. Given this common filter response, the receive filter for each transmitted REC waveform is computed individually. In the determination of the common filter response, the inherent commonality among the REC waveforms stemming from the common phase component, i.e. the base radar waveform, is exploited. Figure 23.7 illustrates the RSM effect within the PCFM-based REC framework. The MF outputs of three REC waveforms are shown for random binary communication sequences of length $N_s = 32$ and a RECT shaping filter for modulation indices $h = 1/4$ and $1/16$. The base radar waveform is a PCFM-implemented LFM with $N_c = 32$, $N_r = 8$ samples/chip, and a RECT shaping filter. The changing sidelobe structures observed across the MF outputs results in residual clutter in the range-Doppler response as previously illustrated in Figures 23.3 and 23.4.

[1]The approaches in [9,33,45–48] can be implemented on a CPI-to-CPI basis. Such an implementation would apply each approach to the set of distinct REC waveforms transmitted within a CPI, which has at most N_p waveforms. When the set of waveforms changes in the next CPI, each approach is applied to the new set and so on. As a result, these approaches can be applied to the PCFM-based REC despite 2^{N_b} being in general very large. However, for practical values of N_p, the resulting computational complexity is still high, and much greater than the methods in [21].

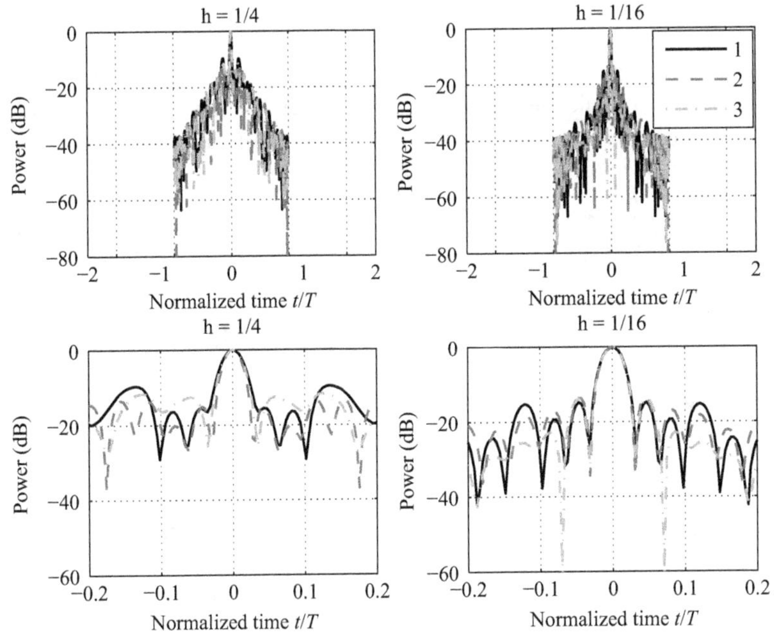

Figure 23.7 Matched filter (MF) outputs of three randomly generated waveforms (top) and close-up view (bottom) for h = 1/4 and h = 1/16

23.1.2.1 Common filter response

Now consider the set of REC waveforms $s_k(t)$ for $k = 0, \ldots, 2^{N_b} - 1$, and the corresponding set of discrete-time representations $\mathbf{s}_k = [s_{k,0}, \ldots, s_{k,N_c N_r - 1}]^T$ obtained by sampling each waveform at a rate of N_r samples per chip to ensure sufficient "oversampling" relative to 3-dB bandwidth (theoretically, a pulsed waveform has infinite bandwidth). Ideally, the objective for receive filter design is to find the associated discrete-time filter $\mathbf{h}_k = [h_{k,0}, \ldots, h_{k,\widetilde{N}-1}]^T$ of length $\widetilde{N}$ (samples) satisfying the matrix representation of convolution denoted as [9]

$$\mathbf{A}_k \mathbf{h}_k = \mathbf{y}, \tag{23.8}$$

where $\mathbf{A}_k$ is the $(\widetilde{N} + N_c N_r - 1) \times \widetilde{N}$ Toeplitz *delay-shift matrix* for the kth discrete-time REC waveform $\mathbf{s}_k$, given as

$$\mathbf{A}_k = \begin{bmatrix} s_{k,0} & 0 & \cdots & 0 \\ \vdots & s_{k,0} & & \vdots \\ s_{k,N_c N_r - 1} & \vdots & \ddots & 0 \\ 0 & s_{k,N_c N_r - 1} & & s_{k,0} \\ \vdots & & \ddots & \vdots \\ 0 & \cdots & & s_{k,N_c N_r - 1} \end{bmatrix}, \tag{23.9}$$

and $\mathbf{y} = [y_0, \ldots, y_{\widetilde{N}+N_c N_r-2}]^T$ is the discrete-time common filter response. However, it was shown in [9] that (23.8) cannot be achieved with finite-length filters for more than two distinct waveforms. Therefore, for an arbitrary number of REC waveforms the objective is to minimize the (average) incoherence among the filter responses. For this purpose, the *mismatch metric* $\Delta_{\text{MM}} = \mathbb{E}\left[\Delta_{k,l}\right]$ is introduced, where $\mathbb{E}[\cdot]$ is the expectation operator and $\Delta_{k,l}$ for $k \neq l$ is defined as

$$\Delta_{k,l} \triangleq \frac{||\mathbf{A}_k \widehat{\mathbf{h}}_k - \mathbf{A}_l \widehat{\mathbf{h}}_l||^2}{\mathbb{E}\left[||\mathbf{A}_k \widehat{\mathbf{h}}_k||^2\right]}, \tag{23.10}$$

for $\widehat{\mathbf{h}}_k$ the estimate of the kth filter. The mismatch metric quantifies the ratio of the pairwise *incoherent* filter response energy (the remaining RSM) to the average filter response energy.

A regularized least-squares (LS) solution [9,35,50,51] to (23.8) is given as

$$\widehat{\mathbf{h}}_k = \left(\mathbf{A}_k^H \mathbf{A}_k + \delta \mathbf{I}_{\widetilde{N}}\right)^{-1} \mathbf{A}_k^H \mathbf{y}, \tag{23.11}$$

where $\mathbf{A}_k^H$ denotes the Hermitian (complex-conjugate transpose) of $\mathbf{A}_k$, $\mathbf{I}_{\widetilde{N}}$ is the $\widetilde{N} \times \widetilde{N}$ identity matrix, and the diagonal loading term $\delta \mathbf{I}_{\widetilde{N}}$ may be invoked to prevent ill-conditioning. The standard LS mismatched filter (MMF) design [50] uses the desired filter response $\mathbf{y} = [0, ..., 0, 1, 0, ..., 0]^T$, with the non-zero element corresponding to the match point. While this filter can minimize the range sidelobes for a single waveform, it does not address the problem of sidelobe incoherence across a set of different waveforms and it can also suffer from severe mismatch loss due to an undesired super-resolution effect caused by "oversampling" with respect to 3 dB bandwidth [52]. Generalizing $\mathbf{y}$ to serve as a desired common filter response provides a framework with which to explore various trade-offs in filter design that are uniquely specified by the common response $\mathbf{y}$.

Another useful metric for receive filter design is the SNR loss due to mismatch [53]. The SNR loss is computed as the expected value of the ratio of the peak SNR at the output of the MF to the peak SNR at the output of the MMF or

$$\text{SNR loss [dB]} = 10 \log_{10} \frac{P_{\text{peak, MF}}/||\mathbf{s}_k||^2}{\mathbb{E}\left[P_{\text{peak},k}/||\widehat{\mathbf{h}}_k||^2\right]}. \tag{23.12}$$

In (23.12), $P_{\text{peak, MF}}$ is the peak power at the output of the MF, which is a constant for all REC waveforms. Likewise, $P_{\text{peak},k}$ is the peak power at the output of the kth MMF. Note that the norm of the MF for waveform k, denoted by $||\mathbf{s}_k||$, is also a constant for all k.

23.1.2.2 Matched filter (MF) formulation

Embedding communication symbols into radar emissions in the PCFM-based REC formulation causes small phase perturbations *relative* to the phase trajectory of the base radar waveform. Accordingly, the MF output of each waveform is similar to the MF output of the base radar waveform. In addition, it intuitively makes sense to choose a common filter response that can be generated by the base radar waveform. Hence, the MF output of $s_r(t)$ can be used to establish the common response via

$$\mathbf{y}_{\text{MF}} = \mathbf{A}_r \, \mathbf{s}_{r,\text{MF}}, \tag{23.13}$$

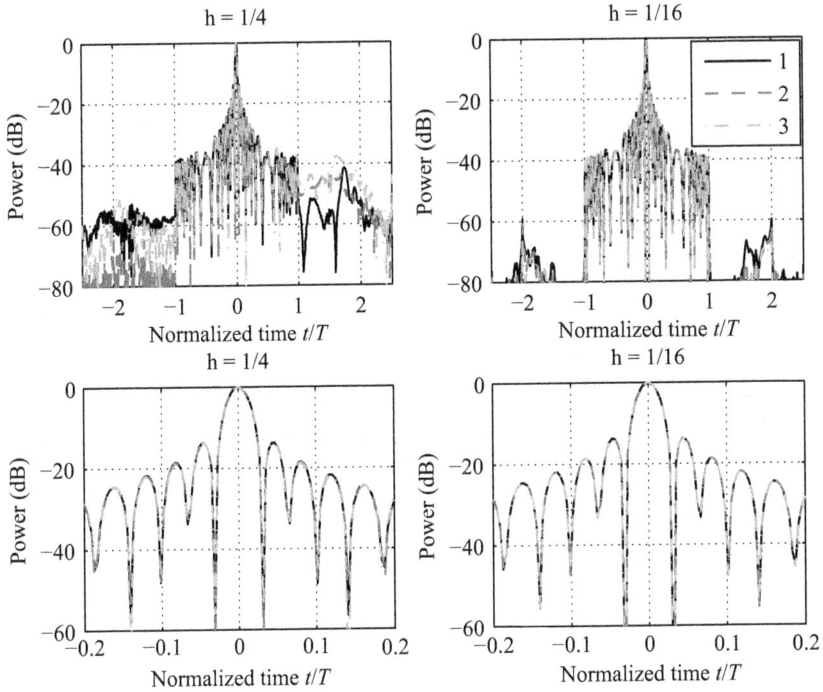

Figure 23.8 Common MF filter responses using (23.11) and (23.13) of three randomly generated REC waveforms (top) and close-up view (bottom) for h = 1/4 and h = 1/16

where $\mathbf{A}_\mathrm{r}$ is the delay-shift matrix containing the discretized base radar waveform like (23.9), and $\mathbf{s}_{\mathrm{r,MF}} = [s^*_{\mathrm{r},N_\mathrm{c}N_\mathrm{r}}, \ldots, s^*_{\mathrm{r},1}]^T$ is the MF of the discrete-time base radar waveform. Then, for each waveform $\mathbf{s}_k$, the filter $\widehat{\mathbf{h}}_k$ is computed from (23.11) using $\underline{\mathbf{y}} = \mathbf{y}_{\mathrm{MF}}$. The filter $\widehat{\mathbf{h}}_k$ is not constrained to have the same length as the MF, so when $\widetilde{N} > N_\mathrm{c}N_\mathrm{r}$, the common response $\mathbf{y}_{\mathrm{MF}}$ is zero-padded accordingly.

In Figure 23.8, the filter responses using (23.11) and (23.13) for three randomly generated REC waveforms are plotted for $h = 1/4$ and $1/16$. The receive filter length is $\widetilde{N} = 1024$ samples (four times the length of the REC waveforms), and all other parameters are the same as those used for Figure 23.7. Diagonal loading in (23.11) is not used, so $\delta = 0$. The filter responses in Figure 23.8 exhibit sufficient coherence for both h values, with the $h = 1/16$ case being superior as expected. The resulting mismatch metric from (23.10) is shown in Table 23.1 (first row) for modulation indices $h = 1/2, 1/4, 1/8$, and $1/16$. The mismatch metric value for the MF, referred to as standard MF (fifth row), is also included for reference. The mismatch metric value decreases with decreasing h as expected. Per Table 23.2 (first row), the SNR loss likewise decreases with decreasing h.

Table 23.1 Mismatch Metric

	$h = 1/2$	$h = 1/4$	$h = 1/8$	$h = 1/16$
MF via (23.11), (23.13)	3×10^{-2}	6×10^{-3}	4×10^{-5}	2×10^{-6}
MMF via (23.11), (23.14)	4×10^{-2}	7×10^{-3}	5×10^{-4}	2×10^{-4}
Cascade via (23.16), (23.19)	6×10^{-2}	1×10^{-2}	3×10^{-4}	4×10^{-5}
Standard LS	1×10^{-1}	8×10^{-2}	1×10^{-2}	3×10^{-3}
Standard MF	9×10^{-1}	7×10^{-1}	3×10^{-1}	1×10^{-1}

Table 23.2 SNR Loss

	$h = 1/2$	$h = 1/4$	$h = 1/8$	$h = 1/16$
MF via (23.11), (23.13)	6.9 dB	5.4 dB	1.3 dB	0.3 dB
MMF via (23.11), (23.14)	7.9 dB	6.9 dB	5.4 dB	4.5 dB
Cascade via (23.16), (23.19)	6.0 dB	4.7 dB	1.6 dB	0.7 dB
Standard LS	4.0 dB	3.4 dB	1.5 dB	0.8 dB

23.1.2.3 Mismatched filter (MMF) formulation

Now consider the case in which the common response is itself constructed from an MMF formulation via

$$\mathbf{y}_{\mathrm{MMF}} = \mathbf{A}_{\mathrm{r}} \left[(\mathbf{A}_{\mathrm{r}}^H \mathbf{A}_{\mathrm{r}} + \delta \mathbf{I}_{\tilde{N}})^{-1} \mathbf{A}_{\mathrm{r}}^H \mathbf{y}_{\mathrm{MF,\,main}} \right], \tag{23.14}$$

where

$$\mathbf{y}_{\mathrm{MF,\,main}} = [0, \ldots, 0, y_{\mathrm{MF},\, N_{\mathrm{c}}N_{\mathrm{r}}-N_{\mathrm{r}}}, \ldots, y_{\mathrm{MF},\, N_{\mathrm{c}}N_{\mathrm{r}}+N_{\mathrm{r}}}, 0, \ldots, 0]^T, \tag{23.15}$$

in which the nonzero terms comprise only the main lobe of the base radar MF response $\mathbf{y}_{\mathrm{MF}}$. In a similar manner to the spectrum-shaping LS formulation in [54], the objective of this approach is to trade the prospect of enhanced range resolution (i.e. we wish to maintain nominal resolution [52]) for a less constrained LS problem.

Using the same parameters as the previous example, the filter response powers using (23.11) and (23.14) are shown in Figure 23.9 for the same three REC waveforms used in Figure 23.8. Diagonal loading is not used because it has been observed to increase the mismatch metric. The filter outputs match more closely as the modulation index decreases. In addition, the range sidelobes are much lower than those obtained with the previous MF formulation. Per Tables 23.1 and 23.2 (second row) the mismatch metric values and SNR loss values are both greater than those obtained with the MF formulation. The difference between the two approaches in terms of SNR loss and the mismatch metric is more pronounced for $h = 1/8$ and $h = 1/16$. The numerical results suggest that the range sidelobes can be reduced at the expense of an increased mismatch metric (i.e. increased RSM).

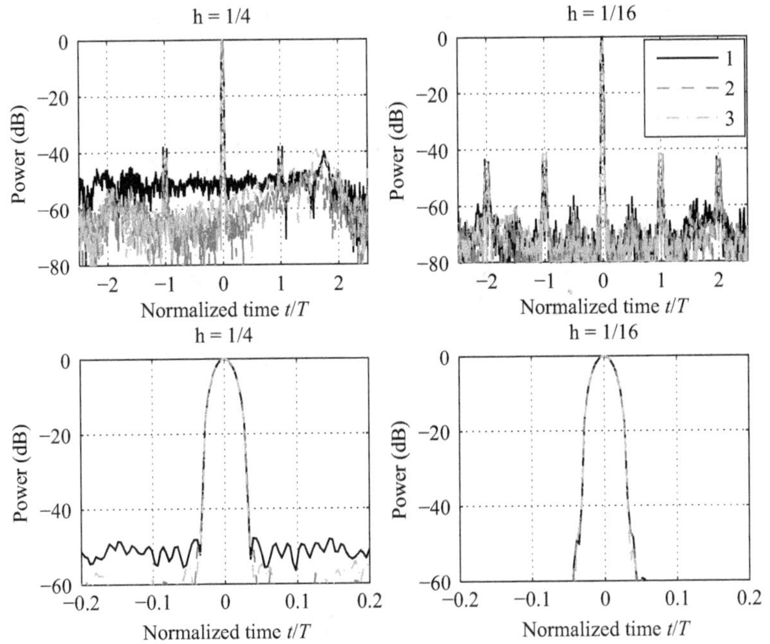

Figure 23.9 Common MMF filter responses using (23.14) and (23.11) for three randomly generated REC waveforms (top) and close-up view (bottom) for h = 1/4 and h = 1/16

For completeness, Tables 23.1 and 23.2 (fourth row) provide the mismatch metric and SNR loss for the case in which the standard LS MMF with $\mathbf{y} = [0, \ldots, 0, 1, 0, \ldots, 0]^T$ is used in (23.11). When the standard LS MMF is used, the "beam-spoiling" approach described in [36,51] is also employed to prevent an undesirable superresolution condition [52]. Diagonal loading is used with $\delta = 0.1$, which minimizes the mismatch metric. It is observed in Table 23.1 that the mismatch metric values are greater than those obtained for the MMF formulation, and much greater for $h = 1/8$ and $h = 1/16$. According to Table 23.2, the SNR loss values are lower than those obtained for the MMF formulation, while they are higher than those obtained for the MF formulation for $h = 1/8$ and $1/16$ (but still lower than MF for $h = 1/2$ and $1/4$).

23.1.2.4 Cascaded filters formulation

Now consider the case where the two goals of (a) maximizing the similarity among the filter responses and (b) minimizing the range sidelobes are *decoupled* by a cascade of two receive filters in series. The first filter (MF formulation) improves the similarity of the filter responses, while the second filter (MMF formulation) subsequently reduces the sidelobes. The motivation here is driven by the success of the MF formulation using (23.11) and (23.13) to produce good similarity among the filter responses.

The first filter has length $\widetilde{N}_1$ and is computed for each waveform $\mathbf{s}_k$ from (23.11) using the MF formulation common response $\mathbf{y}_{\text{MF}}$ from (23.13) as

$$\widehat{\mathbf{h}}_{\text{CMF},k} = \left(\mathbf{A}_k^H \mathbf{A}_k + \delta \mathbf{I}_{\widetilde{N}_1}\right)^{-1} \mathbf{A}_k^H \mathbf{y}_{\text{MF}}, \tag{23.16}$$

where $\mathbf{A}_k$ is $(\widetilde{N}_1 + N_c N_r - 1) \times \widetilde{N}_1$. The output of the first filter for the kth REC waveform is the length $(\widetilde{N}_1 + N_c N_r - 1)$ vector

$$\widehat{\mathbf{y}}_{\text{CMF},k} = \mathbf{A}_k \widehat{\mathbf{h}}_{\text{CMF},k}. \tag{23.17}$$

Using $\widehat{\mathbf{y}}_{\text{CMF},k}$, now construct another delay-shift matrix $\mathbf{B}_k$ of size $(\widetilde{N}_2 + \widetilde{N}_1 + N_c N_r - 2) \times \widetilde{N}_2$, for $\widetilde{N}_2$ the length of the second filter, given as

$$\mathbf{B}_k = \begin{bmatrix} \widehat{y}_{\text{CMF},k,0} & 0 & \cdots & 0 \\ \vdots & \widehat{y}_{\text{CMF},k,0} & & \vdots \\ \widehat{y}_{\text{CMF},k,N_c N_r - 1} & \vdots & \ddots & 0 \\ 0 & \widehat{y}_{\text{CMF},k,N_c N_r - 1} & & \widehat{y}_{\text{CMF},k,0} \\ \vdots & & \ddots & \vdots \\ 0 & \cdots & & \widehat{y}_{\text{CMF},k,N_c N_r - 1} \end{bmatrix}. \tag{23.18}$$

The second filter is then computed based on the MMF formulation from (23.11) as

$$\widehat{\mathbf{h}}_{\text{CMMF},k} = (\mathbf{B}_k^H \mathbf{B}_k + \delta \mathbf{I}_{\widetilde{N}_2})^{-1} \mathbf{B}_k^H \mathbf{y}_{\text{CMMF}}, \tag{23.19}$$

where the common filter response $\mathbf{y}_{\text{CMMF}}$ is

$$\mathbf{y}_{\text{CMMF}} = \mathbf{B}_r[(\mathbf{B}_r^H \mathbf{B}_r + \delta \mathbf{I}_{\widetilde{N}_2})^{-1} \mathbf{B}_r^H \mathbf{y}_{\text{MF, main}}], \tag{23.20}$$

for $\mathbf{y}_{\text{MF, main}}$ defined in (23.15), and $\mathbf{B}_r$ the delay-shift matrix of size $(\widetilde{N}_2 + \widetilde{N}_1 + N_c N_r - 2) \times \widetilde{N}_2$ constructed from $\mathbf{y}_{\text{MF}}$ given in (23.13) like (23.18).

The cascade filtering formulation for waveform $\mathbf{s}_k$ can be represented by a single filter $\widehat{\mathbf{h}}_{\text{C},k}$ of length $\widetilde{N}_1 + \widetilde{N}_2 - 1$ that is the convolution of $\widehat{\mathbf{h}}_{\text{CMF},k}$ and $\widehat{\mathbf{h}}_{\text{CMMF},k}$. A fair comparison between the cascade and previously described formulations requires the (total) filter lengths to be the same, i.e. $\widetilde{N} = \widetilde{N}_1 + \widetilde{N}_2 - 1$. In addition, each filter is constrained to be no shorter than that the length of its input, so $\widetilde{N}_1 \geq N_c N_r$ and $\widetilde{N}_2 \geq \widetilde{N}_1 + N_c N_r - 1$, which together imply $(\widetilde{N} - N_c N_r)/2 + 1 \geq \widetilde{N}_1 \geq N_c N_r$. Note that when the output of the first filter for each waveform satisfies $\widehat{\mathbf{y}}_{\text{CMF},k} \approx \mathbf{y}_{\text{MF}}$, the second filter can be computed based on the (approximately) constant input $\mathbf{y}_{\text{MF}}$. In that case, the second filter does not depend on the particular REC waveform in a given pulse; it can be computed once and used for all REC waveforms. As a result, the filter lengths can be increased, i.e. $\widetilde{N}_1 = \widetilde{N}$ and $\widetilde{N}_2$ arbitrary with $\widetilde{N}_2 \geq \widetilde{N}_1 + N_c N_r - 1$, without increasing the complexity of computing the receive filters compared to the MF and MMF formulations.

The filter lengths used for Figure 23.10 are $\widetilde{N}_1 = 384$ and $\widetilde{N}_2 = 641$. Diagonal loading of $\delta = 0.25$ is used in both (23.19) and (23.20) to reduce the mismatch metric value. Diagonal loading also raises the range sidelobe level at the output of the second filter, so $\delta = 0.25$ is chosen judiciously to balance between the mismatch metric

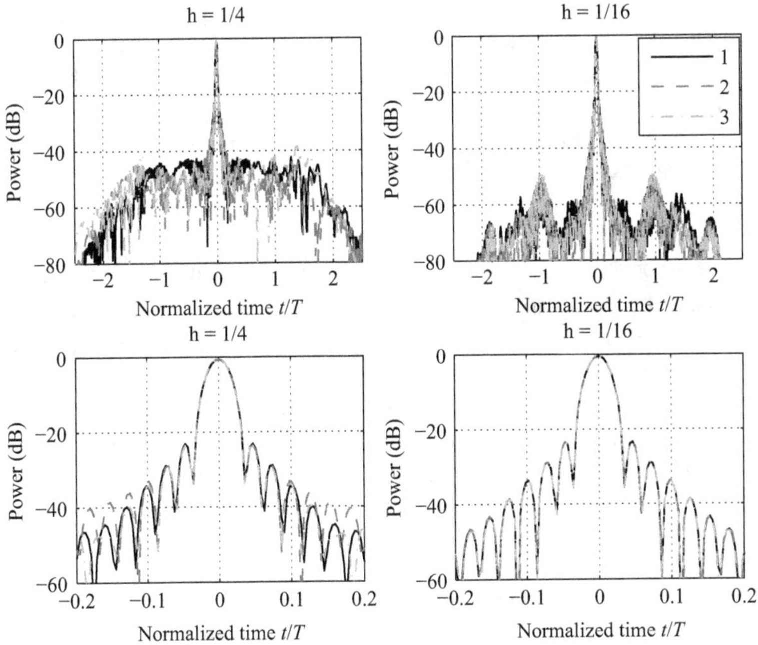

Figure 23.10 Cascaded filter responses using (23.16) and (23.19) for three randomly generated REC waveforms (top) and close-up view (bottom) for h = 1/4 and h = 1/16

performance and range sidelobe level. All other parameters are the same as those used previously. The filter responses using (23.16) and (23.19) for three randomly generated REC waveforms are shown in Figure 23.10. The range sidelobes are reduced compared to the MF formulation, while they are higher than those obtained with the MMF formulation. According to Table 23.1 (third row), the mismatch metric values are lower (higher) than they are with the MMF (MF) formulation for $h = 1/8$ and $1/16$. It can therefore be concluded that the cascade filtering approach provides a further trade-off between coherence among the filter responses and reduced range sidelobes. Per Table 23.2 (third row), the SNR loss values for the cascade formulation are smaller than (comparable) to those observed for the MMF (MF) formulation.

23.1.3 Characterization of RSM

As discussed in the previous sections, RSM [9] is an important design consideration in coding diversity REC systems. It has been addressed by a variety of receive processing techniques [9,21,33,45–48] with varying degrees of performance and computational cost. However, until recently [22], there had not been a theoretical characterization of the impact of RSM on clutter cancelation. Specifically, a closed-form expression for the (average) RSM and residual clutter power after pulse compression and

Doppler processing is derived for a mathematically tractable special case of coding diversity REC. In doing so, the dependence on the various radar/communication parameters are assessed to provide insight into the impact of RSM for arbitrary radar and communication waveforms.

Rewriting (23.1), the REC waveform at the n_{p}th pulse is the product of the base radar waveform and the communication component as

$$s_{n_{\mathrm{p}}}(t;\overline{\boldsymbol{\beta}},\boldsymbol{\alpha}_{n_{\mathrm{p}}}) = s_{\mathrm{r}}(t;\overline{\boldsymbol{\beta}})s_{\mathrm{c},n_{\mathrm{p}}}(t;\boldsymbol{\alpha}_{n_{\mathrm{p}}}), \tag{23.21}$$

which have associated discrete-time representations denoted as $\mathbf{s}_{\mathrm{r}}(\overline{\boldsymbol{\beta}})$ and $\mathbf{s}_{\mathrm{c},n_{\mathrm{p}}}(\boldsymbol{\alpha}_{n_{\mathrm{p}}})$. For this analysis, the radar code $\overline{\boldsymbol{\beta}} = [\overline{\beta}_0,\ldots,\overline{\beta}_{N_{\mathrm{c}}-1}]$ is a random biphase code, i.e. $\overline{\beta}_{n_{\mathrm{c}}} \in \{-1,1\}$ for $n_{\mathrm{c}} = 0,\ldots,N_{\mathrm{c}}-1$, that changes independently from CPI to CPI but remains constant during each CPI. Here, we model the discrete-time base radar waveform $\mathbf{s}_{\mathrm{r}}(\overline{\boldsymbol{\beta}})$ using a zeroth-order representation [55] (i.e. as a discrete code with abrupt phase changes and N_{r} repeated samples/chip), resulting in $\mathbf{s}_{\mathrm{r}}(\overline{\boldsymbol{\beta}}) = [s_{\mathrm{r},0}(\overline{\boldsymbol{\beta}}),\ldots,s_{\mathrm{r},N_{\mathrm{c}}N_{\mathrm{r}}-1}(\overline{\boldsymbol{\beta}})]$, where $\mathbf{s}_{\mathrm{r},n_{\mathrm{c}}N_{\mathrm{r}}+n_{\mathrm{r}}}(\overline{\boldsymbol{\beta}}) = \overline{\beta}_{n_{\mathrm{c}}}$ for all $n_{\mathrm{c}} = 0,\ldots,N_{\mathrm{c}}-1$ and $n_{\mathrm{r}} = 0,\ldots,N_{\mathrm{r}}-1$. The random biphase code assumption and the zeroth-order waveform structure are invoked for the sake of mathematical tractability of the subsequent analysis.

The communication symbol sequence $\boldsymbol{\alpha}_{n_{\mathrm{p}}}$ for the n_{p}th pulse is binary and the number of communication symbols is equal to the number of chips, i.e. $\boldsymbol{\alpha}_{n_{\mathrm{p}}} = [\alpha_{n_{\mathrm{p}},0},\ldots,\alpha_{n_{\mathrm{p}},N_{\mathrm{c}}-1}]$ and $\alpha_{n_{\mathrm{p}},n_{\mathrm{c}}} \in \{0,1\}$ for $n_{\mathrm{c}} = 0,\ldots,N_{\mathrm{c}}-1$. The communication sequences are modulated with PSK [42] having an adjustable phase-shift parameter [10,11] that is the same as h. The relationship between the communication symbols and samples of the discrete-time communication component is given as

$$s_{\mathrm{c},n_{\mathrm{p}},n_{\mathrm{c}}N_{\mathrm{r}}+n_{\mathrm{r}}}(\boldsymbol{\alpha}_{n_{\mathrm{p}}}) = \begin{cases} \exp\{-jh\pi\} & \text{if } \alpha_{n_{\mathrm{p}},n_{\mathrm{c}}} = 0 \\ \exp\{jh\pi\} & \text{if } \alpha_{n_{\mathrm{p}},n_{\mathrm{c}}} = 1, \end{cases} \tag{23.22}$$

for all $n_{\mathrm{c}} = 0,\ldots,N_{\mathrm{c}}-1$, $n_{\mathrm{r}} = 0,\ldots,N_{\mathrm{r}}-1$, and $n_{\mathrm{p}} = 0,\ldots,N_{\mathrm{p}}-1$. The modulation index $0 < h \leq 1/2$ is a rational number that controls the amount of phase shift introduced relative to the base radar waveform when embedding a communication symbol. The binary values $\alpha_{n_{\mathrm{p}},n_{\mathrm{c}}}$ are independent and identically distributed (i.i.d.), i.e. $\alpha_{n_{\mathrm{p}},n_{\mathrm{c}}}$ and $\alpha_{m_{\mathrm{p}},m_{\mathrm{c}}}$ are independent whenever $n_{\mathrm{p}} \neq m_{\mathrm{p}}$ or $n_{\mathrm{c}} \neq m_{\mathrm{c}}$. It follows that the communication waveform samples $s_{\mathrm{c},n_{\mathrm{p}},n_{\mathrm{c}}N_{\mathrm{r}}+n_{\mathrm{r}}}(\boldsymbol{\alpha}_{n_{\mathrm{p}}})$ and $s_{\mathrm{c},m_{\mathrm{p}},m_{\mathrm{c}}N_{\mathrm{r}}+m_{\mathrm{r}}}(\boldsymbol{\alpha}_{m_{\mathrm{p}}})$ are independent for all $n_{\mathrm{r}}, m_{\mathrm{r}} = 0,\ldots,N_{\mathrm{r}}-1$ whenever $n_{\mathrm{p}} \neq m_{\mathrm{p}}$ or $n_{\mathrm{c}} \neq m_{\mathrm{c}}$. Since both $s_{\mathrm{r}}(t;\overline{\boldsymbol{\beta}})$ and $s_{\mathrm{c},n_{\mathrm{p}}}(t;\boldsymbol{\alpha}_{n_{\mathrm{p}}})$ are phase-modulated, their product $s_{n_{\mathrm{p}}}(t;\overline{\boldsymbol{\beta}},\boldsymbol{\alpha}_{n_{\mathrm{p}}})$ is also phase-modulated. This PSK-based REC, which is a special case of the more general PCFM-based REC in which the shaping filter $q(t)$ is simply an impulse, is considered here for mathematical tractability. Because of the memory of CPM, closed-form solutions are often either not obtainable or do not explicitly show the dependence on CPM parameters.

The clutter return for the n_{p}th pulse repetition interval (PRI) is modeled as the convolution of a complex-valued white Gaussian random sequence, denoted as $w = [w_{-N},\ldots,w_0,\ \ldots,w_N]$ for $N \gg N_{\mathrm{c}}N_{\mathrm{r}}$, and the discrete-time representation of the n_{p}th REC waveform $\mathbf{s}_{n_{\mathrm{p}}}(\overline{\boldsymbol{\beta}},\boldsymbol{\alpha}_{n_{\mathrm{p}}})$. Each random sample w_n has variance $\sigma_w^2/N_{\mathrm{r}}$ to ensure that

σ_w^2 is *the average clutter power per range bin*, which does not scale with the sampling rate N_r. The nth clutter response sample in the n_pth PRI, for $-N \leq n \leq N$, is thus

$$c_{n_p,n}(w, \overline{\boldsymbol{\beta}}, \boldsymbol{\alpha}_{n_p}) = \sum_{l=0}^{N_c N_r - 1} w_{n-l} s_{r,l}(\overline{\boldsymbol{\beta}}) s_{c,n_p,l}(\boldsymbol{\alpha}_{n_p}). \tag{23.23}$$

The clutter return in the n_pth PRI is dependent on $\overline{\boldsymbol{\beta}}$, w, and $\boldsymbol{\alpha}_{n_p}$. However, the Gaussian sequence w used to drive the clutter response remains constant within the CPI, while the communication sequence $\boldsymbol{\alpha}_{n_p}$ varies from pulse to pulse. This modeling approach is based on the assumption that clutter does not appreciably change in a time scale smaller than the CPI (i.e. negligible internal clutter motion) and that the radar platform is static.

Because the random process used to drive the clutter return w is stationary, the statistics of the clutter return samples $c_{n_p,n}(w, \overline{\boldsymbol{\beta}}, \boldsymbol{\alpha}_{n_p})$ do not change within the PRI. Therefore, the subscript n can be omitted or simply set to $n = 0$. The clutter return samples after MF of the n_pth PRI can be written as

$$\widehat{c}_{n_p}(w, \overline{\boldsymbol{\beta}}, \boldsymbol{\alpha}_{n_p}) = \frac{1}{N_c N_r} \sum_{l=-N_c N_r+1}^{N_c N_r - 1} w_{-l} \sum_{i=0}^{N_c N_r - 1} s_{r,i}(\overline{\boldsymbol{\beta}}) s^*_{r,i-l}(\overline{\boldsymbol{\beta}}) s_{c,n_p,i}(\boldsymbol{\alpha}_{n_p}) s^*_{c,n_p,i-l}(\boldsymbol{\alpha}_{n_p}), \tag{23.24}$$

where the scaling by $1/N_c N_r$ normalizes the MF energy to unity. The expression in (23.24) is the convolution of w and the autocorrelation of the discrete-time REC waveform $\mathbf{s}_{n_p}(\overline{\boldsymbol{\beta}}, \boldsymbol{\alpha}_{n_p})$.

The clutter power as a function of normalized Doppler frequency $-0.5 \leq \phi \leq 0.5$ and the Doppler windowing function $\boldsymbol{\Gamma} = [\Gamma_0, \ldots, \Gamma_{N_p-1}]$ is then

$$C_{\mathrm{POWER}}(\phi, \boldsymbol{\Gamma}) = \mathbb{E}\left[\left|\sum_{n_p=0}^{N_p-1} \Gamma_{n_p} \widehat{c}_{n_p}(w, \overline{\boldsymbol{\beta}}, \boldsymbol{\alpha}_{n_p}) \exp\{-j2\pi\phi n_p\}\right|^2\right], \tag{23.25}$$

where $\widehat{c}_{n_p}(w, \overline{\boldsymbol{\beta}}, \boldsymbol{\alpha}_{n_p})$ is given in (23.24), the windowing function is normalized such that $\sum_{n_p=1}^{N_p-1} \Gamma_{n_p} = 1$, and the expected value is computed with respect to w, $\overline{\boldsymbol{\beta}}$, and $\boldsymbol{\alpha}_0, \ldots, \boldsymbol{\alpha}_{N_p-1}$. For the REC system under consideration, the clutter power as a function of the parameters $\phi, \boldsymbol{\Gamma}, h, \sigma_w^2, N_p, N_c$, and N_r can be expressed by the closed-form expression [22]

$$\begin{aligned} C_{\mathrm{POWER}}(\phi, \boldsymbol{\Gamma}, h, \sigma_w^2, N_p, N_c, N_r) &= \sigma_w^2 \left(\frac{2}{3} + \frac{1}{3N_r^2}\right) \\ &\times \Bigg[\sin^2(h\pi)(1 + \cos^2(h\pi))\frac{N_c - 1}{N_c}\left(\sum_{n_p=0}^{N_p-1} \Gamma_{n_p}^2\right) \\ &+ \left(1 + \cos^4(h\pi)\frac{N_c - 1}{N_c}\right)|\mathrm{DFT}_{\boldsymbol{\Gamma}}(\phi)|^2\Bigg], \end{aligned} \tag{23.26}$$

where $\text{DFT}_{\mathbf{\Gamma}}(\phi) = \sum_{n_p=0}^{N_p-1} \Gamma_{n_p} \exp\left\{-j2\pi\phi n_p\right\}$ is the DFT of $\mathbf{\Gamma}$ computed at ϕ. In the derivation of (23.26) only the i.i.d. property of w is used, while the Gaussian property is not. The first term inside the square brackets in (23.26) is due to the terms in the expected value expression of (23.25) that vary from pulse to pulse during the CPI; therefore, it quantifies the RSM effect and is referred to as the *RSM power*. Conversely, the second term in (23.26) is due to the terms in (23.25) that remain constant during the CPI and is thus referred to as the *coherent clutter power*. Note that the RSM power does not depend on the normalized Doppler frequency ϕ, while the coherent clutter power term does depend on ϕ through the DFT of the Doppler window at ϕ.

Because we are considering a stationary platform, the main lobe of the coherent clutter component is centered at $\phi = 0$, while its sidelobes decay as $|\phi|$ increases by a pattern and rate specified by the Doppler window and number of pulses in the CPI. By the appropriate choice of the Doppler window, the coherent clutter power can be made arbitrarily small—compared to the RSM power—for Doppler frequency values greater than some threshold $\overline{\phi} > 0$. Therefore, for $|\phi| \geq \overline{\phi}$, the clutter power can be tightly bounded below by the RSM power as

$$C_{\text{POWER}}(\phi, \mathbf{\Gamma}, h, \sigma_w^2, N_p, N_c, N_r) \geq \underbrace{\sigma_w^2\left(\frac{2}{3} + \frac{1}{3N_r^2}\right) \times \sin^2(h\pi)(1+\cos^2(h\pi))\frac{N_c-1}{N_c}\left(\sum_{n_p=0}^{N_p-1}\Gamma_{n_p}^2\right)}_{RSM_{\text{POWER}}(\mathbf{\Gamma}, h, \sigma_w^2, N_p, N_c, N_r)}, \tag{23.27}$$

where the right-hand side is denoted by $RSM_{\text{POWER}}(\mathbf{\Gamma}, h, \sigma_w^2, N_p, N_c, N_r)$. For any target associated with a Doppler frequency greater than $\overline{\phi}$, i.e. $\phi_{\text{trg}} \geq \overline{\phi}$, the detection performance will strongly depend on the RSM power. For a target return with power P_r at Doppler frequency ϕ_{trg} (the return from the target in the n_pth PRI is $\sqrt{P_r}\exp\left\{j2\pi\phi_{\text{trg}}n_p\right\}s_{n_p}(t; \overline{\boldsymbol{\beta}}, \boldsymbol{\alpha}_{n_p})$), the peak signal power after MF-DFT processing (without straddling or scalloping losses) is P_r. Unlike the RSM power, the peak signal power does not depend on $\mathbf{\Gamma}, h, \sigma_w^2, N_p, N_c$, and N_r. The signal-to-clutter ratio (SCR) is defined as P_r/σ_w^2. The SCR is a measure of the strength of the (peak) target return power relative to the peak clutter power after MF-DFT processing, which occurs at zero Doppler and—for small h, and large N_c and N_r—is approximately equal to $4\sigma_w^2/3$.

It can be shown that $\sin^2(h\pi)\left(1+\cos^2(h\pi)\right)$, and hence, the RSM power is a decreasing function of h on $(0, 1/2]$. For $h \ll 1/2$, the small angle approximation (i.e. $\sin(x) \approx x$ and $\cos(x) \approx 1$) can be invoked, thereby resulting in $\sin^2(h\pi)\left(1+\cos^2(h\pi)\right) \approx 2h^2\pi^2$. It follows that for small values of h, further reducing h by a factor of two results in approximately 6 dB reduction in the RSM power. The term $\sum_{n_p=0}^{N_p-1}\Gamma_{n_p}^2$ encapsulates the dependence of the RSM power on both N_p and $\mathbf{\Gamma}$. For a given type of Doppler window (e.g. Taylor, Hamming, etc.), given the normalization $\sum_{n_p=1}^{N_p-1}\Gamma_{n_p} = 1$, this term can be written as the product of $1/N_p$ and a *Doppler windowing loss* term—with respect to the RECT Doppler window—which does not significantly vary with N_p [53]. The dependence on the sampling rate N_r is

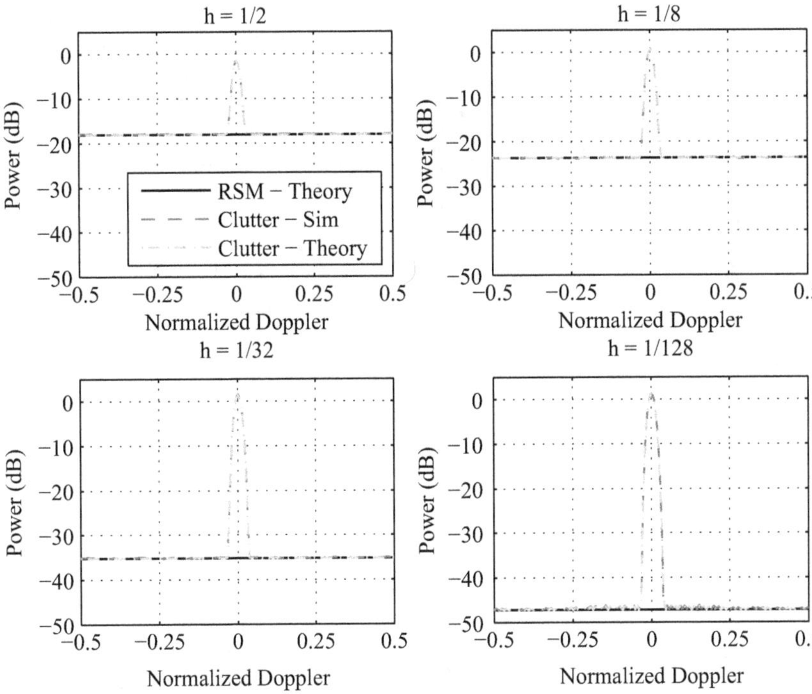

Figure 23.11 Clutter power computed by (23.26) via simulation and the RSM power computed by (23.27) plotted as a function of the normalized Doppler frequency ϕ for h = 1/2, 1/8, 1/32, and 1/128

rather weak since setting $N_r \geq 2$ results in $2/3 < 2/3 + 1/3N_r^2 \leq 3/4$. The dependence on the time-bandwidth product N_c is also weak as practical values of N_c lead to $(N_c - 1)/N_c \approx 1$.

In Figure 23.11, the theoretical clutter power computed by (23.26) via simulation and the theoretical RSM power computed by (23.27) are shown as a function of the normalized Doppler frequency ϕ for modulation index values $h = 1/2, 1/8, 1/32$, and $1/128$. The remaining parameter values are $\sigma_w^2 = 1$, $N_p = 64$, $N_c = 64$, and $N_r = 4$. The DFT is computed with a Taylor window having -55 dB peak-to-sidelobe ratio. It is observed that the theoretical clutter power and the clutter power computed by simulation match for all h values. This observation is expected because the clutter model used for simulations is the same model used in the derivation of (23.26). In addition, the RSM power term from (23.27) establishes a very tight lower bound on the clutter power. Lastly, it is observed that the RSM power decreases by approximately 12 dB when h is divided by four, as predicted by the theoretical analysis. In Figure 23.12, the same three quantities are plotted for $N_p = 16, 32, 128$, and 256 pulses per CPI. The modulation index is fixed at $h = 1/16$, and all other parameters are the same

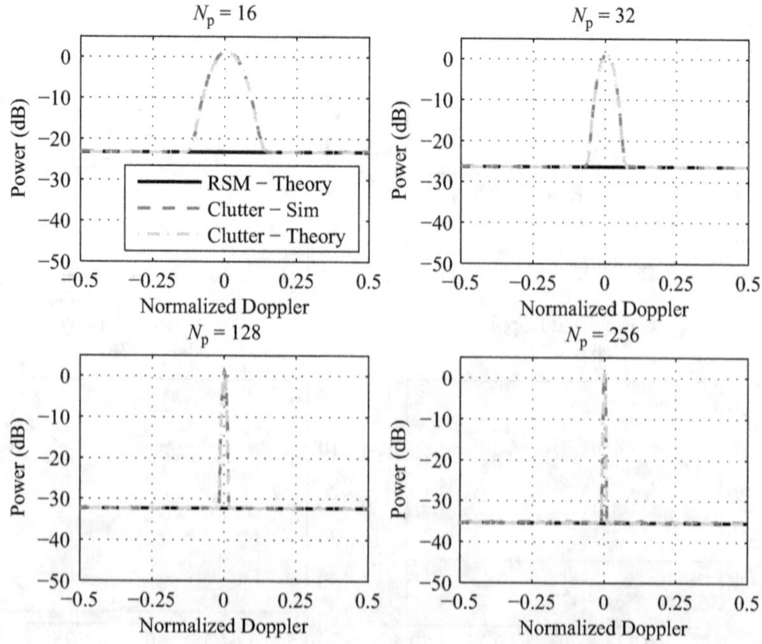

Figure 23.12 Clutter power computed by (23.26) via simulation and the RSM power computed by (23.27) plotted as a function of the normalized Doppler frequency ϕ for $N_p = 16$, 32, 128, and 256

as those used for Figure 23.11. It is observed that the RSM power decreases by 3 dB as N_p is reduced by a factor of 2, which is consistent with the theoretical analysis.

In Figure 23.13, the range-Doppler response (in dB) to a point target at the normalized Doppler frequency $\phi_{trg} = -1/4$ and normalized delay 0 is shown for the modulation indices $h = 1/16, 1/32, 1/64$, and $1/128$, with $N_p = 64$. The clutter power parameter is $\sigma_w^2 = 30$ dB and the peak target return power is $P_r = 0$ dB (i.e. the SCR is −30 dB). The clutter-to-noise ratio (CNR) is $\sigma_w^2/\sigma_n^2 = 50$ dB, where σ_n^2 is the variance of additive Gaussian noise samples after MF-DFT processing with a RECT Doppler window. An ideal "brickwall" filter is used to cancel the clutter response at the normalized Doppler values $\phi \in [-0.1, 0.1]$. The biphase radar code of length $N_c = 65$ is obtained by the Kronecker product of a length-13 and a length-5 Barker code as described in [56]. The remaining parameters have the same values as those used for Figures 23.11 and 23.12. The target response location is indicated by the white circles. It is observed that relative to the surrounding residual clutter, the target visibility is clearly enhanced as the modulation index decreases.

Next, REC system operating curves are presented to illustrate the relationship between the SNR at the communication receiver and the signal-to-RSM ratio (S-RSM) at the radar receiver, defined as $P_r/\text{RSM}_{\text{POWER}}(\mathbf{\Gamma}, h, \sigma_w^2, N_p, N_c, N_r)$, as a

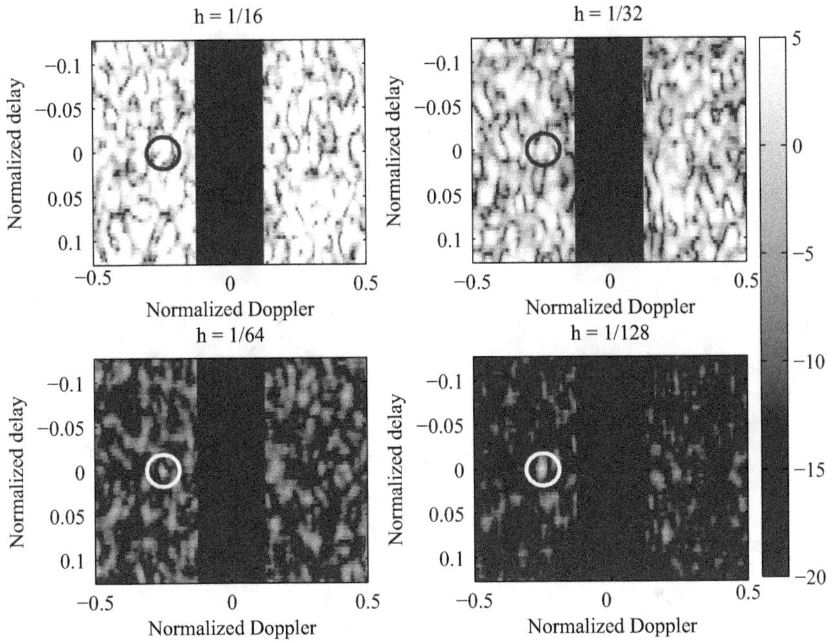

Figure 23.13 The range-Doppler response (in dB) to a target at Doppler frequency $\phi = -\frac{1}{4}$ and normalized delay 0 for modulation indices h = 1/8, 1/16, 1/64, and 1/128 and with $N_p = 64$

function of the modulation index h. Since the clutter power experienced for normalized Doppler values greater than some threshold is tightly approximated by the RSM power, the S-RSM quantifies the effective clutter interference power from an MTI point of view. In Figure 23.14, each curve is generated with the modulation indices $h = 1/2, 1/4, \ldots, 1/1,024$ and a fixed number of pulses per CPI N_p. The Doppler window is RECT, with $N_c = 64$ and $N_r = 4$. The horizontal axis represents the *normalized S-RSM* (dB), while the vertical axis represents the *normalized effective SNR* at the communication receiver (dB). In the computation of the normalized S-RSM, the value of P_r/σ_w^2 is set to unity. For an arbitrary negative value of $P_r/\sigma_w^2(\text{dB}) = -X(\text{dB})$, the respective system operating curves can be obtained by shifting those in Figure 23.14 to the left by $X(\text{dB})$.

From a communication system perspective, the PSK modulation for a given SNR value E_b/N_0, where E_b is the energy per bit and N_0 is the (one-sided) additive white Gaussian noise PSD at the communication receiver, is equivalent to BPSK with an SNR value of $\sin(h\pi)^2 E_b/N_0$ [42]. The BPSK-equivalent SNR $\sin(h\pi)^2 E_b/N_0$ is referred to as the effective SNR, and the normalized effective SNR value is obtained by setting $E_b/N_0 = 1$. For an arbitrary positive value of $E_b/N_0 = Y(\text{dB})$, the respective system operating curves can be obtained by shifting those in Figure 23.14 upward by

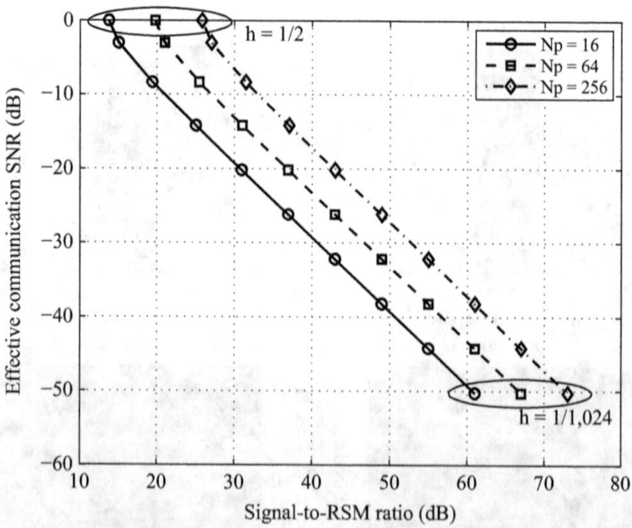

Figure 23.14 *Normalized effective communication SNR versus normalized signal-to-RSM ratio as a function of h, with N_p fixed for each curve and varied across curves with values N_p =16,64, and 256*

Y(dB). The modulation index value $h = 1/2$ leads to the best case scenario from a communication system perspective (i.e. no SNR loss), while it leads to the worst case scenario from a radar system perspective. Because the communication SNR is not a function of N_p, increasing/decreasing N_p shifts the system operating curves right/left as well.

23.2 Spatial waveform diversity

Where Section 23.1 addressed the use of coding to incorporate communications into radar, now consider how spatial degrees-of-freedom can be employed to realize a different dual-function mode. The ability to independently control the waveforms emitted from the elements comprising an antenna array enables the design of fast-time emissions that are spatially dependent by virtue of a significant increase in design degrees-of-freedom. This array structure, known as a fully digital array, facilitates the execution of multiple functions concurrently by the same system. An early example of this idea was the advanced multifunction RF concept [57] that introduced an adaptable subarrayed arrangements to achieve multiple functions simultaneously. The capability of the fully digital array expands on this notion by allowing control over the waveforms emitted by every antenna element so as to achieve a desired emission structure through the coherent combination of the transmitted waveforms in the far-field of the antenna array.

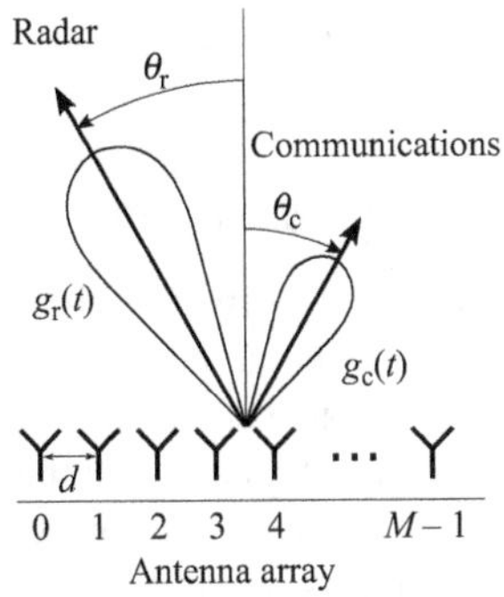

Figure 23.15 Simultaneous emission of radar and communications signals from a common antenna array

Through careful design of these waveforms, the fully digital array can be utilized to simultaneously transmit spatially separated radar and communication signals that have the same time and spectral support (see Figure 23.15). This type of dual-function emission is a particular form of MIMO beampattern design (e.g. [23–27]). In [20], an iterative waveform optimization method denoted as FFRED was introduced that generates pulsed MIMO waveforms, the far-field combination of which realizes simultaneous radar and communications beams, along with the particular constraint that the waveforms are physically realizable and amenable to a high-power amplifier (HPA), i.e. constant amplitude and well contained spectrally (see [35,54,58]). In [28], this method was experimentally demonstrated on the Air Force Research Lab's four-channel software-defined radar testbed BEEMER [59].

In [20], it was shown that to achieve simultaneous emission of two distinct signals in different spatial directions through the use of constant amplitude MIMO waveforms, one must also consider the totality of signals in the orthogonal complement directions. In other words, a portion of the overall energy must be emitted into the spatial regions that are not the desired radar or communication directions (beyond the amount one would normally associate with sidelobes). An interesting trade-off then emerges between the power efficiency gained by enforcing the constant amplitude constraint (and associated use of efficient power amplifiers) and the efficiency lost due to allocation of power in the orthogonal space. This trade is explored in this section by examining the percentage of power emitted into the orthogonal space as well as through the definition of a total directed power (TDP) metric that quantifies the power efficiency of a set of waveforms given a simple amplifier model. Furthermore, how this orthogonal complement power manifests itself in the spatial and spectral domains is also examined. As a result of this orthogonal power, it was observed in [20] that a side effect of the design procedure is that the time-domain signals in the directions away from the communication direction are naturally uncorrelated with the intended communication signal. This behavior is examined further by forming BER versus spatial angle plots for various relative radar/communications beam spacings and power levels.

23.2.1 Far-field radiated emission design

The FFRED approach generates a set of constant-amplitude FM waveforms that, when emitted from a digital array, coherently combine in the far-field to realize the intended radar and communication signals that are separated in space. Here, it is assumed that a pulsed radar emission is the primary function and thus the communications beam is transmitted at a lower power and must exist entirely within the pulse duration and spectral content of the radar emission. The embedding of the communication function into the radar emission represents a loss in transmit power (and therefore loss in received target SNR) for the radar. However, this formulation allows for the data throughput of the communication link to scale with the available bandwidth. It should be noted that this concept can easily revert to a radar-only mission depending on the priority level of the radar transmit power.

23.2.1.1 Signal model

Consider an M element uniform linear array (ULA) with interelement spacing d. If the emission is sufficiently narrowband, the far-field radar and communication signals must respectively satisfy the constraints

$$\sum_{m=0}^{M-1} s_m(t) \exp\left\{ jm\frac{2\pi}{\lambda} d \sin\theta_{\mathrm{r}} \right\} = g_{\mathrm{r}}(t), \tag{23.28}$$

and

$$\sum_{m=0}^{M-1} s_m(t) \exp\left\{ jm\frac{2\pi}{\lambda} d \sin\theta_{\mathrm{c}} \right\} = g_{\mathrm{c}}(t), \tag{23.29}$$

where λ is the free-space wavelength of the carrier frequency f_{c}, the waveform emitted by the mth antenna element is $s_m(t)$, and $g_{\mathrm{r}}(t)$ and $g_{\mathrm{c}}(t)$ are the desired far-field radar and communication signals to be emitted in directions θ_{r} and θ_{c}, respectively. Discretizing $g_{\mathrm{r}}(t)$, $g_{\mathrm{c}}(t)$, and $s_m(t)$ into $N \times 1$ vectors, (23.28) and (23.29) can be rewritten as

$$\mathbf{v}^H(\theta_{\mathrm{r}})\,\mathbf{S} = \mathbf{g}_{\mathrm{r}}^T, \tag{23.30}$$

and

$$\mathbf{v}^H(\theta_{\mathrm{c}})\,\mathbf{S} = \mathbf{g}_{\mathrm{c}}^T, \tag{23.31}$$

where $\mathbf{g}_{\mathrm{r}}$ and $\mathbf{g}_{\mathrm{c}}$ are the $N \times 1$ vectors of the desired radar and communication signals, $\mathbf{S} = [\mathbf{s}_0\ \mathbf{s}_1\ \cdots\ \mathbf{s}_{M-1}]^T$ is an $M \times N$ matrix comprised of the M discretized waveforms, and $\mathbf{v}(\theta)$ is the $M \times 1$ spatial steering vector for direction θ defined as

$$\mathbf{v}(\theta) = \frac{1}{\sqrt{M}}[1\ \exp\{-j\tfrac{2\pi}{\lambda} d \sin\theta\}\ \cdots\ \exp\{-j(M-1)\tfrac{2\pi}{\lambda} d \sin\theta\}]^T. \tag{23.32}$$

The $N \times 1$ vectors $\mathbf{g}_\mathrm{r}$, $\mathbf{g}_\mathrm{c}$, and $\mathbf{s}_m$ are each sufficiently "oversampled" with respect to 3 dB bandwidth (see [35,54,58]) to maintain sufficient fidelity for physical realization (if the desired $\mathbf{g}_\mathrm{r}$ and $\mathbf{g}_\mathrm{c}$ possess good spectral containment then, by extension, so will the waveforms $\mathbf{s}_m$ used to construct them). The discretized constraints of (23.30) and (23.31) can be combined as

$$\mathbf{C}^H\mathbf{S} = \mathbf{G}, \tag{23.33}$$

where the $M \times 2$ matrix

$$\mathbf{C} = [\mathbf{v}\,(\theta_\mathrm{r})\;\;\mathbf{v}\,(\theta_\mathrm{c})] \tag{23.34}$$

contains the spatial steering vectors, and the $N \times 2$ matrix

$$\mathbf{G} = [\mathbf{g}_\mathrm{r}\;\mathbf{g}_\mathrm{c}]^T \tag{23.35}$$

contains the discretized far-field signals.

23.2.1.2 Optimality of waveforms

The optimal waveform matrix (in a minimum-norm sense) can be found by solving the optimization problem

$$\begin{aligned} &\underset{\mathbf{S}}{\text{minimize}} \quad \|\mathbf{S}\|_F^2 \\ &\text{subject to} \quad \mathbf{C}^H\mathbf{S} = \mathbf{G}, \end{aligned} \tag{23.36}$$

where $\|\bullet\|_F^2$ is the squared-Frobenius norm [60]. This problem is convex and has the closed-form solution [60]

$$\mathbf{S}_\star = \mathbf{C}(\mathbf{C}^H\mathbf{C})^{-1}\mathbf{G}. \tag{23.37}$$

This formulation of the waveform matrix is optimal in the sense that all of the energy in $\mathbf{S}_\star$ is used to achieve the emission constraint in (23.33). However, the waveforms in $\mathbf{S}_\star$ are typically not constant amplitude, and thus, they preclude the use of an HPA operating in saturation (which is associated with high power efficiency). From this perspective, it is instructive to reconsider the definition of optimal within the context of a physical system.

Instead, optimality can be viewed in terms of power efficiency, which supports the maximization of energy emitted in the directions θ_r and θ_c. To achieve the peak power requirement of many high-power radar applications, constant amplitude is a necessity [8]. Therefore, under this new interpretation, the constraints of the optimization problem in (23.36) are modified to produce constant amplitude FM waveforms that are amenable to a HPA in saturation. The modified optimization problem is thus

$$\begin{aligned} &\underset{\widetilde{\mathbf{S}}}{\text{minimize}} \quad \|\widetilde{\mathbf{S}}\|_F^2 \\ &\text{subject to} \quad \mathbf{C}^H\widetilde{\mathbf{S}} = \mathbf{G} \\ &\qquad\qquad |\tilde{s}_m(n)| = |\tilde{s}_p(q)| \quad \text{for} \quad n,q = 0,\ldots,N-1 \\ &\qquad\qquad\qquad\qquad\qquad\qquad\qquad\quad m,p = 0,\ldots,M-1, \end{aligned} \tag{23.38}$$

where $\tilde{s}_m(n)$ is the nth time sample of the mth waveform and the tilde $(\tilde{\bullet})$ indicates that $\widetilde{\mathbf{S}}$ is a constant amplitude waveform matrix. An equivalent formulation of (23.38) is

$$\begin{aligned} &\underset{\widetilde{\mathbf{S}},\gamma}{\text{minimize}} \quad \gamma^2 \\ &\text{subject to} \quad \mathbf{C}^H\widetilde{\mathbf{S}} = \mathbf{G} \\ &\qquad\qquad |\tilde{s}_m(n)| = \gamma \quad \text{for} \quad n = 0, \ldots, N-1 \\ &\qquad\qquad\qquad\qquad\qquad\qquad m = 0, \ldots, M-1, \end{aligned} \tag{23.39}$$

where γ is the real, positive-valued amplitude of the elements in waveform matrix $\widetilde{\mathbf{S}}$, and thus $\|\widetilde{\mathbf{S}}\|_F^2 \propto \gamma^2$. Therefore, the optimal waveform matrix [according to optimization problem (23.38)] is one that minimizes amplitude γ while maintaining the constraint $\mathbf{C}^H\widetilde{\mathbf{S}} = \mathbf{G}$. Note that the amplitude γ is dependent on the relative amplitudes in $\mathbf{C}$ and $\mathbf{G}$ and thus can be scaled after the optimization process without affecting the desired signal structure. Unlike (23.36), this problem does not have a closed-form solution, so we shall consider an iterative approach.

23.2.1.3 Utilization of the null space and the total directed power

To realize a waveform matrix that achieves the constraints in (23.39), the null space of the steering vector matrix $\mathbf{C}$ must be leveraged to design a spatial orthogonal complement matrix $\mathbf{S}_\perp$ (where $\mathbf{S}_\star^H\mathbf{S}_\perp = \mathbf{0}_{N\times N}$) such that the summation

$$\widetilde{\mathbf{S}} = \mathbf{S}_\star + \mathbf{S}_\perp \tag{23.40}$$

produces the constant amplitude waveform matrix $\widetilde{\mathbf{S}}$. Further, the columns of $\mathbf{S}_\perp$ are orthogonal to $\mathbf{v}(\theta_\mathrm{r})$ and $\mathbf{v}(\theta_\mathrm{c})$ or simply

$$\mathbf{C}^H\mathbf{S}_\perp = \mathbf{0}_{2\times N}, \tag{23.41}$$

thus maintaining the constraint $\mathbf{C}^H\widetilde{\mathbf{S}} = \mathbf{C}^H(\mathbf{S}_\star + \mathbf{S}_\perp) = \mathbf{G}$. Due to the orthogonality between $\mathbf{S}_\star$ and $\mathbf{S}_\perp$, the average power in $\widetilde{\mathbf{S}}$ is represented as

$$\gamma^2 = \rho_\star + \rho_\perp \tag{23.42}$$

or alternatively as

$$\gamma^2 = \frac{1}{MN}\|\widetilde{\mathbf{S}}\|_F^2, \tag{23.43}$$

where

$$\rho_\star = \frac{1}{MN}\|\mathbf{S}_\star\|_F^2 \tag{23.44}$$

and

$$\rho_\perp = \frac{1}{MN}\|\mathbf{S}_\perp\|_F^2 \tag{23.45}$$

are the average powers in $\mathbf{S}_\star$ and $\mathbf{S}_\perp$, respectively. These average powers are found using the identity $\|\widetilde{\mathbf{S}}\|_F^2 = \mathrm{Tr}\{\widetilde{\mathbf{S}}\widetilde{\mathbf{S}}^H\}$, then substituting in (23.40) and simplifying.

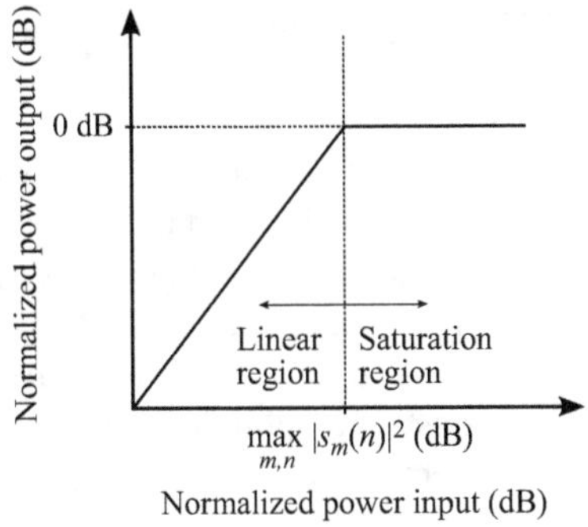

Figure 23.16 Simple amplifier model defining linear and saturation regions

The average power of the minimum-norm solution $\rho_\star$ is fixed for a given $\mathbf{C}$ and $\mathbf{G}$; hence by minimizing amplitude γ [via optimization problem (23.39)], we are effectively minimizing the orthogonal power $\rho_\perp$ necessary to achieve the desired emission using constant amplitude waveforms. The inclusion of the orthogonal emission requires some percentage of the total average power from (23.42). Therefore, there exists a trade-off between the power efficiency lost by placing power into the orthogonal space (via $\mathbf{S}_\perp$) and the efficiency gained by using constant amplitude waveforms that are amenable to an efficient HPA. In Section 23.2.3, we show that for a range of parameterizations, the power efficiency always increases when forming the set of waveforms in this manner.

To quantify the power efficiency of a given waveform matrix, we must first establish an amplifier model to ascertain the transmitter performance given waveforms of different peak-to-average power ratios (PAPRs). A simple amplifier model is shown in Figure 23.16, where the maximum output power is normalized to 0 dB and the power transfer curve has two distinct regions: a linear region and a fully saturated region (in reality, a transition region would exist between these but is omitted here). The amplifier model is enforced by scaling the waveform matrix such that the peak amplitude is unity (thus establishing a maximum instantaneous normalized output power of 0 dB). As such, any amplitude values less than the peak represent a loss relative to constant amplitude waveforms. This notion of "power back-off" is commonly used to prevent amplifier distortion when linear operation is required [61].

Using the amplifier model in Figure 23.16, a normalized TDP metric can be defined that calculates the average power contained in waveform matrix $\mathbf{S}$ that contributes to the desired emission. The normalized TDP for $\mathbf{S}$ satisfying constraint (23.33) is thus

$$\mathrm{TDP}(\mathbf{S}) = \frac{1}{MN}\left\| \frac{\mathbf{P}_\star \mathbf{S}}{\max\limits_{m,n} |s_m(n)|} \right\|_F^2 = \frac{\rho_\star}{\max\limits_{m,n} |s_m(n)|^2}, \tag{23.46}$$

using (23.37) and where $\mathbf{P}_\star = \mathbf{C}(\mathbf{C}^H\mathbf{C})^{-1}\mathbf{C}^H$ is a projection matrix that selects only the portion of the waveform matrix that emits in the desired directions specified in $\mathbf{C}$.

Because the maximum amplitude of the waveform matrix is normalized to unity, the possible values of (23.46) are $0 \leq \text{TDP} \leq 1$, with TDP $= 1$ representing maximum array power efficiency. For constant amplitude waveform matrix $\widetilde{\mathbf{S}}$, the TDP simplifies to $\rho_\star/\gamma^2$. Thus, the minimization of amplitude γ [via optimization problem (23.39)] produces a constant modulus waveform matrix $\widetilde{\mathbf{S}}$ that maximizes TDP. Equivalently, the TDP for waveform matrix $\widetilde{\mathbf{S}}$ can be expressed as $\rho_\star/(\rho_\star + \rho_\perp)$, illustrating that when the orthogonal power is minimized, the TDP is maximized. Maximum array power efficiency (TDP $= 1$) can only occur when $\rho_\perp = 0$, implying that the minimum-norm solution $\mathbf{S}_\star$ is constant amplitude, which can only be achieved when $\mathbf{C}$ contains a single emission direction and $\mathbf{G}$ contains a single constant amplitude waveform.

23.2.2 Realization of emission constraints using error reduction algorithm

The optimization problem in (23.39) does not have a closed-form solution and is nonconvex due to the inclusion of the constant amplitude constraint. A fast and useful approach to perform optimization involving two distinct constraints is to alternate between projecting onto the sets satisfying each constraint. If these sets are convex, this procedure is known as projection onto convex sets (POCS) [62,63]. If one or more of the sets is nonconvex, the procedure is known as the error reduction algorithm (ERA) [64–67]. While the procedure remains the same for these two general methods, the difference lies in the convergence claims. The convergence to an intersection point (or minimum distance point) between the sets can be proven for POCS, but cannot be proven for ERA. However, the error in ERA can be shown to be a nonincreasing sequence [64–67]. Since the constant amplitude constraint in (23.38) represents a nonconvex set, the convergence properties of ERA apply here.

23.2.2.1 FFRED implementation

For a given constant amplitude initialization $\widetilde{\mathbf{S}}_0$, this iterative design procedure alternates between two stages until sufficient convergence is achieved. The first stage projects onto some set A that satisfies a constraint using projection operator $P_A(\bullet)$. Subsequently, the second stage projects onto some set B satisfying the other constraint using projection operator $P_B(\bullet)$. For this problem, set A satisfies the constraint $\mathbf{C}^H\mathbf{S} = \mathbf{G}$ from (23.33) and set B satisfies the constant amplitude constraint from (23.39). Here, set A is convex and set B is nonconvex. This procedure is summarized as follows.

Alternating Projections
Initialize: $\widetilde{\mathbf{S}}_0$, $i = 0$
repeat
  $\mathbf{S}_i = P_A\left(\widetilde{\mathbf{S}}_i\right)$
  $\widetilde{\mathbf{S}}_{i+1} = P_B\left(\mathbf{S}_i\right)$
  $i = i + 1$
until sufficiently converged
return $\widetilde{\mathbf{S}}_i$

The projections $P_A(\bullet)$ and $P_B(\bullet)$ can be defined as the minimum change one could impart to $\widetilde{\mathbf{S}}_i$ or $\mathbf{S}_i$, respectively, such that the corresponding constraints are met. Thus, the projections themselves can be viewed as optimization problems. The first problem is formulated such that the squared distance is minimized between the waveform matrix $\widetilde{\mathbf{S}}_i$ obtained in the previous iteration and the projected waveform matrix $\mathbf{S}_i$, while maintaining the form of constraint (23.33). The first stage projection $P_A(\widetilde{\mathbf{S}}_i)$ is therefore defined as

$$\begin{aligned} &\underset{\mathbf{S}_i}{\text{minimize}} \quad \left\|\widetilde{\mathbf{S}}_i - \mathbf{S}_i\right\|_F^2 \\ &\text{subject to} \quad \mathbf{C}^H\mathbf{S}_i = \mathbf{G}, \end{aligned} \tag{23.47}$$

which has the closed-form solution

$$\mathbf{S}_i = \mathbf{P}_\perp\widetilde{\mathbf{S}}_i + \mathbf{S}_\star \tag{23.48}$$

for projection matrix

$$\mathbf{P}_\perp = \mathbf{I}_M - \mathbf{C}\left(\mathbf{C}^H\mathbf{C}\right)^{-1}\mathbf{C}^H, \tag{23.49}$$

with $\mathbf{I}_M$ the $M \times M$ identity matrix.

Likewise, the second stage projection $P_B(\mathbf{S}_i)$ can be expressed as

$$\begin{aligned} &\underset{\widetilde{\mathbf{S}}_{i+1}}{\text{minimize}} \quad \left\|\widetilde{\mathbf{S}}_{i+1} - \mathbf{S}_i\right\|_F^2 \\ &\text{subject to} \quad \left|\tilde{s}_{m,i+1}(n)\right| = \gamma \quad \text{for} \quad n = 0, \ldots, N-1 \\ &\qquad\qquad\qquad\qquad\qquad\qquad\quad m = 0, \ldots, M-1, \end{aligned} \tag{23.50}$$

which has the closed-form solution

$$\widetilde{\mathbf{S}}_{i+1} = \gamma \exp\{j\angle(\mathbf{S}_i)\}, \tag{23.51}$$

where $\angle(\bullet)$ extracts the phase of the argument. After incorporating the projections from (23.48) and (23.51), the FFRED procedure can be stated as follows for given spatial steering vectors in $\mathbf{C}$ and desired far-field emissions in $\mathbf{G}$.

Far-Field Radiated Emission Design
Initialize: $\widetilde{\mathbf{S}}_0$, γ, $i = 0$
repeat
  $\mathbf{S}_i = \mathbf{P}_\perp\widetilde{\mathbf{S}}_i + \mathbf{S}_\star$
  $\widetilde{\mathbf{S}}_{i+1} = \gamma \exp\{j\angle(\mathbf{S}_i)\}$
  $i = i + 1$
until sufficiently converged
return $\widetilde{\mathbf{S}}_i$

Instead of minimizing amplitude γ [as in problem (23.39)], it is incorporated as a predefined parameter for the FFRED algorithm. We can find a value of γ close to the minimum through a sequence of feasibility problems where, for a given amplitude γ, the intersection between sets A and B is tested (i.e. when both constraints are satisfied). If the intersection exists, the amplitude is reduced, if not, the amplitude is increased. This process can be implemented via a simple bisection procedure and iterated until γ

is within some tolerance. As previously stated, the convergence properties of ERA do not guarantee that for a given initialization $\widetilde{\mathbf{S}}$ and amplitude γ, a solution that intersects the sets A and B will be found. Instead, ERA has the error reduction property

$$\cdots \geq \left\|\widetilde{\mathbf{S}}_i - \mathbf{S}_i\right\|_F^2 \geq \left\|\widetilde{\mathbf{S}}_{i+1} - \mathbf{S}_i\right\|_F^2 \geq \left\|\widetilde{\mathbf{S}}_{i+1} - \mathbf{S}_{i+1}\right\|_F^2 \geq \cdots \tag{23.52}$$

which shows that the squared-distance between the solutions of problems (23.47) and (23.50) forms a nonincreasing sequence. However, in [20] it was shown that by increasing the amount of orthogonal power $\rho_\perp$ (through increase of amplitude γ) the convergence behavior improves by enhancing the overall solution's ability to satisfy the constraints of (23.39). The feasibility procedure to find amplitude γ as described above is obviously not conducive for real-time application. However, prior knowledge of a feasible γ for a given transmission scenario could be used to bypass this process.

The computational complexity of one iteration of FFRED is $8M^2N + 7MN$ floating point operations or "flops," with $8M^2N$ operations arising from (23.48) and $7MN$ operations for (23.51). However, a majority of these operations are readily parallelizable. The number of iterations required to reach convergence depends on the parameter initializations and the convergence criterion, though it has been found to average on the order of 10s to 100s of iterations. Clearly, more effort is needed to guarantee convergence (to a high certainty) given a requirement to operate in real time.

In regards to the initialization $\widetilde{\mathbf{S}}_0$, it has been found thus far that randomly coded PCFM waveforms [35] produce good, consistent results. However, depending on the particular initialization, it is known that ERA-based algorithms may encounter a "tunnel" or "trap" that can slow convergence or prevent the algorithm from finding a solution (see [64] for more details on ERA). Because the foresight to predict these "traps" and "tunnels" does not exist, it is impossible to guarantee that one will not be encountered for a given initialization. Further investigation is necessary to understand how to intelligently choose the initialization of $\widetilde{\mathbf{S}}_0$ so as to best avoid these obstacles. While not elegant, a "brute force" option is simply to employ multiple different initializations and select the best result after each has attained a sufficient degree of convergence.

23.2.3 Simulated characterization of FFRED emissions

To evaluate the performance of FFREDs emission, we examine how multiple factors behave under different parameter initializations. The factors considered here include the TDP, the spatial beampattern, the emission spectral content, and BER for the communications component. The parameter initializations that dictate these factors are relative radar/communication powers, relative radar/communication beam spacing, and the structure of the communication modulation. The results that follow are not to be considered as a comprehensive examination into the performance of these emissions after FFRED optimization. Instead, the results provide insight into the benefits and potential obstacles that can emerge from this waveform optimization technique, with particular attention paid to power efficiency and how the orthogonal power manifests itself in space and frequency.

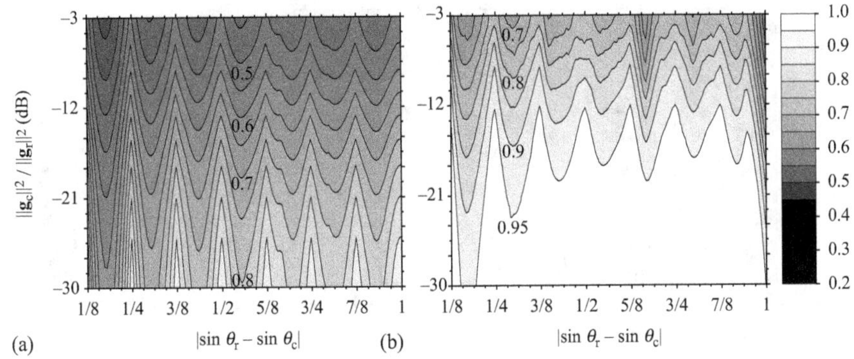

Figure 23.17 TDP *via (23.46) of (a) minimum-norm solution* $\mathbf{S}_\star$ *and (b) FFRED optimized solution* $\tilde{\mathbf{S}}$ *versus* $|\sin\theta_r - \sin\theta_c|$ *and* $\|\mathbf{g}_c\|^2/\|\mathbf{g}_r\|^2$ *using RECT-filtered QPSK*

For the results that follow consider an $M = 16$ element ULA with interelement spacing $d = \lambda/2$. Assume that the elements are omni-directional in the plane in which the array lies and that mutual coupling is negligible. The radar waveform $g_r(t)$ is an up-chirped LFM waveform of time-bandwidth $BT = 64$. Three different communication modulations are considered for signal $g_c(t)$: QPSK with either RECT or square-root-raised-cosine (SRRC) shaping filters [42], and 4-ary CPM with a RECT shaping filter and modulation index $h = 1/2$ [34]. The RECT shaping filter spans one symbol duration, while the SRRC filter spans 10 symbols. Both the RECT-filtered QPSK and CPM communication signals contain 128 information bits per pulse within the given time-bandwidth product of the radar emission, while the SRRC-filtered communication emission contains 108 bits per pulse (due to the temporal extent of the shaping filter). The discretized versions $\mathbf{g}_r$ and $\mathbf{g}_c$ are "oversampled" with respect to bandwidth B such that the final optimized waveform matrix can be readily converted into a set of FM waveforms [35]. The FFRED algorithm is terminated when the result from (23.48) produces a waveform matrix that has a PAPR of 1.01 (thus, the subsequent enforcement of constant amplitude produces a negligible deviation from this result).

Figures 23.17–23.19 show the TDP via (23.46) of the minimum-norm waveform matrices and FFRED optimized waveform matrices for the RECT-filtered QPSK, SRRC-filtered QSPK, and RECT-filtered CPM communications scenarios, respectively. Taking advantage of the spatial invariance of a ULA-containing elements with identical patterns, only the relative spacing between the beams needs to be considered. The relative spacing between the radar and communication beams was varied from $|\sin\theta_r - \sin\theta_c| = 1/8$ (representing the first null of the 16 element ULA beampattern) to $|\sin\theta_r - \sin\theta_c| = 1$ in 1/160 increments. The relative power of the communication signal to the radar signal, computed via $\|\mathbf{g}_c\|^2/\|\mathbf{g}_r\|^2$, was varied from −30 to −3 dB in 1 dB increments. Each evaluation was averaged over 50 Monte Carlo

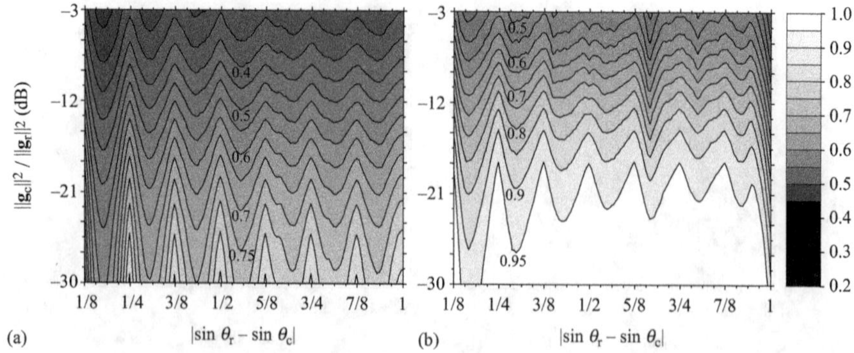

Figure 23.18 TDP *via (23.46) of (a) minimum-norm solution* $\mathbf{S}_\star$ *and (b) FFRED optimized solution* $\widetilde{\mathbf{S}}$ *versus* $|\sin\theta_\mathrm{r} - \sin\theta_\mathrm{c}|$ *and* $\|\mathbf{g}_\mathrm{c}\|^2/\|\mathbf{g}_\mathrm{r}\|^2$ *using SRRC-filtered QPSK*

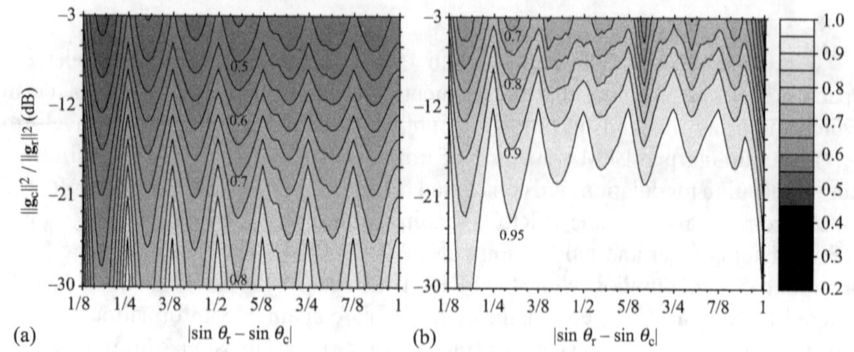

Figure 23.19 TDP *via (23.46) of (a) minimum-norm solution* $\mathbf{S}_\star$ *and (b) FFRED optimized solution* $\widetilde{\mathbf{S}}$ *versus* $|\sin\theta_\mathrm{r} - \sin\theta_\mathrm{c}|$ *and* $\|\mathbf{g}_\mathrm{c}\|^2/\|\mathbf{g}_\mathrm{r}\|^2$ *using RECT-filtered 4-ary CPM with h = 1/2*

trials with independent, randomly generated bit sequences and initialized with a set of $M = 16$ randomly coded PCFM-implemented waveforms.

For all three scenarios in Figures 23.17–23.19, the TDP for the FFRED optimized waveforms in $\widetilde{\mathbf{S}}$ is strictly greater than the TDP for the minimum-norm solution $\mathbf{S}_\star$ for every parameterization, confirming that the FFRED solution is in fact a more power efficient emission. The difference in TDP between $\widetilde{\mathbf{S}}$ and $\mathbf{S}_\star$ for all three cases reaches a maximum of ~35% and occurs at beam spacing $|\sin\theta_\mathrm{r} - \sin\theta_\mathrm{c}| \approx 3/16$. Note that the TDP for the RECT-filtered QPSK and RECT-filtered CPM scenarios in Figures 23.17 and 23.19 have similar outcomes, while a significant decrease in TDP is observed for the SRRC-filtered QPSK case. This result is likely due to the amplitude variation of the communication signal $\mathbf{g}_\mathrm{c}$ for the latter scenario. Since both

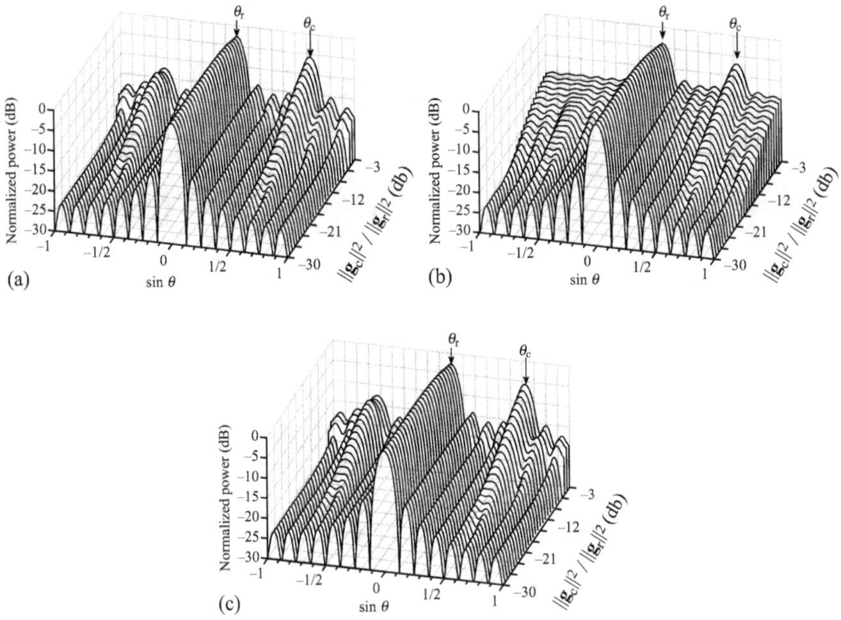

Figure 23.20 Average beam patterns for $\widetilde{\mathbf{S}}$ *versus* $\|\mathbf{g}_c\|^2/\|\mathbf{g}_r\|^2$ *using (a) RECT-filtered QPSK, (b) SRRC-filtered QPSK, and (c) RECT-filtered 4-ary CPM for* $\sin\theta_r = 0$ *and* $\sin\theta_c = 5/8$

the RECT-filtered QPSK and CPM signals have unity PAPR, their TDP efficiencies are likewise similar. In contrast, the SRRC-filtered QPSK signal has an average PAPR of ~2.2, and thus, the emission requires more power to be placed in the orthogonal complement region to attain a solution satisfying the constraints in (23.39), thereby reducing the overall power efficiency of the emission.

Observe in Figures 23.17–23.19 that when the beam separation corresponds to the nulls of the $M = 16$ ULA beampattern as

$$|\sin\theta_r - \sin\theta_c| = \frac{u}{8} \quad \text{for } u = 1, \ldots, 7, \tag{23.53}$$

the TDP increases for both $\widetilde{\mathbf{S}}$ and $\mathbf{S}_\star$ (indicated by the lighter regions in the plots). At these angles, the radar and communication beams are orthogonal, and therefore, no extra power is needed to achieve the spatial orthogonality, hence an increase in TDP.

Figure 23.20 shows the beampatterns of the converged solutions over the 50 Monte Carlo trials for the beam separation case of $|\sin\theta_r - \sin\theta_c| = 5/8$ in each of the three scenarios. Here, the absolute directions are $\sin\theta_r = 0$ and $\sin\theta_c = 5/8$. The beampatterns are normalized such that 0 dB represents the maximum power that the array can emit in any direction. As the relative power $\|\mathbf{g}_c\|^2/\|\mathbf{g}_r\|^2$ increases, the communication beam emerges from the location of the fifth null of the radar

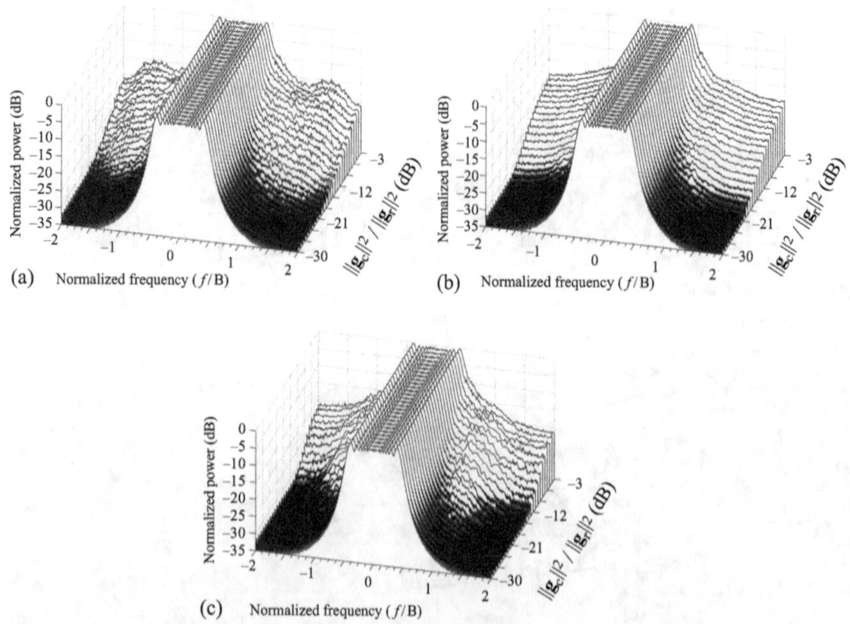

Figure 23.21 Maximum spectral content for $\widetilde{\mathbf{S}}$ *versus* $\|\mathbf{g}_c\|^2/\|\mathbf{g}_r\|^2$ *using (a) RECT-filtered QPSK, (b) SRRC-filtered QPSK, and (c) RECT-filtered 4-ary CPM for* $\sin\theta_r = 0$ *and* $\sin\theta_c = 5/8$

beampattern. As before, the average beampatterns of the RECT-filtered QPSK and the CPM scenarios are nearly identical. It should be noted that for these two scenarios, an additional spurious beam has emerged in the direction $\sin\theta = -5/8$. While symmetry may seem to imply that this spurious beam and the communication beam possess a similar time-domain structure, there is in fact no clear presence of the transmitted information in this direction (as will be shown in the BER plots). In contrast, the average beampattern of the SRRC-filtered QPSK scenario has a much flatter spatial response in the orthogonal complement directions, albeit with a noticeable peak power loss compared to the other cases. The peak power loss of the radar beam (compared to maximum transmit power) is 0.12 dB at a relative beam power of -30 dB for all three scenarios. However, at a relative beam power of -3 dB, the losses of the RECT-filtered QPSK and CPM cases rise to 5.56 and 5.53 dB, respectively, and 8.82 dB for the SRRC-filtered QPSK scenario. Again, this loss in overall power efficiency for the SRRC-filtered QPSK case is a direct result of requiring a larger amount of orthogonal power to satisfy the constraints of (23.39).

Figure 23.21 shows the maximum spectral response over $\sin\theta$ of the converged emissions for all three scenarios when the relative beam separation is 5/8 with $\sin\theta_r = 0$ and $\sin\theta_c = 5/8$ (same arrangement as Figure 23.20 and likewise averaged over the 50 Monte Carlo trials). As there is no spectral containment constraint applied during optimization, the orthogonal emission (dictated by $\mathbf{S}_{\perp}$) is allowed to spread into the spectral roll-off region. For the RECT-filtered QPSK scenario, the sinc structure of the communication spectrum starts to emerge at a relative beam power of $\|\mathbf{g}_c\|^2/\|\mathbf{g}_r\|^2$ (dB) $= -3$ dB. Interestingly, the SRRC-filtered QPSK scenario (which is implemented for its spectral containment properties) also suffers from spectral regrowth in the FFRED optimized solution. The CPM scenario has the least amount of power increase because the communication signal $\mathbf{g}_c$ is both spectrally well-contained and constant modulus [34].

To evaluate the BER performance of the communication link, a maximum likelihood symbol detector [42] is applied for the QPSK cases and the Viterbi algorithm [44] is applied for CPM [34]. Specifically, these symbol decoders are applied to the associated fast-time emission across all spatial angles. The decoding of a signal that is transmitted in a direction other than the communication direction θ_c is expected to produce increased communication errors due to the fast-time/spatial coupling of the emission. Quantifying the expected BER versus spatial angle under different scenarios provides an effective "coherence beamwidth" of spatial angles around direction θ_c that are amenable for communications. To isolate the effects of the spatial dependence of the emission, no additional noise is added to the simulation. The signals received at a hypothetical receiver as a function of spatial angle were phase rotated before symbol estimation so as to achieve the minimum BER possible for the given constellation.

In [20], it was shown that increasing the relative amount of orthogonal power by increasing γ above the minimum feasible amplitude (Section 23.2.2.1), the BER is increased in directions other than θ_c. Here, the amplitude γ is not inflated beyond the minimum. Instead, consider three different relative beam powers $\|\mathbf{g}_c\|^2/\|\mathbf{g}_r\|^2$ (dB) $= \{-20, -13, -6\}$ dB with the radar beam directed toward $\sin\theta_r = 0$ and the communication beam varied from $\sin\theta_c = 1/8$ to $\sin\theta_c = 1$ in $1/80$ increments. Here, the BER of the RECT-filtered QPSK and CPM communication scenarios are examined as they have similar TDP responses (and thus similar orthogonal power) for the parameterizations considered. The same emission configuration is used from the above examples (up-chirped LFM for $\mathbf{g}_r$, $BT = 64$, 16 element ULA, etc.), though the number of Monte Carlo trials is increased to 200 to better estimate the BER. Recall that both the RECT-filtered QPSK and the CPM signals could transmit 128 information bits per pulses with this parameterization, and therefore, the BER at each relative power level and beam separation is based on a total of 25,600 bits.

Figures 23.22–23.24 show the BER versus spatial angle $\sin\theta$ and communication direction $\sin\theta_c$ for relative beam powers of -20, -13, and -6 dB, respectively. Each figure displays the BER for the RECT-filtered QPSK and the RECT-filtered 4-ary CPM communication scenarios. The communication direction is illustrated by the diagonal white bar on each image. The width of this bar signifies the "coherence beamwidth" within which communication symbols can be detected successfully. Thus, by inspection, this coherence beamwidth becomes wider as the relative beam power $\|\mathbf{g}_c\|^2/\|\mathbf{g}_r\|^2$ increases, which may or may not be desirable. Recall in Figure 23.20 that an apparent symmetric beam appeared opposite to the communication direction. Here, this beam can be clearly observed as a diagonal black stripe to the

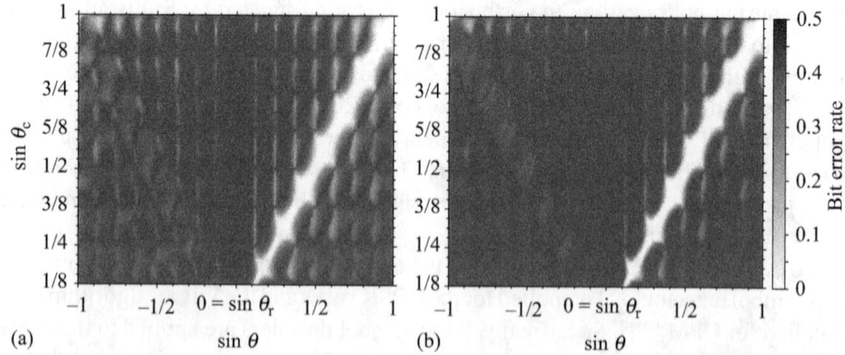

Figure 23.22 Average BER versus $\sin\theta$ *and* $\sin\theta_c$ *for* $\|\mathbf{g}_c\|^2/\|\mathbf{g}_r\|^2$ *(dB)* $= -20$ *dB and* $\sin\theta_r = 0$ *using (a) RECT-filtered QPSK and (b) RECT-filtered 4-ary CPM*

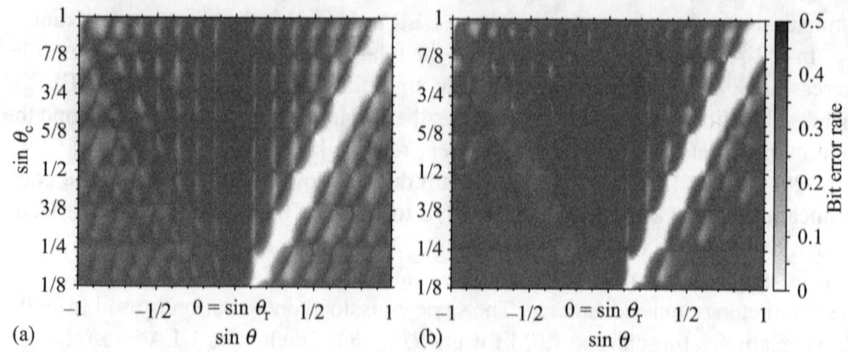

Figure 23.23 Average BER versus $\sin\theta$ *and* $\sin\theta_c$ *for* $\|\mathbf{g}_c\|^2/\|\mathbf{g}_r\|^2$ *(dB)* $= -13$ *dB and* $\sin\theta_r = 0$ *using (a) RECT-filtered QPSK and (b) RECT-filtered 4-ary CPM*

left of the radar emission direction. The high BER of this stripe indicates that the beam does not transmit communication information in the symmetric direction.

Another point of interest is the distinction in the BER performance between the RECT-filtered QPSK and CPM communication scenarios. In Figures 23.22–23.24, the emission containing the CPM signal results in a higher BER in directions other than $\sin\theta_c$ compared to the QPSK-based emission. Thus, the modulation scheme also has an effect on the resulting BER performance in unintended directions.

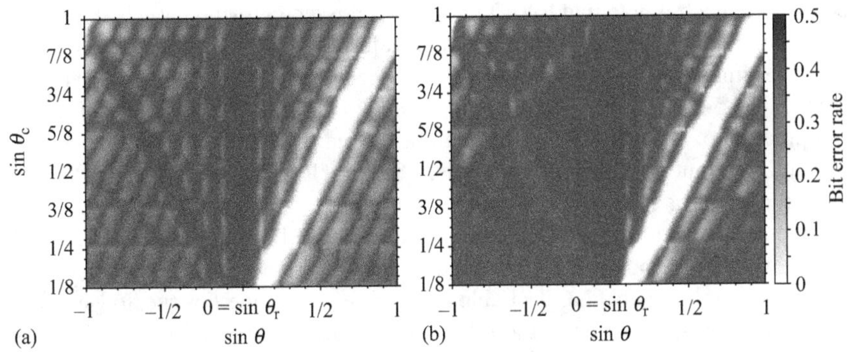

Figure 23.24 Average BER versus $\sin\theta$ *and* $\sin\theta_c$ *for* $\|\mathbf{g}_c\|^2/\|\mathbf{g}_r\|^2$ *(dB) = −6 dB and* $\sin\theta_r = 0$ *using (a) RECT-filtered QPSK and (b) RECT-filtered 4-ary CPM*

23.3 Spectral waveform diversity

Where the previous two sections leveraged coding and spatial degrees-of-freedom, respectively, in this final section we examine multifunction radar/communication spectrum sharing from the most fundamental perspective, that of the frequency domain. Clearly, assigning different frequency channels to the radar and communication aspects of a multifunction system is a straightforward way in which to mitigate their mutual interference (spectral spreading issues notwithstanding). That said, this frequency allocation framework does not really constitute "sharing" nor does it facilitate the inherent flexibility of operation that spectral sharing may ultimately provide. However, if the assigned radar/communication frequency channels are made to vary with time in a cooperative manner, then such flexibility can be achieved.

Here, we describe just such a form of "tandem hopped" radar and communications (or THoRaCs for short [19]) in which a communication signal is generated to reside within the spectral notch of the wider bandwidth emission generated by a collocated radar system. The centralized control of both radar and communication functions thereby permits their frequency hopping to be performed synchronously. Furthermore, the communication signal can be tailored to conform to the envelope of the radar emission's power spectrum such that pulse compression performed in the radar receiver can actually utilize the "composite" radar/communication emission for match filtering.

Since the spectral relationship between the radar and communication components is being made to change with time, it is therefore rather advantageous to use a radar waveform that (a) can readily accommodate changes to its physical structure, (b) can be designed "on-the-fly" in the frequency domain to provide flexibility for spectral notch width and location, (c) is amenable to the rigors of a high-power radar transmitter, and (d) is inherently robust to the sidelobe degradation that is known to accompany spectral notching [68]. To address these requirements, we shall leverage the pseudorandom optimized (PRO) FM waveform structure that has been demonstrated in simulation and experimentally to realize a rather useful form of FM noise radar [29–31,54]. The inherently changing nature of the PRO-FMCW emission allows incorporation of spectral notches that can be moved within the band at an arbitrary rate. Both continuous-wave (CW) and pulsed forms of PRO have been realized, though we shall only focus here on the former.

A communication signal that occupies the hopped spectral notch of a PRO-FMCW waveform, and blends well with the local spectrum around the notch, creates a composite emission that has desirable radar waveform properties (i.e. low autocorrelation sidelobes) by virtue of the PRO design process which produces a desirable power spectrum. A communication signal structure that is well suited for this spectral shaping strategy is that of orthogonal frequency-division multiplexing (OFDM) [42], which is itself the basis for latest generation communication systems [2,69]. That said, OFDM signals do possess a PAPR that can be much greater than 1, which effectively excludes them from high-power amplification and limits their utility as a radar waveform [70]. However, OFDM waveforms allow easy spectral shaping, making them ideal for this emission strategy [19]. It remains a topic of ongoing investigation to determine how to facilitate the practical emission of this communication component at high power.

23.3.1 Designing an FM noise waveform with spectral notches

Each segment of the PRO-FMCW waveform is designed independently, thereby realizing an overall waveform that contains no repeating structure [29]. Each segment is first initialized with a random FM waveform via the PCFM implementation [35]. Subsequent optimization of each segment then forces a desired amplitude envelope (here just a constant) and a desired power spectrum. For the latter, we enforce an approximate Gaussian power spectrum, which in turn provides a Gaussian-like autocorrelation response with corresponding low range sidelobes. The random nature of each segment (due to each being independently initialized) ensures that the range sidelobes do not combine coherently when performing Doppler processing across the segments. As a result of this incoherent combining, Doppler processing serves to further reduce the range sidelobes relative to what is achieved for a given waveform segment.

Designing the power spectrum to conform to a particular shape is performed through the use of a spectral template denoted as $|G(f)|^2$. Spectral notches can be inserted into this template [30] through the use of the null constraint

$$|G(f)| = 0 \text{ for } f \in \Omega, \tag{23.54}$$

where Ω indicates the range(s) of frequencies for the spectral notch(es). It should be noted that forcing a rectangular notch shape in the power spectrum induces a $\sin(x)/x$ roll-off in the autocorrelation sidelobes. While tapering of the notch edges can mitigate this effect [31], it is not necessary when another signal is to be used to "fill in" the spectral notch in such a manner so as to maintain the original (notch-free) power spectrum.

The mth segment of the waveform is initialized with a length T random FM signal, denoted as $p_{0,m}(t)$, here generated by a random instantiation of the PCFM implementation [35] with the "phase change" parameters independently drawn from a uniform distribution on $[-\pi, +\pi]$. The method of alternating projections [29–31,54] is then used to optimize each segment of the waveform via the application of K iterations of

$$r_{k+1,m}(t) = F^{-1}\{|G(f)|\exp\{j\angle F\{p_{k,m}(t)\}\}\} \tag{23.55}$$

and

$$p_{k+1,m}(t) = u(t)\exp\{j\angle r_{k+1,m}(t)\}, \tag{23.56}$$

where $u(t)$ is the amplitude envelope (here a length T rectangular window), and $F\{\cdot\}$ and $F^{-1}\{\cdot\}$ are the Fourier and inverse Fourier transforms, respectively. Through this alternating projection structure, (23.55) produces a time-domain waveform with a power spectrum matching that of $|G(f)|^2$, while (23.56) subsequently removes the amplitude variation, thus yielding a constant amplitude waveform. After optimization of the mth segment, it is phase rotated so that its initial phase value matches the final phase value of the $(m-1)$th segment, thereby avoiding phase discontinuities.

Clearly, the enforcement of constant amplitude causes the waveform power spectrum to deviate from $|G(f)|^2$. While increasing the number of iterations, K realizes a closer match to the desired $|G(f)|^2$, beyond a certain point the diminishing improvement one obtains from each additional iteration becomes negligible. It should also be noted that enforcement of the spectral notches defined in (23.54) via the projection in (23.55) has been found to produce notch depths of about 20 dB relative to the peak. If a greater notch depth is required, the reiterative uniform weighting optimization method [71] can additionally be applied as described in [30]. Finally, because this FM noise waveform structure is designed on a segment-wise basis, it is clear that the rate at which the spectral notch location/width is changed (or "hopped") can be as fast as the segment rate $1/T$, which itself is rather arbitrary, particularly for CW operation.

23.3.2 Embedding communications into radar spectral notches

A cooperative, dual-function arrangement in which a communication signal is designed to occupy the spectral notch in the radar emission can now be considered. Shaping the spectrum of the communication signal to conform to the original radar power spectrum (had the notch been absent) creates a composite spectrum that, when emitted into the far-field, has essentially the same range sidelobe properties as a radar-only emission.

Here, an OFDM implementation of a 4-quadrature amplitude modulation (QAM) (or a $\pi/4$ phase-rotated QPSK) constellation is used for the communication signal. Each OFDM subcarrier is multiplied by an arbitrary 4-QAM symbol that has a narrow frequency and constant amplitude, thereby readily permitting shaping of the communication signal spectrum without affecting the information contained within.

While higher order constellations (e.g. 16-QAM or 64-QAM) could be used to achieve a higher data rate, the resulting per-subcarrier amplitude modulation would detract from our ability to match the desired radar power spectrum. For the sake of convenience, the length of the OFDM symbols are set to match the length of the PRO-FMCW segments, though this arrangement is not required. For the mth segment/symbol interval, the communication signal is represented as

$$r_m(t) = \sum_{n=0}^{N-1} a_{m,n} \exp\{j(2\pi f_{m,n} t + \phi_{m,n})\}, \tag{23.57}$$

where N is the number of OFDM subcarriers, $f_{m,n}$ is the frequency of the nth sub-carrier for the mth waveform segment, $\phi_{m,n}$ is the phase value of the 4-QAM symbol encoded onto the nth subcarrier, and $a_{m,n}$ is an additional amplitude scaling applied to the nth subcarrier that is incorporated here such that the composite (radar + communication) emission realizes the same power spectrum as would otherwise have been achieved by the notch-free radar emission.

23.3.3 Performance assessment of THoRaCs

The efficacy of this tandem-hopped radar and communications approach is validated through simulation measurements for three test cases. The first case involves a spectral notch and OFDM subcarriers that are static during the CPI of the radar emission. The second and third cases allow the radar spectral notch and OFDM subcarriers to hop randomly in tandem around the band to 10 and 100 possible locations, respectively, within the same overall available spectrum. For all cases, the PRO-FMCW waveform has a length of $T_w = 200$ ms partitioned into $M = 10^4$ segments. Thus, each segment has a length of $T = 20\,\mu$s with a postoptimization 3 dB bandwidth of 80 MHz, yielding a time-bandwidth product of $BT \approx 1,600$ per segment such that collectively $BT \approx 1.6 \times 10^7$ across all M segments.

23.3.3.1 Performance evaluation

The three test cases are simulated with high fidelity to mimic physical transmission. In the first case, a single notch is placed in the spectrum as shown in Figure 23.25. The notch has a width of $B/10$ and can accommodate $N = 162$ OFDM subcarriers. An OFDM signal with arbitrary symbols is then spectrally shaped according to (23.57) and superimposed with the radar emission to form the composite spectrum shown in the figure. Since there is no constraint on the amplitude of the overall OFDM signal $r_m(t)$, the PAPR is found to be ~11.7 dB.

It is observed that the composite spectrum matches well with the notch-free spectrum, aside from small deviations at the notch edges. Ample signal-to-interference ratio (SIR) for the OFDM subcarriers is provided by the 65-dB notch depth using [72]. Thus, one can expect good decoding performance at an intended communication receiver that possesses knowledge of the tandem hopping pattern.

Now consider the autocorrelation response achieved by matched filtering each of the M waveform segments followed by coherent integration over all segments (the latter step is simply the processing for the zero Doppler bin). This integrated

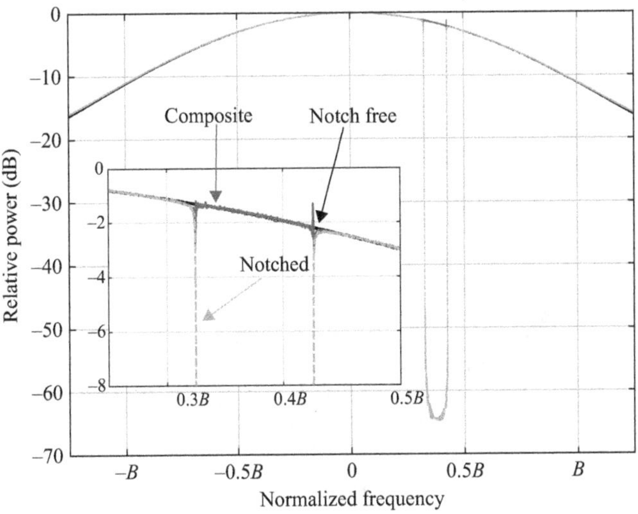

Figure 23.25 Simulated mean power spectrum (with detail inset) for fixed notch/OFDM location

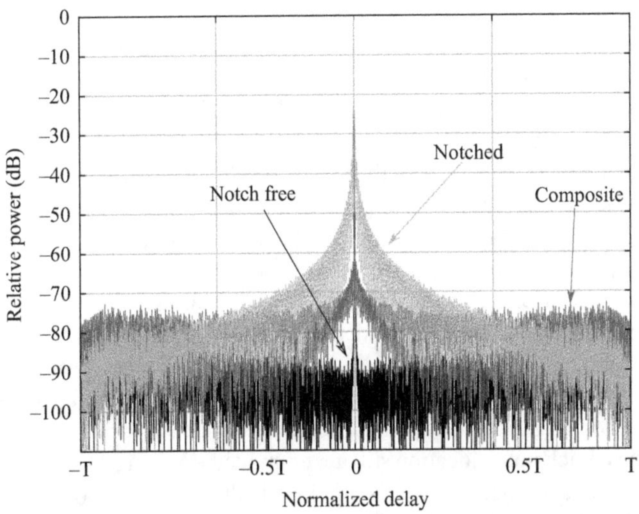

Figure 23.26 Simulated integrated autocorrelation for fixed notch/OFDM location

autocorrelation response is shown in Figure 23.26, with a detailed view in Figure 23.27, for each of the three spectra from Figure 23.25.

The notch-free PRO-FMCW waveform has a peak sidelobe response below −80 dB followed by a very rapid decay into the sidelobe floor. In contrast, the notched waveform has a peak sidelobe at around −37 dB and does not decay into the sidelobe

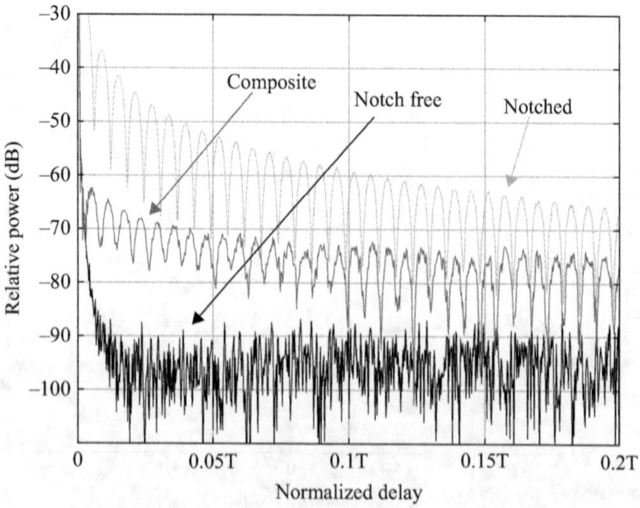

Figure 23.27 Simulated integrated autocorrelation for fixed notch/OFDM location (detail view)

floor until after a delay of $\pm 0.5T$. Between these two extremes, the composite waveform has a peak sidelobe at around -63 dB (or -57 dB depending on the significance of a very narrow shoulder lobe) with a roll-off reaching the floor by $\pm 0.3T$. It can therefore be observed that filling the spectral notch with shaped OFDM subcarriers can create a composite waveform with an autocorrelation response far more favorable than the notched waveform alone. If the small deviations around the notch edges in Figure 23.25 were eliminated, it is expected that the composite autocorrelation would be even closer to that of the notch-free waveform.

In the second case, the notch is hopped randomly around the $\pm 0.5B$ band (without overlapping notch locations) in tandem with the OFDM subcarriers. The $M = 10^4$ segments are partitioned into 10 sets of 10^3 segments, with each set corresponding to a different spectral notch location of width $B/10$. Each location is occupied only once and the hopping order is randomly selected, with each location being occupied for $T = 20$ ms. Each notch location still accommodates $N = 162$ OFDM subcarriers. In the third case, a similar strategy is employed, albeit with the $M = 10^4$ total segments partitioned into 100 sets of 10^2 waveform segments. Each segment is randomly assigned one of the 100 different notch locations without repeat over the band $\pm 0.5B$, and thus, these notch locations do possess some degree of overlap with one another. Each notch persists for $T = 2$ ms in this case. The PAPR of the OFDM waveform in the second and third cases are 13.8 and 13.6 dB, respectively.

The power spectra of the composite waveforms for these tandem hopped cases are shown in Figure 23.28, including a detailed view. The spectra match well to the notch-free case throughout the band. For the case of 10-hopped notches, small perturbations in the spectrum are observed at intervals of $0.1B$, which corresponds to the locations

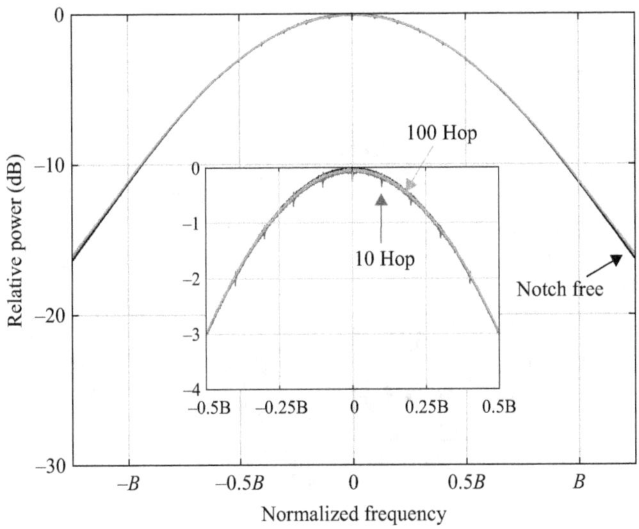

Figure 23.28 Simulated mean power spectrum with detail for tandem hopping

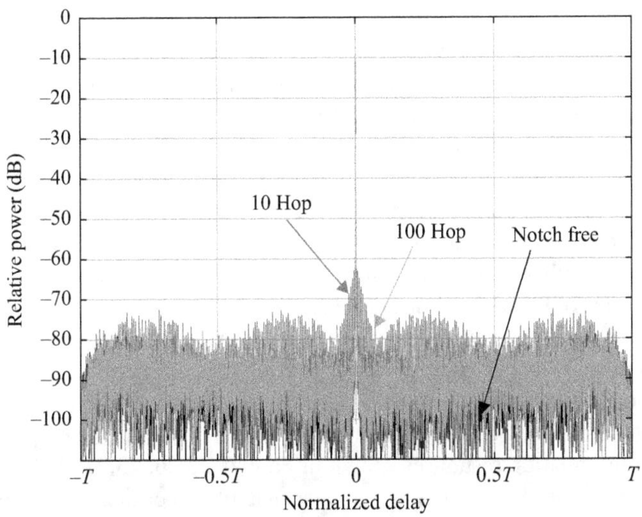

Figure 23.29 Simulated integrated autocorrelation with randomized tandem hopping

of notch edges. In the 100-hopped notch case, the notches are allowed to overlap and the perturbations are smoothed out.

The integrated autocorrelation responses for the hopped cases are shown in Figure 23.29 along with a detailed view in Figure 23.30. The 10-hop autocorrelation

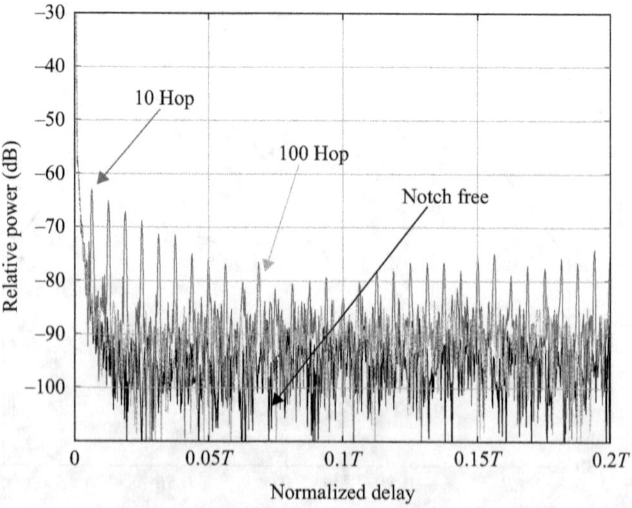

Figure 23.30 Simulated integrated autocorrelation with randomized tandem hopping (detail view)

response exhibits a peak sidelobe at about −63 dB occurring at a delay of $\pm 0.006T$. Successive sidelobes occur at integer multiples of this delay that decay into the floor by $\pm 0.1T$. This repeated sidelobe structure is caused by the regular nonoverlapping notch locations in the spectrum.

The autocorrelation response of the 100-hop case has a peak sidelobe level of about −78 dB that occurs at a delay of $\pm 0.06T$. This factor of 10 increase in delay of the peak sidelobe compared to the 10-hop case is commensurate with the tenfold increase in the number of notch locations. Successive sidelobes once again appear at integer multiples of this delay. Further increasing the spectral overlap of the notches, randomly perturbing their locations, or any other strategy to break up the repetitive structure in the power spectrum will serve to push the commensurately smaller peak sidelobes out further in delay, thus producing an autocorrelation response nearly identical to that of the notch-free waveform.

The autocorrelation responses shown in Figures 23.26, 23.27, 23.29, and 23.30 assumed a stationary point target. It is also useful to consider the response of a moving target during the CPI (i.e. nonzero Doppler). It was shown in [31] that hopping the spectral notch during the CPI caused a significant degradation in the Doppler processed response. In order to determine the impact on Doppler processing of filling the spectral notch in the radar waveform with OFDM subcarriers, the range-Doppler point spread function is analyzed for the 10-hop and 100-hop cases. All segments of each waveform are pulse compressed and then subsequently Doppler processed by weighting across the segments with a Taylor window and then performing a Fourier transform across all 10^4 segments in the waveform. A postoptimization 3 dB bandwidth of 80 MHz is still retained.

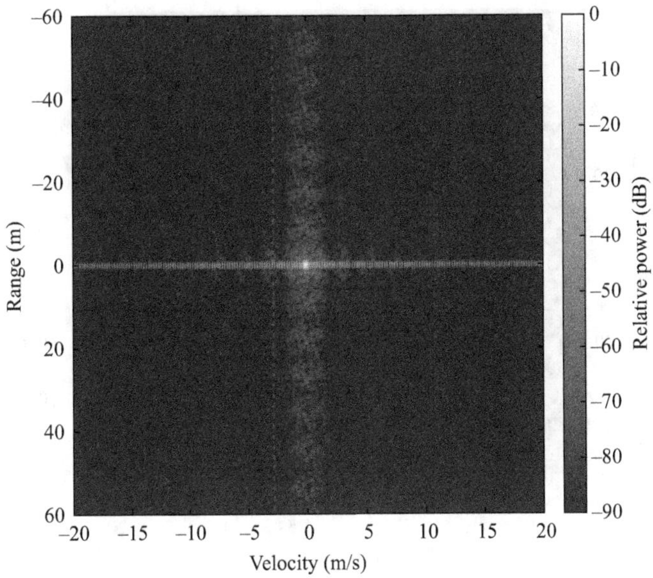

Figure 23.31 Simulated range-Doppler point spread function with 10 randomly hopped notches

Figures 23.31 and 23.32 shown zoomed-in (near the main lobe) regions of the range-Doppler point spread function for the 10-hop and 100-hop composite waveforms, respectively. Both figures show significant energy spread in range and Doppler that is not suppressed by the use of a Taylor window. This spread is more pronounced for the 100-hop case, which is expected since the hopping rate is increased tenfold. Even with this energy spread present, it is much less severe than what was noticed in [31] for the case of an unfilled, randomly hopped notch (with tapering) allowed to occupy 10 different spectral locations during the CPI. Thus, filling the spectral notches with OFDM subcarriers has improved the range-Doppler response of the composite waveform over the case of an unfilled spectral notch in a radar emission.

23.3.4 Commensal radar with spectrally shaped OFDM

Having demonstrated that spectrally shaped OFDM can provide the means to insert a communication function into a spectral notch of a radar emission and in doing so realize a composite emission that retains a favorable autocorrelation, it is reasonable to consider the case in which the entire radar emission is constructed of spectrally shaped OFDM subcarriers with arbitrary symbols. In other words, allow the number of subcarriers N defined in (23.57) to cover the entire radar band while imposing, via the subcarrier amplitude shaping $a_{m,n}$, a power spectrum that corresponds to good autocorrelation sidelobes (e.g. Gaussian). Thus, the spectrally shaped OFDM waveform will serve its primary function as a communication signal, since the 4-QAM

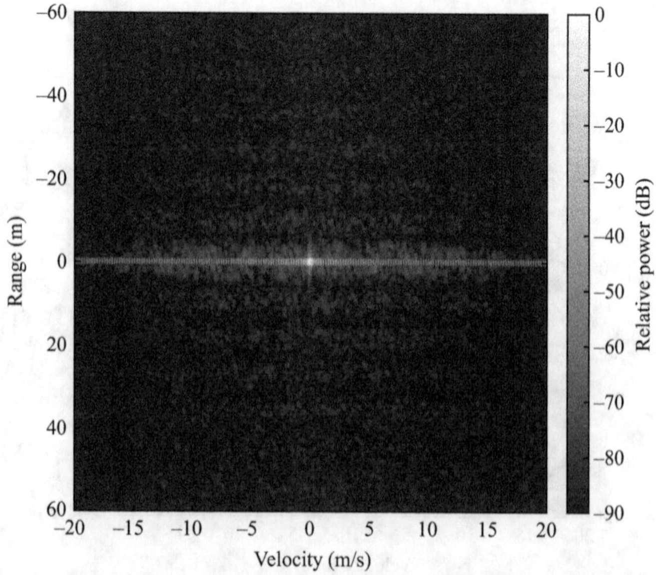

Figure 23.32 Simulated range-Doppler point spread function with 100 randomly hopped notches

symbols are still completely arbitrary, while having properties desirable for a radar waveform, making it useful for passive radar operation [72–74]. In a more general sense, Griffiths, Darwazeh, and Inggs proposed this type of operation as "commensal radar" in [32], where the communication emission is co-designed to also serve as a useful illuminator of opportunity for passive radar.

As an example of this manner of commensal radar waveform, a spectrally shaped OFDM signal is generated via (23.57) with $N = 3,000$ OFDM subcarriers. Figure 23.33 shows the simulated power spectrum of the commensal waveform compared to that of the notch-free PRO-FMCW waveform. Since it is not constrained to constant amplitude, the commensal waveform exhibits a PAPR of 12 dB. However, this amplitude freedom also yields a power spectrum with a sharper roll-off than that of the PRO-FMCW waveform which does have a constant amplitude.

Figure 23.34 shows the integrated autocorrelation response of the commensal waveform compared to that of the PRO-FMCW waveform, and a detailed view is shown in Figure 23.35. Compared to the PRO-FMCW waveform, the commensal waveform has a sidelobe floor that is about 10 dB higher, though the peak sidelobe response is still very low with a maximum value of -75 dB.

Clearly, an OFDM signal with a power spectrum shaped like that in Figure 23.33 deviates from the optimum information capacity for the communication application. However, the point in this demonstration is that there are possible performance trades that could be made so that both radar and communication modes could leverage the same signals. Of course, this example is just one of an infinite variety of

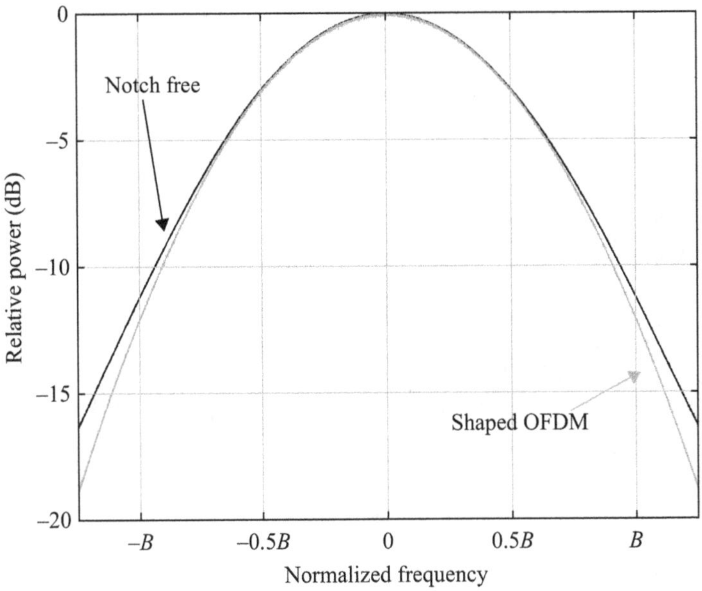

Figure 23.33 Simulated mean power spectrum of spectrally shaped OFDM used as a commensal waveform

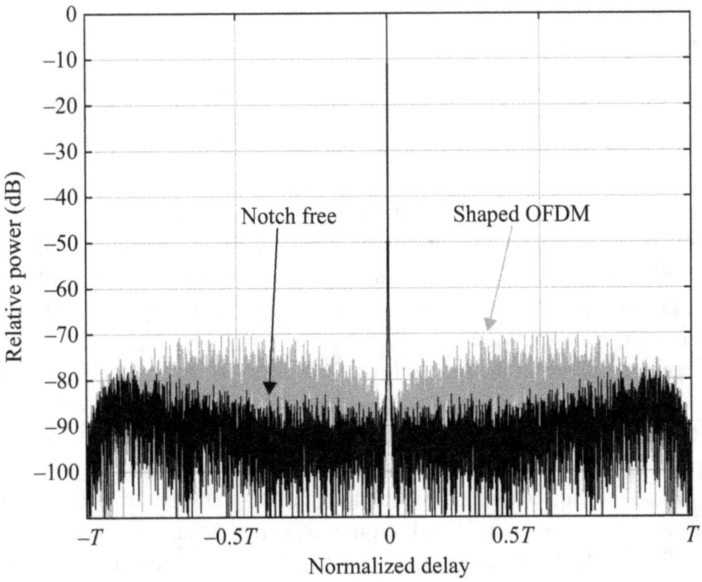

Figure 23.34 Simulated integrated autocorrelation of spectrally shaped OFDM used as a commensal waveform

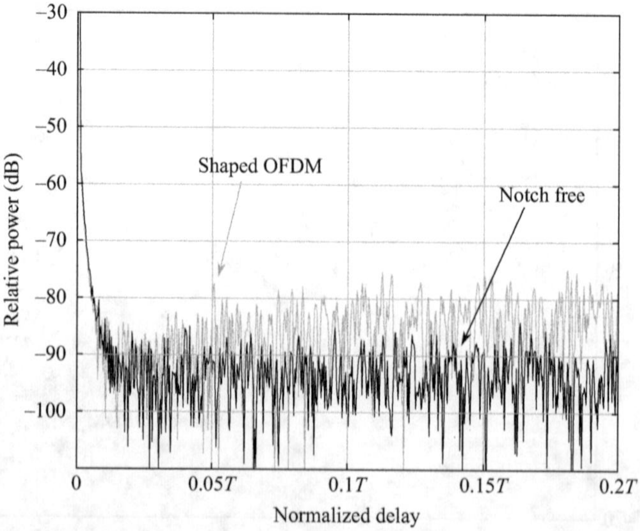

Figure 23.35 Simulated integrated autocorrelation of spectrally shaped OFDM used as a commensal waveform (detail view)

spectral shapes, and it remains to be seen whether an emission can be realized that is sufficiently compatible with both modes.

23.4 Conclusions

The challenge posed by the ever-increasing RF spectrum demand for communication applications is in part met with radar and communication spectrum sharing by co-design of dual-function systems. Co-design of joint radar/communication systems requires the use of some form of waveform diversity via exploitation of the available time, frequency, coding, spatial, or polarization degrees-of-freedom. In this chapter, we have focused on methods for coding, spatial, and spectral forms of waveform diversity, presenting particular dual-function design strategies for each along with practical implementation issues.

In Section 23.1, it was shown that the CPM-based coding diversity REC approach generates constant-envelope information-carrying radar waveforms with good spectral containment, and enables direct control of the degree of RSM by trading-off data throughput and/or BER. For chirp-like waveforms, the spectral broadening is eliminated by the use of guard symbols which translate to guard bands at the edges of the base radar waveform spectrum. In addition, it was shown that with appropriate design of receive filters RSM can be further mitigated. Simulation results demonstrate that coherence across the pulse compression responses can be enhanced, and hence RSM reduced, at the expense of higher range sidelobes. Finally, a closed-form expression was derived that characterizes the dependence of RSM, and thus residual clutter power, on REC system parameters. It is shown that RSM power does not depend on Doppler frequency or time-bandwidth product, but is inversely proportional to the number of pulses in the CPI and is approximately quadratic with the CPM modulation index.

In Section 23.2, the iterative optimization method known as FFRED, which generates constant amplitude, physically realizable waveforms that emit simultaneous radar and communications beams, was discussed. A particularly important attribute for such an arrangement is power efficiency, which was addressed by consideration of the minimum percent orthogonal power needed for convergence and an energy-on-target metric derived from a simple amplifier model. It was found that the converged constant amplitude solutions generated using FFRED produce waveforms that can have as high as a 50% increase in efficiency compared to the minimum-norm solution. It was also discovered that a constant amplitude communication signal yields a higher overall power efficiency in this arrangement compared to a communication signal that requires amplitude modulation (e.g. SRRC-filtered QPSK). How the orthogonal power manifests itself under different scenarios was illustrated by observing the average spatial beampatterns and maximum spectral content. The communication performance was evaluated via a BER simulation under different communication directions and power levels. It was found that the relative communication power level controls the "swath" over which the communication link can be reliably achieved and that the receiver BER is highly dependent on the separation from the desired communication direction.

In Section 23.3, we demonstrated the efficacy of time-varying frequency diversity to facilitate spectral co-design of a radar/communication system. By placement of a spectral notch in an FM noise radar waveform, shaped OFDM subcarriers of a communication waveform can reside at these vacated frequencies with sufficient SIR to allow proper demodulation at the receiver. Use of OFDM subcarriers readily allows for shaping of the communication spectrum to blend in with the surrounding radar waveform spectrum, yielding a favorable autocorrelation response for the composite waveform. Simulation results demonstrate that the autocorrelation response of the composite waveform compares favorably to the response of a notch-free radar waveform. Hence, careful placement and shaping of the OFDM subcarriers in the spectral notch of the radar waveform create a composite emission that serves both radar and communication functions. It was also shown that creating a spectrally shaped waveform composed entirely of OFDM subcarriers, thereby realizing a form of "commensal" waveform, can provide this dual functionality as well.

23.5 Looking ahead

It is anticipated that numerous new approaches for multifunction (particularly dual radar/communication) systems will be proposed in the coming years as spectrum continues to become more congested and as more novel capabilities of these joint modes is revealed. While it is rather daunting to make specific predictions about this field in general, the following discusses possible research paths for the methods presented in this chapter.

Potential future research regarding the CPM-based REC approach branches out in a few directions. It is worth considering the design and optimization of the base radar waveform to maximize radar performance beyond what LFM can achieve. The properties of the ambiguity function of the combined REC waveforms may also be explored. In addition, design of communication strategies that fully exploit the temporal and spectral structure of chirp-like base radar waveforms may be considered to enhance communication performance. Separately, more advanced (i.e. adaptive) radar signal processing techniques to address RSM within this REC framework may be examined to further explore the design trade space (e.g. RSM degree versus range sidelobe level). Further, the trade-offs between communication performance and radar performance can be better understood by extending the theoretical RSM power analysis to the physical CPM-based framework and by invoking a more phenomenologically representative clutter model that accounts for spatial correlation, internal clutter motion, etc. Finally, issues such as synchronization (i.e. timing, phase offset, and frequency offset estimation) at the communication receiver and detection of the presence of communication symbols embedded in the radar waveform by the communication receiver are of practical interest for adoption in real systems.

The spatial waveform diversity approach presented here serves as an illustration of the breadth of research areas in the co-design of dual-function radar/communication systems. Important factors influencing future research topics include, but are not limited to, resource management, power efficiency, physical realizability, real-time waveform optimization, communication methods, and function priority levels. The orthogonal power needed to create a constant-amplitude solution currently has no restrictions on where it is placed, spectrally or spatially. Therefore, additional constraints could be imposed to prevent this power from appearing in preselected sub-bands and/or directions. The initialization to the FFRED algorithm uniquely specifies the final solution and how the algorithm behaves. Obstacles such as "traps" and "tunnels" discussed in [64] need to be avoided to ensure timely convergence. Through intelligent initializations and implementation, these shortcomings may be minimized to ensure the algorithm provides a solution of sufficient quality within a certain time requirement.

Finally, for joint radar/communication using spectral waveform diversity, we have considered the use of OFDM with small signal constellations, for spectral waveform diversity radar/communications. While OFDM is readily amenable to spectrum shaping, it also has a high PAPR, a significant shortcoming that leads to poor power efficiency. It is therefore reasonable to examine constant-envelope communication strategies, such as CPM [34] or constant-envelope OFDM. Higher order constellations that have more than four-phase values may also be considered. Adaptive approaches that vary the constellation size based upon the portion of the spectrum in which the communication signal resides are also of interest. For example, higher power signals could employ larger constellations than lower power signals, thus taking advantage of the improved SNR. The effect of pilot signal ambiguities on the composite autocorrelation response is another important consideration, as it was shown that the presence of a structure in the waveform spectrum creates a cyclostationarity effect on the autocorrelation response. Finally, while the simulation results depicted here and the loopback measurements presented in [19] are promising, it will be interesting to see how this scheme will perform in free-space experiments.

References

[1] Griffiths H, Cohen L, Watts S, *et al.* Radar spectrum engineering and management: technical and regulatory issues. Proceedings of the IEEE. 2015 January; 103(1):85–102.

[2] Andrews JG, Buzzi S, Choi W, *et al.* What will 5G be? IEEE Journal on Selected Areas in Communications. 2014 June; 32(6):1065–1082.

[3] Akyildiz IF, Lee WY, Vuran MC, Mohanty S. NeXt generation/dynamic spectrum access/cognitive radio wireless networks: a survey. Computer Networks. 2006; 50(13):2127–2159.

[4] Strategic Technology Office and Defense Advanced Research Projects Agency. Shared spectrum access for radar and communications (SSPARC). [Online]. Available: http://www.darpa.mil/program/shared-spectrum-access-for-radar-and-communications, accessed January 2016.

[5] Wicks M, Mokole E, Blunt S, Schneible R, Amuso V, editors. Principles of Waveform Diversity and Design. Raleigh, NC: SciTech Publishing; 2010.

[6] Pillai S, Li KY, Himed B. Waveform Diversity: Theory & Applications. New York, NY: McGraw-Hill; 2011.

[7] Gini F, Maio AD, Patton LK, editors. Waveform Design and Diversity for Advanced Radar Systems. Croydon, UK: Institution of Engineering and Technology; 2012.

[8] Blunt SD, Mokole EL. Overview of radar waveform diversity. IEEE Aerospace and Electronic Systems Magazine. 2016 November; 31(11):2–42.

[9] Blunt SD, Cook MR, Stiles J. Embedding information into radar emissions via waveform implementation. In: International Waveform Diversity and Design Conference. Niagara Falls, Canada; 2010. p. 195–199.

[10] Sahin C, Jakabosky J, McCormick PM, Metcalf JG, Blunt SD. A novel approach for embedding communication symbols into physical radar waveforms. In: IEEE Radar Conference. Seattle, WA; 2017.

[11] Nowak MJ, Zhang Z, Qu Y, Dessources DA, Wicks M, Wu Z. Co-designed radar-communication using linear frequency modulation waveform. In: IEEE Military Communications Conference. Baltimore, MD; 2016. p. 918–923.

[12] Chen X, Wang X, Xu S, Zhang J. A novel radar waveform compatible with communication. In: International Conference on Computational Problem-Solving. Chengdu, China; 2011. p. 177–181.

[13] Sturm C, Wiesbeck W. Waveform design and signal processing aspects for fusion of wireless communications and radar sensing. Proceedings of the IEEE. 2011 July; 99(7):1236–1259.

[14] Hassanien A, Amin MG, Zhang YD, Ahmad F. Dual-function radar-communications: information embedding using sidelobe control and waveform diversity. IEEE Transactions on Signal Processing. 2016 April; 64(8):2168–2181.

[15] Metcalf JG, Sahin C, Blunt SD, Rangaswamy M. Analysis of symbol-design strategies for intrapulse radar-embedded communications. IEEE Transactions on Aerospace and Electronic Systems. 2015 October; 51(4):2914–2931.

[16] Huang Y, Piezzo M, Carotenuto V, De Maio A. Radar waveform design under similarity, bandwidth priority, and spectral coexistence constraints. In: IEEE Radar Conference. Seattle, WA; 2017.

[17] Mehrnoush M, Roy S. Interference mitigation in coexistence of WLAN network with radar. In: IEEE Radar Conference. Seattle, WA; 2017.

[18] Chiriyath A, Paul B, Bliss D. Simultaneous radar detection and communications performance with clutter mitigation. In: IEEE Radar Conference. Seattle, WA; 2017.

[19] Ravenscroft B, McCormick PM, Blunt SD, Jakabosky J, Metcalf JG. Tandem-hopped OFDM communications in spectral gaps of FM noise radar. In: IEEE Radar Conference. Seattle, WA; 2017.

[20] McCormick PM, Blunt SD, Metcalf JG. Simultaneous radar and communications emission from a common aperture, Part I: theory. In: IEEE Radar Conference. Seattle, WA; 2017.

[21] Sahin C, Metcalf JG, Blunt SD. Filter design to address range sidelobe modulation in transmit-encoded radar-embedded communications. In: IEEE Radar Conference. Seattle, WA; 2017.

[22] Sahin C, Metcalf JG, Blunt SD. Characterization of range sidelobe modulation arising from radar embedded-communications. In: IET International Conference on Radar Systems. Belfast, UK; 2017.

[23] Fuhrmann DR, Antonio GS. Transmit beamforming for MIMO radar systems using signal cross-correlation. IEEE Transactions on Aerospace and Electronic Systems. 2008 January; 44(1):171–186.

[24] Stoica P, Li J, Zhu X. Waveform synthesis for diversity-based transmit beampattern design. IEEE Transactions on Signal Processing. 2008 June; 56(6):2593–2598.

[25] Friedlander B. On transmit beamforming for MIMO radar. IEEE Transactions on Aerospace and Electronic Systems. 2012 October; 48(4): 3376–3388.

[26] He H, Stoica P, Li J. Wideband MIMO systems: signal design for transmit beampattern synthesis. IEEE Transactions on Signal Processing. 2011 February; 59(2):618–628.

[27] Aldayel O, Monga V, Rangaswamy M. Tractable MIMO beampattern design under constant modulus waveform constraint. In: IEEE Radar Conference. Philadelphia, PA; 2016.

[28] McCormick PM, Duly AJ, Ravenscroft B, Metcalf JG, Blunt SD. Simultaneous radar and communications emission from a common aperture, Part II: experimentation. In: IEEE Radar Conference. Seattle, WA; 2017.

[29] Jakabosky J, Blunt SD, Himed B. Waveform design and receive processing for nonrecurrent nonlinear FMCW radar. In: IEEE Radar Conference. Arlington, VA; 2015. p. 1376–1381.

[30] Jakabosky J, Blunt SD, Martone A. Incorporating hopped spectral gaps into nonrecurrent nonlinear FMCW radar emissions. In: IEEE International Workshop on Computational Advances in Multi-Sensor Adaptive Processing. Cancun, Mexico; 2015. p. 281–284.

[31] Jakabosky J, Ravenscroft B, Blunt SD, Martone A. Gapped spectrum shaping for tandem-hopped radar/communications cognitive sensing. In: IEEE Radar Conference. Philadelphia, PA; 2016.

[32] Griffiths H, Darwazeh I, Inggs M. Waveform design for commensal radar. In: IEEE Radar Conference. Arlington, VA; 2015. p. 1456–1460.

[33] O'Connor AC, Kantor JM, Jakabosky J. Joint equalization filters that mitigate waveform-diversity modulation of clutter. In: IEEE Radar Conference. Philadelphia, PA; 2016.

[34] Anderson JB, Aulin T, Sundberg CE. Digital Phase Modulation. New York, NY: Plenum Press; 1986.

[35] Blunt SD, Cook M, Jakabosky J, de Graaf J, Perrins E. Polyphase-coded FM (PCFM) radar waveforms, Part I: Implementation. IEEE Transactions on Aerospace and Electronic Systems. 2014 July; 50(3):2218–2229.

[36] Blunt SD, Jakabosky J, Cook MR, Stiles J, Seguin S, Mokole E. Polyphase-coded FM (PCFM) radar waveforms, Part II: Optimization. IEEE Transactions on Aerospace and Electronic Systems. 2014 July; 50(3):2230–2241.

[37] Bluetooth Special Interest Group. Specifications of the Bluetooth System, Version 1.2; 2003.

[38] IRIG Standard 106-04: Telemetry Standards. Range Commanders Council, White Sands Missile Range, New Mexico; 2004. Available online at http://www.ntia.doc.gov/osmhome/106.pdf.

[39] Agency DIS. Department of Defense Interface Standard, Interoperability Standard for Single-Access 5-kHz and 25-kHz UHF Satellite Communications Channels. Tech. Rep. MIL-STD-188-181B, Department of Defense, March 1999.

[40] Sahin C, Perrins E. On the symmetric information rate of CPM in the finite blocklength regime. In: IEEE Military Communications Conference. Tampa, FL; 2015. p. 815–819.

[41] Polyanskiy Y, Poor HV, Verdu S. Channel coding rate in the finite blocklength regime. IEEE Transactions on Information Theory. 2010 May; 56(5):2307–2359.

[42] Proakis JG. Digital Communications. New York, NY: McGraw-Hill Education; 2008.

[43] Simon M. Bandwidth-Efficient Digital Modulation with Application to Deep-Space Communication. Hoboken, NJ: Wiley; 2003.

[44] Viterbi A. Error bounds for convolutional codes and an asymptotically optimum decoding algorithm. IEEE Transactions on Information Theory. 1967 April; 13(2):260–269.

[45] Higgins T, Blunt SD, Shackelford AK. Time-range adaptive processing for pulse agile radar. In: International Waveform Diversity and Design Conference. Niagara Falls, Canada; 2010. p. 115–120.

[46] Higgins T, Gerlach K, Shackelford AK, Blunt SD. Aspects of non-identical multiple pulse compression. In: IEEE Radar Conference. Kansas City, MO; 2011. p. 895–900.

[47] Scholnik D. Range-ambiguous clutter suppression with pulse-diverse waveforms. In: IEEE Radar Conference. Kansas City, MO; 2011. p. 336–341.

[48] O'Connor AC, Kantor JM, Jakabosky J. Space-time adaptive mismatch processing. In: IEEE Radar Conference. Philadelphia, PA; 2016.

[49] Cook MR, Blunt SD, Jakabosky J. Optimization of waveform diversity and performance for pulse-agile radar. In: IEEE Radar Conference. Kansas City, MO; 2011. p. 812–8127.

[50] Ackroyd MH, Ghani F. Optimum mismatched filters for sidelobe suppression. IEEE Transactions on Aerospace and Electronic Systems. 1973 March; AES-9(2):214–218.

[51] Henke D, McCormick PM, Blunt SD, Higgins T. Practical aspects of optimal mismatch filtering and adaptive pulse compression for FM waveforms. In: IEEE Radar Conference. Arlington, VA; 2015. p. 1149–1155.

[52] Blunt SD, Gerlach K, Higgins T. Aspects of radar range super-resolution. In: IEEE Radar Conference. Waltham, MA; 2007. p. 683–687.

[53] Richards MA, Scheer JA, Holm WA. Principles of Modern Radar, Vol. I: Basic Principles. Raleigh, NC: SciTech Publishing; 2010.

[54] Jakabosky J, Blunt SD, Himed B. Spectral-shape optimized FM noise radar for pulse agility. In: IEEE Radar Conference. Philadelphia, PA; 2016.

[55] Tan PS, Jakabosky J, Stiles JM, Blunt SD. On higher-order representations of polyphase-coded FM radar waveforms. In: IEEE Radar Conference. Arlington, VA; 2015.

[56] Richards MA. Fundamentals of Radar Signal Processing. New York City, NY: McGraw-Hill Education; 2014.

[57] Tavik GC, Hilterbrick CL, Evins JB, *et al.* The advanced multifunction RF concept. IEEE Transactions on Microwave Theory and Techniques. 2005 March; 53(3):1009–1020.

[58] McCormick PM, Blunt SD, Metcalf JG. Wideband MIMO frequency modulated emission design with space-frequency nulling. IEEE Journal of Selected Topics in Signal Processing. 2017 March; 11(2):363–378.

[59] Mealey TC, Duly AJ. BEEMER: a firmware-tuned, software-defined MIMO radar testbed. In: IEEE International Symposium on Phased Array Systems and Technology. Waltham, MA; 2016.

[60] Boyd S, Vandenberghe L. Convex Optimization. Cambridge, UK: Cambridge University Press; 2004.

[61] Walker JL. Handbook of RF and microwave power amplifiers. Cambridge, UK: Cambridge University Press; 2011.

[62] Youla D. Generalized image restoration by the method of alternating orthogonal projections. IEEE Transactions on Circuits and Systems. 1978 September; 25(9):694–702.

[63] Youla DC, Webb H. Image restoration by the method of convex projections, Part 1: theory. IEEE Transactions on Medical Imaging. 1982 October; 1(2): 81–94.
[64] Levi A, Stark H. Image restoration by the method of generalized projections with application to restoration from magnitude. In: IEEE International Conference on Acoustics, Speech, and Signal Processing. vol. 9. San Diego, CA; 1984. p. 88–91.
[65] Bauschke HH, Combettes PL, Luke DR. Phase retrieval, error reduction algorithm, and Fienup variants: a view from convex optimization. Journal of the Optical Society of America A. 2002; 19(7):1334–1345.
[66] Patton LK, Rigling BD. Phase retrieval for radar waveform optimization. IEEE Transactions on Aerospace and Electronic Systems. 2012 October; 48(4):3287–3302.
[67] Fienup JR. Phase retrieval algorithms: a comparison. Applied Optics. 1982; 21(15):2758–2769.
[68] Frost SW, Rigling B. Sidelobe predictions for spectrally-disjoint radar waveforms. In: 2012 IEEE Radar Conference. Atlanta, GA; 2012. p. 0247–0252.
[69] Ergen M. Mobile Broadband – Including WiMAX and LTE. Berkeley, CA: Springer; 2009.
[70] Jakabosky J, Ryan L, Blunt S. Transmitter-in-the-loop optimization of distorted OFDM radar emissions. In: IEEE Radar Conference. Ottawa, Canada; 2013.
[71] Higgins T, Webster T, Shackelford AK. Mitigating interference via spatial and spectral nulling. IET Radar, Sonar and Navigation. 2014 February; 8(2):84–93.
[72] Poullin D. Passive detection using digital broadcasters (DAB, DVB) with COFDM modulation. IEE Proceedings – Radar, Sonar and Navigation. 2005 June; 152(3):143–152.
[73] Berger CR, Demissie B, Heckenbach J, Willett P, Zhou S. Signal processing for passive radar using OFDM waveforms. IEEE Journal of Selected Topics in Signal Processing. 2010 February; 4(1):226–238.
[74] Palmer JE, Harms HA, Searle SJ, Davis L. DVB-T passive radar signal processing. IEEE Transactions on Signal Processing. 2013 April; 61(8):2116–2126.

Index

www.ingramcontent.com/pod-product-compliance
Lightning Source LLC
LaVergne TN
LVHW011852200926
848484LV00002B/444
* 9 7 8 1 7 8 5 6 1 3 5 7 9 *